Win the complete library o ... a!

ACCURATE EFFICIEN. CUSTOMIZABLE

RSMeansOnline
FROM THE GORDIAN GROUP®

Register your book below to receive your **free** quarterly updates.
Plus you will be entered into a quarterly drawing to win the **complete** RSMeans Online library of 2016 data!

Be sure to keep up-to-date in 2016!

Fill out the card below for RSMeans' free quarterly updates, as well as a chance to win the complete RSMeans Online library. Please provide your name, address, and email below and return this card by mail, or register online:

info.thegordiangroup.com/2016updates.html

Name _____

Email_____ Title_____

Company _____

Street _____

City/Town _____ State/Prov. _____ Zip/Postal Code _____

RSMeans
FROM THE GORDIAN GROUP®

Win the complete library of RSMeans Online data!

ACCURATE

EFFICIENT

CUSTOMIZABLE

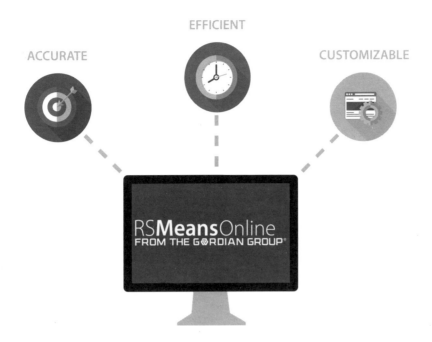

Register your book below to receive your **free** quarterly updates.
Plus you will be entered into a quarterly drawing to win the **complete** RSMeans Online library of 2016 data!

RSMeans
FROM THE GORDIAN GROUP®

Square Foot Costs

Marilyn Phelan, AIA, Senior Editor

2016
37th annual edition

Engineering Director
Bob Mewis, CCP *(1, 2, 4, 13, 14, 31, 32, 33, 34, 35, 41, 44, 46)*

Contributing Editors
Christopher Babbitt
Adrian C. Charest, PE
Cheryl Elsmore
Wafaa Hamitou
Joseph Kelble
Charles Kibbee
Robert J. Kuchta *(8)*
Michael Landry
Thomas Lane *(6, 7)*
Genevieve Medeiros
Elisa Mello
Chris Morris *(26, 27, 28, 48)*

Melville J. Mossman, PE *(21, 22, 23)*
Marilyn Phelan, AIA *(9, 10, 11, 12)*
Stephen C. Plotner *(3, 5)*
Stephen Rosenberg
Kevin Souza
Keegan Spraker
Tim Tonello
David Yazbek

Vice President Data & Engineering
Chris Anderson

Product Manager
Andrea Sillah

Production Manager
Debbie Panarelli

Production
Sharon Larsen
Jonathan Forgit
Sheryl Rose
Mary Lou Geary

Technical Support
Gary L. Hoitt
Kathryn S. Rodriguez

Cover Design
Blaire Gaddis

Numbers in italics are the divisional responsibilities for each editor. Please contact the designated editor directly with any questions.

RSMeans
Construction Publishers & Consultants
1099 Hingham Street, Suite 201
Rockland, MA 02370
United States of America
1-877-756-2789
www.RSMeans.com

Copyright 2015 by RSMeans
All rights reserved.
Cover photo © nikonaft/iStock/Thinkstock

Printed in the United States of America
ISSN 1540-6326
ISBN 978-1-943215-18-8

 $299.99 per copy (in United States)
Price is subject to change without prior notice.

Related RSMeans Products and Services

The residential section of this data set is aimed at new residential construction projects valued at up to $1,000,000. The labor rate used in the calculations is for residential construction. The commercial/industrial/institutional section is aimed primarily at new construction projects valued at $1,500,00 and up. The labor rate in the calculations is union.

The engineers at RSMeans suggest the following products and services as companion information resources to *RSMeans Square Foot Costs*:

Construction Cost Data
Assemblies Cost Data 2016
Building Construction Cost Data 2016
Residential Cost Data 2016
Light Commercial Cost Data 2016

Reference Books
Estimating Building Costs
RSMeans Estimating Handbook
Green Building: Project Planning & Estimating
How to Estimate with RSMeans Data
Plan Reading & Material Takeoff
Project Scheduling & Management for Construction
Universal Design for Style, Comfort & Safety

Seminars and In-House Training
Conceptual Estimating Using RSMeans CostWorks® CD
RSMeans Online® Training
Plan Reading & Material Take-Off

RSMeans Online Store
Visit RSMeans at www.RSMeans.com for the most reliable and current resources available on the market. Learn more about our more than 20 data sets available in Online, Book, eBook, and CD formats. Our library of reference books is also available, along with professional development seminars aimed at improving cost estimating, project management, administration, and facilities management skills.

RSMeans Electronic Data
Receive the most up-to-date cost information with RSMeans Online. This web-based service is quick, intuitive, and easy to use, giving you instant access to RSMeans' comprehensive database. Learn more at: **www.RSMeans.com/Online**.

RSMeans Custom Solutions
Building owners, facility managers, building product manufacturers, attorneys, and even insurance firms across the public and private sectors have engaged in RSMeans' custom solutions to solve their estimating needs including:

Custom Cost Engineering Solutions
Knowing the lifetime maintenance cost of your building, how much it will cost to build a specific building type in different locations across the country and globe, and if the estimate from which you are basing a major decision is accurate are all imperative for building owners and their facility managers.

Market & Custom Analytics
Taking a construction product to market requires accurate market research in order to make critical business development decisions. Once the product goes to market, one way to differentiate your product from the competition is to make any cost saving claims you have to offer.

Third-Party Legal Resources
Natural disasters can lead to reconstruction and repairs just as new developments and expansions can lead to issues of eminent domain. In these cases where one party must pay, it is plausible for legal disputes over costs and estimates to arise.

Construction Costs for Software Applications
More than 25 unit price and assemblies cost databases are available through a number of leading estimating and facilities management software partners. For more information, see "Other RSMeans Products and Services" at the back of this publication.

RSMeans data is also available to federal, state, and local government agencies as multi-year, multi-seat licenses.

For procurement construction cost data such as Job Order Contracting and Change Order Management please refer to The Gordian Group's procurement solutions.

For information on our current partners, call 1-877-756-2789.

Table of Contents

Foreword

Who We Are

Since 1942, RSMeans has delivered construction cost estimating information and consulting throughout North America. In 2014, RSMeans was acquired by The Gordian Group, combining two industry-leading construction cost databases. Through the RSMeans line of products and services, The Gordian Group provides innovative construction cost estimating data to organizations pursuing efficient and effective construction planning, estimating, procurement, and information solutions.

Our Offerings

With RSMeans' construction cost estimating data, contractors, architects, engineers, facility owners, and managers can utilize the most up-to-date data available in many formats to meet their specific planning and estimating needs.

When you purchase information from RSMeans, you are, in effect, hiring the services of a full-time staff of construction and engineering professionals.

Our thoroughly-experienced and highly-qualified staff works daily to collect, analyze, and disseminate comprehensive cost information to meet your needs.

These staff members have years of practical construction experience and engineering training prior to joining the firm. As a result, you can count on them not only for accurate cost figures, but also for additional background reference information that will help you create a realistic estimate.

The RSMeans organization is equipped to help you solve construction problems through its variety of data solutions.

Access our comprehensive database electronically with RSMeans Online. Quick, intuitive, easy to use, and updated continuously throughout the year, RSMeans Online is the most accurate and up-to-date cost estimating data.

This up-to-date, accurate, and localized data can be leveraged for a broad range of applications, such as Custom Cost Engineering solutions, Market and Custom Analytics, and Third-Party Legal Resources.

To ensure you are getting the most from your cost estimating data, we also offer a myriad of reference guides, training, and professional seminars (learn more at **www.RSMeans.com/Learn**).

In short, RSMeans can provide you with the tools and expertise for developing accurate and dependable construction estimates and budgets in a variety of ways.

Our Commitment

Today at RSMeans, we do more than talk about the quality of our data and the usefulness of the information. We stand behind all of our data—from historical cost indexes to construction materials and techniques—to craft current costs and predict future trends.

If you have any questions about our products or services, please call us toll-free at 1-877-756-2789. You can also visit our website at: **www.RSMeans.com**.

How the Cost Data Is Built: An Overview

A Powerful Construction Tool

You now have one of the most powerful construction tools available today. A successful project is built on the foundation of an accurate and dependable estimate. This tool will enable you to construct such an estimate.

For the casual user the information is designed to be:

- quickly and easily understood so you can get right to your estimate.
- filled with valuable information so you can understand the necessary factors that go into a cost estimate.

For the professional user, the information is designed to be:

- a handy reference that can be quickly referred to for key costs.
- a comprehensive, fully reliable source of current construction costs and productivity rates so you'll be prepared to estimate any project.
- a source for preliminary project costs, product selections, and alternate materials and methods.

To meet all of these requirements, we have organized the information into the following clearly defined sections.

Estimating with RSMeans Unit Price Cost Data*

Please refer to these steps for guidance on completing an estimate using RSMeans unit price cost data.

How to Use the Information: The Details

This section contains an in-depth explanation of how the information is arranged and how you can use it to determine a reliable construction cost estimate. It includes how we develop our cost figures and how to prepare your estimate.

Unit Prices*

All cost data has been divided into 50 divisions according to the MasterFormat system of classification and numbering.

Assemblies*

The cost data in this section has been organized in an "Assemblies" format. These assemblies are the functional elements of a building and are arranged according to the 7 elements of the UNIFORMAT II classification system. For a complete explanation of a typical "Assembly", see "How RSMeans Assemblies Data Works."

Residential Models*

Model buildings for four classes of construction—economy, average, custom, and luxury—are developed and shown with complete costs per square foot.

Commercial/Industrial/ Institutional Models*

This section contains complete costs for 77 typical model buildings expressed as costs per square foot.

Green Commercial/Industrial/ Institutional Models*

This section contains complete costs for 25 green model buildings expressed as costs per square foot.

References*

This section includes information on Equipment Rental Costs, Crew Listings, Historical Cost Indexes, City Cost Indexes, Location Factors, Reference Tables, and Change Orders, as well as a listing of abbreviations.

- **Equipment Rental Costs:** Included are the average costs to rent and operate hundreds of pieces of construction equipment.
- **Crew Listings:** This section lists all the crews referenced in the cost data. A crew is composed of more than one trade classification and/or the addition of power equipment to any trade classification. Power equipment is included in the cost of the crew. Costs are shown both with bare labor rates and with the installing contractor's overhead and profit added. For each, the total crew cost per eight-hour day and the composite cost per labor-hour are listed.

- **Historical Cost Indexes:** These indexes provide you with data to adjust construction costs over time.
- **City Cost Indexes:** All costs in this data set are U.S. national averages. Costs vary by region. You can adjust for this by CSI Division to over 700 locations throughout the U.S. and Canada by using this data.
- **Location Factors:** You can adjust total project costs to over 900 locations throughout the U.S. and Canada by using the weighted number, which applies across all divisions.
- **Reference Tables:** At the beginning of selected major classifications in the Unit Prices are reference numbers indicators. These numbers refer you to related information in the Reference Section. In this section, you'll find reference tables, explanations, and estimating information that support how we develop the unit price data, technical data, and estimating procedures.
- **Change Orders:** This section includes information on the factors that influence the pricing of change orders.
- **Abbreviations:** A listing of abbreviations used throughout this information, along with the terms they represent, is included.

Index (printed versions only)

A comprehensive listing of all terms and subjects will help you quickly find what you need when you are not sure where it occurs in MasterFormat.

Conclusion

This information is designed to be as comprehensive and easy to use as possible.

The Construction Specifications Institute (CSI) and Construction Specifications Canada (CSC) have produced the 2014 edition of MasterFormat®, a system of titles and numbers used extensively to organize construction information.

All unit prices in the RSMeans cost data are now arranged in the 50-division MasterFormat® 2014 system.

---

* Not all information is available in all data sets

Note: The material prices in RSMeans cost data are "contractor's prices." They are the prices that contractors can expect to pay at the lumberyards, suppliers'/distributors' warehouses, etc. Small orders of specialty items would be higher than the costs shown, while very large orders, such as truckload lots, would be less. The variation would depend on the size, timing, and negotiating power of the contractor. The labor costs are primarily for new construction or major renovation rather than repairs or minor alterations. With reasonable exercise of judgment, the figures can be used for any building work.

Residential Section

Table of Contents

Introduction to the Square Foot Cost Section

The Square Foot Cost Section contains costs per square foot for four classes of construction in seven building types. Costs are listed for various exterior wall materials which are typical of the class and building type. There are cost tables for wings and ells with modification tables to adjust the base cost of each class of building. Non-standard items can easily be added to the standard structures.

Cost estimating for a residence is a three-step process:
1. Identification
2. Listing dimensions
3. Calculations

Guidelines and a sample cost estimating procedure are shown on the following pages.

Identification

To properly identify a residential building, the class of construction, type, and exterior wall material must be determined. The "Building Classes" information has drawings and guidelines for determining the class of construction. There are also detailed specifications and additional drawings at the beginning of each set of tables to further aid in proper building class and type identification.

Sketches for eight types of residential buildings and their configurations follow. Definitions of living area are next to each sketch. Sketches and definitions of garage types follow.

Living Area

Base cost tables are prepared as costs per square foot of living area. The living area of a residence is that area which is suitable and normally designed for full time living. It does not include basement recreation rooms or finished attics, although these areas are often considered full time living areas by the owners.

Living area is calculated from the exterior dimensions without the need to adjust for exterior wall thickness. When calculating the living area of a 1-1/2 story, two story, three story or tri-level residence, overhangs and other differences in size and shape between floors must be considered.

Only the floor area with a ceiling height of seven feet or more in a 1-1/2 story residence is considered living area. In bi-levels and tri-levels, the areas that are below grade are considered living area, even when these areas may not be completely finished.

Base Tables and Modifications

Base cost tables show the base cost per square foot without a basement, with one full bath and one full kitchen for economy and average homes, and an additional half bath for custom and luxury models. Adjustments for finished and unfinished basements are part of the base cost tables. Adjustments for multi-family residences, additional bathrooms, townhouses, alternative roofs, and air conditioning and heating systems are listed in Modifications, Adjustments, and Alternatives tables below the base cost tables.

Costs for other modifications, adjustments, and alternatives, including garages, breezeways, and site improvements, follow the base tables.

Listing of Dimensions

To use this section, only the dimensions used to calculate the horizontal area of the building and additions, modifications, adjustments, and alternatives are needed. The dimensions, normally the length and width, can come from drawings or field measurements. For ease in calculation, consider measuring in tenths of feet, i.e., 9 ft. 6 in. = 9.5 ft., 9 ft. 4 in. = 9.3 ft.

In all cases, make a sketch of the building. Any protrusions or other variations in shape should be noted on the sketch with dimensions.

Calculations

The calculations portion of the estimate is a two-step activity:
1. The selection of appropriate costs from the tables
2. Computations

Selection of Appropriate Costs

To select the appropriate cost from the base tables, the following information is needed:
1. Class of construction
   - Economy
   - Average
   - Custom
   - Luxury
2. Type of residence
   - 1 story
   - 1-1/2 story
   - 2 story
   - 3 story
   - Bi-level
   - Tri-level
3. Occupancy
   - One family
   - Two family
   - Three family
4. Building configuration
   - Detached
   - Town/Row house
   - Semi-detached
5. Exterior wall construction
   - Wood frame
   - Brick veneer
   - Solid masonry
6. Living areas

Modifications are classified by class, type, and size.

Computations

The computation process should take the following sequence:
1. Multiply the base cost by the area.
2. Add or subtract the modifications, adjustments, and alternatives.
3. Apply the location modifier.

When selecting costs, interpolate or use the cost that most nearly matches the structure under study. This applies to size, exterior wall construction, and class.

How to Use the Residential Square Foot Cost Section

The following is a detailed explanation of a sample entry in the Residential Square Foot Cost Section. Each bold number below corresponds to the item being described in the following list with the appropriate component of the sample entry in parentheses. Prices listed are costs that include overhead and profit of the installing contractor. Total model costs include an additional markup for the general contractor's overhead and profit and fees specific to class of construction.

RESIDENTIAL	Average ❶	2 Story ❷

- Simple design from standard plans
- Single family — 1 full bath, 1 kitchen
- **❸** No basement
- Asphalt shingles on roof
- Hot air heat
- Gypsum wallboard interior finishes
- Materials and workmanship are average
- Detail specifications on p. 27

Note: The illustration shown may contain some optional components (for example: garages and/or fireplaces) whose costs are shown in the modifications, adjustments, & alternatives below or at the end of the square foot section.

Base cost per square foot of living area

Exterior Wall **❹**	Living Area **❺**										
	1000	1200	1400	1600	1800	2000	2200	2600	3000	3400	3800
Wood Siding - Wood Frame	142.45	128.80	122.15	117.55	112.85 **❻**	107.95	104.65	98.35	92.35	89.65	87.15
Brick Veneer - Wood Frame	149.30	135.15	128.05	123.25	118.20	113.05	109.55	102.75	96.45	93.50	90.80
Stucco on Wood Frame	137.75	124.40	118.00	113.65	109.20	104.40	101.25	95.30	89.55	86.90	84.55
Solid Masonry	163.05	147.85	139.95	134.60	128.90	123.40	119.30	111.65	104.60	101.25	98.15
Finished Basement, Add **❼**	22.70	22.45	21.70	21.15	20.65	20.30 **❽**	19.85	19.15	18.60	18.25	17.95
Unfinished Basement, Add	9.00	8.40	7.90	7.60	7.25	7.00	6.75	6.35	6.00	5.75	5.60

Modifications

Add to the total cost

Upgrade Kitchen Cabinets **❾**	$	+ 5525
Solid Surface Countertops (Included)		
Full Bath - including plumbing, wall and floor finishes		+ 7405
Half Bath - including plumbing, wall and floor finishes		+ 4331
One Car Attached Garage		+ 14,471
One Car Detached Garage		+ 19,004
Fireplace & Chimney		+ 7319

Adjustments

For multi family - add to total cost

Additional Kitchen	$	+ 9600
Additional Bath		+ 7405
Additional Entry & Exit		+ 1676
Separate Heating		+ 1336
Separate Electric		+ 1945

For Townhouse/Rowhouse - Multiply cost per square foot by

Inner Unit **❿**	.90
End Unit	.95

Alternatives

Add to or deduct from the cost per square foot of living area

Cedar Shake Roof	+ 1.65
Clay Tile Roof	+ 3.40
Slate Roof	+ 3.75
Upgrade Walls to Skim Coat Plaster **⓫**	+ .53
Upgrade Ceilings to Textured Finish	+ .58
Air Conditioning, in Heating Ductwork	+ 2.96
In Separate Ductwork	+ 5.70
Heating Systems, Hot Water	+ 1.49
Heat Pump	+ 1.61
Electric Heat	– .63
Not Heated	– 3.28

Additional upgrades or components

Kitchen Cabinets & Countertops	Page 58
Bathroom Vanities	59
Fireplaces & Chimneys	59
Windows, Skylights & Dormers	59
Appliances	60
Breezeways & Porches	60
Finished Attic	60
Garages	61
Site Improvements **⓬**	61
Wings & Ells	37

1 Class of Construction (Average)

The class of construction depends upon the design and specifications of the plan. The four classes are economy, average, custom, and luxury.

2 Type of Residence (2 Story)

The building type describes the number of stories or levels in the model. The seven building types are 1 story, 1-1/2 story, 2 story, 2-1/2 story, 3 story, bi-level, and tri-level.

3 Specification Highlights (Hot Air Heat)

These specifications include information concerning the components of the model, including the number of baths, roofing types, HVAC systems, materials, and workmanship. If the components listed are not appropriate, modifications can be made by consulting the information shown below or in the Assemblies Section.

4 Exterior Wall System (Wood Siding–Wood Frame)

This section includes the types of exterior wall systems and the structural frames used. The exterior wall systems shown are typical of the class of construction and building type shown.

5 Living Areas (2000 SF)

The living area is that area of the residence which is suitable and normally designed for full time living. It does not include basement recreation rooms or finished attics. Living area is calculated from the exterior dimensions without the need to adjust for exterior wall thickness. When calculating the living area of a 1-1/2 story, 2 story, 3 story, or tri-level residence, overhangs and other differences in size and shape between floors must be considered. Only the floor area with a ceiling height of seven feet or more in a 1-1/2 story residence is considered living area. In bi-levels and tri-levels, the areas that are below grade are considered living area, even when these areas may not be completely finished. A range of various living areas for the residential model is shown to aid in the selection of values from the matrix.

6 Base Costs per Square Foot of Living Area ($107.95)

Base cost tables show the cost per square foot of living area without a basement, with one full bath and one full kitchen for economy and average homes, and an additional half bath for custom and luxury models. When selecting costs, interpolate or use the cost that most nearly matches the residence under consideration for size, exterior wall system, and class of construction. Prices listed are costs that include overhead and profit of the installing contractor, a general contractor markup, and an allowance for plans that vary by class of construction. For additional information on contractor overhead and architectural fees, see the Reference Section.

7 Basement Types (Finished)

The two types of basements are finished and unfinished. The specifications and components for both are shown under Building Classes in the Introduction to this section.

8 Additional Costs for Basements ($20.30 or $7.00)

These values indicate the additional cost per square foot of living area for either a finished or an unfinished basement.

9 Modifications and Adjustments (Upgrade Kitchen Cabinets $5525)

Modifications and Adjustments are costs added to or subtracted from the total cost of the residence. The total cost of the residence is equal to the cost per square foot of living area times the living area. Typical modifications and adjustments include kitchens, baths, garages, and fireplaces.

10 Multiplier for Townhouse/Rowhouse (Inner Unit .90)

The multipliers shown adjust the base costs per square foot of living area for the common wall condition encountered in townhouses or rowhouses.

11 Alternatives (Skim Coat Plaster $.53)

Alternatives are costs added to or subtracted from the base cost per square foot of living area. Typical alternatives include variations in kitchens, baths, roofing, and air conditioning and heating systems.

12 Additional Upgrades or Components (Wings & Ells)

Costs for additional upgrades or components, including wings or ells, breezeways, porches, finished attics, and site improvements, are shown at the end of each quality section and at the end of the Square Foot Section.

Building Classes

Economy Class

An economy class residence is usually built from stock plans. The materials and workmanship are sufficient to satisfy building codes. Low construction cost is more important than distinctive features. The overall shape of the foundation and structure is seldom other than square or rectangular.

An unfinished basement includes a 7' high, 8" thick foundation wall composed of either concrete block or cast-in-place concrete.

Included in the finished basement cost are inexpensive paneling or drywall as the interior finish on the foundation walls, a low cost sponge backed carpeting adhered to the concrete floor, a drywall ceiling, and overhead lighting.

Custom Class

A custom class residence is usually built from plans and specifications with enough features to give the building a distinction of design. Materials and workmanship are generally above average with obvious attention given to construction details. Construction normally exceeds building code requirements.

An unfinished basement includes a 7'-6" high, 10" thick cast-in-place concrete foundation wall or a 7'-6" high, 12" thick concrete block foundation wall.

A finished basement includes painted drywall on insulated 2" x 4" wood furring as the interior finish to the concrete walls, a suspended ceiling, carpeting adhered to the concrete floor, overhead lighting, and heating.

Average Class

An average class residence is a simple design and built from standard plans. Materials and workmanship are average but often exceed minimum building codes. There are frequently special features that give the residence some distinctive characteristics.

An unfinished basement includes a 7'-6" high, 8" thick foundation wall composed of either cast-in-place concrete or concrete block.

Included in the finished basement are plywood paneling or drywall on furring that is fastened to the foundation walls, sponge backed carpeting adhered to the concrete floor, a suspended ceiling, overhead lighting, and heating.

Luxury Class

A luxury class residence is built from an architect's plan for a specific owner. It is unique both in design and workmanship. There are many special features, and construction usually exceeds all building codes. It is obvious that primary attention is placed on the owner's comfort and pleasure. Construction is supervised by an architect.

An unfinished basement includes an 8' high, 12" thick foundation wall that is composed of cast-in-place concrete or concrete block.

A finished basement includes painted drywall on 2" x 4" wood furring as the interior finish, suspended ceiling, tackless carpet on wood subfloor with sleepers, overhead lighting, and heating.

Configurations

Detached House

This category of residence is a free-standing separate building with or without an attached garage. It has four complete walls.

Semi-Detached House

This category of residence has two living units side-by-side. The common wall is fireproof. Semi-detached residences can be treated as a row house with two end units. Semi-detached residences can be any of the building types.

Town/Row House

This category of residence has a number of attached units made up of inner units and end units. The units are joined by common walls. The inner units have only two exterior walls. The common walls are fireproof. The end units have three walls and a common wall. Town houses/row houses can be any of the building types.

Building Types

One Story

This is an example of a one-story dwelling. The living area of this type of residence is confined to the ground floor. The headroom in the attic is usually too low for use as a living area.

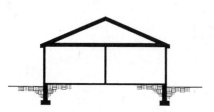

One-and-one-half Story

The living area in the upper level of this type of residence is 50% to 90% of the ground floor. This is made possible by a combination of this design's high-peaked roof and/or dormers. Only the upper level area with a ceiling height of seven feet or more is considered living area. The living area of this residence is the sum of the ground floor area plus the area on the second level with a ceiling height of seven feet or more.

One Story with Finished Attic

The main living area in this type of residence is the ground floor. The upper level or attic area has sufficient headroom for use as a living area. This is made possible by a high peaked roof. The living area in the attic is less than 50% of the ground floor. The living area of this type of residence is the ground floor area only. The finished attic is considered an adjustment.

Two Story

This type of residence has a second floor or upper level area which is equal or nearly equal to the ground floor area. The upper level of this type of residence can range from 90% to 110% of the ground floor area, depending on setbacks or overhangs. The living area is the sum of the ground floor area and the upper level floor area.

Two-and-one-half Story

This type of residence has two levels of equal or nearly equal area and a third level which has a living area that is 50% to 90% of the ground floor. This is made possible by a high peaked roof, extended wall heights, and/or dormers. Only the upper level area with a ceiling height of seven feet or more is considered living area. The living area of this residence is the sum of the ground floor area, the second floor area, and the area on the third level with a ceiling height of seven feet or more.

Bi-level

This type of residence has two living areas, one above the other. One area is about four feet below grade and the second is about four feet above grade. Both areas are equal in size. The lower level in this type of residence is designed and built to serve as a living area and not as a basement. Both levels have full ceiling heights. The living area is the sum of the lower level area and the upper level area.

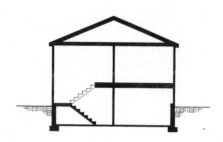

Three Story

This type of residence has three levels which are equal or nearly equal. As in the two story residence, the second and third floor areas may vary slightly depending on setbacks or overhangs. The living area is the sum of the ground floor area and the two upper level floor areas.

Tri-level

This type of residence has three levels of living area: one at grade level, one about four feet below grade, and one about four feet above grade. All levels are designed to serve as living areas. All levels have full ceiling heights. The living area is a sum of the areas of each of the three levels.

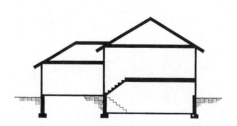

Garage Types

Attached Garage

Shares a common wall with the dwelling. Access is typically through a door between the dwelling and garage.

Basement Garage

Constructed under the roof of the dwelling but below the living area.

Built-In Garage

Constructed under the second floor living space and above the basement level of the dwelling. Reduces gross square feet of the living area.

Detached Garage

Constructed apart from the main dwelling. Shares no common area or wall with the dwelling.

Building Components

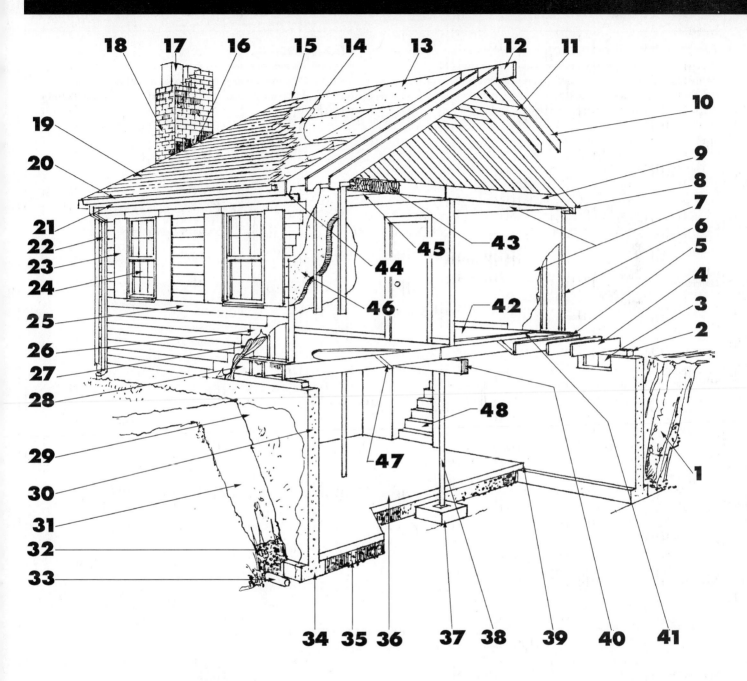

1. Excavation
2. Sill Plate
3. Basement Window
4. Floor Joist
5. Shoe Plate
6. Studs
7. Drywall
8. Plate
9. Ceiling Joists
10. Rafters
11. Collar Ties
12. Ridge Rafter
13. Roof Sheathing
14. Roof Felt
15. Roof Shingles
16. Flashing
17. Flue Lining
18. Chimney
19. Roof Shingles
20. Gutter
21. Fascia
22. Downspout
23. Shutter
24. Window
25. Wall Shingles
26. Weather Barrier
27. Wall Sheathing
28. Fire Stop
29. Dampproofing
30. Foundation Wall
31. Backfill
32. Drainage Stone
33. Drainage Tile
34. Wall Footing
35. Gravel
36. Concrete Slab
37. Column Footing
38. Pipe Column
39. Expansion Joint
40. Girder
41. Sub-floor
42. Finish Floor
43. Attic Insulation
44. Soffit
45. Ceiling Strapping
46. Wall Insulation
47. Cross Bridging
48. Bulkhead Stairs

Exterior Wall Construction

Typical Frame Construction

Typical wood frame construction consists of wood studs with insulation between them. A typical exterior surface is made up of sheathing, building paper, and exterior siding consisting of wood, vinyl, aluminum, or stucco over the wood sheathing.

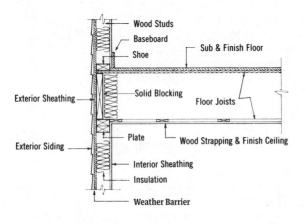

Brick Veneer

Typical brick veneer construction consists of wood studs with insulation between them. A typical exterior surface is sheathing, building paper, and an exterior of brick tied to the sheathing, with metal strips.

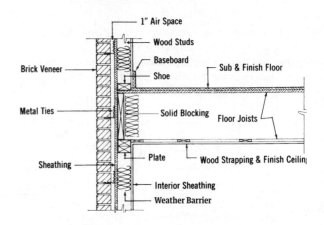

Stone

Typical solid masonry construction consists of a stone or block wall covered on the exterior with brick, stone, or other masonry.

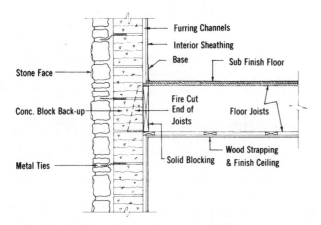

Residential Cost Estimate Worksheet

Worksheet Instructions

The residential cost estimate worksheet can be used as an outline for developing a residential construction or replacement cost. It is also useful for insurance appraisals. The design of the worksheet helps eliminate errors and omissions. To use the worksheet, follow the example below.

1. Fill out the owner's name, residence address, the estimator or appraiser's name, some type of project identifying number or code, and the date.

2. Determine from the plans, specifications, owner's description, photographs, or any other means possible the class of construction. The models in this data set use economy, average, custom, and luxury as classes. Fill in the appropriate box.

3. Fill in the appropriate box for the residence type, configuration, occupancy, and exterior wall. If you require clarification, the pages preceding this worksheet describe each of these.

4. Next, the living area of the residence must be established. The heated or air conditioned space of the residence, not including the basement, should be measured. It is easiest to break the structure up into separate components as shown in the example: the main house (A), a one-and-one-half story wing (B), and a one story wing (C). The breezeway (D), garage (E), and open covered porch (F) will be treated differently. Data entry blocks for the living area are included on the worksheet for your use. Keep each level of each component separate, and fill out the blocks as shown.

5. By using the information on the worksheet, find the model, wing, or ell in the following square foot cost pages that best matches the class, type, exterior finish, and size of the residence being estimated. Use the Modifications, Adjustments, and Alternatives

to determine the adjusted cost per square foot of living area for each component.

6. For each component, multiply the cost per square foot by the living area square footage. If the residence is a townhouse/ rowhouse, a multiplier should be applied based upon the configuration.

7. The second page of the residential cost estimate worksheet has space for the additional components of a house. The cost for additional bathrooms, finished attic space, breezeways, porches, fireplaces, appliance or cabinet upgrades, and garages should be added on this page. The information for each of these components is found with the model being used or under Modifications, Adjustments, and Alternatives.

8. Add the total from page one of the estimate worksheet and the items listed on page two. The sum is the adjusted total building cost.

9. Depending on the use of the final estimated cost, one of the remaining two boxes should be filled out. Any additional items or exclusions should be added or subtracted at this time. The data contained in this data set is a national average. Construction costs are different throughout the country. To allow for this difference, a location factor based upon the first three digits of the residence's zip code must be applied. The location factor is a multiplier that increases or decreases the adjusted total building cost. Find the appropriate location factor and calculate the local cost. If depreciation is a concern, a dollar figure should be subtracted at this point.

10. No residence will match a model exactly. Many differences will be found. At this level of estimating, a variation of plus or minus 10% should be expected.

Adjustments Instructions

No residence matches a model exactly in shape, material, or specifications. The common differences are:

1. Two or more exterior wall systems:
   - Partial basement
   - Partly finished basement
2. Specifications or features that are between two classes
3. Crawl space instead of a basement

Examples

Below are quick examples. See pages 15-17 for complete examples of cost adjustments for these differences:

1. Residence "A" is an average one-story structure with 1,600 S.F. of living area and no basement. Three walls are wood siding on wood frame, and the fourth wall is brick veneer on wood frame. The brick veneer wall is 35% of the exterior wall area.

   Use page 28 to calculate the Base Cost per S.F. of Living Area. Wood Siding for 1,600 S.F. = $111.55 per S.F. Brick Veneer for 1,600 S.F. = $115.65 per S.F.

   .65 ($111.55) + .35 ($115.65) = $112.99 per S.F. of Living Area.

2a. Residence "B" is the same as Residence "A"; However, it has an unfinished basement under 50% of the building. To adjust the $112.99 per S.F. of living area for this partial basement, use page 28.

   $112.99 + .5($11.40) = $118.69 per S.F. of Living Area.

2b. Residence "C" is the same as Residence "A"; However, it has a full basement under the entire building. 640 S.F. or 40% of the basement area is finished.

   Using Page 28:

   $112.99 + .40 ($32.70) + .60 ($11.40) =
   $132.91 per S.F. of Living Area.

3. When specifications or features of a building are between classes, estimate the percent deviation, and use two tables to calculate the cost per S.F.

   A two-story residence with wood siding and 1,800 S.F. of living area has features 30% better than Average, but 70% less than Custom.

   From pages 30 and 42:

Custom 1,800 S.F. Base Cost	=	**$145.25 per S.F.**
Average 1,800 S.F. Base Cost	=	**$112.85 per S.F.**
DIFFERENCE	=	**$32.40 per S.F.**
Cost is $112.85 + .30 ($32.40)	=	**$122.57 per S.F. of Living Area.**

4. To add the cost of a crawl space, use the cost of an unfinished basement as a maximum. For specific costs of components to be added or deducted, such as vapor barrier, underdrain, and floor, see the "Assemblies" section, pages 285 to 518.

Model Residence Example

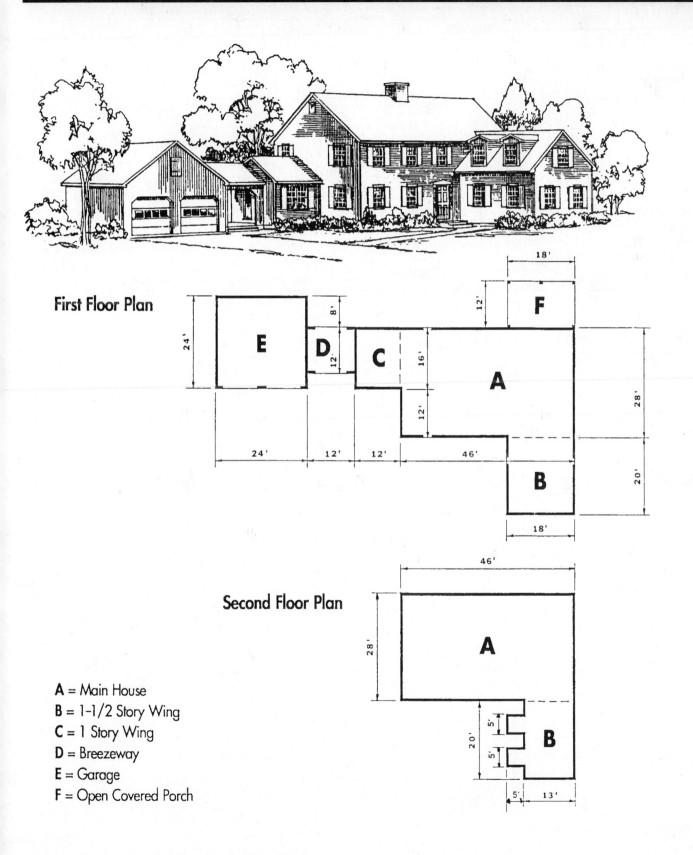

First Floor Plan

E = Garage

D = Breezeway

C = 1 Story Wing

Second Floor Plan

A = Main House
B = 1-1/2 Story Wing
C = 1 Story Wing
D = Breezeway
E = Garage
F = Open Covered Porch

RESIDENTIAL
COST ESTIMATE

OWNERS NAME:	**Albert Westenberg**	APPRAISER:	**Nicole Wojtowicz**
RESIDENCE ADDRESS:	**300 Sygiel Road**	PROJECT:	**# 55**
CITY, STATE, ZIP CODE:	**Three Rivers, MA 01080**	DATE:	**Jan. 1, 2016**

CLASS OF CONSTRUCTION
- ☐ ECONOMY
- ☑ AVERAGE
- ☐ CUSTOM
- ☐ LUXURY

RESIDENCE TYPE
- ☐ 1 STORY
- ☐ 1 1/2 STORY
- ☑ 2 STORY
- ☐ 2 1/2 STORY
- ☐ 3 STORY
- ☐ BI-LEVEL
- ☐ TRI-LEVEL

CONFIGURATION
- ☑ DETACHED
- ☐ TOWN/ROW HOUSE
- ☐ SEMI-DETACHED

OCCUPANCY
- ☑ ONE FAMILY
- ☐ TWO FAMILY
- ☐ THREE FAMILY
- ☐ OTHER

EXTERIOR WALL SYSTEM
- ☑ WOOD SIDING - WOOD FRAME
- ☐ BRICK VENEER - WOOD FRAME
- ☐ STUCCO ON WOOD FRAME
- ☐ PAINTED CONCRETE BLOCK
- ☐ SOLID MASONRY (AVERAGE & CUSTOM)
- ☐ STONE VENEER - WOOD FRAME
- ☐ SOLID BRICK (LUXURY)
- ☐ SOLID STONE (LUXURY)

* LIVING AREA (Main Building)		
First Level	**1288**	S.F.
Second level	**1288**	S.F.
Third Level		S.F.
Total	**2576**	S.F.

* LIVING AREA (Wing or Ell)	(**B**)	
First Level	**360**	S.F.
Second level	**310**	S.F.
Third Level		S.F.
Total	**670**	S.F.

* LIVING AREA (Wing or Ell)	(**C**)	
First Level	**192**	S.F.
Second level		S.F.
Third Level		S.F.
Total	**192**	S.F.

* Basement Area is not part of living area.

MAIN BUILDING		COSTS PER S.F. LIVING AREA	
Cost per Square Foot of Living Area, from Page	**30**	$	**107.95**
Basement Addition:	**100** % Finished, % Unfinished	+	**7.00**
Roof Cover Adjustment: **Cedar Shake**	Type, Page **30** (Add or Deduct)	( )	**1.65**
Central Air Conditioning: ☐ Separate Ducts ☑ Heating Ducts, Page	**30**	+	**2.96**
Heating System Adjustment:	Type, Page (Add or Deduct)	( )	
Main Building: Adjusted Cost per S.F. of Living Area		$	# **119.56**

MAIN BUILDING TOTAL COST	$ **119.56** /S.F.	x **2,576** S.F.	x	= $ **307,987**
	Cost per S.F. Living Area	Living Area	Town/Row House Multiplier (Use 1 for Detached)	TOTAL COST

WING OR ELL (**B**)	**1 - 1/2** STORY	COSTS PER S.F. LIVING AREA	
Cost per Square Foot of Living Area, from Page	**37**	$	**97.10**
Basement Addition: **100** % Finished,	% Unfinished	+	**29.60**
Roof Cover Adjustment:	Type, Page (Add or Deduct)	( )	
Central Air Conditioning: ☐ Separate Ducts ☑ Heating Ducts, Page	**30**	+	**2.96**
Heating System Adjustment:	Type, Page (Add or Deduct)	( )	
Wing or Ell (**B**): Adjusted Cost per S.F. of Living Area		$	# **129.66**

WING OR ELL (**B**) TOTAL COST	$ **129.66** /S.F.	x **670** S.F.	= $ **86,872**
	Cost per S.F. Living Area	Living Area	TOTAL COST

WING OR ELL (**C**)	**1** STORY	COSTS PER S.F. LIVING AREA	
Cost per Square Foot of Living Area, from Page	# **37** **(WOOD SIDING)**	$	# **164.10**
Basement Addition: % Finished,	% Unfinished	+	
Roof Cover Adjustment:	Type, Page (Add or Deduct)	( )	
Central Air Conditioning: ☐ Separate Ducts ☐ Heating Ducts, Page		+	
Heating System Adjustment:	Type, Page (Add or Deduct)	( )	
Wing or Ell (**C**) Adjusted Cost per S.F. of Living Area		$	# **164.1**

WING OR ELL (**C**) TOTAL COST	$ **164.10** /S.F.	x **192** S.F.	= $ **31,507**
	Cost per S.F. Living Area	Living Area	TOTAL COST

TOTAL THIS PAGE	**426,366**

RESIDENTIAL COST ESTIMATE

				QUANTITY	UNIT COST		
Total Page 1						$	426,366
Additional Bathrooms: __2__ Full, __1__ Half	2 @ 7405	1 @ 4331				+	19,141
Finished Attic: __N/A__ Ft. x _____ Ft.					S.F.		
Breezeway: ☑ Open ☐ closed __12__ Ft. x __12__ Ft.				144	S.F. 38.44	+	5,535
Covered Porch: ☑ Open ☐ Enclosed __18__ Ft. x __12__ Ft.				216	S.F. 37.09	+	8,011
Fireplace: ☑ Interior Chimney ☐ Exterior Chimney							
☑ No. of Flues (__2__) ☑ Additional Fireplaces	1 - 2nd Story					+	12,534
Appliances:						+	—
Kitchen Cabinets Adjustments:		(±)					—
☑ Garage ☐ Carport: __2__ Car(s) Description __Wood, Attached__	(±)						24,571
Miscellaneous:						+	
			ADJUSTED TOTAL BUILDING COST			$	496,158

REPLACEMENT COST			INSURANCE COST		
ADJUSTED TOTAL BUILDING COST	$ 496,158		ADJUSTED TOTAL BUILDING COST	$	
Site Improvements			Insurance Exclusions		
(A) Paving & Sidewalks	$		(A) Footings, sitework, Underground Piping	-$	
(B) Landscaping	$		(B) Architects Fees	-$	
(C) Fences	$		Total Building Cost Less Exclusion	$	
(D) Swimming Pools	$		Location Factor	x	
(E) Miscellaneous	$				
TOTAL	$ 496,158		LOCAL INSURABLE REPLACEMENT COST	$	
Location Factor	x 1.060				
Location Replacement Cost	$ 525,927				
Depreciation	-$ 52,592				
LOCAL DEPRECIATED COST	$ 473,335				

SKETCH AND ADDITIONAL CALCULATIONS

17

1 Story

1-1/2 Story

2 Story

Bi-Level

Tri-Level

		Components
1	**Site Work**	Site preparation for slab or excavation for lower level; 4' deep trench excavation for foundation wall.
2	**Foundations**	Continuous reinforced concrete footing, 8" deep x 18" wide; dampproofed and insulated 8" thick reinforced concrete block foundation wall, 4' deep; 4" concrete slab on 4" crushed stone base and polyethylene vapor barrier, trowel finish.
3	**Framing**	Exterior walls—2 x 4 wood studs, 16" O.C.; 1/2" insulation board sheathing; wood truss roof 24" O.C. or 2 x 6 rafters 16" O.C. with 1/2" plywood sheathing, 4 in 12 or 8 in 12 roof pitch.
4	**Exterior Walls**	Beveled wood siding and #15 felt building paper on insulated wood frame walls; sliding sash or double hung wood windows; 2 flush solid core wood exterior doors. **Alternates:** • Brick veneer on wood frame, has 4" veneer of common brick. • Stucco on wood frame has 1" thick colored stucco finish. • Painted concrete block has 8" concrete block sealed and painted on exterior and furring on the interior for the drywall.
5	**Roofing**	20 year asphalt shingles; #15 felt building paper; aluminum gutters, downspouts, drip edge and flashings.
6	**Interiors**	Walls and ceilings—1/2" taped and finished drywall, primed and painted with 2 coats; painted baseboard and trim; rubber backed carpeting 80%, asphalt tile 20%; hollow core wood interior doors.
7	**Specialties**	Economy grade kitchen cabinets—6 L.F. wall and base with plastic laminate counter top and kitchen sink; 30 gallon electric water heater.
8	**Mechanical**	1 lavatory, white, wall hung; 1 water closet, white; 1 bathtub, enameled steel, white; gas fired warm air heat.
9	**Electrical**	100 Amp. service; romex wiring; incandescent lighting fixtures, switches, receptacles.
10	**Overhead and Profit**	General Contractor overhead and profit.

Adjustments

Unfinished Basement:
7' high 8" concrete block or cast-in-place concrete walls.

Finished Basement:
Includes inexpensive paneling or drywall on foundation walls. Inexpensive sponge backed carpeting on concrete floor, drywall ceiling, and lighting.

- **Mass produced from stock plans**
- **Single family — 1 full bath, 1 kitchen**
- **No basement**
- **Asphalt shingles on roof**
- **Hot air heat**
- **Gypsum wallboard interior finishes**
- **Materials and workmanship are sufficient to meet codes**
- **Detail specifications on page 19**

Note: The illustration shown may contain some optional components (for example: garages and/or fireplaces) whose costs are shown in the modifications, adjustments, & alternatives below or at the end of the square foot section.

©Home Planners, Inc.

Base cost per square foot of living area

Exterior Wall	Living Area										
	600	800	1000	1200	1400	1600	1800	2000	2400	2800	3200
Wood Siding - Wood Frame	140.00	126.30	116.00	107.80	100.55	96.05	93.70	90.60	84.50	80.05	77.00
Brick Veneer - Wood Frame	146.40	132.05	121.25	112.50	104.85	100.10	97.55	94.25	87.80	83.05	79.85
Stucco on Wood Frame	131.90	119.10	109.45	101.95	95.25	91.10	88.85	86.05	80.30	76.20	73.45
Painted Concrete Block	138.15	124.70	114.55	106.45	99.40	94.95	92.70	89.60	83.55	79.15	76.15
Finished Basement, Add	32.80	30.90	29.50	28.30	27.25	26.60	26.25	25.70	25.00	24.45	23.90
Unfinished Basement, Add	14.50	12.95	11.90	10.85	10.10	9.55	9.25	8.85	8.25	7.80	7.40

Modifications

Add to the total cost

Upgrade Kitchen Cabinets	$ + 1205
Solid Surface Countertops	+ 849
Full Bath - including plumbing, wall and floor finishes	+ 5924
Half Bath - including plumbing, wall and floor finishes	+ 3465
One Car Attached Garage	+ 13,419
One Car Detached Garage	+ 17,321
Fireplace & Chimney	+ 6285

Adjustments

For multi family - add to total cost

Additional Kitchen	$ + 5288
Additional Bath	+ 5924
Additional Entry & Exit	+ 1676
Separate Heating	+ 1336
Separate Electric	+ 966

For Townhouse/Rowhouse -
Multiply cost per square foot by

Inner Unit	.95
End Unit	.97

Alternatives

Add to or deduct from the cost per square foot of living area

Composition Roll Roofing	– 1.15
Cedar Shake Roof	+ 3.90
Upgrade Walls and Ceilings to Skim Coat Plaster	+ .72
Upgrade Ceilings to Textured Finish	+ .58
Air Conditioning, in Heating Ductwork	+ 4.73
In Separate Ductwork	+ 6.92
Heating Systems, Hot Water	+ 1.57
Heat Pump	+ 1.33
Electric Heat	– 1.47
Not Heated	– 4.21

Additional upgrades or components

Kitchen Cabinets & Countertops	Page 58
Bathroom Vanities	59
Fireplaces & Chimneys	59
Windows, Skylights & Dormers	59
Appliances	60
Breezeways & Porches	60
Finished Attic	60
Garages	61
Site Improvements	61
Wings & Ells	25

- Mass produced from stock plans
- Single family — 1 full bath, 1 kitchen
- No basement
- Asphalt shingles on roof
- Hot air heat
- Gypsum wallboard interior finishes
- Materials and workmanship are sufficient to meet codes
- Detail specifications on page 19

Note: The illustration shown may contain some optional components (for example: garages and/or fireplaces) whose costs are shown in the modifications, adjustments, & alternatives below or at the end of the square foot section.

Base cost per square foot of living area

Exterior Wall	Living Area										
	600	800	1000	1200	1400	1600	1800	2000	2400	2800	3200
Wood Siding - Wood Frame	164.15	136.20	122.35	115.65	110.80	103.50	99.85	96.20	88.50	85.55	82.35
Brick Veneer - Wood Frame	173.40	142.90	128.50	121.50	116.35	108.55	104.65	100.75	92.50	89.30	85.85
Stucco on Wood Frame	152.60	127.75	114.60	108.35	103.90	97.20	93.90	90.55	83.50	80.80	77.95
Painted Concrete Block	161.30	134.10	120.40	113.90	109.15	101.95	98.40	94.85	87.30	84.35	81.25
Finished Basement, Add	25.20	21.40	20.45	19.75	19.25	18.45	18.05	17.65	16.90	16.55	16.10
Unfinished Basement, Add	12.80	9.80	9.00	8.45	8.00	7.45	7.10	6.80	6.15	5.90	5.55

Modifications

Add to the total cost

Upgrade Kitchen Cabinets	$ + 1205
Solid Surface Countertops	+ 849
Full Bath - including plumbing, wall and floor finishes	+ 5924
Half Bath - including plumbing, wall and floor finishes	+ 3465
One Car Attached Garage	+ 13,419
One Car Detached Garage	+ 17,321
Fireplace & Chimney	+ 6285

Adjustments

For multi family - add to total cost

Additional Kitchen	$ + 5288
Additional Bath	+ 5924
Additional Entry & Exit	+ 1676
Separate Heating	+ 1336
Separate Electric	+ 966

For Townhouse/Rowhouse -
Multiply cost per square foot by

Inner Unit	.95
End Unit	.97

Alternatives

Add to or deduct from the cost per square foot of living area

Composition Roll Roofing	– .80
Cedar Shake Roof	+ 2.85
Upgrade Walls and Ceilings to Skim Coat Plaster	+ .73
Upgrade Ceilings to Textured Finish	+ .58
Air Conditioning, in Heating Ductwork	+ 3.52
In Separate Ductwork	+ 6.07
Heating Systems, Hot Water	+ 1.49
Heat Pump	+ 1.46
Electric Heat	– 1.17
Not Heated	– 3.87

Additional upgrades or components

Kitchen Cabinets & Countertops	Page 58
Bathroom Vanities	59
Fireplaces & Chimneys	59
Windows, Skylights & Dormers	59
Appliances	60
Breezeways & Porches	60
Finished Attic	60
Garages	61
Site Improvements	61
Wings & Ells	25

Important: See the Reference Section for Location Factors (to adjust for your city) and Estimating Forms.

- **Mass produced from stock plans**
- **Single family — 1 full bath, 1 kitchen**
- **No basement**
- **Asphalt shingles on roof**
- **Hot air heat**
- **Gypsum wallboard interior finishes**
- **Materials and workmanship are sufficient to meet codes**
- **Detail specifications on page 19**

Note: The illustration shown may contain some optional components (for example: garages and/or fireplaces) whose costs are shown in the modifications, adjustments, & alternatives below or at the end of the square foot section.

Base cost per square foot of living area

Exterior Wall	Living Area										
	1000	1200	1400	1600	1800	2000	2200	2600	3000	3400	3800
Wood Siding - Wood Frame	124.95	113.20	107.50	103.65	99.70	95.35	92.45	87.00	81.65	79.30	77.10
Brick Veneer - Wood Frame	131.60	119.35	113.25	109.25	104.85	100.35	97.20	91.25	85.60	83.00	80.70
Stucco on Wood Frame	116.60	105.50	100.30	96.85	93.20	89.10	86.50	81.65	76.70	74.55	72.65
Painted Concrete Block	123.35	111.70	106.10	102.40	98.45	94.15	91.35	85.95	80.70	78.35	76.25
Finished Basement, Add	17.15	16.45	15.85	15.45	15.10	14.80	14.45	13.90	13.55	13.25	13.00
Unfinished Basement, Add	7.85	7.25	6.80	6.50	6.15	5.90	5.70	5.25	4.95	4.70	4.50

Modifications

Add to the total cost

Upgrade Kitchen Cabinets	$ + 1205
Solid Surface Countertops	+ 849
Full Bath - including plumbing, wall and floor finishes	+ 5924
Half Bath - including plumbing, wall and floor finishes	+ 3465
One Car Attached Garage	+ 13,419
One Car Detached Garage	+ 17,321
Fireplace & Chimney	+ 6945

Adjustments

For multi family - add to total cost

Additional Kitchen	$ + 5288
Additional Bath	+ 5924
Additional Entry & Exit	+ 1676
Separate Heating	+ 1336
Separate Electric	+ 966

For Townhouse/Rowhouse -
Multiply cost per square foot by

Inner Unit	.93
End Unit	.96

Alternatives

Add to or deduct from the cost per square foot of living area

Composition Roll Roofing	– .55
Cedar Shake Roof	+ 1.95
Upgrade Walls and Ceilings to Skim Coat Plaster	+ .74
Upgrade Ceilings to Textured Finish	+ .58
Air Conditioning, in Heating Ductwork	+ 2.87
In Separate Ductwork	+ 5.57
Heating Systems, Hot Water	+ 1.45
Heat Pump	+ 1.55
Electric Heat	– 1.02
Not Heated	– 3.66

Additional upgrades or components

Kitchen Cabinets & Countertops	Page 58
Bathroom Vanities	59
Fireplaces & Chimneys	59
Windows, Skylights & Dormers	59
Appliances	60
Breezeways & Porches	60
Finished Attic	60
Garages	61
Site Improvements	61
Wings & Ells	25

- **Mass produced from stock plans**
- **Single family — 1 full bath, 1 kitchen**
- **No basement**
- **Asphalt shingles on roof**
- **Hot air heat**
- **Gypsum wallboard interior finishes**
- **Materials and workmanship are sufficient to meet codes**
- **Detail specifications on page 19**

Note: The illustration shown may contain some optional components (for example: garages and/or fireplaces) whose costs are shown in the modifications, adjustments, & alternatives below or at the end of the square foot section.

Base cost per square foot of living area

Exterior Wall	Living Area										
	1000	1200	1400	1600	1800	2000	2200	2600	3000	3400	3800
Wood Siding - Wood Frame	115.80	104.70	99.65	96.10	92.55	88.55	85.95	81.10	76.15	74.05	72.15
Brick Veneer - Wood Frame	120.80	109.40	104.00	100.30	96.55	92.30	89.55	84.35	79.25	76.95	74.85
Stucco on Wood Frame	109.45	98.85	94.15	90.85	87.70	83.75	81.40	77.00	72.45	70.50	68.75
Painted Concrete Block	114.50	103.50	98.50	95.05	91.55	87.55	85.00	80.20	75.40	73.35	71.50
Finished Basement, Add	17.15	16.45	15.85	15.45	15.10	14.80	14.45	13.90	13.55	13.25	13.00
Unfinished Basement, Add	7.85	7.25	6.80	6.50	6.15	5.90	5.70	5.25	4.95	4.70	4.50

Modifications

Add to the total cost

Upgrade Kitchen Cabinets	$ + 1205
Solid Surface Countertops	+ 849
Full Bath - including plumbing, wall and floor finishes	+ 5924
Half Bath - including plumbing, wall and floor finishes	+ 3465
One Car Attached Garage	+ 13,419
One Car Detached Garage	+ 17,321
Fireplace & Chimney	+ 6285

Adjustments

For multi family - add to total cost

Additional Kitchen	$ + 5288
Additional Bath	+ 5924
Additional Entry & Exit	+ 1676
Separate Heating	+ 1336
Separate Electric	+ 966

For Townhouse/Rowhouse -
Multiply cost per square foot by

Inner Unit	.94
End Unit	.97

Alternatives

Add to or deduct from the cost per square foot of living area

Composition Roll Roofing	– .55
Cedar Shake Roof	+ 1.95
Upgrade Walls and Ceilings to Skim Coat Plaster	+ .70
Upgrade Ceilings to Textured Finish	+ .58
Air Conditioning, in Heating Ductwork	+ 2.87
In Separate Ductwork	+ 5.57
Heating Systems, Hot Water	+ 1.45
Heat Pump	+ 1.55
Electric Heat	– 1.02
Not Heated	– 3.66

Additional upgrades or components

Kitchen Cabinets & Countertops	Page 58
Bathroom Vanities	59
Fireplaces & Chimneys	59
Windows, Skylights & Dormers	59
Appliances	60
Breezeways & Porches	60
Finished Attic	60
Garages	61
Site Improvements	61
Wings & Ells	25

Important: See the Reference Section for Location Factors (to adjust for your city) and Estimating Forms.

- Mass produced from stock plans
- Single family — 1 full bath, 1 kitchen
- No basement
- Asphalt shingles on roof
- Hot air heat
- Gypsum wallboard interior finishes
- Materials and workmanship are sufficient to meet codes
- Detail specifications on page 19

Note: The illustration shown may contain some optional components (for example: garages and/or fireplaces) whose costs are shown in the modifications, adjustments, & alternatives below or at the end of the square foot section.

©Design Basics, Inc.

Base cost per square foot of living area

Exterior Wall	Living Area										
	1200	1500	1800	2000	2200	2400	2800	3200	3600	4000	4400
Wood Siding - Wood Frame	107.90	99.00	92.40	89.75	85.90	82.80	80.30	77.05	73.25	71.85	68.85
Brick Veneer - Wood Frame	112.60	103.30	96.25	93.40	89.40	86.10	83.50	79.95	75.95	74.45	71.35
Stucco on Wood Frame	102.00	93.65	87.55	85.10	81.50	78.70	76.35	73.40	69.85	68.55	65.80
Solid Masonry	106.65	97.90	91.35	88.75	84.95	81.90	79.50	76.25	72.45	71.15	68.25
Finished Basement, Add*	20.60	19.65	18.85	18.50	18.15	17.75	17.45	17.05	16.65	16.45	16.20
Unfinished Basement, Add*	8.60	7.90	7.25	6.95	6.70	6.40	6.15	5.80	5.50	5.35	5.15

*Basement under middle level only.

Modifications

Add to the total cost

Upgrade Kitchen Cabinets	$ + 1205
Solid Surface Countertops	+ 849
Full Bath - including plumbing, wall and floor finishes	+ 5924
Half Bath - including plumbing, wall and floor finishes	+ 3465
One Car Attached Garage	+ 13,419
One Car Detached Garage	+ 17,321
Fireplace & Chimney	+ 6285

Adjustments

For multi family - add to total cost

Additional Kitchen	$ + 5288
Additional Bath	+ 5924
Additional Entry & Exit	+ 1676
Separate Heating	+ 1336
Separate Electric	+ 966

For Townhouse/Rowhouse -
Multiply cost per square foot by

Inner Unit	.93
End Unit	.96

Alternatives

Add to or deduct from the cost per square foot of living area

Composition Roll Roofing	– .80
Cedar Shake Roof	+ 2.85
Upgrade Walls and Ceilings to Skim Coat Plaster	+ .63
Upgrade Ceilings to Textured Finish	+ .58
Air Conditioning, in Heating Ductwork	+ 2.45
In Separate Ductwork	+ 5.18
Heating Systems, Hot Water	+ 1.39
Heat Pump	+ 1.61
Electric Heat	– .88
Not Heated	– 3.55

Additional upgrades or components

Kitchen Cabinets & Countertops	Page 58
Bathroom Vanities	59
Fireplaces & Chimneys	59
Windows, Skylights & Dormers	59
Appliances	60
Breezeways & Porches	60
Finished Attic	60
Garages	61
Site Improvements	61
Wings & Ells	25

1 Story — Base cost per square foot of living area

Exterior Wall	Living Area							
	50	100	200	300	400	500	600	700
Wood Siding - Wood Frame	186.75	142.15	122.85	101.25	94.95	91.15	88.65	89.10
Brick Veneer - Wood Frame	201.60	152.75	131.70	107.20	100.25	96.15	93.30	93.65
Stucco on Wood Frame	168.20	128.90	111.80	93.90	88.35	84.95	82.75	83.50
Painted Concrete Block	183.85	140.05	121.15	100.10	93.90	90.20	87.70	88.20
Finished Basement, Add	50.30	41.05	37.15	30.70	29.45	28.70	28.15	27.80
Unfinished Basement, Add	27.95	20.75	17.75	12.70	11.70	11.10	10.70	10.40

1-1/2 Story — Base cost per square foot of living area

Exterior Wall	Living Area							
	100	200	300	400	500	600	700	800
Wood Siding - Wood Frame	147.45	118.70	101.00	89.60	84.35	81.70	78.40	77.45
Brick Veneer - Wood Frame	160.70	129.25	109.80	96.55	90.70	87.65	84.00	83.05
Stucco on Wood Frame	130.85	105.40	89.95	81.00	76.40	74.15	71.25	70.40
Painted Concrete Block	144.80	116.55	99.25	88.25	83.05	80.50	77.20	76.30
Finished Basement, Add	33.60	29.75	27.15	24.35	23.55	23.05	22.55	22.50
Unfinished Basement, Add	17.20	14.20	12.15	9.95	9.35	8.95	8.55	8.55

2 Story — Base cost per square foot of living area

Exterior Wall	Living Area							
	100	200	400	600	800	1000	1200	1400
Wood Siding - Wood Frame	150.80	111.95	95.05	77.40	71.95	68.60	66.40	67.15
Brick Veneer - Wood Frame	165.60	122.55	103.90	83.35	77.20	73.55	71.10	71.70
Stucco on Wood Frame	132.20	98.70	84.00	70.05	65.30	62.40	60.50	61.45
Painted Concrete Block	147.85	109.85	93.30	76.30	70.90	67.65	65.45	66.25
Finished Basement, Add	25.20	20.55	18.60	15.40	14.80	14.40	14.15	13.90
Unfinished Basement, Add	13.95	10.35	8.85	6.35	5.85	5.55	5.35	5.20

Base costs do not include bathroom or kitchen facilities. Use Modifications/Adjustments/Alternatives on pages 58-61 where appropriate.

1 Story

1-1/2 Story

2 Story

2-1/2 Story

Bi-Level

Tri-Level

	Components
1 Site Work	Site preparation for slab or excavation for lower level; 4' deep trench excavation for foundation wall.
2 Foundations	Continuous reinforced concrete footing, 8" deep x 18" wide; dampproofed and insulated 8" thick reinforced concrete foundation wall, 4' deep; 4" concrete slab on 4" crushed stone base and polyethylene vapor barrier, trowel finish.
3 Framing	Exterior walls—2 x 4 wood studs, 16" O.C.; 1/2" plywood sheathing; 2 x 6 rafters 16" O.C. with 1/2" plywood sheathing, 4 in 12 or 8 in 12 roof pitch, 2 x 6 ceiling joists; 1/2" plywood subfloor on 1 x 2 wood sleepers 16" O.C.; 2 x 8 floor joists with 5/8" plywood subfloor on models with more than one level.
4 Exterior Walls	Beveled wood siding and #15 felt building paper on insulated wood frame walls; double hung windows; 3 flush solid core wood exterior doors with storms. **Alternates:** • Brick veneer on wood frame, has 4" veneer of common brick. • Stucco on wood frame has 1" thick colored stucco finish. • Solid masonry has a 6" concrete block load bearing wall with insulation and a brick or stone exterior. Also solid brick or stone walls.
5 Roofing	25 year asphalt shingles; #15 felt building paper; aluminum gutters, downspouts, drip edge and flashings.
6 Interiors	Walls and ceilings—1/2" taped and finished drywall, primed and painted with 2 coats; painted baseboard and trim, finished hardwood floor 40%, carpet with 1/2" underlayment 40%, vinyl tile with 1/2" underlayment 15%, ceramic tile with 1/2" underlayment 5%; hollow core and louvered interior doors.
7 Specialties	Average grade kitchen cabinets—14 L.F. wall and base with plastic laminate counter top and kitchen sink; 40 gallon electric water heater.
8 Mechanical	1 lavatory, white, wall hung; 1 water closet, white; 1 bathtub, enameled steel, white; gas fired warm air heat.
9 Electrical	100 Amp. service; romex wiring; incandescent lighting fixtures, switches, receptacles.
10 Overhead and Profit	General Contractor overhead and profit.

Adjustments

Unfinished Basement:
7'-6" high cast-in-place concrete walls 8" thick or 8" concrete block.

Finished Basement:
Includes plywood paneling or drywall on furring on foundation walls, sponge backed carpeting on concrete floor, suspended ceiling, lighting and heating.

- **Simple design from standard plans**
- **Single family — 1 full bath, 1 kitchen**
- **No basement**
- **Asphalt shingles on roof**
- **Hot air heat**
- **Gypsum wallboard interior finishes**
- **Materials and workmanship are average**
- **Detail specifications on p. 27**

Note: The illustration shown may contain some optional components (for example: garages and/or fireplaces) whose costs are shown in the modifications, adjustments, & alternatives below or at the end of the square foot section.

eHome Planners, Inc.

Base cost per square foot of living area

Exterior Wall	Living Area										
	600	800	1000	1200	1400	1600	1800	2000	2400	2800	3200
Wood Siding - Wood Frame	164.45	147.45	135.05	125.20	116.95	111.55	108.60	105.00	97.95	92.85	89.45
Brick Veneer - Wood Frame	171.10	153.35	140.45	130.05	121.35	115.65	112.60	108.75	101.40	96.00	92.30
Stucco on Wood Frame	159.80	143.30	131.25	121.80	113.80	108.65	105.80	102.35	95.50	90.60	87.35
Solid Masonry	183.50	164.45	150.45	139.05	129.60	123.40	120.05	115.75	107.75	101.90	97.80
Finished Basement, Add	39.65	38.30	36.55	34.95	33.55	32.70	32.25	31.55	30.55	29.80	29.15
Unfinished Basement, Add	16.45	14.85	13.75	12.75	11.90	11.40	11.05	10.65	10.05	9.55	9.15

Modifications

Add to the total cost

Upgrade Kitchen Cabinets	$ + 5525
Solid Surface Countertops (Included)	
Full Bath - including plumbing, wall and floor finishes	+ 7405
Half Bath - including plumbing, wall and floor finishes	+ 4331
One Car Attached Garage	+ 14,471
One Car Detached Garage	+ 19,004
Fireplace & Chimney	+ 6627

Adjustments

For multi family - add to total cost

Additional Kitchen	$ + 9600
Additional Bath	+ 7405
Additional Entry & Exit	+ 1676
Separate Heating	+ 1336
Separate Electric	+ 1945

For Townhouse/Rowhouse -
Multiply cost per square foot by

Inner Unit	.92
End Unit	.96

Alternatives

Add to or deduct from the cost per square foot of living area

Cedar Shake Roof	+ 3.35
Clay Tile Roof	+ 6.80
Slate Roof	+ 7.50
Upgrade Walls to Skim Coat Plaster	+ .45
Upgrade Ceilings to Textured Finish	+ .58
Air Conditioning, in Heating Ductwork	+ 4.89
In Separate Ductwork	+ 7.18
Heating Systems, Hot Water	+ 1.60
Heat Pump	+ 1.38
Electric Heat	– .81
Not Heated	– 3.52

Additional upgrades or components

Kitchen Cabinets & Countertops	Page 58
Bathroom Vanities	59
Fireplaces & Chimneys	59
Windows, Skylights & Dormers	59
Appliances	60
Breezeways & Porches	60
Finished Attic	60
Garages	61
Site Improvements	61
Wings & Ells	37

- Simple design from standard plans
- Single family — 1 full bath, 1 kitchen
- No basement
- Asphalt shingles on roof
- Hot air heat
- Gypsum wallboard interior finishes
- Materials and workmanship are average
- Detail specifications on p. 27

Note: The illustration shown may contain some optional components (for example: garages and/or fireplaces) whose costs are shown in the modifications, adjustments, & alternatives below or at the end of the square foot section.

©By Designer

Base cost per square foot of living area

Exterior Wall	Living Area										
	600	800	1000	1200	1400	1600	1800	2000	2400	2800	3200
Wood Siding - Wood Frame	187.40	155.40	139.00	130.90	125.15	116.80	112.65	108.30	99.70	96.20	92.50
Brick Veneer - Wood Frame	196.85	162.35	145.35	136.90	130.85	122.00	117.55	113.00	103.80	100.10	96.05
Stucco on Wood Frame	180.55	150.45	134.40	126.60	121.10	113.10	109.05	104.95	96.75	93.40	89.90
Solid Masonry	213.85	174.65	156.75	147.55	141.00	131.20	126.25	121.30	111.10	107.05	102.50
Finished Basement, Add	32.35	28.15	27.00	26.00	25.35	24.35	23.80	23.30	22.25	21.80	21.20
Unfinished Basement, Add	14.25	11.15	10.35	9.80	9.35	8.70	8.40	8.05	7.40	7.10	6.80

Modifications

Add to the total cost

Upgrade Kitchen Cabinets	$ + 5525
Solid Surface Countertops (Included)	
Full Bath - including plumbing, wall and floor finishes	+ 7405
Half Bath - including plumbing, wall and floor finishes	+ 4331
One Car Attached Garage	+ 14,471
One Car Detached Garage	+ 19,004
Fireplace & Chimney	+ 6627

Adjustments

For multi family - add to total cost

Additional Kitchen	$ + 9600
Additional Bath	+ 7405
Additional Entry & Exit	+ 1676
Separate Heating	+ 1336
Separate Electric	+ 1945

For Townhouse/Rowhouse - Multiply cost per square foot by

Inner Unit	.92
End Unit	.96

Alternatives

Add to or deduct from the cost per square foot of living area

Cedar Shake Roof	+ 2.40
Clay Tile Roof	+ 4.95
Slate Roof	+ 5.40
Upgrade Walls to Skim Coat Plaster	+ .52
Upgrade Ceilings to Textured Finish	+ .58
Air Conditioning, in Heating Ductwork	+ 3.71
In Separate Ductwork	+ 6.28
Heating Systems, Hot Water	+ 1.52
Heat Pump	+ 1.53
Electric Heat	– .74
Not Heated	– 3.38

Additional upgrades or components

Kitchen Cabinets & Countertops	Page 58
Bathroom Vanities	59
Fireplaces & Chimneys	59
Windows, Skylights & Dormers	59
Appliances	60
Breezeways & Porches	60
Finished Attic	60
Garages	61
Site Improvements	61
Wings & Ells	37

- Simple design from standard plans
- Single family — 1 full bath, 1 kitchen
- No basement
- Asphalt shingles on roof
- Hot air heat
- Gypsum wallboard interior finishes
- Materials and workmanship are average
- Detail specifications on p. 27

Note: The illustration shown may contain some optional components (for example: garages and/or fireplaces) whose costs are shown in the modifications, adjustments, & alternatives below or at the end of the square foot section.

Base cost per square foot of living area

Exterior Wall	Living Area										
	1000	1200	1400	1600	1800	2000	2200	2600	3000	3400	3800
Wood Siding - Wood Frame	142.45	128.80	122.15	117.55	112.85	107.95	104.65	98.35	92.35	89.65	87.15
Brick Veneer - Wood Frame	149.30	135.15	128.05	123.25	118.20	113.05	109.55	102.75	96.45	93.50	90.80
Stucco on Wood Frame	137.75	124.40	118.00	113.65	109.20	104.40	101.25	95.30	89.55	86.90	84.55
Solid Masonry	163.05	147.85	139.95	134.60	128.90	123.40	119.30	111.65	104.60	101.25	98.15
Finished Basement, Add	22.70	22.45	21.70	21.15	20.65	20.30	19.85	19.15	18.60	18.25	17.95
Unfinished Basement, Add	9.00	8.40	7.90	7.60	7.25	7.00	6.75	6.35	6.00	5.75	5.60

Modifications

Add to the total cost

Upgrade Kitchen Cabinets	$ + 5525
Solid Surface Countertops (Included)	
Full Bath - including plumbing, wall and floor finishes	+ 7405
Half Bath - including plumbing, wall and floor finishes	+ 4331
One Car Attached Garage	+ 14,471
One Car Detached Garage	+ 19,004
Fireplace & Chimney	+ 7319

Adjustments

For multi family - add to total cost

Additional Kitchen	$ + 9600
Additional Bath	+ 7405
Additional Entry & Exit	+ 1676
Separate Heating	+ 1336
Separate Electric	+ 1945

*For Townhouse/Rowhouse -
Multiply cost per square foot by*

Inner Unit	.90
End Unit	.95

Alternatives

Add to or deduct from the cost per square foot of living area

Cedar Shake Roof	+ 1.65
Clay Tile Roof	+ 3.40
Slate Roof	+ 3.75
Upgrade Walls to Skim Coat Plaster	+ .53
Upgrade Ceilings to Textured Finish	+ .58
Air Conditioning, in Heating Ductwork	+ 2.96
In Separate Ductwork	+ 5.70
Heating Systems, Hot Water	+ 1.49
Heat Pump	+ 1.61
Electric Heat	- .63
Not Heated	- 3.28

Additional upgrades or components

Kitchen Cabinets & Countertops	Page 58
Bathroom Vanities	59
Fireplaces & Chimneys	59
Windows, Skylights & Dormers	59
Appliances	60
Breezeways & Porches	60
Finished Attic	60
Garages	61
Site Improvements	61
Wings & Ells	37

- **Simple design from standard plans**
- **Single family — 1 full bath, 1 kitchen**
- **No basement**
- **Asphalt shingles on roof**
- **Hot air heat**
- **Gypsum wallboard interior finishes**
- **Materials and workmanship are average**
- **Detail specifications on p. 27**

Note: The illustration shown may contain some optional components (for example: garages and/or fireplaces) whose costs are shown in the modifications, adjustments, & alternatives below or at the end of the square foot section.

Base cost per square foot of living area

Exterior Wall	Living Area										
	1200	1400	1600	1800	2000	2400	2800	3200	3600	4000	4400
Wood Siding - Wood Frame	142.15	133.10	121.45	119.05	114.50	107.40	101.85	96.20	93.35	88.20	86.55
Brick Veneer - Wood Frame	149.65	139.85	127.60	125.30	120.30	112.65	106.90	100.80	97.65	92.25	90.45
Stucco on Wood Frame	136.85	128.35	117.05	114.65	110.35	103.70	98.30	93.00	90.30	85.35	83.80
Solid Masonry	164.00	152.85	139.60	137.25	131.55	122.85	116.60	109.60	106.00	100.00	97.95
Finished Basement, Add	19.20	18.80	18.05	18.00	17.50	16.85	16.45	15.90	15.60	15.20	15.05
Unfinished Basement, Add	7.50	6.95	6.45	6.40	6.10	5.65	5.45	5.10	4.90	4.70	4.60

Modifications

Add to the total cost

Upgrade Kitchen Cabinets	$ + 5525
Solid Surface Countertops (Included)	
Full Bath - including plumbing, wall and floor finishes	+ 7405
Half Bath - including plumbing, wall and floor finishes	+ 4331
One Car Attached Garage	+ 14,471
One Car Detached Garage	+ 19,004
Fireplace & Chimney	+ 8029

Adjustments

For multi family - add to total cost

Additional Kitchen	$ + 9600
Additional Bath	+ 7405
Additional Entry & Exit	+ 1676
Separate Heating	+ 1336
Separate Electric	+ 1945

**For Townhouse/Rowhouse -
Multiply cost per square foot by**

Inner Unit	.90
End Unit	.95

Alternatives

Add to or deduct from the cost per square foot of living area

Cedar Shake Roof	+ 1.45
Clay Tile Roof	+ 2.95
Slate Roof	+ 3.25
Upgrade Walls to Skim Coat Plaster	+ .51
Upgrade Ceilings to Textured Finish	+ .58
Air Conditioning, in Heating Ductwork	+ 2.69
In Separate Ductwork	+ 5.51
Heating Systems, Hot Water	+ 1.35
Heat Pump	+ 1.65
Electric Heat	− 1.14
Not Heated	− 3.83

Additional upgrades or components

Kitchen Cabinets & Countertops	Page 58
Bathroom Vanities	59
Fireplaces & Chimneys	59
Windows, Skylights & Dormers	59
Appliances	60
Breezeways & Porches	60
Finished Attic	60
Garages	61
Site Improvements	61
Wings & Ells	37

Important: See the Reference Section for Location Factors (to adjust for your city) and Estimating Forms.

- **Simple design from standard plans**
- **Single family — 1 full bath, 1 kitchen**
- **No basement**
- **Asphalt shingles on roof**
- **Hot air heat**
- **Gypsum wallboard interior finishes**
- **Materials and workmanship are average**
- **Detail specifications on p. 27**

Note: The illustration shown may contain some optional components (for example: garages and/or fireplaces) whose costs are shown in the modifications, adjustments, & alternatives below or at the end of the square foot section.

Base cost per square foot of living area

Exterior Wall	Living Area										
	1500	1800	2100	2500	3000	3500	4000	4500	5000	5500	6000
Wood Siding - Wood Frame	129.00	116.90	111.25	106.80	98.85	95.30	90.35	85.10	83.40	81.50	79.40
Brick Veneer - Wood Frame	135.70	123.10	117.05	112.40	103.90	100.05	94.65	89.10	87.30	85.20	82.95
Stucco on Wood Frame	124.35	112.55	107.25	102.95	95.35	92.00	87.30	82.30	80.70	78.95	77.00
Solid Masonry	149.55	135.95	129.15	123.85	114.25	109.85	103.60	97.35	95.30	92.90	90.20
Finished Basement, Add	16.45	16.25	15.75	15.40	14.85	14.50	14.10	13.75	13.60	13.40	13.20
Unfinished Basement, Add	6.10	5.70	5.35	5.15	4.80	4.55	4.35	4.10	4.00	3.90	3.80

Modifications

Add to the total cost

Upgrade Kitchen Cabinets	$ + 5525
Solid Surface Countertops (Included)	
Full Bath - including plumbing, wall and floor finishes	+ 7405
Half Bath - including plumbing, wall and floor finishes	+ 4331
One Car Attached Garage	+ 14,471
One Car Detached Garage	+ 19,004
Fireplace & Chimney	+ 8029

Adjustments

For multi family - add to total cost

Additional Kitchen	$ + 9600
Additional Bath	+ 7405
Additional Entry & Exit	+ 1676
Separate Heating	+ 1336
Separate Electric	+ 1945

For Townhouse/Rowhouse -
Multiply cost per square foot by

Inner Unit	.88
End Unit	.94

Alternatives

Add to or deduct from the cost per square foot of living area

Cedar Shake Roof	+ 1.10
Clay Tile Roof	+ 2.25
Slate Roof	+ 2.50
Upgrade Walls to Skim Coat Plaster	+ .53
Upgrade Ceilings to Textured Finish	+ .58
Air Conditioning, in Heating Ductwork	+ 2.69
In Separate Ductwork	+ 5.51
Heating Systems, Hot Water	+ 1.35
Heat Pump	+ 1.65
Electric Heat	− .88
Not Heated	− 3.57

Additional upgrades or components

Kitchen Cabinets & Countertops	Page 58
Bathroom Vanities	59
Fireplaces & Chimneys	59
Windows, Skylights & Dormers	59
Appliances	60
Breezeways & Porches	60
Finished Attic	60
Garages	61
Site Improvements	61
Wings & Ells	37

- **Simple design from standard plans**
- **Single family — 1 full bath, 1 kitchen**
- **No basement**
- **Asphalt shingles on roof**
- **Hot air heat**
- **Gypsum wallboard interior finishes**
- **Materials and workmanship are average**
- **Detail specifications on p. 27**

Note: The illustration shown may contain some optional components (for example: garages and/or fireplaces) whose costs are shown in the modifications, adjustments, & alternatives below or at the end of the square foot section.

Base cost per square foot of living area

Exterior Wall	Living Area										
	1000	1200	1400	1600	1800	2000	2200	2600	3000	3400	3800
Wood Siding - Wood Frame	132.95	119.90	113.90	109.70	105.40	100.75	97.85	92.20	86.70	84.25	82.05
Brick Veneer - Wood Frame	138.15	124.75	118.40	114.00	109.50	104.70	101.55	95.55	89.80	87.20	84.80
Stucco on Wood Frame	129.30	116.55	110.75	106.70	102.60	98.00	95.20	89.85	84.55	82.20	80.10
Solid Masonry	148.40	134.20	127.25	122.40	117.45	112.35	108.85	102.20	95.90	92.95	90.30
Finished Basement, Add	22.70	22.45	21.70	21.15	20.65	20.30	19.85	19.15	18.60	18.25	17.95
Unfinished Basement, Add	9.00	8.40	7.90	7.60	7.25	7.00	6.75	6.35	6.00	5.75	5.60

Modifications

Add to the total cost

Upgrade Kitchen Cabinets	$ + 5525
Solid Surface Countertops (Included)	
Full Bath - including plumbing, wall and floor finishes	+ 7405
Half Bath - including plumbing, wall and floor finishes	+ 4331
One Car Attached Garage	+ 14,471
One Car Detached Garage	+ 19,004
Fireplace & Chimney	+ 6627

Adjustments

For multi family - add to total cost

Additional Kitchen	$ + 9600
Additional Bath	+ 7405
Additional Entry & Exit	+ 1676
Separate Heating	+ 1336
Separate Electric	+ 1945

For Townhouse/Rowhouse - Multiply cost per square foot by

Inner Unit	.91
End Unit	.96

Alternatives

Add to or deduct from the cost per square foot of living area

Cedar Shake Roof	+ 1.65
Clay Tile Roof	+ 3.40
Slate Roof	+ 3.75
Upgrade Walls to Skim Coat Plaster	+ .49
Upgrade Ceilings to Textured Finish	+ .58
Air Conditioning, in Heating Ductwork	+ 2.96
In Separate Ductwork	+ 5.70
Heating Systems, Hot Water	+ 1.49
Heat Pump	+ 1.61
Electric Heat	− .63
Not Heated	− 3.28

Additional upgrades or components

Kitchen Cabinets & Countertops	Page 58
Bathroom Vanities	59
Fireplaces & Chimneys	59
Windows, Skylights & Dormers	59
Appliances	60
Breezeways & Porches	60
Finished Attic	60
Garages	61
Site Improvements	61
Wings & Ells	37

- **Simple design from standard plans**
- **Single family — 1 full bath, 1 kitchen**
- **No basement**
- **Asphalt shingles on roof**
- **Hot air heat**
- **Gypsum wallboard interior finishes**
- **Materials and workmanship are average**
- **Detail specifications on p. 27**

Note: The illustration shown may contain some optional components (for example: garages and/or fireplaces) whose costs are shown in the modifications, adjustments, & alternatives below or at the end of the square foot section.

©Design Basics, Inc.

Base cost per square foot of living area

Exterior Wall	Living Area										
	1200	1500	1800	2100	2400	2700	3000	3400	3800	4200	4600
Wood Siding - Wood Frame	123.90	113.55	105.75	99.10	94.85	92.35	89.65	87.10	82.95	79.55	77.90
Brick Veneer - Wood Frame	128.75	117.90	109.75	102.70	98.30	95.60	92.70	90.05	85.75	82.20	80.45
Stucco on Wood Frame	120.50	110.45	102.95	96.55	92.45	90.05	87.50	85.00	81.05	77.80	76.20
Solid Masonry	138.05	126.40	117.35	109.65	104.75	101.85	98.55	95.75	91.05	87.10	85.20
Finished Basement, Add*	26.15	25.70	24.60	23.75	23.15	22.85	22.30	22.05	21.60	21.20	21.00
Unfinished Basement, Add*	10.05	9.25	8.55	8.05	7.70	7.50	7.20	7.05	6.70	6.45	6.35

*Basement under middle level only.

Modifications

Add to the total cost

Upgrade Kitchen Cabinets	$ + 5525
Solid Surface Countertops (Included)	
Full Bath - including plumbing, wall and floor finishes	+ 7405
Half Bath - including plumbing, wall and floor finishes	+ 4331
One Car Attached Garage	+ 14,471
One Car Detached Garage	+ 19,004
Fireplace & Chimney	+ 6627

Adjustments

For multi family - add to total cost

Additional Kitchen	$ + 9600
Additional Bath	+ 7405
Additional Entry & Exit	+ 1676
Separate Heating	+ 1336
Separate Electric	+ 1945

For Townhouse/Rowhouse -
Multiply cost per square foot by

Inner Unit	.90
End Unit	.95

Alternatives

Add to or deduct from the cost per square foot of living area

Cedar Shake Roof	+ 2.40
Clay Tile Roof	+ 4.95
Slate Roof	+ 5.40
Upgrade Walls to Skim Coat Plaster	+ .43
Upgrade Ceilings to Textured Finish	+ .58
Air Conditioning, in Heating Ductwork	+ 2.49
In Separate Ductwork	+ 5.37
Heating Systems, Hot Water	+ 1.44
Heat Pump	+ 1.68
Electric Heat	– .54
Not Heated	– 3.16

Additional upgrades or components

Kitchen Cabinets & Countertops	Page 58
Bathroom Vanities	59
Fireplaces & Chimneys	59
Windows, Skylights & Dormers	59
Appliances	60
Breezeways & Porches	60
Finished Attic	60
Garages	61
Site Improvements	61
Wings & Ells	37

Important: See the Reference Section for Location Factors (to adjust for your city) and Estimating Forms.

- Post and beam frame
- Log exterior walls
- Simple design from standard plans
- Single family — 1 full bath, 1 kitchen
- No basement
- Asphalt shingles on roof
- Hot air heat
- Gypsum wallborad interior finishes
- Materials and workmanship are average
- Detail specification on page 27

Note: The illustration shown may contain some optional components (for example: garages and/or fireplaces) whose costs are shown in the modifications, adjustments, & alternatives below or at the end of the square foot section.

Base cost per square foot of living area

Exterior Wall	Living Area										
	600	800	1000	1200	1400	1600	1800	2000	2400	2800	3200
6" Log - Solid Wall	185.20	167.15	153.95	143.30	134.40	128.65	125.50	121.55	114.05	108.50	104.70
8" Log - Solid Wall	173.00	156.25	144.05	134.40	126.30	121.10	118.15	114.65	107.75	102.75	99.35
Finished Basement, Add	39.65	38.30	36.55	34.95	33.55	32.70	32.25	31.55	30.55	29.80	29.15
Unfinished Basement, Add	16.45	14.85	13.75	12.75	11.90	11.40	11.05	10.65	10.05	9.55	9.15

Modifications

Add to the total cost

Upgrade Kitchen Cabinets	$ + 5525
Solid Surface Countertops (Included)	
Full Bath - including plumbing, wall and floor finishes	+ 7405
Half Bath - including plumbing, wall and floor finishes	+ 4331
One Car Attached Garage	+ 14,471
One Car Detached Garage	+ 19,004
Fireplace & Chimney	+ 6627

Adjustments

For multi family - add to total cost

Additional Kitchen	$ + 9600
Additional Bath	+ 7405
Additional Entry & Exit	+ 1676
Separate Heating	+ 1336
Separate Electric	+ 1945

For Townhouse/Rowhouse - Multiply cost per square foot by

Inner Unit	.92
End Unit	.96

Alternatives

Add to or deduct from the cost per square foot of living area

Cedar Shake Roof	+ 3.35
Air Conditioning, in Heating Ductwork	+ 4.89
In Separate Ductwork	+ 7.17
Heating Systems, Hot Water	+ 1.60
Heat Pump	+ 1.37
Electric Heat	– .85
Not Heated	– 3.52

Additional upgrades or components

Kitchen Cabinets & Countertops	Page 58
Bathroom Vanities	59
Fireplaces & Chimneys	59
Windows, Skylights & Dormers	59
Appliances	60
Breezeways & Porches	60
Finished Attic	60
Garages	61
Site Improvements	61
Wings & Ells	37

- **Post and beam frame**
- **Log exterior walls**
- **Simple design from standard plans**
- **Single family — 1 full bath, 1 kitchen**
- **No basement**
- **Asphalt shingles on roof**
- **Hot air heat**
- **Gypsum wallboard interior finishes**
- **Materials and workmanship are average**
- **Detail specification on page 27**

Note: The illustration shown may contain some optional components (for example: garages and/or fireplaces) whose costs are shown in the modifications, adjustments, & alternatives below or at the end of the square foot section.

Base cost per square foot of living area

Exterior Wall	Living Area										
	1000	1200	1400	1600	1800	2000	2200	2600	3000	3400	3800
6" Log - Solid Wall	157.80	143.40	136.15	131.25	126.05	120.85	117.15	110.25	103.80	100.80	98.00
8" Log - Solid Wall	145.20	131.70	125.25	120.80	116.25	111.40	108.25	102.10	96.25	93.65	91.20
Finished Basement, Add	22.70	22.45	21.70	21.15	20.65	20.30	19.85	19.15	18.60	18.25	17.95
Unfinished Basement, Add	9.00	8.40	7.90	7.60	7.25	7.00	6.75	6.35	6.00	5.75	5.60

Modifications

Add to the total cost

Upgrade Kitchen Cabinets	$ + 5525
Solid Surface Countertops (Included)	
Full Bath - including plumbing, wall and floor finishes	+ 7405
Half Bath - including plumbing, wall and floor finishes	+ 4331
One Car Attached Garage	+ 14,471
One Car Detached Garage	+ 19,004
Fireplace & Chimney	+ 7319

Adjustments

For multi family - add to total cost

Additional Kitchen	$ + 9600
Additional Bath	+ 7405
Additional Entry & Exit	+ 1676
Separate Heating	+ 1336
Separate Electric	+ 1945

For Townhouse/Rowhouse -
Multiply cost per square foot by

Inner Unit	.92
End Unit	.96

Alternatives

Add to or deduct from the cost per square foot of living area

Cedar Shake Roof	+ 1.65
Air Conditioning, in Heating Ductwork	+ 2.96
In Separate Ductwork	+ 5.70
Heating Systems, Hot Water	+ 1.49
Heat Pump	+ 1.61
Electric Heat	– .63
Not Heated	– 3.28

Additional upgrades or components

Kitchen Cabinets & Countertops	Page 58
Bathroom Vanities	59
Fireplaces & Chimneys	59
Windows, Skylights & Dormers	59
Appliances	60
Breezeways & Porches	60
Finished Attic	60
Garages	61
Site Improvements	61
Wings & Ells	37

1 Story   Base cost per square foot of living area

Exterior Wall	Living Area							
	50	100	200	300	400	500	600	700
Wood Siding - Wood Frame	212.80	164.10	143.05	120.15	113.20	109.10	106.30	107.05
Brick Veneer - Wood Frame	215.40	162.30	139.45	113.50	105.95	101.50	98.50	99.05
Stucco on Wood Frame	202.30	156.50	136.65	115.85	109.25	105.35	102.80	103.65
Solid Masonry	263.55	200.30	173.20	140.30	131.30	125.95	124.10	124.00
Finished Basement, Add	63.75	52.80	47.45	38.65	36.85	35.80	35.05	34.55
Unfinished Basement, Add	30.25	22.85	19.75	14.60	13.55	12.95	12.55	12.25

1-1/2 Story   Base cost per square foot of living area

Exterior Wall	Living Area							
	100	200	300	400	500	600	700	800
Wood Siding - Wood Frame	170.75	137.90	118.20	105.95	100.00	97.10	93.45	92.35
Brick Veneer - Wood Frame	226.25	169.75	141.20	123.50	114.95	110.25	105.20	103.35
Stucco on Wood Frame	203.50	151.50	126.00	111.70	104.00	99.95	95.45	93.65
Solid Masonry	257.90	195.05	162.30	140.00	130.10	124.65	118.80	116.85
Finished Basement, Add	42.95	38.75	35.25	31.35	30.30	29.60	28.95	28.80
Unfinished Basement, Add	18.95	15.85	13.80	11.55	10.90	10.50	10.15	10.05

2 Story   Base cost per square foot of living area

Exterior Wall	Living Area							
	100	200	400	600	800	1000	1200	1400
Wood Siding - Wood Frame	170.95	128.30	109.75	91.00	84.95	81.20	78.85	79.80
Brick Veneer - Wood Frame	230.15	161.10	129.80	104.40	95.85	90.75	87.35	87.60
Stucco on Wood Frame	204.60	142.85	114.60	94.25	86.75	82.20	79.25	79.85
Solid Masonry	265.60	186.50	150.95	118.50	108.50	102.55	98.60	98.50
Finished Basement, Add	33.90	28.40	25.75	21.35	20.45	19.95	19.55	19.30
Unfinished Basement, Add	15.30	11.55	10.05	7.45	6.95	6.65	6.45	6.30

Base costs do not include bathroom or kitchen facilities. Use Modifications/Adjustments/Alternatives on pages 58-61 where appropriate.

Important: See the Reference Section for Location Factors (to adjust for your city) and Estimating Forms.

1 Story

1-1/2 Story

2 Story

2-1/2 Story

Bi-Level

Tri-Level

	Components
1 Site Work	Site preparation for slab or excavation for lower level; 4' deep trench excavation for foundation wall.
2 Foundations	Continuous reinforced concrete footing, 8" deep x 18" wide; dampproofed and insulated 8" thick reinforced concrete foundation wall, 4' deep; 4" concrete slab on 4" crushed stone base and polyethylene vapor barrier, trowel finish.
3 Framing	Exterior walls—2 x 6 wood studs, 16" O.C.; 1/2" plywood sheathing; 2 x 8 rafters 16" O.C. with 1/2" plywood sheathing, 4 in 12, 6 in 12 or 8 in 12 roof pitch, 2 x 6 or 2 x 8 ceiling joists; 1/2" plywood subfloor on 1 x 3 wood sleepers 16" O.C.; 2 x 10 floor joists with 5/8" plywood subfloor on models with more than one level.
4 Exterior Walls	Horizontal beveled wood siding and #15 felt building paper on insulated wood frame walls; double hung windows; 3 flush solid core wood exterior doors with storms. **Alternates:** • Brick veneer on wood frame, has 4" veneer high quality face brick or select common brick. • Stone veneer on wood frame has exterior veneer of field stone or 2" thick limestone. • Solid masonry has an 8" concrete block wall with insulation and a brick or stone facing. It may be a solid brick or stone structure.
5 Roofing	30 year asphalt shingles; #15 felt building paper; aluminum gutters, downspouts and drip edge; copper flashings.
6 Interiors	Walls and ceilings—5/8" drywall, skim coat plaster, primed and painted with 2 coats; hardwood baseboard and trim; sanded and finished, hardwood floor 70%, ceramic tile with 1/2" underlayment 20%, vinyl tile with 1/2" underlayment 10%; wood panel interior doors, primed and painted with 2 coats.
7 Specialties	Custom grade kitchen cabinets—20 L.F. wall and base with plastic laminate counter top and kitchen sink; 4 L.F. bathroom vanity; 75 gallon electric water heater; medicine cabinet.
8 Mechanical	Gas fired warm air heat/air conditioning; one full bath including bathtub, corner shower, built in lavatory and water closet; one 1/2 bath including built in lavatory and water closet.
9 Electrical	100 Amp. service; romex wiring; incandescent lighting fixtures, switches, receptacles.
10 Overhead and Profit	General Contractor overhead and profit.

Adjustments

Unfinished Basement:
7'-6" high cast-in-place concrete walls 10" thick or 12" concrete block.

Finished Basement:
Includes painted drywall on 2 x 4 wood furring with insulation, suspended ceiling, carpeting on concrete floor, heating and lighting.

- **A distinct residence from designer's plans**
- **Single family — 1 full bath, 1 half bath, 1 kitchen**
- **No basement**
- **Asphalt shingles on roof**
- **Forced hot air heat/air conditioning**
- **Gypsum wallboard interior finishes**
- **Materials and workmanship are above average**
- **Detail specifications on page 39**

Note: The illustration shown may contain some optional components (for example: garages and/or fireplaces) whose costs are shown in the modifications, adjustments, & alternatives below or at the end of the square foot section.

©Design Basics, Inc.

Base cost per square foot of living area

Exterior Wall	Living Area										
	800	1000	1200	1400	1600	1800	2000	2400	2800	3200	3600
Wood Siding - Wood Frame	206.30	185.95	170.25	157.60	148.95	143.95	138.30	128.10	120.55	115.35	109.90
Brick Veneer - Wood Frame	215.75	194.60	178.00	164.65	155.60	150.35	144.35	133.50	125.65	120.05	114.30
Stone Veneer - Wood Frame	225.60	203.50	186.05	171.95	162.40	156.95	150.55	139.20	130.85	124.95	118.75
Solid Masonry	226.30	204.10	186.65	172.55	162.95	157.45	151.00	139.60	131.25	125.25	119.15
Finished Basement, Add	57.50	57.30	54.85	52.85	51.60	50.85	49.80	48.30	47.15	46.15	45.25
Unfinished Basement, Add	25.40	23.95	22.70	21.65	20.95	20.55	20.00	19.25	18.60	18.10	17.65

Modifications

Add to the total cost

Upgrade Kitchen Cabinets	$ + 1736
Solid Surface Countertops (Included)	
Full Bath - including plumbing, wall and floor finishes	+ 8886
Half Bath - including plumbing, wall and floor finishes	+ 5197
Two Car Attached Garage	+ 28,478
Two Car Detached Garage	+ 32,548
Fireplace & Chimney	+ 6949

Adjustments

For multi family - add to total cost

Additional Kitchen	$ + 20,958
Additional Full Bath & Half Bath	+ 14,083
Additional Entry & Exit	+ 1676
Separate Heating & Air Conditioning	+ 7087
Separate Electric	+ 1945

*For Townhouse/Rowhouse -
Multiply cost per square foot by*

Inner Unit	.90
End Unit	.95

Alternatives

Add to or deduct from the cost per square foot of living area

Cedar Shake Roof	+ 2.35
Clay Tile Roof	+ 5.85
Slate Roof	+ 6.55
Upgrade Ceilings to Textured Finish	+ .58
Air Conditioning, in Heating Ductwork	Base System
Heating Systems, Hot Water	+ 1.64
Heat Pump	+ 1.37
Electric Heat	− 2.38
Not Heated	− 4.41

Additional upgrades or components

Kitchen Cabinets & Countertops	Page 58
Bathroom Vanities	59
Fireplaces & Chimneys	59
Windows, Skylights & Dormers	59
Appliances	60
Breezeways & Porches	60
Finished Attic	60
Garages	61
Site Improvements	61
Wings & Ells	47

- **A distinct residence from designer's plans**
- **Single family — 1 full bath, 1 half bath, 1 kitchen**
- **No basement**
- **Asphalt shingles on roof**
- **Forced hot air heat/air conditioning**
- **Gypsum wallboard interior finishes**
- **Materials and workmanship are above average**
- **Detail specifications on page 39**

Note: The illustration shown may contain some optional components (for example: garages and/or fireplaces) whose costs are shown in the modifications, adjustments, & alternatives below or at the end of the square foot section.

© Donald A. Gardner Architects, Inc.

Base cost per square foot of living area

Exterior Wall	Living Area										
	1000	1200	1400	1600	1800	2000	2400	2800	3200	3600	4000
Wood Siding - Wood Frame	186.50	173.10	163.45	152.15	145.60	139.35	127.45	121.90	116.80	112.95	107.65
Brick Veneer - Wood Frame	196.80	182.70	172.60	160.45	153.50	146.80	134.10	128.15	122.55	118.55	112.85
Stone Veneer - Wood Frame	207.35	192.65	182.05	169.05	161.65	154.55	140.95	134.65	128.55	124.30	118.20
Solid Masonry	208.05	193.25	182.65	169.60	162.15	155.00	141.35	135.00	128.90	124.65	118.55
Finished Basement, Add	38.15	38.45	37.40	35.85	35.00	34.25	32.70	32.00	31.15	30.70	30.05
Unfinished Basement, Add	17.05	16.30	15.75	15.00	14.55	14.10	13.30	12.90	12.45	12.30	11.90

Modifications

Add to the total cost

Upgrade Kitchen Cabinets	$ + 1736
Solid Surface Countertops (Included)	
Full Bath - including plumbing, wall and floor finishes	+ 8886
Half Bath - including plumbing, wall and floor finishes	+ 5197
Two Car Attached Garage	+ 28,478
Two Car Detached Garage	+ 32,548
Fireplace & Chimney	+ 6949

Adjustments

For multi family - add to total cost

Additional Kitchen	$ + 20,958
Additional Full Bath & Half Bath	+ 14,083
Additional Entry & Exit	+ 1676
Separate Heating & Air Conditioning	+ 7087
Separate Electric	+ 1945

For Townhouse/Rowhouse -
Multiply cost per square foot by

Inner Unit	.90
End Unit	.95

Alternatives

Add to or deduct from the cost per square foot of living area

Cedar Shake Roof	+ 1.70
Clay Tile Roof	+ 4.25
Slate Roof	+ 4.70
Upgrade Ceilings to Textured Finish	+ .58
Air Conditioning, in Heating Ductwork	Base System
Heating Systems, Hot Water	+ 1.57
Heat Pump	+ 1.43
Electric Heat	– 2.10
Not Heated	– 4.07

Additional upgrades or components

- A distinct residence from designer's plans
- Single family — 1 full bath, 1 half bath, 1 kitchen
- No basement
- Asphalt shingles on roof
- Forced hot air heat/air conditioning
- Gypsum wallboard interior finishes
- Materials and workmanship are above average
- Detail specifications on page 39

Note: The illustration shown may contain some optional components (for example: garages and/or fireplaces) whose costs are shown in the modifications, adjustments, & alternatives below or at the end of the square foot section.

Base cost per square foot of living area

Exterior Wall	Living Area										
	1200	1400	1600	1800	2000	2400	2800	3200	3600	4000	4400
Wood Siding - Wood Frame	170.35	159.60	152.00	145.25	138.25	128.15	119.60	114.00	110.60	107.00	104.05
Brick Veneer - Wood Frame	180.50	169.15	161.05	153.80	146.50	135.55	126.35	120.30	116.75	112.75	109.60
Stone Veneer - Wood Frame	191.00	178.95	170.50	162.60	155.05	143.30	133.35	126.85	123.05	118.60	115.35
Solid Masonry	192.10	180.00	171.40	163.50	155.90	144.05	134.05	127.55	123.65	119.25	115.95
Finished Basement, Add	30.70	30.85	30.10	29.25	28.70	27.45	26.50	25.80	25.40	24.90	24.60
Unfinished Basement, Add	13.70	13.10	12.70	12.30	12.00	11.30	10.80	10.45	10.30	10.00	9.85

Modifications

Add to the total cost

Upgrade Kitchen Cabinets	$ + 1736
Solid Surface Countertops (Included)	
Full Bath - including plumbing, wall and floor finishes	+ 8886
Half Bath - including plumbing, wall and floor finishes	+ 5197
Two Car Attached Garage	+ 28,478
Two Car Detached Garage	+ 32,548
Fireplace & Chimney	+ 7842

Adjustments

For multi family - add to total cost

Additional Kitchen	$ + 20,958
Additional Full Bath & Half Bath	+ 14,083
Additional Entry & Exit	+ 1676
Separate Heating & Air Conditioning	+ 7087
Separate Electric	+ 1945

For Townhouse/Rowhouse - Multiply cost per square foot by

Inner Unit	.87
End Unit	.93

Alternatives

Add to or deduct from the cost per square foot of living area

Cedar Shake Roof	+ 1.20
Clay Tile Roof	+ 2.90
Slate Roof	+ 3.25
Upgrade Ceilings to Textured Finish	+ .58
Air Conditioning, in Heating Ductwork	Base System
Heating Systems, Hot Water	+ 1.52
Heat Pump	+ 1.60
Electric Heat	− 2.10
Not Heated	− 3.84

Additional upgrades or components

Kitchen Cabinets & Countertops	Page 58
Bathroom Vanities	59
Fireplaces & Chimneys	59
Windows, Skylights & Dormers	59
Appliances	60
Breezeways & Porches	60
Finished Attic	60
Garages	61
Site Improvements	61
Wings & Ells	47

- **A distinct residence from designer's plans**
- **Single family — 1 full bath, 1 half bath, 1 kitchen**
- **No basement**
- **Asphalt shingles on roof**
- **Forced hot air heat/air conditioning**
- **Gypsum wallboard interior finishes**
- **Materials and workmanship are above average**
- **Detail specifications on page 39**

Note: The illustration shown may contain some optional components (for example: garages and/or fireplaces) whose costs are shown in the modifications, adjustments, & alternatives below or at the end of the square foot section.

Base cost per square foot of living area

Exterior Wall	Living Area										
	1500	1800	2100	2400	2800	3200	3600	4000	4500	5000	5500
Wood Siding - Wood Frame	167.55	150.75	140.70	134.60	126.75	119.35	115.20	108.90	105.45	102.35	99.35
Brick Veneer - Wood Frame	178.45	160.75	149.65	143.05	134.85	126.70	122.15	115.35	111.60	108.20	104.85
Stone Veneer - Wood Frame	189.65	171.15	158.90	151.80	143.25	134.30	129.35	122.00	117.95	114.25	110.65
Solid Masonry	190.60	172.05	159.70	152.55	143.95	134.95	129.95	122.55	118.45	114.75	111.10
Finished Basement, Add	24.40	24.30	23.10	22.50	21.95	21.10	20.60	20.10	19.65	19.30	19.00
Unfinished Basement, Add	10.95	10.40	9.70	9.45	9.15	8.70	8.50	8.15	7.95	7.75	7.60

Modifications

Add to the total cost

Upgrade Kitchen Cabinets	$ + 1736
Solid Surface Countertops (Included)	
Full Bath - including plumbing, wall and floor finishes	+ 8886
Half Bath - including plumbing, wall and floor finishes	+ 5197
Two Car Attached Garage	+ 28,478
Two Car Detached Garage	+ 32,548
Fireplace & Chimney	+ 8856

Adjustments

For multi family - add to total cost

Additional Kitchen	$ + 20,958
Additional Full Bath & Half Bath	+ 14,083
Additional Entry & Exit	+ 1676
Separate Heating & Air Conditioning	+ 7087
Separate Electric	+ 1945

For Townhouse/Rowhouse - Multiply cost per square foot by

Inner Unit	.87
End Unit	.94

Alternatives

Add to or deduct from the cost per square foot of living area

Cedar Shake Roof	+ 1.05
Clay Tile Roof	+ 2.55
Slate Roof	+ 2.85
Upgrade Ceilings to Textured Finish	+ .58
Air Conditioning, in Heating Ductwork	Base System
Heating Systems, Hot Water	+ 1.37
Heat Pump	+ 1.65
Electric Heat	− 3.69
Not Heated	− 3.84

Additional upgrades or components

Kitchen Cabinets & Countertops	Page 58
Bathroom Vanities	59
Fireplaces & Chimneys	59
Windows, Skylights & Dormers	59
Appliances	60
Breezeways & Porches	60
Finished Attic	60
Garages	61
Site Improvements	61
Wings & Ells	47

Important: See the Reference Section for Location Factors (to adjust for your city) and Estimating Forms.

- A distinct residence from designer's plans
- Single family — 1 full bath, 1 half bath, 1 kitchen
- No basement
- Asphalt shingles on roof
- Forced hot air heat/air conditioning
- Gypsum wallboard interior finishes
- Materials and workmanship are above average
- Detail specifications on page 39

Note: The illustration shown may contain some optional components (for example: garages and/or fireplaces) whose costs are shown in the modifications, adjustments, & alternatives below or at the end of the square foot section.

Base cost per square foot of living area

Exterior Wall	Living Area										
	1500	1800	2100	2500	3000	3500	4000	4500	5000	5500	6000
Wood Siding - Wood Frame	164.40	147.95	139.65	132.60	122.30	116.95	110.65	104.15	101.75	99.15	96.50
Brick Veneer - Wood Frame	175.25	157.95	149.00	141.60	130.40	124.65	117.70	110.65	107.95	105.20	102.20
Stone Veneer - Wood Frame	186.45	168.35	158.75	150.85	138.85	132.60	124.90	117.35	114.45	111.40	108.05
Solid Masonry	187.65	169.45	159.80	151.85	139.70	133.45	125.70	118.10	115.10	112.05	108.65
Finished Basement, Add	21.35	21.35	20.55	20.00	19.15	18.60	17.90	17.40	17.15	16.90	16.60
Unfinished Basement, Add	9.65	9.15	8.70	8.45	8.00	7.70	7.35	7.10	7.00	6.80	6.65

Modifications

Add to the total cost

Upgrade Kitchen Cabinets	$ + 1736
Solid Surface Countertops (Included)	
Full Bath - including plumbing, wall and floor finishes	+ 8886
Half Bath - including plumbing, wall and floor finishes	+ 5197
Two Car Attached Garage	+ 28,478
Two Car Detached Garage	+ 32,548
Fireplace & Chimney	+ 8856

Adjustments

For multi family - add to total cost

Additional Kitchen	$ + 20,958
Additional Full Bath & Half Bath	+ 14,083
Additional Entry & Exit	+ 1676
Separate Heating & Air Conditioning	+ 7087
Separate Electric	+ 1945

For Townhouse/Rowhouse -
Multiply cost per square foot by

Inner Unit	.85
End Unit	.93

Alternatives

Add to or deduct from the cost per square foot of living area

Cedar Shake Roof	+ .80
Clay Tile Roof	+ 1.95
Slate Roof	+ 2.20
Upgrade Ceilings to Textured Finish	+ .58
Air Conditioning, in Heating Ductwork	Base System
Heating Systems, Hot Water	+ 1.37
Heat Pump	+ 1.65
Electric Heat	– 3.69
Not Heated	– 3.72

Additional upgrades or components

Kitchen Cabinets & Countertops	Page 58
Bathroom Vanities	59
Fireplaces & Chimneys	59
Windows, Skylights & Dormers	59
Appliances	60
Breezeways & Porches	60
Finished Attic	60
Garages	61
Site Improvements	61
Wings & Ells	47

- **A distinct residence from designer's plans**
- **Single family — 1 full bath, 1 half bath, 1 kitchen**
- **No basement**
- **Asphalt shingles on roof**
- **Forced hot air heat/air conditioning**
- **Gypsum wallboard interior finishes**
- **Materials and workmanship are above average**
- **Detail specifications on page 39**

Note: The illustration shown may contain some optional components (for example: garages and/or fireplaces) whose costs are shown in the modifications, adjustments, & alternatives below or at the end of the square foot section.

Base cost per square foot of living area

Exterior Wall	Living Area										
	1200	1400	1600	1800	2000	2400	2800	3200	3600	4000	4400
Wood Siding - Wood Frame	161.20	151.00	143.80	137.45	130.85	121.45	113.45	108.20	105.10	101.80	99.05
Brick Veneer - Wood Frame	168.90	158.30	150.75	144.00	137.10	127.10	118.65	113.10	109.75	106.15	103.30
Stone Veneer - Wood Frame	176.95	165.80	157.90	150.80	143.60	132.95	123.95	118.05	114.55	110.65	107.65
Solid Masonry	177.65	166.45	158.55	151.35	144.20	133.50	124.40	118.55	115.00	111.15	108.10
Finished Basement, Add	30.70	30.85	30.10	29.25	28.70	27.45	26.50	25.80	25.40	24.90	24.60
Unfinished Basement, Add	13.70	13.10	12.70	12.30	12.00	11.30	10.80	10.45	10.30	10.00	9.85

Modifications

Add to the total cost

Upgrade Kitchen Cabinets	$ + 1736
Solid Surface Countertops (Included)	
Full Bath - including plumbing, wall and floor finishes	+ 8886
Half Bath - including plumbing, wall and floor finishes	+ 5197
Two Car Attached Garage	+ 28,478
Two Car Detached Garage	+ 32,548
Fireplace & Chimney	+ 6949

Adjustments

For multi family - add to total cost

Additional Kitchen	$ + 20,958
Additional Full Bath & Half Bath	+ 14,083
Additional Entry & Exit	+ 1676
Separate Heating & Air Conditioning	+ 7087
Separate Electric	+ 1945

For Townhouse/Rowhouse -
Multiply cost per square foot by

Inner Unit	.89
End Unit	.95

Alternatives

Add to or deduct from the cost per square foot of living area

Cedar Shake Roof	+ 1.20
Clay Tile Roof	+ 2.90
Slate Roof	+ 3.25
Upgrade Ceilings to Textured Finish	+ .58
Air Conditioning, in Heating Ductwork	Base System
Heating Systems, Hot Water	+ 1.52
Heat Pump	+ 1.60
Electric Heat	– 2.10
Not Heated	– 3.72

Additional upgrades or components

Important: See the Reference Section for Location Factors (to adjust for your city) and Estimating Forms.

- **A distinct residence from designer's plans**
- **Single family — 1 full bath, 1 half bath, 1 kitchen**
- **No basement**
- **Asphalt shingles on roof**
- **Forced hot air heat/air conditioning**
- **Gypsum wallboard interior finishes**
- **Materials and workmanship are above average**
- **Detail specifications on page 39**

Note: The illustration shown may contain some optional components (for example: garages and/or fireplaces) whose costs are shown in the modifications, adjustments, & alternatives below or at the end of the square foot section.

©Design Basics, Inc.

Base cost per square foot of living area

Exterior Wall	Living Area										
	1200	1500	1800	2100	2400	2800	3200	3600	4000	4500	5000
Wood Siding - Wood Frame	166.60	150.65	138.90	129.20	122.70	117.90	112.35	106.60	104.05	98.60	95.60
Brick Veneer - Wood Frame	174.35	157.65	145.20	134.95	128.15	123.15	117.20	111.10	108.40	102.60	99.35
Stone Veneer - Wood Frame	182.40	164.95	151.80	140.95	133.70	128.50	122.15	115.75	112.95	106.75	103.25
Solid Masonry	183.05	165.60	152.35	141.50	134.25	128.95	122.60	116.10	113.30	107.10	103.60
Finished Basement, Add*	38.40	38.30	36.60	35.30	34.40	33.80	32.95	32.15	31.85	31.10	30.60
Unfinished Basement, Add*	16.90	16.00	15.15	14.45	13.95	13.65	13.25	12.80	12.60	12.25	12.00

*Basement under middle level only.

Modifications

Add to the total cost

Upgrade Kitchen Cabinets	$ + 1736
Solid Surface Countertops (Included)	
Full Bath - including plumbing, wall and floor finishes	+ 8886
Half Bath - including plumbing, wall and floor finishes	+ 5197
Two Car Attached Garage	+ 28,478
Two Car Detached Garage	+ 32,548
Fireplace & Chimney	+ 6949

Adjustments

For multi family - add to total cost

Additional Kitchen	$ + 20,958
Additional Full Bath & Half Bath	+ 14,083
Additional Entry & Exit	+ 1676
Separate Heating & Air Conditioning	+ 7087
Separate Electric	+ 1945

For Townhouse/Rowhouse -
Multiply cost per square foot by

Inner Unit	.87
End Unit	.94

Alternatives

Add to or deduct from the cost per square foot of living area

Cedar Shake Roof	+ 1.70
Clay Tile Roof	+ 4.25
Slate Roof	+ 4.70
Upgrade Ceilings to Textured Finish	+ .58
Air Conditioning, in Heating Ductwork	Base System
Heating Systems, Hot Water	+ 1.47
Heat Pump	+ 1.67
Electric Heat	− 1.87
Not Heated	− 3.72

Additional upgrades or components

Kitchen Cabinets & Countertops	Page 58
Bathroom Vanities	59
Fireplaces & Chimneys	59
Windows, Skylights & Dormers	59
Appliances	60
Breezeways & Porches	60
Finished Attic	60
Garages	61
Site Improvements	61
Wings & Ells	47

Important: See the Reference Section for Location Factors (to adjust for your city) and Estimating Forms.

1 Story — Base cost per square foot of living area

Exterior Wall	Living Area							
	50	100	200	300	400	500	600	700
Wood Siding - Wood Frame	244.80	190.45	166.90	141.20	133.45	128.80	125.75	126.60
Brick Veneer - Wood Frame	269.30	207.95	181.45	150.85	142.20	137.00	133.50	134.05
Stone Veneer - Wood Frame	294.60	226.00	196.50	160.95	151.25	145.40	141.55	141.80
Solid Masonry	300.65	230.35	200.15	163.35	153.40	147.45	143.45	143.65
Finished Basement, Add	93.20	79.40	71.85	59.30	56.80	55.25	54.25	53.55
Unfinished Basement, Add	69.05	47.80	38.00	29.25	26.90	25.50	24.55	23.90

1-1/2 Story — Base cost per square foot of living area

Exterior Wall	Living Area							
	100	200	300	400	500	600	700	800
Wood Siding - Wood Frame	194.60	159.85	138.60	125.65	119.30	116.15	112.20	111.10
Brick Veneer - Wood Frame	216.50	177.35	153.20	136.95	129.75	126.05	121.60	120.35
Stone Veneer - Wood Frame	239.10	195.40	168.25	148.70	140.65	136.30	131.25	129.95
Solid Masonry	244.45	199.75	171.90	151.55	143.20	138.75	133.55	132.20
Finished Basement, Add	62.25	57.50	52.50	47.00	45.50	44.45	43.50	43.40
Unfinished Basement, Add	41.35	31.55	26.90	22.90	21.55	20.60	19.80	19.50

2 Story — Base cost per square foot of living area

Exterior Wall	Living Area							
	100	200	400	600	800	1000	1200	1400
Wood Siding - Wood Frame	193.85	148.00	127.95	108.20	101.60	97.65	95.05	96.20
Brick Veneer - Wood Frame	218.40	165.50	142.55	117.90	110.35	105.80	102.80	103.70
Stone Veneer - Wood Frame	243.70	183.60	157.60	127.95	119.40	114.25	110.85	111.45
Solid Masonry	249.75	187.90	161.25	130.30	121.55	116.25	112.75	113.30
Finished Basement, Add	46.70	39.70	35.95	29.65	28.40	27.65	27.20	26.75
Unfinished Basement, Add	34.55	23.90	19.05	14.60	13.45	12.80	12.35	11.95

Base costs do not include bathroom or kitchen facilities. Use Modifications/Adjustments/Alternatives on pages 58-61 where appropriate.

Important: See the Reference Section for Location Factors (to adjust for your city) and Estimating Forms.

1 Story

1-1/2 Story

2 Story

2-1/2 Story

Bi-Level

Tri-Level

	Components
1 Site Work	Site preparation for slab or excavation for lower level; 4' deep trench excavation for foundation wall.
2 Foundations	Continuous reinforced concrete footing, 8" deep x 18" wide; dampproofed and insulated 12" thick reinforced concrete foundation wall, 4' deep; 4" concrete slab on 4" crushed stone base and polyethylene vapor barrier, trowel finish.
3 Framing	Exterior walls—2 x 6 wood studs, 16" O.C.; 1/2" plywood sheathing; 2 x 8 rafters 16" O.C. with 1/2" plywood sheathing, 4 in 12, 6 in 12 or 8 in 12 roof pitch, 2 x 6 or 2 x 8 ceiling joists; 1/2" plywood subfloor on 1 x 3 wood sleepers 16" O.C.; 2 x 10 floor joists with 5/8" plywood subfloor on models with more than one level.
4 Exterior Walls	Face brick veneer and #15 felt building paper on insulated wood frame walls; double hung windows; 3 flush solid core wood exterior doors with storms. **Alternates:** • Wood siding on wood frame, has top quality cedar or redwood siding or hand split cedar shingles or shakes. • Solid brick may have solid brick exterior wall or brick on concrete block. • Solid stone concrete block with selected fieldstone or limestone exterior.
5 Roofing	Red cedar shingles; #15 felt building paper; aluminum gutters, downspouts and drip edge; copper flashings.
6 Interiors	Walls and ceilings—5/8" drywall, skim coat plaster, primed and painted with 2 coats; hardwood baseboard and trim; sanded and finished, hardwood floor 70%, ceramic tile with 1/2" underlayment 20%, vinyl tile with 1/2" underlayment 10%; wood panel interior doors, primed and painted with 2 coats.
7 Specialties	Luxury grade kitchen cabinets—25 L.F. wall and base with plastic laminate counter top and kitchen sink; 6 L.F. bathroom vanity; 75 gallon electric water heater; medicine cabinet.
8 Mechanical	Gas fired warm air heat/air conditioning; one full bath including bathtub, corner shower, built in lavatory and water closet; one 1/2 bath including built in lavatory and water closet.
9 Electrical	100 Amp. service; romex wiring; incandescent lighting fixtures, switches, receptacles.
10 Overhead and Profit	General Contractor overhead and profit.

Adjustments

Unfinished Basement:
8'-0" high cast-in-place concrete walls 12" thick or 12" concrete block.

Finished Basement:
Includes painted drywall on 2 x 4 wood furring with insulation, suspended ceiling, carpeting on subfloor with sleepers, heating and lighting.

- **Unique residence built from an architect's plan**
- **Single family — 1 full bath, 1 half bath, 1 kitchen**
- **No basement**
- **Cedar shakes on roof**
- **Forced hot air heat/air conditioning**
- **Gypsum wallboard interior finishes**
- **Many special features**
- **Extraordinary materials and workmanship**
- **Detail specifications on page 49**

Note: The illustration shown may contain some optional components (for example: garages and/or fireplaces) whose costs are shown in the modifications, adjustments, & alternatives below or at the end of the square foot section.

eHome Planners, Inc.

Base cost per square foot of living area

Exterior Wall	Living Area										
	1000	1200	1400	1600	1800	2000	2400	2800	3200	3600	4000
Wood Siding - Wood Frame	220.40	201.30	186.05	175.60	169.55	162.65	150.40	141.40	135.15	128.75	123.55
Brick Veneer - Wood Frame	230.05	209.90	193.95	183.00	176.65	169.35	156.45	147.05	140.35	133.60	128.10
Solid Brick	242.05	220.75	203.80	192.20	185.60	177.70	164.10	154.05	146.85	139.65	133.75
Solid Stone	250.70	228.60	210.95	198.90	192.00	183.75	169.60	159.15	151.60	144.05	137.85
Finished Basement, Add	57.80	62.25	59.80	58.15	57.25	55.90	54.05	52.65	51.35	50.25	49.35
Unfinished Basement, Add	26.00	24.40	23.15	22.35	21.85	21.15	20.20	19.45	18.85	18.25	17.80

Modifications

Add to the total cost

Upgrade Kitchen Cabinets	$ + 2445
Solid Surface Countertops (Included)	
Full Bath - including plumbing, wall and floor finishes	+ 10,663
Half Bath - including plumbing, wall and floor finishes	+ 6237
Two Car Attached Garage	+ 32,582
Two Car Detached Garage	+ 36,973
Fireplace & Chimney	+ 10,012

Adjustments

For multi family - add to total cost

Additional Kitchen	$ + 28,496
Additional Full Bath & Half Bath	+ 16,900
Additional Entry & Exit	+ 2306
Separate Heating & Air Conditioning	+ 7087
Separate Electric	+ 1945

For Townhouse/Rowhouse -
Multiply cost per square foot by

Inner Unit	.90
End Unit	.95

Alternatives

Add to or deduct from the cost per square foot of living area

Heavyweight Asphalt Shingles	– 2.35
Clay Tile Roof	+ 3.50
Slate Roof	+ 4.15
Upgrade Ceilings to Textured Finish	+ .58
Air Conditioning, in Heating Ductwork	Base System
Heating Systems, Hot Water	+ 1.77
Heat Pump	+ 1.47
Electric Heat	– 2.10
Not Heated	– 4.80

Additional upgrades or components

Kitchen Cabinets & Countertops	Page 58
Bathroom Vanities	59
Fireplaces & Chimneys	59
Windows, Skylights & Dormers	59
Appliances	60
Breezeways & Porches	60
Finished Attic	60
Garages	61
Site Improvements	61
Wings & Ells	57

- **Unique residence built from an architect's plan**
- **Single family — 1 full bath, 1 half bath, 1 kitchen**
- **No basement**
- **Cedar shakes on roof**
- **Forced hot air heat/air conditioning**
- **Gypsum wallboard interior finishes**
- **Many special features**
- **Extraordinary materials and workmanship**
- **Detail specifications on page 49**

Note: The illustration shown may contain some optional components (for example: garages and/or fireplaces) whose costs are shown in the modifications, adjustments, & alternatives below or at the end of the square foot section.

©Larry E. Belk Designs

Base cost per square foot of living area

Exterior Wall	Living Area										
	1000	1200	1400	1600	1800	2000	2400	2800	3200	3600	4000
Wood Siding - Wood Frame	221.85	205.10	193.30	179.85	171.80	164.15	150.00	143.20	137.05	132.45	126.15
Brick Veneer - Wood Frame	233.25	215.85	203.50	189.10	180.60	172.55	157.40	150.20	143.55	138.65	131.95
Solid Brick	246.85	228.65	215.65	200.15	191.00	182.45	166.20	158.55	151.20	146.10	138.80
Solid Stone	257.65	238.75	225.30	208.90	199.30	190.30	173.20	165.10	157.30	151.95	144.30
Finished Basement, Add	40.55	44.05	42.80	40.90	39.80	38.90	36.95	36.00	35.00	34.45	33.65
Unfinished Basement, Add	18.70	17.80	17.20	16.25	15.70	15.25	14.25	13.75	13.25	12.95	12.55

Modifications

Add to the total cost

Upgrade Kitchen Cabinets	$ + 2445
Solid Surface Countertops (Included)	
Full Bath - including plumbing, wall and floor finishes	+ 10,663
Half Bath - including plumbing, wall and floor finishes	+ 6237
Two Car Attached Garage	+ 32,582
Two Car Detached Garage	+ 36,973
Fireplace & Chimney	+ 10,012

Adjustments

For multi family - add to total cost

Additional Kitchen	$ + 28,496
Additional Full Bath & Half Bath	+ 16,900
Additional Entry & Exit	+ 2306
Separate Heating & Air Conditioning	+ 7087
Separate Electric	+ 1945

For Townhouse/Rowhouse -
Multiply cost per square foot by

Inner Unit	.90
End Unit	.95

Alternatives

Add to or deduct from the cost per square foot of living area

Heavyweight Asphalt Shingles	– 1.70
Clay Tile Roof	+ 2.50
Slate Roof	+ 3
Upgrade Ceilings to Textured Finish	+ .58
Air Conditioning, in Heating Ductwork	Base System
Heating Systems, Hot Water	+ 1.69
Heat Pump	+ 1.64
Electric Heat	– 2.10
Not Heated	– 4.43

Additional upgrades or components

Important: See the Reference Section for Location Factors (to adjust for your city) and Estimating Forms.

- **Unique residence built from an architect's plan**
- **Single family — 1 full bath, 1 half bath, 1 kitchen**
- **No basement**
- **Cedar shakes on roof**
- **Forced hot air heat/air conditioning**
- **Gypsum wallboard interior finishes**
- **Many special features**
- **Extraordinary materials and workmanship**
- **Detail specifications on page 49**

Note: The illustration shown may contain some optional components (for example: garages and/or fireplaces) whose costs are shown in the modifications, adjustments, & alternatives below or at the end of the square foot section.

Base cost per square foot of living area

Exterior Wall	Living Area										
	1200	1400	1600	1800	2000	2400	2800	3200	3600	4000	4400
Wood Siding - Wood Frame	201.55	188.55	179.20	171.15	162.75	150.60	140.45	133.70	129.70	125.35	121.85
Brick Veneer - Wood Frame	212.95	199.15	189.30	180.65	171.95	158.95	147.95	140.80	136.50	131.75	128.05
Solid Brick	228.10	213.35	202.85	193.45	184.25	170.05	158.05	150.25	145.70	140.35	136.40
Solid Stone	237.60	222.25	211.40	201.45	191.95	177.00	164.40	156.15	151.35	145.70	141.55
Finished Basement, Add	32.55	35.40	34.45	33.40	32.70	31.20	29.95	29.15	28.60	28.00	27.60
Unfinished Basement, Add	15.00	14.35	13.85	13.35	13.00	12.20	11.55	11.20	10.90	10.60	10.40

Modifications

Add to the total cost

Upgrade Kitchen Cabinets	$ + 2445
Solid Surface Countertops (Included)	
Full Bath - including plumbing, wall and floor finishes	+ 10,663
Half Bath - including plumbing, wall and floor finishes	+ 6237
Two Car Attached Garage	+ 32,582
Two Car Detached Garage	+ 36,973
Fireplace & Chimney	+ 10,974

Adjustments

For multi family - add to total cost

Additional Kitchen	$ + 28,496
Additional Full Bath & Half Bath	+ 16,900
Additional Entry & Exit	+ 2306
Separate Heating & Air Conditioning	+ 7087
Separate Electric	+ 1945

For Townhouse/Rowhouse -
Multiply cost per square foot by

Inner Unit	.86
End Unit	.93

Alternatives

Add to or deduct from the cost per square foot of living area

Heavyweight Asphalt Shingles	– 1.20
Clay Tile Roof	+ 1.75
Slate Roof	+ 2.10
Upgrade Ceilings to Textured Finish	+ .58
Air Conditioning, in Heating Ductwork	Base System
Heating Systems, Hot Water	+ 1.64
Heat Pump	+ 1.73
Electric Heat	– 1.90
Not Heated	– 4.19

Additional upgrades or components

Kitchen Cabinets & Countertops	Page 58
Bathroom Vanities	59
Fireplaces & Chimneys	59
Windows, Skylights & Dormers	59
Appliances	60
Breezeways & Porches	60
Finished Attic	60
Garages	61
Site Improvements	61
Wings & Ells	57

- Unique residence built from an architect's plan
- Single family — 1 full bath, 1 half bath, 1 kitchen
- No basement
- Cedar shakes on roof
- Forced hot air heat/air conditioning
- Gypsum wallboard interior finishes
- Many special features
- Extraordinary materials and workmanship
- Detail specifications on page 49

Note: The illustration shown may contain some optional components (for example: garages and/or fireplaces) whose costs are shown in the modifications, adjustments, & alternatives below or at the end of the square foot section.

©Larry W. Garnett & Associates, Inc

Base cost per square foot of living area

Exterior Wall	Living Area										
	1500	1800	2100	2500	3000	3500	4000	4500	5000	5500	6000
Wood Siding - Wood Frame	197.10	177.35	165.15	156.25	144.25	135.60	127.45	123.35	119.75	116.15	112.30
Brick Veneer - Wood Frame	209.20	188.55	175.20	165.70	152.80	143.45	134.70	130.20	126.30	122.35	118.30
Solid Brick	224.75	202.85	188.00	177.80	163.80	153.40	143.95	138.95	134.60	130.30	125.95
Solid Stone	235.40	212.75	196.80	186.05	171.30	160.20	150.25	145.00	140.35	135.70	131.15
Finished Basement, Add	26.00	28.00	26.45	25.60	24.40	23.40	22.65	22.15	21.70	21.35	21.00
Unfinished Basement, Add	12.10	11.45	10.65	10.20	9.60	9.10	8.70	8.45	8.25	8.05	7.90

Modifications

Add to the total cost

Upgrade Kitchen Cabinets	$ + 2445
Solid Surface Countertops (Included)	
Full Bath - including plumbing, wall and floor finishes	+ 10,663
Half Bath - including plumbing, wall and floor finishes	+ 6237
Two Car Attached Garage	+ 32,582
Two Car Detached Garage	+ 36,973
Fireplace & Chimney	+ 12,011

Adjustments

For multi family - add to total cost

Additional Kitchen	$ + 28,496
Additional Full Bath & Half Bath	+ 16,900
Additional Entry & Exit	+ 2306
Separate Heating & Air Conditioning	+ 7087
Separate Electric	+ 1945

For Townhouse/Rowhouse - Multiply cost per square foot by

Inner Unit	.86
End Unit	.93

Alternatives

Add to or deduct from the cost per square foot of living area

Heavyweight Asphalt Shingles	– 1.05
Clay Tile Roof	+ 1.50
Slate Roof	+ 1.80
Upgrade Ceilings to Textured Finish	+ .58
Air Conditioning, in Heating Ductwork	Base System
Heating Systems, Hot Water	+ 1.48
Heat Pump	+ 1.78
Electric Heat	– 3.73
Not Heated	– 4.19

Additional upgrades or components

- **Unique residence built from an architect's plan**
- **Single family — 1 full bath, 1 half bath, 1 kitchen**
- **No basement**
- **Cedar shakes on roof**
- **Forced hot air heat/air conditioning**
- **Gypsum wallboard interior finishes**
- **Many special features**
- **Extraordinary materials and workmanship**
- **Detail specifications on page 49**

Note: The illustration shown may contain some optional components (for example: garages and/or fireplaces) whose costs are shown in the modifications, adjustments, & alternatives below or at the end of the square foot section.

Base cost per square foot of living area

Exterior Wall	Living Area										
	1500	1800	2100	2500	3000	3500	4000	4500	5000	5500	6000
Wood Siding - Wood Frame	193.60	174.25	164.20	155.70	143.40	137.05	129.55	122.05	119.10	116.05	112.90
Brick Veneer - Wood Frame	205.65	185.40	174.60	165.70	152.50	145.65	137.35	129.30	126.00	122.80	119.20
Solid Brick	222.20	200.75	188.90	179.35	164.90	157.40	148.05	139.15	135.60	131.95	127.85
Solid Stone	232.05	209.90	197.45	187.45	172.25	164.35	154.45	145.05	141.30	137.40	133.00
Finished Basement, Add	22.70	24.55	23.55	22.85	21.80	21.15	20.30	19.65	19.35	19.10	18.65
Unfinished Basement, Add	10.60	10.00	9.55	9.25	8.70	8.35	7.95	7.60	7.45	7.25	7.05

Modifications

Add to the total cost

Upgrade Kitchen Cabinets	$ + 2445
Solid Surface Countertops (Included)	
Full Bath - including plumbing, wall and floor finishes	+ 10,663
Half Bath - including plumbing, wall and floor finishes	+ 6237
Two Car Attached Garage	+ 32,582
Two Car Detached Garage	+ 36,973
Fireplace & Chimney	+ 12,011

Adjustments

For multi family - add to total cost

Additional Kitchen	$ + 28,496
Additional Full Bath & Half Bath	+ 16,900
Additional Entry & Exit	+ 2306
Separate Heating & Air Conditioning	+ 7087
Separate Electric	+ 1945

For Townhouse/Rowhouse -
Multiply cost per square foot by

Inner Unit	.84
End Unit	.92

Alternatives

Add to or deduct from the cost per square foot of living area

Heavyweight Asphalt Shingles	– .80
Clay Tile Roof	+ 1.15
Slate Roof	+ 1.40
Upgrade Ceilings to Textured Finish	+ .58
Air Conditioning, in Heating Ductwork	Base System
Heating Systems, Hot Water	+ 1.48
Heat Pump	+ 1.78
Electric Heat	– 3.73
Not Heated	– 4.07

Additional upgrades or components

Kitchen Cabinets & Countertops	Page 58
Bathroom Vanities	59
Fireplaces & Chimneys	59
Windows, Skylights & Dormers	59
Appliances	60
Breezeways & Porches	60
Finished Attic	60
Garages	61
Site Improvements	61
Wings & Ells	57

- **Unique residence built from an architect's plan**
- **Single family — 1 full bath, 1 half bath, 1 kitchen**
- **No basement**
- **Cedar shakes on roof**
- **Forced hot air heat/air conditioning**
- **Gypsum wallboard interior finishes**
- **Many special features**
- **Extraordinary materials and workmanship**
- **Detail specifications on page 49**

Note: The illustration shown may contain some optional components (for example: garages and/or fireplaces) whose costs are shown in the modifications, adjustments, & alternatives below or at the end of the square foot section.

Base cost per square foot of living area

Exterior Wall	Living Area										
	1200	1400	1600	1800	2000	2400	2800	3200	3600	4000	4400
Wood Siding - Wood Frame	190.85	178.55	169.65	162.15	154.10	142.85	133.35	127.05	123.30	119.30	116.05
Brick Veneer - Wood Frame	199.50	186.70	177.40	169.45	161.10	149.15	139.15	132.40	128.45	124.25	120.80
Solid Brick	210.80	197.20	187.50	178.90	170.25	157.40	146.60	139.45	135.25	130.55	126.95
Solid Stone	218.20	204.20	194.15	185.15	176.30	162.85	151.55	144.10	139.70	134.80	131.05
Finished Basement, Add	32.55	35.40	34.45	33.40	32.70	31.20	29.95	29.15	28.60	28.00	27.60
Unfinished Basement, Add	15.00	14.35	13.85	13.35	13.00	12.20	11.55	11.20	10.90	10.60	10.40

Modifications

Add to the total cost

Upgrade Kitchen Cabinets	$ + 2445
Solid Surface Countertops (Included)	
Full Bath - including plumbing, wall and floor finishes	+ 10,663
Half Bath - including plumbing, wall and floor finishes	+ 6237
Two Car Attached Garage	+ 32,582
Two Car Detached Garage	+ 36,973
Fireplace & Chimney	+ 10,012

Adjustments

For multi family - add to total cost

Additional Kitchen	$ + 28,496
Additional Full Bath & Half Bath	+ 16,900
Additional Entry & Exit	+ 2306
Separate Heating & Air Conditioning	+ 7087
Separate Electric	+ 1945

For Townhouse/Rowhouse -
Multiply cost per square foot by

Inner Unit	.89
End Unit	.94

Alternatives

Add to or deduct from the cost per square foot of living area

Heavyweight Asphalt Shingles	– 1.20
Clay Tile Roof	+ 1.75
Slate Roof	+ 2.10
Upgrade Ceilings to Textured Finish	+ .58
Air Conditioning, in Heating Ductwork	Base System
Heating Systems, Hot Water	+ 1.64
Heat Pump	+ 1.73
Electric Heat	– 1.90
Not Heated	– 4.19

Additional upgrades or components

Kitchen Cabinets & Countertops	Page 58
Bathroom Vanities	59
Fireplaces & Chimneys	59
Windows, Skylights & Dormers	59
Appliances	60
Breezeways & Porches	60
Finished Attic	60
Garages	61
Site Improvements	61
Wings & Ells	57

Important: See the Reference Section for Location Factors (to adjust for your city) and Estimating Forms.

- **Unique residence built from an architect's plan**
- **Single family — 1 full bath, 1 half bath, 1 kitchen**
- **No basement**
- **Cedar shakes on roof**
- **Forced hot air heat/air conditioning**
- **Gypsum wallboard interior finishes**
- **Many special features**
- **Extraordinary materials and workmanship**
- **Detail specifications on page 49**

Note: The illustration shown may contain some optional components (for example: garages and/or fireplaces) whose costs are shown in the modifications, adjustments, & alternatives below or at the end of the square foot section.

©Home Planners, Inc.

Base cost per square foot of living area

Exterior Wall	Living Area										
	1500	1800	2100	2400	2800	3200	3600	4000	4500	5000	5500
Wood Siding - Wood Frame	177.85	163.65	152.00	144.30	138.45	131.85	125.00	121.80	115.55	111.95	108.00
Brick Veneer - Wood Frame	185.75	170.75	158.50	150.35	144.20	137.20	130.05	126.65	120.00	116.25	112.05
Solid Brick	195.85	179.90	166.75	158.10	151.70	144.10	136.40	132.90	125.75	121.60	117.15
Solid Stone	202.70	186.15	172.50	163.45	156.80	148.85	140.80	137.15	129.65	125.35	120.60
Finished Basement, Add*	38.55	41.50	39.85	38.80	38.10	37.00	36.05	35.65	34.70	34.05	33.50
Unfinished Basement, Add*	17.30	16.30	15.40	14.85	14.50	14.00	13.50	13.25	12.80	12.45	12.20

*Basement under middle level only.

Modifications

Add to the total cost

Upgrade Kitchen Cabinets	$ + 2445
Solid Surface Countertops (Included)	
Full Bath - including plumbing, wall and floor finishes	+ 10,663
Half Bath - including plumbing, wall and floor finishes	+ 6237
Two Car Attached Garage	+ 32,582
Two Car Detached Garage	+ 36,973
Fireplace & Chimney	+ 10,012

Adjustments

For multi family - add to total cost

Additional Kitchen	$ + 28,496
Additional Full Bath & Half Bath	+ 16,900
Additional Entry & Exit	+ 2306
Separate Heating & Air Conditioning	+ 7087
Separate Electric	+ 1945

**For Townhouse/Rowhouse -
Multiply cost per square foot by**

Inner Unit	.86
End Unit	.93

Alternatives

Add to or deduct from the cost per square foot of living area

Heavyweight Asphalt Shingles	– 1.70
Clay Tile Roof	+ 2.50
Slate Roof	+ 3
Upgrade Ceilings to Textured Finish	+ .58
Air Conditioning, in Heating Ductwork	Base System
Heating Systems, Hot Water	+ 1.59
Heat Pump	+ 1.80
Electric Heat	– 1.67
Not Heated	– 4.07

Additional upgrades or components

Kitchen Cabinets & Countertops	Page 58
Bathroom Vanities	59
Fireplaces & Chimneys	59
Windows, Skylights & Dormers	59
Appliances	60
Breezeways & Porches	60
Finished Attic	60
Garages	61
Site Improvements	61
Wings & Ells	57

Important: See the Reference Section for Location Factors (to adjust for your city) and Estimating Forms.

1 Story — Base cost per square foot of living area

Exterior Wall	Living Area							
	50	100	200	300	400	500	600	700
Wood Siding - Wood Frame	270.40	209.55	183.20	154.30	145.70	140.45	137.00	137.95
Brick Veneer - Wood Frame	298.50	229.60	199.90	165.50	155.70	149.80	145.95	146.50
Solid Brick	341.85	260.60	225.70	182.70	171.15	164.25	159.70	159.80
Solid Stone	362.85	275.55	238.25	191.00	178.75	171.30	166.30	166.20
Finished Basement, Add	103.40	93.80	84.25	68.25	65.05	63.15	61.85	60.95
Unfinished Basement, Add	54.05	41.70	36.55	28.05	26.30	25.30	24.60	24.15

1-1/2 Story — Base cost per square foot of living area

Exterior Wall	Living Area							
	100	200	300	400	500	600	700	800
Wood Siding - Wood Frame	214.50	175.55	151.75	137.20	130.05	126.50	122.15	120.85
Brick Veneer - Wood Frame	239.60	195.65	168.55	150.25	142.10	137.90	132.90	131.55
Solid Brick	278.30	226.60	194.35	170.40	160.60	155.45	149.50	148.00
Solid Stone	297.05	241.60	206.85	180.15	169.60	163.90	157.55	155.95
Finished Basement, Add	68.15	67.45	60.95	54.00	52.05	50.75	49.60	49.45
Unfinished Basement, Add	34.40	29.25	25.80	22.05	21.05	20.35	19.75	19.65

2 Story — Base cost per square foot of living area

Exterior Wall	Living Area							
	100	200	400	600	800	1000	1200	1400
Wood Siding - Wood Frame	210.85	159.40	136.95	114.75	107.35	103.00	100.00	101.30
Brick Veneer - Wood Frame	238.95	179.45	153.70	125.90	117.45	112.35	108.95	109.90
Solid Brick	282.30	210.45	179.55	143.10	132.90	126.80	122.70	123.15
Solid Stone	303.30	225.50	192.05	151.40	140.45	133.75	129.30	129.55
Finished Basement, Add	51.70	47.00	42.20	34.15	32.60	31.60	30.95	30.55
Unfinished Basement, Add	27.00	20.85	18.30	14.00	13.15	12.65	12.30	12.05

Base costs do not include bathroom or kitchen facilities. Use Modifications/Adjustments/Alternatives on pages 58-61 where appropriate.

Important: See the Reference Section for Location Factors (to adjust for your city) and Estimating Forms.

Kitchen cabinets - Base units, hardwood (Cost per Unit)

	Economy	Average	Custom	Luxury
24" deep, 35" high,				
One top drawer,				
One door below				
12" wide	$ 251	$ 335	$ 446	$ 586
15" wide	263	350	466	613
18" wide	285	380	505	665
21" wide	293	390	519	683
24" wide	338	450	599	788
Four drawers				
12" wide	263	350	466	613
15" wide	270	360	479	630
18" wide	293	390	519	683
24" wide	319	425	565	744
Two top drawers,				
Two doors below				
27" wide	360	480	638	840
30" wide	394	525	698	919
33" wide	409	545	725	954
36" wide	424	565	751	989
42" wide	443	590	785	1033
48" wide	476	635	845	1111
Range or sink base				
(Cost per unit)				
Two doors below				
30" wide	334	445	592	779
33" wide	356	475	632	831
36" wide	371	495	658	866
42" wide	386	515	685	901
48" wide	409	545	725	954
Corner Base Cabinet				
(Cost per unit)				
36" wide	593	790	1051	1383
Lazy Susan *(Cost per unit)*				
With revolving door	750	1000	1330	1750

Kitchen cabinets - Wall cabinets, hardwood (Cost per Unit)

	Economy	Average	Custom	Luxury
12" deep, 2 doors				
12" high				
30" wide	$ 229	$ 305	$ 406	$ 534
36" wide	270	360	479	630
15" high				
30" wide	233	310	412	543
33" wide	285	380	505	665
36" wide	278	370	492	648
24" high				
30" wide	304	405	539	709
36" wide	334	445	592	779
42" wide	375	500	665	875
30" high, 1 door				
12" wide	225	300	399	525
15" wide	236	315	419	551
18" wide	255	340	452	595
24" wide	296	395	525	691
30" high, 2 doors				
27" wide	323	430	572	753
30" wide	338	450	599	788
36" wide	383	510	678	893
42" wide	413	550	732	963
48" wide	461	615	818	1076
Corner wall, 30" high				
24" wide	334	445	592	779
30" wide	356	475	632	831
36" wide	401	535	712	936
Broom closet				
84" high, 24" deep				
18" wide	623	830	1104	1453
Oven Cabinet				
84" high, 24" deep				
27" wide	938	1250	1663	2188

Kitchen countertops (Cost per L.F.)

	Economy	Average	Custom	Luxury
Solid Surface				
24" wide, no backsplash	$ 112	$ 149	$ 198	$ 261
with backsplash	121	161	214	282
Stock plastic laminate, 24" wide				
with backsplash	26	34	45	60
Custom plastic laminate, no splash				
7/8" thick, alum. molding	36	48	65	85
1-1/4" thick, no splash	42	55	74	97
Marble				
1/2" - 3/4" thick w/splash	54	72	96	126
Maple, laminated				
1-1/2" thick w/splash	87	116	154	203
Stainless steel				
(per S.F.)	143	190	253	333
Cutting blocks, recessed				
16" x 20" x 1" (each)	120	160	213	280

Vanity bases *(Cost per Unit)*

	Economy	Average	Custom	Luxury
2 door, 30" high, 21" deep				
24" wide	$ 293	$ 390	$ 519	$ 683
30" wide	353	470	625	823
36" wide	353	470	625	823
48" wide	454	605	805	1059

Solid surface vanity tops *(Cost Each)*

	Economy	Average	Custom	Luxury
Center bowl				
22" x 25"	$ 425	$ 459	$ 496	$ 536
22" x 31"	490	529	571	617
22" x 37"	565	610	659	712
22" x 49"	700	756	816	881

Fireplaces & Chimneys *(Cost per Unit)*

	1-1/2 Story	2 Story	3 Story
Economy (prefab metal)			
Exterior chimney & 1 fireplace	$ 6284	$ 6944	$ 7617
Interior chimney & 1 fireplace	6022	6695	7004
Average (masonry)			
Exterior chimney & 1 fireplace	6625	7320	8029
Interior chimney & 1 fireplace	6348	7057	7384
For more than 1 flue, add	451	767	1286
For more than 1 fireplace, add	4445	4445	4445
Custom (masonry)			
Exterior chimney & 1 fireplace	6947	7842	8855
Interior chimney & 1 fireplace	6514	7374	7947
For more than 1 flue, add	545	943	1579
For more than 1 fireplace, add	4990	4990	4990
Luxury (masonry)			
Exterior chimney & 1 fireplace	10012	10975	12010
Interior chimney & 1 fireplace	9552	10443	11054
For more than 1 flue, add	826	1380	1926
For more than 1 fireplace, add	7891	7891	7891

Windows and Skylights *(Cost Each)*

	Economy	Average	Custom	Luxury
Fixed Picture Windows				
3'-6" x 4'-0"	$ 364	$ 484	$ 645	$ 849
4'-0" x 6'-0"	634	845	1125	1480
5'-0" x 6'-0"	719	958	1275	1678
6'-0" x 6'-0"	761	1015	1350	1776
Bay/Bow Windows				
8'-0" x 5'-0"	874	1165	1550	2039
10'-0" x 5'-0"	930	1240	1650	2171
10'-0" x 6'-0"	1466	1954	2600	3421
12'-0" x 6'-0"	1903	2537	3375	4441
Palladian Windows				
3'-2" x 6'-4"		1484	1975	2599
4'-0" x 6'-0"		1729	2300	3026
5'-5" x 6'-10"		2255	3000	3947
8'-0" x 6'-0"		2669	3550	4671
Skylights				
46" x 21-1/2"	611	940	1140	1254
46" x 28"	627	965	1180	1298
57" x 44"	692	1065	1195	1315

Dormers *(Cost/S.F. of plan area)*

	Economy	Average	Custom	Luxury
Framing and Roofing Only				
Gable dormer, 2" x 6" roof frame	$ 32	$ 36	$ 39	$ 64
2" x 8" roof frame	33	37	40	67
Shed dormer, 2" x 6" roof frame	20	24	27	41
2" x 8" roof frame	22	25	28	42
2" x 10" roof frame	24	27	29	43

Appliances *(Cost per Unit)*

	Economy	Average	Custom	Luxury
Range				
30" free standing, 1 oven	$ 580	$ 1415	$ 1833	$ 2250
2 oven	1450	2300	2725	3150
30" built-in, 1 oven	1075	1625	1900	2175
2 oven	1700	1955	2083	2210
21" free standing				
1 oven	620	713	759	805
Counter Top Ranges				
4 burner standard	465	1245	1635	2025
As above with griddle	1750	3100	3775	4450
Microwave Oven	252	526	663	800
Compactor				
4 to 1 compaction	850	1138	1282	1425
Deep Freeze				
15 to 23 C.F.	690	870	960	1050
30 C.F.	825	1013	1107	1200
Dehumidifier, portable, auto.				
15 pint	340	391	417	442
30 pint	365	420	448	475
Washing Machine, automatic	770	1560	1955	2350
Water Heater				
Electric, glass lined				
30 gal.	1025	1238	1344	1450
80 gal.	1950	2413	2644	2875
Water Heater, Gas, glass lined				
30 gal.	1800	2188	2382	2575
50 gal.	2075	2550	2788	3025
Dishwasher, built-in				
2 cycles	490	643	719	795
4 or more cycles	630	785	1218	1650
Dryer, automatic	790	1470	1810	2150
Garage Door Opener	525	628	679	730
Garbage Disposal	166	248	289	330
Heater, Electric, built-in				
1250 watt ceiling type	249	317	351	385
1250 watt wall type	330	353	364	375
Wall type w/blower				
1500 watt	310	357	380	403
3000 watt	590	679	723	767
Hood For Range, 2 speed				
30" wide	207	666	896	1125
42" wide	297	1336	1856	2375
Humidifier, portable				
7 gal. per day	167	192	205	217
15 gal. per day	244	281	299	317
Ice Maker, automatic				
13 lb. per day	1275	1467	1563	1658
51 lb. per day	2000	2300	2450	2600
Refrigerator, no frost				
10-12 C.F.	585	655	690	725
14-16 C.F.	710	778	812	845
18-20 C.F.	895	1473	1762	2050
21-29 C.F.	1350	2900	3675	4450
Sump Pump, 1/3 H.P.	325	455	520	585

Breezeway *(Cost per S.F.)*

Class	Type	Area (S.F.)			
		50	100	150	200
Economy	Open	$ 45.56	$ 37.94	$ 33.37	$ 31.09
	Enclosed	169.12	120.38	99.08	85.31
Average	Open	53.56	43.74	38.44	35.79
	Enclosed	173.08	125.37	103.98	93.05
Custom	Open	64.49	53.03	46.74	43.60
	Enclosed	229.22	165.75	137.08	122.50
Luxury	Open	72.27	59.64	52.96	49.62
	Enclosed	271.61	193.74	158.78	141.05

Porches *(Cost per S.F.)*

Class	Type	Area (S.F.)				
		25	50	100	200	300
Economy	Open	$ 82.76	$ 62.05	$ 46.70	$ 35.93	$ 32.36
	Enclosed	179.86	142.87	94.32	71.52	62.33
Average	Open	87.04	66.68	49.01	37.09	33.13
	Enclosed	194.30	154.71	99.66	73.60	63.33
Custom	Open	147.77	106.05	81.46	60.76	59.16
	Enclosed	237.28	191.29	126.52	95.90	91.70
Luxury	Open	165.33	118.86	90.28	65.50	59.44
	Enclosed	275.50	248.06	160.22	116.92	97.01

Finished attic *(Cost per S.F.)*

Class	Area (S.F.)				
	400	500	600	800	1000
Economy	$ 22.01	$ 21.27	$ 20.39	$ 20.05	$ 19.30
Average	34.12	33.38	32.56	32.15	31.25
Custom	43.75	42.77	41.79	41.15	40.31
Luxury	56.15	54.79	53.51	52.23	51.37

Alarm system *(Cost per System)*

	Burglar Alarm	Smoke Detector
Economy	$ 425	$ 70
Average	485	81
Custom	851	199
Luxury	1500	258

Sauna, prefabricated
(Cost per unit, including heater and controls—7' high)

Size	Cost
6' x 4'	$ 5900
6' x 5'	6575
6' x 6'	7050
6' x 9'	9175
8' x 10'	11,700
8' x 12'	13,800
10' x 12'	14,000

Garages *

(Costs include exterior wall systems comparable with the quality of the residence. Included in the cost is an allowance for one personnel door, manual overhead door(s) and electrical fixture.)

Class	Type									
	Detached			Attached			Built-in		Basement	
Economy	One Car	Two Car	Three Car	One Car	Two Car	Three Car	One Car	Two Car	One Car	Two Car
Wood	$17,321	$26,477	$35,632	$13,419	$23,095	$32,251	$-1909	$-3819	$1811	$2520
Masonry	23,431	34,123	44,814	17,242	28,455	39,147	-2660	-5319		
Average										
Wood	19,004	28,582	38,161	14,471	24,571	34,150	-2116	-4232	2060	3018
Masonry	23,406	34,091	44,776	17,226	28,432	39,118	-2657	-4222		
Custom										
Wood	21,204	32,548	43,892	16,454	28,478	39,822	-3390	-3269	2955	4808
Masonry	25,570	38,012	50,454	19,186	32,307	44,749	-3926	-4341		
Luxury										
Wood	23,737	36,973	50,208	18,667	32,582	45,817	-3495	-3479	3997	6393
Masonry	30,739	45,736	60,733	23,048	38,724	53,721	-4355	-5199		

*See the Introduction to this section for definitions of garage types.

Swimming pools *(Cost per S.F.)*

Residential		
In-ground	$	39.50 - 94.00
Deck equipment		1.30
Paint pool, preparation & 3 coats (epoxy)		5.25
Rubber base paint		4.59
Pool Cover		1.86
Swimming Pool Heaters		
(not including wiring, external piping, base or pad)		
Gas		
155 MBH	$	3250.00
190 MBH		3825.00
500 MBH		13,700.
Electric		
15 KW 7200 gallon pool		2850.00
24 KW 9600 gallon pool		3325.00
54 KW 24,000 gallon pool		5200.00

Wood and coal stoves

Wood Only		
Free Standing (minimum)	$	2125
Fireplace Insert (minimum)		1849
Coal Only		
Free Standing	$	2090
Fireplace Insert		2288
Wood and Coal		
Free Standing	$	4299
Fireplace Insert		4405

Sidewalks *(Cost per S.F.)*

Concrete, 3000 psi with wire mesh	4" thick	$	3.96
	5" thick		4.82
	6" thick		5.41
Precast concrete patio blocks (natural)	2" thick		6.60
Precast concrete patio blocks (colors)	2" thick		6.95
Flagstone, bluestone	1" thick		19.60
Flagstone, bluestone	1-1/2" thick		27.25
Slate (natural, irregular)	3/4" thick		18.45
Slate (random rectangular)	1/2" thick		29.00
Seeding			
Fine grading & seeding includes lime, fertilizer & seed per S.Y.			3.03
Lawn Sprinkler System	per S.F.		1.10

Fencing *(Cost per L.F.)*

Chain Link, 4' high, galvanized	$	15.60
Gate, 4' high (each)		212.00
Cedar Picket, 3' high, 2 rail		15.15
Gate (each)		234.00
3 Rail, 4' high		18.80
Gate (each)		252.00
Cedar Stockade, 3 Rail, 6' high		18.25
Gate (each)		255.00
Board & Battens, 2 sides 6' high, pine		28.00
6' high, cedar		36.00
No. 1 Cedar, basketweave, 6' high		37.00
Gate, 6' high (each)		288.00

Carport *(Cost per S.F.)*

Economy	$	10.07
Average		15.07
Custom		22.15
Luxury		25.89

Insurance Exclusions

Insurance exclusions are a matter of policy coverage and are not standard or universal. When making an appraisal for insurance purposes, it is recommended that some time be taken in studying the insurance policy to determine the specific items to be excluded. Most homeowners insurance policies have, as part of their provisions, statements that read "the policy permits the insured to exclude from value, in determining whether the amount of the insurance equals or exceeds 80% of its replacement cost, such items as building excavation, basement footings, underground piping and wiring, piers and other supports which are below the undersurface of the lowest floor or where there is no basement below the ground." Loss to any of these items, however, is covered.

Costs for Excavation, Spread and Strip Footings and Underground Piping

This chart shows excluded items expressed as a percentage of total building cost.

Class	Number of Stories			
	1	1-1/2	2	3
Luxury	3.4%	2.7%	2.4%	1.9%
Custom	3.5%	2.7%	2.4%	1.9%
Average	5.0%	4.3%	3.8%	3.3%
Economy	5.0%	4.4%	4.1%	3.5%

Architect/Designer Fees

Architect/designer fees as presented in the following chart are typical ranges for 4 classes of residences. Factors affecting these ranges include economic conditions, size and scope of project, site selection, and standardization. Where superior quality and detail is required or where closer supervision is required, fees may run higher than listed. Lower quality or simplicity may dictate lower fees.

The editors have included architect/designer fees from the "mean" column in calculating costs.

Listed below are average costs expressed as a range and mean of total building cost for 4 classes of residences.

Class		Range	Mean
Luxury	— Architecturally designed and supervised	5%–15%	10%
Custom	— Architecturally modified designer plans	$2000–$3500	$2500
Average	— Designer plans	$800–$1500	$1100
Economy	— Stock plans	$100–$500	$300

Depreciation as generally defined is "the loss of value due to any cause." Specifically, depreciation can be broken down into three categories: **Physical Depreciation, Functional Obsolescence,** and **Economic Obsolescence.**

Physical Depreciation is the loss of value from the wearing out of the building's components. Such causes may be decay, dry rot, cracks, or structural defects.

Functional Obsolescence is the loss of value from features which render the residence less useful or desirable. Such features may be higher than normal ceilings, design and/or style changes, or outdated mechanical systems.

Economic Obsolescence is the loss of value from external features which render the residence less useful or desirable. These features may be changes in neighborhood socio-economic grouping, zoning changes, legislation, etc.

Depreciation as it applies to the residential portion of this manual deals with the observed physical condition of the residence being appraised. It is in essence *"the cost to cure."*

The ultimate method to arrive at *"the cost to cure"* is to analyze each component of the residence.

For example:

    Component, Roof Covering – Cost = $2400
    Average Life = 15 years
    Actual Life = 5 years
    Depreciation = 5 ÷ 15 = .33 or 33%

    $2400 × 33% = $792 = Amount of Depreciation

The following table, however, can be used as a guide in estimating the % depreciation using the actual age and general condition of the residence.

Depreciation Table — Residential

Age in Years	Good	Average	Poor
2	2%	3%	10%
5	4	6	20
10	7	10	25
15	10	15	30
20	15	20	35
25	18	25	40
30	24	30	45
35	28	35	50
40	32	40	55
45	36	45	60
50	40	50	65

Commercial/Industrial/ Institutional Models

Table of Contents

Introduction to the Commercial/ Industrial/Institutional Section

General

The Commercial/Industrial/Institutional section contains base building costs per square foot of floor area for 77 model buildings. Each model has a table of square foot costs for combinations of exterior wall and framing systems. This table is supplemented by a list of common additives and their unit costs. In printed versions, a breakdown of all componenet costs used to develop the base cost for the model is included. In electronic products, the cost breakdown is available as a printable report. Modifications to the standard models can be performed manually in printed products and automatically using the "swapper" feature in electronic products.

This data may be used directly to estimate the construction cost of most types of buildings knowing only the floor area, exterior wall construction and framing systems. To adjust the base cost for components which are different than the model, use the tables from the Assemblies Section.

Building Identification & Model Selection

The building models in this section represent structures by use. Occupancy, however, does not necessarily identify the building, i.e., a restaurant could be a converted warehouse. In all instances, the building should be described and identified by its own physical characteristics. The model selection should also be guided by comparing specifications with the model. In the case of converted use, data from one model may be used to supplement data from another.

Green Models

Consistent with expanding green trends in the design and construction industry, RSMeans introduced 25 new green building models. Although similar to our standard models in building type and structural system, the new green models meet or exceed Energy Star requirements and address many of the items necessary to obtain LEED certification. Although our models do not include site-specific information, our design assumption is that they are located in climate zone 5. DOE's eQuest software was used to perform an energy analysis of the model buildings. By reducing energy use, we were able to reduce the size of the service entrances, switchgear, power feeds, and generators. Each of the following building systems was researched and analyzed: building envelope, HVAC, plumbing fixtures, lighting, and electrical service. These systems were targeted because of their impact on energy usage and green building.

Wall and roof insulation was increased to reduce heat loss and a recycled vapor barrier was added to the foundation. White roofs were specified for models with flat roofs. Interior finishes include materials with recycled content and low VOC paint. Stainless steel toilet partitions were selected because it is a recycled material and is easy to maintain. Plumbing fixtures were selected to conserve water. Water closets are low-flow and equipped with auto-sensor valves. Waterless urinals are specified throughout. Faucets use auto-sensor flush valves and are powered by a hydroelectric power unit inside the faucet. Energy efficient, low flow water coolers were specified to reduce energy consumption and water usage throughout each building model.

Lighting efficiency was achieved by specifying LED fixtures in place of standard fluorescents. Daylight on/off lighting control systems were incorporated into the models. These controls are equipped with sensors that automatically turn off the lights when sufficient daylight is available. Energy monitoring systems were also included in all green models.

Adjustments

The base cost tables represent the base cost per square foot of floor area for buildings without a basement and without unusual special features. Basement costs and other common additives are available. Cost adjustments can also be made to the model by using the tables from the Assemblies Section. .

Dimensions

All base cost tables are developed so that measurement can be readily made during the inspection process. Areas are calculated from exterior dimensions and story heights are measured from the top surface of one floor to the top surface of the floor above. Roof areas are measured by horizontal area covered and costs related to inclines are converted with appropriate factors. The precision of measurement is a matter of the user's choice and discretion. For ease in calculation, consideration should be given to measuring in tenths of a foot, i.e., 9 ft. 6 in. = 9.5 ft., 9 ft. 4 in. = 9.3 ft.

Floor Area

The term "Floor Area" as used in this section includes the sum of floor plate at grade level and above. This dimension is measured from the outside face of the foundation wall. Basement costs are calculated separately. The user must exercise his/her own judgment, where the lowest level floor is slightly below grade, whether to consider it at grade level or make the basement adjustment.

How to Use the Commercial/Industrial/Institutional Section

The following is a detailed explanation of a sample entry in the Commercial/Industrial/Institutional Square Foot Cost Section. Each bold number below corresponds to the described item on the following page with the appropriate component or cost of the sample entry following in parentheses.

Prices listed are costs that include overhead and profit of the installing contractor and additional markups for General Conditions and Architects' Fees.

COMMERCIAL/INDUSTRIAL/INSTITUTIONAL	M.010 ❶	Apartment, 1-3 Story ❷

Costs per square foot of floor area

Exterior Wall ❸	S.F. Area	8000	12000	15000	19000	22500	25000 ❹	29000	32000	36000
	L.F. Perimeter	213	280	330	350	400	433	442	480	520
Fiber Cement Siding	Wood Frame	179.90	172.20	169.10	163.65	162.10	161.15	158.30	157.65	156.70
Stone Veneer	Wood Frame	199.65	189.55	185.50	177.40	175.40 ❺	174.15	169.75	168.95	167.55
Brick Veneer	Steel Frame	180.70	171.55	167.95	160.90	159.10	157.95	154.15	153.45	152.25
EIFS on Metal Studs	Steel Frame	162.80	156.55	154.15	150.25	149.00	148.25	146.30	145.75	145.05
Brick Veneer	Reinforced Concrete	198.60	187.90	183.65	174.90	172.80	171.45	166.70	165.80	164.35
Stucco on Concrete Block	Reinforced Concrete	187.20	177.95	174.25	167.10	165.20	164.05	160.20	159.40	158.15
Perimeter Adj., Add or Deduct ❻	Per 100 L.F.	14.50	9.70	7.70	6.15	5.15 ❼	4.70	3.95	3.65	3.20
Story Hgt. Adj., Add or Deduct	Per 1 Ft.	2.25	1.95	1.85	1.55	1.50	1.45	1.30	1.25	1.20
❽ For Basement, add $ 37.40 per square foot of basement area										

The above costs were calculated using the basic specifications shown on the facing page. These costs should be adjusted where necessary for design alternatives and owner's requirements. Reported completed project costs, for this type of structure, range from $ 63.55 to $ 236.40 per S.F. ❾

Common additives ❿

Description	Unit	$ Cost
Appliances		
Cooking range, 30" free standing		
1 oven	Ea.	610 - 2300
2 oven	Ea.	1475 - 3175
30" built-in		
1 oven	Ea.	1100 - 2325
2 oven	Ea.	1775 - 3375
Counter top cook tops, 4 burner	Ea.	495 - 2100
Microwave oven	Ea.	296 - 890
Combination range, refrig. & sink, 30" wide	Ea.	1900 - 2750
72" wide	Ea.	2800
Combination range, refrigerator, sink,		
microwave oven & icemaker	Ea.	6725
Compactor, residential, 4-1 compaction	Ea.	880 - 1475
Dishwasher, built-in, 2 cycles	Ea.	675 - 1175
4 cycles	Ea.	815 - 2000
Garbage disposer, sink type	Ea.	239 - 405
Hood for range, 2 speed, vented, 30" wide	Ea.	345 - 1375
42" wide	Ea.	435 - 2600
Refrigerator, no frost 10-12 C.F.	Ea.	615 - 770
18-20 C.F.	Ea.	925 - 2100

Description	Unit	$ Cost
Closed Circuit Surveillance, One station		
Camera and monitor	Ea.	2025
For additional camera stations, add	Ea.	1100
Elevators, Hydraulic passenger, 2 stops		
2000# capacity	Ea.	68,400
2500# capacity	Ea.	70,400
3500# capacity	Ea.	75,400
Additional stop, add	Ea.	8075
Emergency Lighting, 25 watt, battery operated		
Lead battery	Ea.	345
Nickel cadmium	Ea.	685
Laundry Equipment		
Dryer, gas, 16 lb. capacity	Ea.	995
30 lb. capacity	Ea.	3950
Washer, 4 cycle	Ea.	1225
Commercial	Ea.	1625
Smoke Detectors		
Ceiling type	Ea.	227
Duct type	Ea.	565

1 Model Number (M.010)

"M" distinguishes this section of the data and stands for model. The number designation is a sequential number. Our newest models are green and are designated with the letter "G" followed by the building type number.

2 Type of Building (Apartment, 1–3 Story)

There are 50 different types of commercial/industrial/institutional buildings highlighted in this section.

3 Exterior Wall Construction and Building Framing Options (Stone Veneer with Wood Frame)

Three or more commonly used exterior walls and, in most cases, two typical building framing systems are presented for each type of building. The model selected should be based on the actual characteristics of the building being estimated.

4 Total Square Foot of Floor Area and Base Perimeter Used to Compute Base Costs (22,500 Square Feet and 400 Linear Feet)

Square foot of floor area is the total gross area of all floors at grade, and above, and does not include a basement. The perimeter in linear feet used for the base cost is generally for a rectangular, economical building shape.

5 Cost per Square Foot of Floor Area ($175.40)

The highlighted cost is for a building of the selected exterior wall and framing system and floor area. Costs for buildings with floor areas other than those calculated may be interpolated between the costs shown.

6 Building Perimeter and Story Height Adjustments

Square-foot costs for a building with a perimeter or floor-to-floor story height significantly different from the model used to calculate the base cost may be adjusted (add or deduct) to reflect the actual building geometry.

7 Cost per Square Foot of Floor Area for the Perimeter and/or Height Adjustment ($5.15 for Perimeter Difference and $1.50 for Story Height Difference)

Add (or deduct) $5.15 to the base square-foot cost for each 100 feet of perimeter difference between the model and the actual building. Add (or deduct) $1.50 to the base square-foot cost for each 1 foot of story height difference between the model and the actual building.

8 Optional Cost per Square Foot of Basement Floor Area ($37.40)

The cost of an unfinished basement for the building being estimated is $37.40 times the gross floor area of the basement.

9 Range of Cost per Square Foot of Floor Area for Similar Buildings ($63.55 to $236.40)

Many different buildings of the same type have been built using similar materials and systems. RSMeans historical cost data of actual construction projects indicates a range of $63.55 to $236.40 for this type of building.

10 Common Additives

Common components and/or systems used in this type of building are listed. These costs should be added to the total building cost. Additional selections may be found in the Assemblies Section.

How to Use the Commercial/Industrial/Institutional Section (Cont.)

The following is a detailed explanation of the specification and costs for a model building in the Commercial/Industrial/Institutional Square Foot Cost Section. Each bold number below corresponds to the described item on the following page with the appropriate component of the sample entry following in parentheses.

Prices listed are costs that include overhead and profit of the installing contractor.

Model costs calculated for a 3 story building with 10' story height and 22,500 square feet of floor area ❶

❷ **Apartment, 1-3 Story**

				Unit	Unit Cost	Cost Per S.F.	% Of Sub-Total
A. SUBSTRUCTURE							
1010	Standard Foundations	Poured concrete; strip and spread footings		S.F. Ground	5.13	1.71	
1020	Special Foundations	N/A		—	—	—	
1030	Slab on Grade	4" reinforced concrete		S.F. Slab	5.56	1.85	4.0%
2010	Basement Excavation	Site preparation for slab and trench for foundation wall and footing		S.F. Ground	.32	.11	
2020	Basement Walls	4' Foundation wall		L.F. Wall	42.75	1.52	
B. SHELL							
B10 Superstructure							
1010	Floor Construction	Wood Beam and joist on wood columns, fireproofed		S.F. Floor	24.83	16.55	14.5%
1020	Roof Construction	Wood roof, truss, 4/12 slope		S.F. Roof	7.41	2.47	
B20 Exterior Enclosure							
2010	Exterior Walls	Ashlar stone veneer on wood stud back up, insulated	80% of wall	S.F. Wall	39.12	16.69	
2020	Exterior Windows	Aluminum horizontal sliding	20% of wall	Each	525	3.74	15.9%
2030	Exterior Doors	Steel door, hollow metal with frame		Each	2025	.36	
B30 Roofing							
3010	Roof Coverings	Asphalt roofing, strip shingles, flashing		S.F. Roof	7.47	2.49	1.9%
3020	Roof Openings	N/A		—	—	—	
C. INTERIORS ❸		❹					
1010	Partitions	Gypsum board on wood studs	10 S.F. Floor/L.F. Partition	S.F. Partition	7.37	6.55	
1020	Interior Doors	Single leaf solid and hollow core, fire doors	130 S.F. Floor/Door ❺ ❻	Each ❼	632 ❽	.79	
1030	Fittings	Residential wall and base wood cabinets, laminate counters		S.F. Floor	4.33	4.33	
2010	Stair Construction	Wood Stairs with wood rails		Flight	2745	.73	21.6% ❾
3010	Wall Finishes	95% paint, 5% ceramic wall tile		S.F. Surface	1.39	2.47	
3020	Floor Finishes	80% Carpet tile, 10% vinyl composition tile, 10% ceramic tile		S.F. Floor	5.11	5.11	
3030	Ceiling Finishes	Painted gypsum board ceiling on resilient channels		S.F. Ceiling	4.37	4.37	
D. SERVICES							
D10 Conveying							
1010	Elevators & Lifts	Hydraulic passenger elevator		Each	111,600	4.96	3.8%
1020	Escalators & Moving Walks	N/A		—	—	—	
D20 Plumbing							
2010	Plumbing Fixtures	Kitchen, bath, laundry and service fixtures, supply and drainage	1 Fixture/285 S.F. Floor	Each	1918	6.73	
2020	Domestic Water Distribution	Electric water heater		S.F. Floor	7.65	7.65	11.4%
2040	Rain Water Drainage	Roof drains		S.F. Roof	1.53	.51	
D30 HVAC							
3010	Energy Supply	Oil fired hot water, baseboard radiation		S.F. Floor	8.93	8.93	
3020	Heat Generating Systems	N/A		—	—	—	
3030	Cooling Generating Systems	Chilled water, air cooled condenser system		S.F. Floor	9.27	9.27	13.9%
3050	Terminal & Package Units	N/A		—	—	—	
3090	Other HVAC Sys. & Equipment	N/A		—	—	—	
D40 Fire Protection							
4010	Sprinklers	Wet pipe spinkler system, light hazard		S.F. Floor	3.58	3.58	2.7%
4020	Standpipes	N/A		—	—	—	
D50 Electrical							
5010	Electrical Service/Distribution	800 ampere service, panel board and feeders		S.F. Floor	2.67	2.67	
5020	Lighting & Branch Wiring	Incandescent fixtures, receptacles, switches, A.C. and misc. power		S.F. Floor	7.64	7.64	9.2%
5030	Communications & Security	Addressable alarm systems, emergency lighting, internet and phone wiring		S.F. Floor	1.78	1.78	
5090	Other Electrical Systems	N/A		—	—	—	
E. EQUIPMENT & FURNISHINGS							
1010	Commercial Equipment	N/A		—	—	—	
1020	Institutional Equipment	N/A		—	—	—	
1030	Vehicular Equipment	N/A		—	—	—	1.2%
1090	Other Equipment	Residential freestanding gas ranges, dishwashers		S.F. Floor	1.57	1.57	
F. SPECIAL CONSTRUCTION							
1020	Integrated Construction	N/A		—	—	—	0.0%
1040	Special Facilities	N/A		—	—	—	
G. BUILDING SITEWORK	**N/A**						

❿ **Sub-Total**		131.13	100%
CONTRACTOR FEES (General Requirements: 10%, Overhead: 5%, Profit: 10%) ⓫		25%	32.80
ARCHITECT FEES		7%	11.47
⓬ **Total Building Cost**		**175.40**	

70

1 Building Description (Model costs are calculated for a 3-story apartment building with 10' story height and 22,500 square feet of floor area)
The model highlighted is described in terms of building type, number of stories, typical story height, and square footage.

2 Type of Building
(Apartment, 1–3 Story)

3 Division C Interiors
(C1020 Interior Doors)
System costs are presented in divisions according to the 7-element UNIFORMAT II classifications. Each of the component systems is listed.

4 Specification Highlights
(Single leaf solid and hollow core, fire doors)
All systems in each subdivision are described with the material and proportions used.

5 Quantity Criteria (130 S.F. Floor/Door)
The criteria used in determining quantities for the calculations are shown.

6 Unit (Each)
The unit of measure shown in this column is the unit of measure of the particular system shown that corresponds to the unit cost.

7 Unit Cost ($632)
The cost per unit of measure of each system subdivision.

8 Cost per Square Foot ($4.79)
The cost per square foot for each system is the unit cost of the system times the total number of units, divided by the total square feet of building area.

9 % of Sub-Total (21.6%)
The percent of sub-total is the total cost per square foot of all systems in the division divided by the sub-total cost per square foot of the building.

10 Sub-Total ($131.13)
The sub-total is the total of all the system costs per square foot.

11 Project Fees
(Contractor Fees) (25%);
(Architects' Fees) (7%)
Contractor Fees to cover the general requirements, overhead, and profit of the General Contractor are added as a percentage of the sub-total. An Architect's Fee, also as a percentage of the sub-total, is also added. These values vary with the building type.

12 Total Building Cost ($175.40)
The total building cost per square foot of building area is the sum of the square foot costs of all the systems, the General Contractor's general requirements, overhead and profit, and the Architect's fee. The total building cost is the amount which appears shaded in the Cost per Square Foot of Floor Area table shown previously.

Examples

Example 1

This example illustrates the use of the base cost tables. The base cost is adjusted for different exterior wall systems, different story height and a partial basement.

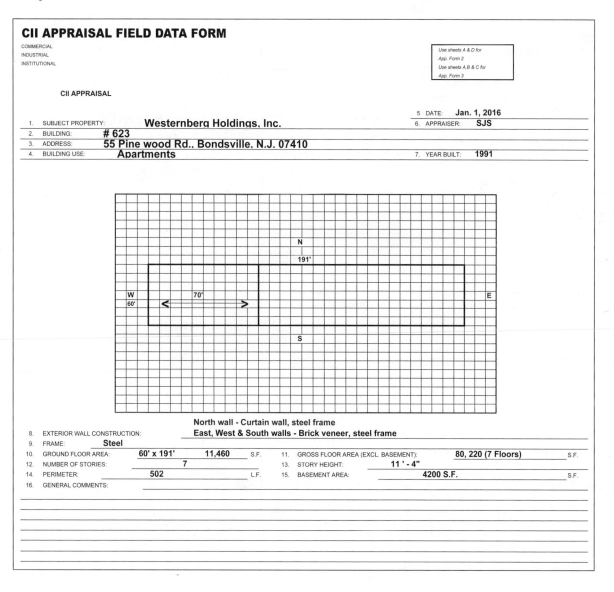

CII APPRAISAL FIELD DATA FORM

COMMERCIAL
INDUSTRIAL
INSTITUTIONAL

Use sheets A & D for
App. Form 2
Use sheets A,B & C for
App. Form 3

CII APPRAISAL

			5 DATE:	**Jan. 1, 2016**
1.	SUBJECT PROPERTY:	**Westernberg Holdings, Inc.**	6. APPRAISER:	**SJS**
2.	BUILDING:	**# 623**		
3.	ADDRESS:	**55 Pine wood Rd., Bondsville, N.J. 07410**		
4.	BUILDING USE:	**Apartments**	7. YEAR BUILT:	**1991**

North wall - Curtain wall, steel frame

8. EXTERIOR WALL CONSTRUCTION: **East, West & South walls - Brick veneer, steel frame**

9. FRAME: **Steel**

10. GROUND FLOOR AREA:	**60' x 191'**	**11,460**	S.F.	11. GROSS FLOOR AREA (EXCL. BASEMENT):	**80, 220 (7 Floors)**	S.F.
12. NUMBER OF STORIES:	**7**			13. STORY HEIGHT:	**11 ' - 4"**	
14. PERIMETER:	**502**		L.F.	15. BASEMENT AREA:	**4200 S.F.**	S.F.

16. GENERAL COMMENTS:

Example 1 (continued)

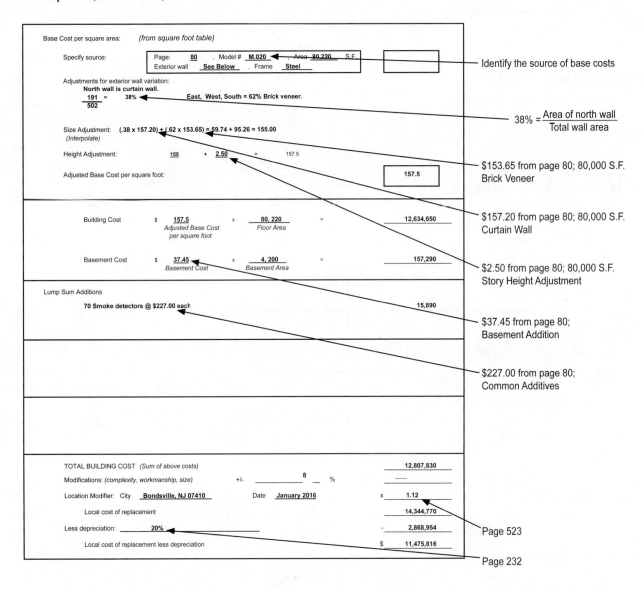

Base Cost per square area: *(from square foot table)*

Specify source: Page: **80** , Model # **M.020** , Area **80,220** S.F.
Exterior wall **See Below** , Frame **Steel**

— Identify the source of base costs

Adjustments for exterior wall variation:
North wall is curtain wall.

$\frac{191}{502}$ = **38%** ← **East, West, South = 62% Brick veneer.**

38% = $\dfrac{\text{Area of north wall}}{\text{Total wall area}}$

Size Adjustment: **(.38 x 157.20) + (.62 x 153.65) = 59.74 + 95.26 = 155.00**
(Interpolate)

Height Adjustment: **155** + **2.50** = 157.5

$153.65 from page 80; 80,000 S.F.
Brick Veneer

Adjusted Base Cost per square foot: **157.5**

$157.20 from page 80; 80,000 S.F.
Curtain Wall

Building Cost $ **157.5** x **80, 220** = 12,634,650
Adjusted Base Cost *Floor Area*
per square foot

$2.50 from page 80; 80,000 S.F.
Story Height Adjustment

Basement Cost $ **37.45** x **4, 200** = 157,290
Basement Cost *Basement Area*

$37.45 from page 80;
Basement Addition

Lump Sum Additions

70 Smoke detectors @ $227.00 each 15,890

$227.00 from page 80;
Common Additives

TOTAL BUILDING COST *(Sum of above costs)* 12,807,830

Modifications: *(complexity, workmanship, size)* +/- **8** % —

Location Modifier: City **Bondsville, NJ 07410** Date **January 2016** x **1.12**

Local cost of replacement 14,344,770

— Page 523

Less depreciation: **20%** - 2,868,954

Local cost of replacement less depreciation $ 11,475,816

— Page 232

74

Example 2

This example shows how to modify a model building.

Model #M.020, 4–7 Story Apartment, page 80, matches the example quite closely. The model specifies a 6-story building with a 10' story height.

The following adjustments must be made:

- add stone ashlar wall
- adjust model to a 5-story building
- change partitions
- change floor finish

Two exterior wall types

CII APPRAISAL

1. SUBJECT PROPERTY: **Westernberg Holdings, Inc.**	5. DATE: **Jan. 2, 2016**
2. BUILDING: **# 623**	6. APPRAISER: **SJS**
3. ADDRESS: **55 Pinewood Rd., Bondsville, N.J. 07410**	
4. BUILDING USE: **Apartments**	7. YEAR BUILT: **1992**

```
                    N
                   170'

        W                              E
       64.7'

                    S
```

North wall - 4" Stone ashlar with 8" block back up (split face)
East, West & South walls - 12" Decorative concrete block

8. EXTERIOR WALL CONSTRUCTION:

9. FRAME: **Steel**

10. GROUND FLOOR AREA: **11,000**	11. GROSS FLOOR AREA (EXCL. BASEMENT): **55,000**	S.F.
12. NUMBER OF STORIES: **5**	13. STORY HEIGHT: **11 Ft.**	
14. PERIMETER: **469.4**	15. BASEMENT AREA: **None**	S.F.

16. GENERAL COMMENTS:

Example 2 (continued)

Square foot area (excluding basement) from item 11 ___55,000___ S.F.
Perimeter from item 14 ___469.4___ Lin. Ft.
Item 17-Model square foot costs (from model sub-total) ___$123.80___

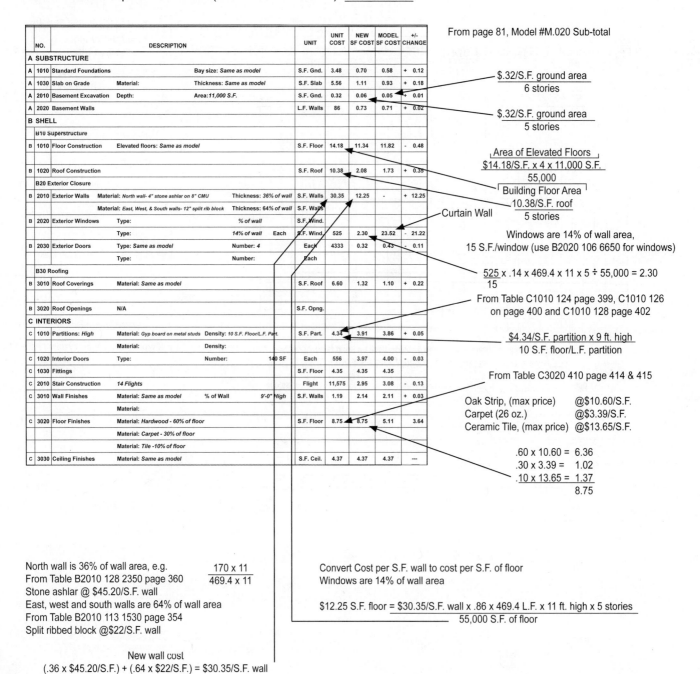

NO.	DESCRIPTION			UNIT	UNIT COST	NEW SF COST	MODEL SF COST	+/- CHANGE
A SUBSTRUCTURE								
A 1010	Standard Foundations	Bay size: *Same as model*		S.F. Gnd.	3.48	0.70	0.58	+ 0.12
A 1030	Slab on Grade	Material:	Thickness: *Same as model*	S.F. Slab	5.56	1.11	0.93	+ 0.18
A 2010	Basement Excavation	Depth:	Area:*11,000 S.F.*	S.F. Gnd.	0.32	0.06	0.05	+ 0.01
A 2020	Basement Walls			L.F. Walls	86	0.73	0.71	+ 0.02
B SHELL								
B10 Superstructure								
B 1010	Floor Construction	Elevated floors: *Same as model*		S.F. Floor	14.18	11.34	11.82	- 0.48
B 1020	Roof Construction			S.F. Roof	10.38	2.08	1.73	+ 0.35
B20 Exterior Closure								
B 2010	Exterior Walls	Material: *North wall- 4" stone ashlar on 8" CMU*	Thickness: *36% of wall*	S.F. Walls	30.35	12.25	-	+ 12.25
		Material: *East, West, & South walls- 12" split rib block*	Thickness: *64% of wall*	S.F. Walls				
B 2020	Exterior Windows	Type:	% of wall	S.F. Wind.				
		Type:	14% of wall Each	S.F. Wind.	525	2.30	23.52	- 21.22
B 2030	Exterior Doors	Type: *Same as model*	Number: 4	Each	4333	0.32	0.43	- 0.11
		Type:	Number:	Each				
B30 Roofing								
B 3010	Roof Coverings	Material: *Same as model*		S.F. Roof	6.60	1.32	1.10	+ 0.22
B 3020	Roof Openings	N/A		S.F. Opng.				
C INTERIORS								
C 1010	Partitions: *High*	Material: *Gyp board on metal studs* Density: *10 S.F. Floor/L.F. Part.*		S.F. Part.	4.34	3.91	3.86	+ 0.05
		Material:	Density:					
C 1020	Interior Doors	Type:	Number: 140 SF	Each	556	3.97	4.00	- 0.03
C 1030	Fittings			S.F. Floor	4.35	4.35	4.35	
C 2010	Stair Construction	*14 Flights*		Flight	11,575	2.95	3.08	- 0.13
C 3010	Wall Finishes	Material: *Same as model* % of Wall 9'-0" High		S.F. Walls	1.19	2.14	2.11	+ 0.03
		Material:						
C 3020	Floor Finishes	Material: *Hardwood - 60% of floor*		S.F. Floor	8.75	8.75	5.11	3.64
		Material: *Carpet - 30% of floor*						
		Material: *Tile -10% of floor*						
C 3030	Ceiling Finishes	Material: *Same as model*		S.F. Ceil.	4.37	4.37	4.37	---

From page 81, Model #M.020 Sub-total

$.32/S.F. ground area
6 stories

$.32/S.F. ground area
5 stories

Area of Elevated Floors
$14.18/S.F. x 4 x 11,000 S.F.
55,000

Building Floor Area
10.38/S.F. roof
5 stories

Curtain Wall

Windows are 14% of wall area,
15 S.F./window (use B2020 106 6650 for windows)

525 x .14 x 469.4 x 11 x 5 ÷ 55,000 = 2.30
15

From Table C1010 124 page 399, C1010 126
on page 400 and C1010 128 page 402

$4.34/S.F. partition x 9 ft. high
10 S.F. floor/L.F. partition

From Table C3020 410 page 414 & 415

Oak Strip, (max price)   @$10.60/S.F.
Carpet (26 oz.)   @$3.39/S.F.
Ceramic Tile, (max price)   @$13.65/S.F.

.60 x 10.60 = 6.36
.30 x 3.39 =   1.02
.10 x 13.65 = 1.37
              8.75

North wall is 36% of wall area, e.g.   170 x 11
From Table B2010 128 2350 page 360   ————————
Stone ashlar @ $45.20/S.F. wall   469.4 x 11
East, west and south walls are 64% of wall area
From Table B2010 113 1530 page 354
Split ribbed block @$22/S.F. wall

New wall cost
(.36 x $45.20/S.F.) + (.64 x $22/S.F.) = $30.35/S.F. wall

Convert Cost per S.F. wall to cost per S.F. of floor
Windows are 14% of wall area

$12.25 S.F. floor = $30.35/S.F. wall x .86 x 469.4 L.F. x 11 ft. high x 5 stories
                    ——
                                    55,000 S.F. of floor

Example 2 (continued)

NO.	SYSTEM/COMPONENT	DESCRIPTION	UNIT	UNIT COST	NEW S.F. COST	MODEL S.F. COST	+/- CHANGE
D SERVICES							
D10 Conveying							
D 1010	Elevators & Lifts 2	Type: *Same as model* Capacity: Stops:	Each	235,500	8.56	7.85	+ 0.71
D 1020	Escalators & Moving Walks	Type:	Each				
D20 Plumbing							
D 2010	Plumbing *Kitchen, bath, + service fixtures*	*1 Fixture/285 S.F. Floor*	Each	1813	6.36	6.34	+ 0.02
D 2020	Domestic Water Distribution		S.F. Floor	7.69	7.69	7.69	---
D 2040	Rain Water Drainage		S.F. Roof	1.68	0.34	0.28	+ 0.06
D30 HVAC							
D 3010	Energy Supply		S.F. Floor	7.87	7.87	7.87	---
D 3020	Heat Generating Systems	Type: *Same as model*	S.F. Floor				
D 3030	Cooling Generating Systems	Type: *Same as model*	S.F. Floor	9.54	9.54	9.54	---
D 3090	Other HVAC Sys. & Equipment		Each				
D40 Fire Protection							
D 4010	Sprinklers	*Same as model*	S.F. Floor	2.78	2.78	2.78	---
D 4020	Standpipes	*Same as model*	S.F. Floor	0.95	0.95	0.95	---
D50 Electrical							
D 5010	Electrical Service/Distribution	*Same as model*	S.F. Floor	2.45	2.45	2.45	---
D 5020	Lighting & Branch Wiring		S.F. Floor	7.59	7.59	7.59	---
D 5030	Communications & Security		S.F. Floor	1.70	1.70	1.70	---
D 5090	Other Electrical Systems		S.F. Floor				
E EQUIPMENT & FURNISHINGS							
E 1010	Commercial Equipment		Each				
E 1020	Institutional Equipment		Each				
E 1030	Vehicular Equipment		Each				
E 1090	Other Equipment		Each	1.01	1.01	1.01	
F SPECIAL CONSTRUCTION							
F 1020	Integrated Construction N/A		S.F.				
F 1040	Special Facilities N/A		S.F.				
G BUILDING SITEWORK							

ITEM						Total Change	- 4.31
17	Model 020 total $ **123.80**						
18	Adjusted S.F. cost $ **119.49** item 17 +/- changes						
19	Building area - from item 11 **55,000** S.F. x adjusted S.F. cost					$	6,571,950
20	Basement area - from item 15 S.F. x S.F. cost $					$	
21	Base building sub-total - item 20 + item 19					$	6,571,950
22	Miscellaneous addition (quality, etc.)					$	
23	Sub-total - item 22 + 21					$	6,571,950
24	General conditions -25 % of item 23					$	1,642,988
25	Sub-total - item 24 + item 23					$	8,214,938
26	Architects fees **5.5** % of item 25					$	451,822
27	Sub-total - item 26 + item 25					$	8,666,760
28	Location modifier .... zip code **07410**					x	1.12
29	Local replacement cost - item 28 x item 27					$	9,706,771
30	Depreciation **20** % of item 29					$	1,941,354
31	Depreciated local replacement cost - item 29 less item 30					$	7,765,417
32	Exclusions					$	
33	Net depreciated replacement cost - item 31 less item 32					$	7,765,417

← From page 523

── Use table on page 232

Costs per square foot of floor area

Exterior Wall	S.F. Area	8000	12000	15000	19000	22500	25000	29000	32000	36000
	L.F. Perimeter	213	280	330	350	400	433	442	480	520
Fiber Cement Siding	Wood Frame	179.90	172.20	169.10	163.65	162.10	161.15	158.30	157.65	156.70
Stone Veneer	Wood Frame	199.65	189.55	185.50	177.40	175.40	174.15	169.75	168.95	167.55
Brick Veneer	Steel Frame	180.70	171.55	167.95	160.90	159.10	157.95	154.15	153.45	152.25
EIFS on Metal Studs	Steel Frame	162.80	156.55	154.15	150.25	149.00	148.25	146.30	145.75	145.05
Brick Veneer	Reinforced Concrete	198.60	187.90	183.65	174.90	172.80	171.45	166.70	165.80	164.35
Stucco on Concrete Block	Reinforced Concrete	187.20	177.95	174.25	167.10	165.20	164.05	160.20	159.40	158.15
Perimeter Adj., Add or Deduct	Per 100 L.F.	14.50	9.70	7.70	6.15	5.15	4.70	3.95	3.65	3.20
Story Hgt. Adj., Add or Deduct	Per 1 Ft.	2.25	1.95	1.85	1.55	1.50	1.45	1.30	1.25	1.20
	For Basement, add $37.40 per square foot of basement area									

The above costs were calculated using the basic specifications shown on the facing page. These costs should be adjusted where necessary for design alternatives and owner's requirements. Reported completed project costs, for this type of structure, range from $63.55 to $236.40 per S.F.

Common additives

Description	Unit	$ Cost	Description	Unit	$ Cost
Appliances			Closed Circuit Surveillance, One station		
Cooking range, 30" free standing			Camera and monitor	Ea.	2025
1 oven	Ea.	610 - 2300	For additional camera stations, add	Ea.	1100
2 oven	Ea.	1475 - 3175	Elevators, Hydraulic passenger, 2 stops		
30" built-in			2000# capacity	Ea.	68,400
1 oven	Ea.	1100 - 2325	2500# capacity	Ea.	70,400
2 oven	Ea.	1775 - 3375	3500# capacity	Ea.	75,400
Counter top cook tops, 4 burner	Ea.	495 - 2100	Additional stop, add	Ea.	8075
Microwave oven	Ea.	296 - 890	Emergency Lighting, 25 watt, battery operated		
Combination range, refrig. & sink, 30" wide	Ea.	1900 - 2750	Lead battery	Ea.	345
72" wide	Ea.	2800	Nickel cadmium	Ea.	685
Combination range, refrigerator, sink,			Laundry Equipment		
microwave oven & icemaker	Ea.	6725	Dryer, gas, 16 lb. capacity	Ea.	995
Compactor, residential, 4-1 compaction	Ea.	880 - 1475	30 lb. capacity	Ea.	3950
Dishwasher, built-in, 2 cycles	Ea.	675 - 1175	Washer, 4 cycle	Ea.	1225
4 cycles	Ea.	815 - 2000	Commercial	Ea.	1625
Garbage disposer, sink type	Ea.	239 - 405	Smoke Detectors		
Hood for range, 2 speed, vented, 30" wide	Ea.	345 - 1375	Ceiling type	Ea.	227
42" wide	Ea.	435 - 2600	Duct type	Ea.	565
Refrigerator, no frost 10-12 C.F.	Ea.	615 - 770			
18-20 C.F.	Ea.	925 - 2100			

Important: See the Reference Section for Location Factors.

Model costs calculated for a 3 story building with 10' story height and 22,500 square feet of floor area

			Unit	Unit Cost	Cost Per S.F.	% Of Sub-Total
A. SUBSTRUCTURE						
1010	Standard Foundations	Poured concrete; strip and spread footings	S.F. Ground	5.13	1.71	
1020	Special Foundations	N/A	—	—	—	
1030	Slab on Grade	4" reinforced concrete	S.F. Slab	5.56	1.85	4.0%
2010	Basement Excavation	Site preparation for slab and trench for foundation wall and footing	S.F. Ground	.32	.11	
2020	Basement Walls	4' Foundation wall	L.F. Wall	42.75	1.52	
B. SHELL						
B10 Superstructure						
1010	Floor Construction	Wood Beam and joist on wood columns, fireproofed	S.F. Floor	24.83	16.55	14.5%
1020	Roof Construction	Wood roof, truss, 4/12 slope	S.F. Roof	7.41	2.47	
B20 Exterior Enclosure						
2010	Exterior Walls	Ashlar stone veneer on wood stud back up, insulated 80% of wall	S.F. Wall	39.12	16.69	
2020	Exterior Windows	Aluminum horizontal sliding 20% of wall	Each	525	3.74	15.9%
2030	Exterior Doors	Steel door , hollow metal with frame	Each	2025	.36	
B30 Roofing						
3010	Roof Coverings	Asphalt roofing, strip shingles, flashing	S.F. Roof	7.47	2.49	1.9%
3020	Roof Openings	N/A	—	—	—	
C. INTERIORS						
1010	Partitions	Gypsum board on wood studs 10 S.F. Floor/L.F. Partition	S.F. Partition	7.37	6.55	
1020	Interior Doors	Single leaf solid and hollow core, fire doors 130 S.F. Floor/Door	Each	632	4.79	
1030	Fittings	Residential wall and base wood cabinets, laminate counters	S.F. Floor	4.33	4.33	
2010	Stair Construction	Wood Stairs with wood rails	Flight	2745	.73	21.6%
3010	Wall Finishes	95% paint, 5% ceramic wall tile	S.F. Surface	1.39	2.47	
3020	Floor Finishes	80% Carpet tile, 10% vinyl composition tile, 10% ceramic tile	S.F. Floor	5.11	5.11	
3030	Ceiling Finishes	Painted gypsum board ceiling on resilient channels	S.F. Ceiling	4.37	4.37	
D. SERVICES						
D10 Conveying						
1010	Elevators & Lifts	Hydraulic passenger elevator	Each	111,600	4.96	3.8%
1020	Escalators & Moving Walks	N/A	—	—	—	
D20 Plumbing						
2010	Plumbing Fixtures	Kitchen, bath, laundry and service fixtures, supply and drainage 1 Fixture/285 S.F. Floor	Each	1918	6.73	
2020	Domestic Water Distribution	Electric water heater	S.F. Floor	7.65	7.65	11.4%
2040	Rain Water Drainage	Roof drains	S.F. Roof	1.53	.51	
D30 HVAC						
3010	Energy Supply	Oil fired hot water, baseboard radiation	S.F. Floor	8.93	8.93	
3020	Heat Generating Systems	N/A	—	—	—	
3030	Cooling Generating Systems	Chilled water, air cooled condenser system	S.F. Floor	9.27	9.27	13.9%
3050	Terminal & Package Units	N/A	—	—	—	
3090	Other HVAC Sys. & Equipment	N/A	—	—	—	
D40 Fire Protection						
4010	Sprinklers	Wet pipe spinkler system, light hazard	S.F. Floor	3.58	3.58	2.7%
4020	Standpipes	N/A	—	—	—	
D50 Electrical						
5010	Electrical Service/Distribution	800 ampere service, panel board and feeders	S.F. Floor	2.67	2.67	
5020	Lighting & Branch Wiring	Incandescent fixtures, receptacles, switches, A.C. and misc. power	S.F. Floor	7.64	7.64	9.2%
5030	Communications & Security	Addressable alarm systems, emergency lighting, internet and phone wiring	S.F. Floor	1.78	1.78	
5090	Other Electrical Systems	N/A	—	—	—	
E. EQUIPMENT & FURNISHINGS						
1010	Commercial Equipment	N/A	—	—	—	
1020	Institutional Equipment	N/A	—	—	—	
1030	Vehicular Equipment	N/A	—	—	—	1.2 %
1090	Other Equipment	Residential freestanding gas ranges, dishwashers	S.F. Floor	1.57	1.57	
F. SPECIAL CONSTRUCTION						
1020	Integrated Construction	N/A	—	—	—	0.0 %
1040	Special Facilities	N/A	—	—	—	
G. BUILDING SITEWORK N/A						

Sub-Total		131.13	**100%**
CONTRACTOR FEES (General Requirements: 10%, Overhead: 5%, Profit: 10%)	25%	32.80	
ARCHITECT FEES	7%	11.47	
Total Building Cost		175.40	

For customer support on your Square Foot Cost Data, call 877.756.2789.

79

Costs per square foot of floor area

Exterior Wall	S.F. Area	40000	45000	50000	55000	60000	70000	80000	90000	100000
	L.F. Perimeter	366	400	433	466	500	510	530	550	594
Curtain Wall	Steel Frame	171.65	169.65	168.00	166.70	165.60	160.55	157.20	154.65	153.60
Brick Veneer	Steel Frame	165.70	163.95	162.45	161.25	160.30	156.35	153.65	151.60	150.75
EIFS on Metal Studs	Steel Frame	160.05	158.40	157.10	156.00	155.15	151.80	149.50	147.80	146.95
Precast Concrete Panel	Reinforced Concrete	189.65	187.45	185.50	184.00	182.80	176.85	172.95	169.95	168.75
Brick Veneer	Reinforced Concrete	183.60	181.50	179.80	178.40	177.25	172.05	168.55	165.90	164.75
Stucco	R/Conc. Frame	152.95	151.15	149.65	148.50	147.55	143.40	140.65	138.50	137.65
Perimeter Adj., Add or Deduct	Per 100 L.F.	10.50	9.30	8.40	7.60	7.00	5.95	5.25	4.65	4.20
Story Hgt. Adj., Add or Deduct	Per 1 Ft.	3.40	3.30	3.25	3.20	3.15	2.75	2.50	2.25	2.25

For Basement, add $37.45 per square foot of basement area.

The above costs were calculated using the basic specifications shown on the facing page. These costs should be adjusted where necessary for design alternatives and owner's requirements. Reported completed project costs, for this type of structure, range from $76.15 to $216.45 per S.F.

Common additives

Description	Unit	$ Cost
Appliances		
Cooking range, 30" free standing		
1 oven	Ea.	610 - 2300
2 oven	Ea.	1475 - 3175
30" built-in		
1 oven	Ea.	1100 - 2325
2 oven	Ea.	1775 - 3375
Counter top cook tops, 4 burner	Ea.	495 - 2100
Microwave oven	Ea.	296 - 890
Combination range, refrig. & sink, 30" wide	Ea.	1900 - 2750
72" wide	Ea.	2800
Combination range, refrigerator, sink,		
microwave oven & icemaker	Ea.	6725
Compactor, residential, 4-1 compaction	Ea.	880 - 1475
Dishwasher, built-in, 2 cycles	Ea.	675 - 1175
4 cycles	Ea.	815 - 2000
Garbage disposer, sink type	Ea.	239 - 405
Hood for range, 2 speed, vented, 30" wide	Ea.	345 - 1375
42" wide	Ea.	435 - 2600
Refrigerator, no frost 10-12 C.F.	Ea.	615 - 770
18-20 C.F.	Ea.	925 - 2100

Description	Unit	$ Cost
Closed Circuit Surveillance, One station		
Camera and monitor	Ea.	2025
For additional camera stations, add	Ea.	1100
Elevators, Electric passenger, 5 stops		
2000# capacity	Ea.	171,100
3500# capacity	Ea.	177,600
5000# capacity	Ea.	184,100
Additional stop, add	Ea.	10,500
Emergency Lighting, 25 watt, battery operated		
Lead battery	Ea.	345
Nickel cadmium	Ea.	685
Laundry Equipment		
Dryer, gas, 16 lb. capacity	Ea.	995
30 lb. capacity	Ea.	3950
Washer, 4 cycle	Ea.	1225
Commercial	Ea.	1625
Smoke Detectors		
Ceiling type	Ea.	227
Duct type	Ea.	565

Important: See the Reference Section for Location Factors.

Model costs calculated for a 6 story building with 10'-4" story height and 60,000 square feet of floor area

				Unit	Unit Cost	Cost Per S.F.	% Of Sub-Total
A. SUBSTRUCTURE							
1010	Standard Foundations	Poured concrete: strip and spread footings		S.F. Ground	3.48	.58	
1020	Special Foundations	N/A		—	—	—	
1030	Slab on Grade	4" reinforced slab on grade		S.F. Slab	5.56	.93	1.8%
2010	Basement Excavation	Site preparation for slab and trench for foundation wall and footing		S.F. Ground	.32	.05	
2020	Basement Walls	4' Foundation wall		L.F. Wall	86	.71	
B. SHELL							
	B10 Superstructure						
1010	Floor Construction	Open web steel joists, slab form, concrete, fireproofed steel columns		S.F. Floor	14.18	11.82	10.9%
1020	Roof Construction	Metal deck, open web steel joists, beams, columns		S.F. Roof	10.38	1.73	
	B20 Exterior Enclosure						
2010	Exterior Walls	N/A		—	—	—	
2020	Exterior Windows	Curtain wall glazing system, thermo-break frame	100% of wall	S.F. Wall	47.50	23.52	19.3 %
2030	Exterior Doors	Aluminum and glass single and double doors		Each	4333	.43	
	B30 Roofing						
3010	Roof Coverings	Single ply membrane, stone ballast, rigid insulation		S.F. Roof	6.60	1.10	0.9%
3020	Roof Openings	N/A		—	—	—	
C. INTERIORS							
1010	Partitions	Gypsum board on metal studs	10 S.F. Floor/L.F. Partition	S.F. Partition	4.34	3.86	
1020	Interior Doors	Single leaf solid and hollow core	140 S.F. Floor/Door	Each	556	4	
1030	Fittings	Residential wall and base wood cabinets, laminate counters		S.F. Floor	4.35	4.35	
2010	Stair Construction	Concrete filled metal pan		Flight	11,575	3.08	21.7%
3010	Wall Finishes	95% paint, 5% ceramic wall tile		S.F. Surface	1.19	2.11	
3020	Floor Finishes	80% Carpet tile, 10% vinyl composition tile, 10% ceramic tile		S.F. Floor	5.11	5.11	
3030	Ceiling Finishes	Painted gypsum board ceiling on resilient channels		S.F. Ceiling	4.37	4.37	
D. SERVICES							
	D10 Conveying						
1010	Elevators & Lifts	Two geared passenger elevators		Each	235,500	7.85	6.3%
1020	Escalators & Moving Walks	N/A		—	—	—	
	D20 Plumbing						
2010	Plumbing Fixtures	Kitchen, bath, laundry and service fixtures, supply and drainage	1 Fixture/285 S.F. Floor	Each	1813	6.34	
2020	Domestic Water Distribution	Electric water heater		S.F. Floor	7.69	7.69	11.6%
2040	Rain Water Drainage	Roof drains		S.F. Roof	1.68	.28	
	D30 HVAC						
3010	Energy Supply	Oil fired hot water, baseboard radiation		S.F. Floor	7.87	7.87	
3020	Heat Generating Systems	N/A		—	—	—	
3030	Cooling Generating Systems	Chilled water, air cooled condenser system		S.F. Floor	9.54	9.54	14.1%
3050	Terminal & Package Units	N/A		—	—	—	
3090	Other HVAC Sys. & Equipment	N/A		—	—	—	
	D40 Fire Protection						
4010	Sprinklers	Wet pipe spinkler system, light hazard		S.F. Floor	2.78	2.78	3.0%
4020	Standpipes	Standpipes and hose systems, with pumps		S.F. Floor	.95	.95	
	D50 Electrical						
5010	Electrical Service/Distribution	1600 Ampere service, panel boards and feeders		S.F. Floor	2.45	2.45	
5020	Lighting & Branch Wiring	Incandescent fixtures, receptacles, switches, A.C. and misc. power		S.F. Floor	7.59	7.59	9.5%
5030	Communications & Security	Addressable alarm systems, emergency lighting, internet and phone wiring		S.F. Floor	1.70	1.70	
5090	Other Electrical Systems	N/A		—	—	—	
E. EQUIPMENT & FURNISHINGS							
1010	Commercial Equipment	N/A		—	—	—	
1020	Institutional Equipment	N/A		—	—	—	
1030	Vehicular Equipment	N/A		—	—	—	0.8 %
1090	Other Equipment	Residential freestanding gas ranges, dishwashers		S.F. Floor	1.01	1.01	
F. SPECIAL CONSTRUCTION							
1020	Integrated Construction	N/A		—	—	—	0.0 %
1040	Special Facilities	N/A		—	—	—	
G. BUILDING SITEWORK	**N/A**						

				Sub-Total	123.80	**100%**
	CONTRACTOR FEES (General Requirements: 10%, Overhead: 5%, Profit: 10%)			25%	30.97	
	ARCHITECT FEES			7%	10.83	

Total Building Cost 165.60

For customer support on your Square Foot Cost Data, call 877.756.2789.

81

Costs per square foot of floor area

Exterior Wall	S.F. Area	95000	112000	129000	145000	170000	200000	275000	400000	600000
	L.F. Perimeter	345	386	406	442	480	510	530	660	800
Curtain Wall	Steel Frame	213.05	210.10	206.30	204.70	201.80	198.40	191.05	187.50	183.70
Brick Veneer	Steel Frame	194.40	191.75	188.55	187.05	184.60	181.85	176.00	173.05	170.00
E.I.F.S.	Steel Frame	196.35	194.00	191.20	189.85	187.70	185.35	180.45	177.90	175.25
Precast Concrete Panel	Reinforced Concrete	219.85	216.40	211.95	210.10	206.70	202.80	194.35	190.25	185.85
Brick Veneer	Reinforced Concrete	216.45	213.20	209.15	207.40	204.30	200.70	193.00	189.20	185.20
Stucco	Reinforced Concrete	198.35	195.65	192.50	191.05	188.55	185.80	180.05	177.05	173.95
Perimeter Adj., Add or Deduct	Per 100 L.F.	10.85	9.30	8.10	7.20	6.10	5.15	3.85	2.55	1.75
Story Hgt. Adj., Add or Deduct	Per 1 Ft.	3.35	3.25	2.90	2.90	2.65	2.35	1.80	1.50	1.30

For Basement, add $38.35 per square foot of basement area

The above costs were calculated using the basic specifications shown on the facing page. These costs should be adjusted where necessary for design alternatives and owner's requirements. Reported completed project costs, for this type of structure, range from $97.45 to $228.70 per S.F.

Common additives

Description	Unit	$ Cost
Appliances		
Cooking range, 30" free standing		
1 oven	Ea.	610 - 2300
2 oven	Ea.	1475 - 3175
30" built-in		
1 oven	Ea.	1100 - 2325
2 oven	Ea.	1775 - 3375
Counter top cook tops, 4 burner	Ea.	495 - 2100
Microwave oven	Ea.	296 - 890
Combination range, refrig. & sink, 30" wide	Ea.	1900 - 2750
72" wide	Ea.	2800
Combination range, refrigerator, sink,		
microwave oven & icemaker	Ea.	6725
Compactor, residential, 4-1 compaction	Ea.	880 - 1475
Dishwasher, built-in, 2 cycles	Ea.	675 - 1175
4 cycles	Ea.	815 - 2000
Garbage disposer, sink type	Ea.	239 - 405
Hood for range, 2 speed, vented, 30" wide	Ea.	345 - 1375
42" wide	Ea.	435 - 2600
Refrigerator, no frost 10-12 C.F.	Ea.	615 - 770
18-20 C.F.	Ea.	925 - 2100

Description	Unit	$ Cost
Closed Circuit Surveillance, One station		
Camera and monitor	Ea.	2025
For additional camera stations, add	Ea.	1100
Elevators, Electric passenger, 10 stops		
3000# capacity	Ea.	411,500
4000# capacity	Ea.	414,000
5000# capacity	Ea.	419,500
Additional stop, add	Ea.	10,500
Emergency Lighting, 25 watt, battery operated		
Lead battery	Ea.	345
Nickel cadmium	Ea.	685
Laundry Equipment		
Dryer, gas, 16 lb. capacity	Ea.	995
30 lb. capacity	Ea.	3950
Washer, 4 cycle	Ea.	1225
Commercial	Ea.	1625
Smoke Detectors		
Ceiling type	Ea.	227
Duct type	Ea.	565

Important: See the Reference Section for Location Factors.

Model costs calculated for a 15 story building with 10'-6" story height and 145,000 square feet of floor area

Apartment, 8-24 Story

				Unit	Unit Cost	Cost Per S.F.	% Of Sub-Total
A. SUBSTRUCTURE							
1010	Standard Foundations	CIP concrete pile caps		S.F. Ground	17.70	1.18	
1020	Special Foundations	Steel H piles, concrete grade beams		S.F. Ground	277	18.46	
1030	Slab on Grade	4" thick reinforced		S.F. Slab	5.56	.37	13.1%
2010	Basement Excavation	Site preparation for slab, piles and grade beams		S.F. Ground	.32	.02	
2020	Basement Walls	4' Foundation wall		L.F. Wall	42.75	.27	
B. SHELL							
	B10 Superstructure						
1010	Floor Construction	Open web steel joists, slab form, concrete, fireproofed steel columns		S.F. Floor	14.34	13.38	9.1%
1020	Roof Construction	Metal deck, open web steel joists, beams, columns		S.F. Roof	10.50	.70	
	B20 Exterior Enclosure						
2010	Exterior Walls	N/A		—	—	—	
2020	Exterior Windows	Curtain wall glazing system, thermo-break frame	100% of wall	S.F. Wall	47.50	21.50	15.8 %
2030	Exterior Doors	Aluminum and glass single and double doors		Each	3248	2.89	
	B30 Roofing						
3010	Roof Coverings	Single ply membrane, stone ballast, rigid insulation		S.F. Roof	6.30	.42	0.3%
3020	Roof Openings	N/A		—	—	—	
C. INTERIORS							
1010	Partitions	Gypsum board on metal studs	10 S.F. Floor/L.F. Partition	S.F. Partition	9.90	8.80	
1020	Interior Doors	Single leaf solid and hollow core	140 S.F. Floor/Door	Each	553	3.87	
1030	Fittings	Residential wall and base wood cabinets, laminate counters		S.F. Floor	4.35	4.35	
2010	Stair Construction	Cement filled metal pan		Flight	11,575	3.43	20.5%
3010	Wall Finishes	95% paint, 5% ceramic wall tile		S.F. Surface	.98	1.74	
3020	Floor Finishes	80% Carpet tile, 10% vinyl composition tile, 10% ceramic tile		S.F. Floor	5.11	5.11	
3030	Ceiling Finishes	Painted gypsum board ceiling on resilient channels		S.F. Ceiling	4.37	4.37	
D. SERVICES							
	D10 Conveying						
1010	Elevators & Lifts	Four geared passenger elevators		Each	475,963	13.13	8.5%
1020	Escalators & Moving Walks	N/A		—	—	—	
	D20 Plumbing						
2010	Plumbing Fixtures	Kitchen, bath, laundry and service fixtures, supply and drainage	1 Fixture/280 S.F. Floor	Each	1770	6.32	
2020	Domestic Water Distribution	Electric water heater		S.F. Floor	7.68	7.68	9.2%
2040	Rain Water Drainage	Roof drains		S.F. Roof	3.75	.25	
	D30 HVAC						
3010	Energy Supply	Oil fired hot water, baseboard radiation		S.F. Floor	7.87	7.87	
3020	Heat Generating Systems	N/A		—	—	—	
3030	Cooling Generating Systems	Chilled water, air cooled condenser system		S.F. Floor	9.54	9.54	11.3%
3050	Terminal & Package Units	N/A		—	—	—	
3090	Other HVAC Sys. & Equipment	N/A		—	—	—	
	D40 Fire Protection						
4010	Sprinklers	Wet pipe spinkler system, light hazard and accessory package		S.F. Floor	2.85	2.85	3.0%
4020	Standpipes	Standpipes and hose systems, with pumps		S.F. Floor	1.85	1.85	
	D50 Electrical						
5010	Electrical Service/Distribution	2000 Ampere service, panel boards and feeders		S.F. Floor	1.16	1.16	
5020	Lighting & Branch Wiring	Incandescent fixtures, receptacles, switches, A.C. and misc. power		S.F. Floor	8.20	8.20	8.1%
5030	Communications & Security	Addressable alarm systems, emergency lighting, internet and phone wiring		S.F. Floor	3.21	3.21	
5090	Other Electrical Systems	N/A		—	—	—	
E. EQUIPMENT & FURNISHINGS							
1010	Commercial Equipment	N/A		—	—	—	
1020	Institutional Equipment	N/A		—	—	—	
1030	Vehicular Equipment	N/A		—	—	—	1.0 %
1090	Other Equipment	Residential freestanding gas ranges, dishwashers		S.F. Floor	1.58	1.58	
F. SPECIAL CONSTRUCTION							
1020	Integrated Construction	N/A		—	—	—	0.0 %
1040	Special Facilities	N/A		—	—	—	
G. BUILDING SITEWORK	**N/A**						

			Sub-Total	154.50	100%
CONTRACTOR FEES (General Requirements: 10%, Overhead: 5%, Profit: 10%)		25%	38.61		
ARCHITECT FEES		6%	11.59		

Total Building Cost	**204.70**

For customer support on your Square Foot Cost Data, call 877.756.2789.

83

Costs per square foot of floor area

Exterior Wall	S.F. Area	2000	3000	4000	5000	7500	10000	15000	25000	50000
	L.F. Perimeter	188	224	260	300	356	400	504	700	1200
Brick Veneer	Wood Frame	247.40	219.55	205.70	197.90	183.75	175.90	168.70	162.70	158.25
	Steel Frame	274.65	247.20	233.55	225.85	211.95	204.25	197.15	191.30	186.95
Brick Veneer on Block	Wood Truss	250.90	221.70	207.20	199.05	184.00	175.65	168.05	161.65	156.95
	Bearing Wall	286.25	257.05	242.50	234.35	219.30	210.95	203.25	196.85	192.15
Wood Siding	Wood Frame	232.35	207.30	194.75	187.70	175.25	168.35	162.05	156.75	152.90
Vinyl Siding	Wood Frame	237.60	211.80	198.95	191.70	178.80	171.75	165.25	159.75	155.80
Perimeter Adj., Add or Deduct	Per 100 L.F.	31.70	21.20	15.85	12.65	8.45	6.35	4.25	2.55	1.35
Story Hgt. Adj., Add or Deduct	Per 1 Ft.	3.90	3.05	2.70	2.50	2.00	1.65	1.40	1.10	1.05

For Basement, add $33.70 per square foot of basement area

The above costs were calculated using the basic specifications shown on the facing page. These costs should be adjusted where necessary for design alternatives and owner's requirements. Reported completed project costs, for this type of structure, range from $95.75 to $240.05 per S.F.

Common additives

Description	Unit	$ Cost
Appliances		
Cooking range, 30" free standing		
1 oven	Ea.	610 - 2300
2 oven	Ea.	1475 - 3175
30" built-in		
1 oven	Ea.	1100 - 2325
2 oven	Ea.	1775 - 3375
Counter top cook tops, 4 burner	Ea.	495 - 2100
Microwave oven	Ea.	296 - 890
Combination range, refrig. & sink, 30" wide	Ea.	1900 - 2750
60" wide	Ea.	2550
72" wide	Ea.	2800
Combination range, refrigerator, sink,		
microwave oven & icemaker	Ea.	6725
Compactor, residential, 4-1 compaction	Ea.	880 - 1475
Dishwasher, built-in, 2 cycles	Ea.	675 - 1175
4 cycles	Ea.	815 - 2000
Garbage disposer, sink type	Ea.	239 - 405
Hood for range, 2 speed, vented, 30" wide	Ea.	345 - 1375
42" wide	Ea.	435 - 2600

Description	Unit	$ Cost
Appliances, cont.		
Refrigerator, no frost 10-12 C.F.	Ea.	615 - 770
14-16 C.F.	Ea.	740 - 900
18-20 C.F.	Ea.	925 - 2100
Laundry Equipment		
Dryer, gas, 16 lb. capacity	Ea.	995
30 lb. capacity	Ea.	3950
Washer, 4 cycle	Ea.	1225
Commercial	Ea.	1625
Sound System		
Amplifier, 250 watts	Ea.	2200
Speaker, ceiling or wall	Ea.	217
Trumpet	Ea.	415

Important: See the Reference Section for Location Factors.

Model costs calculated for a 1 story building with 10' story height and 10,000 square feet of floor area

Assisted - Senior Living

				Unit	Unit Cost	Cost Per S.F.	% Of Sub-Total
A. SUBSTRUCTURE							
1010	Standard Foundations	Poured concrete strip footings		S.F. Ground	2.68	2.68	
1020	Special Foundations	N/A		—	—	—	
1030	Slab on Grade	4" reinforced concrete with vapor barrier and granular base		S.F. Slab	5.56	5.56	9.6%
2010	Basement Excavation	Site preparation for slab and trench for foundation wall and footing		S.F. Ground	.57	.57	
2020	Basement Walls	4' foundation wall		L.F. Wall	86	3.42	
B. SHELL							
B10 Superstructure							
1010	Floor Construction	Steel column		S.F. Floor	.60	.60	6.4%
1020	Roof Construction	Plywood on wood trusses (pitched); fiberglass batt insulation		S.F. Roof	7.50	7.50	
B20 Exterior Enclosure							
2010	Exterior Walls	Brick veneer on wood studs, insulated	80% of wall	S.F. Wall	24.91	7.97	
2020	Exterior Windows	Double hung wood	20% of wall	Each	646	3.23	10.0%
2030	Exterior Doors	Aluminum and glass; 18 ga. steel flush		Each	2521	1.52	
B30 Roofing							
3010	Roof Coverings	Asphalt shingles with flashing (pitched); al. gutters & downspouts		S.F. Roof	3.03	3.03	2.4%
3020	Roof Openings	N/A		—	—	—	
C. INTERIORS							
1010	Partitions	Gypsum board on wood studs	6 S.F. Floor/L.F. Partition	S.F. Partition	5.48	7.30	
1020	Interior Doors	Single leaf wood	80 S.F. Floor/Door	Each	626	7.36	
1030	Fittings	N/A		—	—	—	
2010	Stair Construction	N/A		—	—	—	26.3%
3010	Wall Finishes	Paint on drywall		S.F. Surface	.85	2.27	
3020	Floor Finishes	80% carpet, 20% ceramic tile		S.F. Floor	7.09	7.09	
3030	Ceiling Finishes	Gypsum board on wood furring, painted		S.F. Ceiling	9.30	9.30	
D. SERVICES							
D10 Conveying							
1010	Elevators & Lifts	N/A		—	—	—	0.0%
1020	Escalators & Moving Walks	N/A		—	—	—	
D20 Plumbing							
2010	Plumbing Fixtures	Kitchen, toilet and service fixtures, supply and drainage	1 Fixture/200 S.F. Floor	Each	2958	14.79	
2020	Domestic Water Distribution	Gas water heater		S.F. Floor	2.32	2.32	13.5%
2040	Rain Water Drainage	N/A		—	—	—	
D30 HVAC							
3010	Energy Supply	N/A		—	—	—	
3020	Heat Generating Systems	N/A		—	—	—	
3030	Cooling Generating Systems	N/A		—	—	—	3.5%
3050	Terminal & Package Units	Gas fired hot water boiler; terminal package heat/A/C units		S.F. Floor	4.39	4.39	
3090	Other HVAC Sys. & Equipment	N/A		—	—	—	
D40 Fire Protection							
4010	Sprinklers	Wet pipe sprinkler system		S.F. Floor	3.65	3.65	3.6%
4020	Standpipes	Standpipes		S.F. Floor	.92	.92	
D50 Electrical							
5010	Electrical Service/Distribution	600 ampere service, panel board, feeders and branch wiring		S.F. Floor	8.88	8.88	
5020	Lighting & Branch Wiring	Fuorescent & incandescent fixtures, receptacles, switches, A.C. and misc. power		S.F. Floor	11.83	11.83	23.3%
5030	Communications & Security	Addressable alarm, fire detection and command center, communication system		S.F. Floor	8.67	8.67	
5090	Other Electrical Systems	Emergency generator, 7.5 kW		S.F. Floor	.22	.22	
E. EQUIPMENT & FURNISHINGS							
1010	Commercial Equipment	Residential washers and dryers		S.F. Floor	.45	.45	
1020	Institutional Equipment	N/A		—	—	—	1.3%
1030	Vehicular Equipment	N/A		—	—	—	
1090	Other Equipment	Residential ranges & refrigerators		S.F. Floor	1.26	1.26	
F. SPECIAL CONSTRUCTION							
1020	Integrated Construction	N/A		—	—	—	0.0%
1040	Special Facilities	N/A		—	—	—	
G. BUILDING SITEWORK	**N/A**						

		Sub-Total	126.78	**100%**
CONTRACTOR FEES (General Requirements: 10%, Overhead: 5%, Profit: 10%)		25%	31.69	
ARCHITECT FEES		11%	17.43	
		Total Building Cost	**175.90**	

For customer support on your Square Foot Cost Data, call 877.756.2789.

Costs per square foot of floor area

Exterior Wall	S.F. Area	12000	15000	18000	21000	24000	27000	30000	33000	36000
	L.F. Perimeter	440	500	540	590	640	665	700	732	770
Face Brick with Concrete Block Back-up	Steel Frame	205.95	200.15	194.45	191.20	188.65	185.20	182.95	181.10	179.65
	Bearing Wall	198.90	193.75	188.60	185.70	183.45	180.40	178.40	176.65	175.45
Precast Concrete	Steel Frame	210.55	204.30	198.15	194.60	191.95	188.20	185.80	183.75	182.20
Decorative Concrete Block	Bearing Wall	186.00	182.00	178.20	175.95	174.20	171.85	170.35	169.05	168.10
Concrete Block	Steel Frame	189.25	186.40	183.60	182.05	180.80	179.15	178.10	177.15	176.45
	Bearing Wall	181.80	178.20	174.70	172.70	171.15	169.00	167.65	166.50	165.60
Perimeter Adj., Add or Deduct	Per 100 L.F.	13.85	11.05	9.25	7.85	6.95	6.20	5.60	4.95	4.65
Story Hgt. Adj., Add or Deduct	Per 1 Ft.	2.15	2.00	1.75	1.60	1.65	1.45	1.40	1.30	1.25
For Basement, add $30.60 per square foot of basement area										

The above costs were calculated using the basic specifications shown on the facing page. These costs should be adjusted where necessary for design alternatives and owner's requirements. Reported completed project costs, for this type of structure, range from $96.50 to $244.20 per S.F.

Common additives

Description	Unit	$ Cost
Closed Circuit Surveillance, One station		
Camera and monitor	Ea.	2025
For additional camera stations, add	Ea.	1100
Emergency Lighting, 25 watt, battery operated		
Lead battery	Ea.	345
Nickel cadmium	Ea.	685
Seating		
Auditorium chair, all veneer	Ea.	315
Veneer back, padded seat	Ea.	325
Upholstered, spring seat	Ea.	320
Classroom, movable chair & desk	Set	81 - 171
Lecture hall, pedestal type	Ea.	275 - 620
Smoke Detectors		
Ceiling type	Ea.	227
Duct type	Ea.	565
Sound System		
Amplifier, 250 watts	Ea.	2200
Speaker, ceiling or wall	Ea.	217
Trumpet	Ea.	415

Model costs calculated for a 1 story building with 24' story height and 24,000 square feet of floor area

				Unit	Unit Cost	Cost Per S.F.	% Of Sub-Total
A. SUBSTRUCTURE							
1010	Standard Foundations	Poured concrete; strip and spread footings		S.F. Ground	1.51	1.51	
1020	Special Foundations	N/A		—	—	—	
1030	Slab on Grade	6" reinforced concrete with vapor barrier and granular base		S.F. Slab	6.63	6.63	7.6%
2010	Basement Excavation	Site preparation for slab and trench for foundation wall and footing		S.F. Ground	.18	.18	
2020	Basement Walls	4' foundation wall		L.F. Wall	95	2.52	
B. SHELL							
B10 Superstructure							
1010	Floor Construction	Open web steel joists, slab form, concrete	(balcony)	S.F. Floor	21.76	2.72	8.7%
1020	Roof Construction	Metal deck on steel truss		S.F. Roof	9.83	9.83	
B20 Exterior Enclosure							
2010	Exterior Walls	Precast concrete panel	80% of wall (adjusted for end walls)	S.F. Wall	45.76	23.43	
2020	Exterior Windows	Glass curtain wall	20% of wall	Each	45	5.76	21.8%
2030	Exterior Doors	Double aluminum and glass and hollow metal		Each	4235	2.13	
B30 Roofing							
3010	Roof Coverings	Built-up tar and gravel with flashing; perlite/EPS composite insulation		S.F. Roof	6.36	6.36	4.6%
3020	Roof Openings	Roof hatches		S.F. Roof	.22	.22	
C. INTERIORS							
1010	Partitions	Concrete Block and toilet partitions	40 S.F. Floor/L.F. Partition	S.F. Partition	9.73	3.89	
1020	Interior Doors	Single leaf hollow metal	400 S.F. Floor/Door	Each	1168	2.92	
1030	Fittings	N/A		—	—	—	
2010	Stair Construction	Concrete filled metal pan		Flight	17,275	2.16	18.4%
3010	Wall Finishes	70% paint, 30% epoxy coating		S.F. Surface	5.40	4.32	
3020	Floor Finishes	70% vinyl tile, 30% carpet		S.F. Floor	7.32	7.32	
3030	Ceiling Finishes	Fiberglass board, suspended		S.F. Ceiling	5.78	5.78	
D. SERVICES							
D10 Conveying							
1010	Elevators & Lifts	One hydraulic passenger elevator		Each	94,800	3.95	2.8%
1020	Escalators & Moving Walks	N/A		—	—	—	
D20 Plumbing							
2010	Plumbing Fixtures	Toilet and service fixtures, supply and drainage	1 Fixture/800 S.F. Floor	Each	6784	8.48	
2020	Domestic Water Distribution	Gas fired water heater		S.F. Floor	3.87	3.87	9.4%
2040	Rain Water Drainage	Roof drains		S.F. Roof	1.14	1.14	
D30 HVAC							
3010	Energy Supply	N/A		—	—	—	
3020	Heat Generating Systems	Included in D3050		—	—	—	
3030	Cooling Generating Systems	N/A		—	—	—	9.8 %
3050	Terminal & Package Units	Single zone rooftop unit, gas heating, electric cooling		S.F. Floor	14.05	14.05	
3090	Other HVAC Sys. & Equipment	N/A		—	—	—	
D40 Fire Protection							
4010	Sprinklers	Wet pipe sprinkler system		S.F. Floor	3.65	3.65	2.8%
4020	Standpipes	Standpipe		S.F. Floor	.39	.39	
D50 Electrical							
5010	Electrical Service/Distribution	800 ampere service, panel board and feeders		S.F. Floor	1.91	1.91	
5020	Lighting & Branch Wiring	Fluorescent fixtures, receptacles, switches, A.C. and misc. power		S.F. Floor	12.76	12.76	14.1%
5030	Communications & Security	Addressable alarm systems, emergency lighting and public address system		S.F. Floor	4.24	4.24	
5090	Other Electrical Systems	Emergency generator, 100 kW		S.F. Floor	1.38	1.38	
E. EQUIPMENT & FURNISHINGS							
1010	Commercial Equipment	N/A		—	—	—	
1020	Institutional Equipment	N/A		—	—	—	0.0 %
1030	Vehicular Equipment	N/A		—	—	—	
1090	Other Equipment	N/A		—	—	—	
F. SPECIAL CONSTRUCTION							
1020	Integrated Construction	N/A		—	—	—	0.0 %
1040	Special Facilities	N/A		—	—	—	
G. BUILDING SITEWORK	**N/A**						

		Sub-Total	143.50	100%
CONTRACTOR FEES (General Requirements: 10%, Overhead: 5%, Profit: 10%)		25%	35.89	
ARCHITECT FEES		7%	12.56	

Total Building Cost	**191.95**

Costs per square foot of floor area

Exterior Wall	S.F. Area	2000	2700	3400	4100	4800	5500	6200	6900	7600
	L.F. Perimeter	180	208	236	256	280	303	317	337	357
Face Brick with Concrete Block Back-up	Steel Frame	270.65	254.75	245.50	237.30	232.45	228.60	224.20	221.50	219.35
	R/Conc. Frame	287.15	271.25	261.90	253.80	248.95	245.05	240.65	238.00	235.80
Precast Concrete Panel	Steel Frame	276.05	259.25	249.45	240.85	235.70	231.65	227.00	224.15	221.85
	R/Conc. Frame	297.65	278.60	267.40	257.90	252.10	247.60	242.55	239.45	236.90
Limestone with Concrete Block Back-up	Steel Frame	300.60	280.40	268.50	258.10	251.80	246.90	241.15	237.75	234.95
	R/Conc. Frame	317.05	296.85	285.00	274.60	268.30	263.40	257.65	254.20	251.45
Perimeter Adj., Add or Deduct	Per 100 L.F.	49.65	36.80	29.15	24.20	20.60	18.00	15.95	14.40	13.05
Story Hgt. Adj., Add or Deduct	Per 1 Ft.	4.70	4.05	3.55	3.25	3.00	2.90	2.60	2.55	2.45
For Basement, add $35.40 per square foot of basement area										

The above costs were calculated using the basic specifications shown on the facing page. These costs should be adjusted where necessary for design alternatives and owner's requirements. Reported completed project costs, for this type of structure, range from $147.10 to $362.05 per S.F.

Common additives

Description	Unit	$ Cost	Description	Unit	$ Cost
Bulletproof Teller Window, 44" x 60"	Ea.	5600	Service Windows, Pass thru, steel		
60" x 48"	Ea.	7425	24" x 36"	Ea.	3975
Closed Circuit Surveillance, One station			48" x 48"	Ea.	4175
Camera and monitor	Ea.	2025	72" x 40"	Ea.	6600
For additional camera stations, add	Ea.	1100	Smoke Detectors		
Counters, Complete	Station	6350	Ceiling type	Ea.	227
Door & Frame, 3' x 6'-8", bullet resistant steel			Duct type	Ea.	565
with vision panel	Ea.	7425 - 9600	Twenty-four Hour Teller		
Drive-up Window, Drawer & micr., not incl. glass	Ea.	9175 - 13,000	Automatic deposit cash & memo	Ea.	55,000
Emergency Lighting, 25 watt, battery operated			Vault Front, Door & frame		
Lead battery	Ea.	345	1 hour test, 32"x 78"	Opng.	8250
Nickel cadmium	Ea.	685	2 hour test, 32" door	Opng.	10,100
Night Depository	Ea.	9750 - 14,700	40" door	Opng.	11,700
Package Receiver, painted	Ea.	1975	4 hour test, 32" door	Opng.	10,700
stainless steel	Ea.	3025	40" door	Opng.	12,600
Partitions, Bullet resistant to 8' high	L.F.	340 - 540	Time lock, two movement, add	Ea.	2300
Pneumatic Tube Systems, 2 station	Ea.	33,900			
With TV viewer	Ea.	62,000			

Important: See the Reference Section for Location Factors.

Model costs calculated for a 1 story building with 14' story height and 4,100 square feet of floor area

				Unit	Unit Cost	Cost Per S.F.	% Of Sub-Total
A.	**SUBSTRUCTURE**						
1010	Standard Foundations	Poured concrete; strip and spread footings		S.F. Ground	4.41	4.41	
1020	Special Foundations	N/A		—	—	—	
1030	Slab on Grade	4" reinforced concrete with vapor barrier and granular base		S.F. Slab	5.56	5.56	8.9%
2010	Basement Excavation	Site preparation for slab and trench for foundation wall and footing		S.F. Ground	.32	.32	
2020	Basement Walls	4' Foundation wall		L.F. Wall	86	6.41	
B.	**SHELL**						
	B10 Superstructure						
1010	Floor Construction	Cast-in-place columns		L.F. Column	126	6.14	12.0%
1020	Roof Construction	Cast-in-place concrete flat plate		S.F. Roof	16.45	16.45	
	B20 Exterior Enclosure						
2010	Exterior Walls	Face brick with concrete block backup	80% of wall	S.F. Wall	33.39	23.35	
2020	Exterior Windows	Horizontal aluminum sliding	20% of wall	Each	585	6.81	16.9%
2030	Exterior Doors	Double aluminum and glass and hollow metal		Each	3393	1.66	
	B30 Roofing						
3010	Roof Coverings	Built-up tar and gravel with flashing; perlite/EPS composite insulation		S.F. Roof	7.94	7.94	4.2%
3020	Roof Openings	N/A		—	—	—	
C.	**INTERIORS**						
1010	Partitions	Gypsum board on metal studs	20 S.F. of Floor/L.F. Partition	S.F. Partition	11.26	5.63	
1020	Interior Doors	Single leaf hollow metal	200 S.F. Floor/Door	Each	1168	5.85	
1030	Fittings	N/A		—	—	—	
2010	Stair Construction	N/A		—	—	—	13.6%
3010	Wall Finishes	50% vinyl wall covering, 50% paint		S.F. Surface	1.44	1.44	
3020	Floor Finishes	50% carpet tile, 40% vinyl composition tile, 10% quarry tile		S.F. Floor	5.45	5.45	
3030	Ceiling Finishes	Mineral fiber tile on concealed zee bars		S.F. Ceiling	7.22	7.22	
D.	**SERVICES**						
	D10 Conveying						
1010	Elevators & Lifts	N/A		—	—	—	0.0 %
1020	Escalators & Moving Walks	N/A		—	—	—	
	D20 Plumbing						
2010	Plumbing Fixtures	Toilet and service fixtures, supply and drainage	1 Fixture/580 S.F. Floor	Each	5527	9.53	
2020	Domestic Water Distribution	Gas fired water heater		S.F. Floor	1.64	1.64	6.8%
2040	Rain Water Drainage	Roof drains		S.F. Roof	1.61	1.61	
	D30 HVAC						
3010	Energy Supply	N/A		—	—	—	
3020	Heat Generating Systems	Included in D3050		—	—	—	
3030	Cooling Generating Systems	N/A		—	—	—	7.4 %
3050	Terminal & Package Units	Single zone rooftop unit, gas heating, electric cooling		S.F. Floor	13.95	13.95	
3090	Other HVAC Sys. & Equipment	N/A		—	—	—	
	D40 Fire Protection						
4010	Sprinklers	Wet pipe sprinkler system		S.F. Floor	5.02	5.02	4.3%
4020	Standpipes	Standpipe		S.F. Floor	3.14	3.14	
	D50 Electrical						
5010	Electrical Service/Distribution	200 ampere service, panel board and feeders		S.F. Floor	2.76	2.76	
5020	Lighting & Branch Wiring	Fluorescent fixtures, receptacles, switches, A.C. and misc. power		S.F. Floor	9.86	9.86	12.9%
5030	Communications & Security	Alarm systems, internet and phone wiring, emergency lighting, and security television		S.F. Floor	11	11	
5090	Other Electrical Systems	Emergency generator, 15 kW, Uninterruptible power supply		S.F. Floor	.68	.68	
E.	**EQUIPMENT & FURNISHINGS**						
1010	Commercial Equipment	Automatic teller, drive up window, night depository		S.F. Floor	8.10	8.10	
1020	Institutional Equipment	Closed circuit TV monitoring system		S.F. Floor	2.93	2.93	5.9%
1030	Vehicular Equipment	N/A		—	—	—	
1090	Other Equipment	N/A		—	—	—	
F.	**SPECIAL CONSTRUCTION**						
1020	Integrated Construction	N/A		—	—	—	7.0 %
1040	Special Facilities	Security vault door		S.F. Floor	13.14	13.14	
G.	**BUILDING SITEWORK**	**N/A**					

				Sub-Total	188	100%
CONTRACTOR FEES (General Requirements: 10%, Overhead: 5%, Profit: 10%)				25%	47	
ARCHITECT FEES				8%	18.80	

Total Building Cost 253.80

For customer support on your Square Foot Cost Data, call 877.756.2789.

89

Costs per square foot of floor area

Exterior Wall	S.F. Area	12000	14000	16000	18000	20000	22000	24000	26000	28000
	L.F. Perimeter	460	491	520	540	566	593	620	645	670
Concrete Block	Steel Roof Deck	176.35	173.10	170.65	168.35	166.80	165.50	164.40	163.45	162.60
Decorative Concrete Block	Steel Roof Deck	180.95	177.40	174.60	172.00	170.20	168.75	167.55	166.50	165.50
Face Brick on Concrete Block	Steel Roof Deck	189.10	184.85	181.50	178.35	176.25	174.45	173.00	171.75	170.60
Jumbo Brick on Concrete Block	Steel Roof Deck	185.50	181.55	178.50	175.60	173.60	171.95	170.60	169.45	168.35
Stucco on Concrete Block	Steel Roof Deck	179.90	176.40	173.70	171.15	169.45	168.00	166.80	165.80	164.85
Precast Concrete Panel	Steel Roof Deck	184.95	181.05	177.95	175.15	173.15	171.55	170.25	169.10	168.00
Perimeter Adj., Add or Deduct	Per 100 L.F.	5.40	4.70	4.00	3.60	3.25	2.95	2.70	2.55	2.35
Story Hgt. Adj., Add or Deduct	Per 1 Ft.	1.20	1.10	1.00	0.95	0.90	0.80	0.80	0.80	0.75

For Basement, add $34.90 per square foot of basement area

The above costs were calculated using the basic specifications shown on the facing page. These costs should be adjusted where necessary for design alternatives and owner's requirements. Reported completed project costs, for this type of structure, range from $65.05 to $171.35 per S.F.

Common additives

Description	Unit	$ Cost
Bowling Alleys, Incl. alley, pinsetter,		
scorer, counter & misc. supplies, average	Lane	63,500
For automatic scorer, add (maximum)	Lane	11,000
Emergency Lighting, 25 watt, battery operated		
Lead battery	Ea.	345
Nickel cadmium	Ea.	685
Lockers, Steel, Single tier, 60" or 72"	Opng.	231 - 380
2 tier, 60" or 72" total	Opng.	121 - 170
5 tier, box lockers	Opng.	74 - 84
Locker bench, lam., maple top only	L.F.	34.50
Pedestals, steel pipe	Ea.	76.50
Seating		
Auditorium chair, all veneer	Ea.	315
Veneer back, padded seat	Ea.	325
Upholstered, spring seat	Ea.	320
Sound System		
Amplifier, 250 watts	Ea.	2200
Speaker, ceiling or wall	Ea.	217
Trumpet	Ea.	415

Important: See the Reference Section for Location Factors.

Model costs calculated for a 1 story building with 14' story height and 20,000 square feet of floor area

				Unit	Unit Cost	Cost Per S.F.	% Of Sub-Total
A. SUBSTRUCTURE							
1010	Standard Foundations	Poured concrete; strip and spread footings		S.F. Ground	1.30	1.30	
1020	Special Foundations	N/A		—	—	—	
1030	Slab on Grade	4" reinforced concrete with vapor barrier and granular base		S.F. Slab	5.56	5.56	7.8%
2010	Basement Excavation	Site preparation for slab and trench for foundation wall and footing		S.F. Ground	.57	.57	
2020	Basement Walls	4' Foundation wall		L.F. Wall	86	2.42	
B. SHELL							
B10 Superstructure							
1010	Floor Construction	N/A		—	—	—	7.8 %
1020	Roof Construction	Metal deck on open web steel joists with columns and beams		S.F. Roof	9.86	9.86	
B20 Exterior Enclosure							
2010	Exterior Walls	Concrete block	90% of wall	S.F. Wall	13.46	4.80	
2020	Exterior Windows	Horizontal pivoted	10% of wall	Each	2050	3.38	7.0%
2030	Exterior Doors	Double aluminum and glass and hollow metal		Each	2518	.64	
B30 Roofing							
3010	Roof Coverings	Built-up tar and gravel with flashing; perlite/EPS composite insulation		S.F. Roof	6.09	6.09	5.0%
3020	Roof Openings	Roof hatches		S.F. Roof	.16	.16	
C. INTERIORS							
1010	Partitions	Concrete block	50 S.F. Floor/L.F. Partition	S.F. Partition	10.85	2.17	
1020	Interior Doors	Hollow metal single leaf	1000 S.F. Floor/Door	Each	1168	1.17	
1030	Fittings	N/A		—	—	—	
2010	Stair Construction	N/A		—	—	—	6.3%
3010	Wall Finishes	Paint and block filler		S.F. Surface	6.28	2.51	
3020	Floor Finishes	Vinyl tile	25% of floor	S.F. Floor	2.52	.63	
3030	Ceiling Finishes	Suspended fiberglass board	25% of area	S.F. Ceiling	5.78	1.45	
D. SERVICES							
D10 Conveying							
1010	Elevators & Lifts	N/A		—	—	—	0.0 %
1020	Escalators & Moving Walks	N/A		—	—	—	
D20 Plumbing							
2010	Plumbing Fixtures	Toilet and service fixtures, supply and drainage	1 Fixture/2200 S.F. Floor	Each	3520	1.60	
2020	Domestic Water Distribution	Gas fired water heater		S.F. Floor	.30	.30	2.1%
2040	Rain Water Drainage	Roof drains		S.F. Roof	.71	.71	
D30 HVAC							
3010	Energy Supply	N/A		—	—	—	
3020	Heat Generating Systems	Included in D3050		—	—	—	
3030	Cooling Generating Systems	N/A		—	—	—	12.7 %
3050	Terminal & Package Units	Single zone rooftop unit, gas heating, electric cooling		S.F. Floor	15.95	15.95	
3090	Other HVAC Sys. & Equipment	N/A		—	—	—	
D40 Fire Protection							
4010	Sprinklers	Sprinklers, light hazard		S.F. Floor	3.65	3.65	3.4%
4020	Standpipes	Standpipe, wet, Class III		S.F. Floor	.65	.65	
D50 Electrical							
5010	Electrical Service/Distribution	800 ampere service, panel board and feeders		S.F. Floor	2.30	2.30	
5020	Lighting & Branch Wiring	High efficiency fluorescent fixtures, receptacles, switches, A.C. and misc. power		S.F. Floor	11.43	11.43	12.1%
5030	Communications & Security	Addressable alarm systems and emergency lighting		S.F. Floor	1.37	1.37	
5090	Other Electrical Systems	Emergency generator, 7.5 kW		S.F. Floor	.09	.09	
E. EQUIPMENT & FURNISHINGS							
1010	Commercial Equipment	N/A		—	—	—	
1020	Institutional Equipment	N/A		—	—	—	
1030	Vehicular Equipment	N/A		—	—	—	35.8 %
1090	Other Equipment	Bowling Alley, ballrack, automatic scorers		S.F. Floor	45.12	45.12	
F. SPECIAL CONSTRUCTION							
1020	Integrated Construction	N/A		—	—	—	0.0 %
1040	Special Facilities	N/A		—	—	—	
G. BUILDING SITEWORK	**N/A**						

				Sub-Total	125.88	**100%**
CONTRACTOR FEES (General Requirements: 10%, Overhead: 5%, Profit: 10%)			25%	31.48	
ARCHITECT FEES			6%	9.44	

Total Building Cost   166.80

For customer support on your Square Foot Cost Data, call 877.756.2789.

91

Costs per square foot of floor area

Exterior Wall	S.F. Area	6000	8000	10000	12000	14000	16000	18000	20000	22000
	L.F. Perimeter	320	386	453	475	505	530	570	600	630
Face Brick with Concrete Block Back-up	Bearing Walls	184.10	177.80	174.10	168.35	164.80	161.85	160.20	158.50	157.10
	Steel Frame	187.20	181.00	177.30	171.65	168.10	165.25	163.60	161.95	160.55
Decorative Concrete Block	Bearing Walls	175.35	169.90	166.65	161.90	158.85	156.40	155.00	153.60	152.40
	Steel Frame	178.45	173.05	169.85	165.15	162.15	159.80	158.40	157.05	155.90
Precast Concrete Panels	Bearing Walls	183.25	177.05	173.35	167.75	164.20	161.30	159.65	158.10	156.65
	Steel Frame	186.35	180.20	176.60	171.05	167.50	164.70	163.10	161.50	160.10
Perimeter Adj., Add or Deduct	Per 100 L.F.	14.45	10.85	8.65	7.25	6.10	5.35	4.85	4.35	3.95
Story Hgt. Adj., Add or Deduct	Per 1 Ft.	2.35	2.20	2.00	1.75	1.55	1.45	1.45	1.35	1.30
Basement—Not Applicable										

The above costs were calculated using the basic specifications shown on the facing page. These costs should be adjusted where necessary for design alternatives and owner's requirements. Reported completed project costs, for this type of structure, range from $78.60 to $189.50 per S.F.

Common additives

Description	Unit	$ Cost
Directory Boards, Plastic, glass covered		
30" x 20"	Ea.	610
36" x 48"	Ea.	1525
Aluminum, 24" x 18"	Ea.	640
36" x 24"	Ea.	765
48" x 32"	Ea.	1075
48" x 60"	Ea.	2275
Emergency Lighting, 25 watt, battery operated		
Lead battery	Ea.	345
Nickel cadmium	Ea.	685
Benches, Hardwood	L.F.	141 - 234
Ticket Printer	Ea.	8125
Turnstiles, One way		
4 arm, 46" dia., manual	Ea.	2550
Electric	Ea.	3525
High security, 3 arm		
65" dia., manual	Ea.	9625
Electric	Ea.	12,200
Gate with horizontal bars		
65" dia., 7' high transit type	Ea.	7450

Important: See the Reference Section for Location Factors.

Model costs calculated for a 1 story building with 14' story height and 12,000 square feet of floor area

				Unit	Unit Cost	Cost Per S.F.	% Of Sub-Total
A.	**SUBSTRUCTURE**						
1010	Standard Foundations	Poured concrete; strip and spread footings		S.F. Ground	2.52	2.52	
1020	Special Foundations	N/A		—	—	—	
1030	Slab on Grade	6" reinforced concrete with vapor barrier and granular base		S.F. Slab	8.31	8.31	11.5%
2010	Basement Excavation	Site preparation for slab and trench for foundation wall and footing		S.F. Ground	.32	.32	
2020	Basement Walls	4' Foundation wall		L.F. Wall	88	3.47	
B.	**SHELL**						
	B10 Superstructure						
1010	Floor Construction	N/A		—	—	—	5.3 %
1020	Roof Construction	Metal deck on open web steel joists with columns and beams		S.F. Roof	6.78	6.78	
	B20 Exterior Enclosure						
2010	Exterior Walls	Face brick with concrete block backup	70% of wall	S.F. Wall	33.38	12.95	
2020	Exterior Windows	Store front	30% of wall	Each	32.40	5.39	16.0%
2030	Exterior Doors	Double aluminum and glass		Each	5875	1.96	
	B30 Roofing						
3010	Roof Coverings	Built-up tar and gravel with flashing; perlite/EPS composite insulation		S.F. Roof	6.92	6.92	5.4%
3020	Roof Openings	N/A		—	—	—	
C.	**INTERIORS**						
1010	Partitions	Lightweight concrete block	15 S.F. Floor/L.F. Partition	S.F. Partition	10.83	7.22	
1020	Interior Doors	Hollow metal	150 S.F. Floor/Door	Each	1168	7.78	
1030	Fittings	N/A		—	—	—	
2010	Stair Construction	N/A		—	—	—	27.2%
3010	Wall Finishes	Glazed coating		S.F. Surface	1.98	2.64	
3020	Floor Finishes	Quarry tile and vinyl composition tile		S.F. Floor	9.67	9.67	
3030	Ceiling Finishes	Mineral fiber tile on concealed zee bars		S.F. Ceiling	7.22	7.22	
D.	**SERVICES**						
	D10 Conveying						
1010	Elevators & Lifts	N/A		—	—	—	0.0 %
1020	Escalators & Moving Walks	N/A		—	—	—	
	D20 Plumbing						
2010	Plumbing Fixtures	Toilet and service fixtures, supply and drainage	1 Fixture/850 S.F. Floor	Each	7217	8.49	
2020	Domestic Water Distribution	Electric water heater		S.F. Floor	2.14	2.14	9.3%
2040	Rain Water Drainage	Roof drains		S.F. Roof	1.25	1.25	
	D30 HVAC						
3010	Energy Supply	N/A		—	—	—	
3020	Heat Generating Systems	Included in D3050		—	—	—	
3030	Cooling Generating Systems	N/A		—	—	—	9.6 %
3050	Terminal & Package Units	Single zone rooftop unit, gas heating, electric cooling		S.F. Floor	12.15	12.15	
3090	Other HVAC Sys. & Equipment	N/A		—	—	—	
	D40 Fire Protection						
4010	Sprinklers	Wet pipe sprinkler system		S.F. Floor	4.72	4.72	4.6%
4020	Standpipes	Standpipe		S.F. Floor	1.07	1.07	
	D50 Electrical						
5010	Electrical Service/Distribution	400 ampere service, panel board and feeders		S.F. Floor	1.89	1.89	
5020	Lighting & Branch Wiring	Fluorescent fixtures, receptacles, switches, A.C. and misc. power		S.F. Floor	9.69	9.69	11.1%
5030	Communications & Security	Alarm systems and emergency lighting		S.F. Floor	2.08	2.08	
5090	Other Electrical Systems	Emergency generator, 7.5 kW		S.F. Floor	.44	.44	
E.	**EQUIPMENT & FURNISHINGS**						
1010	Commercial Equipment	N/A		—	—	—	
1020	Institutional Equipment	N/A		—	—	—	0.0 %
1030	Vehicular Equipment	N/A		—	—	—	
1090	Other Equipment	N/A		—	—	—	
F.	**SPECIAL CONSTRUCTION**						
1020	Integrated Construction	N/A		—	—	—	0.0 %
1040	Special Facilities	N/A		—	—	—	
G.	**BUILDING SITEWORK**	**N/A**					

	Sub-Total	127.07	100%
CONTRACTOR FEES (General Requirements: 10%, Overhead: 5%, Profit: 10%)	25%	31.75	
ARCHITECT FEES	6%	9.53	
Total Building Cost		**168.35**	

Costs per square foot of floor area

Exterior Wall	S.F. Area	600	800	1000	1200	1600	2000	2400	3000	4000
	L.F. Perimeter	100	114	128	139	164	189	210	235	280
Brick with Concrete Block Back-up	Steel Frame	351.55	336.80	327.90	320.55	312.40	307.50	303.25	297.65	292.60
	Bearing Walls	344.45	329.75	320.85	313.45	305.30	300.45	296.15	290.60	285.45
Concrete Block	Steel Frame	307.20	298.75	293.70	289.40	284.75	281.95	279.50	276.30	273.45
	Bearing Walls	301.05	292.60	287.50	283.25	278.65	275.80	273.35	270.15	267.25
Galvanized Steel Siding	Steel Frame	311.70	302.55	297.05	292.55	287.50	284.50	281.85	278.45	275.25
Metal Sandwich Panel	Steel Frame	306.30	297.00	291.40	286.70	281.60	278.45	275.80	272.30	269.05
Perimeter Adj., Add or Deduct	Per 100 L.F.	101.65	76.30	61.05	50.80	38.10	30.50	25.40	20.35	15.20
Story Hgt. Adj., Add or Deduct	Per 1 Ft.	6.20	5.30	4.80	4.30	3.80	3.50	3.25	2.95	2.60
Basement—Not Applicable										

The above costs were calculated using the basic specifications shown on the facing page. These costs should be adjusted where necessary for design alternatives and owner's requirements. Reported completed project costs, for this type of structure, range from $141.05 to $400.70 per S.F.

Common additives

Description	Unit	$ Cost	Description	Unit	$ Cost
Air Compressors, Electric			Lockers, Steel, single tier, 60" or 72"	Opng.	231 - 380
1-1/2 H.P., standard controls	Ea.	1650	2 tier, 60" or 72" total	Opng.	121 - 170
Dual controls	Ea.	1975	5 tier, box lockers	Opng.	74 - 84
5 H.P., 115/230 volt, standard control	Ea.	3700	Locker bench, lam. maple top only	L.F.	34.50
Dual controls	Ea.	4775	Pedestals, steel pipe	Ea.	76.50
Emergency Lighting, 25 watt, battery operated			Paving, Bituminous		
Lead battery	Ea.	345	Wearing course plus base course	S.Y.	6.70
Nickel cadmium	Ea.	685	Safe, Office type, 1 hour rating		
Fence, Chain link, 6' high			30" x 18" x 18"	Ea.	2500
9 ga. wire, galvanized	L.F.	28.50	60" x 36" x 18", double door	Ea.	9350
6 ga. wire	L.F.	32	Sidewalks, Concrete 4" thick	S.F.	4.63
Gate	Ea.	400	Yard Lighting,		
Product Dispenser with vapor recovery			20' aluminum pole		
for 6 nozzles	Ea.	27,200	with 400 watt		
Laundry Equipment			high pressure sodium		
Dryer, gas, 16 lb. capacity	Ea.	995	fixture	Ea.	3225
30 lb. capacity	Ea.	3950			
Washer, 4 cycle	Ea.	1225			
Commercial	Ea.	1625			

Important: See the Reference Section for Location Factors.

Model costs calculated for a 1 story building with 12' story height and 800 square feet of floor area

				Unit	Unit Cost	Cost Per S.F.	% Of Sub-Total
A.	**SUBSTRUCTURE**						
1010	Standard Foundations	Poured concrete; strip and spread footings		S.F. Ground	3.37	3.37	
1020	Special Foundations	N/A		—	—	—	
1030	Slab on Grade	5" reinforced concrete with vapor barrier and granular base		S.F. Slab	7.35	7.35	9.0%
2010	Basement Excavation	Site preparation for slab and trench for foundation wall and footing		S.F. Ground	1.08	1.08	
2020	Basement Walls	4' Foundation wall		L.F. Wall	76	10.83	
B.	**SHELL**						
	B10 Superstructure						
1010	Floor Construction	N/A		—	—	—	3.6 %
1020	Roof Construction	Metal deck, open web steel joists, beams, columns		S.F. Roof	9.12	9.12	
	B20 Exterior Enclosure						
2010	Exterior Walls	Face brick with concrete block backup	70% of wall	S.F. Wall	33.40	39.98	
2020	Exterior Windows	Horizontal pivoted steel	5% of wall	Each	807	7.66	26.5%
2030	Exterior Doors	Steel overhead hollow metal		Each	3825	19.13	
	B30 Roofing						
3010	Roof Coverings	Built-up tar and gravel with flashing; perlite/EPS composite insulation		S.F. Roof	9.04	9.04	3.6%
3020	Roof Openings	N/A		—	—	—	
C.	**INTERIORS**						
1010	Partitions	Concrete block	20 S.F. Floor/S.F. Partition	S.F. Partition	10.24	5.12	
1020	Interior Doors	Hollow metal	600 S.F. Floor/Door	Each	1168	1.95	
1030	Fittings	N/A		—	—	—	
2010	Stair Construction	N/A		—	—	—	3.3%
3010	Wall Finishes	Paint		S.F. Surface	1.32	1.32	
3020	Floor Finishes	N/A		—	—	—	
3030	Ceiling Finishes	N/A		—	—	—	
D.	**SERVICES**						
	D10 Conveying						
1010	Elevators & Lifts	N/A		—	—	—	0.0 %
1020	Escalators & Moving Walks	N/A		—	—	—	
	D20 Plumbing						
2010	Plumbing Fixtures	Toilet and service fixtures, supply and drainage	1 Fixture/160 S.F. Floor	Each	2819	17.62	
2020	Domestic Water Distribution	Gas fired water heater		S.F. Floor	36.07	36.07	23.1%
2040	Rain Water Drainage	Roof drains		S.F. Roof	4.54	4.54	
	D30 HVAC						
3010	Energy Supply	N/A		—	—	—	
3020	Heat Generating Systems	N/A		—	—	—	
3030	Cooling Generating Systems	N/A		—	—	—	3.7 %
3050	Terminal & Package Units	Single zone rooftop unit, gas heating, electric cooling		S.F. Floor	9.33	9.33	
3090	Other HVAC Sys. & Equipment	N/A		—	—	—	
	D40 Fire Protection						
4010	Sprinklers	N/A		—	—	—	0.0 %
4020	Standpipes	N/A		—	—	—	
	D50 Electrical						
5010	Electrical Service/Distribution	400 ampere service, panel board and feeders		S.F. Floor	25.41	25.41	
5020	Lighting & Branch Wiring	Fluorescent fixtures, receptacles, switches and misc. power		S.F. Floor	40.97	40.97	27.1%
5030	Communications & Security	N/A		—	—	—	
5090	Other Electrical Systems	Emergency generator, 7.5 kW		S.F. Floor	1.92	1.92	
E.	**EQUIPMENT & FURNISHINGS**						
1010	Commercial Equipment	N/A		—	—	—	
1020	Institutional Equipment	N/A		—	—	—	
1030	Vehicular Equipment	N/A		—	—	—	0.0 %
1090	Other Equipment	N/A		—	—	—	
F.	**SPECIAL CONSTRUCTION**						
1020	Integrated Construction	N/A		—	—	—	
1040	Special Facilities	N/A		—	—	—	0.0 %
G.	**BUILDING SITEWORK**	**N/A**					

		Sub-Total	251.81	100%
CONTRACTOR FEES (General Requirements: 10%, Overhead: 5%, Profit: 10%)		25%	62.96	
ARCHITECT FEES		7%	22.03	
		Total Building Cost	**336.80**	

For customer support on your Square Foot Cost Data, call 877.756.2789.

95

Costs per square foot of floor area

Exterior Wall	S.F. Area	2000	7000	12000	17000	22000	27000	32000	37000	42000
	L.F. Perimeter	180	340	470	540	640	740	762	793	860
Decorative Concrete Block	Wood Arch	268.00	197.95	182.90	171.80	167.60	165.05	159.85	156.45	155.00
	Steel Truss	268.55	198.55	183.50	172.35	168.20	165.60	160.45	157.05	155.60
Stone with Concrete Block Back-up	Wood Arch	331.75	232.40	210.65	194.30	188.20	184.45	176.75	171.60	169.55
	Steel Truss	325.05	225.75	204.00	187.65	181.55	177.80	170.05	165.00	162.85
Face Brick with Concrete Block Back-up	Wood Arch	298.45	214.40	196.15	182.55	177.45	174.30	167.95	163.70	161.95
	Steel Truss	291.80	207.80	189.55	175.90	170.85	167.70	161.25	157.05	155.30
Perimeter Adj., Add or Deduct	Per 100 L.F.	68.60	19.60	11.45	8.15	6.30	5.05	4.25	3.65	3.25
Story Hgt. Adj., Add or Deduct	Per 1 Ft.	3.90	2.15	1.70	1.40	1.30	1.20	1.10	0.90	0.90

For Basement, add $36.50 per square foot of basement area

The above costs were calculated using the basic specifications shown on the facing page. These costs should be adjusted where necessary for design alternatives and owner's requirements. Reported completed project costs, for this type of structure, range from $90.05 to $368.20 per S.F.

Common additives

Description	Unit	$ Cost
Altar, Wood, custom design, plain	Ea.	3300
Deluxe	Ea.	16,800
Granite or marble, average	Ea.	17,200
Deluxe	Ea.	46,100
Ark, Prefabricated, plain	Ea.	11,800
Deluxe	Ea.	152,500
Baptistry, Fiberglass, incl. plumbing	Ea.	7725 - 12,600
Bells & Carillons, 48 bells	Ea.	1,095,500
24 bells	Ea.	515,500
Confessional, Prefabricated wood		
Single, plain	Ea.	4575
Deluxe	Ea.	11,000
Double, plain	Ea.	8475
Deluxe	Ea.	23,900
Emergency Lighting, 25 watt, battery operated		
Lead battery	Ea.	345
Nickel cadmium	Ea.	685
Lecterns, Wood, plain	Ea.	980
Deluxe	Ea.	6725

Description	Unit	$ Cost
Pews/Benches, Hardwood	L.F.	141 - 234
Pulpits, Prefabricated, hardwood	Ea.	2025 - 11,800
Railing, Hardwood	L.F.	258
Steeples, translucent fiberglass		
30" square, 15' high	Ea.	11,300
25' high	Ea.	12,700
Painted fiberglass, 24" square, 14' high	Ea.	9500
28' high	Ea.	8675
Aluminum		
20' high, 3'- 6" base	Ea.	11,000
35' high, 8'- 0" base	Ea.	40,400
60' high, 14'- 0" base	Ea.	80,000

Important: See the Reference Section for Location Factors.

Model costs calculated for a 1 story building with 24' story height and 17,000 square feet of floor area

Church

				Unit	Unit Cost	Cost Per S.F.	% Of Sub-Total
A.	**SUBSTRUCTURE**						
1010	Standard Foundations	Poured concrete; strip and spread footings		S.F. Ground	4.18	4.18	
1020	Special Foundations	N/A		—	—	—	
1030	Slab on Grade	4" reinforced concrete with vapor barrier and granular base		S.F. Slab	5.56	5.56	10.5%
2010	Basement Excavation	Site preparation for slab and trench for foundation wall and footing		S.F. Ground	.18	.18	
2020	Basement Walls	4' Foundation wall		L.F. Wall	86	4.07	
B.	**SHELL**						
	B10 Superstructure						
1010	Floor Construction	N/A		—	—	—	15.8 %
1020	Roof Construction	Wood deck on laminated wood arches		S.F. Roof	19.71	21.03	
	B20 Exterior Enclosure						
2010	Exterior Walls	Face brick with concrete block backup	80% of wall (adjusted for end walls)	S.F. Wall	39.66	24.19	
2020	Exterior Windows	Aluminum, top hinged, in-swinging and curtain wall panels	20% of wall	S.F. Window	32.66	4.98	22.4%
2030	Exterior Doors	Double hollow metal swinging, single hollow metal		Each	1475	.52	
	B30 Roofing						
3010	Roof Coverings	Asphalt shingles with flashing; polystyrene insulation		S.F. Roof	6.63	6.63	5.0%
3020	Roof Openings	N/A		—	—	—	
C.	**INTERIORS**						
1010	Partitions	Plaster on metal studs	40 S.F. Floor/L.F. Partitions	S.F. Partition	15.52	9.31	
1020	Interior Doors	Hollow metal	400 S.F. Floor/Door	Each	1168	2.92	
1030	Fittings	N/A		—	—	—	
2010	Stair Construction	N/A		—	—	—	14.1%
3010	Wall Finishes	Paint		S.F. Surface	1.20	1.44	
3020	Floor Finishes	Carpet		S.F. Floor	5.05	5.05	
3030	Ceiling Finishes	N/A		—	—	—	
D.	**SERVICES**						
	D10 Conveying						
1010	Elevators & Lifts	N/A		—	—	—	0.0 %
1020	Escalators & Moving Walks	N/A		—	—	—	
	D20 Plumbing						
2010	Plumbing Fixtures	Kitchen, toilet and service fixtures, supply and drainage	1 Fixture/2430 S.F. Floor	Each	8068	3.32	
2020	Domestic Water Distribution	Gas fired hot water heater		S.F. Floor	.40	.40	2.8%
2040	Rain Water Drainage	N/A		—	—	—	
	D30 HVAC						
3010	Energy Supply	Oil fired hot water, wall fin radiation		S.F. Floor	10.65	10.65	
3020	Heat Generating Systems	N/A		—	—	—	
3030	Cooling Generating Systems	N/A		—	—	—	16.0%
3050	Terminal & Package Units	Split systems with air cooled condensing units		S.F. Floor	10.64	10.64	
3090	Other HVAC Sys. & Equipment	N/A		—	—	—	
	D40 Fire Protection						
4010	Sprinklers	Wet pipe sprinkler system		S.F. Floor	3.65	3.65	3.7%
4020	Standpipes	Standpipe		S.F. Floor	1.30	1.30	
	D50 Electrical						
5010	Electrical Service/Distribution	400 ampere service, panel board and feeders		S.F. Floor	.82	.82	
5020	Lighting & Branch Wiring	Fluorescent fixtures, receptacles, switches, A.C. and misc. power		S.F. Floor	8.67	8.67	9.6%
5030	Communications & Security	Alarm systems, sound system and emergency lighting		S.F. Floor	3.12	3.12	
5090	Other Electrical Systems	Emergency generator, 7.5 kW		S.F. Floor	.12	.12	
E.	**EQUIPMENT & FURNISHINGS**						
1010	Commercial Equipment	N/A		—	—	—	
1020	Institutional Equipment	N/A		—	—	—	
1030	Vehicular Equipment	N/A		—	—	—	0.0 %
1090	Other Equipment	N/A		—	—	—	
F.	**SPECIAL CONSTRUCTION**						
1020	Integrated Construction	N/A		—	—	—	0.0 %
1040	Special Facilities	N/A		—	—	—	
G.	**BUILDING SITEWORK**	**N/A**					

		Sub-Total	132.75	100%
CONTRACTOR FEES (General Requirements: 10%, Overhead: 5%, Profit: 10%)		25%	33.21	
ARCHITECT FEES		10%	16.59	
	Total Building Cost		**182.55**	

For customer support on your Square Foot Cost Data, call 877.756.2789.

97

Costs per square foot of floor area

Exterior Wall	S.F. Area	2000	4000	6000	8000	12000	15000	18000	20000	22000
	L.F. Perimeter	180	280	340	386	460	535	560	600	640
Stone Ashlar Veneer On Concrete Block	Wood Truss	262.95	234.85	219.20	209.75	198.85	195.70	191.00	189.80	188.65
	Steel Joists	267.55	238.85	222.80	213.05	201.85	198.65	193.80	192.55	191.40
Stucco on Concrete Block	Wood Truss	234.50	212.75	201.35	194.50	186.80	184.45	181.20	180.25	179.50
	Steel Joists	239.60	217.20	205.30	198.25	190.25	187.80	184.40	183.45	182.70
Brick Veneer	Wood Frame	265.95	225.35	206.90	196.40	184.75	181.05	176.55	175.15	174.00
Wood Shingles	Wood Frame	254.15	216.10	199.45	190.05	179.75	176.40	172.45	171.20	170.15
Perimeter Adj., Add or Deduct	Per 100 L.F.	47.15	23.60	15.75	11.80	7.85	6.30	5.25	4.70	4.35
Story Hgt. Adj., Add or Deduct	Per 1 Ft.	5.40	4.25	3.40	2.90	2.30	2.15	1.85	1.75	1.80

For Basement, add $31.95 per square foot of basement area

The above costs were calculated using the basic specifications shown on the facing page. These costs should be adjusted where necessary for design alternatives and owner's requirements. Reported completed project costs, for this type of structure, range from $89.05 to $296.35 per S.F.

Common additives

Description	Unit	$ Cost	Description	Unit	$ Cost
Bar, Front bar	L.F.	435	Sauna, Prefabricated, complete, 6' x 4'	Ea.	6750
Back bar	L.F.	345	6' x 9'	Ea.	10,500
Booth, Upholstered, custom, straight	L.F.	252 - 465	8' x 8'	Ea.	12,000
"L" or "U" shaped	L.F.	261 - 440	10' x 12'	Ea.	16,100
Fireplaces, Brick not incl. chimney			Smoke Detectors		
or foundation, 30" x 24" opening	Ea.	3200	Ceiling type	Ea.	227
Chimney, standard brick			Duct type	Ea.	565
Single flue 16" x 20"	V.L.F.	108	Sound System		
20" x 20"	V.L.F.	126	Amplifier, 250 watts	Ea.	2200
2 flue, 20" x 32"	V.L.F.	193	Speaker, ceiling or wall	Ea.	217
Lockers, Steel, single tier, 60" or 72"	Opng.	231 - 380	Trumpet	Ea.	415
2 tier, 60" or 72" total	Opng.	121 - 170	Steam Bath, Complete, to 140 C.F.	Ea.	3000
5 tier, box lockers	Opng.	74 - 84	To 300 C.F.	Ea.	3325
Locker bench, lam. maple top only	L.F.	34.50	To 800 C.F.	Ea.	7100
Pedestals, steel pipe	Ea.	76.50	To 2500 C.F.	Ea.	9325
Refrigerators, Prefabricated, walk-in			Swimming Pool Complete, gunite	S.F.	104 - 129
7'-6" High, 6' x 6'	S.F.	203	Tennis Court, Complete with fence		
10' x 10'	S.F.	159	Bituminous	Ea.	49,000 - 85,500
12' x 14'	S.F.	142	Clay	Ea.	52,500 - 87,000
12' x 20'	S.F.	125			

Important: See the Reference Section for Location Factors.

Model costs calculated for a 1 story building with 12' story height and 6,000 square feet of floor area

					Unit	Unit Cost	Cost Per S.F.	% Of Sub-Total
A. SUBSTRUCTURE								
1010	Standard Foundations	Poured concrete; strip and spread footings			S.F. Ground	2.58	2.58	
1020	Special Foundations	N/A			—	—	—	
1030	Slab on Grade	4" reinforced concrete with vapor barrier and granular base			S.F. Slab	5.56	5.56	9.1%
2010	Basement Excavation	Site preparation for slab and trench for foundation wall and footing			S.F. Ground	.32	.32	
2020	Basement Walls	4' Foundation wall			L.F. Wall	86	6.30	
B. SHELL								
B10 Superstructure								
1010	Floor Construction	6 x 6 wood columns			S.F. Floor	.06	.06	4.6%
1020	Roof Construction	Wood truss with plywood sheathing			S.F. Roof	7.42	7.42	
B20 Exterior Enclosure								
2010	Exterior Walls	Stone ashlar veneer on concrete block	65% of wall		S.F. Wall	45.20	19.98	
2020	Exterior Windows	Aluminum horizontal sliding	35% of wall		Each	525	8.33	19.2%
2030	Exterior Doors	Double aluminum and glass, hollow metal			Each	2907	2.90	
B30 Roofing								
3010	Roof Coverings	Asphalt shingles			S.F. Roof	18.72	3.20	2.0%
3020	Roof Openings	N/A			—	—	—	
C. INTERIORS								
1010	Partitions	Gypsum board on metal studs, load bearing	14 S.F. Floor/L.F. Partition		S.F. Partition	8.53	6.09	
1020	Interior Doors	Single leaf wood	140 S.F. Floor/Door		Each	626	4.47	
1030	Fittings	N/A			—	—	—	
2010	Stair Construction	N/A			—	—	—	16.8%
3010	Wall Finishes	40% vinyl wall covering, 40% paint, 20% ceramic tile			S.F. Surface	2.68	3.83	
3020	Floor Finishes	50% carpet, 30% hardwood tile, 20% ceramic tile			S.F. Floor	8.16	8.16	
3030	Ceiling Finishes	Gypsum plaster on wood furring			S.F. Ceiling	4.66	4.66	
D. SERVICES								
D10 Conveying								
1010	Elevators & Lifts	N/A			—	—	—	0.0 %
1020	Escalators & Moving Walks	N/A			—	—	—	
D20 Plumbing								
2010	Plumbing Fixtures	Kitchen, toilet and service fixtures, supply and drainage	1 Fixture/125 S.F. Floor		Each	2146	17.17	
2020	Domestic Water Distribution	Gas fired hot water			S.F. Floor	8.22	8.22	15.6%
2040	Rain Water Drainage	N/A			—	—	—	
D30 HVAC								
3010	Energy Supply	N/A			—	—	—	
3020	Heat Generating Systems	Included in D3050			—	—	—	
3030	Cooling Generating Systems	N/A			—	—	—	18.8 %
3050	Terminal & Package Units	Multizone rooftop unit, gas heating, electric cooling			S.F. Floor	30.52	30.52	
3090	Other HVAC Sys. & Equipment	N/A			—	—	—	
D40 Fire Protection								
4010	Sprinklers	Wet pipe sprinkler system			S.F. Floor	5.98	5.98	4.8%
4020	Standpipes	Standpipe			S.F. Floor	1.85	1.85	
D50 Electrical								
5010	Electrical Service/Distribution	400 ampere service, panel board and feeders			S.F. Floor	3.51	3.51	
5020	Lighting & Branch Wiring	High efficiency fluorescent fixtures, receptacles, switches, A.C. and misc. power			S.F. Floor	9.02	9.02	9.1%
5030	Communications & Security	Addressable alarm systems and emergency lighting			S.F. Floor	2	2	
5090	Other Electrical Systems	Emergency generator, 11.5 kW			S.F. Floor	.24	.24	
E. EQUIPMENT & FURNISHINGS								
1010	Commercial Equipment	N/A			—	—	—	
1020	Institutional Equipment	N/A			—	—	—	0.0 %
1030	Vehicular Equipment	N/A			—	—	—	
1090	Other Equipment	N/A			—	—	—	
F. SPECIAL CONSTRUCTION								
1020	Integrated Construction	N/A			—	—	—	0.0 %
1040	Special Facilities	N/A			—	—	—	
G. BUILDING SITEWORK	**N/A**							

Sub-Total		162.37	100%
CONTRACTOR FEES (General Requirements: 10%, Overhead: 5%, Profit: 10%)	25%	40.59	
ARCHITECT FEES	8%	16.24	
Total Building Cost		**219.20**	

For customer support on your Square Foot Cost Data, call 877.756.2789.

99

Costs per square foot of floor area

Exterior Wall	S.F. Area	4000	8000	12000	17000	22000	27000	32000	37000	42000
	L.F. Perimeter	280	386	520	585	640	740	840	940	940
Stone Ashlar on Concrete Block	Steel Joists	222.30	188.80	180.00	168.65	161.85	159.40	157.70	156.45	153.05
	Wood Joists	209.80	181.40	174.25	164.05	158.00	155.85	154.40	153.35	150.15
Face Brick on Concrete Block	Steel Joists	206.70	180.10	173.40	163.90	158.30	156.35	155.00	153.95	151.05
	Wood Joists	201.50	175.65	169.15	159.90	154.50	152.60	151.25	150.30	147.50
Decorative Concrete Block	Steel Joists	197.50	173.75	167.70	159.35	154.45	152.75	151.50	150.60	148.05
	Wood Joists	190.30	167.45	161.60	153.60	148.90	147.25	146.05	145.25	142.75
Perimeter Adj., Add or Deduct	Per 100 L.F.	26.05	13.00	8.70	6.05	4.80	3.85	3.30	2.85	2.40
Story Hgt. Adj., Add or Deduct	Per 1 Ft.	4.75	3.25	2.95	2.30	2.00	1.85	1.80	1.70	1.50
For Basement, add $32.60 per square foot of basement area										

The above costs were calculated using the basic specifications shown on the facing page. These costs should be adjusted where necessary for design alternatives and owner's requirements. Reported completed project costs, for this type of structure, range from $91.10 to $261.75 per S.F.

Common additives

Description	Unit	$ Cost	Description	Unit	$ Cost
Bar, Front bar	L.F.	435	Kitchen Equipment, cont.		
Back bar	L.F.	345	Freezer, 44 C.F., reach-in	Ea.	6175
Booth, Upholstered, custom, straight	L.F.	252 - 465	Ice cube maker, 50 lb. per day	Ea.	2000
"L" or "U" shaped	L.F.	261 - 440	Lockers, Steel, single tier, 60" or 72"	Opng.	231 - 380
Emergency Lighting, 25 watt, battery operated			2 tier, 60" or 72" total	Opng.	121 - 170
Lead battery	Ea.	345	5 tier, box lockers	Opng.	74 - 84
Nickel cadmium	Ea.	685	Locker bench, lam. maple top only	L.F.	34.50
Flagpoles, Complete			Pedestals, steel pipe	Ea.	76.50
Aluminum, 20' high	Ea.	1900	Refrigerators, Prefabricated, walk-in		
40' High	Ea.	4425	7'-6" High, 6' x 6'	S.F.	203
70' High	Ea.	11,500	10' x 10'	S.F.	159
Fiberglass, 23' High	Ea.	1350	12' x 14'	S.F.	142
39'-5" High	Ea.	3400	12' x 20'	S.F.	125
59' High	Ea.	6650	Smoke Detectors		
Kitchen Equipment			Ceiling type	Ea.	227
Broiler	Ea.	4075	Duct type	Ea.	565
Coffee urn, twin 6 gallon	Ea.	3025	Sound System		
Cooler, 6 ft. long, reach-in	Ea.	3925	Amplifier, 250 watts	Ea.	2200
Dishwasher, 10-12 racks per hr.	Ea.	4275	Speaker, ceiling or wall	Ea.	217
Food warmer, counter 1.2 kw	Ea.	770	Trumpet	Ea.	415

Important: See the Reference Section for Location Factors.

Model costs calculated for a 1 story building with 12' story height and 22,000 square feet of floor area

				Unit	Unit Cost	Cost Per S.F.	% Of Sub-Total
A. SUBSTRUCTURE							
1010	Standard Foundations	Poured concrete; strip and spread footings		S.F. Ground	1.45	1.45	
1020	Special Foundations	N/A		—	—	—	
1030	Slab on Grade	4" reinforced concrete with vapor barrier and granular base		S.F. Slab	5.56	5.56	8.0%
2010	Basement Excavation	Site preparation for slab and trench for foundation wall and footing		S.F. Ground	.18	.18	
2020	Basement Walls	4' Foundation wall		L.F. Wall	86	2.48	
B. SHELL							
B10 Superstructure							
1010	Floor Construction	Steel column		S.F. Floor	.65	.65	6.3%
1020	Roof Construction	Metal deck on open web steel joists		S.F. Roof	6.96	6.96	
B20 Exterior Enclosure							
2010	Exterior Walls	Stone, ashlar veneer on concrete block	65% of wall	S.F. Wall	45.22	10.26	
2020	Exterior Windows	Window wall	35% of wall	Each	51	6.24	14.3%
2030	Exterior Doors	Double aluminum and glass doors		Each	2907	.80	
B30 Roofing							
3010	Roof Coverings	Built-up tar and gravel with flashing; perlite/EPS composite insulation		S.F. Roof	6.46	6.46	5.3%
3020	Roof Openings	N/A		—	—	—	
C. INTERIORS							
1010	Partitions	Lightweight concrete block	14 S.F. Floor/L.F. Partition	S.F. Partition	11.90	8.50	
1020	Interior Doors	Single leaf wood	140 S.F. Floor/Door	Each	626	4.47	
1030	Fittings	N/A		—	—	—	
2010	Stair Construction	N/A		—	—	—	25.1%
3010	Wall Finishes	65% paint, 25% vinyl wall covering, 10% ceramic tile		S.F. Surface	1.81	2.59	
3020	Floor Finishes	60% carpet, 35% hardwood, 15% ceramic tile		S.F. Floor	7.54	7.54	
3030	Ceiling Finishes	Mineral fiber tile on concealed zee bars		S.F. Ceiling	7.22	7.22	
D. SERVICES							
D10 Conveying							
1010	Elevators & Lifts	N/A		—	—	—	0.0 %
1020	Escalators & Moving Walks	N/A		—	—	—	
D20 Plumbing							
2010	Plumbing Fixtures	Kitchen, toilet and service fixtures, supply and drainage	1 Fixture/1050 S.F. Floor	Each	8799	8.38	
2020	Domestic Water Distribution	Gas fired hot water heater		S.F. Floor	.34	.34	7.9%
2040	Rain Water Drainage	Roof drains		S.F. Roof	.82	.82	
D30 HVAC							
3010	Energy Supply	N/A		—	—	—	
3020	Heat Generating Systems	Included in D3050		—	—	—	
3030	Cooling Generating Systems	N/A		—	—	—	18.9 %
3050	Terminal & Package Units	Multizone rooftop unit, gas heating, electric cooling		S.F. Floor	22.82	22.82	
3090	Other HVAC Sys. & Equipment	N/A		—	—	—	
D40 Fire Protection							
4010	Sprinklers	Wet pipe sprinkler system		S.F. Floor	5.09	5.09	4.6%
4020	Standpipes	Standpipe		S.F. Floor	.50	.50	
D50 Electrical							
5010	Electrical Service/Distribution	400 ampere service, panel board and feeders		S.F. Floor	1.23	1.23	
5020	Lighting & Branch Wiring	High efficiency fluorescent fixtures, receptacles, switches, A.C. and misc. power		S.F. Floor	8.83	8.83	9.7%
5030	Communications & Security	Addressable alarm systems and emergency lighting		S.F. Floor	1.52	1.52	
5090	Other Electrical Systems	Emergency generator, 11.5 kW		S.F. Floor	.13	.13	
E. EQUIPMENT & FURNISHINGS							
1010	Commercial Equipment	N/A		—	—	—	
1020	Institutional Equipment	N/A		—	—	—	0.0 %
1030	Vehicular Equipment	N/A		—	—	—	
1090	Other Equipment	N/A		—	—	—	
F. SPECIAL CONSTRUCTION							
1020	Integrated Construction	N/A		—	—	—	0.0 %
1040	Special Facilities	N/A		—	—	—	
G. BUILDING SITEWORK	**N/A**						

		Sub-Total	121.02	100%
CONTRACTOR FEES (General Requirements: 10%, Overhead: 5%, Profit: 10%)		25%	30.24	
ARCHITECT FEES		7%	10.59	
	Total Building Cost		**161.85**	

For customer support on your Square Foot Cost Data, call 877.756.2789.

101

Costs per square foot of floor area

Exterior Wall	S.F. Area	15000	20000	28000	38000	50000	65000	85000	100000	150000
	L.F. Perimeter	350	400	480	550	630	660	750	825	1035
Fiber Cement Siding	Wood Frame	189.00	180.50	173.25	167.45	163.40	159.20	156.60	155.40	152.85
Stone Veneer	Wood Frame	201.70	191.25	182.25	174.85	169.70	164.05	160.60	159.10	155.75
Brick Veneer	Steel Frame	185.80	176.70	168.75	162.50	158.10	153.45	150.55	149.35	146.55
Curtain Wall	Steel Frame	205.15	193.40	183.25	174.90	169.10	162.55	158.60	156.85	153.05
Brick Veneer	Reinforced Concrete	211.05	200.35	191.10	183.60	178.35	172.45	169.00	167.40	164.00
Stone Veneer	Reinforced Concrete	221.20	208.30	197.10	187.75	181.25	173.80	169.40	167.45	163.10
Perimeter Adj., Add or Deduct	Per 100 L.F.	8.30	6.15	4.30	3.15	2.50	1.90	1.45	1.20	0.85
Story Hgt. Adj., Add or Deduct	Per 1 Ft.	1.20	1.05	0.80	0.70	0.65	0.45	0.45	0.45	0.30

For Basement, add $38.00 per square foot of basement area

The above costs were calculated using the basic specifications shown on the facing page. These costs should be adjusted where necessary for design alternatives and owner's requirements. Reported completed project costs, for this type of structure, range from $131.75 to $309.65 per S.F.

Common additives

Description	Unit	$ Cost
Carrels Hardwood	Ea.	945 - 2400
Clock System		
20 Room	Ea.	20,600
50 Room	Ea.	48,500
Elevators, Hydraulic passenger, 2 stops		
1500# capacity	Ea.	67,400
2500# capacity	Ea.	70,400
3500# capacity	Ea.	75,400
Additional stop, add	Ea.	8075
Emergency Lighting, 25 watt, battery operated		
Lead battery	Ea.	345
Nickel cadmium	Ea.	685
Flagpoles, Complete		
Aluminum, 20' high	Ea.	1900
40' High	Ea.	4425
70' High	Ea.	11,500
Fiberglass, 23' High	Ea.	1350
39'-5" High	Ea.	3400
59' High	Ea.	6650

Description	Unit	$ Cost
Lockers, Steel, single tier, 60" or 72"	Opng.	231 - 380
2 tier, 60" or 72" total	Opng.	121 - 170
5 tier, box lockers	Opng.	74 - 84
Locker bench, lam. maple top only	L.F.	34.50
Pedestals, steel pipe	Ea.	76.50
Seating		
Auditorium chair, all veneer	Ea.	315
Veneer back, padded seat	Ea.	325
Upholstered, spring seat	Ea.	320
Classroom, movable chair & desk	Set	81 - 171
Lecture hall, pedestal type	Ea.	275 - 620
Smoke Detectors		
Ceiling type	Ea.	227
Duct type	Ea.	565
Sound System		
Amplifier, 250 watts	Ea.	2200
Speaker, ceiling or wall	Ea.	217
Trumpet	Ea.	415
TV Antenna, Master system, 12 outlet	Outlet	263
30 outlet	Outlet	225
100 outlet	Outlet	213

Important: See the Reference Section for Location Factors.

Model costs calculated for a 2 story building with 12' story height and 50,000 square feet of floor area

College, Classroom, 2-3 Story

			Unit	Unit Cost	Cost Per S.F.	% Of Sub-Total
A. SUBSTRUCTURE						
1010	Standard Foundations	Poured concrete; strip and spread footings	S.F. Ground	2.02	1.01	
1020	Special Foundations	N/A	—	—	—	
1030	Slab on Grade	4" reinforced slab on grade	S.F. Slab	5.56	2.79	4.0%
2010	Basement Excavation	Site preparation for slab and trench for foundation wall and footing	S.F. Ground	.32	.16	
2020	Basement Walls	4' Foundation wall	L.F. Wall	42.75	1.07	
B. SHELL						
B10 Superstructure						
1010	Floor Construction	Open web steel joists, slab form, concrete, fireproofed steel columns	S.F. Floor	15.56	7.78	10.3%
1020	Roof Construction	Metal deck on open web steel joints, columns	S.F. Roof	10.40	5.20	
B20 Exterior Enclosure						
2010	Exterior Walls	N/A	—	—	—	
2020	Exterior Windows	Curtain wall glazing system, thermo-break frame 100% of wall	S.F. Wall	47.50	14.22	11.8 %
2030	Exterior Doors	Aluminum and glass double doors	Each	5875	.71	
B30 Roofing						
3010	Roof Coverings	Single ply membrane, stone ballast, rigid insulation	S.F. Roof	5.58	2.79	2.3%
3020	Roof Openings	Roof and smoke hatches	S.F. Roof	2060	.08	
C. INTERIORS						
1010	Partitions	Gypsum board on Metal studs 20 S.F. Floor/L.F. Partition	S.F. Partition	6.34	3.17	
1020	Interior Doors	Single leaf hollow metal doors 420 S.F. Floor/Door	Each	1168	2.77	
1030	Fittings	Toilet partitions, chalkboards, classroom cabinetry	S.F. Floor	9.38	9.38	
2010	Stair Construction	Cement filled metal pan	Flight	17,275	1.73	22.8%
3010	Wall Finishes	95% paint, 5% ceramic wall tile	S.F. Surface	1.40	1.40	
3020	Floor Finishes	25% Carpet tile, 35% vinyl composition tile, 5% ceramic tile	S.F. Floor	3.17	3.17	
3030	Ceiling Finishes	Acoustic ceiling tiles on suspended channel grid	S.F. Ceiling	7.22	7.22	
D. SERVICES						
D10 Conveying						
1010	Elevators & Lifts	Two hydraulic passenger elevators	Each	85,250	3.41	2.7%
1020	Escalators & Moving Walks	N/A	—	—	—	
D20 Plumbing						
2010	Plumbing Fixtures	Restroom and service fixtures, supply and drainage 1 Fixture/395 S.F. Floor	Each	2250	5.71	
2020	Domestic Water Distribution	Gas fired water heater	S.F. Floor	2.84	2.84	7.1%
2040	Rain Water Drainage	Roof drains	S.F. Roof	.92	.46	
D30 HVAC						
3010	Energy Supply	N/A	—	—	—	
3020	Heat Generating Systems	Included in D3050	—	—	—	
3030	Cooling Generating Systems	N/A	—	—	—	16.7 %
3050	Terminal & Package Units	Multizone gas heating, electric cooling unit	S.F. Floor	21.15	21.15	
3090	Other HVAC Sys. & Equipment	N/A	—	—	—	
D40 Fire Protection						
4010	Sprinklers	Wet pipe spinkler system, light hazard	S.F. Floor	3.15	3.15	2.8%
4020	Standpipes	Standpipes	S.F. Floor	.39	.39	
D50 Electrical						
5010	Electrical Service/Distribution	2000 ampere service, panel board and feeders	S.F. Floor	3.82	3.82	
5020	Lighting & Branch Wiring	Incandescent fixtures, receptacles, switches, A.C. and misc. power	S.F. Floor	13.56	13.56	19.5%
5030	Communications & Security	Addressable alarm systems, emergency lighting, internet and phone wiring	S.F. Floor	7.29	7.29	
5090	Other Electrical Systems	N/A	—	—	—	
E. EQUIPMENT & FURNISHINGS						
1010	Commercial Equipment	N/A	—	—	—	
1020	Institutional Equipment	N/A	—	—	—	
1030	Vehicular Equipment	N/A	—	—	—	0.0 %
1090	Other Equipment	N/A	—	—	—	
F. SPECIAL CONSTRUCTION						
1020	Integrated Construction	N/A	—	—	—	0.0 %
1040	Special Facilities	N/A	—	—	—	
G. BUILDING SITEWORK	**N/A**					

		Sub-Total	126.43	100%
CONTRACTOR FEES (General Requirements: 10%, Overhead: 5%, Profit: 10%)		25%	31.61	
ARCHITECT FEES		7%	11.06	

Total Building Cost	**169.10**

For customer support on your Square Foot Cost Data, call 877.756.2789.

103

Costs per square foot of floor area

Exterior Wall	S.F. Area	10000	15000	25000	40000	55000	70000	80000	90000	100000
	L.F. Perimeter	260	320	400	476	575	628	684	721	772
Fiber Cement Siding	Wood Frame	214.25	196.85	180.85	170.30	166.05	162.80	161.65	160.40	159.60
Stone Veneer	Wood Frame	240.70	219.10	198.25	184.00	178.50	173.85	172.35	170.65	169.65
Brick Veneer	Steel Frame	220.15	200.30	181.40	168.65	163.65	159.55	158.20	156.65	155.70
Curtain Wall	Steel Frame	245.45	222.35	199.65	184.00	177.95	172.75	171.15	169.25	168.15
Brick Veneer	Reinforced Concrete	247.50	224.75	202.45	187.10	181.10	176.10	174.45	172.60	171.50
Stone Veneer	Reinforced Concrete	281.45	251.65	221.30	199.75	191.65	184.40	182.10	179.40	177.90
Perimeter Adj., Add or Deduct	Per 100 L.F.	13.65	9.20	5.45	3.45	2.60	1.95	1.70	1.55	1.40
Story Hgt. Adj., Add or Deduct	Per 1 Ft.	2.10	1.75	1.25	0.95	0.90	0.75	0.65	0.65	0.70
For Basement, add $37.00 per square foot of basement area										

The above costs were calculated using the basic specifications shown on the facing page. These costs should be adjusted where necessary for design alternatives and owner's requirements. Reported completed project costs, for this type of structure, range from $85.10 to $250.80 per S.F.

Common additives

Description	Unit	$ Cost		Description	Unit	$ Cost
Carrels Hardwood	Ea.	945 - 2400		Kitchen Equipment		
Closed Circuit Surveillance, One station				Broiler	Ea.	4075
Camera and monitor	Ea.	2025		Coffee urn, twin 6 gallon	Ea.	3025
For additional camera stations, add	Ea.	1100		Cooler, 6 ft. long	Ea.	3925
Elevators, Hydraulic passenger, 2 stops				Dishwasher, 10-12 racks per hr.	Ea.	4275
2000# capacity	Ea.	68,400		Food warmer	Ea.	770
2500# capacity	Ea.	70,400		Freezer, 44 C.F., reach-in	Ea.	6175
3500# capacity	Ea.	75,400		Ice cube maker, 50 lb. per day	Ea.	2000
Additional stop, add	Ea.	8075		Range with 1 oven	Ea.	3300
Emergency Lighting, 25 watt, battery operated				Laundry Equipment		
Lead battery	Ea.	345		Dryer, gas, 16 lb. capacity	Ea.	995
Nickel cadmium	Ea.	685		30 lb. capacity	Ea.	3950
Furniture	Student	2800 - 5350		Washer, 4 cycle	Ea.	1225
Intercom System, 25 station capacity				Commercial	Ea.	1625
Master station	Ea.	3175		Smoke Detectors		
Intercom outlets	Ea.	192		Ceiling type	Ea.	227
Handset	Ea.	525		Duct type	Ea.	565
				TV Antenna, Master system, 12 outlet	Outlet	263
				30 outlet	Outlet	225
				100 outlet	Outlet	213

Important: See the Reference Section for Location Factors.

Model costs calculated for a 3 story building with 12' story height and 25,000 square feet of floor area

				Unit	Unit Cost	Cost Per S.F.	% Of Sub-Total
A. SUBSTRUCTURE							
1010	Standard Foundations	Poured concrete; strip and spread footings		S.F. Ground	9.48	3.16	
1020	Special Foundations	N/A		—	—	—	
1030	Slab on Grade	4" reinforced slab on grade		S.F. Slab	5.56	1.85	4.8%
2010	Basement Excavation	Site preparation for slab and trench for foundation wall and footing		S.F. Ground	.18	.06	
2020	Basement Walls	4' Foundation wall		L.F. Wall	86	1.37	
B. SHELL							
B10 Superstructure							
1010	Floor Construction	Wood Beam and Joist on wood columns, fireproofed		S.F. Floor	25.37	16.91	14.3%
1020	Roof Construction	Wood roof, truss, 4/12 slope		S.F. Roof	7.41	2.47	
B20 Exterior Enclosure							
2010	Exterior Walls	Vinyl siding on wood studs, insulated	80% of wall	S.F. Wall	15.69	7.23	
2020	Exterior Windows	Aluminum awning windows	20% of wall	Each	746	3.73	9.2%
2030	Exterior Doors	Aluminum and glass double doors		Each	6125	1.47	
B30 Roofing							
3010	Roof Coverings	Asphalt roofing, strip shingles, flashing, gutters and downspouts		S.F. Roof	4.17	1.39	1.0%
3020	Roof Openings	N/A		—	—	—	
C. INTERIORS							
1010	Partitions	Sound deadening gypsum board on wood studs	10 S.F. Floor/L.F. Partition	S.F. Partition	8.96	9.95	
1020	Interior Doors	Single leaf solid core wood doors	100 S.F. Floor/Door	Each	626	6.46	
1030	Fittings	Toilet partitions and bathroom accessories		S.F. Floor	3.81	3.81	
2010	Stair Construction	Prefabricated wood stairs with rails		Flight	2745	1.21	23.1%
3010	Wall Finishes	95% paint, 5% ceramic wall tile		S.F. Surface	1.32	2.94	
3020	Floor Finishes	80% Carpet tile, 10% vinyl composition tile, 10% ceramic tile		S.F. Floor	6.22	6.22	
3030	Ceiling Finishes	Acoustic ceiling tiles on suspended channel grid		S.F. Ceiling	.58	.58	
D. SERVICES							
D10 Conveying							
1010	Elevators & Lifts	Hydraulic passenger elevator		Each	117,000	4.68	3.5%
1020	Escalators & Moving Walks	N/A		—	—	—	
D20 Plumbing							
2010	Plumbing Fixtures	Restroom and service fixtures, supply and drainage	1 Fixture/300 S.F. Floor	Each	4295	14.27	
2020	Domestic Water Distribution	Electric water heater		S.F. Floor	5.82	5.82	15.5%
2040	Rain Water Drainage	Roof drains		S.F. Roof	2.40	.80	
D30 HVAC							
3010	Energy Supply	N/A		—	—	—	
3020	Heat Generating Systems	Included in D3050		—	—	—	
3030	Cooling Generating Systems	N/A		—	—	—	9.8%
3050	Terminal & Package Units	Multizone gas heating, electric cooling unit		S.F. Floor	13.25	13.25	
3090	Other HVAC Sys. & Equipment	N/A		—	—	—	
D40 Fire Protection							
4010	Sprinklers	Wet pipe spinkler system, light hazard		S.F. Floor	2.96	2.96	2.9%
4020	Standpipes	Standpipes		S.F. Floor	.96	.96	
D50 Electrical							
5010	Electrical Service/Distribution	600 ampere service, panel board and feeders		S.F. Floor	2.42	2.42	
5020	Lighting & Branch Wiring	Fluorescent fixtures, receptacles, switches, A.C. and misc. power		S.F. Floor	10.24	10.24	14.9%
5030	Communications & Security	Addressable alarm systems, emergency lighting, internet and phone wiring		S.F. Floor	7.47	7.47	
5090	Other Electrical Systems	N/A		—	—	—	
E. EQUIPMENT & FURNISHINGS							
1010	Commercial Equipment	N/A		—	—	—	
1020	Institutional Equipment	N/A		—	—	—	
1030	Vehicular Equipment	N/A		—	—	—	1.1%
2020	Moveable Furnishings	Dormitory furniture		S.F. Floor	1.52	1.52	
F. SPECIAL CONSTRUCTION							
1020	Integrated Construction	N/A		—	—	—	0.0%
1040	Special Facilities	N/A		—	—	—	
G. BUILDING SITEWORK	**N/A**						

	Sub-Total	135.20	**100%**
CONTRACTOR FEES (General Requirements: 10%, Overhead: 5%, Profit: 10%)	25%	33.82	
ARCHITECT FEES	7%	11.83	
	Total Building Cost	**180.85**	

For customer support on your Square Foot Cost Data, call 877.756.2789.

105

Costs per square foot of floor area

Exterior Wall	S.F. Area	20000	35000	45000	65000	85000	110000	135000	160000	200000
	L.F. Perimeter	260	340	400	440	500	540	560	590	640
Brick Veneer	Steel Frame	189.00	174.50	170.75	162.30	158.65	154.80	151.80	150.00	148.10
Curtain Wall	Steel Frame	209.05	189.55	184.55	172.95	167.90	162.60	158.45	155.95	153.30
E.I.F.S.	Steel Frame	181.65	169.30	166.15	159.10	156.05	152.85	150.40	148.90	147.30
Precast Concrete Panel	Tilt Up Panel	233.15	208.45	202.15	187.15	180.70	173.85	168.55	165.30	161.90
Brick Veneer	Reinforced Concrete	214.35	195.35	190.45	179.15	174.30	169.15	165.15	162.65	160.10
Stone Veneer	Reinforced Concrete	252.80	222.65	214.95	196.50	188.55	180.10	173.55	169.50	165.30
Perimeter Adj., Add or Deduct	Per 100 L.F.	18.10	10.35	8.05	5.55	4.20	3.30	2.70	2.30	1.80
Story Hgt. Adj., Add or Deduct	Per 1 Ft.	3.20	2.40	2.15	1.70	1.40	1.20	1.05	0.90	0.80

For Basement, add $37.00 per square foot of basement area

The above costs were calculated using the basic specifications shown on the facing page. These costs should be adjusted where necessary for design alternatives and owner's requirements. Reported completed project costs, for this type of structure, range from $120.95 to $291.75 per S.F.

Common additives

Description	Unit	$ Cost		Description	Unit	$ Cost
Carrels Hardwood	Ea.	945 - 2400		Kitchen Equipment		
Closed Circuit Surveillance, One station				Broiler	Ea.	4075
Camera and monitor	Ea.	2025		Coffee urn, twin, 6 gallon	Ea.	3025
For additional camera stations, add	Ea.	1100		Cooler, 6 ft. long	Ea.	3925
Elevators, Electric passenger, 5 stops				Dishwasher, 10-12 racks per hr.	Ea.	4275
2000# capacity	Ea.	171,100		Food warmer	Ea.	770
2500# capacity	Ea.	175,600		Freezer, 44 C.F., reach-in	Ea.	6175
3500# capacity	Ea.	177,600		Ice cube maker, 50 lb. per day	Ea.	2000
Additional stop, add	Ea.	10,500		Range with 1 oven	Ea.	3300
Emergency Lighting, 25 watt, battery operated				Laundry Equipment		
Lead battery	Ea.	345		Dryer, gas, 16 lb. capacity	Ea.	995
Nickel cadmium	Ea.	685		30 lb. capacity	Ea.	3950
Furniture	Student	2800 - 5350		Washer, 4 cycle	Ea.	1225
Intercom System, 25 station capacity				Commercial	Ea.	1625
Master station	Ea.	3175		Smoke Detectors		
Intercom outlets	Ea.	192		Ceiling type	Ea.	227
Handset	Ea.	525		Duct type	Ea.	565
				TV Antenna, Master system, 12 outlet	Outlet	263
				30 outlet	Outlet	225
				100 outlet	Outlet	213

Important: See the Reference Section for Location Factors.

Model costs calculated for a 6 story building with 12' story height and 85,000 square feet of floor area

					Unit	Unit Cost	Cost Per S.F.	% Of Sub-Total
A. SUBSTRUCTURE								
1010	Standard Foundations	Poured concrete; strip and spread footings			S.F. Ground	5.40	.90	
1020	Special Foundations	N/A			—	—	—	
1030	Slab on Grade	4" reinforced slab on grade			S.F. Slab	5.56	.93	1.8%
2010	Basement Excavation	Site preparation for slab and trench for foundation wall and footing			S.F. Ground	.32	.05	
2020	Basement Walls	4' Foundation wall			L.F. Wall	42.75	.50	
B. SHELL								
B10 Superstructure								
1010	Floor Construction	Cast-in-place column, I beam, precast double T beam & topping			S.F. Floor	25.34	21.12	17.5%
1020	Roof Construction	Precast double T beams with 2" topping			S.F. Roof	15.24	2.54	
B20 Exterior Enclosure								
2010	Exterior Walls	Precast concrete panels, insulated	80% of wall		S.F. Wall	48.34	16.38	
2020	Exterior Windows	Aluminum sliding windows	20% of wall		Each	525	2.96	14.6%
2030	Exterior Doors	Aluminum and glass double doors and single doors			Each	3725	.34	
B30 Roofing								
3010	Roof Coverings	Single ply membrane, stone ballast, rigid insulation			S.F. Roof	6	1	0.8%
3020	Roof Openings	Roof and smoke hatches			S.F. Roof	2060	.09	
C. INTERIORS								
1010	Partitions	Sound deadening gypsum board on metal studs, CMU partitions	10 S.F. Floor/L.F. Partition		S.F. Partition	6.14	6.82	
1020	Interior Doors	Single leaf solid core wood doors	100 S.F. Floor/Door		Each	626	6.46	
1030	Fittings	Toilet partitions and bathroom accessories			S.F. Floor	1.68	1.68	
2010	Stair Construction	Cement filled metal pan with rails			Flight	17,275	4.07	23.5%
3010	Wall Finishes	95% paint, 5% ceramic wall tile			S.F. Surface	2.60	5.77	
3020	Floor Finishes	80% Carpet tile, 10% vinyl composition tile, 10% ceramic tile			S.F. Floor	6.36	6.36	
3030	Ceiling Finishes	Acoustic ceiling tiles on suspended channel grid			S.F. Ceiling	.58	.58	
D. SERVICES								
D10 Conveying								
1010	Elevators & Lifts	Traction geared passenger elevators			Each	249,050	11.72	8.7%
1020	Escalators & Moving Walks	N/A			—	—	—	
D20 Plumbing								
2010	Plumbing Fixtures	Restroom and service fixtures, supply and drainage	1 Fixture/300 S.F. Floor		Each	1969	6.54	
2020	Domestic Water Distribution	Electric water heater			S.F. Floor	2.30	2.30	6.7%
2040	Rain Water Drainage	Roof drains			S.F. Roof	1.62	.27	
D30 HVAC								
3010	Energy Supply	Oil fired hot water, wall fin radiation			S.F. Floor	4.57	4.57	
3020	Heat Generating Systems	N/A			—	—	—	
3030	Cooling Generating Systems	Chilled water, air cooled condenser system			S.F. Floor	9.54	9.54	10.4%
3050	Terminal & Package Units	N/A			—	—	—	
3090	Other HVAC Sys. & Equipment	N/A			—	—	—	
D40 Fire Protection								
4010	Sprinklers	Wet pipe spinkler system, light hazard			S.F. Floor	2.89	2.89	2.7%
4020	Standpipes	Standpipes			S.F. Floor	.71	.71	
D50 Electrical								
5010	Electrical Service/Distribution	800 ampere service, panel board and feeders			S.F. Floor	.88	.88	
5020	Lighting & Branch Wiring	Fluorescent fixtures, receptacles, switches, A.C. and misc. power			S.F. Floor	10.06	10.06	12.8%
5030	Communications & Security	Addressable alarm systems, emergency lighting, internet and phone wiring			S.F. Floor	6.42	6.42	
5090	Other Electrical Systems	N/A			—	—	—	
E. EQUIPMENT & FURNISHINGS								
1010	Commercial Equipment	N/A			—	—	—	
1020	Institutional Equipment	N/A			—	—	—	
1030	Vehicular Equipment	N/A			—	—	—	0.5 %
2020	Moveable Furnishings	Dormitory furniture			S.F. Floor	.67	.67	
F. SPECIAL CONSTRUCTION								
1020	Integrated Construction	N/A			—	—	—	0.0 %
1040	Special Facilities	N/A			—	—	—	
G. BUILDING SITEWORK	**N/A**							

		Sub-Total	135.12	100%
CONTRACTOR FEES (General Requirements: 10%, Overhead: 5%, Profit: 10%)			25%	33.76	
ARCHITECT FEES			7%	11.82	

Total Building Cost   **180.70**

Costs per square foot of floor area

Exterior Wall	S.F. Area	12000	20000	28000	37000	45000	57000	68000	80000	92000
	L.F. Perimeter	470	600	698	793	900	1000	1075	1180	1275
Brick Veneer	Steel Frame	219.40	190.90	177.10	168.20	163.90	158.35	154.65	152.20	150.15
Curtain Wall	Steel Frame	234.80	202.20	186.15	175.70	170.75	164.05	159.55	156.65	154.25
EFIS	Steel Frame	217.05	189.80	176.70	168.35	164.20	159.05	155.55	153.25	151.40
Precast Concrete Panel	Reinforced Concrete	242.30	210.15	194.35	184.05	179.20	172.65	168.25	165.40	163.00
Brick Veneer	Reinforced Concrete	237.05	206.10	191.00	181.20	176.55	170.25	166.10	163.40	161.20
Stone Veneer	Reinforced Concrete	267.40	228.75	209.35	196.60	190.75	182.40	176.80	173.25	170.20
Perimeter Adj., Add or Deduct	Per 100 L.F.	11.15	6.65	4.75	3.60	2.95	2.40	1.90	1.60	1.45
Story Hgt. Adj., Add or Deduct	Per 1 Ft.	1.25	0.95	0.80	0.75	0.65	0.55	0.50	0.45	0.45

For Basement, add $23.90 per square foot of basement area

The above costs were calculated using the basic specifications shown on the facing page. These costs should be adjusted where necessary for design alternatives and owner's requirements. Reported completed project costs, for this type of structure, range from $174.20 to $325.35 per S.F.

Common additives

Description	Unit	$ Cost		Description	Unit	$ Cost
Cabinets, Base, door units, metal	L.F.	325		Safety Equipment, Eye wash, hand held	Ea.	445
Drawer units	L.F.	645		Deluge shower	Ea.	850
Tall storage cabinets, open	L.F.	620		Sink, One piece plastic		
With doors	L.F.	900		Flask wash, freestanding	Ea.	2600
Wall, metal 12-1/2" deep, open	L.F.	263		Tables, acid resist. top, drawers	L.F.	221
With doors	L.F.	460		Titration Unit, Four 2000 ml reservoirs	Ea.	6100
Carrels Hardwood	Ea.	945 - 2400				
Countertops, not incl. base cabinets, acid proof	S.F.	61.50 - 74.50				
Stainless steel	S.F.	180				
Fume Hood, Not incl. ductwork	L.F.	780 - 1900				
Ductwork	Hood	5950 - 9775				
Glassware Washer, Distilled water rinse	Ea.	7625 - 16,700				
Seating						
Auditorium chair, all veneer	Ea.	315				
Veneer back, padded seat	Ea.	325				
Upholstered, spring seat	Ea.	320				
Classroom, movable chair & desk	Set	81 - 171				
Lecture hall, pedestal type	Ea.	275 - 620				

Model costs calculated for a 1 story building with 12' story height and 45,000 square feet of floor area

College, Laboratory

				Unit	Unit Cost	Cost Per S.F.	% Of Sub-Total
A.	**SUBSTRUCTURE**						
1010	Standard Foundations	Poured concrete; strip and spread footings		S.F. Ground	2.12	2.12	
1020	Special Foundations	N/A		—	—	—	
1030	Slab on Grade	4" reinforced slab on grade		S.F. Slab	5.56	5.56	8.0%
2010	Basement Excavation	Site preparation for slab and trench for foundation wall and footing		S.F. Ground	.18	.18	
2020	Basement Walls	4' Foundation wall		L.F. Wall	42.75	2.14	
B.	**SHELL**						
	B10 Superstructure						
1010	Floor Construction	Structural fireproofing		S.F. Floor	6.92	.90	9.0%
1020	Roof Construction	Metal deck on open web steel joists, columns		S.F. Roof	10.38	10.38	
	B20 Exterior Enclosure						
2010	Exterior Walls	N/A		—	—	—	
2020	Exterior Windows	Curtain wall glazing system, thermo-break frame	100% of wall	S.F. Wall	47.50	11.30	10.8 %
2030	Exterior Doors	Aluminum and glass single and double doors		Each	5053	2.24	
	B30 Roofing						
3010	Roof Coverings	Single ply membrane, stone ballast, rigid insulation		S.F. Roof	5.33	5.33	4.5%
3020	Roof Openings	Roof and smoke hatches		S.F. Roof	.28	.28	
C.	**INTERIORS**						
1010	Partitions	Gypsum board on metal studs	10 S.F. Floor/L.F. Partition	S.F. Partition	5.56	5.56	
1020	Interior Doors	Single leaf steel doors	820 S.F. Floor/Door	Each	1417	1.73	
1030	Fittings	Toilet partitions and lockers		S.F. Floor	.91	.91	
2010	Stair Construction	N/A		—	—	—	18.5%
3010	Wall Finishes	80% paint, 20% ceramic wall tile		S.F. Surface	.85	1.70	
3020	Floor Finishes	80% Carpet tile, 10% vinyl composition tile, 10% ceramic tile		S.F. Floor	6.05	6.05	
3030	Ceiling Finishes	Acoustic ceiling tiles on suspended channel grid		S.F. Ceiling	7.22	7.22	
D.	**SERVICES**						
	D10 Conveying						
1010	Elevators & Lifts	N/A		—	—	—	0.0 %
1020	Escalators & Moving Walks	N/A		—	—	—	
	D20 Plumbing						
2010	Plumbing Fixtures	Toilet and service fixtures, supply and drainage	1 Fixture/260 S.F. Floor	Each	1391	5.35	
2020	Domestic Water Distribution	Gas fired water heater		S.F. Floor	3.16	3.16	7.1%
2040	Rain Water Drainage	Roof drains		S.F. Roof	.42	.42	
	D30 HVAC						
3010	Energy Supply	N/A		—	—	—	
3020	Heat Generating Systems	Included in D3050		—	—	—	
3030	Cooling Generating Systems	N/A		—	—	—	16.9 %
3050	Terminal & Package Units	Multizone gas heating, electric cooling unit		S.F. Floor	21.15	21.15	
3090	Other HVAC Sys. & Equipment	N/A		—	—	—	
	D40 Fire Protection						
4010	Sprinklers	Wet pipe spinkler system, light hazard		S.F. Floor	2.84	2.84	2.5%
4020	Standpipes	Standpipes		S.F. Floor	.35	.35	
	D50 Electrical						
5010	Electrical Service/Distribution	1000 ampere service, panel board and feeders		S.F. Floor	1.62	1.62	
5020	Lighting & Branch Wiring	Fluorescent fixtures, receptacles, switches, A.C. and misc. power		S.F. Floor	11.67	11.67	13.6%
5030	Communications & Security	Addressable alarm systems, emergency lighting, internet and phone wiring		S.F. Floor	3.50	3.50	
5090	Other Electrical Systems	Emergency generator, 11.5 kW, Uninterruptible power supply		S.F. Floor	.23	.23	
E.	**EQUIPMENT & FURNISHINGS**						
1010	Commercial Equipment	N/A		—	—	—	
1020	Institutional Equipment	Laboratory equipment		S.F. Floor	11.42	11.42	9.1 %
1030	Vehicular Equipment	N/A		—	—	—	
1090	Other Equipment	N/A		—	—	—	
F.	**SPECIAL CONSTRUCTION**						
1020	Integrated Construction	N/A		—	—	—	0.0 %
1040	Special Facilities	N/A		—	—	—	
G.	**BUILDING SITEWORK**	**N/A**					

			Sub-Total	125.31	100%
CONTRACTOR FEES (General Requirements: 10%, Overhead: 5%, Profit: 10%)			25%	31.34	
ARCHITECT FEES			9%	14.10	

Total Building Cost	**170.75**

For customer support on your Square Foot Cost Data, call 877.756.2789.

109

Costs per square foot of floor area

Exterior Wall	S.F. Area	15000	20000	25000	30000	35000	40000	45000	50000	55000
	L.F. Perimeter	354	425	457	513	568	583	629	644	683
Brick Face with Concrete Block Back-up	Steel Frame	194.75	188.90	183.10	180.40	178.45	175.50	174.20	172.30	171.35
	R/Conc. Frame	183.90	178.00	172.20	169.55	167.55	164.65	163.30	161.40	160.45
Precast Concrete Panel	Steel Frame	200.90	194.50	187.95	185.00	182.85	179.50	178.00	175.80	174.75
	R/Conc. Frame	190.70	184.15	177.50	174.50	172.25	168.90	167.35	165.15	164.05
Limestone Face Concrete Block Back-up	Steel Frame	207.25	200.15	192.80	189.45	187.00	183.25	181.60	179.15	177.90
	R/Conc. Frame	196.35	189.25	181.90	178.55	176.15	172.35	170.75	168.25	167.00
Perimeter Adj., Add or Deduct	Per 100 L.F.	9.90	7.45	5.95	5.00	4.20	3.75	3.30	2.95	2.70
Story Hgt. Adj., Add or Deduct	Per 1 Ft.	2.45	2.20	1.95	1.80	1.65	1.55	1.45	1.30	1.30
For Basement, add $39.65 per square foot of basement area										

The above costs were calculated using the basic specifications shown on the facing page. These costs should be adjusted where necessary for design alternatives and owner's requirements. Reported completed project costs, for this type of structure, range from $142.75 to $291.20 per S.F.

Common additives

Description	Unit	$ Cost		Description	Unit	$ Cost
Carrels Hardwood	Ea.	945 - 2400		Lockers, Steel, Single tier, 60" or 72"	Opng.	231 - 380
Elevators, Hydraulic passenger, 2 stops				2 tier, 60" or 72" total	Opng.	121 - 170
2000# capacity	Ea.	68,400		5 tier, box lockers	Opng.	74 - 84
2500# capacity	Ea.	70,400		Locker bench, lam. maple top only	L.F.	34.50
3500# capacity	Ea.	75,400		Pedestals, steel pipe	Ea.	76.50
Emergency Lighting, 25 watt, battery operated				Sound System		
Lead battery	Ea.	345		Amplifier, 250 watts	Ea.	2200
Nickel cadmium	Ea.	685		Speaker, ceiling or wall	Ea.	217
Escalators, Metal				Trumpet	Ea.	415
32" wide, 10' story height	Ea.	155,500				
20' story height	Ea.	189,500				
48" wide, 10' Story height	Ea.	164,000				
20' story height	Ea.	199,000				
Glass						
32" wide, 10' story height	Ea.	147,500				
20' story height	Ea.	180,500				
48" wide, 10' story height	Ea.	155,000				
20' story height	Ea.	189,500				

Important: See the Reference Section for Location Factors.

Model costs calculated for a 2 story building with 12′ story height and 25,000 square feet of floor area

				Unit	Unit Cost	Cost Per S.F.	% Of Sub-Total
A.	**SUBSTRUCTURE**						
1010	Standard Foundations	Poured concrete; strip and spread footings		S.F. Ground	4.82	2.41	
1020	Special Foundations	N/A		—	—	—	
1030	Slab on Grade	4″ reinforced concrete with vapor barrier and granular base		S.F. Slab	5.56	2.79	5.3%
2010	Basement Excavation	Site preparation for slab and trench for foundation wall and footing		S.F. Ground	.18	.09	
2020	Basement Walls	4′ foundation wall		L.F. Wall	86	1.56	
B.	**SHELL**						
	B10 Superstructure						
1010	Floor Construction	Concrete flat plate		S.F. Floor	27.80	13.90	16.9%
1020	Roof Construction	Concrete flat plate		S.F. Roof	15.82	7.91	
	B20 Exterior Enclosure						
2010	Exterior Walls	Face brick with concrete block backup	75% of wall	S.F. Wall	33.40	10.99	
2020	Exterior Windows	Window wall	25% of wall	Each	43.30	4.75	12.6%
2030	Exterior Doors	Double aluminum and glass		Each	2725	.44	
	B30 Roofing						
3010	Roof Coverings	Built-up tar and gravel with flashing; perlite/EPS composite insulation		S.F. Roof	6.82	3.41	2.7%
3020	Roof Openings	Roof hatches		S.F. Roof	.14	.07	
C.	**INTERIORS**						
1010	Partitions	Gypsum board on metal studs	14 S.F. Floor/L.F. Partition	S.F. Partition	6.48	4.63	
1020	Interior Doors	Single leaf hollow metal	140 S.F. Floor/Door	Each	1168	8.34	
1030	Fittings	N/A		—	—	—	
2010	Stair Construction	Cast in place concrete		Flight	8400	1.35	21.9%
3010	Wall Finishes	50% paint, 50% vinyl wall covering		S.F. Surface	2.46	3.51	
3020	Floor Finishes	50% carpet, 50% vinyl composition tile		S.F. Floor	4.61	4.61	
3030	Ceiling Finishes	Suspended fiberglass board		S.F. Ceiling	5.78	5.78	
D.	**SERVICES**						
	D10 Conveying						
1010	Elevators & Lifts	One hydraulic passenger elevator		Each	85,250	3.41	2.6%
1020	Escalators & Moving Walks	N/A		—	—	—	
	D20 Plumbing						
2010	Plumbing Fixtures	Toilet and service fixtures, supply and drainage	1 Fixture/1040 S.F. Floor	Each	3182	3.06	
2020	Domestic Water Distribution	Gas fired water heater		S.F. Floor	.92	.92	3.6%
2040	Rain Water Drainage	Roof drains		S.F. Roof	1.24	.62	
	D30 HVAC						
3010	Energy Supply	N/A		—	—	—	
3020	Heat Generating Systems	Included in D3050		—	—	—	
3030	Cooling Generating Systems	N/A		—	—	—	16.4 %
3050	Terminal & Package Units	Multizone unit, gas heating, electric cooling		S.F. Floor	21.15	21.15	
3090	Other HVAC Sys. & Equipment	N/A		—	—	—	
	D40 Fire Protection						
4010	Sprinklers	Wet pipe sprinkler system		S.F. Floor	3.15	3.15	3.2%
4020	Standpipes	Standpipe, wet, Class III		S.F. Floor	.96	.96	
	D50 Electrical						
5010	Electrical Service/Distribution	600 ampere service, panel board and feeders		S.F. Floor	1.88	1.88	
5020	Lighting & Branch Wiring	High efficiency fluorescent fixtures, receptacles, switches, A.C. and misc. power		S.F. Floor	13.26	13.26	14.7%
5030	Communications & Security	Addressable alarm systems, internet wiring, communications systems and emergency lighting		S.F. Floor	3.63	3.63	
5090	Other Electrical Systems	Emergency generator, 11.5 kW		S.F. Floor	.17	.17	
E.	**EQUIPMENT & FURNISHINGS**						
1010	Commercial Equipment	N/A		—	—	—	
1020	Institutional Equipment	N/A		—	—	—	0.0 %
1030	Vehicular Equipment	N/A		—	—	—	
1090	Other Equipment	N/A		—	—	—	
F.	**SPECIAL CONSTRUCTION**						
1020	Integrated Construction	N/A		—	—	—	0.0 %
1040	Special Facilities	N/A		—	—	—	
G.	**BUILDING SITEWORK**	**N/A**					

			Sub-Total	128.75	100%
CONTRACTOR FEES (General Requirements: 10%, Overhead: 5%, Profit: 10%)		25%	32.18		
ARCHITECT FEES		7%	11.27		
	Total Building Cost		**172.20**		

For customer support on your Square Foot Cost Data, call 877.756.2789.

111

Costs per square foot of floor area

Exterior Wall	S.F. Area	4000	6000	8000	10000	12000	14000	16000	18000	20000
	L.F. Perimeter	260	340	420	453	460	510	560	610	600
Face Brick with Concrete Block Back-up	Bearing Walls	175.05	167.05	163.05	156.85	151.05	149.35	148.05	147.05	143.85
	Steel Frame	179.75	171.55	167.40	161.05	155.05	153.35	151.95	150.90	147.65
Decorative Concrete Block	Bearing Walls	153.45	147.40	144.40	139.80	135.55	134.30	133.35	132.55	130.25
	Steel Frame	165.70	159.30	156.05	151.25	146.75	145.40	144.40	143.60	141.15
Tilt Up Concrete Wall Panels	Bearing Walls	156.40	150.80	148.00	143.90	140.05	138.90	138.05	137.35	135.25
	Steel Frame	161.10	155.30	152.40	148.10	144.05	142.90	141.90	141.20	139.00
Perimeter Adj., Add or Deduct	Per 100 L.F.	19.85	13.25	9.90	8.00	6.60	5.65	5.00	4.45	4.00
Story Hgt. Adj., Add or Deduct	Per 1 Ft.	2.90	2.50	2.35	2.05	1.75	1.65	1.60	1.55	1.35

For Basement, add $34.70 per square foot of basement area

The above costs were calculated using the basic specifications shown on the facing page. These costs should be adjusted where necessary for design alternatives and owner's requirements. Reported completed project costs, for this type of structure, range from $87.10 to $275.20 per S.F.

Common additives

Description	Unit	$ Cost	Description	Unit	$ Cost
Bar, Front bar	L.F.	435	Movie Equipment		
Back bar	L.F.	345	Projector, 35mm	Ea.	12,700 - 17,400
Booth, Upholstered, custom straight	L.F.	252 - 465	Screen, wall or ceiling hung	S.F.	9.45 - 13.70
"L" or "U" shaped	L.F.	261 - 440	Partitions, Folding leaf, wood		
Bowling Alleys, incl. alley, pinsetter			Acoustic type	S.F.	82.50 - 136
Scorer, counter & misc. supplies, average	Lane	63,500	Seating		
For automatic scorer, add	Lane	11,000	Auditorium chair, all veneer	Ea.	315
Emergency Lighting, 25 watt, battery operated			Veneer back, padded seat	Ea.	325
Lead battery	Ea.	345	Upholstered, spring seat	Ea.	320
Nickel cadmium	Ea.	685	Classroom, movable chair & desk	Set	81 - 171
Kitchen Equipment			Lecture hall, pedestal type	Ea.	275 - 620
Broiler	Ea.	4075	Sound System		
Coffee urn, twin 6 gallon	Ea.	3025	Amplifier, 250 watts	Ea.	2200
Cooler, 6 ft. long	Ea.	3925	Speaker, ceiling or wall	Ea.	217
Dishwasher, 10-12 racks per hr.	Ea.	4275	Trumpet	Ea.	415
Food warmer	Ea.	770	Stage Curtains, Medium weight	S.F.	11.05 - 41
Freezer, 44 C.F., reach-in	Ea.	6175	Curtain Track, Light duty	L.F.	90.50
Ice cube maker, 50 lb. per day	Ea.	2000	Swimming Pools, Complete, gunite	S.F.	104 - 129
Range with 1 oven	Ea.	3300			

Important: See the Reference Section for Location Factors.

Model costs calculated for a 1 story building with 12' story height and 10,000 square feet of floor area

Community Center

			Unit	Unit Cost	Cost Per S.F.	% Of Sub-Total
A. SUBSTRUCTURE						
1010	Standard Foundations	Poured concrete; strip and spread footings	S.F. Ground	1.99	1.99	
1020	Special Foundations	N/A	—	—	—	
1030	Slab on Grade	4" reinforced concrete with vapor barrier and granular base	S.F. Slab	5.56	5.56	10.8%
2010	Basement Excavation	Site preparation for slab and trench for foundation wall and footing	S.F. Ground	.32	.32	
2020	Basement Walls	4' foundation wall	L.F. Wall	81	4.73	
B. SHELL						
	B10 Superstructure					
1010	Floor Construction	N/A	—	—	—	
1020	Roof Construction	Metal deck on open web steel joists	S.F. Roof	7.82	7.82	6.7 %
	B20 Exterior Enclosure					
2010	Exterior Walls	Face brick with concrete block backup 80% of wall	S.F. Wall	33.39	14.52	
2020	Exterior Windows	Aluminum sliding 20% of wall	Each	625	2.13	15.2%
2030	Exterior Doors	Double aluminum and glass and hollow metal	Each	2664	1.07	
	B30 Roofing					
3010	Roof Coverings	Built-up tar and gravel with flashing; perlite/EPS composite insulation	S.F. Roof	7.18	7.18	6.3%
3020	Roof Openings	Roof hatches	S.F. Roof	.09	.09	
C. INTERIORS						
1010	Partitions	Gypsum board on metal studs 14 S.F. Floor/L.F. Partition	S.F. Partition	8.78	6.27	
1020	Interior Doors	Single leaf hollow metal 140 S.F. Floor/Door	Each	1168	8.34	
1030	Fittings	Toilet partitions, directory board, mailboxes	S.F. Floor	1.74	1.74	
2010	Stair Construction	N/A	—	—	—	26.4%
3010	Wall Finishes	Paint	S.F. Surface	1.86	2.65	
3020	Floor Finishes	50% carpet, 50% vinyl tile	S.F. Floor	4.49	4.49	
3030	Ceiling Finishes	Mineral fiber tile on concealed zee bars	S.F. Ceiling	7.22	7.22	
D. SERVICES						
	D10 Conveying					
1010	Elevators & Lifts	N/A	—	—	—	0.0 %
1020	Escalators & Moving Walks	N/A	—	—	—	
	D20 Plumbing					
2010	Plumbing Fixtures	Kitchen, toilet and service fixtures, supply and drainage 1 Fixture/910 S.F. Floor	Each	3221	3.54	
2020	Domestic Water Distribution	Electric water heater	S.F. Floor	9.69	9.69	11.9%
2040	Rain Water Drainage	Roof drains	S.F. Roof	.59	.59	
	D30 HVAC					
3010	Energy Supply	N/A	—	—	—	
3020	Heat Generating Systems	Included in D3050	—	—	—	
3030	Cooling Generating Systems	N/A	—	—	—	9.6 %
3050	Terminal & Package Units	Single zone rooftop unit, gas heating, electric cooling	S.F. Floor	11.20	11.20	
3090	Other HVAC Sys. & Equipment	N/A	—	—	—	
	D40 Fire Protection					
4010	Sprinklers	Wet pipe sprinkler system	S.F. Floor	3.65	3.65	3.1%
4020	Standpipes	N/A	—	—	—	
	D50 Electrical					
5010	Electrical Service/Distribution	200 ampere service, panel board and feeders	S.F. Floor	1.14	1.14	
5020	Lighting & Branch Wiring	High efficiency fluorescent fixtures, receptacles, switches, A.C. and misc. power	S.F. Floor	6.26	6.26	7.9%
5030	Communications & Security	Addressable alarm systems and emergency lighting	S.F. Floor	1.55	1.55	
5090	Other Electrical Systems	Emergency generator, 15 kW	S.F. Floor	.19	.19	
E. EQUIPMENT & FURNISHINGS						
1010	Commercial Equipment	Freezer, chest type	S.F. Floor	.95	.95	
1020	Institutional Equipment	N/A	—	—	—	2.0%
1030	Vehicular Equipment	N/A	—	—	—	
1090	Other Equipment	Kitchen equipment, directory board, mailboxes, built-in coat racks	S.F. Floor	1.32	1.32	
F. SPECIAL CONSTRUCTION						
1020	Integrated Construction	N/A	—	—	—	0.0 %
1040	Special Facilities	N/A	—	—	—	
G. BUILDING SITEWORK N/A						

		Sub-Total	116.20	100%
CONTRACTOR FEES (General Requirements: 10%, Overhead: 5%, Profit: 10%)		25%	29.03	
ARCHITECT FEES		8%	11.62	

Total Building Cost	**156.85**

For customer support on your Square Foot Cost Data, call 877.756.2789.

113

Costs per square foot of floor area

Exterior Wall	S.F. Area	10000	12500	15000	17500	20000	22500	25000	30000	40000
	L.F. Perimeter	400	450	500	550	600	625	650	700	800
Fiber Cement	Wood Frame	326.25	323.30	321.30	319.95	318.90	317.30	316.00	314.00	311.55
Stone Veneer	Wood Frame	344.60	339.40	335.95	333.50	331.60	328.75	326.50	323.10	318.80
Curtain Wall	Steel Frame	353.80	348.05	344.20	341.45	339.45	336.30	333.80	330.10	325.40
EIFS	Steel Frame	331.35	328.15	325.95	324.40	323.25	321.55	320.15	318.10	315.45
Brick Veneer	Reinforced Concrete	358.10	352.65	349.05	346.45	344.50	341.55	339.15	335.50	331.00
Metal Panel	Reinforced Concrete	354.20	349.25	345.95	343.60	341.85	339.10	336.95	333.65	329.55
Perimeter Adj., Add or Deduct	Per 100 L.F.	7.35	5.85	4.95	4.20	3.65	3.20	2.90	2.45	1.90
Story Hgt. Adj., Add or Deduct	Per 1 Ft.	1.00	0.90	0.90	0.80	0.75	0.70	0.65	0.60	0.50

For Basement, add $38.10 per square foot of basement area

The above costs were calculated using the basic specifications shown on the facing page. These costs should be adjusted where necessary for design alternatives and owner's requirements. Reported completed project costs, for this type of structure, range from $179.75 to $364.75 per S.F.

Common additives

Description	Unit	$ Cost
Clock System		
20 room	Ea.	20,600
50 room	Ea.	48,500
Closed Circuit Surveillance, one station		
Camera and monitor	Total	2025
For additional camera stations, add	Ea.	1100
For zoom lens - remote control, add	Ea.	3175 - 10,700
For automatic iris for low light, add	Ea.	2825
Directory Boards, plastic, glass covered		
30" x 20"	Ea.	610
36" x 48"	Ea.	1525
Aluminum, 24" x 18"	Ea.	640
36" x 24"	Ea.	765
48" x 32"	Ea.	1075
48" x 60"	Ea.	2275
Emergency Lighting, 25 watt, battery operated		
Lead battery	Ea.	345
Nickel cadmium	Ea.	685

Description	Unit	$ Cost
Smoke Detectors		
Ceiling type	Ea.	227
Duct type	Ea.	565
Sound System		
Amplifier, 250 watts	Ea.	2200
Speakers, ceiling or wall	Ea.	217
Trumpets	Ea.	415

Model costs calculated for a 1 story building with 16'-6" story height and 22,500 square feet of floor area

				Unit	Unit Cost	Cost Per S.F.	% Of Sub-Total
A. SUBSTRUCTURE							
1010	Standard Foundations	Poured concrete; strip and spread footings		S.F. Ground	1.66	1.66	
1020	Special Foundations	N/A		—	—	—	
1030	Slab on Grade	4" reinforced concrete		S.F. Slab	5.56	5.56	4.1%
2010	Basement Excavation	Site preparation for slab and trench for foundation wall and footing		S.F. Ground	.18	.20	
2020	Basement Walls	4' Foundation wall		L.F. Wall	86	2.38	
B. SHELL							
B10 Superstructure							
1010	Floor Construction	Wood columns, fireproofed		S.F. Floor	3.16	3.16	4.6%
1020	Roof Construction	Wood roof truss 4:12 pitch		S.F. Roof	7.85	7.85	
B20 Exterior Enclosure							
2010	Exterior Walls	Fiber cement siding on wood studs, insulated	95% of wall	S.F. Wall	18.81	8.19	
2020	Exterior Windows	Aluminum horizontal sliding	5% of wall	Each	585	.89	4.3%
2030	Exterior Doors	Aluminum and glass single and double doors		Each	3942	1.05	
B30 Roofing							
3010	Roof Coverings	Asphalt roofing, strip shingles, gutters and downspouts		S.F. Roof	2.30	2.30	1.1%
3020	Roof Openings	Roof hatch		S.F. Roof	.20	.20	
C. INTERIORS							
1010	Partitions	Sound deadening gypsum board on wood studs	15 S.F. Floor/L.F. Partition	S.F. Partition	9.07	8.37	
1020	Interior Doors	Solid core wood doors in metal frames	370 S.F. Floor/Door	Each	776	2.10	
1030	Fittings	Plastic laminate toilet partitions		S.F. Floor	.30	.30	
2010	Stair Construction	N/A		—	—	—	11.5%
3010	Wall Finishes	90% Paint, 10% ceramic wall tile		S.F. Surface	1.73	3.20	
3020	Floor Finishes	65% Carpet, 10% porcelain tile, 10% quarry tile		S.F. Floor	6.17	6.17	
3030	Ceiling Finishes	Acoustic ceiling tiles on suspended channel grid		S.F. Ceiling	7.22	7.22	
D. SERVICES							
D10 Conveying							
1010	Elevators & Lifts	N/A		—	—	—	0.0 %
1020	Escalators & Moving Walks	N/A		—	—	—	
D20 Plumbing							
2010	Plumbing Fixtures	Restroom and service fixtures, supply and drainage	1 Fixture/1180 S.F. Floor	Each	2065	1.75	
2020	Domestic Water Distribution	Gas fired water heater		S.F. Floor	.81	.81	1.1%
2040	Rain Water Drainage	N/A		—	—	—	
D30 HVAC							
3010	Energy Supply	Hot water reheat system		S.F. Floor	6.57	6.57	
3020	Heat Generating Systems	Oil fired hot water boiler and pumps		Each	53,025	10.30	
3030	Cooling Generating Systems	Cooling tower and chiller		S.F. Floor	9.58	9.58	42.0%
3050	Terminal & Package Units	N/A		—	—	—	
3090	Other HVAC Sys. & Equipment	Plate heat exchanger, ductwork, AHUs and VAV terminals		S.F. Floor	73	73.13	
D40 Fire Protection							
4010	Sprinklers	85% light hazard wet pipe sprinkler system, 15% preaction system		S.F. Floor	4.37	4.37	2.1%
4020	Standpipes	Standpipes and hose systems		S.F. Floor	.70	.70	
D50 Electrical							
5010	Electrical Service/Distribution	1200 ampere service, panel boards and feeders		S.F. Floor	3.73	3.73	
5020	Lighting & Branch Wiring	Fluorescent fixtures, receptacles, switches, A.C. and misc. power		S.F. Floor	17.21	17.21	27.3%
5030	Communications & Security	Telephone systems, internet wiring, and addressable alarm systems		S.F. Floor	30.79	30.79	
5090	Other Electrical Systems	Emergency generator, UPS system with 15 minute pack		S.F. Floor	12.94	12.94	
E. EQUIPMENT & FURNISHINGS							
1010	Commercial Equipment	N/A		—	—	—	
1020	Institutional Equipment	N/A		—	—	—	0.0 %
1030	Vehicular Equipment	N/A		—	—	—	
1090	Other Equipment	N/A		—	—	—	
F. SPECIAL CONSTRUCTION & DEMOLITION							
1020	Integrated Construction	Pedestal access floor		S.F. Floor	4.53	4.53	1.9%
1040	Special Facilities	N/A		—	—	—	
G. BUILDING SITEWORK	**N/A**						

		Sub-Total	237.21	100%
CONTRACTOR FEES (General Requirements: 10%, Overhead: 5%, Profit: 10%)		25%	59.33	
ARCHITECT FEES		7%	20.76	
		Total Building Cost	**317.30**	

For customer support on your Square Foot Cost Data, call 877.756.2789.

115

Costs per square foot of floor area

Exterior Wall	S.F. Area	16000	23000	30000	37000	44000	51000	58000	65000	72000
	L.F. Perimeter	597	763	821	968	954	1066	1090	1132	1220
Limestone with Concrete Block Back-up	R/Conc. Frame	236.10	225.95	215.55	212.45	205.25	203.50	200.00	197.70	196.70
	Steel Frame	237.45	227.20	216.85	213.70	206.55	204.80	201.25	198.95	198.00
Face Brick with Concrete Block Back-up	R/Conc. Frame	219.05	210.75	203.05	200.45	195.35	193.90	191.40	189.70	188.95
	Steel Frame	220.35	212.00	204.35	201.70	196.65	195.20	192.65	191.00	190.25
Stone with Concrete Block Back-up	R/Conc. Frame	225.20	216.25	207.60	204.80	198.95	197.40	194.50	192.60	191.75
	Steel Frame	226.50	217.50	208.85	206.05	200.20	198.65	195.80	193.85	193.00
Perimeter Adj., Add or Deduct	Per 100 L.F.	8.55	5.95	4.60	3.70	3.20	2.65	2.35	2.10	1.90
Story Hgt. Adj., Add or Deduct	Per 1 Ft.	3.10	2.70	2.25	2.15	1.80	1.70	1.50	1.40	1.40
For Basement, add $32.75 per square foot of basement area										

The above costs were calculated using the basic specifications shown on the facing page. These costs should be adjusted where necessary for design alternatives and owner's requirements. Reported completed project costs, for this type of structure, range from $153.35 to $285.70 per S.F.

Common additives

Description	Unit	$ Cost	Description	Unit	$ Cost
Benches, Hardwood	L.F.	141 - 234	Flagpoles, Complete		
Clock System			Aluminum, 20' high	Ea.	1900
20 room	Ea.	20,600	40' high	Ea.	4425
50 room	Ea.	48,500	70' high	Ea.	11,500
Closed Circuit Surveillance, One station			Fiberglass, 23' high	Ea.	1350
Camera and monitor	Ea.	2025	39'-5" high	Ea.	3400
For additional camera stations, add	Ea.	1100	59' high	Ea.	6650
Directory Boards, Plastic, glass covered			Intercom System, 25 station capacity		
30" x 20"	Ea.	610	Master station	Ea.	3175
36" x 48"	Ea.	1525	Intercom outlets	Ea.	192
Aluminum, 24" x 18"	Ea.	640	Handset	Ea.	525
36" x 24"	Ea.	765	Safe, Office type, 1 hour rating		
48" x 32"	Ea.	1075	30" x 18" x 18"	Ea.	2500
48" x 60"	Ea.	2275	60" x 36" x 18", double door	Ea.	9350
Emergency Lighting, 25 watt, battery operated			Smoke Detectors		
Lead battery	Ea.	345	Ceiling type	Ea.	227
Nickel cadmium	Ea.	685	Duct type	Ea.	565

Model costs calculated for a 1 story building with 14′ story height and 30,000 square feet of floor area

				Unit	Unit Cost	Cost Per S.F.	% Of Sub-Total
A.	**SUBSTRUCTURE**						
1010	Standard Foundations	Poured concrete; strip and spread footings		S.F. Ground	1.91	1.91	
1020	Special Foundations	N/A		—	—	—	
1030	Slab on Grade	4″ reinforced concrete with vapor barrier and granular base		S.F. Slab	5.56	5.56	6.3%
2010	Basement Excavation	Site preparation for slab and trench for foundation wall and footing		S.F. Ground	.18	.18	
2020	Basement Walls	4′ foundation wall		L.F. Wall	86	2.50	
B.	**SHELL**						
	B10 Superstructure						
1010	Floor Construction	Cast-in-place columns		L.F. Column	76	1.70	13.9%
1020	Roof Construction	Cast-in-place concrete waffle slab		S.F. Roof	20.70	20.70	
	B20 Exterior Enclosure						
2010	Exterior Walls	Limestone panels with concrete block backup	75% of wall	S.F. Wall	66	18.97	
2020	Exterior Windows	Aluminum with insulated glass	25% of wall	Each	746	3.11	14.0%
2030	Exterior Doors	Double wood		Each	1921	.44	
	B30 Roofing						
3010	Roof Coverings	Built-up tar and gravel with flashing; perlite/EPS composite insulation		S.F. Roof	6.07	6.07	3.8%
3020	Roof Openings	Roof hatches		S.F. Roof	.09	.09	
C.	**INTERIORS**						
1010	Partitions	Plaster on metal studs	10 S.F. Floor/L.F. Partition	S.F. Partition	14.15	16.98	
1020	Interior Doors	Single leaf wood	100 S.F. Floor/Door	Each	626	6.26	
1030	Fittings	Toilet partitions		S.F. Floor	.51	.51	
2010	Stair Construction	N/A		—	—	—	33.2%
3010	Wall Finishes	70% paint, 20% wood paneling, 10% vinyl wall covering		S.F. Surface	2.19	5.26	
3020	Floor Finishes	60% hardwood, 20% carpet, 20% terrazzo		S.F. Floor	13.71	13.71	
3030	Ceiling Finishes	Gypsum plaster on metal lath, suspended		S.F. Ceiling	10.77	10.77	
D.	**SERVICES**						
	D10 Conveying						
1010	Elevators & Lifts	N/A		—	—	—	0.0 %
1020	Escalators & Moving Walks	N/A		—	—	—	
	D20 Plumbing						
2010	Plumbing Fixtures	Toilet and service fixtures, supply and drainage	1 Fixture/1110 S.F. Floor	Each	4573	4.12	
2020	Domestic Water Distribution	Electric hot water heater		S.F. Floor	3.48	3.48	5.5%
2040	Rain Water Drainage	Roof drains		S.F. Roof	1.21	1.21	
	D30 HVAC						
3010	Energy Supply	N/A		—	—	—	
3020	Heat Generating Systems	Included in D3050		—	—	—	
3030	Cooling Generating Systems	N/A		—	—	—	13.1 %
3050	Terminal & Package Units	Multizone unit, gas heating, electric cooling		S.F. Floor	21.15	21.15	
3090	Other HVAC Sys. & Equipment	N/A		—	—	—	
	D40 Fire Protection						
4010	Sprinklers	Wet pipe sprinkler system		S.F. Floor	2.84	2.84	2.1%
4020	Standpipes	Standpipe, wet, Class III		S.F. Floor	.53	.53	
	D50 Electrical						
5010	Electrical Service/Distribution	400 ampere service, panel board and feeders		S.F. Floor	1.03	1.03	
5020	Lighting & Branch Wiring	High efficiency fluorescent fixtures, receptacles, switches, A.C. and misc. power		S.F. Floor	10.24	10.24	8.1%
5030	Communications & Security	Addressable alarm systems, internet wiring, and emergency lighting		S.F. Floor	1.72	1.72	
5090	Other Electrical Systems	Emergency generator, 11.5 kW		S.F. Floor	.14	.14	
E.	**EQUIPMENT & FURNISHINGS**						
1010	Commercial Equipment	N/A		—	—	—	
1020	Institutional Equipment	N/A		—	—	—	0.0 %
1030	Vehicular Equipment	N/A		—	—	—	
1090	Other Equipment	N/A		—	—	—	
F.	**SPECIAL CONSTRUCTION**						
1020	Integrated Construction	N/A		—	—	—	0.0 %
1040	Special Facilities	N/A		—	—	—	
G.	**BUILDING SITEWORK**	**N/A**					

		Sub-Total	161.18	**100%**
CONTRACTOR FEES (General Requirements: 10%, Overhead: 5%, Profit: 10%)		25%	40.27	
ARCHITECT FEES		7%	14.10	
	Total Building Cost		**215.55**	

For customer support on your Square Foot Cost Data, call 877.756.2789.

117

Costs per square foot of floor area

Exterior Wall	S.F. Area	30000	40000	45000	50000	60000	70000	80000	90000	100000
	L.F. Perimeter	400	466	500	533	600	666	733	800	867
Limestone with Concrete Block Back-up	R/Conc. Frame	254.95	242.35	238.15	234.75	229.75	226.20	223.45	221.35	219.70
	Steel Frame	260.05	247.40	243.30	239.90	234.85	231.25	228.60	226.45	224.85
Face Brick with Concrete Block Back-up	R/Conc. Frame	239.30	228.65	225.10	222.30	218.00	215.00	212.75	210.95	209.50
	Steel Frame	244.50	233.85	230.30	227.45	223.20	220.15	217.90	216.10	214.70
Stone with Concrete Block Back-up	R/Conc. Frame	245.00	233.65	229.90	226.85	222.30	219.05	216.65	214.75	213.25
	Steel Frame	250.15	238.80	235.05	231.95	227.45	224.20	221.75	219.85	218.40
Perimeter Adj., Add or Deduct	Per 100 L.F.	12.70	9.45	8.45	7.65	6.40	5.35	4.75	4.25	3.85
Story Hgt. Adj., Add or Deduct	Per 1 Ft.	3.95	3.45	3.30	3.20	3.00	2.85	2.75	2.65	2.55

For Basement, add $33.45 per square foot of basement area

The above costs were calculated using the basic specifications shown on the facing page. These costs should be adjusted where necessary for design alternatives and owner's requirements. Reported completed project costs, for this type of structure, range from $154.05 to $287.35 per S.F.

Common additives

Description	Unit	$ Cost
Benches, Hardwood	L.F.	141 - 234
Clock System		
20 room	Ea.	20,600
50 room	Ea.	48,500
Closed Circuit Surveillance, One station		
Camera and monitor	Ea.	2025
For additional camera stations, add	Ea.	1100
Directory Boards, Plastic, glass covered		
30" x 20"	Ea.	610
36" x 48"	Ea.	1525
Aluminum, 24" x 18"	Ea.	640
36" x 24"	Ea.	765
48" x 32"	Ea.	1075
48" x 60"	Ea.	2275
Elevators, Hydraulic passenger, 2 stops		
1500# capacity	Ea.	67,400
2500# capacity	Ea.	70,400
3500# capacity	Ea.	75,400
Additional stop, add	Ea.	8075

Description	Unit	$ Cost
Emergency Lighting, 25 watt, battery operated		
Lead battery	Ea.	345
Nickel cadmium	Ea.	685
Flagpoles, Complete		
Aluminum, 20' high	Ea.	1900
40' high	Ea.	4425
70' high	Ea.	11,500
Fiberglass, 23' high	Ea.	1350
39'-5" high	Ea.	3400
59' high	Ea.	6650
Intercom System, 25 station capacity		
Master station	Ea.	3175
Intercom outlets	Ea.	192
Handset	Ea.	525
Safe, Office type, 1 hour rating		
30" x 18" x 18"	Ea.	2500
60" x 36" x 18", double door	Ea.	9350
Smoke Detectors		
Ceiling type	Ea.	227
Duct type	Ea.	565

Important: See the Reference Section for Location Factors.

Model costs calculated for a 3 story building with 12' story height and 60,000 square feet of floor area

				Unit	Unit Cost	Cost Per S.F.	% Of Sub-Total
A. SUBSTRUCTURE							
1010	Standard Foundations	Poured concrete; strip and spread footings		S.F. Ground	2.04	.68	
1020	Special Foundations	N/A		—	—	—	
1030	Slab on Grade	4" reinforced concrete with vapor barrier and granular base		S.F. Slab	5.56	1.85	2.1%
2010	Basement Excavation	Site preparation for slab and trench for foundation wall and footing		S.F. Ground	.18	.06	
2020	Basement Walls	4' foundation wall		L.F. Wall	81	.98	
B. SHELL							
B10 Superstructure							
1010	Floor Construction	Concrete slab with metal deck and beams		S.F. Floor	36.27	24.18	18.5%
1020	Roof Construction	Concrete slab with metal deck and beams		S.F. Roof	20.82	6.94	
B20 Exterior Enclosure							
2010	Exterior Walls	Face brick with concrete block backup	75% of wall	S.F. Wall	36.04	9.73	
2020	Exterior Windows	Horizontal pivoted steel	25% of wall	Each	770	7.70	10.6%
2030	Exterior Doors	Double aluminum and glass and hollow metal		Each	3640	.37	
B30 Roofing							
3010	Roof Coverings	Built-up tar and gravel with flashing; perlite/EPS composite insulation		S.F. Roof	6.48	2.16	1.3%
3020	Roof Openings	N/A		—	—	—	
C. INTERIORS							
1010	Partitions	Plaster on metal studs	10 S.F. Floor/L.F. Partition	S.F. Partition	14.30	14.30	
1020	Interior Doors	Single leaf wood	100 S.F. Floor/Door	Each	626	6.26	
1030	Fittings	Toilet partitions		S.F. Floor	.25	.25	
2010	Stair Construction	Concrete filled metal pan		Flight	20,150	1.68	30.5%
3010	Wall Finishes	70% paint, 20% wood paneling, 10% vinyl wall covering		S.F. Surface	2.19	4.38	
3020	Floor Finishes	60% hardwood, 20% terrazzo, 20% carpet		S.F. Floor	13.71	13.71	
3030	Ceiling Finishes	Gypsum plaster on metal lath, suspended		S.F. Ceiling	10.77	10.77	
D. SERVICES							
D10 Conveying							
1010	Elevators & Lifts	Five hydraulic passenger elevators		Each	158,520	13.21	7.8%
1020	Escalators & Moving Walks	N/A		—	—	—	
D20 Plumbing							
2010	Plumbing Fixtures	Toilet and service fixtures, supply and drainage	1 Fixture/665 S.F. Floor	Each	3265	4.91	
2020	Domestic Water Distribution	Electric water heater		S.F. Floor	5.81	5.81	6.8%
2040	Rain Water Drainage	Roof drains		S.F. Roof	2.19	.73	
D30 HVAC							
3010	Energy Supply	N/A		—	—	—	
3020	Heat Generating Systems	Included in D3050		—	—	—	
3030	Cooling Generating Systems	N/A		—	—	—	12.6 %
3050	Terminal & Package Units	Multizone unit, gas heating, electric cooling		S.F. Floor	21.15	21.15	
3090	Other HVAC Sys. & Equipment	N/A		—	—	—	
D40 Fire Protection							
4010	Sprinklers	Wet pipe sprinkler system		S.F. Floor	3.05	3.05	2.1%
4020	Standpipes	Standpipe, wet, Class III		S.F. Floor	.48	.48	
D50 Electrical							
5010	Electrical Service/Distribution	800 ampere service, panel board and feeders		S.F. Floor	.95	.95	
5020	Lighting & Branch Wiring	High efficiency fluorescent fixtures, receptacles, switches, A.C. and misc. power		S.F. Floor	10.37	10.37	7.8%
5030	Communications & Security	Addressable alarm systems, internet wiring, and emergency lighting		S.F. Floor	1.56	1.56	
5090	Other Electrical Systems	Emergency generator, 15 kW		S.F. Floor	.24	.24	
E. EQUIPMENT & FURNISHINGS							
1010	Commercial Equipment	N/A		—	—	—	
1020	Institutional Equipment	N/A		—	—	—	0.0 %
1030	Vehicular Equipment	N/A		—	—	—	
1090	Other Equipment	N/A		—	—	—	
F. SPECIAL CONSTRUCTION							
1020	Integrated Construction	N/A		—	—	—	0.0 %
1040	Special Facilities	N/A		—	—	—	
G. BUILDING SITEWORK	**N/A**						

		Sub-Total	168.46	100%
CONTRACTOR FEES (General Requirements: 10%, Overhead: 5%, Profit: 10%)		25%	42.11	
ARCHITECT FEES		6%	12.63	
	Total Building Cost		**223.20**	

For customer support on your Square Foot Cost Data, call 877.756.2789.

119

Costs per square foot of floor area

Exterior Wall	S.F. Area	2000	5000	7000	10000	12000	15000	18000	21000	25000
	L.F. Perimeter	200	310	360	440	480	520	560	600	660
Tiltup Concrete Panel	Steel Joists	215.50	184.40	176.60	171.05	168.30	164.75	162.40	160.65	159.30
Decorative Concrete Block	Bearing Walls	221.35	187.35	178.70	172.60	169.55	165.60	163.05	161.10	159.50
Brick on Block	Steel Joists	254.85	208.35	196.45	188.00	183.65	178.15	174.50	171.85	169.60
Stucco on Concrete Block	Wood Truss	214.60	185.20	177.60	172.25	169.55	166.00	163.65	161.95	160.55
Brick Veneer	Wood Frame	224.60	191.60	183.00	176.95	173.85	169.80	167.10	165.20	163.60
Wood Siding	Wood Frame	205.85	180.20	173.65	169.05	166.70	163.70	161.70	160.20	159.00
Perimeter Adj., Add or Deduct	Per 100 L.F.	27.90	11.15	7.90	5.65	4.60	3.75	3.15	2.75	2.15
Story Hgt. Adj., Add or Deduct	Per 1 Ft.	2.70	1.70	1.40	1.20	1.05	0.95	0.90	0.80	0.65
	For Basement, add $32.55 per square foot of basement area									

The above costs were calculated using the basic specifications shown on the facing page. These costs should be adjusted where necessary for design alternatives and owner's requirements. Reported completed project costs, for this type of structure, range from $66.45 to $242.10 per S.F.

Common additives

Description	Unit	$ Cost		Description	Unit	$ Cost
Emergency Lighting, 25 watt, battery operated				Lockers, Steel, single tier, 60" to 72"	Opng.	231 - 380
Lead battery	Ea.	345		2 tier, 60" to 72" total	Opng.	121 - 170
Nickel cadmium	Ea.	685		5 tier, box lockers	Opng.	74 - 84
Flagpoles, Complete				Locker bench, lam. maple top only	L.F.	34.50
Aluminum, 20' high	Ea.	1900		Pedestals, steel pipe	Ea.	76.50
40' high	Ea.	4425		Smoke Detectors		
70' high	Ea.	11,500		Ceiling type	Ea.	227
Fiberglass, 23' high	Ea.	1350		Duct type	Ea.	565
39'-5" high	Ea.	3400		Sound System		
59' high	Ea.	6650		Amplifier, 250 watts	Ea.	2200
Gym Floor, Incl. sleepers and finish, maple	S.F.	14.75		Speaker, ceiling or wall	Ea.	217
Intercom System, 25 Station capacity				Trumpet	Ea.	415
Master station	Ea.	3175				
Intercom outlets	Ea.	192				
Handset	Ea.	525				

Important: See the Reference Section for Location Factors.

Model costs calculated for a 1 story building with 12' story height and 10,000 square feet of floor area

				Unit	Unit Cost	Cost Per S.F.	% Of Sub-Total
A. SUBSTRUCTURE							
1010	Standard Foundations	Poured concrete; strip and spread footings		S.F. Ground	2.21	2.21	
1020	Special Foundations	N/A		—	—	—	
1030	Slab on Grade	4" concrete with vapor barrier and granular base		S.F. Slab	5.03	5.03	8.7%
2010	Basement Excavation	Site preparation for slab and trench for foundation wall and footing		S.F. Ground	.32	.32	
2020	Basement Walls	4' foundation wall		L.F. Wall	86	3.76	
B. SHELL							
	B10 Superstructure						
1010	Floor Construction	Wood beams on columns		S.F. Floor	.22	.22	6.2%
1020	Roof Construction	Wood trusses		S.F. Roof	7.85	7.85	
	B20 Exterior Enclosure						
2010	Exterior Walls	Brick veneer on wood studs	85% of wall	S.F. Wall	25.36	11.38	
2020	Exterior Windows	Window wall	15% of wall	Each	51	4.05	15.6%
2030	Exterior Doors	Aluminum and glass; steel		Each	2558	4.87	
	B30 Roofing						
3010	Roof Coverings	Asphalt shingles, 9" fiberglass batt insulation, gutters and downspouts		S.F. Roof	4.68	4.68	3.6%
3020	Roof Openings	N/A		—	—	—	
C. INTERIORS							
1010	Partitions	Gypsum board on wood studs	8 S.F. Floor/S.F. Partition	S.F. Partition	8.09	5.39	
1020	Interior Doors	Single leaf hollow metal	700 S.F. Floor/Door	Each	1168	3.11	
1030	Fittings	Toilet partitions		S.F. Floor	.46	.46	
2010	Stair Construction	N/A		—	—	—	15.3%
3010	Wall Finishes	Paint		S.F. Surface	1.60	2.13	
3020	Floor Finishes	5% quarry tile, 95% vinyl composition tile		S.F. Floor	3.02	3.02	
3030	Ceiling Finishes	Fiberglass tile on tee grid		S.F. Ceiling	5.78	5.78	
D. SERVICES							
	D10 Conveying						
1010	Elevators & Lifts	N/A		—	—	—	0.0 %
1020	Escalators & Moving Walks	N/A		—	—	—	
	D20 Plumbing						
2010	Plumbing Fixtures	Toilet and service fixtures, supply and drainage	1 Fixture/455 S.F. Floor	Each	2982	16.75	
2020	Domestic Water Distribution	Electric water heater		S.F. Floor	2.79	2.79	15.0%
2040	Rain Water Drainage	N/A		—	—	—	
	D30 HVAC						
3010	Energy Supply	Oil fired hot water, wall fin radiation		S.F. Floor	11.72	11.72	
3020	Heat Generating Systems	N/A		—	—	—	
3030	Cooling Generating Systems	N/A		—	—	—	19.5%
3050	Terminal & Package Units	Split systems with air cooled condensing units		S.F. Floor	13.61	13.61	
3090	Other HVAC Sys. & Equipment	N/A		—	—	—	
	D40 Fire Protection						
4010	Sprinklers	Sprinkler, light hazard		S.F. Floor	3.65	3.65	3.5%
4020	Standpipes	Standpipe		S.F. Floor	.92	.92	
	D50 Electrical						
5010	Electrical Service/Distribution	200 ampere service, panel board and feeders		S.F. Floor	1.14	1.14	
5020	Lighting & Branch Wiring	High efficiency fluorescent fixtures, receptacles, switches, A.C. and misc. power		S.F. Floor	8.53	8.53	8.7%
5030	Communications & Security	Addressable alarm systems and emergency lighting		S.F. Floor	1.57	1.57	
5090	Other Electrical Systems	Emergency generator, 15 kW		S.F. Floor	.09	.09	
E. EQUIPMENT & FURNISHINGS							
1010	Commercial Equipment	N/A		—	—	—	
1020	Institutional Equipment	Cabinets and countertop		S.F. Floor	4.85	4.85	3.7 %
1030	Vehicular Equipment	N/A		—	—	—	
1090	Other Equipment	N/A		—	—	—	
F. SPECIAL CONSTRUCTION							
1020	Integrated Construction	N/A		—	—	—	0.0 %
1040	Special Facilities	N/A		—	—	—	
G. BUILDING SITEWORK	**N/A**						

				Sub-Total	129.88	**100%**
CONTRACTOR FEES (General Requirements: 10%, Overhead: 5%, Profit: 10%)			25%	32.46		
ARCHITECT FEES			9%	14.61		

Total Building Cost	**176.95**

For customer support on your Square Foot Cost Data, call 877.756.2789.

121

Costs per square foot of floor area

Exterior Wall	S.F. Area	12000	18000	24000	30000	36000	42000	48000	54000	60000
	L.F. Perimeter	460	580	713	730	826	880	965	1006	1045
Concrete Block	Steel Frame	141.05	133.70	130.35	125.80	124.20	122.30	121.40	120.15	119.10
	Bearing Walls	138.70	131.30	127.95	123.40	121.80	119.90	119.00	117.75	116.65
Precast Concrete Panels	Steel Frame	148.70	140.10	136.25	130.60	128.75	126.50	125.40	123.85	122.55
Insulated Metal Panels	Steel Frame	143.45	135.70	132.20	127.25	125.60	123.60	122.65	121.30	120.20
Face Brick on Common Brick	Steel Frame	159.80	149.45	144.85	137.65	135.40	132.55	131.25	129.25	127.55
Tilt-up Concrete Panel	Steel Frame	142.85	135.20	131.75	126.90	125.25	123.30	122.35	121.05	119.90
Perimeter Adj., Add or Deduct	Per 100 L.F.	5.75	3.85	2.90	2.25	1.90	1.65	1.45	1.25	1.10
Story Hgt. Adj., Add or Deduct	Per 1 Ft.	0.85	0.70	0.65	0.45	0.50	0.45	0.45	0.40	0.40
For Basement, add $33.15 per square foot of basement area										

The above costs were calculated using the basic specifications shown on the facing page. These costs should be adjusted where necessary for design alternatives and owner's requirements. Reported completed project costs, for this type of structure, range from $49.30 to $190.85 per S.F.

Common additives

Description	Unit	$ Cost	Description	Unit	$ Cost
Clock System			Dock Levelers, Hinged 10 ton cap.		
20 room	Ea.	20,600	6' x 8'	Ea.	6150
50 room	Ea.	48,500	7' x 8'	Ea.	8025
Dock Bumpers, Rubber blocks			Partitions, Woven wire, 10 ga., 1-1/2" mesh		
4-1/2" thick, 10" high, 14" long	Ea.	89	4' wide x 7' high	Ea.	206
24" long	Ea.	132	8' high	Ea.	232
36" long	Ea.	179	10' High	Ea.	277
12" high, 14" long	Ea.	106	Platform Lifter, Portable, 6'x 6'		
24" long	Ea.	159	3000# cap.	Ea.	10,200
36" long	Ea.	189	4000# cap.	Ea.	12,500
6" thick, 10" high, 14" long	Ea.	104	Fixed, 6' x 8', 5000# cap.	Ea.	12,700
24" long	Ea.	159			
36" long	Ea.	203			
20" high, 11" long	Ea.	183			
Dock Boards, Heavy					
60" x 60" Aluminum, 5,000# cap.	Ea.	1500			
9000# cap.	Ea.	1500			
15,000# cap.	Ea.	1725			

Important: See the Reference Section for Location Factors.

Model costs calculated for a 1 story building with 20' story height and 30,000 square feet of floor area

				Unit	Unit Cost	Cost Per S.F.	% Of Sub-Total
A. SUBSTRUCTURE							
1010	Standard Foundations	Poured concrete; strip and spread footings		S.F. Ground	1.59	1.59	
1020	Special Foundations	N/A		—	—	—	
1030	Slab on Grade	4" reinforced concrete with vapor barrier and granular base		S.F. Slab	7.35	7.35	12.4%
2010	Basement Excavation	Site preparation for slab and trench for foundation wall and footing		S.F. Ground	.18	.18	
2020	Basement Walls	4' foundation wall		L.F. Wall	86	2.50	
B. SHELL							
	B10 Superstructure						
1010	Floor Construction	N/A		—	—	—	10.4 %
1020	Roof Construction	Metal deck, open web steel joists, beams and columns		S.F. Roof	9.79	9.79	
	B20 Exterior Enclosure						
2010	Exterior Walls	Concrete block	75% of wall	S.F. Wall	8.99	3.28	
2020	Exterior Windows	Industrial horizontal pivoted steel	25% of wall	Each	860	3.27	8.4%
2030	Exterior Doors	Double aluminum and glass, hollow metal, steel overhead		Each	2610	1.32	
	B30 Roofing						
3010	Roof Coverings	Built-up tar and gravel with flashing; perlite/EPS composite insulation		S.F. Roof	6.25	6.25	7.1%
3020	Roof Openings	Roof hatches		S.F. Roof	.38	.38	
C. INTERIORS							
1010	Partitions	Concrete block	60 S.F. Floor/L.F. Partition	S.F. Partition	9.10	1.82	
1020	Interior Doors	Single leaf hollow metal and fire doors	600 S.F. Floor/Door	Each	1168	1.95	
1030	Fittings	Toilet partitions		S.F. Floor	1.02	1.02	
2010	Stair Construction	N/A		—	—	—	8.3%
3010	Wall Finishes	Paint		S.F. Surface	4.90	1.96	
3020	Floor Finishes	Vinyl composition tile	10% of floor	S.F. Floor	2.80	.28	
3030	Ceiling Finishes	Fiberglass board on exposed grid system	10% of area	S.F. Ceiling	7.22	.73	
D. SERVICES							
	D10 Conveying						
1010	Elevators & Lifts	N/A		—	—	—	0.0 %
1020	Escalators & Moving Walks	N/A		—	—	—	
	D20 Plumbing						
2010	Plumbing Fixtures	Toilet and service fixtures, supply and drainage	1 Fixture/1000 S.F. Floor	Each	5020	5.02	
2020	Domestic Water Distribution	Gas fired water heater		S.F. Floor	.62	.62	7.1%
2040	Rain Water Drainage	Roof drains		S.F. Roof	1.03	1.03	
	D30 HVAC						
3010	Energy Supply	Oil fired hot water, unit heaters		S.F. Floor	10.08	10.08	
3020	Heat Generating Systems	N/A		—	—	—	
3030	Cooling Generating Systems	Chilled water, air cooled condenser system		S.F. Floor	11.88	11.88	23.4%
3050	Terminal & Package Units	N/A		—	—	—	
3090	Other HVAC Sys. & Equipment	N/A		—	—	—	
	D40 Fire Protection						
4010	Sprinklers	Sprinklers, ordinary hazard		S.F. Floor	4.15	4.15	5.1%
4020	Standpipes	Standpipe, wet, Class III		S.F. Floor	.67	.67	
	D50 Electrical						
5010	Electrical Service/Distribution	600 ampere service, panel board and feeders		S.F. Floor	1.12	1.12	
5020	Lighting & Branch Wiring	High intensity discharge fixtures, receptacles, switches, A.C. and misc. power		S.F. Floor	13.75	13.75	17.4%
5030	Communications & Security	Addressable alarm systems and emergency lighting		S.F. Floor	1.52	1.52	
5090	Other Electrical Systems	N/A		—	—	—	
E. EQUIPMENT & FURNISHINGS							
1010	Commercial Equipment	N/A		—	—	—	
1020	Institutional Equipment	N/A		—	—	—	0.6 %
1030	Vehicular Equipment	Dock shelters		Each	.53	.53	
1090	Other Equipment	N/A		—	—	—	
F. SPECIAL CONSTRUCTION							
1020	Integrated Construction	N/A		—	—	—	0.0 %
1040	Special Facilities	N/A		—	—	—	
G. BUILDING SITEWORK	N/A						

		Sub-Total	94.04	100%
CONTRACTOR FEES (General Requirements: 10%, Overhead: 5%, Profit: 10%)		25%	23.53	
ARCHITECT FEES		7%	8.23	
		Total Building Cost	125.80	

For customer support on your Square Foot Cost Data, call 877.756.2789.

123

Costs per square foot of floor area

Exterior Wall	S.F. Area	20000	30000	40000	50000	60000	70000	80000	90000	100000
	L.F. Perimeter	625	750	850	950	1015	1095	1150	1275	1300
Face Brick Common Brick Back-up	Steel Frame	206.65	187.30	175.95	169.15	163.05	159.30	155.70	154.80	151.55
	Concrete Frame	197.10	177.75	166.40	159.50	153.45	149.70	146.10	145.25	141.95
Face Brick Concrete Block Back-up	Steel Frame	204.90	185.85	174.75	168.00	162.10	158.40	154.85	154.00	150.80
	Concrete Frame	195.30	176.30	165.15	158.45	152.45	148.85	145.30	144.45	141.25
Stucco on Concrete Block	Steel Frame	185.90	170.70	161.85	156.50	151.80	148.90	146.15	145.40	142.90
	Concrete Frame	176.35	161.10	152.25	146.90	142.25	139.35	136.55	135.85	133.35
Perimeter Adj., Add or Deduct	Per 100 L.F.	13.10	8.70	6.50	5.15	4.35	3.70	3.30	2.95	2.65
Story Hgt. Adj., Add or Deduct	Per 1 Ft.	6.25	5.00	4.20	3.75	3.35	3.10	2.85	2.90	2.60

For Basement, add $36.15 per square foot of basement area

The above costs were calculated using the basic specifications shown on the facing page. These costs should be adjusted where necessary for design alternatives and owner's requirements. Reported completed project costs, for this type of structure, range from $49.30 to $190.90 per S.F.

Common additives

Description	Unit	$ Cost
Clock System		
20 room	Ea.	20,600
50 room	Ea.	48,500
Dock Bumpers, Rubber blocks		
4-1/2" thick, 10" high, 14" long	Ea.	89
24" long	Ea.	132
36" long	Ea.	179
12" high, 14" long	Ea.	106
24" long	Ea.	159
36" long	Ea.	189
6" thick, 10" high, 14" long	Ea.	104
24" long	Ea.	159
36" long	Ea.	203
20" high, 11" long	Ea.	183
Dock Boards, Heavy		
60" x 60" Aluminum, 5,000# cap.	Ea.	1500
9000# cap.	Ea.	1500
15,000# cap.	Ea.	1725

Description	Unit	$ Cost
Dock Levelers, Hinged 10 ton cap.		
6' x 8'	Ea.	6150
7' x 8'	Ea.	8025
Elevator, Hydraulic freight, 2 stops		
3500# capacity	Ea.	145,100
4000# capacity	Ea.	150,100
Additional stop, add	Ea.	9400
Partitions, Woven wire, 10 ga., 1-1/2" mesh		
4' Wide x 7" high	Ea.	206
8' High	Ea.	232
10' High	Ea.	277
Platform Lifter, Portable, 6'x 6'		
3000# cap.	Ea.	10,200
4000# cap.	Ea.	12,500
Fixed, 6'x 8', 5000# cap.	Ea.	12,700

Important: See the Reference Section for Location Factors.

Model costs calculated for a 3 story building with 12' story height and 90,000 square feet of floor area

					Unit	Unit Cost	Cost Per S.F.	% Of Sub-Total

A. SUBSTRUCTURE

				Unit	Unit Cost	Cost Per S.F.	% Of Sub-Total
1010	Standard Foundations	Poured concrete; strip and spread footings		S.F. Ground	2.31	.77	
1020	Special Foundations	N/A		—	—	—	
1030	Slab on Grade	4" reinforced concrete with vapor barrier and granular base		S.F. Slab	11.29	3.77	5.3%
2010	Basement Excavation	Site preparation for slab and trench for foundation wall and footing		S.F. Ground	.18	.06	
2020	Basement Walls	4' foundation wall		L.F. Wall	86	1.21	

B. SHELL

B10 Superstructure

				Unit	Unit Cost	Cost Per S.F.	% Of Sub-Total
1010	Floor Construction	Concrete flat slab		S.F. Floor	22.64	15.09	18.7%
1020	Roof Construction	Concrete flat slab		S.F. Roof	16.29	5.43	

B20 Exterior Enclosure

				Unit	Unit Cost	Cost Per S.F.	% Of Sub-Total
2010	Exterior Walls	Face brick with common brick backup	70% of wall	S.F. Wall	35.10	12.53	
2020	Exterior Windows	Industrial, horizontal pivoted steel	30% of wall	Each	1366	13.06	23.8%
2030	Exterior Doors	Double aluminum & glass, hollow metal, overhead doors		Each	2470	.46	

B30 Roofing

				Unit	Unit Cost	Cost Per S.F.	% Of Sub-Total
3010	Roof Coverings	Built-up tar and gravel with flashing; perlite/EPS composite insulation		S.F. Roof	6.84	2.28	2.2%
3020	Roof Openings	Roof hatches		S.F. Roof	.42	.14	

C. INTERIORS

				Unit	Unit Cost	Cost Per S.F.	% Of Sub-Total
1010	Partitions	Gypsum board on metal studs	50 S.F. Floor/L.F. Partition	S.F. Partition	4.35	.87	
1020	Interior Doors	Single leaf fire doors	500 S.F. Floor/Door	Each	1417	2.83	
1030	Fittings	Toilet partitions		S.F. Floor	.51	.51	
2010	Stair Construction	Concrete		Flight	8400	.37	8.7%
3010	Wall Finishes	Paint		S.F. Surface	.85	.34	
3020	Floor Finishes	90% metallic hardener, 10% vinyl composition tile		S.F. Floor	4.08	4.08	
3030	Ceiling Finishes	Fiberglass board on exposed grid systems	10% of area	S.F. Ceiling	5.78	.58	

D. SERVICES

D10 Conveying

				Unit	Unit Cost	Cost Per S.F.	% Of Sub-Total
1010	Elevators & Lifts	Two hydraulic freight elevators		Each	210,600	4.68	4.3%
1020	Escalators & Moving Walks	N/A		—	—	—	

D20 Plumbing

				Unit	Unit Cost	Cost Per S.F.	% Of Sub-Total
2010	Plumbing Fixtures	Toilet and service fixtures, supply and drainage	1 Fixture/1345 S.F. Floor	Each	8474	6.30	
2020	Domestic Water Distribution	Gas fired water heater		S.F. Floor	.37	.37	6.5%
2040	Rain Water Drainage	Roof drains		S.F. Roof	1.35	.45	

D30 HVAC

				Unit	Unit Cost	Cost Per S.F.	% Of Sub-Total
3010	Energy Supply	Oil fired hot water, unit heaters		S.F. Floor	4.82	4.82	
3020	Heat Generating Systems	N/A		—	—	—	
3030	Cooling Generating Systems	Chilled water, air cooled condenser system		S.F. Floor	11.88	11.88	15.2%
3050	Terminal & Package Units	N/A		—	—	—	
3090	Other HVAC Sys. & Equipment	N/A		—	—	—	

D40 Fire Protection

				Unit	Unit Cost	Cost Per S.F.	% Of Sub-Total
4010	Sprinklers	Wet pipe sprinkler system		S.F. Floor	3.63	3.63	3.9%
4020	Standpipes	Standpipe, wet, Class III		S.F. Floor	.64	.64	

D50 Electrical

				Unit	Unit Cost	Cost Per S.F.	% Of Sub-Total
5010	Electrical Service/Distribution	800 ampere service, panel board and feeders		S.F. Floor	.63	.63	
5020	Lighting & Branch Wiring	High intensity discharge fixtures, switches, A.C. and misc. power		S.F. Floor	10.21	10.21	11.2%
5030	Communications & Security	Addressable alarm systems and emergency lighting		S.F. Floor	1.25	1.25	
5090	Other Electrical Systems	Emergency generator, 30 kW		S.F. Floor	.21	.21	

E. EQUIPMENT & FURNISHINGS

				Unit	Unit Cost	Cost Per S.F.	% Of Sub-Total
1010	Commercial Equipment	N/A		—	—	—	
1020	Institutional Equipment	N/A		—	—	—	
1030	Vehicular Equipment	Dock Shelters		Each	.17	.17	0.2%
1090	Other Equipment	N/A		—	—	—	

F. SPECIAL CONSTRUCTION

				Unit	Unit Cost	Cost Per S.F.	% Of Sub-Total
1020	Integrated Construction	N/A		—	—	—	
1040	Special Facilities	N/A		—	—	—	0.0%

G. BUILDING SITEWORK   N/A

	Sub-Total	109.62	100%
CONTRACTOR FEES (General Requirements: 10%, Overhead: 5%, Profit: 10%)	25%	27.41	
ARCHITECT FEES	6%	8.22	

Total Building Cost	**145.25**

For customer support on your Square Foot Cost Data, call 877.756.2789.

125

Costs per square foot of floor area

Exterior Wall	S.F. Area	4000	4500	5000	5500	6000	6500	7000	7500	8000
	L.F. Perimeter	260	280	300	310	320	336	353	370	386
Face Brick Concrete Block Back-up	Steel Joists	193.75	189.95	187.00	183.00	179.70	177.60	176.05	174.65	173.30
	Bearing Walls	179.15	175.45	172.45	168.45	165.15	163.05	161.50	160.10	158.75
Decorative Concrete Block	Steel Joists	178.60	175.50	173.05	169.90	167.40	165.65	164.40	163.20	162.15
	Bearing Walls	164.10	160.95	158.50	155.35	152.80	151.10	149.85	148.65	147.60
Limestone with Concrete Block Back-up	Steel Joists	214.25	209.65	206.00	200.85	196.60	194.05	192.05	190.30	188.65
	Bearing Walls	199.75	195.10	191.45	186.30	182.05	179.50	177.50	175.75	174.05
Perimeter Adj., Add or Deduct	Per 100 L.F.	21.15	18.85	16.90	15.40	14.10	13.05	12.10	11.30	10.60
Story Hgt. Adj., Add or Deduct	Per 1 Ft.	2.85	2.75	2.60	2.50	2.30	2.30	2.20	2.15	2.10

For Basement, add $40.55 per square foot of basement area

The above costs were calculated using the basic specifications shown on the facing page. These costs should be adjusted where necessary for design alternatives and owner's requirements. Reported completed project costs, for this type of structure, range from $79.80 to $236.10 per S.F.

Common additives

Description	Unit	$ Cost
Appliances		
Cooking range, 30" free standing		
1 oven	Ea.	610 - 2300
2 oven	Ea.	1475 - 3175
30" built-in		
1 oven	Ea.	1100 - 2325
2 oven	Ea.	1775 - 3375
Counter top cook tops, 4 burner	Ea.	495 - 2100
Microwave oven	Ea.	296 - 890
Combination range, refrig. & sink, 30" wide	Ea.	1900 - 2750
60" wide	Ea.	2550
72" wide	Ea.	2800
Combination range refrigerator, sink		
microwave oven & icemaker	Ea.	6725
Compactor, residential, 4-1 compaction	Ea.	880 - 1475
Dishwasher, built-in, 2 cycles	Ea.	675 - 1175
4 cycles	Ea.	815 - 2000
Garbage disposer, sink type	Ea.	239 - 405
Hood for range, 2 speed, vented, 30" wide	Ea.	345 - 1375
42" wide	Ea.	435 - 2600

Description	Unit	$ Cost
Appliances, cont.		
Refrigerator, no frost 10-12 C.F.	Ea.	615 - 770
14-16 C.F.	Ea.	740 - 900
18-20 C.F.	Ea.	925 - 2100
Lockers, Steel, single tier, 60" or 72"	Opng.	231 - 380
2 tier, 60" or 72" total	Opng.	121 - 170
5 tier, box lockers	Opng.	74 - 84
Locker bench, lam. maple top only	L.F.	34.50
Pedestals, steel pipe	Ea.	76.50
Sound System		
Amplifier, 250 watts	Ea.	2200
Speaker, ceiling or wall	Ea.	217
Trumpet	Ea.	415

Important: See the Reference Section for Location Factors.

Fire Station, 1 Story

Model costs calculated for a 1 story building with 14' story height and 6,000 square feet of floor area

					Unit	Unit Cost	Cost Per S.F.	% Of Sub-Total
A.	**SUBSTRUCTURE**							
1010	Standard Foundations	Poured concrete; strip and spread footings			S.F. Ground	2.79	2.79	
1020	Special Foundations	N/A			—	—	—	
1030	Slab on Grade	6" reinforced concrete with vapor barrier and granular base			S.F. Slab	7.35	7.35	11.5%
2010	Basement Excavation	Site preparation for slab and trench for foundation wall and footing			S.F. Ground	.57	.57	
2020	Basement Walls	4' foundation wall			L.F. Wall	86	4.56	
B.	**SHELL**							
	B10 Superstructure							
1010	Floor Construction	N/A			—	—	—	
1020	Roof Construction	Metal deck, open web steel joists, beams on columns			S.F. Roof	10.86	10.86	8.2 %
	B20 Exterior Enclosure							
2010	Exterior Walls	Face brick with concrete block backup		75% of wall	S.F. Wall	33.39	18.70	
2020	Exterior Windows	Aluminum insulated glass		10% of wall	Each	860	2.01	19.2%
2030	Exterior Doors	Single aluminum and glass, overhead, hollow metal		15% of wall	S.F. Door	43.04	4.82	
	B30 Roofing							
3010	Roof Coverings	Built-up tar and gravel with flashing; perlite/EPS composite insulation			S.F. Roof	7.54	7.54	5.8%
3020	Roof Openings	Skylights, roof hatches			S.F. Roof	.16	.16	
C.	**INTERIORS**							
1010	Partitions	Concrete block	17 S.F. Floor/L.F. Partition		S.F. Partition	9.72	5.72	
1020	Interior Doors	Single leaf hollow metal	500 S.F. Floor/Door		Each	1168	2.34	
1030	Fittings	Toilet partitions			S.F. Floor	.51	.51	
2010	Stair Construction	N/A			—	—	—	12.8%
3010	Wall Finishes	Paint			S.F. Surface	2.16	2.54	
3020	Floor Finishes	50% vinyl tile, 50% paint			S.F. Floor	2.32	2.32	
3030	Ceiling Finishes	Fiberglass board on exposed grid, suspended	50% of area		S.F. Ceiling	7.22	3.62	
D.	**SERVICES**							
	D10 Conveying							
1010	Elevators & Lifts	N/A			—	—	—	0.0 %
1020	Escalators & Moving Walks	N/A			—	—	—	
	D20 Plumbing							
2010	Plumbing Fixtures	Kitchen, toilet and service fixtures, supply and drainage	1 Fixture/375 S.F. Floor		Each	3881	10.35	
2020	Domestic Water Distribution	Gast fired water heater			S.F. Floor	3.09	3.09	11.1%
2040	Rain Water Drainage	Roof drains			S.F. Roof	1.40	1.40	
	D30 HVAC							
3010	Energy Supply	N/A			—	—	—	
3020	Heat Generating Systems	Included in D3050			—	—	—	
3030	Cooling Generating Systems	N/A			—	—	—	18.5 %
3050	Terminal & Package Units	Rooftop multizone unit system			S.F. Floor	24.59	24.59	
3090	Other HVAC Sys. & Equipment	N/A			—	—	—	
	D40 Fire Protection							
4010	Sprinklers	Wet pipe sprinkler system			S.F. Floor	5.02	5.02	4.9%
4020	Standpipes	Standpipe, wet, Class III			S.F. Floor	1.54	1.54	
	D50 Electrical							
5010	Electrical Service/Distribution	200 ampere service, panel board and feeders			S.F. Floor	1.66	1.66	
5020	Lighting & Branch Wiring	High efficiency fluorescent fixtures, receptacles, switches, A.C. and misc. power			S.F. Floor	7.06	7.06	8.1%
5030	Communications & Security	Addressable alarm systems			S.F. Floor	2	2	
5090	Other Electrical Systems	N/A			—	—	—	
E.	**EQUIPMENT & FURNISHINGS**							
1010	Commercial Equipment	N/A			—	—	—	
1020	Institutional Equipment	N/A			—	—	—	
1030	Vehicular Equipment	N/A			—	—	—	0.0 %
1090	Other Equipment	N/A			—	—	—	
F.	**SPECIAL CONSTRUCTION**							
1020	Integrated Construction	N/A			—	—	—	0.0 %
1040	Special Facilities	N/A			—	—	—	
G.	**BUILDING SITEWORK**	**N/A**						

		Sub-Total	133.12	100%
CONTRACTOR FEES (General Requirements: 10%, Overhead: 5%, Profit: 10%)		25%	33.27	
ARCHITECT FEES		8%	13.31	

Total Building Cost 179.70

For customer support on your Square Foot Cost Data, call 877.756.2789.

127

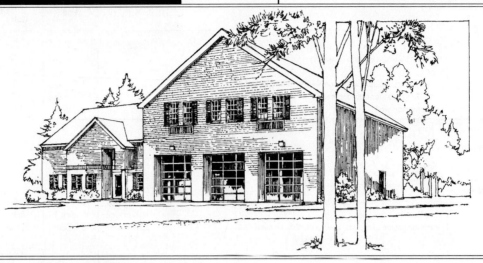

Costs per square foot of floor area

Exterior Wall	S.F. Area	6000	7000	8000	9000	10000	11000	12000	13000	14000
	L.F. Perimeter	220	240	260	280	286	303	320	336	353
Face Brick with Concrete Block Back-up	Steel Joists	204.15	197.70	192.85	189.10	184.00	181.30	179.05	177.05	175.45
	Precast Conc.	211.40	205.00	200.15	196.40	191.30	188.60	186.45	184.35	182.80
Decorative Concrete Block	Steel Joists	188.55	182.95	178.75	175.45	171.30	168.90	167.10	165.30	163.95
	Precast Conc.	198.25	192.70	188.50	185.25	181.10	178.70	176.90	175.10	173.75
Limestone with Concrete Block Back-up	Steel Joists	224.60	216.65	210.65	206.05	199.40	195.95	193.25	190.70	188.70
	Precast Conc.	234.05	226.20	220.25	215.60	208.95	205.55	202.85	200.30	198.30
Perimeter Adj., Add or Deduct	Per 100 L.F.	24.90	21.25	18.65	16.50	14.90	13.50	12.50	11.50	10.65
Story Hgt. Adj., Add or Deduct	Per 1 Ft.	3.25	3.05	2.90	2.75	2.55	2.45	2.40	2.30	2.25

For Basement, add $39.35 per square foot of basement area

The above costs were calculated using the basic specifications shown on the facing page. These costs should be adjusted where necessary for design alternatives and owner's requirements. Reported completed project costs, for this type of structure, range from $79.60 to $236.10 per S.F.

Common additives

Description	Unit	$ Cost
Appliances		
Cooking range, 30" free standing		
1 oven	Ea.	610 - 2300
2 oven	Ea.	1475 - 3175
30" built-in		
1 oven	Ea.	1100 - 2325
2 oven	Ea.	1775 - 3375
Counter top cook tops, 4 burner	Ea.	495 - 2100
Microwave oven	Ea.	296 - 890
Combination range, refrig. & sink, 30" wide	Ea.	1900 - 2750
60" wide	Ea.	2550
72" wide	Ea.	2800
Combination range, refrigerator, sink,		
microwave oven & icemaker	Ea.	6725
Compactor, residential, 4-1 compaction	Ea.	880 - 1475
Dishwasher, built-in, 2 cycles	Ea.	675 - 1175
4 cycles	Ea.	815 - 2000
Garbage diposer, sink type	Ea.	239 - 405
Hood for range, 2 speed, vented, 30" wide	Ea.	345 - 1375
42" wide	Ea.	435 - 2600

Description	Unit	$ Cost
Appliances, cont.		
Refrigerator, no frost 10-12 C.F.	Ea.	615 - 770
14-16 C.F.	Ea.	740 - 900
18-20 C.F.	Ea.	925 - 2100
Elevators, Hydraulic passenger, 2 stops		
1500# capacity	Ea.	67,400
2500# capacity	Ea.	70,400
3500# capacity	Ea.	75,400
Lockers, Steel, single tier, 60" or 72"	Opng.	231 - 380
2 tier, 60" or 72" total	Opng.	121 - 170
5 tier, box lockers	Opng.	74 - 84
Locker bench, lam. maple top only	L.F.	34.50
Pedestals, steel pipe	Ea.	76.50
Sound System		
Amplifier, 250 watts	Ea.	2200
Speaker, ceiling or wall	Ea.	217
Trumpet	Ea.	415

Important: See the Reference Section for Location Factors.

Model costs calculated for a 2 story building with 14' story height and 10,000 square feet of floor area

Fire Station, 2 Story

				Unit	Unit Cost	Cost Per S.F.	% Of Sub-Total
A. SUBSTRUCTURE							
1010	Standard Foundations	Poured concrete; strip and spread footings		S.F. Ground	2.72	1.36	
1020	Special Foundations	N/A		—	—	—	
1030	Slab on Grade	6" reinforced concrete with vapor barrier and granular base		S.F. Slab	7.35	3.68	6.3%
2010	Basement Excavation	Site preparation for slab and trench for foundation wall and footing		S.F. Ground	.57	.29	
2020	Basement Walls	4' foundation wall		L.F. Wall	95	2.71	
B. SHELL							
B10 Superstructure							
1010	Floor Construction	Open web steel joists, slab form, concrete		S.F. Floor	15.88	7.94	8.3%
1020	Roof Construction	Metal deck on open web steel joists		S.F. Roof	5.14	2.57	
B20 Exterior Enclosure							
2010	Exterior Walls	Decorative concrete block	75% of wall	S.F. Wall	20.75	12.46	
2020	Exterior Windows	Aluminum insulated glass	10% of wall	Each	746	2.60	15.0%
2030	Exterior Doors	Single aluminum and glass, steel overhead, hollow metal	15% of wall	S.F. Door	33.55	4.03	
B30 Roofing							
3010	Roof Coverings	Built-up tar and gravel with flashing; perlite/EPS composite insulation		S.F. Roof	7.74	3.87	3.1%
3020	Roof Openings	N/A		—	—	—	
C. INTERIORS							
1010	Partitions	Concrete block	10 S.F. Floor/L.F. Partition	S.F. Partition	9.72	5.72	
1020	Interior Doors	Single leaf hollow metal	500 S.F. Floor/Door	Each	1417	2.83	
1030	Fittings	Toilet partitions		S.F. Floor	.46	.46	
2010	Stair Construction	Concrete filled metal pan		Flight	20,150	4.03	17.8%
3010	Wall Finishes	Paint		S.F. Surface	3.66	4.30	
3020	Floor Finishes	50% vinyl tile, 50% paint		S.F. Floor	2.32	2.32	
3030	Ceiling Finishes	Fiberglass board on exposed grid, suspended	50% of area	S.F. Ceiling	5.78	2.89	
D. SERVICES							
D10 Conveying							
1010	Elevators & Lifts	One hydraulic passenger elevator		Each	85,300	8.53	6.7%
1020	Escalators & Moving Walks	N/A		—	—	—	
D20 Plumbing							
2010	Plumbing Fixtures	Kitchen toilet and service fixtures, supply and drainage	1 Fixture/400 S.F. Floor	Each	4004	10.01	
2020	Domestic Water Distribution	Gas fired water heater		S.F. Floor	2.47	2.47	10.6%
2040	Rain Water Drainage	Roof drains		S.F. Roof	1.82	.91	
D30 HVAC							
3010	Energy Supply	N/A		—	—	—	
3020	Heat Generating Systems	Included in D3050		—	—	—	
3030	Cooling Generating Systems	N/A		—	—	—	19.4 %
3050	Terminal & Package Units	Rooftop multizone unit system		S.F. Floor	24.59	24.59	
3090	Other HVAC Sys. & Equipment	N/A		—	—	—	
D40 Fire Protection							
4010	Sprinklers	Wet pipe sprinkler system		S.F. Floor	3.95	3.95	4.4%
4020	Standpipes	Standpipe, wet, Class III		S.F. Floor	1.57	1.57	
D50 Electrical							
5010	Electrical Service/Distribution	400 ampere service, panel board and feeders		S.F. Floor	1.66	1.66	
5020	Lighting & Branch Wiring	High efficiency fluorescent fixtures, receptacles, switches, A.C. and misc. power		S.F. Floor	7.30	7.30	8.5%
5030	Communications & Security	Addressable alarm systems and emergency lighting		S.F. Floor	1.57	1.57	
5090	Other Electrical Systems	Emergency generator, 15 kW		S.F. Floor	.26	.26	
E. EQUIPMENT & FURNISHINGS							
1010	Commercial Equipment	N/A		—	—	—	
1020	Institutional Equipment	N/A		—	—	—	
1030	Vehicular Equipment	N/A		—	—	—	0.0 %
1090	Other Equipment	N/A		—	—	—	
F. SPECIAL CONSTRUCTION							
1020	Integrated Construction	N/A		—	—	—	
1040	Special Facilities	N/A		—	—	—	0.0 %
G. BUILDING SITEWORK	**N/A**						

	Sub-Total	126.88	**100%**
CONTRACTOR FEES (General Requirements: 10%, Overhead: 5%, Profit: 10%)	25%	31.73	
ARCHITECT FEES	8%	12.69	
	Total Building Cost	**171.30**	

For customer support on your Square Foot Cost Data, call 877.756.2789.

129

Costs per square foot of floor area

Exterior Wall	S.F. Area	4000	5000	6000	8000	10000	12000	14000	16000	18000
	L.F. Perimeter	180	205	230	260	300	340	353	386	420
Cedar Beveled Siding	Wood Frame	220.15	211.50	205.70	196.70	192.00	188.85	185.30	183.50	182.20
Aluminum Siding	Wood Frame	217.10	208.70	203.10	194.50	190.00	186.95	183.60	181.90	180.65
Board and Batten	Wood Frame	214.95	207.80	203.00	195.35	191.45	188.80	185.60	184.15	183.05
Face Brick on Block	Bearing Wall	222.55	212.20	205.25	193.85	188.10	184.25	179.40	177.15	175.50
Stucco on Block	Bearing Wall	205.60	196.70	190.75	181.60	176.70	173.45	169.80	167.95	166.60
Decorative Block	Wood Joists	224.85	215.75	209.70	200.10	195.15	191.80	187.95	186.05	184.65
Perimeter Adj., Add or Deduct	Per 100 L.F.	16.95	13.50	11.25	8.45	6.80	5.65	4.85	4.30	3.75
Story Hgt. Adj., Add or Deduct	Per 1 Ft.	2.20	2.00	1.85	1.60	1.50	1.40	1.30	1.20	1.15
For Basement, add $25.05 per square foot of basement area										

The above costs were calculated using the basic specifications shown on the facing page. These costs should be adjusted where necessary for design alternatives and owner's requirements. Reported completed project costs, for this type of structure, range from $115.75 to $253.10 per S.F.

Common additives

Description	Unit	$ Cost
Appliances		
Cooking range, 30" free standing		
1 oven	Ea.	610 - 2300
2 oven	Ea.	1475 - 3175
30" built-in		
1 oven	Ea.	1100 - 2325
2 oven	Ea.	1775 - 3375
Counter top cook tops, 4 burner	Ea.	495 - 2100
Microwave oven	Ea.	296 - 890
Combination range, refrig. & sink, 30" wide	Ea.	1900 - 2750
60" wide	Ea.	2550
72" wide	Ea.	2800
Combination range, refrigerator, sink,		
microwave oven & icemaker	Ea.	6725
Compactor, residential, 4-1 compaction	Ea.	880 - 1475
Dishwasher, built-in, 2 cycles	Ea.	675 - 1175
4 cycles	Ea.	815 - 2000
Garbage disposer, sink type	Ea.	239 - 405
Hood for range, 2 speed, vented, 30" wide	Ea.	345 - 1375
42" wide	Ea.	435 - 2600

Description	Unit	$ Cost
Appliances, cont.		
Refrigerator, no frost 10-12 C.F.	Ea.	615 - 770
14-16 C.F.	Ea.	740 - 900
18-20 C.F.	Ea.	925 - 2100
Elevators, Hydraulic passenger, 2 stops		
1500# capacity	Ea.	67,400
2500# capacity	Ea.	70,400
3500# capacity	Ea.	75,400
Laundry Equipment		
Dryer, gas, 16 lb. capacity	Ea.	995
30 lb. capacity	Ea.	3950
Washer, 4 cycle	Ea.	1225
Commercial	Ea.	1625
Sound System		
Amplifier, 250 watts	Ea.	2200
Speaker, ceiling or wall	Ea.	217
Trumpet	Ea.	415

Important: See the Reference Section for Location Factors.

Model costs calculated for a 2 story building with 10' story height and 10,000 square feet of floor area

Fraternity/Sorority House

				Unit	Unit Cost	Cost Per S.F.	% Of Sub-Total
A.	**SUBSTRUCTURE**						
1010	Standard Foundations	Poured concrete; strip and spread footings		S.F. Ground	3.02	1.51	
1020	Special Foundations	N/A		—			
1030	Slab on Grade	4" reinforced concrete with vapor barrier and granular base		S.F. Slab	5.56	2.79	4.9%
2010	Basement Excavation	Site preparation for slab and trench for foundation wall and footing		S.F. Ground	.57	.29	
2020	Basement Walls	4' foundation wall		L.F. Wall	76	2.29	
B.	**SHELL**						
	B10 Superstructure						
1010	Floor Construction	Plywood on wood joists		S.F. Floor	41.84	20.92	16.4%
1020	Roof Construction	Plywood on wood rafters (pitched)		S.F. Roof	3.80	2.13	
	B20 Exterior Enclosure						
2010	Exterior Walls	Cedar bevel siding on wood studs, insulated	80% of wall	S.F. Wall	13.67	6.56	
2020	Exterior Windows	Double hung wood	20% of wall	Each	646	3.10	8.0%
2030	Exterior Doors	Solid core wood		Each	2388	1.68	
	B30 Roofing						
3010	Roof Coverings	Asphalt shingles with flashing (pitched); rigid fiberglass insulation		S.F. Roof	5.36	2.68	1.9%
3020	Roof Openings	N/A		—	—	—	
C.	**INTERIORS**						
1010	Partitions	Gypsum board on wood studs	25 S.F. Floor/L.F. Partition	S.F. Partition	8.56	2.74	
1020	Interior Doors	Single leaf wood	200 S.F. Floor/Door	Each	626	3.14	
1030	Fittings	N/A		—	—	—	
2010	Stair Construction	Wood		Flight	2745	.83	13.8%
3010	Wall Finishes	Paint		S.F. Surface	.84	.54	
3020	Floor Finishes	10% hardwood, 70% carpet, 20% ceramic tile		S.F. Floor	7.59	7.59	
3030	Ceiling Finishes	Gypsum board on wood furring		S.F. Ceiling	4.66	4.66	
D.	**SERVICES**						
	D10 Conveying						
1010	Elevators & Lifts	One hydraulic passenger elevator		Each	85,300	8.53	6.1%
1020	Escalators & Moving Walks	N/A		—			
	D20 Plumbing						
2010	Plumbing Fixtures	Kitchen toilet and service fixtures, supply and drainage	1 Fixture/150 S.F. Floor	Each	2217	14.78	
2020	Domestic Water Distribution	Gas fired water heater		S.F. Floor	3.32	3.32	12.8%
2040	Rain Water Drainage	N/A		—			
	D30 HVAC						
3010	Energy Supply	Oil fired hot water, baseboard radiation		S.F. Floor	8.05	8.05	
3020	Heat Generating Systems	N/A		—	—	—	
3030	Cooling Generating Systems	N/A		—	—	—	11.4%
3050	Terminal & Package Units	Split system with air cooled condensing unit		S.F. Floor	8.04	8.04	
3090	Other HVAC Sys. & Equipment	N/A		—	—	—	
	D40 Fire Protection						
4010	Sprinklers	Wet pipe sprinkler system		S.F. Floor	3.95	3.95	3.6%
4020	Standpipes	Standpipe		S.F. Floor	1.13	1.13	
	D50 Electrical						
5010	Electrical Service/Distribution	600 ampere service, panel board and feeders		S.F. Floor	4.16	4.16	
5020	Lighting & Branch Wiring	High efficiency fluorescent fixtures, receptacles, switches, A.C. and misc. power		S.F. Floor	10.54	10.54	21.1%
5030	Communications & Security	Addressable alarm, communication system, internet and phone wiring, and generator set		S.F. Floor	14.76	14.76	
5090	Other Electrical Systems	Emergency generator, 7.5 kW		S.F. Floor	.22	.22	
E.	**EQUIPMENT & FURNISHINGS**						
1010	Commercial Equipment	N/A		—	—	—	
1020	Institutional Equipment	N/A		—	—	—	
1030	Vehicular Equipment	N/A		—	—	—	0.0%
1090	Other Equipment	N/A		—	—	—	
F.	**SPECIAL CONSTRUCTION**						
1020	Integrated Construction	N/A		—	—	—	
1040	Special Facilities	N/A		—	—	—	0.0%
G.	**BUILDING SITEWORK**	**N/A**					

		Sub-Total	140.93	100%
CONTRACTOR FEES (General Requirements: 10%, Overhead: 5%, Profit: 10%)		25%	35.22	
ARCHITECT FEES		9%	15.85	

Total Building Cost	**192**

For customer support on your Square Foot Cost Data, call 877.756.2789.

131

Costs per square foot of floor area

Exterior Wall	S.F. Area	4000	6000	8000	10000	12000	14000	16000	18000	20000
	L.F. Perimeter	260	320	384	424	460	484	510	540	576
Vertical Redwood Siding	Wood Frame	166.45	153.00	146.55	141.55	138.05	135.05	133.00	131.40	130.35
Brick Veneer	Wood Frame	214.15	186.50	173.00	163.40	156.75	151.45	147.55	144.65	142.55
Aluminum Siding	Wood Frame	199.50	174.50	162.20	153.80	148.10	143.65	140.40	137.90	136.05
Brick on Block	Wood Truss	222.00	193.05	178.90	168.70	161.60	155.85	151.70	148.60	146.35
Limestone on Block	Wood Truss	244.25	211.20	195.25	183.10	174.60	167.55	162.40	158.65	156.00
Stucco on Block	Wood Truss	205.10	179.15	166.45	157.65	151.70	146.85	143.40	140.80	138.85
Perimeter Adj., Add or Deduct	Per 100 L.F.	11.85	7.95	5.90	4.75	3.90	3.45	2.95	2.65	2.40
Story Hgt. Adj., Add or Deduct	Per 1 Ft.	1.60	1.35	1.15	1.05	0.95	0.90	0.75	0.80	0.75

For Basement, add $33.40 per square foot of basement area

The above costs were calculated using the basic specifications shown on the facing page. These costs should be adjusted where necessary for design alternatives and owner's requirements. Reported completed project costs, for this type of structure, range from $113.30 to $348.40 per S.F.

Common additives

Description	Unit	$ Cost
Autopsy Table, Standard	Ea.	11,400
Deluxe	Ea.	19,100
Directory Boards, Plastic, glass covered		
30" x 20"	Ea.	610
36" x 48"	Ea.	1525
Aluminum, 24" x 18"	Ea.	640
36" x 24"	Ea.	765
48" x 32"	Ea.	1075
48" x 60"	Ea.	2275
Emergency Lighting, 25 watt, battery operated		
Lead battery	Ea.	345
Nickel cadmium	Ea.	685
Mortuary Refrigerator, End operated		
Two capacity	Ea.	10,800
Six capacity	Ea.	16,300

Description	Unit	$ Cost
Planters, Precast concrete		
48" diam., 24" high	Ea.	705
7" diam., 36" high	Ea.	1825
Fiberglass, 36" diam., 24" high	Ea.	785
60" diam., 24" high	Ea.	1325
Smoke Detectors		
Ceiling type	Ea.	227
Duct type	Ea.	565

Important: See the Reference Section for Location Factors.

Model costs calculated for a 1 story building with 12' story height and 10,000 square feet of floor area

				Unit	Unit Cost	Cost Per S.F.	% Of Sub-Total
A. SUBSTRUCTURE							
1010	Standard Foundations	Poured concrete; strip and spread footings		S.F. Ground	2.69	2.69	
1020	Special Foundations	N/A		—			
1030	Slab on Grade	4" reinforced concrete with vapor barrier and granular base		S.F. Slab	5.56	5.56	11.4%
2010	Basement Excavation	Site preparation for slab and trench for foundation wall and footing		S.F. Ground	.32	.32	
2020	Basement Walls	4' foundation wall		L.F. Wall	76	3.22	
B. SHELL							
	B10 Superstructure						
1010	Floor Construction	Wood beams on columns		S.F. Floor	.85	.85	6.0%
1020	Roof Construction	Plywood on wood truss		S.F. Roof	5.39	5.39	
	B20 Exterior Enclosure						
2010	Exterior Walls	1" x 4" vertical T & G redwood siding on wood studs	90% of wall	S.F. Wall	13.85	6.34	
2020	Exterior Windows	Double hung wood	10% of wall	Each	521	2.21	9.4%
2030	Exterior Doors	Wood swinging double doors, single leaf hollow metal		Each	1995	1.20	
	B30 Roofing						
3010	Roof Coverings	Single ply membrane, fully adhered; polyisocyanurate sheets		S.F. Roof	7.43	7.43	7.2%
3020	Roof Openings	N/A		—	—	—	
C. INTERIORS							
1010	Partitions	Gypsum board on wood studs with sound deadening board	15 S.F. Floor/L.F. Partition	S.F. Partition	9.73	5.19	
1020	Interior Doors	Single leaf wood	150 S.F. Floor/Door	Each	626	4.17	
1030	Fittings	N/A		—	—	—	
2010	Stair Construction	N/A		—	—	—	26.2%
3010	Wall Finishes	50% wallpaper, 25% wood paneling, 25% paint		S.F. Surface	3.23	3.45	
3020	Floor Finishes	70% carpet, 30% ceramic tile		S.F. Floor	8.60	8.60	
3030	Ceiling Finishes	Fiberglass board on exposed grid, suspended		S.F. Ceiling	5.78	5.78	
D. SERVICES							
	D10 Conveying						
1010	Elevators & Lifts	N/A		—	—	—	0.0 %
1020	Escalators & Moving Walks	N/A		—	—	—	
	D20 Plumbing						
2010	Plumbing Fixtures	Toilet and service fixtures, supply and drainage	1 Fixture/770 S.F. Floor	Each	2895	3.76	
2020	Domestic Water Distribution	Electric water heater		S.F. Floor	4.95	4.95	8.4%
2040	Rain Water Drainage	Roof drain		S.F. Roof			
	D30 HVAC						
3010	Energy Supply	N/A		—	—	—	
3020	Heat Generating Systems	Included in D3030		—	—	—	
3030	Cooling Generating Systems	Multizone rooftop unit, gas heating, electric cooling		S.F. Floor	17.30	17.30	16.7 %
3050	Terminal & Package Units	N/A		—	—	—	
3090	Other HVAC Sys. & Equipment	N/A		—	—	—	
	D40 Fire Protection						
4010	Sprinklers	Wet pipe sprinkler system		S.F. Floor	3.65	3.65	4.4%
4020	Standpipes	Standpipe, wet, Class III		S.F. Floor	.92	.92	
	D50 Electrical						
5010	Electrical Service/Distribution	400 ampere service, panel board and feeders		S.F. Floor	2	2	
5020	Lighting & Branch Wiring	High efficiency fluorescent fixtures, receptacles, switches, A.C. and misc. power		S.F. Floor	7.23	7.23	10.5%
5030	Communications & Security	Addressable alarm systems and emergency lighting		S.F. Floor	1.57	1.57	
5090	Other Electrical Systems	Emergency generator, 15 kW		S.F. Floor	.09	.09	
E. EQUIPMENT & FURNISHINGS							
1010	Commercial Equipment	N/A		—	—	—	
1020	Institutional Equipment	N/A		—	—	—	0.0 %
1030	Vehicular Equipment	N/A		—	—	—	
1090	Other Equipment	N/A		—	—	—	
F. SPECIAL CONSTRUCTION							
1020	Integrated Construction	N/A		—	—	—	0.0 %
1040	Special Facilities	N/A		—	—	—	
G. BUILDING SITEWORK	**N/A**						

	Sub-Total	103.87	100%
CONTRACTOR FEES (General Requirements: 10%, Overhead: 5%, Profit: 10%)	25%	25.99	
ARCHITECT FEES	9%	11.69	
	Total Building Cost	**141.55**	

For customer support on your Square Foot Cost Data, call 877.756.2789.

Costs per square foot of floor area

Exterior Wall	S.F. Area	12000	14000	16000	19000	21000	23000	26000	28000	30000
	L.F. Perimeter	440	474	510	556	583	607	648	670	695
E.I.F.S. on Concrete Block	Steel Frame	134.80	131.95	129.90	127.35	125.90	124.60	123.10	122.15	121.40
Tilt-up Concrete Wall	Steel Frame	132.25	129.60	127.70	125.30	123.95	122.70	121.40	120.45	119.80
Face Brick with Concrete Block Back-up	Bearing Walls	134.45	131.10	128.65	125.60	123.85	122.25	120.55	119.40	118.50
	Steel Frame	141.45	138.15	135.65	132.60	130.85	129.25	127.55	126.40	125.55
Stucco on Concrete Block	Bearing Walls	125.95	123.30	121.30	118.75	117.40	116.15	114.75	113.85	113.15
	Steel Frame	133.35	130.65	128.65	126.10	124.75	123.50	122.10	121.20	120.50
Perimeter Adj., Add or Deduct	Per 100 L.F.	7.10	6.15	5.40	4.45	4.00	3.65	3.30	3.10	2.90
Story Hgt. Adj., Add or Deduct	Per 1 Ft.	1.55	1.40	1.40	1.20	1.10	1.10	1.00	1.00	0.95

For Basement, add $37.95 per square foot of basement area

The above costs were calculated using the basic specifications shown on the facing page. These costs should be adjusted where necessary for design alternatives and owner's requirements. Reported completed project costs, for this type of structure, range from $60.30 to $175.35 per S.F.

Common additives

Description	Unit	$ Cost
Emergency Lighting, 25 watt, battery operated		
Lead battery	Ea.	345
Nickel cadmium	Ea.	685
Smoke Detectors		
Ceiling type	Ea.	227
Duct type	Ea.	565
Sound System		
Amplifier, 250 watts	Ea.	2200
Speaker, ceiling or wall	Ea.	217
Trumpet	Ea.	415

Important: See the Reference Section for Location Factors.

Model costs calculated for a 1 story building with 14' story height and 21,000 square feet of floor area

					Unit	Unit Cost	Cost Per S.F.	% Of Sub-Total
A.	**SUBSTRUCTURE**							
1010	Standard Foundations	Poured concrete; strip and spread footings			S.F. Ground	1.51	1.51	
1020	Special Foundations	N/A			—	—	—	
1030	Slab on Grade	4" reinforced concrete with vapor barrier and granular base			S.F. Slab	8.31	8.31	13.3%
2010	Basement Excavation	Site preparation for slab and trench for foundation wall and footing			S.F. Ground	.32	.32	
2020	Basement Walls	4' foundation wall			L.F. Wall	86	2.38	
B.	**SHELL**							
	B10 Superstructure							
1010	Floor Construction	N/A			—	—	—	12.8 %
1020	Roof Construction	Metal deck, open web steel joists, beams, columns			S.F. Roof	12.02	12.02	
	B20 Exterior Enclosure							
2010	Exterior Walls	E.I.F.S.		70% of wall	S.F. Wall	18.30	4.98	
2020	Exterior Windows	Window wall		30% of wall	Each	58	6.75	16.4%
2030	Exterior Doors	Double aluminum and glass, hollow metal, steel overhead			Each	4302	3.70	
	B30 Roofing							
3010	Roof Coverings	Built-up tar and gravel with flashing; perlite/EPS composite insulation			S.F. Roof	7.10	7.10	7.5%
3020	Roof Openings	N/A			—	—	—	
C.	**INTERIORS**							
1010	Partitions	Gypsum board on metal studs	28 S.F. Floor/L.F. Partition		S.F. Partition	5.11	2.19	
1020	Interior Doors	Hollow metal	280 S.F. Floor/Door		Each	1168	4.17	
1030	Fittings	N/A			—	—	—	
2010	Stair Construction	N/A			—	—	—	13.3%
3010	Wall Finishes	Paint			S.F. Surface	1.12	.96	
3020	Floor Finishes	50% vinyl tile, 50% paint			S.F. Floor	2.32	2.32	
3030	Ceiling Finishes	Fiberglass board on exposed grid, suspended	50% of area		S.F. Ceiling	5.78	2.89	
D.	**SERVICES**							
	D10 Conveying							
1010	Elevators & Lifts	N/A			—	—	—	0.0 %
1020	Escalators & Moving Walks	N/A			—	—	—	
	D20 Plumbing							
2010	Plumbing Fixtures	Toilet and service fixtures, supply and drainage	1 Fixture/1500 S.F. Floor		Each	2955	1.97	
2020	Domestic Water Distribution	Gas fired water heater			S.F. Floor	2.06	2.06	6.9%
2040	Rain Water Drainage	Roof drains			S.F. Roof	2.42	2.42	
	D30 HVAC							
3010	Energy Supply	N/A			—	—	—	
3020	Heat Generating Systems	N/A			—	—	—	
3030	Cooling Generating Systems	N/A			—	—	—	11.2 %
3050	Terminal & Package Units	Single zone rooftop unit, gas heating, electric cooling			S.F. Floor	9.84	9.84	
3090	Other HVAC Sys. & Equipment	Underfloor garage exhaust system			S.F. Floor	.67	.67	
	D40 Fire Protection							
4010	Sprinklers	Wet pipe sprinkler system			S.F. Floor	4.72	4.72	5.5%
4020	Standpipes	Standpipe			S.F. Floor	.48	.48	
	D50 Electrical							
5010	Electrical Service/Distribution	200 ampere service, panel board and feeders			S.F. Floor	.49	.49	
5020	Lighting & Branch Wiring	T-8 fluorescent fixtures, receptacles, switches, A.C. and misc. power			S.F. Floor	6.94	6.94	11.6%
5030	Communications & Security	Addressable alarm systems, partial internet wiring and emergency lighting			S.F. Floor	3.41	3.41	
5090	Other Electrical Systems	Emergency generator, 7.5 kW			S.F. Floor	.07	.07	
E.	**EQUIPMENT & FURNISHINGS**							
1010	Commercial Equipment	N/A			—	—	—	
1020	Institutional Equipment	N/A			—	—	—	1.5 %
1030	Vehicular Equipment	Hoists, compressor, fuel pump			S.F. Floor	1.43	1.43	
1090	Other Equipment	N/A			—	—	—	
F.	**SPECIAL CONSTRUCTION**							
1020	Integrated Construction	N/A			—	—	—	0.0 %
1040	Special Facilities	N/A			—	—	—	
G.	**BUILDING SITEWORK**	**N/A**						

	Sub-Total	94.10	100%
CONTRACTOR FEES (General Requirements: 10%, Overhead: 5%, Profit: 10%)	25%	23.57	
ARCHITECT FEES	7%	8.23	
Total Building Cost		**125.90**	

For customer support on your Square Foot Cost Data, call 877.756.2789.

135

Costs per square foot of floor area

Exterior Wall	S.F. Area	85000	115000	145000	175000	205000	235000	265000	295000	325000
	L.F. Perimeter	529	638	723	823	875	910	975	1027	1075
Face Brick with Concrete Block Back-up	Steel Frame	82.30	80.60	79.35	78.70	77.85	77.05	76.65	76.30	75.95
	Precast Concrete	71.30	69.60	68.35	67.75	66.85	66.10	65.70	65.35	64.95
Precast Concrete	Steel Frame	90.70	88.05	86.05	85.05	83.55	82.25	81.65	80.95	80.45
	Precast Concrete	78.25	75.85	74.10	73.10	71.80	70.65	70.00	69.45	68.95
Reinforced Concrete	Steel Frame	79.70	78.30	77.30	76.75	76.05	75.50	75.10	74.85	74.55
	R/Conc. Frame	71.55	70.05	69.10	68.55	67.85	67.25	66.95	66.65	66.40
Perimeter Adj., Add or Deduct	Per 100 L.F.	1.75	1.30	1.05	0.85	0.70	0.65	0.60	0.50	0.50
Story Hgt. Adj., Add or Deduct	Per 1 Ft.	0.80	0.70	0.65	0.60	0.50	0.50	0.45	0.40	0.45
Basement—Not Applicable										

The above costs were calculated using the basic specifications shown on the facing page. These costs should be adjusted where necessary for design alternatives and owner's requirements. Reported completed project costs, for this type of structure, range from $37.00 to $143.15 per S.F.

Common additives

Description	Unit	$ Cost
Elevators, Electric passenger, 5 stops		
2000# capacity	Ea.	171,100
3500# capacity	Ea.	177,600
5000# capacity	Ea.	184,100
Barrier gate w/programmable controller	Ea.	4225
Booth for attendant, average	Ea.	13,400
Fee computer	Ea.	13,600
Ticket spitter with time/date stamp	Ea.	8125
Mag strip encoding	Ea.	21,400
Collection station, pay on foot	Ea.	128,500
Parking control software	Ea.	26,200 - 105,000
Painting, Parking stalls	Stall	8.85
Parking Barriers		
Timber with saddles, 4" x 4"	L.F.	7.75
Precast concrete, 6" x 10" x 6'	Ea.	63.50
Traffic Signs, directional, 12" x 18", high density	Ea.	99

Important: See the Reference Section for Location Factors.

Model costs calculated for a 5 story building with 10' story height and 145,000 square feet of floor area

				Unit	Unit Cost	Cost Per S.F.	% Of Sub-Total
A.	**SUBSTRUCTURE**						
1010	Standard Foundations	Poured concrete; strip and spread footings		S.F. Ground	2.10	.42	
1020	Special Foundations	N/A		—	—	—	
1030	Slab on Grade	6" reinforced concrete with vapor barrier and granular base		S.F. Slab	7.25	1.45	4.6%
2010	Basement Excavation	Site preparation for slab and trench for foundation wall and footing		S.F. Ground	.18	.04	
2020	Basement Walls	4' foundation wall		L.F. Wall	86	.47	
B.	**SHELL**						
	B10 Superstructure						
1010	Floor Construction	Double tee precast concrete slab, precast concrete columns		S.F. Floor	29.23	23.38	45.3%
1020	Roof Construction	N/A		—	—	—	
	B20 Exterior Enclosure						
2010	Exterior Walls	Face brick with concrete block backup	40% of story height	S.F. Wall	50	5	
2020	Exterior Windows	N/A		—	—	—	9.7%
2030	Exterior Doors	N/A		—	—	—	
	B30 Roofing						
3010	Roof Coverings	N/A		—	—	—	0.0%
3020	Roof Openings	N/A		—	—	—	
C.	**INTERIORS**						
1010	Partitions	Concrete block		S.F. Partition	34.32	1.32	
1020	Interior Doors	Hollow metal		Each	23,360	.16	
1030	Fittings	N/A		—	—	—	
2010	Stair Construction	Concrete		Flight	11,575	1.27	14.9%
3010	Wall Finishes	Paint		S.F. Surface	1.82	.14	
3020	Floor Finishes	Parking deck surface coating		S.F. Floor	4.82	4.82	
3030	Ceiling Finishes	N/A		—	—	—	
D.	**SERVICES**						
	D10 Conveying						
1010	Elevators & Lifts	Two hydraulic passenger elevators		Each	168,200	2.32	4.5%
1020	Escalators & Moving Walks	N/A		—	—	—	
	D20 Plumbing						
2010	Plumbing Fixtures	Toilet and service fixtures, supply and drainage	1 Fixture/18,125 S.F. Floor	Each	725	.04	
2020	Domestic Water Distribution	Electric water heater		S.F. Floor	.10	.10	3.1%
2040	Rain Water Drainage	Roof drains		S.F. Roof	7.20	1.44	
	D30 HVAC						
3010	Energy Supply	N/A		—	—	—	
3020	Heat Generating Systems	N/A		—	—	—	
3030	Cooling Generating Systems	N/A		—	—	—	0.0%
3050	Terminal & Package Units	N/A		—	—	—	
3090	Other HVAC Sys. & Equipment	N/A		—	—	—	
	D40 Fire Protection						
4010	Sprinklers	Dry pipe sprinkler system		S.F. Floor	4.29	4.29	8.5%
4020	Standpipes	Standpipes and hose systems		S.F. Floor	.10	.10	
	D50 Electrical						
5010	Electrical Service/Distribution	400 ampere service, panel board and feeders		S.F. Floor	.25	.25	
5020	Lighting & Branch Wiring	T-8 fluorescent fixtures, receptacles, switches and misc. power		S.F. Floor	3.18	3.18	7.0%
5030	Communications & Security	Addressable alarm systems and emergency lighting		S.F. Floor	.13	.13	
5090	Other Electrical Systems	Emergency generator, 7.5 kW		S.F. Floor	.06	.06	
E.	**EQUIPMENT & FURNISHINGS**						
1010	Commercial Equipment	N/A		—	—	—	
1020	Institutional Equipment	N/A		—	—	—	2.4%
1030	Vehicular Equipment	Ticket dispensers, booths, automatic gates		S.F. Floor	1.24	1.24	
1090	Other Equipment	N/A		—	—	—	
F.	**SPECIAL CONSTRUCTION**						
1020	Integrated Construction	N/A		—	—	—	0.0%
1040	Special Facilities	N/A		—	—	—	
G.	**BUILDING SITEWORK**	**N/A**					
				Sub-Total		51.62	100%
	CONTRACTOR FEES (General Requirements: 10%, Overhead: 5%, Profit: 10%)				25%	12.86	
	ARCHITECT FEES				6%	3.87	
				Total Building Cost		68.35	

For customer support on your Square Foot Cost Data, call 877.756.2789.

137

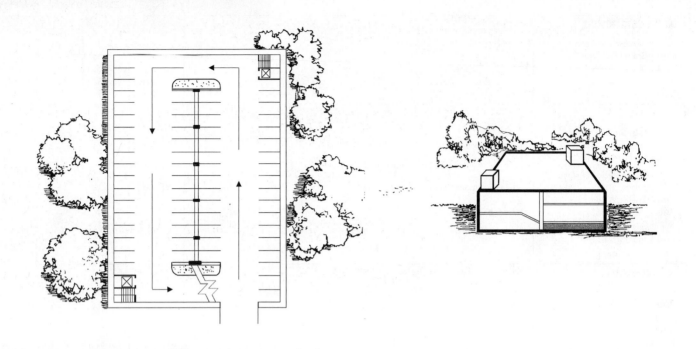

Costs per square foot of floor area

Exterior Wall	S.F. Area	20000	30000	40000	50000	75000	100000	125000	150000	175000
	L.F. Perimeter	400	500	600	650	775	900	1000	1100	1185
Reinforced Concrete	R/Conc. Frame	105.20	98.30	94.90	91.55	87.20	85.00	83.45	82.35	81.55
Perimeter Adj., Add or Deduct	Per 100 L.F.	6.10	4.10	3.00	2.45	1.65	1.20	0.95	0.85	0.70
Story Hgt. Adj., Add or Deduct	Per 1 Ft.	2.30	1.95	1.75	1.55	1.20	1.05	0.90	0.90	0.75
Basement—Not Applicable										

The above costs were calculated using the basic specifications shown on the facing page. These costs should be adjusted where necessary for design alternatives and owner's requirements. Reported completed project costs, for this type of structure, range from $53.80 to $128.15 per S.F.

Common additives

Description	Unit	$ Cost
Elevators, Hydraulic passenger, 2 stops		
1500# capacity	Ea.	67,400
2500# capacity	Ea.	70,400
3500# capacity	Ea.	75,400
Barrier gate w/programmable controller	Ea.	4225
Booth for attendant, average	Ea.	13,400
Fee computer	Ea.	13,600
Ticket spitter with time/date stamp	Ea.	8125
Mag strip encoding	Ea.	21,400
Collection station, pay on foot	Ea.	128,500
Parking control software	Ea.	26,200 - 105,000
Painting, Parking stalls	Stall	8.85
Parking Barriers		
Timber with saddles, 4" x 4"	L.F.	7.75
Precast concrete, 6" x 10" x 6'	Ea.	63.50
Traffic Signs, directional, 12" x 18"	Ea.	99

Important: See the Reference Section for Location Factors.

Garage, Underground Parking

Model costs calculated for a 2 story building with 10' story height and 100,000 square feet of floor area

			Unit	Unit Cost	Cost Per S.F.	% Of Sub-Total
A. SUBSTRUCTURE						
1010	Standard Foundations	Poured concrete; strip and spread footings and waterproofing	S.F. Ground	8.18	4.09	
1020	Special Foundations	N/A	—	—	—	
1030	Slab on Grade	5" reinforced concrete with vapor barrier and granular base	S.F. Slab	7.35	3.68	21.3%
2010	Basement Excavation	Excavation 24' deep	S.F. Ground	11.30	5.65	
2020	Basement Walls	N/A	—	—	—	
B. SHELL						
	B10 Superstructure					
1010	Floor Construction	Cast-in-place concrete beam and slab, concrete columns	S.F. Floor	31.72	15.86	46.2%
1020	Roof Construction	Cast-in-place concrete beam and slab, concrete columns	S.F. Roof	26.40	13.20	
	B20 Exterior Enclosure					
2010	Exterior Walls	Cast-in place concrete	S.F. Wall	23.17	4.17	
2020	Exterior Windows	N/A	—	—	—	6.9%
2030	Exterior Doors	Steel overhead, hollow metal	Each	4543	.19	
	B30 Roofing					
3010	Roof Coverings	Neoprene membrane traffic deck	S.F. Roof	5.50	2.75	4.4%
3020	Roof Openings	N/A	—	—	—	
C. INTERIORS						
1010	Partitions	Concrete block	S.F. Partition	44.72	.86	
1020	Interior Doors	Hollow metal	Each	9344	.09	
1030	Fittings	N/A	—	—	—	
2010	Stair Construction	Concrete	Flight	6950	.34	2.2%
3010	Wall Finishes	Paint	S.F. Surface	2.60	.10	
3020	Floor Finishes	N/A	—	—	—	
3030	Ceiling Finishes	N/A	—	—	—	
D. SERVICES						
	D10 Conveying					
1010	Elevators & Lifts	Two hydraulic passenger elevators	Each	85,500	1.71	2.7%
1020	Escalators & Moving Walks	N/A	—	—	—	
	D20 Plumbing					
2010	Plumbing Fixtures	Drainage in parking areas, toilets, & service fixtures 1 Fixture/5000 S.F. Floor	Each	.05	.05	
2020	Domestic Water Distribution	Electric water heater	S.F. Floor	.15	.15	2.2%
2040	Rain Water Drainage	Roof drains	S.F. Roof	2.42	1.21	
	D30 HVAC					
3010	Energy Supply	N/A	—	—	—	
3020	Heat Generating Systems	N/A	—	—	—	
3030	Cooling Generating Systems	N/A	—	—	—	0.3 %
3050	Terminal & Package Units	Exhaust fans	S.F. Floor	.17	.17	
3090	Other HVAC Sys. & Equipment	N/A	—	—	—	
	D40 Fire Protection					
4010	Sprinklers	Dry pipe sprinkler system	S.F. Floor	4.32	4.32	7.1%
4020	Standpipes	Dry standpipe system, class III	S.F. Floor	.17	.17	
	D50 Electrical					
5010	Electrical Service/Distribution	200 ampere service, panel board and feeders	S.F. Floor	.13	.13	
5020	Lighting & Branch Wiring	T-8 fluorescent fixtures, receptacles, switches and misc. power	S.F. Floor	3.42	3.42	6.0%
5030	Communications & Security	Addressable alarm systems and emergency lighting	S.F. Floor	.18	.18	
5090	Other Electrical Systems	Emergency generator, 11.5 kW	S.F. Floor	.07	.07	
E. EQUIPMENT & FURNISHINGS						
1010	Commercial Equipment	N/A	—	—	—	
1020	Institutional Equipment	N/A	—	—	—	
1030	Vehicular Equipment	Ticket dispensers, booths, automatic gates	S.F. Floor	.40	.40	0.6 %
1090	Other Equipment	N/A	—	—	—	
F. SPECIAL CONSTRUCTION						
1020	Integrated Construction	N/A	—	—	—	0.0 %
1040	Special Facilities	N/A	—	—	—	
G. BUILDING SITEWORK	**N/A**					

		Sub-Total	62.96	100%
CONTRACTOR FEES (General Requirements: 10%, Overhead: 5%, Profit: 10%)		25%	15.74	
ARCHITECT FEES		8%	6.30	
	Total Building Cost		**85**	

For customer support on your Square Foot Cost Data, call 877.756.2789.

139

Costs per square foot of floor area

Exterior Wall	S.F. Area	2000	4000	6000	8000	10000	12000	14000	16000	18000
	L.F. Perimeter	180	260	340	420	450	500	550	575	600
Concrete Block	Wood Joists	165.60	145.85	139.25	135.95	131.30	129.05	127.50	125.45	123.90
	Steel Joists	172.35	150.20	142.80	139.10	134.00	131.50	129.80	127.50	125.80
Cast in Place Concrete	Wood Joists	177.15	154.45	146.95	143.15	137.45	134.80	132.90	130.40	128.50
	Steel Joists	185.40	159.65	151.00	146.70	140.50	137.55	135.45	132.70	130.60
Stucco	Wood Frame	171.55	148.55	140.80	137.00	131.15	128.40	126.50	123.90	122.00
Insulated Metal Panels	Steel Frame	188.05	161.25	152.25	147.80	141.45	138.40	136.25	133.45	131.30
Perimeter Adj., Add or Deduct	Per 100 L.F.	26.90	13.45	9.05	6.75	5.35	4.50	3.85	3.40	3.00
Story Hgt. Adj., Add or Deduct	Per 1 Ft.	2.15	1.55	1.35	1.25	1.05	1.05	0.95	0.90	0.80
For Basement, add $35.70 per square foot of basement area										

The above costs were calculated using the basic specifications shown on the facing page. These costs should be adjusted where necessary for design alternatives and owner's requirements. Reported completed project costs, for this type of structure, range from $77.10 to $231.70 per S.F.

Common additives

Description	Unit	$ Cost
Air Compressors		
Electric 1-1/2 H.P., standard controls	Ea.	1650
Dual controls	Ea.	1975
5 H.P. 115/230 Volt, standard controls	Ea.	3700
Dual controls	Ea.	4775
Product Dispenser		
with vapor recovery for 6 nozzles	Ea.	27,200
Lifts, Single post		
8000# cap., swivel arm	Ea.	10,300
Two post, adjustable frames, 12,000# cap.	Ea.	5650
24,000# cap.	Ea.	11,300
30,000# cap.	Ea.	51,500
Four post, roll on ramp, 25,000# cap.	Ea.	20,600
Lockers, Steel, single tier, 60" or 72"	Opng.	231 - 380
2 tier, 60" or 72" total	Opng.	121 - 170
5 tier, box lockers	Opng.	74 - 84
Locker bench, lam. maple top only	L.F.	34.50
Pedestals, steel pipe	Ea.	76.50
Lube Equipment		
3 reel type, with pumps, no piping	Ea.	13,000
Spray Painting Booth, 26' long, complete	Ea.	15,700

Important: See the Reference Section for Location Factors.

Model costs calculated for a 1 story building with 14' story height and 10,000 square feet of floor area

				Unit	Unit Cost	Cost Per S.F.	% Of Sub-Total
A. SUBSTRUCTURE							
1010	Standard Foundations	Poured concrete; strip and spread footings		S.F. Ground	2.78	2.78	
1020	Special Foundations	N/A		—	—	—	
1030	Slab on Grade	6" reinforced concrete with vapor barrier and granular base		S.F. Slab	8.31	8.31	15.4%
2010	Basement Excavation	Site preparation for slab and trench for foundation wall and footing		S.F. Ground	.32	.32	
2020	Basement Walls	4' foundation wall		L.F. Wall	86	3.85	
B. SHELL							
	B10 Superstructure						
1010	Floor Construction	N/A		—	—	—	
1020	Roof Construction	Metal deck on open web steel joists		S.F. Roof	5.71	5.71	5.8 %
	B20 Exterior Enclosure						
2010	Exterior Walls	Concrete block	80% of wall	S.F. Wall	13.45	6.78	
2020	Exterior Windows	Hopper type commercial steel	5% of wall	Each	525	1.11	10.1%
2030	Exterior Doors	Steel overhead and hollow metal	15% of wall	S.F. Door	22.96	2.17	
	B30 Roofing						
3010	Roof Coverings	Built-up tar and gravel; perlite/EPS composite insulation		S.F. Roof	6.93	6.93	7.0%
3020	Roof Openings	Skylight		S.F. Roof	.02	.02	
C. INTERIORS							
1010	Partitions	Concrete block	50 S.F. Floor/L.F. Partition	S.F. Partition	23.55	4.71	
1020	Interior Doors	Single leaf hollow metal	3000 S.F. Floor/Door	Each	1168	.39	
1030	Fittings	Toilet partitions		S.F. Floor	.15	.15	
2010	Stair Construction	N/A		—	—	—	10.2%
3010	Wall Finishes	Paint		S.F. Surface	7.53	3.01	
3020	Floor Finishes	90% metallic floor hardener, 10% vinyl composition tile		S.F. Floor	1.41	1.41	
3030	Ceiling Finishes	Gypsum board on wood joists in office and washrooms	10% of area	S.F. Ceiling	4.23	.43	
D. SERVICES							
	D10 Conveying						
1010	Elevators & Lifts	N/A		—	—	—	0.0 %
1020	Escalators & Moving Walks	N/A		—	—	—	
	D20 Plumbing						
2010	Plumbing Fixtures	Toilet and service fixtures, supply and drainage	1 Fixture/500 S.F. Floor	Each	1605	3.21	
2020	Domestic Water Distribution	Gas fired water heater		S.F. Floor	.67	.67	6.3%
2040	Rain Water Drainage	Roof drains		S.F. Roof	2.34	2.34	
	D30 HVAC						
3010	Energy Supply	N/A		—	—	—	
3020	Heat Generating Systems	N/A		—	—	—	
3030	Cooling Generating Systems	N/A		—	—	—	11.1 %
3050	Terminal & Package Units	Single zone AC unit		S.F. Floor	10.09	10.09	
3090	Other HVAC Sys. & Equipment	Garage exhaust system		S.F. Floor	.95	.95	
	D40 Fire Protection						
4010	Sprinklers	Sprinklers, ordinary hazard		S.F. Floor	4.72	4.72	5.8%
4020	Standpipes	Standpipe		S.F. Floor	1.01	1.01	
	D50 Electrical						
5010	Electrical Service/Distribution	200 ampere service, panel board and feeders		S.F. Floor	.47	.47	
5020	Lighting & Branch Wiring	T-8 fluorescent fixtures, receptacles, switches, A.C. and misc. power		S.F. Floor	8.65	8.65	12.9%
5030	Communications & Security	Addressable alarm systems, partial internet wiring and emergency lighting		S.F. Floor	3.59	3.59	
5090	Other Electrical Systems	Emergency generator, 15 kW		S.F. Floor	.09	.09	
E. EQUIPMENT & FURNISHINGS							
1010	Commercial Equipment	N/A		—	—	—	
1020	Institutional Equipment	N/A		—	—	—	
1030	Vehicular Equipment	Hoists		S.F. Floor	15.38	15.38	15.5 %
1090	Other Equipment	N/A		—	—	—	
F. SPECIAL CONSTRUCTION							
1020	Integrated Construction	N/A		—	—	—	0.0 %
1040	Special Facilities	N/A		—	—	—	
G. BUILDING SITEWORK	**N/A**						

			Sub-Total	99.25	100%
CONTRACTOR FEES (General Requirements: 10%, Overhead: 5%, Profit: 10%)			25%	24.83	
ARCHITECT FEES			8%	9.92	
		Total Building Cost	**134**		

For customer support on your Square Foot Cost Data, call 877.756.2789.

141

Costs per square foot of floor area

Exterior Wall	S.F. Area	600	800	1000	1200	1400	1600	1800	2000	2200
	L.F. Perimeter	100	120	126	140	153	160	170	180	190
Face Brick with Concrete Block Back-up	Wood Truss	290.05	271.80	250.30	241.05	233.80	225.65	220.55	216.45	213.00
	Steel Joists	310.15	290.95	268.25	258.45	250.90	242.25	236.85	232.50	228.95
Enameled Sandwich Panel	Steel Frame	274.15	256.65	236.55	227.75	220.95	213.30	208.50	204.65	201.45
Tile on Concrete Block	Steel Joists	328.15	307.15	281.80	271.05	262.65	253.00	247.05	242.20	238.25
Aluminum Siding	Wood Frame	248.45	233.65	217.40	210.10	204.45	198.35	194.45	191.25	188.70
Wood Siding	Wood Frame	249.95	235.00	218.55	211.10	205.45	199.20	195.25	192.05	189.45
Perimeter Adj., Add or Deduct	Per 100 L.F.	125.85	94.45	75.55	62.95	54.00	47.20	41.95	37.70	34.35
Story Hgt. Adj., Add or Deduct	Per 1 Ft.	8.15	7.35	6.15	5.70	5.35	4.90	4.55	4.40	4.20
Basement—Not Applicable										

The above costs were calculated using the basic specifications shown on the facing page. These costs should be adjusted where necessary for design alternatives and owner's requirements. Reported completed project costs, for this type of structure, range from $55.55 to $280.65 per S.F.

Common additives

Description	Unit	$ Cost
Air Compressors		
Electric 1-1/2 H.P., standard controls	Ea.	1650
Dual controls	Ea.	1975
5 H.P. 115/230 volt, standard controls	Ea.	3700
Dual controls	Ea.	4775
Product Dispenser		
with vapor recovery for 6 nozzles	Ea.	27,200
Lifts, Single post		
8000# cap. swivel arm	Ea.	10,300
Two post, adjustable frames, 12,000# cap.	Ea.	5650
24,000# cap.	Ea.	11,300
30,000# cap.	Ea.	51,500
Four post, roll on ramp, 25,000# cap.	Ea.	20,600
Lockers, Steel, single tier, 60" or 72"	Opng.	231 - 380
2 tier, 60" or 72" total	Opng.	121 - 170
5 tier, box lockers	Ea.	74 - 84
Locker bench, lam. maple top only	L.F.	34.50
Pedestals, steel pipe	Ea.	76.50
Lube Equipment		
3 reel type, with pumps, no piping	Ea.	13,000

Important: See the Reference Section for Location Factors.

Model costs calculated for a 1 story building with 12' story height and 1,400 square feet of floor area

					Unit	Unit Cost	Cost Per S.F.	% Of Sub-Total
A.	**SUBSTRUCTURE**							
1010	Standard Foundations	Poured concrete; strip and spread footings			S.F. Ground	4.49	4.49	
1020	Special Foundations	N/A			–	–	–	
1030	Slab on Grade	6" reinforced concrete with vapor barrier and granular base			S.F. Slab	8.31	8.31	12.8%
2010	Basement Excavation	Site preparation for slab and trench for foundation wall and footing			S.F. Ground	1.08	1.08	
2020	Basement Walls	4' foundation wall			L.F. Wall	76	8.31	
B.	**SHELL**							
	B10 Superstructure							
1010	Floor Construction	N/A			–	–	–	4.5 %
1020	Roof Construction	Plywood on wood trusses			S.F. Roof	7.82	7.82	
	B20 Exterior Enclosure							
2010	Exterior Walls	Face brick with concrete block backup	60% of wall		S.F. Wall	33.40	26.28	
2020	Exterior Windows	Store front and metal top hinged outswinging	20% of wall		Each	43.30	11.36	28.9%
2030	Exterior Doors	Steel overhead, aluminum & glass and hollow metal	20% of wall		Each	47.47	12.45	
	B30 Roofing							
3010	Roof Coverings	Asphalt shingles with flashing; perlite/EPS composite insulation			S.F. Roof	5.50	5.50	3.2%
3020	Roof Openings	N/A			–	–	–	
C.	**INTERIORS**							
1010	Partitions	Concrete block	25 S.F. Floor/L.F. Partition		S.F. Partition	7.97	2.55	
1020	Interior Doors	Single leaf hollow metal	700 S.F. Floor/Door		Each	1168	1.67	
1030	Fittings	Toilet partitions			S.F. Floor	2.17	2.17	
2010	Stair Construction	N/A			–	–	–	7.4%
3010	Wall Finishes	Paint			S.F. Surface	5.95	3.81	
3020	Floor Finishes	Vinyl composition tile	35% of floor area		S.F. Floor	2.77	.97	
3030	Ceiling Finishes	Painted gypsum board on furring in sales area & washrooms	35% of floor area		S.F. Ceiling	4.66	1.63	
D.	**SERVICES**							
	D10 Conveying							
1010	Elevators & Lifts	N/A			–	–	–	0.0 %
1020	Escalators & Moving Walks	N/A			–	–	–	
	D20 Plumbing							
2010	Plumbing Fixtures	Toilet and service fixtures, supply and drainage	1 Fixture/235 S.F. Floor		Each	2265	9.64	
2020	Domestic Water Distribution	Gas fired water heater			S.F. Floor	4.80	4.80	8.3%
2040	Rain Water Drainage	N/A			–	–	–	
	D30 HVAC							
3010	Energy Supply	N/A			–	–	–	
3020	Heat Generating Systems	N/A			–	–	–	
3030	Cooling Generating Systems	N/A			–	–	–	14.7 %
3050	Terminal & Package Units	Single zone AC unit			S.F. Floor	19.87	19.87	
3090	Other HVAC Sys. & Equipment	Underfloor exhaust system			S.F. Floor	5.64	5.64	
	D40 Fire Protection							
4010	Sprinklers	Wet pipe sprinkler system			S.F. Floor	9.63	9.63	8.1%
4020	Standpipes	Standpipe			S.F. Floor	4.34	4.34	
	D50 Electrical							
5010	Electrical Service/Distribution	200 ampere service, panel board and feeders			S.F. Floor	4.95	4.95	
5020	Lighting & Branch Wiring	High efficiency fluorescent fixtures, receptacles, switches, A.C. and misc. power			S.F. Floor	6.58	6.58	12.1%
5030	Communications & Security	Addressable alarm systems and emergency lighting			S.F. Floor	8.70	8.70	
5090	Other Electrical Systems	Emergency generator			S.F. Floor	.66	.66	
E.	**EQUIPMENT & FURNISHINGS**							
1010	Commercial Equipment	N/A			–	–	–	
1020	Institutional Equipment	N/A			–	–	–	0.0 %
1030	Vehicular Equipment	N/A			–	–	–	
1090	Other Equipment	N/A			–	–	–	
F.	**SPECIAL CONSTRUCTION**							
1020	Integrated Construction	N/A			–	–	–	0.0 %
1040	Special Facilities	N/A			–	–	–	
G.	**BUILDING SITEWORK**	**N/A**						

						Sub-Total	173.21	100%
	CONTRACTOR FEES (General Requirements: 10%, Overhead: 5%, Profit: 10%)					25%	43.27	
	ARCHITECT FEES					8%	17.32	

			Total Building Cost	**233.80**

For customer support on your Square Foot Cost Data, call 877.756.2789.

143

Costs per square foot of floor area

Exterior Wall	S.F. Area	12000	16000	20000	25000	30000	35000	40000	45000	50000
	L.F. Perimeter	440	520	600	700	708	780	841	910	979
Reinforced Concrete Block	Lam. Wood Arches	178.50	173.85	171.05	168.80	164.70	163.35	162.15	161.30	160.65
	Rigid Steel Frame	187.00	182.10	179.20	176.85	172.50	171.15	169.80	168.95	168.25
Face Brick with Concrete Block Back-up	Lam. Wood Arches	202.15	194.80	190.40	186.85	179.95	177.70	175.70	174.35	173.20
	Rigid Steel Frame	209.10	201.75	197.35	193.80	186.90	184.70	182.65	181.30	180.20
Metal Sandwich Panels	Lam. Wood Arches	189.05	183.20	179.70	176.85	171.50	169.75	168.15	167.10	166.25
	Rigid Steel Frame	199.65	193.45	189.70	186.65	180.80	178.95	177.25	176.10	175.20
Perimeter Adj., Add or Deduct	Per 100 L.F.	7.05	5.25	4.20	3.40	2.85	2.45	2.05	1.80	1.65
Story Hgt. Adj., Add or Deduct	Per 1 Ft.	1.00	0.85	0.85	0.80	0.70	0.65	0.55	0.55	0.50
Basement—Not Applicable										

The above costs were calculated using the basic specifications shown on the facing page. These costs should be adjusted where necessary for design alternatives and owner's requirements. Reported completed project costs, for this type of structure, range from $84.30 to $251.50 per S.F.

Common additives

Description	Unit	$ Cost
Bleachers, Telescoping, manual		
To 15 tier	Seat	143 - 199
16-20 tier	Seat	294 - 360
21-30 tier	Seat	310 - 410
For power operation, add	Seat	57.50 - 90
Gym Divider Curtain, Mesh top		
Manual roll-up	S.F.	13.80
Gym Mats		
2" naugahyde covered	S.F.	5.15
2" nylon	S.F.	9.10
1-1/2" wall pads	S.F.	7.10
1" wrestling mats	S.F.	5.15
Scoreboard		
Basketball, one side	Ea.	3625 - 14,200
Basketball Backstop		
Wall mtd., 6' extended, fixed	Ea.	2700 - 3025
Swing up, wall mtd.	Ea.	2800 - 4275

Description	Unit	$ Cost
Lockers, Steel, single tier, 60" or 72"	Opng.	231 - 380
2 tier, 60" or 72" total	Opng.	121 - 170
5 tier, box lockers	Opng.	74 - 84
Locker bench, lam. maple top only	L.F.	34.50
Pedestals, steel pipe	Ea.	76.50
Sound System		
Amplifier, 250 watts	Ea.	2200
Speaker, ceiling or wall	Ea.	217
Trumpet	Ea.	415
Emergency Lighting, 25 watt, battery operated		
Lead battery	Ea.	345
Nickel cadmium	Ea.	685

Model costs calculated for a 1 story building with 25′ story height and 20,000 square feet of floor area

Gymnasium

			Unit	Unit Cost	Cost Per S.F.	% Of Sub-Total
A. SUBSTRUCTURE						
1010	Standard Foundations	Poured concrete; strip and spread footings	S.F. Ground	1.60	1.60	
1020	Special Foundations	N/A	—	—	—	
1030	Slab on Grade	4" reinforced concrete with vapor barrier and granular base	S.F. Slab	5.56	5.56	7.5%
2010	Basement Excavation	Site preparation for slab and trench for foundation wall and footing	S.F. Ground	.18	.18	
2020	Basement Walls	4′ foundation wall	L.F. Wall	76	2.29	
B. SHELL						
	B10 Superstructure					
1010	Floor Construction	Steel column	S.F. Floor	2.33	2.33	17.1%
1020	Roof Construction	Wood deck on laminated wood arches	S.F. Roof	19.48	19.48	
	B20 Exterior Enclosure					
2010	Exterior Walls	Reinforced concrete block (end walls included) 90% of wall	S.F. Wall	11.97	8.08	
2020	Exterior Windows	Metal horizontal pivoted 10% of wall	Each	674	5.06	10.8%
2030	Exterior Doors	Aluminum and glass, hollow metal, steel overhead	Each	2208	.66	
	B30 Roofing					
3010	Roof Coverings	EPDM, 60 mils, fully adhered; polyisocyanurate insulation	S.F. Roof	5.47	5.47	4.3%
3020	Roof Openings	N/A	—	—	—	
C. INTERIORS						
1010	Partitions	Concrete block 50 S.F. Floor/L.F. Partition	S.F. Partition	9.75	1.95	
1020	Interior Doors	Single leaf hollow metal 500 S.F. Floor/Door	Each	1168	2.34	
1030	Fittings	Toilet partitions	S.F. Floor	.31	.31	
2010	Stair Construction	N/A	—	—	—	19.8%
3010	Wall Finishes	50% paint, 50% ceramic tile	S.F. Surface	10.40	4.16	
3020	Floor Finishes	90% hardwood, 10% ceramic tile	S.F. Floor	15.51	15.51	
3030	Ceiling Finishes	Mineral fiber tile on concealed zee bars 15% of area	S.F. Ceiling	7.22	1.08	
D. SERVICES						
	D10 Conveying					
1010	Elevators & Lifts	N/A	—	—	—	0.0 %
1020	Escalators & Moving Walks	N/A	—	—	—	
	D20 Plumbing					
2010	Plumbing Fixtures	Toilet and service fixtures, supply and drainage 1 Fixture/515 S.F. Floor	Each	4661	9.05	
2020	Domestic Water Distribution	Electric water heater	S.F. Floor	4.93	4.93	10.9%
2040	Rain Water Drainage	N/A	—	—	—	
	D30 HVAC					
3010	Energy Supply	N/A	—	—	—	
3020	Heat Generating Systems	Included in D3050	—	—	—	
3030	Cooling Generating Systems	N/A	—	—	—	9.5 %
3050	Terminal & Package Units	Single zone rooftop unit, gas heating, electric cooling	S.F. Floor	12.15	12.15	
3090	Other HVAC Sys. & Equipment	N/A	—	—	—	
	D40 Fire Protection					
4010	Sprinklers	Wet pipe sprinkler system	S.F. Floor	3.65	3.65	3.7%
4020	Standpipes	Standpipe	S.F. Floor	1.10	1.10	
	D50 Electrical					
5010	Electrical Service/Distribution	400 ampere service, panel board and feeders	S.F. Floor	1.14	1.14	
5020	Lighting & Branch Wiring	High efficiency fluorescent fixtures, receptacles, switches, A.C. and misc. power	S.F. Floor	9.65	9.65	10.9%
5030	Communications & Security	Addressable alarm systems, sound system and emergency lighting	S.F. Floor	2.99	2.99	
5090	Other Electrical Systems	Emergency generator, 7.5 kW	S.F. Floor	.22	.22	
E. EQUIPMENT & FURNISHINGS						
1010	Commercial Equipment	N/A	—	—	—	
1020	Institutional Equipment	N/A	—	—	—	
1030	Vehicular Equipment	N/A	—	—	—	5.4 %
1090	Other Equipment	Bleachers, sauna, weight room	S.F. Floor	6.96	6.96	
F. SPECIAL CONSTRUCTION						
1020	Integrated Construction	N/A	—	—	—	0.0 %
1040	Special Facilities	N/A	—	—	—	
G. BUILDING SITEWORK	**N/A**					

				Sub-Total	127.90	100%
	CONTRACTOR FEES (General Requirements: 10%, Overhead: 5%, Profit: 10%)			25%	31.96	
	ARCHITECT FEES			7%	11.19	
				Total Building Cost	**171.05**	

For customer support on your Square Foot Cost Data, call 877.756.2789.

145

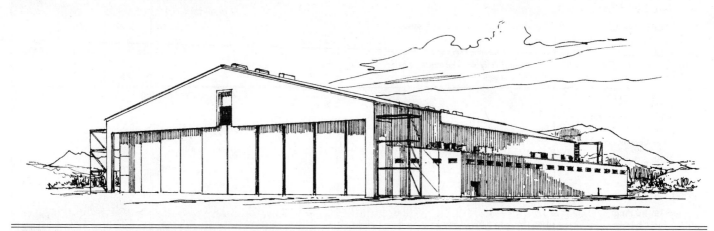

Costs per square foot of floor area

Exterior Wall	S.F. Area	5000	10000	15000	20000	30000	40000	50000	75000	100000
	L.F. Perimeter	300	410	500	580	710	830	930	1150	1300
Concrete Block Reinforced	Steel Frame	187.05	152.35	139.35	132.30	124.30	120.05	117.05	112.60	109.75
	Bearing Walls	195.30	160.35	147.30	140.20	132.10	127.85	124.80	120.35	117.40
Precast Concrete	Steel Frame	235.75	185.60	166.40	155.85	143.50	136.85	132.10	125.05	120.25
	Bearing Walls	238.75	188.30	168.95	158.35	145.90	139.20	134.45	127.35	122.50
Galv. Steel Siding	Steel Frame	181.65	148.65	136.30	129.65	122.10	118.15	115.30	111.25	108.55
Metal Sandwich Panel	Steel Frame	192.75	155.50	141.65	134.10	125.65	121.10	118.00	113.30	110.25
Perimeter Adj., Add or Deduct	Per 100 L.F.	20.80	10.40	6.95	5.25	3.45	2.55	2.05	1.40	0.95
Story Hgt. Adj., Add or Deduct	Per 1 Ft.	2.05	1.35	1.15	1.00	0.85	0.70	0.65	0.55	0.40
Basement—Not Applicable										

The above costs were calculated using the basic specifications shown on the facing page. These costs should be adjusted where necessary for design alternatives and owner's requirements. Reported completed project costs, for this type of structure, range from $ 48.40 to $ 220.65 per S.F.

Common additives

Description	Unit	$ Cost
Closed Circuit Surveillance, One station		
Camera and monitor	Ea.	2025
For additional camera stations, add	Ea.	1100
Emergency Lighting, 25 watt, battery operated		
Lead battery	Ea.	345
Nickel cadmium	Ea.	685
Lockers, Steel, single tier, 60" or 72"	Opng.	231 - 380
2 tier, 60" or 72" total	Opng.	121 - 170
5 tier, box lockers	Opng.	74 - 84
Locker bench, lam. maple top only	L.F.	34.50
Pedestals, steel pipe	Ea.	76.50
Safe, Office type, 1 hour rating		
30" x 18" x 18"	Ea.	2500
60" x 36" x 18", double door	Ea.	9350
Sound System		
Amplifier, 250 watts	Ea.	2200
Speaker, ceiling or wall	Ea.	217
Trumpet	Ea.	415

Important: See the Reference Section for Location Factors.

Model costs calculated for a 1 story building with 24′ story height and 20,000 square feet of floor area

				Unit	Unit Cost	Cost Per S.F.	% Of Sub-Total
A.	**SUBSTRUCTURE**						
1010	Standard Foundations	Poured concrete; strip and spread footings		S.F. Ground	1.48	1.48	
1020	Special Foundations	N/A		—	—	—	
1030	Slab on Grade	6″ reinforced concrete with vapor barrier and granular base		S.F. Slab	8.31	8.31	12.8%
2010	Basement Excavation	Site preparation for slab and trench for foundation wall and footing		S.F. Ground	.18	.18	
2020	Basement Walls	4′ foundation wall		L.F. Wall	86	2.48	
B.	**SHELL**						
	B10 Superstructure						
1010	Floor Construction	N/A		—	—	—	12.6 %
1020	Roof Construction	Metal deck, open web steel joists, beams, columns		S.F. Roof	12.23	12.23	
	B20 Exterior Enclosure						
2010	Exterior Walls	Galvanized steel siding		S.F. Wall	11.49	4	
2020	Exterior Windows	Industrial horizontal pivoted steel	20% of wall	Each	2050	11.89	25.5%
2030	Exterior Doors	Steel overhead and sliding	30% of wall	Each	42.19	8.81	
	B30 Roofing						
3010	Roof Coverings	Elastomeric membrane; fiberboard insulation		S.F. Roof	5.48	5.48	5.8%
3020	Roof Openings	Roof hatches		S.F. Roof	.13	.13	
C.	**INTERIORS**						
1010	Partitions	Concrete block	200 S.F. Floor/L.F. Partition	S.F. Partition	10.20	.51	
1020	Interior Doors	Single leaf hollow metal	5000 S.F. Floor/Door	Each	1168	.24	
1030	Fittings	Toilet partitions		S.F. Floor	.26	.26	
2010	Stair Construction	N/A		—	—	—	1.2%
3010	Wall Finishes	Paint		S.F. Surface	1.30	.13	
3020	Floor Finishes	N/A		—	—	—	
3030	Ceiling Finishes	N/A		—	—	—	
D.	**SERVICES**						
	D10 Conveying						
1010	Elevators & Lifts	N/A		—	—	—	0.0 %
1020	Escalators & Moving Walks	N/A		—	—	—	
	D20 Plumbing						
2010	Plumbing Fixtures	Toilet and service fixtures, supply and drainage	1 Fixture/1000 S.F. Floor	Each	2690	2.69	
2020	Domestic Water Distribution	Gas fired water heater		S.F. Floor	2.47	2.47	6.7%
2040	Rain Water Drainage	Roof drains		S.F. Roof	1.31	1.31	
	D30 HVAC						
3010	Energy Supply	N/A		—	—	—	
3020	Heat Generating Systems	Unit heaters		Each	7.71	7.71	
3030	Cooling Generating Systems	N/A		—	—	—	8.3 %
3050	Terminal & Package Units	Exhaust fan		S.F. Floor	.29	.29	
3090	Other HVAC Sys. & Equipment	N/A		—	—	—	
	D40 Fire Protection						
4010	Sprinklers	Sprinklers, extra hazard, deluge system		S.F. Floor	7.72	7.72	13.3%
4020	Standpipes	Standpipe and fire pump		S.F. Floor	5.22	5.22	
	D50 Electrical						
5010	Electrical Service/Distribution	200 ampere service, panel board and feeders		S.F. Floor	1.01	1.01	
5020	Lighting & Branch Wiring	High intensity discharge fixtures, receptacles, switches and misc. power		S.F. Floor	10.92	10.92	13.8%
5030	Communications & Security	Addressable alarm systems and emergency lighting		S.F. Floor	1.37	1.37	
5090	Other Electrical Systems	Emergency generator, 7.5 kW		S.F. Floor	.12	.12	
E.	**EQUIPMENT & FURNISHINGS**						
1010	Commercial Equipment	N/A		—	—	—	
1020	Institutional Equipment	N/A		—	—	—	0.0 %
1030	Vehicular Equipment	N/A		—	—	—	
1090	Other Equipment	N/A		—	—	—	
F.	**SPECIAL CONSTRUCTION**						
1020	Integrated Construction	N/A		—	—	—	0.0 %
1040	Special Facilities	N/A		—	—	—	
G.	**BUILDING SITEWORK**	N/A					

		Sub-Total	96.96	100%
CONTRACTOR FEES (General Requirements: 10%, Overhead: 5%, Profit: 10%)		25%	24.21	
ARCHITECT FEES		7%	8.48	
		Total Building Cost	**129.65**	

For customer support on your Square Foot Cost Data, call 877.756.2789.

147

Costs per square foot of floor area

Exterior Wall	S.F. Area	25000	40000	55000	70000	85000	100000	115000	130000	145000
	L.F. Perimeter	388	520	566	666	766	866	878	962	1045
Fiber Cement	Wood Frame	314.50	306.70	300.90	298.70	297.30	296.30	294.50	293.85	293.30
Brick Veneer	Wood Frame	458.85	398.30	367.80	351.80	341.50	334.30	327.55	323.35	320.05
Stone Veneer	Steel Frame	468.50	405.15	371.70	354.90	343.95	336.40	328.55	324.05	320.45
Curtain Wall	Steel Frame	455.55	392.65	359.70	343.00	332.20	324.70	316.95	312.55	309.05
EFIS	Reinforced Concrete	493.70	432.75	401.90	385.75	375.30	368.10	361.15	356.95	353.60
Metal Panel	Reinforced Concrete	463.70	400.95	368.25	351.60	340.80	333.30	325.65	321.30	317.80
Perimeter Adj., Add or Deduct	Per 100 L.F.	5.85	3.65	2.70	2.10	1.75	1.45	1.20	1.10	1.00
Story Hgt. Adj., Add or Deduct	Per 1 Ft.	1.30	1.15	0.90	0.85	0.80	0.70	0.60	0.60	0.65

For Basement, add $37.25 per square foot of basement area

The above costs were calculated using the basic specifications shown on the facing page. These costs should be adjusted where necessary for design alternatives and owner's requirements. Reported completed project costs, for this type of structure, range from $186.00 to $469.80 per S.F.

Common additives

Description	Unit	$ Cost
Cabinet, Base, door units, metal	L.F.	325
Drawer units	L.F.	645
Tall storage cabinets, 7' high, open	L.F.	620
With doors	L.F.	900
Wall, metal 12-1/2" deep, open	L.F.	263
With doors	L.F.	460
Closed Circuit TV (Patient monitoring)		
One station camera & monitor	Ea.	2025
For additional camera, add	Ea.	1100
For automatic iris for low light, add	Ea.	2825
Hubbard Tank, with accessories		
Stainless steel, 125 GPM 45 psi	Ea.	13,700
For electric hoist, add	Ea.	2975
Mortuary Refrigerator, End operated		
2 capacity	Ea.	10,800
6 capacity	Ea.	16,300

Description	Unit	$ Cost
Nurses Call Station		
Single bedside call station	Ea.	292
Ceiling speaker station	Ea.	160
Emergency call station	Ea.	166
Pillow speaker	Ea.	287
Double bedside call station	Ea.	330
Duty station	Ea.	320
Standard call button	Ea.	183
Master control station for 20 stations	Ea.	5825
Sound System		
Amplifier, 250 watts	Ea.	2200
Speaker, ceiling or wall	Ea.	217
Trumpet	Ea.	415
Station, Dietary with ice	Ea.	18,400
Sterilizers		
Single door, steam	Ea.	134,500
Double door, steam	Ea.	227,000
Portable, countertop, steam	Ea.	3300 - 5225
Gas	Ea.	43,700
Automatic washer/sterilizer	Ea.	60,500

Important: See the Reference Section for Location Factors.

Model costs calculated for a 3 story building with 12' story height and 55,000 square feet of floor area

				Unit	Unit Cost	Cost Per S.F.	% Of Sub-Total
A. SUBSTRUCTURE							
1010	Standard Foundations	Poured concrete; strip and spread footings		S.F. Ground	2.94	.98	
1020	Special Foundations	N/A		—	—	—	
1030	Slab on Grade	4" reinforced slab on grade		S.F. Slab	5.56	1.85	1.4%
2010	Basement Excavation	Site preparation for slab and trench for foundation wall and footing		S.F. Ground	.18	.06	
2020	Basement Walls	4' Foundation wall		L.F. Wall	86	.88	
B. SHELL							
B10 Superstructure							
1010	Floor Construction	Wood Beam and Joist on wood columns, fireproofed		S.F. Floor	23.84	15.89	6.9%
1020	Roof Construction	Wood roof, truss, 4/12 slope		S.F. Roof	7.86	2.62	
B20 Exterior Enclosure							
2010	Exterior Walls	Brick veneer on wood stud back up, insulated	80% of wall	S.F. Wall	27.09	8.03	
2020	Exterior Windows	Aluminum awning, insulated	20% of wall	Each	746	2.41	4.0%
2030	Exterior Doors	Aluminum and glass double doors, hollow metal egress doors		Each	3308	.37	
B30 Roofing							
3010	Roof Coverings	Asphalt roofing, strip shingles, flashing, gutters and downspouts		S.F. Roof	2.46	.82	0.4%
3020	Roof Openings	Roof hatches		S.F. Roof	.45	.15	
C. INTERIORS							
1010	Partitions	Concrete block partition, gypsum board on wood studs	10 S.F. Floor/L.F. Partition	S.F. Partition	9.63	10.70	
1020	Interior Doors	Hollow metal fire doors	90 S.F. Floor/Door	Each	1218	13.53	
1030	Fittings	Hospital curtains		S.F. Floor	1.10	1.10	
2010	Stair Construction	Cast in place concrete stairs, with landings and rails		Flight	8400	1.23	18.9%
3010	Wall Finishes	65% paint, 35% ceramic tile		S.F. Surface	3.59	7.97	
3020	Floor Finishes	60% vinyl tile, 20% ceramic , 15% terrazzo, 5% epoxy		S.F. Floor	9.21	9.21	
3030	Ceiling Finishes	Acoustic ceiling tiles on suspended channel grid		S.F. Ceiling	7.22	7.22	
D. SERVICES							
D10 Conveying							
1010	Elevators & Lifts	Two hydraulic passenger elevators		Each	130,075	4.73	1.8%
1020	Escalators & Moving Walks	N/A		—	—	—	
D20 Plumbing							
2010	Plumbing Fixtures	Medical, patient & specialty, supply and drainage	1 Fixture/225 S.F. Floor	Each	1727	7.71	
2020	Domestic Water Distribution	Electric water heater		S.F. Floor	17.14	17.14	9.2%
2040	Rain Water Drainage	Roof drains		S.F. Roof			
D30 HVAC							
3010	Energy Supply	Conditioned air with hot water reheat system		S.F. Floor	4.92	4.92	
3020	Heat Generating Systems	Boiler		Each	129,400	3.70	
3030	Cooling Generating Systems	Conditioned air, chillers, cooling towers		S.F. Floor	6.57	6.57	21.4%
3050	Terminal & Package Units	N/A		—	—	—	
3090	Other HVAC Sys. & Equipment	Conditioned air, ductwork, surgey air curtain		S.F. Floor	42.66	42.66	
D40 Fire Protection							
4010	Sprinklers	Wet pipe spinkler system		S.F. Floor	3.05	3.05	1.4%
4020	Standpipes	Standpipe		S.F. Floor	.85	.85	
D50 Electrical							
5010	Electrical Service/Distribution	1200 ampere service, panel board and feeders		S.F. Floor	49.17	49.17	
5020	Lighting & Branch Wiring	Fluorescent fixtures, receptacles, switches, A.C. and misc. power		S.F. Floor	17.59	17.59	28.6%
5030	Communications & Security	Addressable alarm systems, emergency lighting, internet and phone wiring		S.F. Floor	5.59	5.59	
5090	Other Electrical Systems	Emergency generator, 200 Kw with fuel tank, UPS system		S.F. Floor	4.81	4.81	
E. EQUIPMENT & FURNISHINGS							
1010	Commercial Equipment	N/A		—	—	—	
1020	Institutional Equipment	Laboratory and commercial kitchen equipment		S.F. Floor	13.16	13.16	6.1 %
1030	Vehicular Equipment	N/A		—	—	—	
2020	Moveable Furnishings	Patient wall systems		S.F. Floor	3.28	3.28	
F. SPECIAL CONSTRUCTION							
1020	Integrated Construction	N/A		—	—	—	0.0 %
1040	Special Facilities	N/A		—	—	—	
G. BUILDING SITEWORK	**N/A**						

	Sub-Total	269.95	**100%**
CONTRACTOR FEES (General Requirements: 10%, Overhead: 5%, Profit: 10%)	25%	67.48	
ARCHITECT FEES	9%	30.37	
Total Building Cost		**367.80**	

For customer support on your Square Foot Cost Data, call 877.756.2789.

149

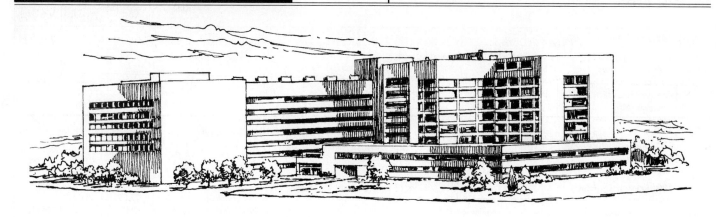

Costs per square foot of floor area

Exterior Wall	S.F. Area	100000	125000	150000	175000	200000	225000	250000	275000	300000
	L.F. Perimeter	594	705	816	783	866	950	1033	1116	1200
Fiber Cement	Steel Frame	285.35	283.05	281.45	277.75	276.80	276.05	275.50	274.95	274.60
Curtain Wall	Steel Frame	301.00	297.85	295.80	289.95	288.60	287.65	286.80	286.05	285.60
Metal Panel	Steel Frame	284.85	282.50	280.85	277.15	276.20	275.45	274.80	274.30	273.90
Brick Veneer	Reinforced Concrete	304.55	301.45	299.40	293.45	292.10	291.15	290.35	289.60	289.05
Stone Veneer	Reinforced Concrete	317.05	313.30	310.80	302.80	301.15	300.00	298.95	298.05	297.40
EIFS	Reinforced Concrete	306.90	303.90	301.85	296.30	295.05	294.10	293.30	292.65	292.15
Perimeter Adj., Add or Deduct	Per 100 L.F.	3.20	2.55	2.10	1.75	1.55	1.40	1.25	1.15	1.05
Story Hgt. Adj., Add or Deduct	Per 1 Ft.	1.15	1.05	1.00	0.80	0.80	0.85	0.75	0.75	0.75

For Basement, add $38.45 per square foot of basement area

The above costs were calculated using the basic specifications shown on the facing page. These costs should be adjusted where necessary for design alternatives and owner's requirements. Reported completed project costs, for this type of structure, range from $190.25 to $463.95 per S.F.

Common additives

Description	Unit	$ Cost	Description	Unit	$ Cost
Cabinets, Base, door units, metal	L.F.	325	Nurses Call Station		
Drawer units	L.F.	645	Single bedside call station	Ea.	292
Tall storage cabinets, 7' high, open	L.F.	620	Ceiling speaker station	Ea.	160
With doors	L.F.	900	Emergency call station	Ea.	166
Wall, metal 12-1/2" deep, open	L.F.	263	Pillow speaker	Ea.	287
With doors	L.F.	460	Double bedside call station	Ea.	330
Closed Circuit TV (Patient monitoring)			Duty station	Ea.	320
One station camera & monitor	Ea.	2025	Standard call button	Ea.	183
For additional camera add	Ea.	1100	Master control station for 20 stations	Ea.	5825
For automatic iris for low light add	Ea.	2825	Sound System		
Hubbard Tank, with accessories			Amplifier, 250 watts	Ea.	2200
Stainless steel, 125 GPM 45 psi	Ea.	13,700	Speaker, ceiling or wall	Ea.	217
For electric hoist, add	Ea.	2975	Trumpet	Ea.	415
Mortuary Refrigerator, End operated			Station, Dietary with ice	Ea.	18,400
2 capacity	Ea.	10,800	Sterilizers		
6 capacity	Ea.	16,300	Single door, steam	Ea.	134,500
			Double door, steam	Ea.	227,000
			Portable, counter top, steam	Ea.	3300 - 5225
			Gas	Ea.	43,700
			Automatic washer/sterilizer	Ea.	60,500

Important: See the Reference Section for Location Factors.

Model costs calculated for a 6 story building with 12' story height and 200,000 square feet of floor area

				Unit	Unit Cost	Cost Per S.F.	% Of Sub-Total
A. SUBSTRUCTURE							
1010	Standard Foundations	Poured concrete; strip and spread footings		S.F. Ground	2.46	.41	
1020	Special Foundations	N/A		—	—	—	
1030	Slab on Grade	4" reinforced slab on grade		S.F. Slab	14.60	2.44	1.6%
2010	Basement Excavation	Site preparation for slab and trench for foundation wall and footing		S.F. Ground	.18	.03	
2020	Basement Walls	4' Foundation wall		L.F. Wall	86	.37	
B. SHELL							
	B10 Superstructure						
1010	Floor Construction	Open web steel joists, slab form, concrete, fireproofed steel columns		S.F. Floor	14.34	11.95	6.7%
1020	Roof Construction	Metal deck on open web steel joints, columns		S.F. Roof	10.38	1.73	
	B20 Exterior Enclosure						
2010	Exterior Walls	Fiber cement siding on metal studs, insulated	70% of wall	S.F. Wall	18.79	4.10	
2020	Exterior Windows	Aluminum sliding windows	30% of wall	Each	585	2.44	3.6%
2030	Exterior Doors	Aluminum and glass double doors, hollow metal egress doors		Each	5981	.84	
	B30 Roofing						
3010	Roof Coverings	Single ply membrane, stone ballast		S.F. Roof	5.46	.91	0.5%
3020	Roof Openings	Roof hatches		S.F. Roof	.18	.03	
C. INTERIORS							
1010	Partitions	Concrete block partitions, gypsum board on metal studs	10 S.F. Floor/L.F. Partition	S.F. Partition	6.74	7.49	
1020	Interior Doors	Single leaf hollow metal doors, fire doors	90 S.F. Floor/Door	Each	1218	13.53	
1030	Fittings	Hospital curtains		S.F. Floor	1.21	1.21	
2010	Stair Construction	Concrete filled metal pan		Flight	11,575	1.50	23.2%
3010	Wall Finishes	65% Paint, 35 % ceramic wall tile		S.F. Surface	3.33	7.39	
3020	Floor Finishes	60% vinyl tile, 20% ceramic , 20% terrazzo		S.F. Floor	8.71	8.71	
3030	Ceiling Finishes	Acoustic ceiling tiles on suspended channel grid		S.F. Ceiling	7.22	7.22	
D. SERVICES							
	D10 Conveying						
1010	Elevators & Lifts	Six traction geared hospital elevators		Each	232,667	6.98	3.4%
1020	Escalators & Moving Walks	N/A		—	—	—	
	D20 Plumbing						
2010	Plumbing Fixtures	Medical, patient & specialty, supply and drainage	1 Fixture/415 S.F. Floor	Each	2991	7.19	
2020	Domestic Water Distribution	Electric water heater		S.F. Floor	10.43	10.43	9.0%
2040	Rain Water Drainage	Roof drains		S.F. Floor	3.54	.59	
	D30 HVAC						
3010	Energy Supply	Conditioned air with hot water reheat system		S.F. Floor	4.25	4.25	
3020	Heat Generating Systems	Steam boiler for services		Each	43,425	.54	
3030	Cooling Generating Systems	Chillers		S.F. Floor	3.06	3.06	23.0%
3050	Terminal & Package Units	N/A		—	—	—	
3090	Other HVAC Sys. & Equipment	Hot water boilers, ductwork, VAV terminals, ventilation system		S.F. Floor	38.93	38.93	
	D40 Fire Protection						
4010	Sprinklers	Wet pipe spinkler system		S.F. Floor	2.85	2.85	1.7%
4020	Standpipes	Standpipe		S.F. Floor	.55	.55	
	D50 Electrical						
5010	Electrical Service/Distribution	2000 ampere service, panel board and feeders		S.F. Floor	2.31	2.31	
5020	Lighting & Branch Wiring	Fluorescent fixtures, receptacles, switches, A.C. and misc. power		S.F. Floor	20.53	20.53	14.8%
5030	Communications & Security	Addressable alarms, emergency lighting, internet and phone wiring		S.F. Floor	2.51	2.51	
5090	Other Electrical Systems	Emergency generator, 400 Kw with fuel tank, UPS system		S.F. Floor	4.73	4.73	
E. EQUIPMENT & FURNISHINGS							
1010	Commercial Equipment	N/A		—	—	—	
1020	Institutional Equipment	Laboratory and kitchen equipment		S.F. Floor	22.11	22.11	12.5 %
1030	Vehicular Equipment	N/A		—	—	—	
2020	Moveable Furnishings	Patient wall systems		S.F. Floor	3.29	3.29	
F. SPECIAL CONSTRUCTION							
1020	Integrated Construction	N/A		—	—	—	0.0 %
1040	Special Facilities	N/A		—	—	—	
G. BUILDING SITEWORK	**N/A**						

			Sub-Total	203.15	**100%**
CONTRACTOR FEES (General Requirements: 10%, Overhead: 5%, Profit: 10%)			25%	50.80	
ARCHITECT FEES			9%	22.85	
		Total Building Cost		**276.80**	

Costs per square foot of floor area

Exterior Wall	S.F. Area	35000	55000	75000	95000	115000	135000	155000	175000	195000
	L.F. Perimeter	314	401	497	555	639	722	754	783	850
Brick Veneer	Reinforced Concrete	205.40	193.00	187.85	182.80	180.60	179.05	176.25	173.90	173.10
Stone Veneer	Steel Frame	200.30	189.15	184.50	180.00	178.05	176.65	174.15	172.20	171.45
EFIS	Reinforced Concrete	185.10	176.50	172.85	169.65	168.10	167.05	165.40	164.05	163.50
Curtain Wall	Steel Frame	199.60	188.15	183.40	178.95	176.90	175.50	173.05	171.10	170.35
Precast Concrete Panel	Reinforced Concrete	194.45	183.50	178.90	174.65	172.70	171.35	169.05	167.20	166.45
Fiber Cement	Steel Frame	176.15	168.30	164.95	162.05	160.70	159.70	158.25	157.10	156.60
Perimeter Adj., Add or Deduct	Per 100 L.F.	14.90	9.50	6.95	5.50	4.50	3.90	3.40	3.00	2.70
Story Hgt. Adj., Add or Deduct	Per 1 Ft.	2.80	2.30	2.10	1.80	1.75	1.70	1.50	1.40	1.40
For Basement, add $36.60 per square foot of basement area										

The above costs were calculated using the basic specifications shown on the facing page. These costs should be adjusted where necessary for design alternatives and owner's requirements. Reported completed project costs, for this type of structure, range from $125.50 to $240.90 per S.F.

Common additives

Description	Unit	$ Cost	Description	Unit	$ Cost
Bar, Front bar	L.F.	435	Laundry Equipment		
Back bar	L.F.	345	Folders, blankets & sheets, king size	Ea.	73,500
Booth, Upholstered, custom, straight	L.F.	252 - 465	Ironers, 110" single roll	Ea.	39,200
"L" or "U" shaped	L.F.	261 - 440	Combination washer extractor 50#	Ea.	13,600
Closed Circuit Surveillance, One station			125#	Ea.	37,000
Camera and monitor	Ea.	2025	Sauna, Prefabricated, complete		
For additional camera stations, add	Ea.	1100	6' x 4'	Ea.	6750
Directory Boards, Plastic, glass covered			6' x 6'	Ea.	8075
30" x 20"	Ea.	610	6' x 9'	Ea.	10,500
36" x 48"	Ea.	1525	8' x 8'	Ea.	12,000
Aluminum, 24" x 18"	Ea.	640	10' x 12'	Ea.	16,100
48" x 32"	Ea.	1075	Smoke Detectors		
48" x 60"	Ea.	2275	Ceiling type	Ea.	227
Elevators, Electric passenger, 5 stops			Duct type	Ea.	565
3500# capacity	Ea.	177,600	Sound System		
5000# capacity	Ea.	184,100	Amplifier, 250 watts	Ea.	2200
Additional stop, add	Ea.	10,500	Speaker, ceiling or wall	Ea.	217
Emergency Lighting, 25 watt, battery operated			Trumpet	Ea.	415
Lead battery	Ea.	345	TV Antenna, Master system, 12 outlet	Outlet	263
Nickel cadmium	Ea.	685	30 outlet	Outlet	225
			100 outlet	Outlet	213

Model costs calculated for a 6 story building with 10' story height and 135,000 square feet of floor area

				Unit	Unit Cost	Cost Per S.F.	% Of Sub-Total
A. SUBSTRUCTURE							
1010	Standard Foundations	Poured concrete; strip and spread footings		S.F. Ground	2.58	.43	
1020	Special Foundations	N/A		—	—	—	
1030	Slab on Grade	4" reinforced slab on grade		S.F. Slab	5.56	.93	1.4%
2010	Basement Excavation	Site preparation for slab and trench for foundation wall and footing		S.F. Ground	.18	.03	
2020	Basement Walls	4' Foundation wall		L.F. Wall	86	.50	
B. SHELL							
B10 Superstructure							
1010	Floor Construction	Cast in place concrete columns, beams, and slab, fireproofed		S.F. Floor	25.09	20.91	15.5%
1020	Roof Construction	Included in B1010		—	—	—	
B20 Exterior Enclosure							
2010	Exterior Walls	Face brick with concrete block backup. Insulated	80% of wall	S.F. Wall	38.80	9.96	
2020	Exterior Windows	Aluminum awning windows	20% of wall	Each	746	2.08	9.1%
2030	Exterior Doors	Aluminum and glass double doors, hollow metal egress doors		Each	3684	.28	
B30 Roofing							
3010	Roof Coverings	Single ply membrane, stone ballast, rigid insulation		S.F. Roof	5.94	.99	0.8%
3020	Roof Openings	Roof hatches		S.F. Roof	.18	.03	
C. INTERIORS							
1010	Partitions	Fire rated gypsum board on metal studs, CMU partitions	10 S.F. Floor/L.F. Partition	S.F. Partition	6.45	5.73	
1020	Interior Doors	Single leaf hollow metal	90 S.F. Floor/Door	Each	1168	12.98	
1030	Fittings	N/A		—	—	—	
2010	Stair Construction	Concrete filled metal pan		Flight	14,450	2.78	26.0%
3010	Wall Finishes	80% paint, 20% ceramic tile		S.F. Surface	2.33	4.14	
3020	Floor Finishes	80% carpet, 10% vinyl composition tile, 10% ceramic tile		S.F. Floor	5.47	5.47	
3030	Ceiling Finishes	Painted gypsum board ceiling on metal furring		S.F. Ceiling	4.03	4.03	
D. SERVICES							
D10 Conveying							
1010	Elevators & Lifts	Four geared passenger elevators		Each	235,238	6.97	5.2%
1020	Escalators & Moving Walks	N/A		—	—	—	
D20 Plumbing							
2010	Plumbing Fixtures	Kitchen, bath, laundry and service fixtures, supply and drainage	1 Fixture/150 S.F. Floor	Each	2835	18.29	
2020	Domestic Water Distribution	Gas fired water heater		S.F. Floor	.64	.64	14.2%
2040	Rain Water Drainage	Roof drains		S.F. Roof	1.92	.32	
D30 HVAC							
3010	Energy Supply	Gas fired hot water, wall fin radiation		S.F. Floor	5.03	5.03	
3020	Heat Generating Systems	N/A		—	—	—	
3030	Cooling Generating Systems	Chilled water, fan coil units		S.F. Floor	14	14	14.1%
3050	Terminal & Package Units	N/A		—	—	—	
3090	Other HVAC Sys. & Equipment	N/A		—	—	—	
D40 Fire Protection							
4010	Sprinklers	Wet pipe spinkler system, light hazard and accessory package		S.F. Floor	2.87	2.87	2.4%
4020	Standpipes	Standpipes and hose systems, with pumps		S.F. Floor	.39	.39	
D50 Electrical							
5010	Electrical Service/Distribution	2000 Ampere service, panel boards and feeders		S.F. Floor	1.48	1.48	
5020	Lighting & Branch Wiring	Fluorescent fixtures, receptacles, switches, A.C. and misc. power		S.F. Floor	9.05	9.05	11.4%
5030	Communications & Security	Addressable alarm systems, emergency lighting, internet and phone wiring		S.F. Floor	4.37	4.37	
5090	Other Electrical Systems	Emergency generator muffler and transfer switch, 250 kW		S.F. Floor	.45	.45	
E. EQUIPMENT & FURNISHINGS							
1010	Commercial Equipment	N/A		—	—	—	
1020	Institutional Equipment	N/A		—	—	—	
1030	Vehicular Equipment	N/A		—	—	—	0.0 %
1090	Other Equipment	N/A		—	—	—	
F. SPECIAL CONSTRUCTION							
1020	Integrated Construction	N/A		—	—	—	0.0 %
1040	Special Facilities	N/A		—	—	—	
G. BUILDING SITEWORK	**N/A**						

				Sub-Total	135.13	100%
CONTRACTOR FEES (General Requirements: 10%, Overhead: 5%, Profit: 10%)				25%	33.79	
ARCHITECT FEES				6%	10.13	

Total Building Cost | **179.05**

For customer support on your Square Foot Cost Data, call 877.756.2789.

153

Costs per square foot of floor area

Exterior Wall	S.F. Area	140000	243000	346000	450000	552000	655000	760000	860000	965000
	L.F. Perimeter	403	587	672	800	936	1073	1213	1195	1312
Brick Veneer	Reinforced Concrete	211.75	203.00	195.30	192.50	191.10	190.05	189.25	186.20	185.65
Stone Veneer	Steel Frame	200.40	193.35	187.35	185.10	184.00	183.20	182.55	180.20	179.75
EFIS	Reinforced Concrete	190.35	185.40	181.55	180.00	179.25	178.70	178.25	176.90	176.55
Curtain Wall	Steel Frame	200.80	194.20	188.60	186.50	185.50	184.70	184.10	182.00	181.55
Precast Concrete Panel	Reinforced Concrete	238.15	218.55	207.35	202.30	199.45	197.45	195.95	192.90	191.90
Fiber Cement	Steel Frame	212.85	196.35	188.15	184.10	181.90	180.15	178.95	177.20	176.40
Perimeter Adj., Add or Deduct	Per 100 L.F.	10.40	6.00	4.20	3.15	2.55	2.20	1.85	1.65	1.50
Story Hgt. Adj., Add or Deduct	Per 1 Ft.	2.25	1.95	1.50	1.40	1.30	1.30	1.30	1.05	1.10

For Basement, add $38.00 per square foot of basement area

The above costs were calculated using the basic specifications shown on the facing page. These costs should be adjusted where necessary for design alternatives and owner's requirements. Reported completed project costs, for this type of structure, range from $139.30 to $243.50 per S.F.

Common additives

Description	Unit	$ Cost	Description	Unit	$ Cost
Bar, Front bar	L.F.	435	Laundry Equipment		
Back bar	L.F.	345	Folders, blankets & sheets, king size	Ea.	73,500
Booth, Upholstered, custom, straight	L.F.	252 - 465	Ironers, 110" single roll	Ea.	39,200
"L" or "U" shaped	L.F.	261 - 440	Combination washer & extractor 50#	Ea.	13,600
Closed Circuit Surveillance, One station			125#	Ea.	37,000
Camera and monitor	Ea.	2025	Sauna, Prefabricated, complete		
For additional camera stations, add	Ea.	1100	6' x 4'	Ea.	6750
Directory Boards, Plastic, glass covered			6' x 6'	Ea.	8075
30" x 20"	Ea.	610	6' x 9'	Ea.	10,500
36" x 48"	Ea.	1525	8' x 8'	Ea.	12,000
Aluminum, 24" x 18"	Ea.	640	10' x 12'	Ea.	16,100
48" x 32"	Ea.	1075	Smoke Detectors		
48" x 60"	Ea.	2275	Ceiling type	Ea.	227
Elevators, Electric passenger, 10 stops			Duct type	Ea.	565
3500# capacity	Ea.	411,500	Sound System		
5000# capacity	Ea.	419,500	Amplifier, 250 watts	Ea.	2200
Additional stop, add	Ea.	10,500	Speaker, ceiling or wall	Ea.	217
Emergency Lighting, 25 watt, battery operated			Trumpet	Ea.	415
Lead battery	Ea.	345	TV Antenna, Master system, 12 outlet	Outlet	263
Nickel cadmium	Ea.	685	30 outlet	Outlet	225
			100 outlet	Outlet	213

Important: See the Reference Section for Location Factors.

Model costs calculated for a 15 story building with 10' story height and 450,000 square feet of floor area

				Unit	Unit Cost	Cost Per S.F.	% Of Sub-Total
A. SUBSTRUCTURE							
1010	Standard Foundations	CIP concrete pile caps		S.F. Ground	12.60	.84	
1020	Special Foundations	Steel H piles, concrete grade beams		S.F. Ground	195	13.01	
1030	Slab on Grade	4" reinforced slab on grade		S.F. Slab	5.56	.37	9.4%
2010	Basement Excavation	Site preparation for slab, piles and grade beams		S.F. Ground	.18	.01	
2020	Basement Walls	4' Foundation wall		L.F. Wall	86	.16	
B. SHELL							
	B10 Superstructure						
1010	Floor Construction	Cast in place concrete columns, beams, and slab		S.F. Floor	20.72	19.34	13.4%
1020	Roof Construction	Precast double T with 2" topping		S.F. Roof	15.90	1.06	
	B20 Exterior Enclosure						
2010	Exterior Walls	Precast concrete panels, insulated	80% of wall	S.F. Wall	48.33	10.31	
2020	Exterior Windows	Aluminum awning type	20% of wall	Each	746	1.73	8.0%
2030	Exterior Doors	Aluminum & glass doors and entrances		Each	3353	.24	
	B30 Roofing						
3010	Roof Coverings	Single ply membrane, stone ballast, rigid insulation		S.F. Roof	5.70	.38	0.3%
3020	Roof Openings	Roof hatches		S.F. Roof	.15	.01	
C. INTERIORS							
1010	Partitions	Gypsum board on metal studs, sound attenuation insulation	10 S.F. Floor/L.F. Partition	S.F. Partition	6.38	5.67	
1020	Interior Doors	Single leaf hollow metal	90 S.F. Floor/Door	Each	1168	12.98	
1030	Fittings	N/A		—	—	—	
2010	Stair Construction	Concrete filled metal pan with rails		Flight	14,450	1.99	22.4%
3010	Wall Finishes	80% Paint, 20% ceramic tile		S.F. Surface	2.31	4.11	
3020	Floor Finishes	80% carpet, 10% vinyl composition tile, 10% ceramic tile		S.F. Floor	5.47	5.47	
3030	Ceiling Finishes	Gypsum board on resilient channel		S.F. Ceiling	4.03	4.03	
D. SERVICES							
	D10 Conveying						
1010	Elevators & Lifts	Five traction geared passenger elevators		Each	495,000	6.60	4.3%
1020	Escalators & Moving Walks	N/A		—	—	—	
	D20 Plumbing						
2010	Plumbing Fixtures	Kitchen, bath, laundry and service fixtures, supply and drainage	1 Fixture/165 S.F. Floor	Each	2978	18.05	
2020	Domestic Water Distribution	Electric water heaters, gas fired water heaters		S.F. Floor	6.86	6.86	16.4%
2040	Rain Water Drainage	Roof drains		S.F. Roof	2.85	.19	
	D30 HVAC						
3010	Energy Supply	Gas fired hot water, wall fin radiation		S.F. Floor	2.65	2.65	
3020	Heat Generating Systems	N/A		—	—	—	
3030	Cooling Generating Systems	Chilled water, fan coil units		S.F. Floor	14	14	10.9%
3050	Terminal & Package Units	N/A		—	—	—	
3090	Other HVAC Sys. & Equipment	N/A		—	—	—	
	D40 Fire Protection						
4010	Sprinklers	Wet pipe spinkler system, light hazard and accessory package		S.F. Floor	4	4	5.3%
4020	Standpipes	Standpipes and hose systems, with pumps		S.F. Floor	4.08	4.08	
	D50 Electrical						
5010	Electrical Service/Distribution	2000 Ampere service, panel boards and feeders		S.F. Floor	.96	.96	
5020	Lighting & Branch Wiring	Fluorescent fixtures, receptacles, switches, A.C. and misc. power		S.F. Floor	8.90	8.90	9.5%
5030	Communications & Security	Addressable alarm systems, emergency lighting, internet and phone wiring		S.F. Floor	4.34	4.34	
5090	Other Electrical Systems	Emergency generator muffler and transfer switch, 250 kW		S.F. Floor	.33	.33	
E. EQUIPMENT & FURNISHINGS							
1010	Commercial Equipment	N/A		—	—	—	
1020	Institutional Equipment	N/A		—	—	—	
1030	Vehicular Equipment	N/A		—	—	—	0.0 %
1090	Other Equipment	N/A		—	—	—	
F. SPECIAL CONSTRUCTION							
1020	Integrated Construction	N/A		—	—	—	0.0 %
1040	Special Facilities	N/A		—	—	—	
G. BUILDING SITEWORK	**N/A**						

		Sub-Total	152.67	100%
CONTRACTOR FEES (General Requirements: 10%, Overhead: 5%, Profit: 10%)		25%	38.18	
ARCHITECT FEES		6%	11.45	
	Total Building Cost		**202.30**	

For customer support on your Square Foot Cost Data, call 877.756.2789.

155

Costs per square foot of floor area

Exterior Wall	S.F. Area	5500	12000	20000	30000	40000	60000	80000	100000	145000
	L.F. Perimeter	310	330	400	490	530	590	680	760	920
Face Brick with Concrete Block Back-up	Steel Frame	541.10	396.10	353.65	332.65	318.60	303.40	296.95	292.80	287.15
	R/Conc. Frame	531.60	386.60	344.15	323.20	309.10	293.90	287.50	283.30	277.65
Stucco on Concrete Block	Steel Frame	523.75	388.05	348.05	328.25	315.10	301.10	295.10	291.20	285.95
	R/Conc. Frame	512.70	377.40	337.45	317.70	304.65	290.60	284.60	280.75	275.50
Reinforced Concrete	Steel Frame	517.20	384.45	345.15	325.75	312.95	299.25	293.35	289.60	284.45
	R/Conc. Frame	507.70	375.00	335.70	316.25	303.45	289.80	283.85	280.10	275.00
Perimeter Adj., Add or Deduct	Per 100 L.F.	52.95	24.25	14.60	9.65	7.25	4.85	3.65	2.90	2.00
Story Hgt. Adj., Add or Deduct	Per 1 Ft.	12.60	6.15	4.50	3.65	2.95	2.25	1.95	1.70	1.45
For Basement, add $52.55 per square foot of basement area										

The above costs were calculated using the basic specifications shown on the facing page. These costs should be adjusted where necessary for design alternatives and owner's requirements. Reported completed project costs, for this type of structure, range from $202.65 to $538.25 per S.F.

Common additives

Description	Unit	$ Cost	Description	Unit	$ Cost
Clock System			Emergency Lighting, 25 watt, battery operated		
20 room	Ea.	20,600	Nickel cadmium	Ea.	685
50 room	Ea.	48,500	Flagpoles, Complete		
Closed Circuit Surveillance, One station			Aluminum, 20' high	Ea.	1900
Camera and monitor	Ea.	2025	40' High	Ea.	4425
For additional camera stations, add	Ea.	1100	70' High	Ea.	11,500
Elevators, Hydraulic passenger, 2 stops			Fiberglass, 23' High	Ea.	1350
1500# capacity	Ea.	67,400	39'-5" High	Ea.	3400
2500# capacity	Ea.	70,400	59' High	Ea.	6650
3500# capacity	Ea.	75,400	Laundry Equipment		
Additional stop, add	Ea.	8075	Folders, blankets, & sheets	Ea.	73,500
Emergency Generators, Complete system, gas			Ironers, 110" single roll	Ea.	39,200
15 KW	Ea.	18,500	Combination washer & extractor, 50#	Ea.	13,600
70 KW	Ea.	35,200	125#	Ea.	37,000
85 KW	Ea.	39,000	Safe, Office type, 1 hour rating		
115 KW	Ea.	75,500	30" x 18" x 18"	Ea.	2500
170 KW	Ea.	98,000	60" x 36" x 18", double door	Ea.	9350
Diesel, 50 KW	Ea.	26,100	Sound System		
100 KW	Ea.	36,200	Amplifier, 250 watts	Ea.	2200
150 KW	Ea.	47,900	Speaker, ceiling or wall	Ea.	217
350 KW	Ea.	72,500	Trumpet	Ea.	415

Important: See the Reference Section for Location Factors.

Model costs calculated for a 3 story building with 12' story height and 40,000 square feet of floor area

Jail

				Unit	Unit Cost	Cost Per S.F.	% Of Sub-Total
A. SUBSTRUCTURE							
1010	Standard Foundations	Poured concrete; strip and spread footings		S.F. Ground	2.94	.98	
1020	Special Foundations	N/A		—	—	—	
1030	Slab on Grade	4" reinforced concrete with vapor barrier and granular base		S.F. Slab	5.56	1.85	1.7%
2010	Basement Excavation	Site preparation for slab and trench for foundation wall and footing		S.F. Ground	.32	.11	
2020	Basement Walls	4' foundation wall		L.F. Wall	86	1.13	
B. SHELL							
B10 Superstructure							
1010	Floor Construction	Concrete slab with metal deck and beams		S.F. Floor	30.66	20.44	12.0%
1020	Roof Construction	Concrete slab with metal deck and beams		S.F. Roof	24.12	8.04	
B20 Exterior Enclosure							
2010	Exterior Walls	Face brick with reinforced concrete block backup	85% of wall	S.F. Wall	33.42	13.55	
2020	Exterior Windows	Bullet resisting and metal horizontal pivoted	15% of wall	Each	164	11.72	10.7%
2030	Exterior Doors	Metal		Each	5590	.27	
B30 Roofing							
3010	Roof Coverings	Built-up tar and gravel with flashing; perlite/EPS composite insulation		S.F. Roof	6.69	2.23	0.9%
3020	Roof Openings	Roof hatches		S.F. Roof	.09	.03	
C. INTERIORS							
1010	Partitions	Concrete block	45 S.F. of Floor/L.F. Partition	S.F. Partition	10.84	2.89	
1020	Interior Doors	Hollow metal fire doors	930 S.F. Floor/Door	Each	1168	1.26	
1030	Fittings	N/A		—	—	—	
2010	Stair Construction	Concrete filled metal pan		Flight	17,275	2.16	6.1%
3010	Wall Finishes	Paint		S.F. Surface	4.37	2.33	
3020	Floor Finishes	70% vinyl composition tile, 20% carpet, 10% ceramic tile		S.F. Floor	3.70	3.70	
3030	Ceiling Finishes	Mineral tile on zee runners	30% of area	S.F. Ceiling	7.22	2.17	
D. SERVICES							
D10 Conveying							
1010	Elevators & Lifts	One hydraulic passenger elevator		Each	168,800	4.22	1.8%
1020	Escalators & Moving Walks	N/A		—	—	—	
D20 Plumbing							
2010	Plumbing Fixtures	Toilet and service fixtures, supply and drainage	1 Fixture/78 S.F. Floor	Each	4224	54.16	
2020	Domestic Water Distribution	Electric water heater		S.F. Floor	8.86	8.86	26.8%
2040	Rain Water Drainage	Roof drains		S.F. Roof	2.70	.90	
D30 HVAC							
3010	Energy Supply	N/A		—	—	—	
3020	Heat Generating Systems	Included in D3050		—	—	—	
3030	Cooling Generating Systems	N/A		—	—	—	7.0 %
3050	Terminal & Package Units	Rooftop multizone unit systems		S.F. Floor	16.56	16.56	
3090	Other HVAC Sys. & Equipment	N/A		—	—	—	
D40 Fire Protection							
4010	Sprinklers	Sprinkler system, light hazard		S.F. Floor	2.97	2.97	1.5%
4020	Standpipes	Standpipes and hose systems		S.F. Floor	.67	.67	
D50 Electrical							
5010	Electrical Service/Distribution	600 ampere service, panel board and feeders		S.F. Floor	1.05	1.05	
5020	Lighting & Branch Wiring	Fluorescent fixtures, receptacles, switches, A.C. and misc. power		S.F. Floor	10.23	10.23	7.0%
5030	Communications & Security	Addressable alarm systems, intercom, and emergency lighting		S.F. Floor	4.60	4.60	
5090	Other Electrical Systems	Emergency generator, 80 kW		S.F. Floor	.77	.77	
E. EQUIPMENT & FURNISHINGS							
1010	Commercial Equipment	N/A		—	—	—	
1020	Institutional Equipment	Prefabricated cells, visitor cubicles		S.F. Floor	58	58.33	24.5 %
1030	Vehicular Equipment	N/A		—	—	—	
1090	Other Equipment	N/A		—	—	—	
F. SPECIAL CONSTRUCTION							
1020	Integrated Construction	N/A		—	—	—	0.0 %
1040	Special Facilities	N/A		—	—	—	
G. BUILDING SITEWORK	**N/A**						
				Sub-Total		238.18	100%
	CONTRACTOR FEES (General Requirements: 10%, Overhead: 5%, Profit: 10%)			25%		59.58	
	ARCHITECT FEES			7%		20.84	
				Total Building Cost		**318.60**	

For customer support on your Square Foot Cost Data, call 877.756.2789.

157

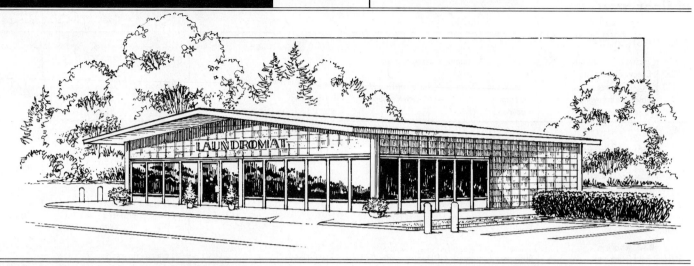

Costs per square foot of floor area

Exterior Wall	S.F. Area	1000	2000	3000	4000	5000	10000	15000	20000	25000
	L.F. Perimeter	126	179	219	253	283	400	490	568	632
Decorative Concrete Block	Steel Frame	303.50	254.55	235.60	225.20	218.50	203.05	196.75	193.30	190.85
	Bearing Walls	300.40	250.95	231.75	221.30	214.45	198.75	192.50	188.95	186.40
Face Brick with Concrete Block Back-up	Steel Frame	333.70	275.90	252.95	240.20	231.90	212.40	204.40	199.90	196.70
	Bearing Walls	330.85	272.50	249.35	236.50	228.10	208.40	200.35	195.75	192.50
Metal Sandwich Panel	Steel Frame	303.25	254.40	235.50	225.20	218.50	203.15	196.95	193.45	191.00
Precast Concrete Panel	Bearing Walls	295.40	247.40	228.85	218.75	212.20	197.20	191.20	187.80	185.40
Perimeter Adj., Add or Deduct	Per 100 L.F.	60.85	30.45	20.25	15.20	12.15	6.05	4.05	3.00	2.45
Story Hgt. Adj., Add or Deduct	Per 1 Ft.	3.95	2.85	2.25	2.00	1.80	1.20	1.05	0.95	0.75
For Basement, add $34.80 per square foot of basement area										

The above costs were calculated using the basic specifications shown on the facing page. These costs should be adjusted where necessary for design alternatives and owner's requirements. Reported completed project costs, for this type of structure, range from $112.85 to $275.80 per S.F.

Common additives

Description	Unit	$ Cost
Closed Circuit Surveillance, One station		
Camera and monitor	Ea.	2025
For additional camera stations, add	Ea.	1100
Emergency Lighting, 25 watt, battery operated		
Lead battery	Ea.	345
Nickel cadmium	Ea.	685
Laundry Equipment		
Dryers, coin operated 30 lb.	Ea.	3950
Double stacked	Ea.	8600
50 lb.	Ea.	4450
Dry cleaner 20 lb.	Ea.	44,400
30 lb.	Ea.	66,500
Washers, coin operated	Ea.	1625
Washer/extractor 20 lb.	Ea.	7325
30 lb.	Ea.	11,600
50 lb.	Ea.	13,600
75 lb.	Ea.	26,200
Smoke Detectors		
Ceiling type	Ea.	227
Duct type	Ea.	565

Important: See the Reference Section for Location Factors.

Model costs calculated for a 1 story building with 12' story height and 3,000 square feet of floor area

Laundromat

			Unit	Unit Cost	Cost Per S.F.	% Of Sub-Total
A. SUBSTRUCTURE						
1010	Standard Foundations	Poured concrete; strip and spread footings	S.F. Ground	3.62	3.62	
1020	Special Foundations	N/A	—	—	—	
1030	Slab on Grade	5" reinforced concrete with vapor barrier and granular base	S.F. Slab	6.01	6.01	8.9%
2010	Basement Excavation	Site preparation for slab and trench for foundation wall and footing	S.F. Ground	.57	.57	
2020	Basement Walls	4' Foundation wall	L.F. Wall	76	5.55	
B. SHELL						
B10 Superstructure						
1010	Floor Construction	Steel column fireproofing	S.F. Floor	.96	.96	6.4%
1020	Roof Construction	Metal deck, open web steel joists, beams, columns	S.F. Roof	10.38	10.38	
B20 Exterior Enclosure						
2010	Exterior Walls	Decorative concrete block	90% of wall	S.F. Wall	16.30	12.85	
2020	Exterior Windows	Store front	10% of wall	Each	45.55	4	10.9%
2030	Exterior Doors	Double aluminum and glass		Each	3605	2.40	
B30 Roofing						
3010	Roof Coverings	Built-up tar and gravel with flashing; perlite/EPS composite insulation	S.F. Roof	9.08	9.08	5.2%
3020	Roof Openings	N/A	—	—	—	
C. INTERIORS						
1010	Partitions	Gypsum board on metal studs	60 S.F. Floor/L.F. Partition	S.F. Partition	28.02	4.67	
1020	Interior Doors	Single leaf wood	750 S.F. Floor/Door	Each	626	.83	
1030	Fittings	N/A	—	—	—	
2010	Stair Construction	N/A	—	—	—	7.0%
3010	Wall Finishes	Paint	S.F. Surface	.84	.28	
3020	Floor Finishes	Vinyl composition tile	S.F. Floor	2.36	2.36	
3030	Ceiling Finishes	Fiberglass board on exposed grid system	S.F. Ceiling	4.23	4.23	
D. SERVICES						
D10 Conveying						
1010	Elevators & Lifts	N/A	—	—	—	0.0%
1020	Escalators & Moving Walks	N/A	—	—	—	
D20 Plumbing						
2010	Plumbing Fixtures	Toilet and service fixtures, supply and drainage	1 Fixture/600 S.F. Floor	Each	6774	11.29	
2020	Domestic Water Distribution	Gas fired hot water heater	S.F. Floor	49.30	49.30	35.6%
2040	Rain Water Drainage	Roof drains	S.F. Roof	2.11	2.11	
D30 HVAC						
3010	Energy Supply	N/A	—	—	—	
3020	Heat Generating Systems	Included in D3050	—	—	—	
3030	Cooling Generating Systems	N/A	—	—	—	5.3%
3050	Terminal & Package Units	Rooftop single zone unit systems	S.F. Floor	9.41	9.41	
3090	Other HVAC Sys. & Equipment	N/A	—	—	—	
D40 Fire Protection						
4010	Sprinklers	Sprinkler, ordinary hazard	S.F. Floor	5.50	5.50	3.1%
4020	Standpipes	N/A	—	—	—	
D50 Electrical						
5010	Electrical Service/Distribution	200 ampere service, panel board and feeders	S.F. Floor	4.17	4.17	
5020	Lighting & Branch Wiring	High efficiency fluorescent fixtures, receptacles, switches, A.C. and misc. power	S.F. Floor	22.87	22.87	17.4%
5030	Communications & Security	Addressable alarm systems and emergency lighting	S.F. Floor	3.16	3.16	
5090	Other Electrical Systems	Emergency generator, 7.5 kW	S.F. Floor	.52	.52	
E. EQUIPMENT & FURNISHINGS						
1010	Commercial Equipment	N/A	—	—	—	
1020	Institutional Equipment	N/A	—	—	—	0.0%
1030	Vehicular Equipment	N/A	—	—	—	
1090	Other Equipment	N/A	—	—	—	
F. SPECIAL CONSTRUCTION						
1020	Integrated Construction	N/A	—	—	—	0.0%
1040	Special Facilities	N/A	—	—	—	
G. BUILDING SITEWORK	N/A					

		Sub-Total	176.12	100%
CONTRACTOR FEES (General Requirements: 10%, Overhead: 5%, Profit: 10%)		25%	44.07	
ARCHITECT FEES		7%	15.41	
	Total Building Cost		**235.60**	

For customer support on your Square Foot Cost Data, call 877.756.2789.

159

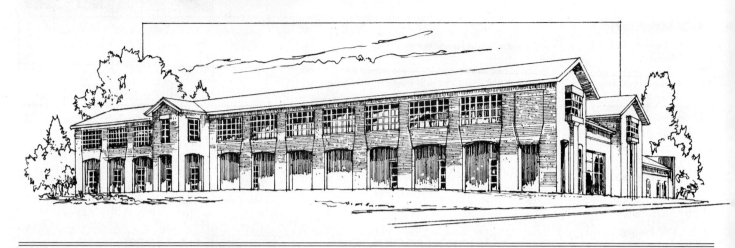

Costs per square foot of floor area

Exterior Wall	S.F. Area	7000	10000	13000	16000	19000	22000	25000	28000	31000
	L.F. Perimeter	240	300	336	386	411	435	472	510	524
Face Brick with Concrete Block Back-up	R/Conc. Frame	198.65	188.90	180.50	176.70	172.00	168.35	166.55	165.15	162.75
	Steel Frame	198.70	188.95	180.50	176.75	172.00	168.35	166.55	165.20	162.75
Limestone with Concrete Block	R/Conc. Frame	231.90	216.30	203.20	197.15	189.80	184.25	181.45	179.20	175.55
	Steel Frame	226.25	213.05	201.35	196.20	189.40	184.25	181.75	179.80	176.35
Precast Concrete Panels	R/Conc. Frame	222.35	207.90	195.90	190.40	183.75	178.70	176.15	174.20	170.85
	Steel Frame	222.35	207.85	195.95	190.40	183.75	178.75	176.20	174.15	170.90
Perimeter Adj., Add or Deduct	Per 100 L.F.	24.35	17.10	13.10	10.70	8.90	7.75	6.80	6.10	5.55
Story Hgt. Adj., Add or Deduct	Per 1 Ft.	3.65	3.20	2.75	2.65	2.25	2.10	2.00	2.00	1.80

For Basement, add $49.80 per square foot of basement area

The above costs were calculated using the basic specifications shown on the facing page. These costs should be adjusted where necessary for design alternatives and owner's requirements. Reported completed project costs, for this type of structure, range from $108.45 to $278.25 per S.F.

Common additives

Description	Unit	$ Cost
Carrels Hardwood	Ea.	945 - 2400
Closed Circuit Surveillance, One station		
Camera and monitor	Ea.	2025
For additional camera stations, add	Ea.	1100
Elevators, Hydraulic passenger, 2 stops		
1500# capacity	Ea.	67,400
2500# capacity	Ea.	70,400
3500# capacity	Ea.	75,400
Emergency Lighting, 25 watt, battery operated		
Lead battery	Ea.	345
Nickel cadmium	Ea.	685
Flagpoles, Complete		
Aluminum, 20' high	Ea.	1900
40' high	Ea.	4425
70' high	Ea.	11,500
Fiberglass, 23' high	Ea.	1350
39'-5" high	Ea.	3400
59' high	Ea.	6650

Description	Unit	$ Cost
Library Furnishings		
Bookshelf, 90" high, 10" shelf double face	L.F.	288
single face	L.F.	184
Charging desk, built-in with counter		
Plastic laminated top	L.F.	420
Reading table, laminated		
top 60" x 36"	Ea.	775

Important: See the Reference Section for Location Factors.

Model costs calculated for a 2 story building with 14' story height and 22,000 square feet of floor area

Library

				Unit	Unit Cost	Cost Per S.F.	% Of Sub-Total
A.	**SUBSTRUCTURE**						
1010	Standard Foundations	Poured concrete; strip and spread footings		S.F. Ground	4.06	2.03	
1020	Special Foundations	N/A		—	—	—	
1030	Slab on Grade	4" reinforced concrete with vapor barrier and granular base		S.F. Slab	5.56	2.79	5.3%
2010	Basement Excavation	Site preparation for slab and trench for foundation wall and footing		S.F. Ground	.32	.16	
2020	Basement Walls	4' foundation wall		L.F. Wall	86	1.69	
B.	**SHELL**						
	B10 Superstructure						
1010	Floor Construction	Concrete waffle slab		S.F. Floor	26.42	13.21	18.7%
1020	Roof Construction	Concrete waffle slab		S.F. Roof	20.26	10.13	
	B20 Exterior Enclosure						
2010	Exterior Walls	Face brick with concrete block backup	90% of wall	S.F. Wall	32.89	16.39	
2020	Exterior Windows	Window wall	10% of wall	Each	56	3.08	16.0%
2030	Exterior Doors	Double aluminum and glass, single leaf hollow metal		Each	5875	.53	
	B30 Roofing						
3010	Roof Coverings	Single ply membrane, EPDM, fully adhered; perlite/EPS composite insulation		S.F. Roof	5.62	2.81	2.3%
3020	Roof Openings	Roof hatches		S.F. Roof	.08	.04	
C.	**INTERIORS**						
1010	Partitions	Gypsum board on metal studs	30 S.F. Floor/L.F. Partition	S.F. Partition	12.63	5.05	
1020	Interior Doors	Single leaf wood	300 S.F. Floor/Door	Each	626	2.09	
1030	Fittings	N/A		—	—	—	
2010	Stair Construction	Concrete filled metal pan		Flight	9850	.89	15.9%
3010	Wall Finishes	Paint		S.F. Surface	.85	.68	
3020	Floor Finishes	50% carpet, 50% vinyl tile		S.F. Floor	3.93	3.93	
3030	Ceiling Finishes	Mineral fiber on concealed zee bars		S.F. Ceiling	7.22	7.22	
D.	**SERVICES**						
	D10 Conveying						
1010	Elevators & Lifts	One hydraulic passenger elevator		Each	88,880	4.04	3.2%
1020	Escalators & Moving Walks	N/A		—	—	—	
	D20 Plumbing						
2010	Plumbing Fixtures	Toilet and service fixtures, supply and drainage	1 Fixture/1835 S.F. Floor	Each	7487	4.08	
2020	Domestic Water Distribution	Gas fired water heater		S.F. Floor	1.31	1.31	4.8%
2040	Rain Water Drainage	Roof drains		S.F. Roof	1.28	.64	
	D30 HVAC						
3010	Energy Supply	N/A		—	—	—	
3020	Heat Generating Systems	Included in D3050		—	—	—	
3030	Cooling Generating Systems	N/A		—	—	—	17.9 %
3050	Terminal & Package Units	Multizone unit, gas heating, electric cooling		S.F. Floor	22.30	22.30	
3090	Other HVAC Sys. & Equipment	N/A		—	—	—	
	D40 Fire Protection						
4010	Sprinklers	Wet pipe sprinkler system		S.F. Floor	3.15	3.15	3.4%
4020	Standpipes	Standpipe		S.F. Floor	1.05	1.05	
	D50 Electrical						
5010	Electrical Service/Distribution	400 ampere service, panel board and feeders		S.F. Floor	1.20	1.20	
5020	Lighting & Branch Wiring	High efficiency fluorescent fixtures, receptacles, switches, A.C. and misc. power		S.F. Floor	11.83	11.83	12.4%
5030	Communications & Security	Addressable alarm systems, internet wiring, and emergency lighting		S.F. Floor	2.11	2.11	
5090	Other Electrical Systems	Emergency generator, 7.5 kW, Uninterruptible power supply		S.F. Floor	.27	.27	
E.	**EQUIPMENT & FURNISHINGS**						
1010	Commercial Equipment	N/A		—	—	—	
1020	Institutional Equipment	N/A		—	—	—	0.0 %
1030	Vehicular Equipment	N/A		—	—	—	
1090	Other Equipment	N/A		—	—	—	
F.	**SPECIAL CONSTRUCTION**						
1020	Integrated Construction	N/A		—	—	—	0.0 %
1040	Special Facilities	N/A		—	—	—	
G.	**BUILDING SITEWORK**	N/A					

				Sub-Total	124.70	**100%**
	CONTRACTOR FEES (General Requirements: 10%, Overhead: 5%, Profit: 10%)			25%	31.18	
	ARCHITECT FEES			8%	12.47	

Total Building Cost 168.35

For customer support on your Square Foot Cost Data, call 877.756.2789.

161

Costs per square foot of floor area

Exterior Wall	S.F. Area	4000	5500	7000	8500	10000	11500	13000	14500	16000
	L.F. Perimeter	280	320	380	440	453	503	510	522	560
Vinyl Siding	Wood Frame	176.30	167.40	164.20	162.10	157.60	156.35	153.15	150.95	150.15
Stone Veneer	Wood Frame	193.70	182.05	177.90	175.25	169.15	167.55	163.35	160.35	159.30
Fiber Cement	Steel Frame	177.20	168.20	164.95	162.90	158.30	157.00	153.80	151.55	150.75
E.I.F.S.	Steel Frame	182.40	172.90	169.45	167.25	162.35	161.00	157.60	155.15	154.35
Precast Concrete Panel	Reinforced Concrete	212.95	199.25	194.45	191.40	184.20	182.30	177.35	173.75	172.60
Brick Veneer	Reinforced Concrete	220.90	205.80	200.50	197.20	189.20	187.10	181.60	177.60	176.35
Perimeter Adj., Add or Deduct	Per 100 L.F.	16.60	12.05	9.50	7.85	6.65	5.75	5.10	4.55	4.10
Story Hgt. Adj., Add or Deduct	Per 1 Ft.	2.25	1.80	1.75	1.65	1.45	1.35	1.25	1.10	1.10

For Basement, add $32.40 per square foot of basement area

The above costs were calculated using the basic specifications shown on the facing page. These costs should be adjusted where necessary for design alternatives and owner's requirements. Reported completed project costs, for this type of structure, range from $91.35 to $235.00 per S.F.

Common additives

Description	Unit	$ Cost	Description	Unit	$ Cost
Cabinets, Hospital, base			Directory Boards, Plastic, glass covered		
Laminated plastic	L.F.	420	30" x 20"	Ea.	610
Stainless steel	L.F.	675	36" x 48"	Ea.	1525
Counter top, laminated plastic	L.F.	81.50	Aluminum, 24" x 18"	Ea.	640
Stainless steel	L.F.	199	36" x 24"	Ea.	765
For drop-in sink, add	Ea.	1100	48" x 32"	Ea.	1075
Nurses station, door type			48" x 60"	Ea.	2275
Laminated plastic	L.F.	470	Heat Therapy Unit		
Enameled steel	L.F.	455	Humidified, 26" x 78" x 28"	Ea.	3450
Stainless steel	L.F.	790	Smoke Detectors		
Wall cabinets, laminated plastic	L.F.	305	Ceiling type	Ea.	227
Enameled steel	L.F.	360	Duct type	Ea.	565
Stainless steel	L.F.	635	Tables, Examining, vinyl top		
			with base cabinets	Ea.	1650 - 6775
			Utensil Washer, Sanitizer	Ea.	9650
			X-Ray, Mobile	Ea.	18,500 - 85,500

Important: See the Reference Section for Location Factors.

Model costs calculated for a 1 story building with 10' story height and 7,000 square feet of floor area

				Unit	Unit Cost	Cost Per S.F.	% Of Sub-Total
A. SUBSTRUCTURE							
1010	Standard Foundations	Poured concrete; strip and spread footings		S.F. Ground	3.74	3.74	
1020	Special Foundations	N/A		—	—	—	
1030	Slab on Grade	4" reinforced slab on grade		S.F. Slab	5.56	5.56	10.9%
2010	Basement Excavation	Site preparation for slab and trench for foundation wall and footing		S.F. Ground	.32	.32	
2020	Basement Walls	4' Foundation wall		L.F. Wall	86	4.64	
B. SHELL							
B10 Superstructure							
1010	Floor Construction	Wood columns, fireproofed		S.F. Floor	3.49	3.49	8.7%
1020	Roof Construction	Wood roof trusses, 4:12 slope		S.F. Roof	7.85	7.85	
B20 Exterior Enclosure							
2010	Exterior Walls	Ashlar stone veneer on wood stud back up	70% of wall	S.F. Wall	41.97	15.95	
2020	Exterior Windows	Double hung, insulated wood windows	30% of wall	Each	646	6.19	17.9%
2030	Exterior Doors	Single leaf solid core wood doors		Each	1475	1.27	
B30 Roofing							
3010	Roof Coverings	Asphalt shingles with flashing, gutters and downspouts		S.F. Roof	2.76	2.76	2.1%
3020	Roof Openings	N/A		—	—	—	
C. INTERIORS							
1010	Partitions	Fire rated gypsum board on wood studs	6 S.F. Floor/L.F. Partition	S.F. Partition	7.73	10.30	
1020	Interior Doors	Single leaf solid core wood doors	240 S.F. Floor/Door	Each	626	2.61	
1030	Fittings	N/A		—	—	—	
2010	Stair Construction	N/A		—	—	—	20.4%
3010	Wall Finishes	Paint on drywall		S.F. Surface	.97	2.59	
3020	Floor Finishes	50% carpet , 50% vinyl composition tile		S.F. Floor	4.61	4.61	
3030	Ceiling Finishes	Acoustic ceiling tiles on suspended support system		S.F. Ceiling	6.56	6.56	
D. SERVICES							
D10 Conveying							
1010	Elevators & Lifts	N/A		—	—	—	0.0 %
1020	Escalators & Moving Walks	N/A		—	—	—	
D20 Plumbing							
2010	Plumbing Fixtures	Restroom, exam room and service fixtures, supply and drainage	1 Fixture/350 S.F. Floor	Each	1967	5.62	
2020	Domestic Water Distribution	Gas fired water heater		S.F. Floor	2.07	2.07	5.9%
2040	Rain Water Drainage	Roof drains		S.F. Roof			
D30 HVAC							
3010	Energy Supply	N/A		—	—	—	
3020	Heat Generating Systems	Included in D3050		—	—	—	
3030	Cooling Generating Systems	N/A		—	—	—	12.2 %
3050	Terminal & Package Units	Multizone unit, gas heating , electric cooling		S.F. Floor	15.90	15.90	
3090	Other HVAC Sys. & Equipment	N/A		—	—	—	
D40 Fire Protection							
4010	Sprinklers	Wet pipe spinkler system		S.F. Floor	5.02	5.02	4.8%
4020	Standpipes	Standpipes and hose systems, with pumps		S.F. Floor	1.31	1.31	
D50 Electrical							
5010	Electrical Service/Distribution	200 Ampere service, panel boards and feeders		S.F. Floor	1.65	1.65	
5020	Lighting & Branch Wiring	Fluorescent fixtures, receptacles, switches, A.C. and misc. power		S.F. Floor	8.14	8.14	13.7%
5030	Communications & Security	Alarm systems, emergency lighting, internet and phone wiring		S.F. Floor	8.04	8.04	
5090	Other Electrical Systems	N/A		—	—	—	
E. EQUIPMENT & FURNISHINGS							
1010	Commercial Equipment	N/A		—	—	—	
1020	Institutional Equipment	Exam room equipment, cabinets and countertops		S.F. Floor	4.38	4.38	3.4 %
1030	Vehicular Equipment	N/A		—	—	—	
1090	Other Equipment	N/A		—	—	—	
F. SPECIAL CONSTRUCTION							
1020	Integrated Construction	N/A		—	—	—	0.0 %
1040	Special Facilities	N/A		—	—	—	
G. BUILDING SITEWORK	**N/A**						

		Sub-Total	130.57	100%
CONTRACTOR FEES (General Requirements: 10%, Overhead: 5%, Profit: 10%)		25%	32.64	
ARCHITECT FEES		9%	14.69	
	Total Building Cost		**177.90**	

For customer support on your Square Foot Cost Data, call 877.756.2789.

163

Costs per square foot of floor area

Exterior Wall	S.F. Area	4000	5500	7000	8500	10000	11500	13000	14500	16000
	L.F. Perimeter	180	210	240	270	286	311	336	361	386
Vinyl Siding	Wood Frame	211.10	201.50	196.00	192.45	188.60	186.50	184.85	183.65	182.60
Stone Veneer	Wood Frame	229.90	217.25	210.00	205.30	200.10	197.30	195.15	193.45	192.10
Fiber Cement Siding	Steel Frame	218.00	207.90	202.05	198.35	194.30	192.05	190.35	189.05	187.95
E.I.F.S.	Steel Frame	209.20	199.30	193.55	189.85	185.85	183.65	182.00	180.75	179.65
Precast Concrete Panel	Reinforced Concrete	255.10	240.65	232.30	227.00	220.90	217.65	215.20	213.25	211.70
Brick Veneer	Reinforced Concrete	216.60	203.75	196.40	191.70	186.30	183.45	181.25	179.60	178.20
Perimeter Adj., Add or Deduct	Per 100 L.F.	24.50	17.85	14.00	11.55	9.80	8.60	7.65	6.70	6.15
Story Hgt. Adj., Add or Deduct	Per 1 Ft.	2.75	2.35	2.10	1.95	1.70	1.70	1.60	1.50	1.50

For Basement, add $36.00 per square foot of basement area

The above costs were calculated using the basic specifications shown on the facing page. These costs should be adjusted where necessary for design alternatives and owner's requirements. Reported completed project costs, for this type of structure, range from $93.80 to $320.35 per S.F.

Common additives

Description	Unit	$ Cost
Cabinets, Hospital, base		
Laminated plastic	L.F.	420
Stainless steel	L.F.	675
Counter top, laminated plastic	L.F.	81.50
Stainless steel	L.F.	199
For drop-in sink, add	Ea.	1100
Nurses station, door type		
Laminated plastic	L.F.	470
Enameled steel	L.F.	455
Stainless steel	L.F.	790
Wall cabinets, laminated plastic	L.F.	305
Enameled steel	L.F.	360
Stainless steel	L.F.	635
Elevators, Hydraulic passenger, 2 stops		
1500# capacity	Ea.	67,400
2500# capacity	Ea.	70,400
3500# capacity	Ea.	75,400

Description	Unit	$ Cost
Directory Boards, Plastic, glass covered		
30" x 20"	Ea.	610
36" x 48"	Ea.	1525
Aluminum, 24" x 18"	Ea.	640
36" x 24"	Ea.	765
48" x 32"	Ea.	1075
48" x 60"	Ea.	2275
Emergency Lighting, 25 watt, battery operated		
Lead battery	Ea.	345
Nickel cadmium	Ea.	685
Heat Therapy Unit		
Humidified, 26" x 78" x 28"	Ea.	3450
Smoke Detectors		
Ceiling type	Ea.	227
Duct type	Ea.	565
Tables, Examining, vinyl top		
with base cabinets	Ea.	1650 - 6775
Utensil Washer, Sanitizer	Ea.	9650
X-Ray, Mobile	Ea.	18,500 - 85,500

Important: See the Reference Section for Location Factors.

Model costs calculated for a 2 story building with 10' story height and 7,000 square feet of floor area

Medical Office, 2 Story

				Unit	Unit Cost	Cost Per S.F.	% Of Sub-Total
A.	**SUBSTRUCTURE**						
1010	Standard Foundations	Poured concrete; strip and spread footings		S.F. Ground	5.32	2.66	
1020	Special Foundations	N/A		—	—	—	
1030	Slab on Grade	4" reinforced slab on grade		S.F. Slab	5.56	2.79	5.8%
2010	Basement Excavation	Site preparation for slab and trench for foundation wall and footing		S.F. Ground	.57	.29	
2020	Basement Walls	4' Foundation wall		L.F. Wall	86	2.93	
B.	**SHELL**						
	B10 Superstructure						
1010	Floor Construction	Open web bar joist, slab form, fireproofed columns		S.F. Floor	15.84	7.92	8.8%
1020	Roof Construction	Metal deck, open web steel joists, beams, columns		S.F. Roof	10.40	5.20	
	B20 Exterior Enclosure						
2010	Exterior Walls	Fiber cement siding on metal studs	70% of wall	S.F. Wall	16.46	7.90	
2020	Exterior Windows	Aluminum sliding windows	30% of wall	Each	525	7.20	12.4%
2030	Exterior Doors	Aluminum & glass entrance doors		Each	5875	3.35	
	B30 Roofing						
3010	Roof Coverings	Single ply membrane, stone ballast, rigid insulation		S.F. Roof	7.58	3.79	3.1%
3020	Roof Openings	Roof and smoke hatches		S.F. Roof	1.76	.88	
C.	**INTERIORS**						
1010	Partitions	Fire rated gypsum board on metal studs	6 S.F. Floor/L.F. Partition	S.F. Partition	5.77	7.69	
1020	Interior Doors	Single leaf solid core wood doors	240 S.F. Floor/Door	Each	626	2.95	
1030	Fittings	N/A		—	—	—	
2010	Stair Construction	Concrete filled metal pan, with rails		Flight	14,450	10.32	23.5%
3010	Wall Finishes	Paint on drywall		S.F. Surface	1.03	2.74	
3020	Floor Finishes	50% carpet , 50% vinyl composition tile		S.F. Floor	4.61	4.61	
3030	Ceiling Finishes	Acoustic ceiling tiles on suspended support system		S.F. Ceiling	6.56	6.56	
D.	**SERVICES**						
	D10 Conveying						
1010	Elevators & Lifts	One hydraulic elevator		Each	110,320	15.76	10.6%
1020	Escalators & Moving Walks	N/A		—	—	—	
	D20 Plumbing						
2010	Plumbing Fixtures	Restroom, exam room and service fixtures, supply and drainage	1 Fixture/350 S.F. Floor	Each	2065	5.90	
2020	Domestic Water Distribution	Gas fired water heater		S.F. Floor	2.07	2.07	6.1%
2040	Rain Water Drainage	Roof drain		S.F. Roof	2.26	1.13	
	D30 HVAC						
3010	Energy Supply	N/A		—	—	—	
3020	Heat Generating Systems	Included in D3050		—	—	—	
3030	Cooling Generating Systems	N/A		—	—	—	10.7 %
3050	Terminal & Package Units	Multizone unit, gas heating , electric cooling		S.F. Floor	15.90	15.90	
3090	Other HVAC Sys. & Equipment	N/A		—	—	—	
	D40 Fire Protection						
4010	Sprinklers	Wet pipe spinkler system		S.F. Floor	3.95	3.95	4.0%
4020	Standpipes	Standpipes and hose systems, with pumps		S.F. Floor	1.98	1.98	
	D50 Electrical						
5010	Electrical Service/Distribution	400 Ampere service, panel boards and feeders		S.F. Floor	2.43	2.43	
5020	Lighting & Branch Wiring	Fluorescent fixtures, receptacles, switches, A.C. and misc. power		S.F. Floor	8.14	8.14	12.5%
5030	Communications & Security	Alarm systems, emergency lighting, internet and phone wiring		S.F. Floor	8.04	8.04	
5090	Other Electrical Systems	N/A		—	—	—	
E.	**EQUIPMENT & FURNISHINGS**						
1010	Commercial Equipment	N/A		—	—	—	
1020	Institutional Equipment	Exam room equipment, cabinets and countertops		S.F. Floor	3.22	3.22	2.2 %
1030	Vehicular Equipment	N/A		—	—	—	
1090	Other Equipment	N/A		—	—	—	
F.	**SPECIAL CONSTRUCTION**						
1020	Integrated Construction	N/A		—	—	—	0.0 %
1040	Special Facilities	N/A		—	—	—	
G.	**BUILDING SITEWORK**	**N/A**					

			Sub-Total	148.30	100%
CONTRACTOR FEES (General Requirements: 10%, Overhead: 5%, Profit: 10%)		25%	37.07		
ARCHITECT FEES		9%	16.68		

Total Building Cost	**202.05**

For customer support on your Square Foot Cost Data, call 877.756.2789.

165

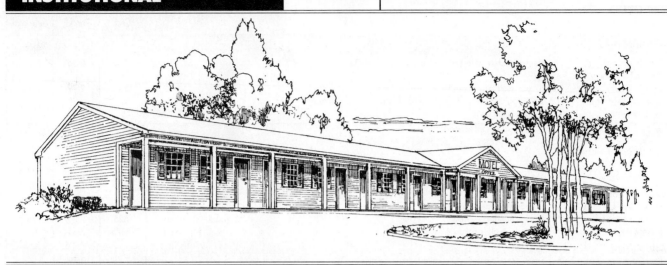

Costs per square foot of floor area

Exterior Wall	S.F. Area	2000	3000	4000	6000	8000	10000	12000	14000	16000
	L.F. Perimeter	240	260	280	380	480	560	580	660	740
Brick Veneer	Wood Frame	199.30	174.90	162.70	156.10	152.80	149.65	144.85	143.70	142.90
Aluminum Siding	Wood Frame	181.55	162.10	152.35	146.75	143.90	141.40	137.65	136.75	136.05
Wood Siding	Wood Frame	181.25	161.90	152.20	146.60	143.75	141.25	137.60	136.65	135.90
Wood Shingles	Wood Frame	184.75	164.35	154.15	148.35	145.45	142.80	138.85	137.95	137.20
Precast Concrete Block	Wood Truss	180.40	161.25	151.70	146.15	143.30	140.80	137.20	136.30	135.55
Brick on Concrete Block	Wood Truss	208.60	181.65	168.15	161.05	157.45	154.00	148.55	147.40	146.45
Perimeter Adj., Add or Deduct	Per 100 L.F.	27.85	18.60	13.95	9.30	6.95	5.60	4.60	4.05	3.55
Story Hgt. Adj., Add or Deduct	Per 1 Ft.	5.10	3.65	3.00	2.70	2.55	2.40	2.05	2.05	1.95

For Basement, add $ 24.75 per square foot of basement area

The above costs were calculated using the basic specifications shown on the facing page. These costs should be adjusted where necessary for design alternatives and owner's requirements. Reported completed project costs, for this type of structure, range from $75.70 to $358.35 per S.F.

Common additives

Description	Unit	$ Cost
Closed Circuit Surveillance, One station		
Camera and monitor	Ea.	2025
For additional camera stations, add	Ea.	1100
Emergency Lighting, 25 watt, battery operated		
Lead battery	Ea.	345
Nickel cadmium	Ea.	685
Laundry Equipment		
Dryer, gas, 16 lb. capacity	Ea.	995
30 lb. capacity	Ea.	3950
Washer, 4 cycle	Ea.	1225
Commercial	Ea.	1625
Sauna, Prefabricated, complete		
6' x 4'	Ea.	6750
6' x 6'	Ea.	8075
6' x 9'	Ea.	10,500
8' x 8'	Ea.	12,000
8' x 10'	Ea.	13,400
10' x 12'	Ea.	16,100
Smoke Detectors		
Ceiling type	Ea.	227
Duct type	Ea.	565

Description	Unit	$ Cost
Swimming Pools, Complete, gunite	S.F.	104 - 129
TV Antenna, Master system, 12 outlet	Outlet	263
30 outlet	Outlet	225
100 outlet	Outlet	213

Important: See the Reference Section for Location Factors.

Model costs calculated for a 1 story building with 9' story height and 8,000 square feet of floor area

Motel, 1 Story

				Unit	Unit Cost	Cost Per S.F.	% Of Sub-Total
A. SUBSTRUCTURE							
1010	Standard Foundations	Poured concrete; strip and spread footings		S.F. Ground	2.47	2.47	
1020	Special Foundations	N/A		—	—	—	
1030	Slab on Grade	4" reinforced concrete with vapor barrier and granular base		S.F. Slab	5.56	5.56	11.5%
2010	Basement Excavation	Site preparation for slab and trench for foundation wall and footing		S.F. Ground	.32	.32	
2020	Basement Walls	4' foundation wall		L.F. Wall	81	4.83	
B. SHELL							
	B10 Superstructure						
1010	Floor Construction	N/A		—	—	—	6.5 %
1020	Roof Construction	Plywood on wood trusses		S.F. Roof	7.42	7.42	
	B20 Exterior Enclosure						
2010	Exterior Walls	Face brick on wood studs with sheathing, insulation and paper	80% of wall	S.F. Wall	25.35	10.95	
2020	Exterior Windows	Wood double hung	20% of wall	Each	521	4.02	16.6%
2030	Exterior Doors	Wood solid core		Each	1475	4.05	
	B30 Roofing						
3010	Roof Coverings	Asphalt shingles with flashing (pitched); rigid fiberglass insulation		S.F. Roof	4.60	4.60	4.0%
3020	Roof Openings	N/A		—	—	—	
C. INTERIORS							
1010	Partitions	Gypsum bd. and sound deadening bd. on wood studs	9 S.F. Floor/L.F. Partition	S.F. Partition	10.51	9.34	
1020	Interior Doors	Single leaf hollow core wood	300 S.F. Floor/Door	Each	542	1.81	
1030	Fittings	N/A		—	—	—	
2010	Stair Construction	N/A		—	—	—	22.3%
3010	Wall Finishes	90% paint, 10% ceramic tile		S.F. Surface	1.54	2.73	
3020	Floor Finishes	85% carpet, 15% ceramic tile		S.F. Floor	6.96	6.96	
3030	Ceiling Finishes	Painted gypsum board on furring		S.F. Ceiling	4.66	4.66	
D. SERVICES							
	D10 Conveying						
1010	Elevators & Lifts	N/A		—	—	—	0.0 %
1020	Escalators & Moving Walks	N/A		—	—	—	
	D20 Plumbing						
2010	Plumbing Fixtures	Toilet and service fixtures, supply and drainage	1 Fixture/90 S.F. Floor	Each	1817	20.19	
2020	Domestic Water Distribution	Gas fired water heater		S.F. Floor	1.94	1.94	19.4%
2040	Rain Water Drainage	N/A		—	—	—	
	D30 HVAC						
3010	Energy Supply	N/A		—	—	—	
3020	Heat Generating Systems	Included in D3050		—	—	—	
3030	Cooling Generating Systems	N/A		—	—	—	3.4 %
3050	Terminal & Package Units	Through the wall electric heating and cooling units		S.F. Floor	3.84	3.84	
3090	Other HVAC Sys. & Equipment	N/A		—	—	—	
	D40 Fire Protection						
4010	Sprinklers	Wet pipe sprinkler system		S.F. Floor	5.02	5.02	5.4%
4020	Standpipes	Standpipe, wet, Class III		S.F. Floor	1.15	1.15	
	D50 Electrical						
5010	Electrical Service/Distribution	200 ampere service, panel board and feeders		S.F. Floor	1.32	1.32	
5020	Lighting & Branch Wiring	Fluorescent fixtures, receptacles, switches and misc. power		S.F. Floor	7.29	7.29	10.0%
5030	Communications & Security	Addressable alarm systems		S.F. Floor	2.78	2.78	
5090	Other Electrical Systems	N/A		—	—	—	
E. EQUIPMENT & FURNISHINGS							
1010	Commercial Equipment	Laundry equipment		S.F. Floor	.99	.99	
1020	Institutional Equipment	N/A		—	—	—	
1030	Vehicular Equipment	N/A		—	—	—	0.9%
1090	Other Equipment	N/A		—	—	—	
F. SPECIAL CONSTRUCTION							
1020	Integrated Construction	N/A		—	—	—	0.0 %
1040	Special Facilities	N/A		—	—	—	
G. BUILDING SITEWORK	**N/A**						

		Sub-Total	114.24	100%
CONTRACTOR FEES (General Requirements: 10%, Overhead: 5%, Profit: 10%)		25%	28.56	
ARCHITECT FEES		7%	10	
	Total Building Cost		**152.80**	

For customer support on your Square Foot Cost Data, call 877.756.2789.

167

Costs per square foot of floor area

Exterior Wall	S.F. Area	25000	37000	49000	61000	73000	81000	88000	96000	104000
	L.F. Perimeter	433	593	606	720	835	911	978	1054	1074
Decorative Concrete Block	Wood Joists	162.40	159.55	154.75	153.65	152.95	152.65	152.40	152.10	151.35
	Precast Conc.	179.75	175.70	170.10	168.65	167.70	167.25	166.90	166.50	165.60
Stucco on Concrete Block	Wood Joists	167.70	163.50	157.45	155.85	154.85	154.40	154.00	153.65	152.65
	Precast Conc.	184.10	179.35	172.80	171.05	169.95	169.40	169.00	168.55	167.50
Wood Siding	Wood Frame	159.35	156.75	152.55	151.60	150.90	150.65	150.40	150.15	149.50
Brick Veneer	Wood Frame	167.60	163.45	157.35	155.80	154.75	154.30	153.90	153.55	152.60
Perimeter Adj., Add or Deduct	Per 100 L.F.	4.45	3.00	2.20	1.85	1.55	1.40	1.20	1.20	1.05
Story Hgt. Adj., Add or Deduct	Per 1 Ft.	1.75	1.65	1.30	1.20	1.20	1.15	1.05	1.15	1.00
For Basement, add $32.50 per square foot of basement area										

The above costs were calculated using the basic specifications shown on the facing page. These costs should be adjusted where necessary for design alternatives and owner's requirements. Reported completed project costs, for this type of structure, range from $69.70 to $358.35 per S.F.

Common additives

Description	Unit	$ Cost		Description	Unit	$ Cost
Closed Circuit Surveillance, One station				Sauna, Prefabricated, complete		
Camera and monitor	Ea.	2025		6' x 4'	Ea.	6750
For additional camera station, add	Ea.	1100		6' x 6'	Ea.	8075
Elevators, Hydraulic passenger, 2 stops				6' x 9'	Ea.	10,500
1500# capacity	Ea.	67,400		8' x 8'	Ea.	12,000
2500# capacity	Ea.	70,400		8' x 10'	Ea.	13,400
3500# capacity	Ea.	75,400		10' x 12'	Ea.	16,100
Additional stop, add	Ea.	8075		Smoke Detectors		
Emergency Lighting, 25 watt, battery operated				Ceiling type	Ea.	227
Lead battery	Ea.	345		Duct type	Ea.	565
Nickel cadmium	Ea.	685		Swimming Pools, Complete, gunite	S.F.	104 - 129
Laundry Equipment				TV Antenna, Master system, 12 outlet	Outlet	263
Dryer, gas, 16 lb. capacity	Ea.	995		30 outlet	Outlet	225
30 lb. capacity	Ea.	3950		100 outlet	Outlet	213
Washer, 4 cycle	Ea.	1225				
Commercial	Ea.	1625				

Model costs calculated for a 3 story building with 9' story height and 49,000 square feet of floor area

			Unit	Unit Cost	Cost Per S.F.	% Of Sub-Total
A. SUBSTRUCTURE						
1010	Standard Foundations	Poured concrete; strip and spread footings	S.F. Ground	2.67	.89	
1020	Special Foundations	N/A	—	—	—	
1030	Slab on Grade	4" reinforced concrete with vapor barrier and granular base	S.F. Slab	5.56	1.85	3.0%
2010	Basement Excavation	Site preparation for slab and trench for foundation wall and footing	S.F. Ground	.32	.11	
2020	Basement Walls	4' foundation wall	L.F. Wall	86	1.06	
B. SHELL						
	B10 Superstructure					
1010	Floor Construction	Precast concrete plank	S.F. Floor	11.04	7.36	8.5%
1020	Roof Construction	Precast concrete plank	S.F. Roof	10.56	3.52	
	B20 Exterior Enclosure					
2010	Exterior Walls	Decorative concrete block 85% of wall	S.F. Wall	18.43	5.23	
2020	Exterior Windows	Aluminum sliding 15% of wall	Each	585	1.95	10.8%
2030	Exterior Doors	Aluminum and glass doors and entrance with transom	Each	2031	6.72	
	B30 Roofing					
3010	Roof Coverings	Built-up tar and gravel with flashing; perlite/EPS composite insulation	S.F. Roof	6.63	2.21	1.8%
3020	Roof Openings	Roof hatches	S.F. Roof	.12	.04	
C. INTERIORS						
1010	Partitions	Concrete block 7 S.F. Floor/L.F. Partition	S.F. Partition	20.38	23.29	
1020	Interior Doors	Wood hollow core 70 S.F. Floor/Door	Each	542	7.74	
1030	Fittings	N/A	—	—	—	
2010	Stair Construction	Concrete filled metal pan	Flight	14,450	3.54	38.7%
3010	Wall Finishes	90% paint, 10% ceramic tile	S.F. Surface	1.54	3.51	
3020	Floor Finishes	85% carpet, 5% vinyl composition tile, 10% ceramic tile	S.F. Floor	6.96	6.96	
3030	Ceiling Finishes	Textured finish	S.F. Ceiling	4.66	4.66	
D. SERVICES						
	D10 Conveying					
1010	Elevators & Lifts	Two hydraulic passenger elevators	Each	110,985	4.53	3.5%
1020	Escalators & Moving Walks	N/A	—	—	—	
	D20 Plumbing					
2010	Plumbing Fixtures	Toilet and service fixtures, supply and drainage 1 Fixture/180 S.F. Floor	Each	4108	22.82	
2020	Domestic Water Distribution	Gas fired water heater	S.F. Floor	1.26	1.26	19.1%
2040	Rain Water Drainage	Roof drains	S.F. Roof	1.50	.50	
	D30 HVAC					
3010	Energy Supply	N/A	—	—	—	
3020	Heat Generating Systems	Included in D3050	—	—	—	
3030	Cooling Generating Systems	N/A	—	—	—	2.7 %
3050	Terminal & Package Units	Through the wall electric heating and cooling units	S.F. Floor	3.46	3.46	
3090	Other HVAC Sys. & Equipment	N/A	—	—	—	
	D40 Fire Protection					
4010	Sprinklers	Sprinklers, wet, light hazard	S.F. Floor	2.97	2.97	2.5%
4020	Standpipes	Standpipe, wet, Class III	S.F. Floor	.26	.26	
	D50 Electrical					
5010	Electrical Service/Distribution	800 ampere service, panel board and feeders	S.F. Floor	1.57	1.57	
5020	Lighting & Branch Wiring	Fluorescent fixtures, receptacles, switches and misc. power	S.F. Floor	8.01	8.01	9.1%
5030	Communications & Security	Addressable alarm systems and emergency lighting	S.F. Floor	1.92	1.92	
5090	Other Electrical Systems	Emergency generator, 7.5 kW	S.F. Floor	.13	.13	
E. EQUIPMENT & FURNISHINGS						
1010	Commercial Equipment	Commercial laundry equipment	S.F. Floor	.32	.32	
1020	Institutional Equipment	N/A	—	—	—	
1030	Vehicular Equipment	N/A	—	—	—	0.2%
1090	Other Equipment	N/A	—	—	—	
F. SPECIAL CONSTRUCTION						
1020	Integrated Construction	N/A	—	—	—	0.0 %
1040	Special Facilities	N/A	—	—	—	
G. BUILDING SITEWORK	**N/A**					

			Sub-Total	128.39	**100%**
CONTRACTOR FEES (General Requirements: 10%, Overhead: 5%, Profit: 10%)			25%	32.08	
ARCHITECT FEES			6%	9.63	
		Total Building Cost		**170.10**	

For customer support on your Square Foot Cost Data, call 877.756.2789.

169

Costs per square foot of floor area

Exterior Wall	S.F. Area	9000	10000	12000	13000	14000	15000	16000	18000	20000
	L.F. Perimeter	385	410	440	460	480	500	510	547	583
Decorative Concrete Block	Steel Joists	190.20	185.25	176.20	173.20	170.50	168.20	165.55	162.10	159.35
Painted Concrete Block	Steel Joists	183.50	178.85	170.55	167.65	165.15	163.00	160.60	157.35	154.80
Face Brick on Conc. Block	Steel Joists	201.80	196.35	186.15	182.75	179.80	177.25	174.15	170.35	167.25
Precast Concrete Panels	Steel Joists	215.45	209.55	197.90	194.10	190.80	187.90	184.40	180.05	176.60
Tilt-up Panels	Steel Joists	187.50	182.75	174.00	170.95	168.35	166.10	163.60	160.20	157.50
Metal Sandwich Panels	Steel Joists	186.30	181.60	172.90	169.95	167.40	165.15	162.65	159.35	156.70
Perimeter Adj., Add or Deduct	Per 100 L.F.	11.60	10.45	8.75	8.05	7.45	7.00	6.55	5.80	5.20
Story Hgt. Adj., Add or Deduct	Per 1 Ft.	1.75	1.75	1.55	1.40	1.35	1.35	1.30	1.25	1.20
Basement—Not Applicable										

The above costs were calculated using the basic specifications shown on the facing page. These costs should be adjusted where necessary for design alternatives and owner's requirements. Reported completed project costs, for this type of structure, range from $89.20 to $233.25 per S.F.

Common additives

Description	Unit	$ Cost
Emergency Lighting, 25 watt, battery operated		
Lead battery	Ea.	345
Nickel cadmium	Ea.	685
Seating		
Auditorium chair, all veneer	Ea.	315
Veneer back, padded seat	Ea.	325
Upholstered, spring seat	Ea.	320
Classroom, movable chair & desk	Set	81 - 171
Lecture hall, pedestal type	Ea.	275 - 620
Smoke Detectors		
Ceiling type	Ea.	227
Duct type	Ea.	565
Sound System		
Amplifier, 250 watts	Ea.	2200
Speaker, ceiling or wall	Ea.	217
Trumpet	Ea.	415

Important: See the Reference Section for Location Factors.

Model costs calculated for a 1 story building with 20' story height and 12,000 square feet of floor area

				Unit	Unit Cost	Cost Per S.F.	% Of Sub-Total
A.	**SUBSTRUCTURE**						
1010	Standard Foundations	Poured concrete; strip and spread footings		S.F. Ground	1.57	1.57	
1020	Special Foundations	N/A		—	—	—	
1030	Slab on Grade	4" reinforced concrete with vapor barrier and granular base		S.F. Slab	5.56	5.56	8.0%
2010	Basement Excavation	Site preparation for slab and trench for foundation wall and footing		S.F. Ground	.32	.32	
2020	Basement Walls	4' foundation wall		L.F. Wall	86	3.14	
B.	**SHELL**						
	B10 Superstructure						
1010	Floor Construction	Open web steel joists, slab form, concrete	mezzanine 2250 S.F.	S.F. Floor	11.82	2.21	9.9%
1020	Roof Construction	Metal deck on open web steel joists		S.F. Roof	10.79	10.79	
	B20 Exterior Enclosure						
2010	Exterior Walls	Decorative concrete block	80% of wall	S.F. Wall	16.60	12.17	
2020	Exterior Windows	Window wall	20% of wall	Each	58	8.48	16.7%
2030	Exterior Doors	Sliding mallfront aluminum and glass and hollow metal		Each	2592	1.30	
	B30 Roofing						
3010	Roof Coverings	Built-up tar and gravel with flashing; perlite/EPS composite insulation		S.F. Roof	6.60	6.60	5.0%
3020	Roof Openings	N/A		—	—	—	
C.	**INTERIORS**						
1010	Partitions	Concrete block	40 S.F. Floor/L.F. Partition	S.F. Partition	9.73	3.89	
1020	Interior Doors	Single leaf hollow metal	705 S.F. Floor/Door	Each	1168	1.66	
1030	Fittings	Toilet partitions		S.F. Floor	.66	.66	
2010	Stair Construction	Concrete filled metal pan		Flight	17,275	2.88	19.9%
3010	Wall Finishes	Paint		S.F. Surface	4.29	3.43	
3020	Floor Finishes	Carpet 50%, ceramic tile 5%	50% of area	S.F. Floor	12.88	6.44	
3030	Ceiling Finishes	Mineral fiber tile on concealed zee runners suspended		S.F. Ceiling	7.22	7.22	
D.	**SERVICES**						
	D10 Conveying						
1010	Elevators & Lifts	N/A		—	—	—	0.0 %
1020	Escalators & Moving Walks	N/A		—	—	—	
	D20 Plumbing						
2010	Plumbing Fixtures	Toilet and service fixtures, supply and drainage	1 Fixture/500 S.F. Floor	Each	4960	9.92	
2020	Domestic Water Distribution	Gas fired water heater		S.F. Floor	.56	.56	9.0%
2040	Rain Water Drainage	Roof drains		S.F. Roof	1.36	1.36	
	D30 HVAC						
3010	Energy Supply	N/A		—	—	—	
3020	Heat Generating Systems	Included in D3050		—	—	—	
3030	Cooling Generating Systems	N/A		—	—	—	8.5 %
3050	Terminal & Package Units	Single zone rooftop unit, gas heating, electric cooling		S.F. Floor	11.20	11.20	
3090	Other HVAC Sys. & Equipment	N/A		—	—	—	
	D40 Fire Protection						
4010	Sprinklers	Wet pipe sprinkler system		S.F. Floor	4.36	4.36	4.0%
4020	Standpipes	Standpipe		S.F. Floor	.94	.94	
	D50 Electrical						
5010	Electrical Service/Distribution	400 ampere service, panel board and feeders		S.F. Floor	1.76	1.76	
5020	Lighting & Branch Wiring	High efficiency fluorescent fixtures, receptacles, switches, A.C. and misc. power		S.F. Floor	6.99	6.99	8.6%
5030	Communications & Security	Addressable alarm systems, sound system and emergency lighting		S.F. Floor	2.36	2.36	
5090	Other Electrical Systems	Emergency generator, 7.5 kW		S.F. Floor	.18	.18	
E.	**EQUIPMENT & FURNISHINGS**						
1010	Commercial Equipment	N/A		—	—	—	
1020	Institutional Equipment	Projection equipment, screen		S.F. Floor	9.36	9.36	10.5 %
1030	Vehicular Equipment	N/A		—	—	—	
2010	Fixed Furnishings	Movie theater seating		S.F. Floor	4.45	4.45	
F.	**SPECIAL CONSTRUCTION**						
1020	Integrated Construction	N/A		—	—	—	0.0 %
1040	Special Facilities	N/A		—	—	—	
G.	**BUILDING SITEWORK**	**N/A**					

			Sub-Total	131.76	**100%**
CONTRACTOR FEES (General Requirements: 10%, Overhead: 5%, Profit: 10%)			25%	32.91	
ARCHITECT FEES			7%	11.53	
		Total Building Cost		**176.20**	

For customer support on your Square Foot Cost Data, call 877.756.2789.

171

Costs per square foot of floor area

Exterior Wall	S.F. Area	10000	15000	20000	25000	30000	35000	40000	45000	50000
	L.F. Perimeter	286	370	453	457	513	568	624	680	735
Precast Concrete Panels	Bearing Walls	223.90	214.95	210.35	202.45	200.05	198.30	197.00	195.95	195.15
	Steel Frame	229.85	220.90	216.35	208.50	206.05	204.35	203.15	202.05	201.25
Face Brick with Concrete Block Back-up	Bearing Walls	214.60	206.90	202.95	196.50	194.45	193.00	191.95	191.00	190.30
	Steel Joists	221.40	213.70	209.80	203.40	201.35	199.90	198.90	198.00	197.25
Stucco on Concrete Block	Bearing Walls	209.60	202.60	199.00	193.25	191.50	190.15	189.20	188.35	187.70
	Steel Joists	216.35	209.40	205.85	200.20	198.40	197.10	196.15	195.30	194.70
Perimeter Adj., Add or Deduct	Per 100 L.F.	16.15	10.70	8.10	6.50	5.40	4.55	4.00	3.60	3.20
Story Hgt. Adj., Add or Deduct	Per 1 Ft.	3.95	3.40	3.15	2.50	2.35	2.25	2.20	2.05	2.00
For Basement, add $33.70 per square foot of basement area										

The above costs were calculated using the basic specifications shown on the facing page. These costs should be adjusted where necessary for design alternatives and owner's requirements. Reported completed project costs, for this type of structure, range from $95.75 to $240.00 per S.F.

Common additives

Description	Unit	$ Cost		Description	Unit	$ Cost
Beds, Manual	Ea.	865 - 2700		Kitchen Equipment, cont.		
Elevators, Hydraulic passenger, 2 stops				Ice cube maker, 50 lb. per day	Ea.	2000
1500# capacity	Ea.	67,400		Range with 1 oven	Ea.	3300
2500# capacity	Ea.	70,400		Laundry Equipment		
3500# capacity	Ea.	75,400		Dryer, gas, 16 lb. capacity	Ea.	995
Emergency Lighting, 25 watt, battery operated				30 lb. capacity	Ea.	3950
Lead battery	Ea.	345		Washer, 4 cycle	Ea.	1225
Nickel cadmium	Ea.	685		Commercial	Ea.	1625
Intercom System, 25 station capacity				Nurses Call System		
Master station	Ea.	3175		Single bedside call station	Ea.	292
Intercom outlets	Ea.	192		Pillow speaker	Ea.	287
Handset	Ea.	525		Refrigerator, Prefabricated, walk-in		
Kitchen Equipment				7'-6" high, 6' x 6'	S.F.	203
Broiler	Ea.	4075		10' x 10'	S.F.	159
Coffee urn, twin 6 gallon	Ea.	3025		12' x 14'	S.F.	142
Cooler, 6 ft. long	Ea.	3925		12' x 20'	S.F.	125
Dishwasher, 10-12 racks per hr.	Ea.	4275		TV Antenna, Master system, 12 outlet	Outlet	263
Food warmer	Ea.	770		30 outlet	Outlet	225
Freezer, 44 C.F., reach-in	Ea.	6175		100 outlet	Outlet	213
				Whirlpool Bath, Mobile, 18" x 24" x 60"	Ea.	6350
				X-Ray, Mobile	Ea.	18,500 - 85,500

Important: See the Reference Section for Location Factors.

Model costs calculated for a 2 story building with 10' story height and 25,000 square feet of floor area

Nursing Home

			Unit	Unit Cost	Cost Per S.F.	% Of Sub-Total
A. SUBSTRUCTURE						
1010	Standard Foundations	Poured concrete; strip and spread footings	S.F. Ground	1.88	.94	
1020	Special Foundations	N/A	—	—	—	
1030	Slab on Grade	4" reinforced concrete with vapor barrier and granular base	S.F. Slab	5.56	2.79	3.6%
2010	Basement Excavation	Site preparation for slab and trench for foundation wall and footing	S.F. Ground	.32	.16	
2020	Basement Walls	4' foundation wall	L.F. Wall	76	1.38	
B. SHELL						
	B10 Superstructure					
1010	Floor Construction	Pre-cast double tees with concrete topping	S.F. Floor	13.34	6.67	8.4%
1020	Roof Construction	Pre-cast double tees	S.F. Roof	11.26	5.63	
	B20 Exterior Enclosure					
2010	Exterior Walls	Precast concrete panels	S.F. Wall	46.76	14.53	
2020	Exterior Windows	Wood double hung *15% of wall*	Each	585	2.13	11.7%
2030	Exterior Doors	Double aluminum & glass doors, single leaf hollow metal	Each	2355	.47	
	B30 Roofing					
3010	Roof Coverings	Built-up tar and gravel with flashing; perlite/EPS composite insulation	S.F. Roof	6.62	3.31	2.3%
3020	Roof Openings	Roof hatches	S.F. Roof	.08	.04	
C. INTERIORS						
1010	Partitions	Gypsum board on metal studs *8 S.F. Floor/L.F. Partition*	S.F. Partition	9.44	9.44	
1020	Interior Doors	Single leaf wood *80 S.F. Floor/Door*	Each	626	7.82	
1030	Fittings	N/A	—	—	—	
2010	Stair Construction	Concrete stair	Flight	6950	1.11	20.2%
3010	Wall Finishes	50% vinyl wall coverings, 45% paint, 5% ceramic tile	S.F. Surface	1.77	3.54	
3020	Floor Finishes	95% vinyl tile, 5% ceramic tile	S.F. Floor	3.20	3.20	
3030	Ceiling Finishes	Painted gypsum board	S.F. Ceiling	4.37	4.37	
D. SERVICES						
	D10 Conveying					
1010	Elevators & Lifts	One hydraulic hospital elevator	Each	110,250	4.41	3.0%
1020	Escalators & Moving Walks	N/A	—	—	—	
	D20 Plumbing					
2010	Plumbing Fixtures	Kitchen, toilet and service fixtures, supply and drainage *1 Fixture/230 S.F. Floor*	Each	8692	37.79	
2020	Domestic Water Distribution	Oil fired water heater	S.F. Floor	1.33	1.33	27.2%
2040	Rain Water Drainage	Roof drains	S.F. Roof	1.22	.61	
	D30 HVAC					
3010	Energy Supply	Oil fired hot water, wall fin radiation	S.F. Floor	8.52	8.52	
3020	Heat Generating Systems	N/A	—	—	—	
3030	Cooling Generating Systems	N/A	—	—	—	11.3%
3050	Terminal & Package Units	Split systems with air cooled condensing units	S.F. Floor	7.93	7.93	
3090	Other HVAC Sys. & Equipment	N/A	—	—	—	
	D40 Fire Protection					
4010	Sprinklers	Sprinkler, light hazard	S.F. Floor	3.15	3.15	2.5%
4020	Standpipes	Standpipe	S.F. Floor	.49	.49	
	D50 Electrical					
5010	Electrical Service/Distribution	800 ampere service, panel board and feeders	S.F. Floor	1.63	1.63	
5020	Lighting & Branch Wiring	High efficiency fluorescent fixtures, receptacles, switches, A.C. and misc. power	S.F. Floor	10.53	10.53	9.7%
5030	Communications & Security	Addressable alarm systems and emergency lighting	S.F. Floor	1.43	1.43	
5090	Other Electrical Systems	Emergency generator, 15 kW	S.F. Floor	.56	.56	
E. EQUIPMENT & FURNISHINGS						
1010	Commercial Equipment	N/A	—	—	—	
1020	Institutional Equipment	N/A	—	—	—	
1030	Vehicular Equipment	N/A	—	—	—	0.0 %
1090	Other Equipment	N/A	—	—	—	
F. SPECIAL CONSTRUCTION						
1020	Integrated Construction	N/A	—	—	—	0.0 %
1040	Special Facilities	N/A	—	—	—	
G. BUILDING SITEWORK	**N/A**					

				Sub-Total	145.91	100%
	CONTRACTOR FEES (General Requirements: 10%, Overhead: 5%, Profit: 10%)			25%	36.48	
	ARCHITECT FEES			11%	20.06	

Total Building Cost	**202.45**

For customer support on your Square Foot Cost Data, call 877.756.2789.

173

Costs per square foot of floor area

Exterior Wall	S.F. Area	2000	3000	5000	7000	9000	12000	15000	20000	25000
	L.F. Perimeter	220	260	320	360	420	480	520	640	700
Vinyl Siding	Wood Frame	213.25	191.20	171.25	161.05	156.70	151.45	147.55	145.15	142.30
Stone Veneer	Wood Frame	252.60	221.65	193.50	178.90	172.75	165.15	159.45	156.10	151.95
Fiber Cement Siding	Steel Frame	174.05	149.60	127.60	116.45	111.55	105.85	101.55	98.85	95.85
E.I.F.S.	Steel Frame	227.15	203.25	181.80	170.85	166.15	160.55	156.35	153.75	150.80
Precast Concrete Panel	Reinforced Concrete	282.25	248.10	216.55	199.95	193.05	184.40	177.80	174.05	169.20
Brick Veneer	Reinforced Concrete	289.95	253.00	218.95	201.10	193.70	184.45	177.30	173.25	168.10
Perimeter Adj., Add or Deduct	Per 100 L.F.	28.35	18.90	11.40	8.15	6.35	4.75	3.75	2.80	2.35
Story Hgt. Adj., Add or Deduct	Per 1 Ft.	3.00	2.35	1.75	1.40	1.25	1.15	0.95	0.85	0.80

For Basement, add $34.15 per square foot of basement area

The above costs were calculated using the basic specifications shown on the facing page. These costs should be adjusted where necessary for design alternatives and owner's requirements. Reported completed project costs, for this type of structure, range from $73.10 to $282.60 per S.F.

Common additives

Description	Unit	$ Cost	Description	Unit	$ Cost
Closed circuit surveillance, one station			Security access systems		
Camera and monitor	Ea.	2025	Metal detectors, wand type	Ea.	138
For additional camera stations, add	Ea.	1100	Walk-through portal type, single-zone	Ea.	5175
Directory boards, plastic, glass covered			Multi-zone	Ea.	6325
30" x 20"	Ea.	610	X-ray equipment		
36" x 48"	Ea.	1525	Desk top, for mail, small packages	Ea.	4000
Aluminum, 24" x 18"	Ea.	640	Conveyer type, including monitor, minimum	Ea.	17,900
36" x 24"	Ea.	765	Maximum	Ea.	31,100
48" x 32"	Ea.	1075	Explosive detection equipment		
48" x 60"	Ea.	2275	Hand held, battery operated	Ea.	28,100
Electronic, wall mounted	S.F.	2900	Walk-through portal type	Ea.	48,900
Free standing	S.F.	4125	Uninterruptible power supply, 15 kVA/12.75 kW	kW	1890
Pedestal access floor system w/ plastic laminate cover					
Computer room, less than 6000 SF	S.F.	21			
Greater than 6000 SF	S.F.	20.50			
Office, greater than 6000 S.F.	S.F.	16.95			

Important: See the Reference Section for Location Factors.

Model costs calculated for a 1 story building with 12' story height and 7,000 square feet of floor area

				Unit	Unit Cost	Cost Per S.F.	% Of Sub-Total
A.	**SUBSTRUCTURE**						
1010	Standard Foundations	Poured concrete; strip and spread footings		S.F. Ground	3.54	3.54	
1020	Special Foundations	N/A		—	—	—	
1030	Slab on Grade	4" reinforced concrete		S.F. Slab	5.56	5.56	9.2%
2010	Basement Excavation	Site preparation for slab and trench for foundation wall and footing		S.F. Ground	.32	.32	
2020	Basement Walls	4' Foundation wall		L.F. Wall	86	4.40	
B.	**SHELL**						
	B10 Superstructure						
1010	Floor Construction	Cast in place concrete columns		S.F. Floor	2.96	2.96	12.6%
1020	Roof Construction	Concrete beam and slab		S.F. Roof	15.80	15.80	
	B20 Exterior Enclosure						
2010	Exterior Walls	Precast concrete panels, insulated	80% of wall	S.F. Wall	48.35	23.87	
2020	Exterior Windows	Aluminum awning windows	20% of wall	Each	746	4	20.7%
2030	Exterior Doors	Aluminum and glass single and double doors, hollow metal doors		Each	3570	3.06	
	B30 Roofing						
3010	Roof Coverings	Single ply membrane, stone ballast, rigid insulation		S.F. Roof	6.91	6.91	4.8%
3020	Roof Openings	Smoke hatches		S.F. Roof	.32	.32	
C.	**INTERIORS**						
1010	Partitions	Gypsum board on metal studs	20 S.F. Floor/L.F. Partition	S.F. Partition	6.28	3.14	
1020	Interior Doors	70% Single leaf solid wood doors, 30% hollow metal single leaf	320 S.F. Floor/Door	Each	789	2.48	
1030	Fittings	Plastic laminate toilet partitions		S.F. Floor	.38	.38	
2010	Stair Construction	N/A		—	—	—	13.5%
3010	Wall Finishes	90% Paint, 10% ceramic wall tile		S.F. Surface	1.27	1.27	
3020	Floor Finishes	65% Carpet, 10% porcelain tile, 10% quarry tile		S.F. Floor	5.68	5.68	
3030	Ceiling Finishes	Acoustic ceiling tiles on suspended channel grid		S.F. Ceiling	7.22	7.22	
D.	**SERVICES**						
	D10 Conveying						
1010	Elevators & Lifts	N/A		—	—	—	0.0%
1020	Escalators & Moving Walks	N/A		—	—	—	
	D20 Plumbing						
2010	Plumbing Fixtures	Restroom and service fixtures, supply and drainage	1 Fixture/700 S.F. Floor	Each	2261	3.23	
2020	Domestic Water Distribution	Gas fired water heater		S.F. Floor	1.55	1.55	3.8%
2040	Rain Water Drainage	Roof drains		S.F. Roof	.84	.84	
	D30 HVAC						
3010	Energy Supply	N/A		—	—	—	
3020	Heat Generating Systems	Included in D3050		—	—	—	
3030	Cooling Generating Systems	N/A		—	—	—	15.0%
3050	Terminal & Package Units	Multizone gas heating, electric cooling unit		S.F. Floor	22.35	22.35	
3090	Other HVAC Sys. & Equipment	N/A		—	—	—	
	D40 Fire Protection						
4010	Sprinklers	Wet pipe light hazard sprinkler system		S.F. Floor	3.65	3.65	3.5%
4020	Standpipes	Standpipes		S.F. Floor	1.57	1.57	
	D50 Electrical						
5010	Electrical Service/Distribution	400 ampere service, panel boards and feeders		S.F. Floor	5.17	5.17	
5020	Lighting & Branch Wiring	Fluorescent fixtures, receptacles, switches, A.C. and misc. power		S.F. Floor	12.90	12.90	17.0%
5030	Communications & Security	Telephone systems, internet wiring, and addressable alarm systems		S.F. Floor	7.30	7.30	
5090	Other Electrical Systems	Emergency generator, 7.5 kW		S.F. Floor			
E.	**EQUIPMENT & FURNISHINGS**						
1010	Commercial Equipment	N/A		—	—	—	
1020	Institutional Equipment	N/A		—	—	—	0.0%
1030	Vehicular Equipment	N/A		—	—	—	
1090	Other Equipment	N/A		—	—	—	
F.	**SPECIAL CONSTRUCTION**						
1020	Integrated Construction	N/A		—	—	—	0.0%
1040	Special Facilities	N/A		—	—	—	
G.	**BUILDING SITEWORK**	**N/A**					

		Sub-Total	149.47	100%
	CONTRACTOR FEES (General Requirements: 10%, Overhead: 5%, Profit: 10%)	25%	37.40	
	ARCHITECT FEES	7%	13.08	
	Total Building Cost		**199.95**	

For customer support on your Square Foot Cost Data, call 877.756.2789.

175

Costs per square foot of floor area

Exterior Wall	S.F. Area	5000	8000	12000	16000	20000	35000	50000	65000	80000
	L.F. Perimeter	220	260	310	330	360	440	550	600	675
Vinyl Siding	Wood Frame	210.10	187.15	174.05	165.00	160.30	150.90	147.95	145.10	143.70
Stone Veneer	Wood Frame	249.45	216.15	197.00	183.25	176.20	161.95	157.60	153.15	151.05
Fiber Cement Siding	Steel Frame	209.05	185.75	172.40	163.20	158.40	148.80	145.80	142.95	141.50
E.I.F.S.	Steel Frame	211.90	187.45	173.45	163.65	158.55	148.35	145.20	142.05	140.55
Precast Concrete Panel	Reinforced Concrete	274.45	236.35	214.40	198.40	190.25	173.65	168.60	163.45	161.05
Brick Veneer	Reinforced Concrete	264.20	225.60	203.35	187.20	178.95	162.05	157.00	151.75	149.30
Perimeter Adj., Add or Deduct	Per 100 L.F.	26.30	16.45	10.90	8.20	6.55	3.70	2.60	1.95	1.65
Story Hgt. Adj., Add or Deduct	Per 1 Ft.	3.60	2.70	2.05	1.65	1.45	1.00	0.85	0.75	0.70
For Basement, add $38.25 per square foot of basement area										

The above costs were calculated using the basic specifications shown on the facing page. These costs should be adjusted where necessary for design alternatives and owner's requirements. Reported completed project costs, for this type of structure, range from $76.50 to $296.35 per S.F.

Common additives

Description	Unit	$ Cost		Description	Unit	$ Cost
Closed circuit surveillance, one station				Security access systems		
Camera and monitor	Ea.	2025		Metal detectors, wand type	Ea.	138
For additional camera stations, add	Ea.	1100		Walk-through portal type, single-zone	Ea.	5175
Directory boards, plastic, glass covered				Multi-zone	Ea.	6325
30" x 20"	Ea.	610		X-ray equipment		
36" x 48"	Ea.	1525		Desk top, for mail, small packages	Ea.	4000
Aluminum, 24" x 18"	Ea.	640		Conveyer type, including monitor, minimum	Ea.	17,900
36" x 24"	Ea.	765		Maximum	Ea.	31,100
48" x 32"	Ea.	1075		Explosive detection equipment		
48" x 60"	Ea.	2275		Hand held, battery operated	Ea.	28,100
Electronic, wall mounted	S.F.	2900		Walk-through portal type	Ea.	48,900
Free standing	S.F.	4125		Uninterruptible power supply, 15 kVA/12.75 kW	kW	1890
Escalators, 10' rise, 32" wide, glass balustrade	Ea.	147,500				
Metal balustrade	Ea.	155,500				
48" wide, glass balustrade	Ea.	155,000				
Metal balustrade	Ea.	164,000				
Pedestal access floor system w/ plastic laminate cover						
Computer room, less than 6000 SF	S.F.	21				
Greater than 6000 SF	S.F.	20.50				
Office, greater than 6000 S.F.	S.F.	16.95				

Important: See the Reference Section for Location Factors.

Model costs calculated for a 3 story building with 12' story height and 20,000 square feet of floor area

				Unit	Unit Cost	Cost Per S.F.	% Of Sub-Total
A. SUBSTRUCTURE							
1010	Standard Foundations	Poured concrete; strip and spread footings		S.F. Ground	4.05	1.35	
1020	Special Foundations	N/A		—	—	—	
1030	Slab on Grade	4" reinforced slab on grade		S.F. Slab	5.56	1.85	3.7%
2010	Basement Excavation	Site preparation for slab and trench for foundation wall and footing		S.F. Ground	.32	.11	
2020	Basement Walls	4' Foundation wall		L.F. Wall	42.75	1.54	
B. SHELL							
	B10 Superstructure						
1010	Floor Construction	Wood Beam and joist on wood columns, fireproofed		S.F. Floor	25.98	17.32	15.1%
1020	Roof Construction	Wood roof, truss, 4/12 slope		S.F. Roof	7.86	2.62	
	B20 Exterior Enclosure						
2010	Exterior Walls	Ashlar stone veneer on wood stud back up, insulated	80% of wall	S.F. Wall	38.89	20.16	
2020	Exterior Windows	Aluminum awning type windows	20% of wall	Each	746	4.21	19.3%
2030	Exterior Doors	Aluminum and glass single and double doors, hollow metal doors		Each	3570	1.08	
	B30 Roofing						
3010	Roof Coverings	Asphalt shingles with flashing, gutters and downspouts		S.F. Roof	2.52	.84	0.6%
3020	Roof Openings	N/A		—	—	—	
C. INTERIORS							
1010	Partitions	Gypsum board on wood studs	20 S.F. Floor/L.F. Partition	S.F. Partition	8.03	3.21	
1020	Interior Doors	Single leaf solid core wood doors, hollow metal egress doors	315 S.F. Floor/Door	Each	789	2.53	
1030	Fittings	Plastic laminate toilet partitions		S.F. Floor	.22	.22	
2010	Stair Construction	Concrete filled metal pan with rails		Flight	2745	.96	15.9%
3010	Wall Finishes	Paint on drywall		S.F. Surface	1.40	1.12	
3020	Floor Finishes	60% Carpet tile, 30% vinyl composition tile, 10% ceramic tile		S.F. Floor	5.68	5.68	
3030	Ceiling Finishes	Acoustic ceiling tiles on suspended support system		S.F. Ceiling	7.22	7.22	
D. SERVICES							
	D10 Conveying						
1010	Elevators & Lifts	Two hydraulic passenger elevators		Each	120,000	12	9.1%
1020	Escalators & Moving Walks	N/A		—	—	—	
	D20 Plumbing						
2010	Plumbing Fixtures	Restroom and service fixtures, supply and drainage	1 Fixture/1050 S.F. Floor	Each	2001	1.90	
2020	Domestic Water Distribution	Gas fired water heater		S.F. Floor	.54	.54	1.9%
2040	Rain Water Drainage	Roof drains		S.F. Roof			
	D30 HVAC						
3010	Energy Supply	N/A		—	—	—	
3020	Heat Generating Systems	Included in D3050		—	—	—	
3030	Cooling Generating Systems	N/A		—	—	—	13.7 %
3050	Terminal & Package Units	Multizone unit, gas heating , electric cooling		S.F. Floor	18.05	18.05	
3090	Other HVAC Sys. & Equipment	N/A		—	—	—	
	D40 Fire Protection						
4010	Sprinklers	Wet pipe spinkler system		S.F. Floor	3.86	3.86	3.7%
4020	Standpipes	Standpipes and hose systems		S.F. Floor	1.02	1.02	
	D50 Electrical						
5010	Electrical Service/Distribution	1000 Ampere service, panel boards and feeders		S.F. Floor	3.66	3.66	
5020	Lighting & Branch Wiring	Fluorescent fixtures, receptacles, switches, A.C. and misc. power		S.F. Floor	12.41	12.41	17.0%
5030	Communications & Security	Addressable alarm systems, emergency lighting, internet and phone wiring		S.F. Floor	6.09	6.09	
5090	Other Electrical Systems	Uninterruptible power supply with battery pack, 12.75kW		S.F. Floor	.19	.19	
E. EQUIPMENT & FURNISHINGS							
1010	Commercial Equipment	N/A		—	—	—	
1020	Institutional Equipment	N/A		—	—	—	
1030	Vehicular Equipment	N/A		—	—	—	0.0 %
1090	Other Equipment	N/A		—	—	—	
F. SPECIAL CONSTRUCTION							
1020	Integrated Construction	N/A		—	—	—	
1040	Special Facilities	N/A		—	—	—	0.0 %
G. BUILDING SITEWORK	**N/A**						

		Sub-Total	131.74	100%
	CONTRACTOR FEES (General Requirements: 10%, Overhead: 5%, Profit: 10%)	25%	32.93	
	ARCHITECT FEES	7%	11.53	

Total Building Cost 176.20

For customer support on your Square Foot Cost Data, call 877.756.2789.

177

Costs per square foot of floor area

Exterior Wall	S.F. Area	20000	40000	60000	80000	100000	150000	200000	250000	300000
	L.F. Perimeter	260	360	400	420	460	520	650	720	800
Stone Veneer	Steel Frame	226.05	192.80	174.60	163.75	158.55	149.85	148.00	145.15	143.50
E.I.F.S.	Steel Frame	181.50	162.80	153.05	147.40	144.55	139.90	138.85	137.40	136.50
Curtain Wall	Steel Frame	224.45	192.50	175.55	165.50	160.65	152.50	150.70	148.15	146.60
Brick Veneer	Reinforced Concrete	285.45	234.15	205.40	188.20	180.05	166.20	163.40	158.85	156.25
Metal Panel	Reinforced Concrete	217.70	189.05	173.65	164.45	160.00	152.60	151.05	148.65	147.20
E.I.F.S.	Reinforced Concrete	267.00	221.30	195.90	180.70	173.45	161.20	158.75	154.70	152.45
Perimeter Adj., Add or Deduct	Per 100 L.F.	35.95	18.05	11.95	8.90	7.30	4.80	3.60	2.90	2.40
Story Hgt. Adj., Add or Deduct	Per 1 Ft.	7.30	5.10	3.75	2.95	2.65	1.90	1.80	1.65	1.55
For Basement, add $41.20 per square foot of basement area										

The above costs were calculated using the basic specifications shown on the facing page. These costs should be adjusted where necessary for design alternatives and owner's requirements. Reported completed project costs, for this type of structure, range from $86.10 to $253.10 per S.F.

Common additives

Description	Unit	$ Cost
Closed circuit surveillance, one station		
Camera and monitor	Ea.	2025
For additional camera stations, add	Ea.	1100
Directory boards, plastic, glass covered		
30" x 20"	Ea.	610
36" x 48"	Ea.	1525
Aluminum, 24" x 18"	Ea.	640
36" x 24"	Ea.	765
48" x 32"	Ea.	1075
48" x 60"	Ea.	2275
Electronic, wall mounted	S.F.	2900
Free standing	S.F.	4125
Escalators, 10' rise, 32" wide, glass balustrade	Ea.	147,500
Metal balustrade	Ea.	155,500
48" wide, glass balustrade	Ea.	155,000
Metal balustrade	Ea.	164,000
Pedestal access floor system w/ plastic laminate cover		
Computer room, less than 6000 SF	S.F.	21
Greater than 6000 SF	S.F.	20.50
Office, greater than 6000 S.F.	S.F.	16.95

Description	Unit	$ Cost
Security access systems		
Metal detectors, wand type	Ea.	138
Walk-through portal type, single-zone	Ea.	5175
Multi-zone	Ea.	6325
X-ray equipment		
Desk top, for mail, small packages	Ea.	4000
Conveyer type, including monitor, minimum	Ea.	17,900
Maximum	Ea.	31,100
Explosive detection equipment		
Hand held, battery operated	Ea.	28,100
Walk-through portal type	Ea.	48,900
Uninterruptible power supply, 15 kVA/12.75 kW	kW	1890

Important: See the Reference Section for Location Factors.

Model costs calculated for a 8 story building with 12' story height and 80,000 square feet of floor area

Office, 5-10 Story

				Unit	Unit Cost	Cost Per S.F.	% Of Sub-Total
A. SUBSTRUCTURE							
1010	Standard Foundations	Poured concrete; strip and spread footings		S.F. Ground	3.68	.46	
1020	Special Foundations	N/A		—	—	—	
1030	Slab on Grade	4" reinforced slab on grade		S.F. Slab	5.56	.70	1.3%
2010	Basement Excavation	Site preparation for slab and trench for foundation wall and footing		S.F. Ground	.32	.04	
2020	Basement Walls	4' Foundation wall		L.F. Wall	86	.45	
B. SHELL							
B10 Superstructure							
1010	Floor Construction	Open web steel joists, slab form, concrete, fireproofed steel columns		S.F. Floor	14.11	12.35	11.0%
1020	Roof Construction	Metal deck, open web steel joists, beams, columns		S.F. Roof	10.40	1.30	
B20 Exterior Enclosure							
2010	Exterior Walls	Limestone stone veneer on metal studs	80% of wall	S.F. Wall	54	21.80	
2020	Exterior Windows	Aluminum sliding windows	20% of wall	Each	585	3.94	21.1%
2030	Exterior Doors	Aluminum and glass double doors, hollow metal egress doors		Each	3905	.29	
B30 Roofing							
3010	Roof Coverings	Single ply membrane, stone ballast, rigid insulation		S.F. Roof	6.40	.80	0.8%
3020	Roof Openings	Roof and smoke hatches		S.F. Roof	2060	.15	
C. INTERIORS							
1010	Partitions	Gypsum board on metal studs	30 S.F. Floor/L.F. Partition	S.F. Partition	9.90	3.30	
1020	Interior Doors	Single leaf solid core wood doors, hollow metal egress doors	350 S.F. Floor/Door	Each	789	2.23	
1030	Fittings	Plastic laminate toilet partitions		S.F. Floor	.18	.18	
2010	Stair Construction	Concrete filled metal pan, with rails		Flight	11,575	3.76	18.8%
3010	Wall Finishes	Paint on drywall		S.F. Surface	1.37	.91	
3020	Floor Finishes	60% Carpet tile, 30% vinyl composition tile, 10% ceramic tile		S.F. Floor	5.68	5.68	
3030	Ceiling Finishes	Acoustic ceiling tiles on suspended support system		S.F. Ceiling	7.22	7.22	
D. SERVICES							
D10 Conveying							
1010	Elevators & Lifts	Traction geared passenger elevators		Each	303,200	15.16	12.3%
1020	Escalators & Moving Walks	N/A		—	—	—	
D20 Plumbing							
2010	Plumbing Fixtures	Restroom and service fixtures, supply and drainage	1 Fixture/1600 S.F. Floor	Each	2336	1.46	
2020	Domestic Water Distribution	Gas fired water heater		S.F. Floor	.58	.58	1.9%
2040	Rain Water Drainage	Roof drains		S.F. Roof	2.08	.26	
D30 HVAC							
3010	Energy Supply	N/A		—	—	—	
3020	Heat Generating Systems	Included in D3050		—	—	—	
3030	Cooling Generating Systems	N/A		—	—	—	14.6 %
3050	Terminal & Package Units	Multizone unit, gas heating , electric cooling		S.F. Floor	18.05	18.05	
3090	Other HVAC Sys. & Equipment	N/A		—	—	—	
D40 Fire Protection							
4010	Sprinklers	Wet pipe spinkler system		S.F. Floor	2.88	2.88	3.3%
4020	Standpipes	Standpipes and hose systems, with pumps		S.F. Floor	1.15	1.15	
D50 Electrical							
5010	Electrical Service/Distribution	800 Ampere service, panel boards and feeders		S.F. Floor	1.08	1.08	
5020	Lighting & Branch Wiring	Fluorescent fixtures, receptacles, switches, A.C. and misc. power		S.F. Floor	12.32	12.32	15.0%
5030	Communications & Security	Alarm systems, emergency lighting, internet and phone wiring		S.F. Floor	4.52	4.52	
5090	Other Electrical Systems	Uninterruptible power supply with battery pack, 12.75kW		S.F. Floor	.59	.59	
E. EQUIPMENT & FURNISHINGS							
1010	Commercial Equipment	N/A		—	—	—	
1020	Institutional Equipment	N/A		—	—	—	0.0 %
1030	Vehicular Equipment	N/A		—	—	—	
1090	Other Equipment	N/A		—	—	—	
F. SPECIAL CONSTRUCTION							
1020	Integrated Construction	N/A		—	—	—	0.0 %
1040	Special Facilities	N/A		—	—	—	
G. BUILDING SITEWORK	**N/A**						

		Sub-Total	123.61	100%
CONTRACTOR FEES (General Requirements: 10%, Overhead: 5%, Profit: 10%)		25%	30.87	
ARCHITECT FEES		6%	9.27	
	Total Building Cost		**163.75**	

For customer support on your Square Foot Cost Data, call 877.756.2789.

179

Costs per square foot of floor area

Exterior Wall	S.F. Area	120000	145000	170000	200000	230000	260000	400000	600000	800000
	L.F. Perimeter	420	450	470	490	510	530	650	800	900
Stone Veneer	Steel Frame	184.50	176.65	170.35	164.60	160.35	157.10	149.20	143.80	140.40
E.I.F.S.	Steel Frame	157.05	152.40	148.85	145.60	143.20	141.35	136.80	133.70	131.95
Curtain Wall	Steel Frame	184.50	176.95	170.90	165.35	161.30	158.15	150.60	145.40	142.05
Brick Veneer	Reinforced Concrete	184.35	175.75	168.85	162.60	157.95	154.30	145.65	139.75	135.90
Metal Panel	Reinforced Concrete	179.10	172.60	167.50	162.85	159.45	156.80	150.40	145.95	143.20
E.I.F.S.	Reinforced Concrete	181.70	174.25	168.40	162.95	158.95	155.85	148.50	143.40	140.10
Perimeter Adj., Add or Deduct	Per 100 L.F.	11.10	9.20	7.90	6.65	5.85	5.15	3.35	2.20	1.55
Story Hgt. Adj., Add or Deduct	Per 1 Ft.	3.65	3.20	2.90	2.55	2.35	2.10	1.70	1.40	1.10

For Basement, add $41.20 per square foot of basement area

The above costs were calculated using the basic specifications shown on the facing page. These costs should be adjusted where necessary for design alternatives and owner's requirements. Reported completed project costs, for this type of structure, range from $107.95 to $263.45 per S.F.

Common additives

Description	Unit	$ Cost
Closed circuit surveillance, one station		
Camera and monitor	Ea.	2025
For additional camera stations, add	Ea.	1100
Directory boards, plastic, glass covered		
30" x 20"	Ea.	610
36" x 48"	Ea.	1525
Aluminum, 24" x 18"	Ea.	640
36" x 24"	Ea.	765
48" x 32"	Ea.	1075
48" x 60"	Ea.	2275
Electronic, wall mounted	S.F.	2900
Free standing	S.F.	4125
Escalators, 10' rise, 32" wide, glass balustrade	Ea.	147,500
Metal balustrade	Ea.	155,500
48" wide, glass balustrade	Ea.	155,000
Metal balustrade	Ea.	164,000
Pedestal access floor system w/ plastic laminate cover		
Computer room, less than 6000 SF	S.F.	21
Greater than 6000 SF	S.F.	20.50
Office, greater than 6000 S.F.	S.F.	16.95

Description	Unit	$ Cost
Security access systems		
Metal detectors, wand type	Ea.	138
Walk-through portal type, single-zone	Ea.	5175
Multi-zone	Ea.	6325
X-ray equipment		
Desk top, for mail, small packages	Ea.	4000
Conveyer type, including monitor, minimum	Ea.	17,900
Maximum	Ea.	31,100
Explosive detection equipment		
Hand held, battery operated	Ea.	28,100
Walk-through portal type	Ea.	48,900
Uninterruptible power supply, 15 kVA/12.75 kW	kW	1890

Model costs calculated for a 16 story building with 12' story height and 260,000 square feet of floor area

				Unit	Unit Cost	Cost Per S.F.	% Of Sub-Total
A. SUBSTRUCTURE							
1010	Standard Foundations	CIP concrete pile caps		S.F. Ground	10.40	.65	
1020	Special Foundations	Steel H-piles, concrete grade beams		S.F. Ground	170	10.65	
1030	Slab on Grade	4" reinforced concrete with vapor barrier and granular base		S.F. Slab	5.56	.35	10.1%
2010	Basement Excavation	Site preparation for slab and trench for foundation wall and footing		S.F. Ground	.32	.02	
2020	Basement Walls	4' Foundation wall		L.F. Wall	86	.34	
B. SHELL							
	B10 Superstructure						
1010	Floor Construction	Open web steel joists, slab form, concrete, fireproofed steel columns		S.F. Floor	13.60	12.75	11.3%
1020	Roof Construction	Metal deck, open web steel joists, beams, columns		S.F. Roof	10.40	.65	
	B20 Exterior Enclosure						
2010	Exterior Walls	Limestone stone veneer on metal studs	80% of wall	S.F. Wall	43.26	16.93	
2020	Exterior Windows	Curtain wall glazing system, thermo-break frame	20% of wall	Each	47.50	1.57	16.2%
2030	Exterior Doors	Double aluminum & glass doors		Each	6521	.71	
	B30 Roofing						
3010	Roof Coverings	Single ply membrane, stone ballast, rigid insulation		S.F. Roof	5.92	.37	0.4%
3020	Roof Openings	Roof and smoke hatches.		S.F. Roof	2060	.06	
C. INTERIORS							
1010	Partitions	Concrete block partitions, gypsum board on metal studs	30 S.F. Floor/L.F. Partition	S.F. Partition	9.42	3.14	
1020	Interior Doors	Single leaf solid core wood doors, hollow metal egress doors	350 S.F. Floor/Door	Each	789	2.22	
1030	Fittings	Plastic laminate toilet partitions		S.F. Floor	.18	.18	
2010	Stair Construction	Concrete filled metal pan, with rails		Flight	17,275	4.39	19.4%
3010	Wall Finishes	Paint on drywall		S.F. Surface	1.26	.84	
3020	Floor Finishes	60% Carpet tile, 30% vinyl composition tile, 10% ceramic tile		S.F. Floor	5.02	5.02	
3030	Ceiling Finishes	Acoustic ceiling tiles on suspended support system		S.F. Ceiling	7.22	7.22	
D. SERVICES							
	D10 Conveying						
1010	Elevators & Lifts	Traction geared passenger elevators		Each	475,800	7.32	6.2%
1020	Escalators & Moving Walks	N/A		—	—	—	
	D20 Plumbing						
2010	Plumbing Fixtures	Restroom and service fixtures, supply and drainage	1 Fixture/1600 S.F. Floor	Each	2089	1.31	
2020	Domestic Water Distribution	Gas fired water heater		S.F. Floor	.33	.33	1.5%
2040	Rain Water Drainage	Roof drains		S.F. Roof	1.28	.08	
	D30 HVAC						
3010	Energy Supply	N/A		—	—	—	
3020	Heat Generating Systems	Boiler, heat exchanger and fans		Each	509,825	2.69	
3030	Cooling Generating Systems	Chiled water, fan coil units		S.F. Floor	16.83	16.83	16.5 %
3050	Terminal & Package Units	N/A		—	—	—	
3090	Other HVAC Sys. & Equipment	N/A		—	—	—	
	D40 Fire Protection						
4010	Sprinklers	Sprinkler system and accessory package		S.F. Floor	2.78	2.78	2.8%
4020	Standpipes	Standpipes, pumps and hose systems		S.F. Floor	.59	.59	
	D50 Electrical						
5010	Electrical Service/Distribution	1200 Ampere service, panel boards and feeders		S.F. Floor	.56	.56	
5020	Lighting & Branch Wiring	Fluorescent fixtures, receptacles, switches, A.C. and misc. power		S.F. Floor	12.20	12.20	15.6%
5030	Communications & Security	Alarm systems, emergency lighting, internet and phone wiring		S.F. Floor	5.43	5.43	
5090	Other Electrical Systems	Uninterruptible power supply with battery pack, 12.75kW		S.F. Floor	.36	.36	
E. EQUIPMENT & FURNISHINGS							
1010	Commercial Equipment	N/A		—	—	—	
1020	Institutional Equipment	N/A		—	—	—	
1030	Vehicular Equipment	N/A		—	—	—	0.0 %
1090	Other Equipment	N/A		—	—	—	
F. SPECIAL CONSTRUCTION							
1020	Integrated Construction	N/A		—	—	—	
1040	Special Facilities	N/A		—	—	—	0.0 %
G. BUILDING SITEWORK	**N/A**						

		Sub-Total	118.54	100%
CONTRACTOR FEES (General Requirements: 10%, Overhead: 5%, Profit: 10%)		25%	29.67	
ARCHITECT FEES		6%	8.89	
	Total Building Cost		**157.10**	

Costs per square foot of floor area

Exterior Wall	S.F. Area	5000	7500	10000	12500	15000	17500	20000	22500	25000
	L.F. Perimeter	284	350	410	450	490	530	570	600	632
Brick Veneer	Wood Truss	458.85	445.50	438.30	432.25	428.30	425.45	423.30	421.15	419.50
Brick Veneer	Steel Frame	458.85	446.10	439.15	433.45	429.70	426.90	424.95	422.90	421.35
Decorative Concrete Block	Steel Frame	461.40	448.20	441.00	435.10	431.15	428.35	426.25	424.10	422.50
Decorative Concrete Block	Bearing Wall	452.30	440.70	434.45	429.30	425.90	423.45	421.60	419.85	418.40
Tiltup Concrete Panel	Steel Frame	453.60	441.85	435.45	430.20	426.75	424.25	422.40	420.60	419.10
EIFS on Concrete Block	Steel Frame	457.35	443.40	435.95	430.10	426.20	423.35	421.30	419.30	417.70
Perimeter Adj., Add or Deduct	Per 100 L.F.	20.40	13.65	10.20	8.15	6.80	5.80	5.10	4.55	4.15
Story Hgt. Adj., Add or Deduct	Per 1 Ft.	3.05	2.50	2.15	1.90	1.75	1.60	1.50	1.45	1.35

For Basement, add $32.40 per square foot of basement area

The above costs were calculated using the basic specifications shown on the facing page. These costs should be adjusted where necessary for design alternatives and owner's requirements. Reported completed project costs, for this type of structure, range from $189.10 to $980.55 per S.F.

Common additives

Description	Unit	$ Cost
Cabinets, base, door units, metal	L.F.	325
Drawer units	L.F.	645
Tall storage cabinets, open, 7' high	L.F.	620
With glazed doors	L.F.	900
Wall cabinets, metal, 12-1/2" deep, open	L.F.	263
With doors	L.F.	460
Counter top, laminated plastic, no backsplash	L.F.	81.50
Stainless steel counter top	L.F.	199
For drop-in stainless 43" x 21" sink, add	Ea.	1100
Nurse Call Station		
Single bedside call station	Ea.	292
Ceiling speaker station	Ea.	160
Emergency call station	Ea.	166
Pillow speaker	Ea.	287
Standard call button	Ea.	183
Master control station for 20 stations	Total	5825

Description	Unit	$ Cost
Directory Boards, Plastic, glass covered		
Plastic, glass covered, 30" x 20"	Ea.	610
36" x 48"	Ea.	1525
Building directory, alum., black felt panels, 1 door, 24" x 18"	Ea.	640
36" x 24"	Ea.	765
48" x 32"	Ea.	1075
48" x 60"	Ea.	2275
Tables, Examining, vinyl top		
with base cabinets	Ea.	1650
Utensil washer-sanitizer	Ea.	9650
X-ray, mobile, minimum	Ea.	18,500

Important: See the Reference Section for Location Factors.

Model costs calculated for a 1 story building with 15'-4" story height and 12,500 square feet of floor area

					Unit	Unit Cost	Cost Per S.F.	% Of Sub-Total

A. SUBSTRUCTURE

				Unit	Unit Cost	Cost Per S.F.	% Of Sub-Total
1010	Standard Foundations	Poured concrete; strip and spread footings		S.F. Ground	1.90	1.90	
1020	Special Foundations	N/A		—	—	—	
1030	Slab on Grade	4" reinforced concrete with vapor barrier and granular base		S.F. Slab	6.90	6.90	3.9%
2010	Basement Excavation	Site preparation for slab and trench for foundation wall and footing		S.F. Ground	.32	.32	
2020	Basement Walls	4' Foundation wall		L.F. Wall	86	3.08	

B. SHELL

B10 Superstructure

				Unit	Unit Cost	Cost Per S.F.	% Of Sub-Total
1010	Floor Construction	N/A		—	—	—	
1020	Roof Construction	Steel Joists and girders on columns		S.F. Roof	7.84	7.84	2.5 %

B20 Exterior Enclosure

				Unit	Unit Cost	Cost Per S.F.	% Of Sub-Total
2010	Exterior Walls	Brick veneer	85% of wall	S.F. Wall	19.48	9.14	
2020	Exterior Windows	Aluminum frame, fixed	15% of wall	Each	850	4.69	6.1%
2030	Exterior Doors	Aluminum and glass entry; steel flush doors		Each	6877	5.50	

B30 Roofing

				Unit	Unit Cost	Cost Per S.F.	% Of Sub-Total
3010	Roof Coverings	Single-ply membrane, loose laid and ballasted; polyisocyanurate insulation		S.F. Roof	6.74	6.74	2.2%
3020	Roof Openings	Roof hatches		S.F. Roof	.10	.10	

C. INTERIORS

				Unit	Unit Cost	Cost Per S.F.	% Of Sub-Total
1010	Partitions	Gypsum board on metal studs	7 S.F. Floor/L.F. Partition	S.F. Partition	8.13	11.61	
1020	Interior Doors	Single and double leaf fire doors	200 S.F. Floor/Door	Each	1397	6.99	
1030	Fittings	Stainless steel toilet partitions, steel lockers		S.F. Floor	1.70	1.70	
2010	Stair Construction	N/A		—	—	—	12.9%
3010	Wall Finishes	80% Paint, 10% vinyl wall covering, 10% ceramic tile		S.F. Surface	2.07	5.90	
3020	Floor Finishes	10% Carpet, 80% vinyl sheet, 10% ceramic tile		S.F. Floor	10.40	10.40	
3030	Ceiling Finishes	Mineral fiber tile and plastic coated tile on suspension system		S.F. Ceiling	4.28	4.28	

D. SERVICES

D10 Conveying

				Unit	Unit Cost	Cost Per S.F.	% Of Sub-Total
1010	Elevators & Lifts	N/A		—	—	—	0.0 %
1020	Escalators & Moving Walks	N/A		—	—	—	

D20 Plumbing

				Unit	Unit Cost	Cost Per S.F.	% Of Sub-Total
2010	Plumbing Fixtures	Medical, patient & specialty fixtures, supply and drainage	1 fixture/350 S.F. Floor	Each	3441	9.83	
2020	Domestic Water Distribution	Electric water heater		S.F. Floor	2.23	2.23	4.6%
2040	Rain Water Drainage	Roof drains		S.F. Roof	2.39	2.39	

D30 HVAC

				Unit	Unit Cost	Cost Per S.F.	% Of Sub-Total
3010	Energy Supply	Hot water reheat for surgery		S.F. Floor	11.50	11.50	
3020	Heat Generating Systems	Boiler		Each	63,800	7.09	
3030	Cooling Generating Systems	N/A		—	—	—	37.6%
3050	Terminal & Package Units	N/A		—	—	—	
3090	Other HVAC Sys. & Equipment	Rooftop AC, CAV, roof vents, heat exchanger, surgical air curtains		S.F. Floor	100	100.27	

D40 Fire Protection

				Unit	Unit Cost	Cost Per S.F.	% Of Sub-Total
4010	Sprinklers	Wet pipe sprinkler system		S.F. Floor	3.65	3.65	1.7%
4020	Standpipes	Standpipe		S.F. Floor	1.72	1.72	

D50 Electrical

				Unit	Unit Cost	Cost Per S.F.	% Of Sub-Total
5010	Electrical Service/Distribution	1200 ampere service, panel board and feeders		S.F. Floor	6.71	6.71	
5020	Lighting & Branch Wiring	T-8 fluorescent light fixtures, receptacles, switches, A.C. and misc. power		S.F. Floor	17.63	17.63	13.0%
5030	Communications & Security	Addressable alarm system, internet wiring, communications system, emergency lighting		S.F. Floor	10.34	10.34	
5090	Other Electrical Systems	Emergency generator, 300 kW with fuel tank		S.F. Floor	6.32	6.32	

E. EQUIPMENT & FURNISHINGS

				Unit	Unit Cost	Cost Per S.F.	% Of Sub-Total
1010	Commercial Equipment	N/A		—	—	—	
1020	Institutional Equipment	Medical gas system, cabinetry, scrub sink, sterilizer		S.F. Floor	47.93	47.93	15.5 %
1030	Vehicular Equipment	N/A		—	—	—	
1090	Other Equipment	Ice maker, washing machine		S.F. Floor	1.05	1.05	

F. SPECIAL CONSTRUCTION & DEMOLITION

				Unit	Unit Cost	Cost Per S.F.	% Of Sub-Total
1020	Integrated Construction	N/A		—	—	—	0.0 %
1040	Special Facilities	N/A		—	—	—	

G. BUILDING SITEWORK   N/A

Sub-Total		315.75	**100%**
CONTRACTOR FEES (General Requirements: 10%, Overhead: 5%, Profit: 5%)	20%	78.93	
ARCHITECT FEES	9%	35.52	
Total Building Cost		**430.20**	

For customer support on your Square Foot Cost Data, call 877.756.2789.

183

Costs per square foot of floor area

Exterior Wall	S.F. Area	7000	9000	11000	13000	15000	17000	19000	21000	23000
	L.F. Perimeter	240	280	303	325	354	372	397	422	447
Limestone with Concrete Block Back-up	Bearing Walls	290.95	272.45	256.65	245.55	238.65	231.65	227.10	223.45	220.40
	R/Conc. Frame	298.60	280.80	265.95	255.50	248.95	242.40	238.15	234.65	231.80
Face Brick with Concrete Block Back-up	Bearing Walls	252.95	237.85	226.00	217.65	212.25	207.15	203.70	200.95	198.60
	R/Conc. Frame	269.35	254.25	242.45	234.20	228.80	223.75	220.30	217.55	215.20
Decorative Concrete Block	Bearing Walls	241.65	227.55	216.85	209.40	204.45	199.95	196.80	194.30	192.20
	R/Conc. Frame	259.20	245.20	234.55	227.15	222.20	217.75	214.60	212.10	210.05
Perimeter Adj., Add or Deduct	Per 100 L.F.	36.75	28.60	23.45	19.85	17.15	15.15	13.55	12.25	11.20
Story Hgt. Adj., Add or Deduct	Per 1 Ft.	6.65	6.00	5.35	4.85	4.55	4.25	4.10	3.90	3.75
For Basement, add $32.40 per square foot of basement area										

The above costs were calculated using the basic specifications shown on the facing page. These costs should be adjusted where necessary for design alternatives and owner's requirements. Reported completed project costs, for this type of structure, range from $124.05 to $323.60 per S.F.

Common additives

Description	Unit	$ Cost
Cells Prefabricated, 5'-6' wide, 7'-8' high, 7'-8' deep	Ea.	12,900
Elevators, Hydraulic passenger, 2 stops		
1500# capacity	Ea.	67,400
2500# capacity	Ea.	70,400
3500# capacity	Ea.	75,400
Emergency Lighting, 25 watt, battery operated		
Lead battery	Ea.	345
Nickel cadmium	Ea.	685
Flagpoles, Complete		
Aluminum, 20' high	Ea.	1900
40' high	Ea.	4425
70' high	Ea.	11,500
Fiberglass, 23' high	Ea.	1350
39'-5" high	Ea.	3400
59' high	Ea.	6650

Description	Unit	$ Cost
Lockers, Steel, Single tier, 60" to 72"	Opng.	231 - 380
2 tier, 60" or 72" total	Opng.	121 - 170
5 tier, box lockers	Opng.	74 - 84
Locker bench, lam. maple top only	L.F.	34.50
Pedestals, steel pipe	Ea.	76.50
Safe, Office type, 1 hour rating		
30" x 18" x 18"	Ea.	2500
60" x 36" x 18", double door	Ea.	9350
Shooting Range, Incl. bullet traps, target provisions, and contols, not incl. structural shell	Ea.	52,500
Smoke Detectors		
Ceiling type	Ea.	227
Duct type	Ea.	565
Sound System		
Amplifier, 250 watts	Ea.	2200
Speaker, ceiling or wall	Ea.	217
Trumpet	Ea.	415

Important: See the Reference Section for Location Factors.

Model costs calculated for a 2 story building with 12' story height and 11,000 square feet of floor area

Police Station

		Description		Unit	Unit Cost	Cost Per S.F.	% Of Sub-Total
A. SUBSTRUCTURE							
1010	Standard Foundations	Poured concrete; strip and spread footings		S.F. Ground	3.02	1.51	
1020	Special Foundations	N/A		—	—	—	
1030	Slab on Grade	4" reinforced concrete with vapor barrier and granular base		S.F. Slab	5.56	2.79	3.6%
2010	Basement Excavation	Site preparation for slab and trench for foundation wall and footing		S.F. Ground	.32	.16	
2020	Basement Walls	4' foundation wall		L.F. Wall	86	2.35	
B. SHELL							
	B10 Superstructure						
1010	Floor Construction	Open web steel joists, slab form, concrete		S.F. Floor	12.56	6.28	4.7%
1020	Roof Construction	Metal deck on open web steel joists		S.F. Roof	4.96	2.48	
	B20 Exterior Enclosure						
2010	Exterior Walls	Limestone with concrete block backup	80% of wall	S.F. Wall	66	34.90	
2020	Exterior Windows	Metal horizontal sliding	20% of wall	Each	1167	10.28	25.1%
2030	Exterior Doors	Hollow metal		Each	2851	2.06	
	B30 Roofing						
3010	Roof Coverings	Built-up tar and gravel with flashing; perlite/EPS composite insulation		S.F. Roof	7.64	3.82	2.0%
3020	Roof Openings	N/A		—	—	—	
C. INTERIORS							
1010	Partitions	Concrete block	20 S.F. Floor/L.F. Partition	S.F. Partition	14.98	7.49	
1020	Interior Doors	Single leaf kalamein fire door	200 S.F. Floor/Door	Each	1168	5.85	
1030	Fittings	Toilet partitions		S.F. Floor	.98	.98	
2010	Stair Construction	Concrete filled metal pan		Flight	17,275	4.71	18.2%
3010	Wall Finishes	90% paint, 10% ceramic tile		S.F. Surface	3.88	3.88	
3020	Floor Finishes	70% vinyl composition tile, 20% carpet tile, 10% ceramic tile		S.F. Floor	4.11	4.11	
3030	Ceiling Finishes	Mineral fiber tile on concealed zee bars		S.F. Ceiling	7.22	7.22	
D. SERVICES							
	D10 Conveying						
1010	Elevators & Lifts	One hydraulic passenger elevator		Each	85,360	7.76	4.1%
1020	Escalators & Moving Walks	N/A		—	—	—	
	D20 Plumbing						
2010	Plumbing Fixtures	Toilet and service fixtures, supply and drainage	1 Fixture/580 S.F. Floor	Each	4565	7.87	
2020	Domestic Water Distribution	Oil fired water heater		S.F. Floor	4.33	4.33	7.4%
2040	Rain Water Drainage	Roof drains		S.F. Roof	3.30	1.65	
	D30 HVAC						
3010	Energy Supply	N/A		—	—	—	
3020	Heat Generating Systems	N/A		—	—	—	
3030	Cooling Generating Systems	N/A		—	—	—	11.9 %
3050	Terminal & Package Units	Multizone HVAC air cooled system		S.F. Floor	22.35	22.35	
3090	Other HVAC Sys. & Equipment	N/A		—	—	—	
	D40 Fire Protection						
4010	Sprinklers	Wet pipe sprinkler system		S.F. Floor	3.95	3.95	2.7%
4020	Standpipes	Standpipe		S.F. Floor	1.11	1.11	
	D50 Electrical						
5010	Electrical Service/Distribution	400 ampere service, panel board and feeders		S.F. Floor	2.05	2.05	
5020	Lighting & Branch Wiring	T-8 fluorescent fixtures, receptacles, switches, A.C. and misc. power		S.F. Floor	12.59	12.59	11.9%
5030	Communications & Security	Addressable alarm systems, internet wiring, intercom and emergency lighting		S.F. Floor	7.46	7.46	
5090	Other Electrical Systems	Emergency generator, 15 kW		S.F. Floor	.24	.24	
E. EQUIPMENT & FURNISHINGS							
1010	Commercial Equipment	N/A		—	—	—	
1020	Institutional Equipment	Lockers, detention rooms, cells, gasoline dispensers		S.F. Floor	13.68	13.68	8.6 %
1030	Vehicular Equipment	Gasoline dispenser system		S.F. Floor	2.47	2.47	
1090	Other Equipment	N/A		—	—	—	
F. SPECIAL CONSTRUCTION							
1020	Integrated Construction	N/A		—	—	—	0.0 %
1040	Special Facilities	N/A		—	—	—	
G. BUILDING SITEWORK	**N/A**						

			Sub-Total	188.38	100%
CONTRACTOR FEES (General Requirements: 10%, Overhead: 5%, Profit: 10%)			25%	47.08	
ARCHITECT FEES			9%	21.19	
		Total Building Cost		**256.65**	

For customer support on your Square Foot Cost Data, call 877.756.2789.

185

Costs per square foot of floor area

Exterior Wall	S.F. Area	600	750	925	1150	1450	1800	2100	2600	3200
	L.F. Perimeter	100	112	128	150	164	182	200	216	240
Steel Siding on Wood Studs	Wood Truss	77.05	71.65	68.15	65.55	60.45	56.90	55.15	51.65	49.30
Aluminum Siding	Wood Truss	75.95	71.05	67.85	65.50	60.95	57.85	56.20	53.10	51.00
Wood Board Siding	Wood Truss	79.55	74.40	71.05	68.60	63.80	60.45	58.80	55.50	53.25
Plywood Siding	Wood Truss	75.80	70.90	67.75	65.40	60.85	57.65	56.10	52.95	50.90
Perimeter Adj., Add or Deduct	Per 100 L.F.	45.10	36.10	29.25	23.50	18.60	15.00	12.85	10.40	8.50
Story Hgt. Adj., Add or Deduct	Per 1 Ft.	3.25	2.85	2.70	2.50	2.20	1.95	1.85	1.65	1.45
Basement—Not Applicable										

The above costs were calculated using the basic specifications shown on the facing page. These costs should be adjusted where necessary for design alternatives and owner's requirements. Reported completed project costs, for this type of structure, range from $39.65 to $158.30 per S.F.

Common additives

Description	Unit	$ Cost
Slab on Grade		
4″ thick, non-industrial, non-reinforced	S.F.	5.03
Reinforced	S.F.	5.56
Light industrial, non-reinforced	S.F.	6.37
Reinforced	S.F.	6.90
Industrial, non-reinforced	S.F.	10.74
Reinforced	S.F.	11.29
6″ thick, non-industrial, non-reinforced	S.F.	5.89
Reinforced	S.F.	6.63
Light industrial, non-reinforced	S.F.	7.25
Reinforced	S.F.	8.31
Heavy industrial, non-reinforced	S.F.	13.85
Reinforced	S.F.	14.60
Insulation		
Fiberboard, low density, 1″ thick, R2.78	S.F.	1.46
Foil on reinforced scrim, single bubble air space, R8.8	C.S.F.	68
Vinyl faced fiberglass, 1-1/2″ thick, R5	S.F.	0.89
Vinyl/scrim/foil (VSF), 1-1/2″ thick, R5	S.F.	1.09

Description	Unit	$ Cost
Roof ventilation, ridge vent, aluminum	L.F.	8
Mushroom vent, aluminum	Ea.	90.50
Gutter, aluminum, 5″ K type, enameled	L.F.	8.65
Steel, half round or box, 5″, enameled	L.F.	8
Wood, clear treated cedar, fir or hemlock, 4″ x 5″	L.F.	24.50
Downspout, aluminum, 3″ x 4″, enameled	L.F.	7.60
Steel, round, 4″ diameter	L.F.	7.60

Important: See the Reference Section for Location Factors.

Model costs calculated for a 1 story building with 14' story height and 1,800 square feet of floor area

Post Frame Barn

			Unit	Unit Cost	Cost Per S.F.	% Of Sub-Total
A. SUBSTRUCTURE						
1010	Standard Foundations	Included in B2010 Exterior Walls	—	—	—	
1020	Special Foundations	N/A	—	—	—	
1030	Slab on Grade	N/A	—	—	—	0.0%
2010	Basement Excavation	N/A	—	—	—	
2020	Basement Walls	N/A	—	—	—	
B. SHELL						
B10 Superstructure						
1010	Floor Construction	N/A	—	—	—	17.6 %
1020	Roof Construction	Wood truss, 4/12 pitch	S.F. Roof	7.42	7.42	
B20 Exterior Enclosure						
2010	Exterior Walls	2 x 6 Wood framing, 6 x 6 wood posts in concrete; steel siding panels	S.F. Wall	11.90	16	
2020	Exterior Windows	Steel windows, fixed	Each	540	4.25	58.4%
2030	Exterior Doors	Steel overhead, hollow metal doors	Each	1973	4.38	
B30 Roofing						
3010	Roof Coverings	Metal panel roof	S.F. Roof	4.33	4.33	10.3%
3020	Roof Openings	N/A	—	—	—	
C. INTERIORS						
1010	Partitions	N/A	—	—	—	
1020	Interior Doors	N/A	—	—	—	
1030	Fittings	N/A	—	—	—	
2010	Stair Construction	N/A	—	—	—	0.0 %
3010	Wall Finishes	N/A	—	—	—	
3020	Floor Finishes	N/A	—	—	—	
3030	Ceiling Finishes	N/A	—	—	—	
D. SERVICES						
D10 Conveying						
1010	Elevators & Lifts	N/A	—	—	—	0.0 %
1020	Escalators & Moving Walks	N/A	—	—	—	
D20 Plumbing						
2010	Plumbing Fixtures	N/A	—	—	—	
2020	Domestic Water Distribution	N/A	—	—	—	0.0 %
2040	Rain Water Drainage	N/A	—	—	—	
D30 HVAC						
3010	Energy Supply	N/A	—	—	—	
3020	Heat Generating Systems	N/A	—	—	—	
3030	Cooling Generating Systems	N/A	—	—	—	0.0 %
3050	Terminal & Package Units	N/A	—	—	—	
3090	Other HVAC Sys. & Equipment	N/A	—	—	—	
D40 Fire Protection						
4010	Sprinklers	N/A	—	—	—	0.0 %
4020	Standpipes	N/A	—	—	—	
D50 Electrical						
5010	Electrical Service/Distribution	60 Ampere service, panel board & feeders	S.F. Floor	1.87	1.87	
5020	Lighting & Branch Wiring	High bay fixtures, receptacles & switches	S.F. Floor	3.91	3.91	13.7%
5030	Communications & Security	N/A	—	—	—	
5090	Other Electrical Systems	N/A	—	—	—	
E. EQUIPMENT & FURNISHINGS						
1010	Commercial Equipment	N/A	—	—	—	
1020	Institutional Equipment	N/A	—	—	—	0.0 %
1030	Vehicular Equipment	N/A	—	—	—	
1090	Other Equipment	N/A	—	—	—	
F. SPECIAL CONSTRUCTION & DEMOLITION						
1020	Integrated Construction	N/A	—	—	—	0.0 %
1040	Special Facilities	N/A	—	—	—	
G. BUILDING SITEWORK	**N/A**					

			Sub-Total	42.16	100%
CONTRACTOR FEES (General Requirements: 10%, Overhead: 5%, Profit: 5%)			20%	10.52	
ARCHITECT FEES			8%	4.22	

Total Building Cost	**56.90**

For customer support on your Square Foot Cost Data, call 877.756.2789.

187

Costs per square foot of floor area

Exterior Wall	S.F. Area	5000	7000	9000	11000	13000	15000	17000	19000	21000
	L.F. Perimeter	300	380	420	468	486	513	540	580	620
Face Brick with Concrete Block Back-up	Steel Frame	176.50	166.60	157.05	151.70	145.75	142.10	139.15	137.55	136.20
	Bearing Walls	177.00	167.05	157.40	151.95	145.95	142.20	139.25	137.65	136.30
Limestone with Concrete Block Back-up	Steel Frame	206.35	193.60	180.25	172.85	164.35	159.10	154.95	152.70	150.90
	Bearing Walls	206.85	194.05	180.60	173.10	164.55	159.20	155.10	152.80	151.00
Decorative Concrete Block	Steel Frame	164.90	156.15	148.05	143.50	138.55	135.45	133.05	131.65	130.50
	Bearing Walls	165.45	156.55	148.35	143.70	138.75	135.55	133.15	131.75	130.60
Perimeter Adj., Add or Deduct	Per 100 L.F.	18.35	13.15	10.25	8.30	7.05	6.05	5.40	4.80	4.35
Story Hgt. Adj., Add or Deduct	Per 1 Ft.	2.95	2.65	2.30	2.05	1.85	1.60	1.55	1.50	1.45

For Basement, add $31.60 per square foot of basement area

The above costs were calculated using the basic specifications shown on the facing page. These costs should be adjusted where necessary for design alternatives and owner's requirements. Reported completed project costs, for this type of structure, range from $102.90 to $267.15 per S.F.

Common additives

Description	Unit	$ Cost	Description	Unit	$ Cost
Closed Circuit Surveillance, One station			Mail Boxes, Horizontal, key lock, 15" x 6" x 5"	Ea.	64.50
Camera and monitor	Ea.	2025	Double 15" x 12" x 5"	Ea.	89.50
For additional camera stations, add	Ea.	1100	Quadruple 15" x 12" x 10"	Ea.	141
Emergency Lighting, 25 watt, battery operated			Vertical, 6" x 5" x 15", aluminum	Ea.	64.50
Lead battery	Ea.	345	Bronze	Ea.	70.50
Nickel cadmium	Ea.	685	Steel, enameled	Ea.	67
Flagpoles, Complete			Scales, Dial type, 5 ton cap.		
Aluminum, 20' high	Ea.	1900	8' x 6' platform	Ea.	12,900
40' high	Ea.	4425	9' x 7' platform	Ea.	15,100
70' high	Ea.	11,500	Smoke Detectors		
Fiberglass, 23' high	Ea.	1350	Ceiling type	Ea.	227
39'-5" high	Ea.	3400	Duct type	Ea.	565
59' high	Ea.	6650			

Important: See the Reference Section for Location Factors.

Model costs calculated for a 1 story building with 14' story height and 13,000 square feet of floor area

Post Office

				Unit	Unit Cost	Cost Per S.F.	% Of Sub-Total
A. SUBSTRUCTURE							
1010	Standard Foundations	Poured concrete; strip and spread footings		S.F. Ground	1.72	1.72	
1020	Special Foundations	N/A		—	—	—	
1030	Slab on Grade	4" reinforced concrete with vapor barrier and granular base		S.F. Slab	5.56	5.56	10.1%
2010	Basement Excavation	Site preparation for slab and trench for foundation wall and footing		S.F. Ground	.32	.32	
2020	Basement Walls	4' foundation wall		L.F. Wall	86	3.20	
B. SHELL							
	B10 Superstructure						
1010	Floor Construction	Steel column fireproofing		L.F. Column	36.45	.27	9.0%
1020	Roof Construction	Metal deck, open web steel joists, columns		S.F. Roof	9.39	9.39	
	B20 Exterior Enclosure						
2010	Exterior Walls	Face brick with concrete block backup	80% of wall	S.F. Wall	33.39	13.98	
2020	Exterior Windows	Double strength window glass	20% of wall	Each	746	3.39	17.4%
2030	Exterior Doors	Double aluminum & glass, single aluminum, hollow metal, steel overhead		Each	2753	1.27	
	B30 Roofing						
3010	Roof Coverings	Built-up tar and gravel with flashing; perlite/EPS composite insulation		S.F. Roof	6.83	6.83	6.4%
3020	Roof Openings	N/A		—	—	—	
C. INTERIORS							
1010	Partitions	Concrete block	15 S.F. Floor/L.F. Partition	S.F. Partition	9.73	7.78	
1020	Interior Doors	Single leaf hollow metal	150 S.F. Floor/Door	Each	1168	7.78	
1030	Fittings	Toilet partitions, cabinets, shelving, lockers		S.F. Floor	1.58	1.58	
2010	Stair Construction	N/A		—	—	—	23.4%
3010	Wall Finishes	Paint		S.F. Surface	2.38	3.80	
3020	Floor Finishes	50% vinyl tile, 50% paint		S.F. Floor	2.32	2.32	
3030	Ceiling Finishes	Mineral fiber tile on concealed zee bars	25% of area	S.F. Ceiling	7.22	1.81	
D. SERVICES							
	D10 Conveying						
1010	Elevators & Lifts	N/A		—	—	—	0.0 %
1020	Escalators & Moving Walks	N/A		—	—	—	
	D20 Plumbing						
2010	Plumbing Fixtures	Toilet and service fixtures, supply and drainage	1 Fixture/1180 S.F. Floor	Each	2289	1.94	
2020	Domestic Water Distribution	Gas fired water heater		S.F. Floor	.51	.51	3.7%
2040	Rain Water Drainage	Roof drains		S.F. Roof	1.48	1.48	
	D30 HVAC						
3010	Energy Supply	N/A		—	—	—	
3020	Heat Generating Systems	Included in D3050		—	—	—	
3030	Cooling Generating Systems	N/A		—	—	—	8.9 %
3050	Terminal & Package Units	Single zone, gas heating, electric cooling		S.F. Floor	9.57	9.57	
3090	Other HVAC Sys. & Equipment	N/A		—	—	—	
	D40 Fire Protection						
4010	Sprinklers	Wet pipe sprinkler system		S.F. Floor	4.72	4.72	5.3%
4020	Standpipes	Standpipe		S.F. Floor	.99	.99	
	D50 Electrical						
5010	Electrical Service/Distribution	600 ampere service, panel board and feeders		S.F. Floor	2.59	2.59	
5020	Lighting & Branch Wiring	T-8 fluorescent fixtures, receptacles, switches, A.C. and misc. power		S.F. Floor	9.36	9.36	15.7%
5030	Communications & Security	Addressable alarm systems, internet wiring and emergency lighting		S.F. Floor	4.69	4.69	
5090	Other Electrical Systems	Emergency generator, 15 kW		S.F. Floor	.14	.14	
E. EQUIPMENT & FURNISHINGS							
1010	Commercial Equipment	N/A		—	—	—	
1020	Institutional Equipment	N/A		—	—	—	0.0 %
1030	Vehicular Equipment	N/A		—	—	—	
1090	Other Equipment	N/A		—	—	—	
F. SPECIAL CONSTRUCTION							
1020	Integrated Construction	N/A		—	—	—	0.0 %
1040	Special Facilities	N/A		—	—	—	
G. BUILDING SITEWORK	**N/A**						

		Sub-Total	106.99	100%
CONTRACTOR FEES (General Requirements: 10%, Overhead: 5%, Profit: 10%)		25%	26.72	
ARCHITECT FEES		9%	12.04	
	Total Building Cost		**145.75**	

Costs per square foot of floor area

Exterior Wall	S.F. Area	5000	10000	15000	21000	25000	30000	40000	50000	60000
	L.F. Perimeter	287	400	500	600	660	700	834	900	1000
Face Brick with Concrete Block Back-up	Steel Frame	249.45	213.70	200.45	191.40	187.40	182.00	177.40	172.60	170.25
	Bearing Walls	247.30	211.05	197.65	188.50	184.40	178.95	174.30	169.40	167.00
Concrete Block	Steel Frame	215.05	189.75	180.50	174.30	171.60	168.05	164.95	161.80	160.25
Brick Veneer	Steel Frame	235.20	203.75	192.15	184.35	180.85	176.25	172.30	168.15	166.10
Galvanized Steel Siding	Steel Frame	205.05	183.20	175.30	170.05	167.75	164.80	162.25	159.70	158.35
Metal Sandwich Panel	Steel Frame	258.80	220.30	205.90	196.10	191.70	185.85	180.80	175.55	172.95
Perimeter Adj., Add or Deduct	Per 100 L.F.	31.10	15.60	10.35	7.40	6.15	5.20	3.90	3.10	2.55
Story Hgt. Adj., Add or Deduct	Per 1 Ft.	6.30	4.45	3.70	3.20	2.90	2.55	2.35	2.00	1.85
For Basement, add $30.40 per square foot of basement area										

The above costs were calculated using the basic specifications shown on the facing page. These costs should be adjusted where necessary for design alternatives and owner's requirements. Reported completed project costs, for this type of structure, range from $97.50 to $304.05 per S.F.

Common additives

Description	Unit	$ Cost	Description	Unit	$ Cost
Bar, Front Bar	L.F.	435	Lockers, Steel, single tier, 60" or 72"	Opng.	231 - 380
Back Bar	L.F.	345	2 tier, 60" or 72" total	Opng.	121 - 170
Booth, Upholstered, custom straight	L.F.	252 - 465	5 tier, box lockers	Opng.	74 - 84
"L" or "U" shaped	L.F.	261 - 440	Locker bench, lam. maple top only	L.F.	34.50
Bleachers, Telescoping, manual			Pedestals, steel pipe	Ea.	76.50
To 15 tier	Seat	143 - 199	Sauna, Prefabricated, complete		
21-30 tier	Seat	310 - 410	6' x 4'	Ea.	6750
Courts			6' x 9'	Ea.	10,500
Ceiling	Court	10,000	8' x 8'	Ea.	12,000
Floor	Court	16,700	8' x 10'	Ea.	13,400
Walls	Court	28,800	10' x 12'	Ea.	16,100
Emergency Lighting, 25 watt, battery operated			Sound System		
Lead battery	Ea.	345	Amplifier, 250 watts	Ea.	2200
Nickel cadmium	Ea.	685	Speaker, ceiling or wall	Ea.	217
Kitchen Equipment			Trumpet	Ea.	415
Broiler	Ea.	4075	Steam Bath, Complete, to 140 C.F.	Ea.	3000
Cooler, 6 ft. long, reach-in	Ea.	3925	To 300 C.F.	Ea.	3325
Dishwasher, 10-12 racks per hr.	Ea.	4275	To 800 C.F.	Ea.	7100
Food warmer, counter 1.2 KW	Ea.	770	To 2500 C.F.	Ea.	9325
Freezer, reach-in, 44 C.F.	Ea.	6175			
Ice cube maker, 50 lb. per day	Ea.	2000			

Important: See the Reference Section for Location Factors.

Model costs calculated for a 2 story building with 12′ story height and 30,000 square feet of floor area

					Unit	Unit Cost	Cost Per S.F.	% Of Sub-Total
A. SUBSTRUCTURE								
1010	Standard Foundations	Poured concrete; strip and spread footings			S.F. Ground	2.28	1.14	
1020	Special Foundations	N/A			—	—	—	
1030	Slab on Grade	5″ reinforced concrete with vapor barrier and granular base			S.F. Slab	7.35	3.68	5.2%
2010	Basement Excavation	Site preparation for slab and trench for foundation wall and footing			S.F. Ground	.32	.21	
2020	Basement Walls	4′ foundation wall			L.F. Wall	86	1.99	
B. SHELL								
	B10 Superstructure							
1010	Floor Construction	Open web steel joists, slab form, concrete, columns	50% of area		S.F. Floor	18.94	9.47	9.9%
1020	Roof Construction	Metal deck on open web steel joists, columns			S.F. Roof	7.86	3.93	
	B20 Exterior Enclosure							
2010	Exterior Walls	Face brick with concrete block backup	95% of wall		S.F. Wall	32.89	17.50	
2020	Exterior Windows	Storefront	5% of wall		S.F. Window	103	2.87	15.4%
2030	Exterior Doors	Aluminum and glass and hollow metal			Each	3151	.43	
	B30 Roofing							
3010	Roof Coverings	Built-up tar and gravel with flashing; perlite/EPS composite insulation			S.F. Roof	7	3.50	2.6%
3020	Roof Openings	Roof hatches			S.F. Roof	.14	.07	
C. INTERIORS								
1010	Partitions	Concrete block, gypsum board on metal studs	25 S.F. Floor/L.F. Partition		S.F. Partition	15.03	6.01	
1020	Interior Doors	Single leaf hollow metal	810 S.F. Floor/Door		Each	1168	1.44	
1030	Fittings	Toilet partitions			S.F. Floor	.41	.41	
2010	Stair Construction	Concrete filled metal pan			Flight	17,275	1.73	13.5%
3010	Wall Finishes	Paint			S.F. Surface	1.45	1.16	
3020	Floor Finishes	80% carpet tile, 20% ceramic tile	50% of floor area		S.F. Floor	6.34	3.17	
3030	Ceiling Finishes	Mineral fiber tile on concealed zee bars	60% of area		S.F. Ceiling	7.22	4.33	
D. SERVICES								
	D10 Conveying							
1010	Elevators & Lifts	N/A			—	—	—	0.0 %
1020	Escalators & Moving Walks	N/A			—	—	—	
	D20 Plumbing							
2010	Plumbing Fixtures	Kitchen, bathroom and service fixtures, supply and drainage	1 Fixture/1000 S.F. Floor		Each	3680	3.68	
2020	Domestic Water Distribution	Gas fired water heater			S.F. Floor	3.28	3.28	5.5%
2040	Rain Water Drainage	Roof drains			S.F. Roof	.80	.40	
	D30 HVAC							
3010	Energy Supply	N/A			—	—	—	
3020	Heat Generating Systems	Included in D3050			—	—	—	
3030	Cooling Generating Systems	N/A			—	—	—	22.5 %
3050	Terminal & Package Units	Multizone unit, gas heating, electric cooling			S.F. Floor	30.35	30.35	
3090	Other HVAC Sys. & Equipment	N/A			—	—	—	
	D40 Fire Protection							
4010	Sprinklers	Sprinklers, light hazard			S.F. Floor	3.40	3.40	2.8%
4020	Standpipes	Standpipe			S.F. Floor	.40	.40	
	D50 Electrical							
5010	Electrical Service/Distribution	400 ampere service, panel board and feeders			S.F. Floor	.90	.90	
5020	Lighting & Branch Wiring	Fluorescent and high intensity discharge fixtures, receptacles, switches, A.C. and misc. power			S.F. Floor	8.48	8.48	7.8%
5030	Communications & Security	Addressable alarm systems and emergency lighting			S.F. Floor	1.12	1.12	
5090	Other Electrical Systems	Emergency generator, 15 kW			S.F. Floor	.06	.06	
E. EQUIPMENT & FURNISHINGS								
1010	Commercial Equipment	N/A			—	—	—	
1020	Institutional Equipment	N/A			—	—	—	14.6 %
1030	Vehicular Equipment	N/A			—	—	—	
1090	Other Equipment	Courts, sauna baths			S.F. Floor	19.71	19.71	
F. SPECIAL CONSTRUCTION								
1020	Integrated Construction	N/A			—	—	—	0.0 %
1040	Special Facilities	N/A			—	—	—	
G. BUILDING SITEWORK	**N/A**							

				Sub-Total	134.82	**100%**
	CONTRACTOR FEES (General Requirements: 10%, Overhead: 5%, Profit: 10%)			25%	33.70	
	ARCHITECT FEES			8%	13.48	
			Total Building Cost		**182**	

For customer support on your Square Foot Cost Data, call 877.756.2789.

191

Costs per square foot of floor area

Exterior Wall	S.F. Area	5000	6000	7000	8000	9000	10000	11000	12000	13000
	L.F. Perimeter	286	320	353	386	397	425	454	460	486
Face Brick with Concrete Block Back-up	Steel Joists	195.05	190.45	187.00	184.55	180.45	178.65	177.30	174.50	173.45
	Wood Joists	197.80	193.15	189.70	187.15	183.05	181.20	179.85	177.05	175.95
Stucco on Concrete Block	Steel Joists	198.10	193.95	190.75	188.45	184.70	183.10	181.80	179.30	178.30
	Wood Joists	192.55	188.35	185.15	182.80	179.10	177.45	176.15	173.55	172.55
Limestone with Concrete Block Back-up	Steel Joists	223.90	217.85	213.30	210.00	204.35	202.00	200.10	196.20	194.80
	Wood Joists	219.15	213.05	208.50	205.15	199.50	197.10	195.25	191.35	189.85
Perimeter Adj., Add or Deduct	Per 100 L.F.	17.10	14.30	12.30	10.70	9.55	8.65	7.80	7.10	6.50
Story Hgt. Adj., Add or Deduct	Per 1 Ft.	3.00	2.85	2.65	2.50	2.35	2.25	2.15	2.00	1.90

For Basement, add $31.95 per square foot of basement area

The above costs were calculated using the basic specifications shown on the facing page. These costs should be adjusted where necessary for design alternatives and owner's requirements. Reported completed project costs, for this type of structure, range from $85.15 to $263.65 per S.F.

Common additives

Description	Unit	$ Cost
Carrels Hardwood	Ea.	945 - 2400
Emergency Lighting, 25 watt, battery operated		
Lead battery	Ea.	345
Nickel cadmium	Ea.	685
Flagpoles, Complete		
Aluminum, 20' high	Ea.	1900
40' high	Ea.	4425
70' high	Ea.	11,500
Fiberglass, 23' high	Ea.	1350
39'-5" high	Ea.	3400
59' high	Ea.	6650
Gym Floor, Incl. sleepers and finish, maple	S.F.	14.75
Intercom System, 25 Station capacity		
Master station	Ea.	3175
Intercom outlets	Ea.	192
Handset	Ea.	525

Description	Unit	$ Cost
Lockers, Steel, single tier, 60" to 72"	Opng.	231 - 380
2 tier, 60" to 72" total	Opng.	121 - 170
5 tier, box lockers	Opng.	74 - 84
Locker bench, lam. maple top only	L.F.	34.50
Pedestals, steel pipe	Ea.	76.50
Seating		
Auditorium chair, all veneer	Ea.	315
Veneer back, padded seat	Ea.	325
Upholstered, spring seat	Ea.	320
Classroom, movable chair & desk	Set	81 - 171
Lecture hall, pedestal type	Ea.	275 - 620
Smoke Detectors		
Ceiling type	Ea.	227
Duct type	Ea.	565
Sound System		
Amplifier, 250 watts	Ea.	2200
Speaker, ceiling or wall	Ea.	217
Trumpet	Ea.	415
Swimming Pools, Complete, gunite	S.F.	104 - 129

Important: See the Reference Section for Location Factors.

Model costs calculated for a 1 story building with 12' story height and 10,000 square feet of floor area

				Unit	Unit Cost	Cost Per S.F.	% Of Sub-Total
A.	**SUBSTRUCTURE**						
1010	Standard Foundations	Poured concrete; strip and spread footings		S.F. Ground	2.31	2.31	
1020	Special Foundations	N/A		—	—	—	
1030	Slab on Grade	4" reinforced concrete with vapor barrier and granular base		S.F. Slab	5.56	5.56	9.0%
2010	Basement Excavation	Site preparation for slab and trench for foundation wall and footing		S.F. Ground	.32	.32	
2020	Basement Walls	4' foundation wall		L.F. Wall	86	3.63	
B.	**SHELL**						
	B10 Superstructure						
1010	Floor Construction	N/A		—	—	—	7.2 %
1020	Roof Construction	Metal deck, open web steel joists, beams, interior columns		S.F. Roof	9.45	9.45	
	B20 Exterior Enclosure						
2010	Exterior Walls	Face brick with concrete block backup	85% of wall	S.F. Wall	27.50	11.92	
2020	Exterior Windows	Window wall	15% of wall	Each	576	6.59	16.1%
2030	Exterior Doors	Double aluminum and glass		Each	6588	2.65	
	B30 Roofing						
3010	Roof Coverings	Built-up tar and gravel with flashing; perlite/EPS composite insulation		S.F. Roof	6.84	6.84	5.2%
3020	Roof Openings	N/A		—	—	—	
C.	**INTERIORS**						
1010	Partitions	Concrete block	8 S.F. Floor/S.F. Partition	S.F. Partition	9.73	12.16	
1020	Interior Doors	Single leaf hollow metal	700 S.F. Floor/Door	Each	1168	1.67	
1030	Fittings	Toilet partitions		S.F. Floor	1.22	1.22	
2010	Stair Construction	N/A		—	—	—	24.8%
3010	Wall Finishes	Paint		S.F. Surface	2.32	5.81	
3020	Floor Finishes	50% vinyl tile, 50% carpet		S.F. Floor	4.49	4.49	
3030	Ceiling Finishes	Mineral fiber tile on concealed zee bars		S.F. Ceiling	7.22	7.22	
D.	**SERVICES**						
	D10 Conveying						
1010	Elevators & Lifts	N/A		—	—	—	0.0 %
1020	Escalators & Moving Walks	N/A		—	—	—	
	D20 Plumbing						
2010	Plumbing Fixtures	Toilet and service fixtures, supply and drainage	1 Fixture/455 S.F. Floor	Each	3822	8.40	
2020	Domestic Water Distribution	Gas fired water heater		S.F. Floor	1.64	1.64	8.5%
2040	Rain Water Drainage	Roof drains		S.F. Roof	1.12	1.12	
	D30 HVAC						
3010	Energy Supply	N/A		—	—	—	
3020	Heat Generating Systems	N/A		—	—	—	
3030	Cooling Generating Systems	N/A		—	—	—	19.8 %
3050	Terminal & Package Units	Multizone HVAC air cooled system		S.F. Floor	26	26	
3090	Other HVAC Sys. & Equipment	N/A		—	—	—	
	D40 Fire Protection						
4010	Sprinklers	Sprinkler, light hazard		S.F. Floor	3.65	3.65	3.2%
4020	Standpipes	Standpipe		S.F. Floor	.61	.61	
	D50 Electrical						
5010	Electrical Service/Distribution	200 ampere service, panel board and feeders		S.F. Floor	1.14	1.14	
5020	Lighting & Branch Wiring	T-8 fluorescent fixtures, receptacles, switches, A.C. and misc. power		S.F. Floor	3.05	3.05	6.0%
5030	Communications & Security	Addressable alarm systems, partial internet wiring and emergency lighting		S.F. Floor	3.59	3.59	
5090	Other Electrical Systems	Emergency generator, 15 kW		S.F. Floor	.09	.09	
E.	**EQUIPMENT & FURNISHINGS**						
1010	Commercial Equipment	N/A		—	—	—	
1020	Institutional Equipment	N/A		—	—	—	0.0 %
1030	Vehicular Equipment	N/A		—	—	—	
1090	Other Equipment	N/A		—	—	—	
F.	**SPECIAL CONSTRUCTION**						
1020	Integrated Construction	N/A		—	—	—	0.0 %
1040	Special Facilities	N/A		—	—	—	
G.	**BUILDING SITEWORK**	**N/A**					

		Sub-Total	131.13	100%
CONTRACTOR FEES (General Requirements: 10%, Overhead: 5%, Profit: 10%)		25%	32.77	
ARCHITECT FEES		9%	14.75	
	Total Building Cost		**178.65**	

For customer support on your Square Foot Cost Data, call 877.756.2789.

193

Costs per square foot of floor area

Exterior Wall	S.F. Area	2000	2800	3500	4200	5000	5800	6500	7200	8000
	L.F. Perimeter	180	212	240	268	300	314	336	344	368
Wood Siding	Wood Frame	222.55	210.85	204.95	201.10	198.00	194.15	192.35	189.90	188.55
Brick Veneer	Wood Frame	227.20	214.30	207.85	203.55	200.10	195.85	193.80	191.05	189.50
Stone Veneer	Steel Frame	263.50	246.00	237.25	231.40	226.80	220.55	217.75	213.75	211.60
Fiber Cement Siding	Steel Frame	229.05	217.15	211.20	207.25	204.10	200.20	198.40	195.90	194.50
E.I.F.S.	Reinforced Concrete	254.35	240.55	233.70	229.10	225.40	220.75	218.55	215.55	213.90
Stucco	Reinforced Concrete	258.25	243.95	236.85	232.10	228.25	223.40	221.15	218.00	216.35
Perimeter Adj., Add or Deduct	Per 100 L.F.	25.15	18.00	14.45	11.95	10.05	8.65	7.75	7.05	6.25
Story Hgt. Adj., Add or Deduct	Per 1 Ft.	2.40	2.00	1.90	1.70	1.55	1.50	1.40	1.35	1.25
For Basement, add $37.00 per square foot of basement area										

The above costs were calculated using the basic specifications shown on the facing page. These costs should be adjusted where necessary for design alternatives and owner's requirements. Reported completed project costs, for this type of structure, range from $132.00 to $307.60 per S.F.

Common additives

Description	Unit	$ Cost
Bar, Front Bar	L.F.	435
Back bar	L.F.	345
Booth, Upholstered, custom straight	L.F.	252 - 465
"L" or "U" shaped	L.F.	261 - 440
Cupola, Stock unit, redwood		
30" square, 37" high, aluminum roof	Ea.	775
Copper roof	Ea.	435
Fiberglass, 5'-0" base, 63" high	Ea.	4025 - 4725
6'-0" base, 63" high	Ea.	5225 - 7550
Decorative Wood Beams, Non load bearing		
Rough sawn, 4" x 6"	L.F.	9
4" x 8"	L.F.	10.10
4" x 10"	L.F.	11.95
4" x 12"	L.F.	13.55
8" x 8"	L.F.	14.45
Emergency Lighting, 25 watt, battery operated		
Lead battery	Ea.	345
Nickel cadmium	Ea.	685

Description	Unit	$ Cost
Fireplace, Brick, not incl. chimney or foundation		
30" x 29" opening	Ea.	3200
Chimney, standard brick		
Single flue, 16" x 20"	V.L.F.	108
20" x 20"	V.L.F.	126
2 Flue, 20" x 24"	V.L.F.	157
20" x 32"	V.L.F.	193
Kitchen Equipment		
Broiler	Ea.	4075
Coffee urn, twin 6 gallon	Ea.	3025
Cooler, 6 ft. long	Ea.	3925
Dishwasher, 10-12 racks per hr.	Ea.	4275
Food warmer, counter, 1.2 KW	Ea.	770
Freezer, 44 C.F., reach-in	Ea.	6175
Ice cube maker, 50 lb. per day	Ea.	2000
Range with 1 oven	Ea.	3300
Refrigerators, Prefabricated, walk-in		
7'-6" high, 6' x 6'	S.F.	203
10' x 10'	S.F.	159
12' x 14'	S.F.	142
12' x 20'	S.F.	125

Important: See the Reference Section for Location Factors.

Model costs calculated for a 1 story building with 12' story height and 5,000 square feet of floor area

				Unit	Unit Cost	Cost Per S.F.	% Of Sub-Total
A.	**SUBSTRUCTURE**						
1010	Standard Foundations	Poured concrete; strip and spread footings		S.F. Ground	4.55	4.55	
1020	Special Foundations	N/A		—	—	—	
1030	Slab on Grade	4" reinforced concrete		S.F. Slab	5.56	5.56	10.7%
2010	Basement Excavation	Site preparation for slab and trench for foundation wall and footing		S.F. Ground	.57	.57	
2020	Basement Walls	4' Foundation wall		L.F. Wall	86	5.13	
B.	**SHELL**						
	B10 Superstructure						
1010	Floor Construction	Wood columns		S.F. Floor	.17	.42	5.3%
1020	Roof Construction	Wood roof truss, 4:12 slope		S.F. Roof	6.63	7.42	
	B20 Exterior Enclosure						
2010	Exterior Walls	Wood siding on wood studs, insulated	70% of wall	S.F. Wall	20.93	10.55	
2020	Exterior Windows	Curtain glazing wall system	30% of wall	Each	46.15	4.45	13.8%
2030	Exterior Doors	Aluminum & glass double doors, hollow metal doors		Each	5472	5.49	
	B30 Roofing						
3010	Roof Coverings	Asphalt strip shingles, gutters and downspouts		S.F. Roof	15.95	2.46	1.7%
3020	Roof Openings	N/A		—	—	—	
C.	**INTERIORS**						
1010	Partitions	Fire rated gypsum board on wood studs	25 S.F. Floor/L.F. Partition	S.F. Partition	7	2.80	
1020	Interior Doors	Single leaf hollow core wood doors	1000 S.F. Floor/Door	Each	542	.54	
1030	Fittings	Plastic laminate toilet partitions		S.F. Floor	.88	.88	
2010	Stair Construction	N/A		—	—	—	13.3%
3010	Wall Finishes	80% Paint, 20% ceramic tile		S.F. Surface	2.75	2.20	
3020	Floor Finishes	70% Carpet tile, 30% quarry tile		S.F. Floor	8.57	8.57	
3030	Ceiling Finishes	Gypsum board ceiling on wood furring		S.F. Ceiling	4.65	4.65	
D.	**SERVICES**						
	D10 Conveying						
1010	Elevators & Lifts	N/A		—	—	—	0.0 %
1020	Escalators & Moving Walks	N/A		—	—	—	
	D20 Plumbing						
2010	Plumbing Fixtures	Restroom, kitchen and service fixtures, supply and drainage	1 Fixture/350 S.F. Floor	Each	3064	8.63	
2020	Domestic Water Distribution	Gas fired water heater		S.F. Floor	5.08	5.08	9.3%
2040	Rain Water Drainage	Roof drains		S.F. Roof			
	D30 HVAC						
3010	Energy Supply	N/A		—	—	—	
3020	Heat Generating Systems	Included in D3050		—	—	—	
3030	Cooling Generating Systems	N/A		—	—	—	25.9 %
3050	Terminal & Package Units	Multizone unit gas heating, electric cooling, kitchen exhaust system		S.F. Floor	38.32	38.32	
3090	Other HVAC Sys. & Equipment	N/A		—	—	—	
	D40 Fire Protection						
4010	Sprinklers	Wet pipe sprinkler system		S.F. Floor	9.19	9.19	7.7%
4020	Standpipes	Standpipe		S.F. Floor	2.21	2.21	
	D50 Electrical						
5010	Electrical Service/Distribution	400 Ampere service, panelboards and feeders		S.F. Floor	4.70	4.70	
5020	Lighting & Branch Wiring	Fluorescent fixtures, receptacles, switches, a.c. and misc. power		S.F. Floor	9.98	9.98	12.4%
5030	Communications & Security	Addressable alarm system		S.F. Floor	3.67	3.67	
5090	Other Electrical Systems	N/A		—	—	—	
E.	**EQUIPMENT & FURNISHINGS**						
1010	Commercial Equipment	N/A		—	—	—	
1020	Institutional Equipment	N/A		—	—	—	0.0 %
1030	Vehicular Equipment	N/A		—	—	—	
1090	Other Equipment	N/A		—	—	—	
F.	**SPECIAL CONSTRUCTION**						
1020	Integrated Construction	N/A		—	—	—	0.0 %
1040	Special Facilities	N/A		—	—	—	
G.	**BUILDING SITEWORK**	**N/A**					

		Sub-Total	148.02	**100%**
CONTRACTOR FEES (General Requirements: 10%, Overhead: 5%, Profit: 10%)		25%	37.03	
ARCHITECT FEES		7%	12.95	
		Total Building Cost	**198**	

For customer support on your Square Foot Cost Data, call 877.756.2789.

195

Costs per square foot of floor area

Exterior Wall	S.F. Area	2000	2800	3500	4000	5000	5800	6500	7200	8000
	L.F. Perimeter	180	212	240	260	300	314	336	344	368
Wood Siding	Wood Frame	216.05	208.85	205.25	203.45	200.95	197.95	196.65	194.80	193.85
Brick Veneer	Wood Frame	225.35	217.15	213.05	211.00	208.10	204.70	203.25	201.10	200.05
Stone Veneer	Steel Frame	253.90	241.85	235.75	232.75	228.50	223.55	221.45	218.20	216.70
Fiber Cement Siding	Steel Frame	219.50	212.50	209.00	207.20	204.75	201.85	200.65	198.80	197.90
E.I.F.S.	Reinforced Concrete	246.90	237.95	233.50	231.25	228.15	224.45	222.90	220.50	219.30
Stucco	Reinforced Concrete	246.90	237.95	233.50	231.30	228.20	224.50	222.95	220.55	219.40
Perimeter Adj., Add or Deduct	Per 100 L.F.	25.25	18.00	14.40	12.60	10.10	8.65	7.85	6.95	6.25
Story Hgt. Adj., Add or Deduct	Per 1 Ft.	2.65	2.20	2.00	1.85	1.65	1.60	1.55	1.35	1.30
Basement—Not Applicable										

The above costs were calculated using the basic specifications shown on the facing page. These costs should be adjusted where necessary for design alternatives and owner's requirements. Reported completed project costs, for this type of structure, range from $131.70 to $256.80 per S.F.

Common additives

Description	Unit	$ Cost	Description	Unit	$ Cost
Bar, Front Bar	L.F.	435	Refrigerators, Prefabricated, walk-in		
Back bar	L.F.	345	7'-6" High, 6' x 6'	S.F.	203
Booth, Upholstered, custom straight	L.F.	252 - 465	10' x 10'	S.F.	159
"L" or "U" shaped	L.F.	261 - 440	12' x 14'	S.F.	142
Drive-up Window	Ea.	9175 - 13,000	12' x 20'	S.F.	125
Emergency Lighting, 25 watt, battery operated			Serving		
Lead battery	Ea.	345	Counter top (Stainless steel)	L.F.	199
Nickel cadmium	Ea.	685	Base cabinets	L.F.	420 - 675
Kitchen Equipment			Sound System		
Broiler	Ea.	4075	Amplifier, 250 watts	Ea.	2200
Coffee urn, twin 6 gallon	Ea.	3025	Speaker, ceiling or wall	Ea.	217
Cooler, 6 ft. long	Ea.	3925	Trumpet	Ea.	415
Dishwasher, 10-12 racks per hr.	Ea.	4275	Storage		
Food warmer, counter, 1.2 KW	Ea.	770	Shelving	S.F.	10.30
Freezer, 44 C.F., reach-in	Ea.	6175	Washing		
Ice cube maker, 50 lb. per day	Ea.	2000	Stainless steel counter	L.F.	199
Range with 1 oven	Ea.	3300			

Important: See the Reference Section for Location Factors.

Model costs calculated for a 1 story building with 10' story height and 4,000 square feet of floor area

				Unit	Unit Cost	Cost Per S.F.	% Of Sub-Total
A. SUBSTRUCTURE							
1010	Standard Foundations	Poured concrete; strip and spread footings		S.F. Ground	3.61	3.61	
1020	Special Foundations	N/A		—	—	—	
1030	Slab on Grade	4" reinforced concrete		S.F. Slab	5.56	5.56	10.2%
2010	Basement Excavation	Site preparation for slab and trench for foundation wall and footing		S.F. Ground	.57	.57	
2020	Basement Walls	4' Foundation wall		L.F. Wall	86	5.56	
B. SHELL							
B10 Superstructure							
1010	Floor Construction	Wood columns		S.F. Floor	.42	.42	3.3%
1020	Roof Construction	Flat wood rafter roof		S.F. Roof	4.50	4.50	
B20 Exterior Enclosure							
2010	Exterior Walls	Wood siding on wood studs, insulated	70% of wall	S.F. Wall	17.87	8.13	
2020	Exterior Windows	Curtain glazing wall system	30% of wall	Each	52	4.73	13.9%
2030	Exterior Doors	Aluminum & glass double doors, hollow metal doors		Each	4055	8.12	
B30 Roofing							
3010	Roof Coverings	Single ply membrane, stone ballast, rigid insulation		S.F. Roof	6.55	6.55	4.7%
3020	Roof Openings	Roof and smoke hatches		S.F. Roof	.52	.52	
C. INTERIORS							
1010	Partitions	Water resistant gypsum board on wood studs	25 S.F. Floor/L.F. Partition	S.F. Partition	6.90	2.76	
1020	Interior Doors	Single leaf hollow core wood doors	1000 S.F. Floor/Door	Each	542	.54	
1030	Fittings	Stainless steel toilet partitions		S.F. Floor	1.53	1.53	
2010	Stair Construction	N/A		—	—	—	20.2%
3010	Wall Finishes	90% Paint, 10% ceramic tile		S.F. Surface	2.01	1.61	
3020	Floor Finishes	Quarry tile		S.F. Floor	16.76	16.76	
3030	Ceiling Finishes	Acoustic ceiling tiles on suspended support grid		S.F. Ceiling	7.22	7.22	
D. SERVICES							
D10 Conveying							
1010	Elevators & Lifts	N/A		—	—	—	0.0 %
1020	Escalators & Moving Walks	N/A		—	—	—	
D20 Plumbing							
2010	Plumbing Fixtures	Restroom, kitchen and service fixtures, supply and drainage	1 Fixture/330 S.F. Floor	Each	2681	8.05	
2020	Domestic Water Distribution	Gas fired water heater		S.F. Floor	3.63	3.63	9.0%
2040	Rain Water Drainage	Roof drains		S.F. Roof	1.84	1.84	
D30 HVAC							
3010	Energy Supply	N/A		—	—	—	
3020	Heat Generating Systems	Included in D3050		—	—	—	
3030	Cooling Generating Systems	N/A		—	—	—	12.7 %
3050	Terminal & Package Units	Single zone unit gas heating, electric cooling		S.F. Floor	19.14	19.14	
3090	Other HVAC Sys. & Equipment	N/A		—	—	—	
D40 Fire Protection							
4010	Sprinklers	Wet pipe sprinkler system, ordinary hazard		S.F. Floor	9.19	9.19	7.6%
4020	Standpipes	Standpipe		S.F. Floor	2.30	2.30	
D50 Electrical							
5010	Electrical Service/Distribution	400 Ampere service, panelboards and feeders		S.F. Floor	4.98	4.98	
5020	Lighting & Branch Wiring	Fluorescent fixtures, receptacles, switches, a.c. and misc. power		S.F. Floor	10.83	10.83	13.5%
5030	Communications & Security	Addressable alarm, emergency lighting		S.F. Floor	4.57	4.57	
5090	Other Electrical Systems	N/A		—	—	—	
E. EQUIPMENT & FURNISHINGS							
1010	Commercial Equipment	N/A		—	—	—	
1020	Institutional Equipment	N/A		—	—	—	
1030	Vehicular Equipment	N/A		—	—	—	5.0 %
1090	Other Equipment	Prefabricated walk in refrigerators		S.F. Floor	7.49	7.49	
F. SPECIAL CONSTRUCTION							
1020	Integrated Construction	N/A		—	—	—	0.0 %
1040	Special Facilities	N/A		—	—	—	
G. BUILDING SITEWORK	**N/A**						

			Sub-Total	150.71	**100%**
CONTRACTOR FEES (General Requirements: 10%, Overhead: 5%, Profit: 10%)		25%	37.67		
ARCHITECT FEES		8%	15.07		

Total Building Cost	**203.45**

For customer support on your Square Foot Cost Data, call 877.756.2789.

197

Costs per square foot of floor area

Exterior Wall	S.F. Area	10000	15000	20000	25000	30000	35000	40000	45000	50000
	L.F. Perimeter	450	500	600	700	740	822	890	920	966
Face Brick with Concrete Block Back-up	Steel Joists	204.30	186.15	180.60	177.20	172.20	170.25	168.35	165.65	163.95
	Lam. Wood Truss	215.90	198.40	193.35	190.25	185.30	183.50	181.70	179.00	177.35
Concrete Block	Steel Frame	199.65	187.70	183.90	181.60	178.35	177.00	175.80	174.05	173.00
	Lam. Wood Truss	190.35	179.40	176.15	174.15	171.10	169.95	168.85	167.10	166.10
Galvanized Steel Siding	Steel Frame	198.50	186.85	183.15	180.90	177.75	176.40	175.25	173.50	172.45
Metal Sandwich Panel	Steel Frame	201.45	189.00	185.05	182.75	179.30	178.00	176.65	174.90	173.70
Perimeter Adj., Add or Deduct	Per 100 L.F.	14.15	9.40	7.05	5.70	4.65	4.05	3.55	3.10	2.85
Story Hgt. Adj., Add or Deduct	Per 1 Ft.	2.25	1.65	1.45	1.45	1.20	1.20	1.10	1.05	1.00
Basement—Not Applicable										

The above costs were calculated using the basic specifications shown on the facing page. These costs should be adjusted where necessary for design alternatives and owner's requirements. Reported completed project costs, for this type of structure, range from $77.90 to $242.45 per S.F.

Common additives

Description	Unit	$ Cost
Bar, Front Bar	L.F.	435
Back bar	L.F.	345
Booth, Upholstered, custom straight	L.F.	252 - 465
"L" or "U" shaped	L.F.	261 - 440
Bleachers, Telescoping, manual		
To 15 tier	Seat	143 - 199
16-20 tier	Seat	294 - 360
21-30 tier	Seat	310 - 410
For power operation, add	Seat	57.50 - 90
Emergency Lighting, 25 watt, battery operated		
Lead battery	Ea.	345
Nickel cadmium	Ea.	685
Lockers, Steel, single tier, 60" or 72"	Opng.	231 - 380
2 tier, 60" or 72" total	Opng.	121 - 170
5 tier, box lockers	Opng.	74 - 84
Locker bench, lam. maple top only	L.F.	34.50
Pedestals, steel pipe	Ea.	76.50

Description	Unit	$ Cost
Rink		
Dasher boards & top guard	Ea.	196,000
Mats, rubber	S.F.	23.50
Score Board	Ea.	17,000 - 29,200

Important: See the Reference Section for Location Factors.

Model costs calculated for a 1 story building with 24' story height and 30,000 square feet of floor area

				Unit	Unit Cost	Cost Per S.F.	% Of Sub-Total
A.	**SUBSTRUCTURE**						
1010	Standard Foundations	Poured concrete; strip and spread footings		S.F. Ground	1.68	1.68	
1020	Special Foundations	N/A		—	—	—	
1030	Slab on Grade	6" reinforced concrete with vapor barrier and granular base		S.F. Slab	6.63	6.63	8.1%
2010	Basement Excavation	Site preparation for slab and trench for foundation wall and footing		S.F. Ground	.18	.18	
2020	Basement Walls	4' foundation wall		L.F. Wall	81	2.30	
B.	**SHELL**						
	B10 Superstructure						
1010	Floor Construction	Wide flange beams and columns		S.F. Floor	43.57	30.50	22.9%
1020	Roof Construction	(incl. in B1010)					
	B20 Exterior Enclosure						
2010	Exterior Walls	Concrete block	95% of wall	S.F. Wall	15.15	8.52	
2020	Exterior Windows	Store front	5% of wall	Each	50	1.50	8.2%
2030	Exterior Doors	Aluminum and glass, hollow metal, overhead		Each	3381	.90	
	B30 Roofing						
3010	Roof Coverings	Elastomeric neoprene membrane with flashing; perlite/EPS composite insulation		S.F. Roof	5.17	5.17	4.1%
3020	Roof Openings	Roof hatches		S.F. Roof	.24	.24	
C.	**INTERIORS**						
1010	Partitions	Concrete block	140 S.F. Floor/L.F. Partition	S.F. Partition	9.80	.84	
1020	Interior Doors	Hollow metal	2500 S.F. Floor/Door	Each	1168	.46	
1030	Fittings	N/A		—	—	—	
2010	Stair Construction	N/A		—	—	—	6.6%
3010	Wall Finishes	Paint		S.F. Surface	12.78	2.19	
3020	Floor Finishes	80% rubber mat, 20% paint	50% of floor area	S.F. Floor	9.08	4.54	
3030	Ceiling Finishes	Mineral fiber tile on concealed zee bar	10% of area	S.F. Ceiling	7.22	.73	
D.	**SERVICES**						
	D10 Conveying						
1010	Elevators & Lifts	N/A		—	—	—	0.0 %
1020	Escalators & Moving Walks	N/A		—	—	—	
	D20 Plumbing						
2010	Plumbing Fixtures	Toilet and service fixtures, supply and drainage	1 Fixture/1070 S.F. Floor	Each	6474	6.05	
2020	Domestic Water Distribution	Oil fired water heater		S.F. Floor	3.28	3.28	7.6%
2040	Rain Water Drainage	Roof drains		S.F. Roof	.87	.87	
	D30 HVAC						
3010	Energy Supply	Oil fired hot water, unit heaters	10% of area	S.F. Floor	.92	.92	
3020	Heat Generating Systems	N/A		—	—	—	
3030	Cooling Generating Systems	N/A		—	—	—	11.5%
3050	Terminal & Package Units	Single zone, electric cooling	90% of area	S.F. Floor	14.36	14.36	
3090	Other HVAC Sys. & Equipment	N/A		—	—	—	
	D40 Fire Protection						
4010	Sprinklers	N/A		—	—	—	0.5 %
4020	Standpipes	Standpipe		S.F. Floor	.72	.72	
	D50 Electrical						
5010	Electrical Service/Distribution	600 ampere service, panel board and feeders		S.F. Floor	1.58	1.58	
5020	Lighting & Branch Wiring	High intensity discharge and fluorescent fixtures, receptacles, switches, A.C. and misc. power		S.F. Floor	8.16	8.16	9.1%
5030	Communications & Security	Addressable alarm systems, emergency lighting and public address		S.F. Floor	2.14	2.14	
5090	Other Electrical Systems	Emergency generator		S.F. Floor	.22	.22	
E.	**EQUIPMENT & FURNISHINGS**						
1010	Commercial Equipment	N/A		—	—	—	
1020	Institutional Equipment	N/A		—	—	—	
1030	Vehicular Equipment	N/A		—	—	—	0.0 %
1090	Other Equipment	N/A		—	—	—	
F.	**SPECIAL CONSTRUCTION**						
1020	Integrated Construction	N/A		—	—	—	21.5 %
1040	Special Facilities	Dasher boards and rink (including ice making system)		S.F. Floor	28.66	28.66	
G.	**BUILDING SITEWORK**	**N/A**					

		Sub-Total	133.34	100%
CONTRACTOR FEES (General Requirements: 10%, Overhead: 5%, Profit: 10%)		25%	33.34	
ARCHITECT FEES		7%	11.67	
	Total Building Cost		**178.35**	

For customer support on your Square Foot Cost Data, call 877.756.2789.

199

Costs per square foot of floor area

Exterior Wall	S.F. Area	25000	30000	35000	40000	45000	50000	55000	60000	65000
	L.F. Perimeter	900	1050	1200	1350	1510	1650	1800	1970	2100
Fiber Cement Siding	Steel Frame	166.50	165.30	164.45	163.75	163.45	162.80	162.40	162.45	161.85
Metal Panel	Steel Frame	167.80	166.50	165.50	164.85	164.50	163.85	163.45	163.50	162.85
E.I.F.S.	Steel Frame	169.25	168.05	167.10	166.45	166.10	165.45	165.15	165.10	164.55
Tilt-up Concrete Panel	Reinforced Concrete	206.05	204.05	202.60	201.55	201.05	200.00	199.45	199.55	198.60
Brick Veneer	Reinforced Concrete	193.40	191.75	190.60	189.75	189.30	188.45	188.05	188.05	187.30
Stone Veneer	Reinforced Concrete	228.70	226.10	224.15	222.80	222.15	220.75	220.05	220.15	218.85
Perimeter Adj., Add or Deduct	Per 100 L.F.	3.30	2.70	2.30	1.95	1.75	1.65	1.50	1.30	1.30
Story Hgt. Adj., Add or Deduct	Per 1 Ft.	1.25	1.20	1.10	1.15	1.10	1.15	1.15	1.15	1.10
For Basement, add $27.40 per square foot of basement area										

The above costs were calculated using the basic specifications shown on the facing page. These costs should be adjusted where necessary for design alternatives and owner's requirements. Reported completed project costs, for this type of structure, range from $91.00 to $231.50 per S.F.

Common additives

Description	Unit	$ Cost
Bleachers, Telescoping, manual		
To 15 tier	Seat	143 - 199
16-20 tier	Seat	294 - 360
21-30 tier	Seat	310 - 410
For power operation, add	Seat	57.50 - 90
Carrels Hardwood	Ea.	945 - 2400
Clock System		
20 room	Ea.	20,600
50 room	Ea.	48,500
Emergency Lighting, 25 watt, battery operated		
Lead battery	Ea.	345
Nickel cadmium	Ea.	685
Flagpoles, Complete		
Aluminum, 20' high	Ea.	1900
40' high	Ea.	4425
Fiberglass, 23' high	Ea.	1350
39'-5" high	Ea.	3400
Kitchen Equipment		
Broiler	Ea.	4075
Cooler, 6 ft. long, reach-in	Ea.	3925

Description	Unit	$ Cost
Kitchen Equipment, cont.		
Dishwasher, 10-12 racks per hr.	Ea.	4275
Food warmer, counter, 1.2 KW	Ea.	770
Freezer, 44 C.F., reach-in	Ea.	6175
Ice cube maker, 50 lb. per day	Ea.	2000
Range with 1 oven	Ea.	3300
Lockers, Steel, single tier, 60" to 72"	Opng.	231 - 380
2 tier, 60" to 72" total	Opng.	121 - 170
5 tier, box lockers	Opng.	74 - 84
Locker bench, lam. maple top only	L.F.	34.50
Pedestals, steel pipe	Ea.	76.50
Seating		
Auditorium chair, all veneer	Ea.	315
Veneer back, padded seat	Ea.	325
Upholstered, spring seat	Ea.	320
Classroom, movable chair & desk	Set	81 - 171
Lecture hall, pedestal type	Ea.	275 - 620
Sound System		
Amplifier, 250 watts	Ea.	2200
Speaker, ceiling or wall	Ea.	217
Trumpet	Ea.	415

Important: See the Reference Section for Location Factors.

Model costs calculated for a 1 story building with 15' story height and 45,000 square feet of floor area

School, Elementary

				Unit	Unit Cost	Cost Per S.F.	% Of Sub-Total
A. SUBSTRUCTURE							
1010	Standard Foundations	Poured concrete; strip and spread footings		S.F. Ground	3.16	3.16	
1020	Special Foundations	N/A		—	—	—	
1030	Slab on Grade	4" reinforced concrete		S.F. Slab	5.56	5.56	9.7%
2010	Basement Excavation	Site preparation for slab and trench for foundation wall and footing		S.F. Ground	.18	.32	
2020	Basement Walls	4' Foundation wall		L.F. Wall	86	2.87	
B. SHELL							
	B10 Superstructure						
1010	Floor Construction	Fireproofing for structural columns		S.F. Floor	1.08	1.08	9.3%
1020	Roof Construction	Metal deck, open web steel joists, beams, columns		S.F. Roof	10.38	10.38	
	B20 Exterior Enclosure						
2010	Exterior Walls	Textured metal panel on metal stud, insulated	70% of wall	S.F. Wall	17.63	6.21	
2020	Exterior Windows	Curtain wall system, thermo-break frame, aluminum awning windows	30% of wall	Each	804	6.18	10.6%
2030	Exterior Doors	Aluminum & glass double doors, hollow metal doors		Each	3905	.69	
	B30 Roofing						
3010	Roof Coverings	Single ply membrane, stone ballast, rigid insulation		S.F. Roof	6.54	6.54	5.5%
3020	Roof Openings	Roof and smoke hatches		S.F. Roof	2060	.19	
C. INTERIORS							
1010	Partitions	Gypsum board on metal studs, sound attentuation insulation	20 S.F. Floor/L.F. Partition	S.F. Partition	5.72	2.86	
1020	Interior Doors	Hollow metal doors	700 S.F. Floor/Door	Each	1168	1.67	
1030	Fittings	Chalkboards, lockers, toilet partitions		S.F. Floor	6.42	6.42	
2010	Stair Construction	N/A		—	—	—	22.4%
3010	Wall Finishes	90% Paint, 10% ceramic tile		S.F. Surface	1.83	1.83	
3020	Floor Finishes	10% Carpet, 10% terrazzo, 60% VCT, 20% wood		S.F. Floor	7.49	7.49	
3030	Ceiling Finishes	Acoustic ceiling tiles on suspended channel grid		S.F. Ceiling	7.22	7.22	
D. SERVICES							
	D10 Conveying						
1010	Elevators & Lifts	N/A		—	—	—	0.0 %
1020	Escalators & Moving Walks	N/A		—	—	—	
	D20 Plumbing						
2010	Plumbing Fixtures	Restroom, kitchen, and service fixtures, supply and drainage	1 Fixture/325 S.F. Floor	Each	2457	7.56	
2020	Domestic Water Distribution	Gas fired water heater		S.F. Floor	.64	.64	7.4%
2040	Rain Water Drainage	Roof drains		S.F. Roof	.92	.92	
	D30 HVAC						
3010	Energy Supply	Forced hot water heating system, fin tube radiation		S.F. Floor	10.65	10.65	
3020	Heat Generating Systems	N/A		—	—	—	
3030	Cooling Generating Systems	N/A		—	—	—	19.1%
3050	Terminal & Package Units	Split system with air cooled condensing unit		S.F. Floor	12.85	12.85	
3090	Other HVAC Sys. & Equipment	N/A		—	—	—	
	D40 Fire Protection						
4010	Sprinklers	Wet pipe spinkler system		S.F. Floor	2.84	2.84	2.7%
4020	Standpipes	Standpipe		S.F. Floor	.44	.44	
	D50 Electrical						
5010	Electrical Service/Distribution	800 Ampere service, panelboards and feeders		S.F. Floor	1.11	1.11	
5020	Lighting & Branch Wiring	Fluorescent fixtures, receptacles, switches, a.c. and misc. power		S.F. Floor	11.18	11.18	13.2%
5030	Communications & Security	Addressable alarm system, emergency lighting, internet & phone wiring		S.F. Floor	3.85	3.85	
5090	Other Electrical Systems	Emergency generator, 15kW		S.F. Floor	.08	.08	
E. EQUIPMENT & FURNISHINGS							
1010	Commercial Equipment	N/A		—	—	—	
1020	Institutional Equipment	Laboratory equipment		S.F. Floor	.20	.20	0.2 %
1030	Vehicular Equipment	N/A		—	—	—	
1090	Other Equipment	N/A		—	—	—	
F. SPECIAL CONSTRUCTION							
1020	Integrated Construction	N/A		—	—	—	0.0 %
1040	Special Facilities	N/A		—	—	—	
G. BUILDING SITEWORK	**N/A**						

		Sub-Total	122.99	100%
CONTRACTOR FEES (General Requirements: 10%, Overhead: 5%, Profit: 10%)		25%	30.75	
ARCHITECT FEES		7%	10.76	
	Total Building Cost		**164.50**	

For customer support on your Square Foot Cost Data, call 877.756.2789.

201

Costs per square foot of floor area

Exterior Wall	S.F. Area	50000	70000	90000	110000	130000	150000	170000	190000	210000
	L.F. Perimeter	850	1140	1420	1700	1980	2280	2560	2840	3120
Fiber Cement Siding	Steel Frame	179.20	176.70	175.10	174.05	173.40	173.10	172.70	172.40	172.10
Metal Panel	Steel Frame	185.95	182.25	180.00	178.60	177.60	177.10	176.55	176.15	175.70
E.I.F.S.	Steel Frame	187.65	183.95	181.70	180.25	179.30	178.80	178.25	177.80	177.40
Tilt-up Concrete Panel	Reinforced Concrete	216.30	211.60	208.55	206.60	205.35	204.80	204.00	203.40	202.85
Brick Veneer	Reinforced Concrete	210.40	206.10	203.35	201.60	200.55	200.00	199.25	198.70	198.25
Stone Veneer	Reinforced Concrete	238.35	232.80	229.10	226.80	225.20	224.65	223.65	222.90	222.25
Perimeter Adj., Add or Deduct	Per 100 L.F.	3.90	2.80	2.15	1.80	1.50	1.30	1.25	0.95	0.90
Story Hgt. Adj., Add or Deduct	Per 1 Ft.	1.55	1.45	1.40	1.35	1.40	1.35	1.40	1.35	1.30

For Basement, add $35.80 per square foot of basement area

The above costs were calculated using the basic specifications shown on the facing page. These costs should be adjusted where necessary for design alternatives and owner's requirements. Reported completed project costs, for this type of structure, range from $105.80 to $249.15 per S.F.

Common additives

Description	Unit	$ Cost
Bleachers, Telescoping, manual		
To 15 tier	Seat	143 - 199
16-20 tier	Seat	294 - 360
21-30 tier	Seat	310 - 410
For power operation, add	Seat	57.50 - 90
Carrels Hardwood	Ea.	945 - 2400
Clock System		
20 room	Ea.	20,600
50 room	Ea.	48,500
Elevators, Hydraulic passenger, 2 stops		
1500# capacity	Ea.	67,400
2500# capacity	Ea.	70,400
Emergency Lighting, 25 watt, battery operated		
Lead battery	Ea.	345
Nickel cadmium	Ea.	685
Flagpoles, Complete		
Aluminum, 20' high	Ea.	1900
40' high	Ea.	4425
Fiberglass, 23' high	Ea.	1350
39'-5" high	Ea.	3400

Description	Unit	$ Cost
Kitchen Equipment		
Broiler	Ea.	4075
Cooler, 6 ft. long, reach-in	Ea.	3925
Dishwasher, 10-12 racks per hr.	Ea.	4275
Food warmer, counter, 1.2 KW	Ea.	770
Freezer, 44 C.F., reach-in	Ea.	6175
Lockers, Steel, single tier, 60" or 72"	Opng.	231 - 380
2 tier, 60" or 72" total	Opng.	121 - 170
5 tier, box lockers	Opng.	74 - 84
Locker bench, lam. maple top only	L.F.	34.50
Pedestals, steel pipe	Ea.	76.50
Seating		
Auditorium chair, all veneer	Ea.	315
Veneer back, padded seat	Ea.	325
Upholstered, spring seat	Ea.	320
Classroom, movable chair & desk	Set	81 - 171
Lecture hall, pedestal type	Ea.	275 - 620
Sound System		
Amplifier, 250 watts	Ea.	2200
Speaker, ceiling or wall	Ea.	217
Trumpet	Ea.	415

Important: See the Reference Section for Location Factors.

Model costs calculated for a 2 story building with 15' story height and 130,000 square feet of floor area

School, High, 2-3 Story

				Unit	Unit Cost	Cost Per S.F.	% Of Sub-Total
A. SUBSTRUCTURE							
1010	Standard Foundations	Poured concrete; strip and spread footings		S.F. Ground	2.98	1.49	
1020	Special Foundations	N/A		—	—	—	
1030	Slab on Grade	4" reinforced concrete		S.F. Slab	5.56	2.79	4.3%
2010	Basement Excavation	Site preparation for slab and trench for foundation wall and footing		S.F. Ground	.18	.14	
2020	Basement Walls	4' Foundation wall		L.F. Wall	86	1.31	
B. SHELL							
	B10 Superstructure						
1010	Floor Construction	Open web steel joists, slab form, concrete, fireproofed steel columns		S.F. Floor	15.62	7.81	9.8%
1020	Roof Construction	Metal deck, open web steel joists, beams, columns		S.F. Roof	10.40	5.20	
	B20 Exterior Enclosure						
2010	Exterior Walls	Textured metal panel on metal stud, insulated	75% of wall	S.F. Wall	17.65	6.05	
2020	Exterior Windows	Curtain wall system, thermo-break frame	25% of wall	Each	77	8.75	11.5%
2030	Exterior Doors	Aluminum & glass double doors, hollow metal doors, overhead doors		Each	2318	.53	
	B30 Roofing						
3010	Roof Coverings	Single ply membrane, stone ballast, rigid insulation		S.F. Roof	5.72	2.86	2.3%
3020	Roof Openings	Roof and smoke hatches		S.F. Roof	.32	.16	
C. INTERIORS							
1010	Partitions	Gypsum board on metal studs, sound attentuation insulation	25 S.F. Floor/L.F. Partition	S.F. Partition	6.69	3.21	
1020	Interior Doors	Hollow metal doors	700 S.F. Floor/Door	Each	1168	1.67	
1030	Fittings	Chalkboards, lockers, toilet partitions		S.F. Floor	10.12	10.12	
2010	Stair Construction	Cement filled metal pan, picket rail		Flight	11,575	.89	24.4%
3010	Wall Finishes	90% Paint, 10% ceramic tile		S.F. Surface	1.82	1.75	
3020	Floor Finishes	10% Carpet, 10% terrazzo, 60% VCT, 20% wood		S.F. Floor	7.49	7.49	
3030	Ceiling Finishes	Acoustic ceiling tiles on suspended channel grid		S.F. Ceiling	7.22	7.22	
D. SERVICES							
	D10 Conveying						
1010	Elevators & Lifts	Hydraulic passenger elevator		Each	85,800	.66	0.5%
1020	Escalators & Moving Walks	N/A		—	—	—	
	D20 Plumbing						
2010	Plumbing Fixtures	Restroom, kitchen, and service fixtures, supply and drainage	1 Fixture/340 S.F. Floor	Each	2054	6.04	
2020	Domestic Water Distribution	Gas fired water heater		S.F. Floor	1.12	1.12	5.7%
2040	Rain Water Drainage	Roof drains		S.F. Roof	.86	.43	
	D30 HVAC						
3010	Energy Supply	Forced hot water heating system, fin tube radiation		S.F. Floor	5.26	5.26	
3020	Heat Generating Systems	N/A		—	—	—	
3030	Cooling Generating Systems	Packaged chiller		S.F. Floor	18.23	18.23	17.7%
3050	Terminal & Package Units	N/A		—	—	—	
3090	Other HVAC Sys. & Equipment	N/A		—	—	—	
	D40 Fire Protection						
4010	Sprinklers	Wet pipe spinkler system		S.F. Floor	2.49	2.49	2.2%
4020	Standpipes	Standpipe		S.F. Floor	.38	.38	
	D50 Electrical						
5010	Electrical Service/Distribution	2000 Ampere service, panelboards and feeders		S.F. Floor	1.09	1.09	
5020	Lighting & Branch Wiring	Fluorescent fixtures, receptacles, switches, a.c. and misc. power		S.F. Floor	10.39	10.39	12.2%
5030	Communications & Security	Addressable alarm system, emergency lighting, internet & phone wiring		S.F. Floor	4.21	4.21	
5090	Other Electrical Systems	Emergency generator, 250kW		S.F. Floor	.46	.46	
E. EQUIPMENT & FURNISHINGS							
1010	Commercial Equipment	N/A		—	—	—	
1020	Institutional Equipment	Laboratory equipment		S.F. Floor	8.92	8.92	9.5 %
1030	Vehicular Equipment	N/A		—	—	—	
1090	Other Equipment	Gym equipment, bleachers, scoreboards		S.F. Floor	3.67	3.67	
F. SPECIAL CONSTRUCTION							
1020	Integrated Construction	N/A		—	—	—	0.0 %
1040	Special Facilities	N/A		—	—	—	
G. BUILDING SITEWORK	**N/A**						

	Sub-Total	132.79	**100%**
CONTRACTOR FEES (General Requirements: 10%, Overhead: 5%, Profit: 10%)	25%	33.19	
ARCHITECT FEES	7%	11.62	
Total Building Cost		**177.60**	

For customer support on your Square Foot Cost Data, call 877.756.2789.

203

Costs per square foot of floor area

Exterior Wall	S.F. Area	50000	65000	80000	95000	110000	125000	140000	155000	170000
	L.F. Perimeter	850	1060	1280	1490	1700	1920	2140	2340	2560
Fiber Cement Siding	Steel Frame	164.10	162.25	161.35	160.50	159.90	159.60	159.35	158.95	158.80
Metal Panel	Steel Frame	166.30	164.15	163.05	162.15	161.45	161.05	160.75	160.30	160.15
E.I.F.S.	Steel Frame	167.00	165.15	164.25	163.40	162.80	162.45	162.20	161.85	161.70
Tilt-up Concrete Panel	Reinforced Concrete	190.55	188.15	186.90	185.80	185.05	184.60	184.30	183.75	183.55
Brick Veneer	Reinforced Concrete	196.55	194.00	192.75	191.60	190.80	190.40	190.05	189.45	189.35
Stone Veneer	Reinforced Concrete	213.20	209.75	208.10	206.55	205.50	204.95	204.45	203.75	203.40
Perimeter Adj., Add or Deduct	Per 100 L.F.	3.20	2.45	1.95	1.75	1.55	1.30	1.15	1.10	1.00
Story Hgt. Adj., Add or Deduct	Per 1 Ft.	1.35	1.25	1.25	1.20	1.20	1.20	1.20	1.15	1.20

For Basement, add $37.25 per square foot of basement area

The above costs were calculated using the basic specifications shown on the facing page. These costs should be adjusted where necessary for design alternatives and owner's requirements. Reported completed project costs, for this type of structure, range from $99.85 to $234.65 per S.F.

Common additives

Description	Unit	$ Cost
Bleachers, Telescoping, manual		
To 15 tier	Seat	143 - 199
16-20 tier	Seat	294 - 360
21-30 tier	Seat	310 - 410
For power operation, add	Seat	57.50 - 90
Carrels Hardwood	Ea.	945 - 2400
Clock System		
20 room	Ea.	20,600
50 room	Ea.	48,500
Elevators, Hydraulic passenger, 2 stops		
1500# capacity	Ea.	67,400
2500# capacity	Ea.	70,400
Emergency Lighting, 25 watt, battery operated		
Lead battery	Ea.	345
Nickel cadmium	Ea.	685
Flagpoles, Complete		
Aluminum, 20' high	Ea.	1900
40' high	Ea.	4425
Fiberglass, 23' high	Ea.	1350
39'-5" high	Ea.	3400

Description	Unit	$ Cost
Kitchen Equipment		
Broiler	Ea.	4075
Cooler, 6 ft. long, reach-in	Ea.	3925
Dishwasher, 10-12 racks per hr.	Ea.	4275
Food warmer, counter, 1.2 KW	Ea.	770
Freezer, 44 C.F., reach-in	Ea.	6175
Lockers, Steel, single tier, 60" to 72"	Opng.	231 - 380
2 tier, 60" to 72" total	Opng.	121 - 170
5 tier, box lockers	Opng.	74 - 84
Locker bench, lam. maple top only	L.F.	34.50
Pedestals, steel pipe	Ea.	76.50
Seating		
Auditorium chair, all veneer	Ea.	315
Veneer back, padded seat	Ea.	325
Upholstered, spring seat	Ea.	320
Classroom, movable chair & desk	Set	81 - 171
Lecture hall, pedestal type	Ea.	275 - 620
Sound System		
Amplifier, 250 watts	Ea.	2200
Speaker, ceiling or wall	Ea.	217
Trumpet	Ea.	415

Important: See the Reference Section for Location Factors.

Model costs calculated for a 2 story building with 15' story height and 110,000 square feet of floor area

School, Jr High, 2-3 Story

				Unit	Unit Cost	Cost Per S.F.	% Of Sub-Total
A. SUBSTRUCTURE							
1010	Standard Foundations	Poured concrete; strip and spread footings		S.F. Ground	1.92	.96	
1020	Special Foundations	N/A		—	—	—	
1030	Slab on Grade	4" reinforced concrete		S.F. Slab	5.56	2.79	4.4%
2010	Basement Excavation	Site preparation for slab and trench for foundation wall and footing		S.F. Ground	.18	.16	
2020	Basement Walls	4' Foundation wall		L.F. Wall	86	1.32	
B. SHELL							
	B10 Superstructure						
1010	Floor Construction	Open web steel joists, slab form, concrete, fireproofed steel columns		S.F. Floor	15.80	7.90	11.0%
1020	Roof Construction	Metal deck, open web steel joists, beams, columns		S.F. Roof	10.40	5.20	
	B20 Exterior Enclosure						
2010	Exterior Walls	Fiber cement siding on wood studs, insulated	75% of wall	S.F. Wall	16.45	5.72	
2020	Exterior Windows	Curtain wall system, thermo-break frame	25% of wall	Each	58	6.68	10.9%
2030	Exterior Doors	Aluminum & glass double doors, hollow metal doors, overhead doors		Each	2452	.65	
	B30 Roofing						
3010	Roof Coverings	Single ply membrane, stone ballast, rigid insulation		S.F. Roof	6.36	3.18	2.8%
3020	Roof Openings	Roof and smoke hatches		S.F. Roof	.30	.15	
C. INTERIORS							
1010	Partitions	Gypsum board on metal studs, sound attentuation insulation	20 S.F. Floor/L.F. Partition	S.F. Partition	6.43	3.86	
1020	Interior Doors	Hollow metal doors	750 S.F. Floor/Door	Each	1168	1.56	
1030	Fittings	Chalkboards, lockers, toilet partitions		S.F. Floor	8.32	8.32	
2010	Stair Construction	Cement filled metal pan, picket rail		Flight	11,575	.84	25.5%
3010	Wall Finishes	90% Paint, 10% ceramic tile		S.F. Surface	1.78	2.13	
3020	Floor Finishes	10% Carpet, 10% terrazzo, 70% VCT, 10% wood		S.F. Floor	6.61	6.61	
3030	Ceiling Finishes	Acoustic ceiling tiles on suspended channel grid		S.F. Ceiling	7.22	7.22	
D. SERVICES							
	D10 Conveying						
1010	Elevators & Lifts	Hydraulic passenger elevator		Each	84,700	.77	0.6%
1020	Escalators & Moving Walks	N/A		—	—	—	
	D20 Plumbing						
2010	Plumbing Fixtures	Restroom, kitchen, and service fixtures, supply and drainage	1 Fixture/350 S.F. Floor	Each	2142	6.12	
2020	Domestic Water Distribution	Gas fired water heater		S.F. Floor	.65	.65	6.0%
2040	Rain Water Drainage	Roof drains		S.F. Roof	.76	.38	
	D30 HVAC						
3010	Energy Supply	N/A		—	—	—	
3020	Heat Generating Systems	Included in D3050		—	—	—	
3030	Cooling Generating Systems	N/A		—	—	—	17.7 %
3050	Terminal & Package Units	Single zone unit gas heating, electric cooling		S.F. Floor	21.15	21.15	
3090	Other HVAC Sys. & Equipment	N/A		—	—	—	
	D40 Fire Protection						
4010	Sprinklers	Wet pipe spinkler system		S.F. Floor	2.49	2.49	2.4%
4020	Standpipes	Standpipe		S.F. Floor	.43	.43	
	D50 Electrical						
5010	Electrical Service/Distribution	1600 Ampere service, panelboards and feeders		S.F. Floor	1.04	1.04	
5020	Lighting & Branch Wiring	Fluorescent fixtures, receptacles, switches, a.c. and misc. power		S.F. Floor	10.39	10.39	13.6%
5030	Communications & Security	Addressable alarm system, emergency lighting, internet & phone wiring		S.F. Floor	4.50	4.50	
5090	Other Electrical Systems	Emergency generator, 100kW		S.F. Floor	.33	.33	
E. EQUIPMENT & FURNISHINGS							
1010	Commercial Equipment	N/A		—	—	—	
1020	Institutional Equipment	Laboratory equipment		S.F. Floor	1.74	1.74	5.1 %
1030	Vehicular Equipment	N/A		—	—	—	
1090	Other Equipment	Gym equipment, bleachers, scoreboards		S.F. Floor	4.34	4.34	
F. SPECIAL CONSTRUCTION							
1020	Integrated Construction	N/A		—	—	—	0.0 %
1040	Special Facilities	N/A		—	—	—	
G. BUILDING SITEWORK	**N/A**						

		Sub-Total	119.58	100%
CONTRACTOR FEES (General Requirements: 10%, Overhead: 5%, Profit: 10%)		25%	29.86	
ARCHITECT FEES		7%	10.46	
	Total Building Cost		**159.90**	

For customer support on your Square Foot Cost Data, call 877.756.2789.

205

Costs per square foot of floor area

Exterior Wall	S.F. Area	20000	30000	40000	50000	60000	70000	80000	90000	100000
	L.F. Perimeter	440	590	740	900	1050	1200	1360	1510	1660
Face Brick with Concrete Block Back-up	Steel Frame	191.30	184.95	181.75	180.25	178.90	178.00	177.50	177.00	176.40
	Bearing Walls	186.65	180.30	177.10	175.60	174.25	173.30	172.85	172.30	171.75
Decorative Concrete Block	Steel Frame	177.95	173.05	170.55	169.30	168.30	167.55	167.20	166.75	166.35
	Bearing Walls	173.00	168.05	165.65	164.40	163.35	162.60	162.25	161.80	161.45
Steel Siding on Steel Studs	Steel Frame	172.10	167.80	165.70	164.55	163.70	163.00	162.70	162.30	161.95
Metal Sandwich Panel	Steel Frame	172.45	168.15	165.95	164.90	163.95	163.30	162.95	162.55	162.20
Perimeter Adj., Add or Deduct	Per 100 L.F.	9.65	6.45	4.85	3.90	3.25	2.75	2.45	2.10	2.00
Story Hgt. Adj., Add or Deduct	Per 1 Ft.	2.35	2.15	2.05	1.95	1.95	1.80	1.90	1.70	1.80
For Basement, add $36.15 per square foot of basement area										

The above costs were calculated using the basic specifications shown on the facing page. These costs should be adjusted where necessary for design alternatives and owner's requirements. Reported completed project costs, for this type of structure, range from $89.65 to $256.10 per S.F.

Common additives

Description	Unit	$ Cost
Carrels Hardwood	Ea.	945 - 2400
Clock System		
20 room	Ea.	20,600
50 room	Ea.	48,500
Directory Boards, Plastic, glass covered		
30" x 20"	Ea.	610
36" x 48"	Ea.	1525
Aluminum, 24" x 18"	Ea.	640
36" x 24"	Ea.	765
48" x 32"	Ea.	1075
48" x 60"	Ea.	2275
Elevators, Hydraulic passenger, 2 stops		
1500# capacity	Ea.	67,400
2500# capacity	Ea.	70,400
3500# capacity	Ea.	75,400
Emergency Lighting, 25 watt, battery operated		
Lead battery	Ea.	345
Nickel cadmium	Ea.	685

Description	Unit	$ Cost
Flagpoles, Complete		
Aluminum, 20' high	Ea.	1900
40' high	Ea.	4425
Fiberglass, 23' high	Ea.	1350
39'-5" high	Ea.	3400
Seating		
Auditorium chair, all veneer	Ea.	315
Veneer back, padded seat	Ea.	325
Upholstered, spring seat	Ea.	320
Classroom, movable chair & desk	Set	81 - 171
Lecture hall, pedestal type	Ea.	275 - 620
Shops & Workroom:		
Benches, metal	Ea.	870
Parts bins 6'-3" high, 3' wide, 12" deep, 72 bins	Ea.	720
Shelving, metal 1' x 3'	S.F.	12
Wide span 6' wide x 24" deep	S.F.	10.30
Sound System		
Amplifier, 250 watts	Ea.	2200
Speaker, ceiling or wall	Ea.	217
Trumpet	Ea.	415

Important: See the Reference Section for Location Factors.

Model costs calculated for a 2 story building with 16' story height and 40,000 square feet of floor area

School, Vocational

				Unit	Unit Cost	Cost Per S.F.	% Of Sub-Total
A. SUBSTRUCTURE							
1010	Standard Foundations	Poured concrete; strip and spread footings		S.F. Ground	3.06	1.53	
1020	Special Foundations	N/A		—	—	—	
1030	Slab on Grade	5" reinforced concrete with vapor barrier and granular base		S.F. Slab	13.81	6.91	7.6%
2010	Basement Excavation	Site preparation for slab and trench for foundation wall and footing		S.F. Ground	.18	.09	
2020	Basement Walls	4' foundation wall		L.F. Wall	86	1.58	
B. SHELL							
	B10 Superstructure						
1010	Floor Construction	Open web steel joists, slab form, concrete, beams, columns		S.F. Floor	21.44	10.72	10.6%
1020	Roof Construction	Metal deck, open web steel joists, beams, columns		S.F. Roof	6.60	3.30	
	B20 Exterior Enclosure						
2010	Exterior Walls	Face brick with concrete block backup	85% of wall	S.F. Wall	33.39	16.80	
2020	Exterior Windows	Tubular aluminum framing with insulated glass	15% of wall	Each	58	5.11	17.0%
2030	Exterior Doors	Metal and glass doors		Each	3054	.61	
	B30 Roofing						
3010	Roof Coverings	Single-ply membrane with polyisocyanurate insulation		S.F. Roof	6.84	3.42	2.7%
3020	Roof Openings	Roof hatches		S.F. Roof	.32	.16	
C. INTERIORS							
1010	Partitions	Concrete block	20 S.F. Floor/L.F. Partition	S.F. Partition	9.73	5.84	
1020	Interior Doors	Single leaf kalamein fire doors	600 S.F. Floor/Door	Each	1168	1.95	
1030	Fittings	Toilet partitions, chalkboards		S.F. Floor	1.57	1.57	
2010	Stair Construction	Concrete filled metal pan		Flight	14,450	2.17	20.9%
3010	Wall Finishes	50% paint, 40% glazed coating, 10% ceramic tile		S.F. Surface	4.76	5.71	
3020	Floor Finishes	70% vinyl composition tile, 20% carpet, 10% terrazzo		S.F. Floor	6.06	6.06	
3030	Ceiling Finishes	Mineral fiber tile on concealed zee bars	60% of area	S.F. Ceiling	7.22	4.33	
D. SERVICES							
	D10 Conveying						
1010	Elevators & Lifts	One hydraulic passenger elevator		Each	85,600	2.14	1.6%
1020	Escalators & Moving Walks	N/A		—	—	—	
	D20 Plumbing						
2010	Plumbing Fixtures	Toilet and service fixtures, supply and drainage	1 Fixture/700 S.F. Floor	Each	3206	4.58	
2020	Domestic Water Distribution	Gas fired water heater		S.F. Floor	1.07	1.07	5.0%
2040	Rain Water Drainage	Roof drains		S.F. Roof	2.02	1.01	
	D30 HVAC						
3010	Energy Supply	Oil fired hot water, wall fin radiation		S.F. Floor	10.05	10.05	
3020	Heat Generating Systems	N/A		—	—	—	
3030	Cooling Generating Systems	Chilled water, cooling tower systems		S.F. Floor	15.50	15.50	19.3%
3050	Terminal & Package Units	N/A		—	—	—	
3090	Other HVAC Sys. & Equipment	N/A		—	—	—	
	D40 Fire Protection						
4010	Sprinklers	Sprinklers, light hazard		S.F. Floor	3.15	3.15	3.2%
4020	Standpipes	Standpipe, wet, Class III		S.F. Floor	1.03	1.03	
	D50 Electrical						
5010	Electrical Service/Distribution	800 ampere service, panel board and feeders		S.F. Floor	1.25	1.25	
5020	Lighting & Branch Wiring	High efficiency fluorescent fixtures, receptacles, switches, A.C. and misc. power		S.F. Floor	11.23	11.23	12.0%
5030	Communications & Security	Addressable alarm systems, internet wiring, communications systems and emergency lighting		S.F. Floor	3.26	3.26	
5090	Other Electrical Systems	Emergency generator, 11.5 kW		S.F. Floor	.12	.12	
E. EQUIPMENT & FURNISHINGS							
1010	Commercial Equipment	N/A		—	—	—	
1020	Institutional Equipment	Stainless steel countertops		S.F. Floor	.18	.18	0.1%
1030	Vehicular Equipment	N/A		—	—	—	
1090	Other Equipment	N/A		—	—	—	
F. SPECIAL CONSTRUCTION							
1020	Integrated Construction	N/A		—	—	—	0.0%
1040	Special Facilities	N/A		—	—	—	
G. BUILDING SITEWORK	**N/A**						

			Sub-Total	132.43	100%
CONTRACTOR FEES (General Requirements: 10%, Overhead: 5%, Profit: 10%)			25%	33.08	
ARCHITECT FEES			7%	11.59	

Total Building Cost   177.10

For customer support on your Square Foot Cost Data, call 877.756.2789.

207

Costs per square foot of floor area

Exterior Wall	S.F. Area	1000	2000	3000	4000	6000	8000	10000	12000	15000
	L.F. Perimeter	126	179	219	253	310	358	400	438	490
Wood Siding	Wood Frame	167.00	140.35	129.55	123.50	116.55	112.65	110.00	108.05	106.00
Face Brick Veneer	Wood Frame	210.90	177.20	163.25	155.25	146.15	140.85	137.30	134.70	131.90
Stucco on Concrete Block	Steel Frame	184.75	154.10	141.50	134.30	126.10	121.40	118.25	115.95	113.45
	Bearing Walls	179.70	149.00	136.40	129.20	121.00	116.30	113.15	110.80	108.35
Metal Sandwich Panel	Steel Frame	197.10	162.85	148.60	140.45	131.10	125.70	122.05	119.40	116.55
Precast Concrete	Steel Frame	234.05	188.60	169.30	158.15	145.25	137.75	132.70	128.90	124.85
Perimeter Adj., Add or Deduct	Per 100 L.F.	44.90	22.45	14.95	11.20	7.50	5.60	4.50	3.80	3.00
Story Hgt. Adj., Add or Deduct	Per 1 Ft.	3.30	2.35	1.95	1.65	1.45	1.15	1.05	1.00	0.90
For Basement, add $27.30 per square foot of basement area										

The above costs were calculated using the basic specifications shown on the facing page. These costs should be adjusted where necessary for design alternatives and owner's requirements. Reported completed project costs, for this type of structure, range from $74.25 to $257.35 per S.F.

Common additives

Description	Unit	$ Cost
Check Out Counter		
Single belt	Ea.	3600
Double belt	Ea.	5125
Emergency Lighting, 25 watt, battery operated		
Lead battery	Ea.	345
Nickel cadmium	Ea.	685
Refrigerators, Prefabricated, walk-in		
7'-6" high, 6' x 6'	S.F.	203
10' x 10'	S.F.	159
12' x 14'	S.F.	142
12' x 20'	S.F.	125
Refrigerated Food Cases		
Dairy, multi deck, 12' long	Ea.	12,800
Delicatessen case, single deck, 12' long	Ea.	9600
Multi deck, 18 S.F. shelf display	Ea.	8875
Freezer, self-contained chest type, 30 C.F.	Ea.	10,300
Glass door upright, 78 C.F.	Ea.	13,600

Description	Unit	$ Cost
Refrigerated Food Cases, cont.		
Frozen food, chest type, 12' long	Ea.	9450
Glass door reach-in, 5 door	Ea.	15,900
Island case 12' long, single deck	Ea.	8425
Multi deck	Ea.	9800
Meat cases, 12' long, single deck	Ea.	8425
Multi deck	Ea.	11,800
Produce, 12' long single deck	Ea.	7950
Multi deck	Ea.	10,000
Safe, Office type, 1 hour rating		
30" x 18" x 18"	Ea.	2500
60" x 36" x 18", double door	Ea.	9350
Smoke Detectors		
Ceiling type	Ea.	227
Duct type	Ea.	565
Sound System		
Amplifier, 250 watts	Ea.	2200
Speaker, ceiling or wall	Ea.	217
Trumpet	Ea.	415

Important: See the Reference Section for Location Factors.

Model costs calculated for a 1 story building with 12' story height and 4,000 square feet of floor area

Store, Convenience

				Unit	Unit Cost	Cost Per S.F.	% Of Sub-Total
A.	**SUBSTRUCTURE**						
1010	Standard Foundations	Poured concrete; strip and spread footings		S.F. Ground	1.18	1.18	
1020	Special Foundations	N/A		—	—	—	
1030	Slab on Grade	4" reinforced concrete with vapor barrier and granular base		S.F. Slab	5.56	5.56	13.1%
2010	Basement Excavation	Site preparation for slab and trench for foundation wall and footing		S.F. Ground	.57	.57	
2020	Basement Walls	4' foundation wall		L.F. Wall	76	4.81	
B.	**SHELL**						
	B10 Superstructure						
1010	Floor Construction	N/A		—	—	—	
1020	Roof Construction	Wood truss with plywood sheathing		S.F. Roof	7.42	7.42	8.0 %
	B20 Exterior Enclosure						
2010	Exterior Walls	Wood siding on wood studs, insulated	80% of wall	S.F. Wall	10.31	6.26	
2020	Exterior Windows	Storefront	20% of wall	Each	47.50	7.21	17.1%
2030	Exterior Doors	Double aluminum and glass, solid core wood		Each	3042	2.29	
	B30 Roofing						
3010	Roof Coverings	Asphalt shingles; rigid fiberglass insulation		S.F. Roof	5.95	5.95	6.4%
3020	Roof Openings	N/A		—	—	—	
C.	**INTERIORS**						
1010	Partitions	Gypsum board on wood studs	60 S.F. Floor/L.F. Partition	S.F. Partition	13.80	2.30	
1020	Interior Doors	Single leaf wood, hollow metal	1300 S.F. Floor/Door	Each	1259	.97	
1030	Fittings	N/A		—	—	—	
2010	Stair Construction	N/A		—	—	—	14.1%
3010	Wall Finishes	Paint		S.F. Surface	1.20	.40	
3020	Floor Finishes	Vinyl composition tile		S.F. Floor	2.76	2.76	
3030	Ceiling Finishes	Mineral fiber tile on wood furring		S.F. Ceiling	6.56	6.56	
D.	**SERVICES**						
	D10 Conveying						
1010	Elevators & Lifts	N/A		—	—	—	0.0 %
1020	Escalators & Moving Walks	N/A		—	—	—	
	D20 Plumbing						
2010	Plumbing Fixtures	Toilet and service fixtures, supply and drainage	1 Fixture/1000 S.F. Floor	Each	3920	3.92	
2020	Domestic Water Distribution	Gas fired water heater		S.F. Floor	1.68	1.68	6.1%
2040	Rain Water Drainage	N/A		—	—	—	
	D30 HVAC						
3010	Energy Supply	N/A		—	—	—	
3020	Heat Generating Systems	Included in D3050		—	—	—	
3030	Cooling Generating Systems	N/A		—	—	—	8.3 %
3050	Terminal & Package Units	Single zone rooftop unit, gas heating, electric cooling		S.F. Floor	7.69	7.69	
3090	Other HVAC Sys. & Equipment	N/A		—	—	—	
	D40 Fire Protection						
4010	Sprinklers	Sprinkler, ordinary hazard		S.F. Floor	5.02	5.02	7.9%
4020	Standpipes	Standpipe		S.F. floor	2.30	2.30	
	D50 Electrical						
5010	Electrical Service/Distribution	200 ampere service, panel board and feeders		S.F. Floor	3.12	3.12	
5020	Lighting & Branch Wiring	High efficiency fluorescent fixtures, receptacles, switches, A.C. and misc. power		S.F. Floor	11.12	11.12	18.9%
5030	Communications & Security	Addressable alarm systems and emergency lighting		S.F. Floor	2.85	2.85	
5090	Other Electrical Systems	Emergency generator, 7.5 kW		S.F. Floor	.38	.38	
E.	**EQUIPMENT & FURNISHINGS**						
1010	Commercial Equipment	N/A		—	—	—	
1020	Institutional Equipment	N/A		—	—	—	0.0 %
1030	Vehicular Equipment	N/A		—	—	—	
1090	Other Equipment	N/A		—	—	—	
F.	**SPECIAL CONSTRUCTION**						
1020	Integrated Construction	N/A		—	—	—	0.0 %
1040	Special Facilities	N/A		—	—	—	
G.	**BUILDING SITEWORK**	**N/A**					

		Sub-Total	92.32	100%
	CONTRACTOR FEES (General Requirements: 10%, Overhead: 5%, Profit: 10%)	25%	23.10	
	ARCHITECT FEES	7%	8.08	
	Total Building Cost		**123.50**	

For customer support on your Square Foot Cost Data, call 877.756.2789.

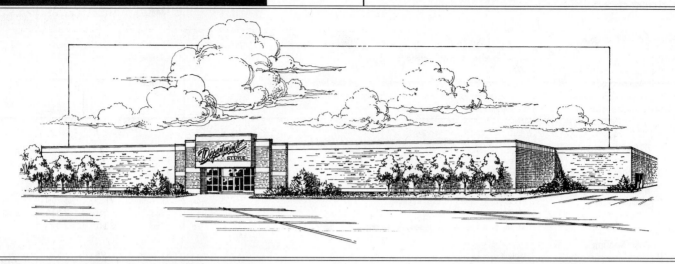

Costs per square foot of floor area

Exterior Wall	S.F. Area	50000	65000	80000	95000	110000	125000	140000	155000	170000
	L.F. Perimeter	920	1065	1167	1303	1333	1433	1533	1633	1733
Face Brick with Concrete Block Back-up	R/Conc. Frame	134.05	131.50	129.40	128.35	126.60	125.85	125.15	124.70	124.30
	Steel Frame	117.05	114.50	112.40	111.30	109.60	108.75	108.15	107.70	107.25
Decorative Concrete Block	R/Conc. Frame	128.60	126.60	125.05	124.20	123.00	122.40	121.90	121.55	121.25
	Steel Joists	112.65	110.60	108.90	108.00	106.65	106.05	105.50	105.15	104.80
Precast Concrete Panels	R/Conc. Frame	132.65	130.35	128.40	127.40	125.85	125.15	124.55	124.15	123.80
	Steel Joists	115.10	112.80	110.85	109.85	108.30	107.55	106.95	106.55	106.15
Perimeter Adj., Add or Deduct	Per 100 L.F.	1.95	1.55	1.30	1.00	0.85	0.70	0.80	0.65	0.55
Story Hgt. Adj., Add or Deduct	Per 1 Ft.	1.00	0.90	0.80	0.75	0.60	0.60	0.65	0.55	0.55

For Basement, add $26.85 per square foot of basement area

The above costs were calculated using the basic specifications shown on the facing page. These costs should be adjusted where necessary for design alternatives and owner's requirements. Reported completed project costs, for this type of structure, range from $64.05 to $159.40 per S.F.

Common additives

Description	Unit	$ Cost
Closed Circuit Surveillance, One station		
Camera and monitor	Ea.	2025
For additional camera stations, add	Ea.	1100
Directory Boards, Plastic, glass covered		
30" x 20"	Ea.	610
36" x 48"	Ea.	1525
Aluminum, 24" x 18"	Ea.	640
36" x 24"	Ea.	765
48" x 32"	Ea.	1075
48" x 60"	Ea.	2275
Emergency Lighting, 25 watt, battery operated		
Lead battery	Ea.	345
Nickel cadmium	Ea.	685
Safe, Office type, 1 hour rating		
30" x 18" x 18"	Ea.	2500
60" x 36" x 18", double door	Ea.	9350
Sound System		
Amplifier, 250 watts	Ea.	2200
Speaker, ceiling or wall	Ea.	217
Trumpet	Ea.	415

Important: See the Reference Section for Location Factors.

Model costs calculated for a 1 story building with 14' story height and 110,000 square feet of floor area

				Unit	Unit Cost	Cost Per S.F.	% Of Sub-Total
A. SUBSTRUCTURE							
1010	Standard Foundations	Poured concrete; strip and spread footings		S.F. Ground	.97	.97	
1020	Special Foundations	N/A		—	—	—	
1030	Slab on Grade	4" reinforced concrete with vapor barrier and granular base		S.F. Slab	5.56	5.56	8.3%
2010	Basement Excavation	Site preparation for slab and trench for foundation wall and footing		S.F. Ground	.32	.32	
2020	Basement Walls	4' foundation wall		L.F. Wall	86	1.04	
B. SHELL							
B10 Superstructure							
1010	Floor Construction	Cast-in-place concrete columns		L.F. Column	76	.97	25.3%
1020	Roof Construction	Precast concrete beam and plank, concrete columns		S.F. Roof	23.19	23.19	
B20 Exterior Enclosure							
2010	Exterior Walls	Face brick with concrete block backup	90% of wall	S.F. Wall	32.94	5.03	
2020	Exterior Windows	Storefront	10% of wall	Each	70	1.18	6.9%
2030	Exterior Doors	Sliding electric operated entrance, hollow metal		Each	7823	.43	
B30 Roofing							
3010	Roof Coverings	Built-up tar and gravel with flashing; perlite/EPS composite insulation		S.F. Roof	5.65	5.65	6.0%
3020	Roof Openings	Roof hatches		S.F. Roof	.05	.05	
C. INTERIORS							
1010	Partitions	Gypsum board on metal studs	60 S.F. Floor/L.F. Partition	S.F. Partition	11.10	1.85	
1020	Interior Doors	Single leaf hollow metal	600 S.F. Floor/Door	Each	1168	1.95	
1030	Fittings	N/A		—	—	—	
2010	Stair Construction	N/A		—	—	—	26.2%
3010	Wall Finishes	Paint		S.F. Surface	.84	.28	
3020	Floor Finishes	50% ceramic tile, 50% carpet tile		S.F. Floor	16.05	16.05	
3030	Ceiling Finishes	Mineral fiber board on exposed grid system, suspended		S.F. Ceiling	4.87	4.87	
D. SERVICES							
D10 Conveying							
1010	Elevators & Lifts	N/A		—	—	—	0.0 %
1020	Escalators & Moving Walks	N/A		—	—	—	
D20 Plumbing							
2010	Plumbing Fixtures	Toilet and service fixtures, supply and drainage	1 Fixture/4075 S.F. Floor	Each	6968	1.71	
2020	Domestic Water Distribution	Gas fired water heater		S.F. Floor	.39	.39	3.1%
2040	Rain Water Drainage	Roof drains		S.F. Roof	.87	.87	
D30 HVAC							
3010	Energy Supply	N/A		—	—	—	
3020	Heat Generating Systems	Included in D3050		—	—	—	
3030	Cooling Generating Systems	N/A		—	—	—	9.2 %
3050	Terminal & Package Units	Single zone unit gas heating, electric cooling		S.F. Floor	8.81	8.81	
3090	Other HVAC Sys. & Equipment	N/A		—	—	—	
D40 Fire Protection							
4010	Sprinklers	Sprinklers, light hazard		S.F. Floor	2.84	2.84	3.3%
4020	Standpipes	Standpipe		S.F. Floor	.30	.30	
D50 Electrical							
5010	Electrical Service/Distribution	1200 ampere service, panel board and feeders		S.F. Floor	.76	.76	
5020	Lighting & Branch Wiring	High efficiency fluorescent fixtures, receptacles, switches, A.C. and misc. power		S.F. Floor	9.17	9.17	11.8%
5030	Communications & Security	Addressable alarm systems, internet wiring and emergency lighting		S.F. Floor	1.26	1.26	
5090	Other Electrical Systems	Emergency generator, 7.5 kW		S.F. Floor	.04	.04	
E. EQUIPMENT & FURNISHINGS							
1010	Commercial Equipment	N/A		—	—	—	
1020	Institutional Equipment	N/A		—	—	—	
1030	Vehicular Equipment	N/A		—	—	—	0.0 %
1090	Other Equipment	N/A		—	—	—	
F. SPECIAL CONSTRUCTION							
1020	Integrated Construction	N/A		—	—	—	0.0 %
1040	Special Facilities	N/A		—	—	—	
G. BUILDING SITEWORK	**N/A**						

		Sub-Total	95.54	100%
CONTRACTOR FEES (General Requirements: 10%, Overhead: 5%, Profit: 10%)		25%	23.89	
ARCHITECT FEES		6%	7.17	

Total Building Cost	**126.60**

For customer support on your Square Foot Cost Data, call 877.756.2789.

211

Costs per square foot of floor area

Exterior Wall	S.F. Area	50000	65000	80000	95000	110000	125000	140000	155000	170000
	L.F. Perimeter	533	593	670	715	778	840	871	923	976
Face Brick with Concrete Block Back-up	Steel Frame	163.45	157.45	154.20	151.15	149.35	147.95	146.30	145.25	144.50
	R/Conc. Frame	162.20	156.15	152.95	149.90	148.10	146.70	145.00	144.05	143.25
Face Brick on Steel Studs	Steel Frame	158.50	153.20	150.35	147.70	146.05	144.85	143.45	142.50	141.85
	R/Conc. Frame	160.05	154.70	151.85	149.20	147.60	146.40	144.95	144.05	143.40
Precast Concrete Panels Exposed Aggregate	Steel Frame	168.90	162.10	158.55	155.05	153.05	151.50	149.55	148.40	147.55
	R/Conc. Frame	169.85	163.05	159.45	156.00	153.95	152.40	150.50	149.35	148.45
Perimeter Adj., Add or Deduct	Per 100 L.F.	5.25	4.00	3.30	2.80	2.35	2.10	1.85	1.75	1.50
Story Hgt. Adj., Add or Deduct	Per 1 Ft.	1.55	1.35	1.25	1.10	1.05	1.00	0.90	0.90	0.85

For Basement, add $45.50 per square foot of basement area

The above costs were calculated using the basic specifications shown on the facing page. These costs should be adjusted where necessary for design alternatives and owner's requirements. Reported completed project costs, for this type of structure, range from $66.20 to $177.15 per S.F.

Common additives

Description	Unit	$ Cost
Closed Circuit Surveillance, One station		
Camera and monitor	Ea.	2025
For additional camera stations, add	Ea.	1100
Directory Boards, Plastic, glass covered		
30" x 20"	Ea.	610
36" x 48"	Ea.	1525
Aluminum, 24" x 18"	Ea.	640
36" x 24"	Ea.	765
48" x 32"	Ea.	1075
48" x 60"	Ea.	2275
Elevators, Hydraulic passenger, 2 stops		
1500# capacity	Ea.	67,400
2500# capacity	Ea.	70,400
3500# capacity	Ea.	75,400
Additional stop, add	Ea.	8075
Emergency Lighting, 25 watt, battery operated		
Lead battery	Ea.	345
Nickel cadmium	Ea.	685

Description	Unit	$ Cost
Escalators, Metal		
32" wide, 10' story height	Ea.	155,500
20' story height	Ea.	189,500
48" wide, 10' story height	Ea.	164,000
20' story height	Ea.	199,000
Glass		
32" wide, 10' story height	Ea.	147,500
20' story height	Ea.	180,500
48" wide, 10' story height	Ea.	155,000
20' story height	Ea.	189,500
Safe, Office type, 1 hour rating		
30" x 18" x 18"	Ea.	2500
60" x 36" x 18", double door	Ea.	9350
Sound System		
Amplifier, 250 watts	Ea.	2200
Speaker, ceiling or wall	Ea.	217
Trumpet	Ea.	415

Important: See the Reference Section for Location Factors.

Model costs calculated for a 3 story building with 16′ story height and 95,000 square feet of floor area

Store, Department, 3 Story

				Unit	Unit Cost	Cost Per S.F.	% Of Sub-Total
A. SUBSTRUCTURE							
1010	Standard Foundations	Poured concrete; strip and spread footings		S.F. Ground	3.87	1.29	
1020	Special Foundations	N/A		—	—	—	
1030	Slab on Grade	4″ reinforced concrete with vapor barrier and granular base		S.F. Slab	5.56	1.85	3.4%
2010	Basement Excavation	Site preparation for slab and trench for foundation wall and footing		S.F. Ground	.18	.06	
2020	Basement Walls	4′ foundation wall		L.F. Wall	81	.68	
B. SHELL							
B10 Superstructure							
1010	Floor Construction	Concrete slab with metal deck and beams, steel columns		S.F. Floor	30.86	20.57	21.1%
1020	Roof Construction	Metal deck, open web steel joists, beams, columns		S.F. Roof	10.62	3.54	
B20 Exterior Enclosure							
2010	Exterior Walls	Face brick with concrete block backup	90% of wall	S.F. Wall	32.91	10.70	
2020	Exterior Windows	Storefront	10% of wall	Each	50	1.81	12.6%
2030	Exterior Doors	Revolving and sliding panel, mall-front		Each	12,250	1.81	
B30 Roofing							
3010	Roof Coverings	Built-up tar and gravel with flashing; perlite/EPS composite insulation		S.F. Roof	6.15	2.05	1.9%
3020	Roof Openings	Roof hatches		S.F. Roof	.21	.07	
C. INTERIORS							
1010	Partitions	Gypsum board on metal studs	60 S.F. Floor/L.F. Partition	S.F. Partition	6.54	1.09	
1020	Interior Doors	Single leaf hollow metal	600 S.F. Floor/Door	Each	1168	1.95	
1030	Fittings	N/A		—	—	—	
2010	Stair Construction	Concrete filled metal pan		Flight	20,150	2.12	26.5%
3010	Wall Finishes	70% paint, 20% vinyl wall covering, 10% ceramic tile		S.F. Surface	5.01	1.67	
3020	Floor Finishes	50% carpet tile, 40% marble tile, 10% terrazzo		S.F. Floor	16.22	16.22	
3030	Ceiling Finishes	Mineral fiber tile on concealed zee bars		S.F. Ceiling	7.22	7.22	
D. SERVICES							
D10 Conveying							
1010	Elevators & Lifts	One hydraulic passenger, one hydraulic freight		Each	311,600	3.28	8.8%
1020	Escalators & Moving Walks	Four escalators		Each	161,025	6.78	
D20 Plumbing							
2010	Plumbing Fixtures	Toilet and service fixtures, supply and drainage	1 Fixture/2570 S.F. Floor	Each	4755	1.85	
2020	Domestic Water Distribution	Gas fired water heater		S.F. Floor	.46	.46	2.5%
2040	Rain Water Drainage	Roof drains		S.F. Roof	1.71	.57	
D30 HVAC							
3010	Energy Supply	N/A		—	—	—	
3020	Heat Generating Systems	Included in D3050		—	—	—	
3030	Cooling Generating Systems	N/A		—	—	—	7.7 %
3050	Terminal & Package Units	Multizone rooftop unit, gas heating, electric cooling		S.F. Floor	8.81	8.81	
3090	Other HVAC Sys. & Equipment	N/A		—	—	—	
D40 Fire Protection							
4010	Sprinklers	Sprinklers, light hazard		S.F. Floor	2.39	2.39	3.7%
4020	Standpipes	Standpipe		S.F. Floor	1.84	1.84	
D50 Electrical							
5010	Electrical Service/Distribution	1200 ampere service, panel board and feeders		S.F. Floor	1.22	1.22	
5020	Lighting & Branch Wiring	High efficiency fluorescent fixtures, receptacles, switches, A.C. and misc. power		S.F. Floor	9.37	9.37	11.7%
5030	Communications & Security	Addressable alarm systems, internet wiring and emergency lighting		S.F. Floor	2.53	2.53	
5090	Other Electrical Systems	Emergency generator, 50 kW		S.F. Floor	.28	.28	
E. EQUIPMENT & FURNISHINGS							
1010	Commercial Equipment	N/A		—	—	—	
1020	Institutional Equipment	N/A		—	—	—	
1030	Vehicular Equipment	N/A		—	—	—	0.0 %
1090	Other Equipment	N/A		—	—	—	
F. SPECIAL CONSTRUCTION							
1020	Integrated Construction	N/A		—	—	—	
1040	Special Facilities	N/A		—	—	—	0.0 %
G. BUILDING SITEWORK	**N/A**						

			Sub-Total	114.08	**100%**
CONTRACTOR FEES (General Requirements: 10%, Overhead: 5%, Profit: 10%)			25%	28.51	
ARCHITECT FEES			6%	8.56	

		Total Building Cost	**151.15**

For customer support on your Square Foot Cost Data, call 877.756.2789.

213

Costs per square foot of floor area

Exterior Wall	S.F. Area	4000	6000	8000	10000	12000	15000	18000	20000	22000
	L.F. Perimeter	260	340	360	410	440	490	540	565	594
Vinyl Siding	Wood Frame	135.95	125.55	116.15	112.20	108.65	105.20	102.95	101.60	100.55
Fiber Cement Siding	Wood Frame	137.70	127.30	118.00	114.05	110.50	107.05	104.85	103.50	102.45
EIFS on Metal Studs	Steel Joists	146.15	135.25	125.15	120.95	117.10	113.50	111.05	109.60	108.50
Stone Veneer	Steel Frame	190.15	174.45	157.70	151.20	144.90	138.95	135.00	132.60	130.75
Brick Veneer	Reinforced Concrete	180.95	166.50	151.50	145.60	139.95	134.60	131.05	128.85	127.20
Stucco	Reinforced Concrete	186.50	170.95	154.45	148.05	141.85	135.95	132.05	129.70	127.85
Perimeter Adj., Add or Deduct	Per 100 L.F.	14.05	9.35	7.05	5.65	4.70	3.80	3.15	2.80	2.55
Story Hgt. Adj., Add or Deduct	Per 1 Ft.	1.55	1.35	1.10	1.00	0.85	0.85	0.75	0.70	0.65
For Basement, add $37.80 per square foot of basement area										

The above costs were calculated using the basic specifications shown on the facing page. These costs should be adjusted where necessary for design alternatives and owner's requirements. Reported completed project costs, for this type of structure, range from $64.60 to $224.60 per S.F.

Common additives

Description	Unit	$ Cost
Emergency Lighting, 25 watt, battery operated		
Lead battery	Ea.	345
Nickel cadmium	Ea.	685
Safe, Office type, 1 hour rating		
30" x 18" x 18"	Ea.	2500
60" x 36" x 18", double door	Ea.	9350
Smoke Detectors		
Ceiling type	Ea.	227
Duct type	Ea.	565
Sound System		
Amplifier, 250 watts	Ea.	2200
Speaker, ceiling or wall	Ea.	217
Trumpet	Ea.	415

Important: See the Reference Section for Location Factors.

Model costs calculated for a 1 story building with 14' story height and 8,000 square feet of floor area

				Unit	Unit Cost	Cost Per S.F.	% Of Sub-Total
A.	**SUBSTRUCTURE**						
1010	Standard Foundations	Poured concrete; strip and spread footings		S.F. Ground	2.16	2.16	
1020	Special Foundations	N/A		—	—	—	
1030	Slab on Grade	4" reinforced concrete		S.F. Slab	5.56	5.56	13.8%
2010	Basement Excavation	Site preparation for slab and trench for foundation wall and footing		S.F. Ground	.32	.32	
2020	Basement Walls	4' Foundation wall		L.F. Wall	86	3.85	
B.	**SHELL**						
	B10 Superstructure						
1010	Floor Construction	Wood columns,		S.F. Floor	.42	.42	9.6%
1020	Roof Construction	Wood roof truss 4:12 pitch		S.F. Roof	7.85	7.85	
	B20 Exterior Enclosure						
2010	Exterior Walls	Vinyl siding on wood studs, insulated	85% of wall	S.F. Wall	14.57	7.80	
2020	Exterior Windows	Storefront glazing	15% of wall	Each	50	2.21	13.1%
2030	Exterior Doors	Aluminum and glass single doors, hollow metal doors		Each	2528	1.27	
	B30 Roofing						
3010	Roof Coverings	Asphalt roofing, strip shingles, gutters and downspouts		S.F. Roof	3.90	3.90	4.7%
3020	Roof Openings	Roof hatches		S.F. Roof	.12	.12	
C.	**INTERIORS**						
1010	Partitions	Sound deadening gypsum board on wood studs	60 S.F. Floor/L.F. Partition	S.F. Partition	10.14	1.69	
1020	Interior Doors	Hollow metal doors	1150 S.F. Floor/Door	Each	1168	1.02	
1030	Fittings	Toilet partitions		S.F. Floor	.19	.19	
2010	Stair Construction	N/A		—	—	—	16.1%
3010	Wall Finishes	90% Paint, 10% ceramic tile		S.F. Surface	2.85	.95	
3020	Floor Finishes	Vinyl composition tile		S.F. Floor	2.76	2.76	
3030	Ceiling Finishes	Acoustic ceiling tiles on suspended channel grid		S.F. Ceiling	7.22	7.22	
D.	**SERVICES**						
	D10 Conveying						
1010	Elevators & Lifts	N/A		—	—	—	0.0 %
1020	Escalators & Moving Walks	N/A		—	—	—	
	D20 Plumbing						
2010	Plumbing Fixtures	Restroom and service fixtures, supply and drainage	1 Fixture/1350 S.F. Floor	Each	2666	2	
2020	Domestic Water Distribution	Gas fired water heater		S.F. Floor	3.18	3.18	7.6%
2040	Rain Water Drainage	Roof Drains		S.F. Roof	1.35	1.35	
	D30 HVAC						
3010	Energy Supply	N/A		—	—	—	
3020	Heat Generating Systems	Included in D3050		—	—	—	
3030	Cooling Generating Systems	N/A		—	—	—	10.2 %
3050	Terminal & Package Units	Single zone unit gas heating, electric cooling		S.F. Floor	8.81	8.81	
3090	Other HVAC Sys. & Equipment	N/A		—	—	—	
	D40 Fire Protection						
4010	Sprinklers	Wet pipe spinkler system		S.F. Floor	4.72	4.72	6.8%
4020	Standpipes	Standpipe		S.F. Floor	1.15	1.15	
	D50 Electrical						
5010	Electrical Service/Distribution	400 Ampere service, panelboards and feeders		S.F. Floor	2.83	2.83	
5020	Lighting & Branch Wiring	Flourescent fixtures, receptacles, a.c. and misc. power		S.F. Floor	10.99	10.99	18.1%
5030	Communications & Security	Addressable alarm system, emergency lighting		S.F. Floor	1.72	1.72	
5090	Other Electrical Systems	N/A		—	—	—	
E.	**EQUIPMENT & FURNISHINGS**						
1010	Commercial Equipment	N/A		—	—	—	
1020	Institutional Equipment	N/A		—	—	—	0.0 %
1030	Vehicular Equipment	N/A		—	—	—	
1090	Other Equipment	N/A		—	—	—	
F.	**SPECIAL CONSTRUCTION**						
1020	Integrated Construction	N/A		—	—	—	0.0 %
1040	Special Facilities	N/A		—	—	—	
G.	**BUILDING SITEWORK**	**N/A**					

		Sub-Total	86.04	100%
	CONTRACTOR FEES (General Requirements: 10%, Overhead: 5%, Profit: 10%)	25%	21.51	
	ARCHITECT FEES	8%	8.60	

Total Building Cost	**116.15**	

For customer support on your Square Foot Cost Data, call 877.756.2789.

215

Costs per square foot of floor area

Exterior Wall	S.F. Area	12000	16000	20000	26000	32000	38000	44000	52000	60000
	L.F. Perimeter	450	510	570	650	716	780	840	920	1020
Stucco Face	Reinforced Concrete	173.30	159.20	150.80	142.25	136.10	131.75	128.40	125.10	123.40
Curtain Wall	Steel Frame	151.25	140.00	133.30	126.55	121.75	118.35	115.75	113.25	111.90
E.I.F.S.	Steel Frame	129.75	121.80	117.05	112.35	109.05	106.75	104.95	103.25	102.30
Metal Panel	Steel Frame	126.80	118.90	114.20	109.50	106.30	104.00	102.20	100.55	99.60
Tilt-up Concrete Panel	Reinforced Concrete	146.85	137.05	131.20	125.35	121.20	118.25	115.95	113.85	112.65
Brick Veneer	Reinforced Concrete	183.50	167.90	158.55	149.05	142.15	137.30	133.55	129.95	128.00
Perimeter Adj., Add or Deduct	Per 100 L.F.	16.00	12.00	9.60	7.30	6.00	5.05	4.40	3.70	3.25
Story Hgt. Adj., Add or Deduct	Per 1 Ft.	1.60	1.35	1.15	1.05	0.90	0.85	0.75	0.80	0.75
For Basement, add $27.45 per square foot of basement area										

The above costs were calculated using the basic specifications shown on the facing page. These costs should be adjusted where necessary for design alternatives and owner's requirements. Reported completed project costs, for this type of structure, range from $75.45 to $216.10 per S.F.

Common additives

Description	Unit	$ Cost
Check Out Counter		
Single belt	Ea.	3600
Double belt	Ea.	5125
Scanner, registers, guns & memory 2 lanes	Ea.	18,300
10 lanes	Ea.	173,500
Power take away	Ea.	6025
Emergency Lighting, 25 watt, battery operated		
Lead battery	Ea.	345
Nickel cadmium	Ea.	685
Refrigerators, Prefabricated, walk-in		
7'-6" High, 6' x 6'	S.F.	203
10' x 10'	S.F.	159
12' x 14'	S.F.	142
12' x 20'	S.F.	125
Refrigerated Food Cases		
Dairy, multi deck, 12' long	Ea.	12,800
Delicatessen case, single deck, 12' long	Ea.	9600
Multi deck, 18 S.F. shelf display	Ea.	8875
Freezer, self-contained chest type, 30 C.F.	Ea.	10,300
Glass door upright, 78 C.F.	Ea.	13,600

Description	Unit	$ Cost
Refrigerated Food Cases, cont.		
Frozen food, chest type, 12' long	Ea.	9450
Glass door reach in, 5 door	Ea.	15,900
Island case 12' long, single deck	Ea.	8425
Multi deck	Ea.	9800
Meat case 12' long, single deck	Ea.	8425
Multi deck	Ea.	11,800
Produce, 12' long single deck	Ea.	7950
Multi deck	Ea.	10,000
Safe, Office type, 1 hour rating		
30" x 18" x 18"	Ea.	2500
60" x 36" x 18", double door	Ea.	9350
Smoke Detectors		
Ceiling type	Ea.	227
Duct type	Ea.	565
Sound System		
Amplifier, 250 watts	Ea.	2200
Speaker, ceiling or wall	Ea.	217
Trumpet	Ea.	415

Important: See the Reference Section for Location Factors.

Model costs calculated for a 1 story building with 18' story height and 44,000 square feet of floor area

Supermarket

				Unit	Unit Cost	Cost Per S.F.	% Of Sub-Total
A. SUBSTRUCTURE							
1010	Standard Foundations	Poured concrete; strip and spread footings		S.F. Ground	1.07	1.07	
1020	Special Foundations	N/A		—	—	—	
1030	Slab on Grade	4" reinforced slab on grade		S.F. Slab	5.56	5.56	9.8%
2010	Basement Excavation	Site preparation for slab and trench for foundation wall and footing		S.F. Ground	.18	.18	
2020	Basement Walls	4' Foundation wall		L.F. Wall	86	1.63	
B. SHELL							
B10 Superstructure							
1010	Floor Construction	N/A		—	—	—	12.0 %
1020	Roof Construction	Metal deck on open web steel joints, columns		S.F. Roof	10.38	10.38	
B20 Exterior Enclosure							
2010	Exterior Walls	N/A		—	—	—	
2020	Exterior Windows	Curtain wall glazing system, thermo-break frame	100% of wall	Each	47.50	16.16	20.3 %
2030	Exterior Doors	Glass sliding entrance doors, overhead doors, hollow metal doors		Each	5612	1.40	
B30 Roofing							
3010	Roof Coverings	Single ply membrane, rigid insulation, gravel stop		S.F. Roof	5.17	5.17	6.1%
3020	Roof Openings	Roof and smoke hatches		S.F. Roof	.14	.14	
C. INTERIORS							
1010	Partitions	Concrete partitions, gypsum board on metal stud	50 S.F. Floor/L.F. Partition	S.F. Partition	9.21	2.21	
1020	Interior Doors	Hollow metal doors	2000 S.F. Floor/Door	Each	1168	.59	
1030	Fittings	Stainless steel toilet partitions		S.F. Floor	.10	.10	
2010	Stair Construction	N/A		—	—	—	16.0%
3010	Wall Finishes	Paint		S.F. Surface	2	.96	
3020	Floor Finishes	Vinyl composition tile		S.F. Floor	2.76	2.76	
3030	Ceiling Finishes	Acoustic ceiling tiles on suspended channel grid		S.F. Ceiling	7.22	7.22	
D. SERVICES							
D10 Conveying							
1010	Elevators & Lifts	N/A		—	—	—	0.0 %
1020	Escalators & Moving Walks	N/A		—	—	—	
D20 Plumbing							
2010	Plumbing Fixtures	Restroom, food prep and service fixtures, supply and drainage	1 Fixture/1750 S.F. Floor	Each	2112	1.20	
2020	Domestic Water Distribution	Gas fired water heater		S.F. Floor	.25	.25	2.9%
2040	Rain Water Drainage	Roof drains		S.F. Roof	1.07	1.07	
D30 HVAC							
3010	Energy Supply	N/A		—	—	—	
3020	Heat Generating Systems	Included in D3050		—	—	—	
3030	Cooling Generating Systems	N/A		—	—	—	5.8 %
3050	Terminal & Package Units	Single zone unit gas heating, electric cooling		S.F. Floor	5	5	
3090	Other HVAC Sys. & Equipment	N/A		—	—	—	
D40 Fire Protection							
4010	Sprinklers	Wet pipe sprinkler system		S.F. Floor	3.65	3.65	4.9%
4020	Standpipes	Standpipe		S.F. Floor	.59	.59	
D50 Electrical							
5010	Electrical Service/Distribution	1600 ampere service, panel board and feeders		S.F. Floor	2.57	2.57	
5020	Lighting & Branch Wiring	Fluorescent fixtures, receptacles, switches, A.C. and misc. power		S.F. Floor	14.36	14.36	22.3%
5030	Communications & Security	Addressable alarm system, emergency lighting, internet & phone wiring		S.F. Floor	2.28	2.28	
5090	Other Electrical Systems	Emergency generator, 15kW		S.F. Floor	.05	.05	
E. EQUIPMENT & FURNISHINGS							
1010	Commercial Equipment	N/A		—	—	—	
1020	Institutional Equipment	N/A		—	—	—	0.0 %
1030	Vehicular Equipment	N/A		—	—	—	
1090	Other Equipment	N/A		—	—	—	
F. SPECIAL CONSTRUCTION							
1020	Integrated Construction	N/A		—	—	—	0.0 %
1040	Special Facilities	N/A		—	—	—	
G. BUILDING SITEWORK	**N/A**						
					Sub-Total	86.55	**100%**
	CONTRACTOR FEES (General Requirements: 10%, Overhead: 5%, Profit: 10%)				25%	21.63	
	ARCHITECT FEES				7%	7.57	
					Total Building Cost	**115.75**	

For customer support on your Square Foot Cost Data, call 877.756.2789.

217

Costs per square foot of floor area

Exterior Wall	S.F. Area	10000	16000	20000	22000	24000	26000	28000	30000	32000
	L.F. Perimeter	420	510	600	640	660	684	706	740	737
Face Brick with Concrete Block Back-up	Wood Truss	277.65	259.05	255.35	253.60	250.80	248.60	246.70	245.70	242.90
	Precast Conc.	318.60	296.35	292.05	289.90	286.50	283.95	281.65	280.45	277.05
Metal Sandwich Panel	Wood Truss	291.75	271.95	267.45	265.40	262.65	260.50	258.60	257.45	254.95
Precast Concrete Panel	Precast Conc.	320.00	293.35	287.60	284.95	281.10	278.20	275.55	274.05	270.35
Painted Concrete Block	Wood Frame	278.00	264.50	261.75	260.45	258.40	256.90	255.50	254.75	252.80
	Precast Conc.	298.60	277.15	272.30	270.15	267.10	264.80	262.70	261.45	258.70
Perimeter Adj., Add or Deduct	Per 100 L.F.	16.55	10.40	8.30	7.50	6.85	6.45	5.95	5.50	5.20
Story Hgt. Adj., Add or Deduct	Per 1 Ft.	2.25	1.65	1.60	1.55	1.45	1.45	1.35	1.25	1.20

For Basement, add $40.70 per square foot of basement area

The above costs were calculated using the basic specifications shown on the facing page. These costs should be adjusted where necessary for design alternatives and owner's requirements. Reported completed project costs, for this type of structure, range from $119.20 to $350.40 per S.F.

Common additives

Description	Unit	$ Cost
Bleachers, Telescoping, manual		
To 15 tier	Seat	143 - 199
16-20 tier	Seat	294 - 360
21-30 tier	Seat	310 - 410
For power operation, add	Seat	57.50 - 90
Emergency Lighting, 25 watt, battery operated		
Lead battery	Ea.	345
Nickel cadmium	Ea.	685
Lockers, Steel, single tier, 60" or 72"	Opng.	231 - 380
2 tier, 60" or 72" total	Opng.	121 - 170
5 tier, box lockers	Opng.	74 - 84
Locker bench, lam. maple top only	L.F.	34.50
Pedestal, steel pipe	Ea.	76.50
Pool Equipment		
Diving stand, 3 meter	Ea.	20,200
1 meter	Ea.	10,900
Diving board, 16' aluminum	Ea.	4925
Fiberglass	Ea.	4075
Lifeguard chair, fixed	Ea.	4200
Portable	Ea.	3050
Lights, underwater, 12 volt, 300 watt	Ea.	1000

Description	Unit	$ Cost
Sauna, Prefabricated, complete		
6' x 4'	Ea.	6750
6' x 6'	Ea.	8075
6' x 9'	Ea.	10,500
8' x 8'	Ea.	12,000
8' x 10'	Ea.	13,400
10' x 12'	Ea.	16,100
Sound System		
Amplifier, 250 watts	Ea.	2200
Speaker, ceiling or wall	Ea.	217
Trumpet	Ea.	415
Steam Bath, Complete, to 140 C.F.	Ea.	3000
To 300 C.F.	Ea.	3325
To 800 C.F.	Ea.	7100
To 2500 C.F.	Ea.	9325

Important: See the Reference Section for Location Factors.

Model costs calculated for a 1 story building with 24' story height and 20,000 square feet of floor area

Swimming Pool, Enclosed

				Unit	Unit Cost	Cost Per S.F.	% Of Sub-Total
A. SUBSTRUCTURE							
1010	Standard Foundations	Poured concrete; strip and spread footings		S.F. Ground	3.99	3.99	
1020	Special Foundations	N/A		—	—	—	
1030	Slab on Grade	4" reinforced concrete with vapor barrier and granular base		S.F. Slab	5.56	5.56	8.1%
2010	Basement Excavation	Site preparation for slab and trench for foundation wall and footing		S.F. Ground	.32	.64	
2020	Basement Walls	4' foundation wall		L.F. Wall	85	5.13	
B. SHELL							
B10 Superstructure							
1010	Floor Construction	N/A		—	—	—	10.4 %
1020	Roof Construction	Wood deck on laminated wood truss		S.F. Roof	19.64	19.64	
B20 Exterior Enclosure							
2010	Exterior Walls	Face brick with concrete block backup	80% of wall	S.F. Wall	32.90	18.95	
2020	Exterior Windows	Outward projecting steel	20% of wall	Each	50	7.21	14.2%
2030	Exterior Doors	Double aluminum and glass, hollow metal		Each	2723	.69	
B30 Roofing							
3010	Roof Coverings	Asphalt shingles with flashing; perlite/EPS composite insulation		S.F. Roof	4.50	4.50	2.4%
3020	Roof Openings	N/A		—	—	—	
C. INTERIORS							
1010	Partitions	Concrete block	100 S.F. Floor/L.F. Partition	S.F. Partition	9.70	.97	
1020	Interior Doors	Single leaf hollow metal	1000 S.F. Floor/Door	Each	1168	1.17	
1030	Fittings	Toilet partitions		S.F. Floor	.18	.18	
2010	Stair Construction	N/A		—	—	—	15.5%
3010	Wall Finishes	Acrylic glazed coating		S.F. Surface	11.90	2.38	
3020	Floor Finishes	70% terrazzo, 30% ceramic tile		S.F. Floor	23.60	23.60	
3030	Ceiling Finishes	Mineral fiber tile on concealed zee bars	20% of area	S.F. Ceiling	4.87	.98	
D. SERVICES							
D10 Conveying							
1010	Elevators & Lifts	N/A		—	—	—	0.0 %
1020	Escalators & Moving Walks	N/A		—	—	—	
D20 Plumbing							
2010	Plumbing Fixtures	Toilet and service fixtures, supply and drainage	1 Fixture/210 S.F. Floor	Each	1266	6.03	
2020	Domestic Water Distribution	Gas fired water heater		S.F. Floor	4.32	4.32	5.5%
2040	Rain Water Drainage	N/A		—	—	—	
D30 HVAC							
3010	Energy Supply	Terminal unit heaters	10% of area	S.F. Floor	.92	.92	
3020	Heat Generating Systems	N/A		—	—	—	
3030	Cooling Generating Systems	N/A		—	—	—	8.1%
3050	Terminal & Package Units	Single zone unit gas heating, electric cooling		S.F. Floor	14.36	14.36	
3090	Other HVAC Sys. & Equipment	N/A		—	—	—	
D40 Fire Protection							
4010	Sprinklers	Sprinklers, light hazard		S.F. Floor	1.83	1.83	1.3%
4020	Standpipes	Standpipe		S.F. Floor	.60	.60	
D50 Electrical							
5010	Electrical Service/Distribution	400 ampere service, panel board and feeders		S.F. Floor	1.18	1.18	
5020	Lighting & Branch Wiring	High efficiency fluorescent fixtures, receptacles, switches, A.C. and misc. power		S.F. Floor	21.72	21.72	12.7%
5030	Communications & Security	Addressable alarm systems and emergency lighting		S.F. Floor	1.06	1.06	
5090	Other Electrical Systems	Emergency generator, 15 kW		S.F. Floor	.09	.09	
E. EQUIPMENT & FURNISHINGS							
1010	Commercial Equipment	N/A		—	—	—	
1020	Institutional Equipment	N/A		—	—	—	
1030	Vehicular Equipment	N/A		—	—	—	21.9 %
1090	Other Equipment	Swimming pool		S.F. Floor	41.47	41.47	
F. SPECIAL CONSTRUCTION							
1020	Integrated Construction	N/A		—	—	—	0.0 %
1040	Special Facilities	N/A		—	—	—	
G. BUILDING SITEWORK	**N/A**						

		Sub-Total	189.17	100%
CONTRACTOR FEES (General Requirements: 10%, Overhead: 5%, Profit: 10%)		25%	47.26	
ARCHITECT FEES		8%	18.92	
Total Building Cost			**255.35**	

For customer support on your Square Foot Cost Data, call 877.756.2789.

219

Costs per square foot of floor area

Exterior Wall	S.F. Area	2000	3000	4000	5000	6000	7000	8000	9000	10000
	L.F. Perimeter	180	220	260	286	320	353	368	397	425
Face Brick with Concrete Block Back-up	Steel Frame	229.80	203.90	191.00	180.45	174.75	170.45	165.05	162.40	160.15
	Bearing Walls	219.65	196.45	184.90	175.15	170.00	166.05	161.00	158.55	156.50
Limestone with Concrete Block Back-up	Steel Frame	264.00	233.65	218.45	205.30	198.45	193.30	186.30	183.00	180.25
	Bearing Walls	263.60	230.45	213.95	200.00	192.65	187.05	179.80	176.30	173.35
Decorative Concrete Block	Steel Frame	214.65	191.55	180.10	170.85	165.75	161.95	157.25	155.00	153.00
	Bearing Walls	209.40	186.30	174.75	165.55	160.45	156.65	152.05	149.75	147.75
Perimeter Adj., Add or Deduct	Per 100 L.F.	50.70	33.80	25.30	20.25	16.85	14.50	12.65	11.25	10.15
Story Hgt. Adj., Add or Deduct	Per 1 Ft.	5.85	4.80	4.30	3.70	3.50	3.30	3.00	2.90	2.80

For Basement, add $39.45 per square foot of basement area

The above costs were calculated using the basic specifications shown on the facing page. These costs should be adjusted where necessary for design alternatives and owner's requirements. Reported completed project costs, for this type of structure, range from $105.35 to $380.10 per S.F.

Common additives

Description	Unit	$ Cost
Emergency Lighting, 25 watt, battery operated		
Lead battery	Ea.	345
Nickel cadmium	Ea.	685
Smoke Detectors		
Ceiling type	Ea.	227
Duct type	Ea.	565
Emergency Generators, complete system, gas		
15 kw	Ea.	18,500
85 kw	Ea.	39,000
170 kw	Ea.	98,000
Diesel, 50 kw	Ea.	26,100
150 kw	Ea.	47,900
350 kw	Ea.	72,500

Telephone Exchange

Model costs calculated for a 1 story building with 12' story height and 5,000 square feet of floor area

				Unit	Unit Cost	Cost Per S.F.	% Of Sub-Total
A.	**SUBSTRUCTURE**						
1010	Standard Foundations	Poured concrete; strip and spread footings		S.F. Ground	2.65	2.65	
1020	Special Foundations	N/A		—	—	—	
1030	Slab on Grade	4" reinforced concrete with vapor barrier and granular base		S.F. Slab	5.56	5.56	10.6%
2010	Basement Excavation	Site preparation for slab and trench for foundation wall and footing		S.F. Ground	.57	.71	
2020	Basement Walls	4' foundation wall		L.F. Wall	86	4.89	
B.	**SHELL**						
	B10 Superstructure						
1010	Floor Construction	Steel column fireproofing		S.F. Floor	.85	.85	8.6%
1020	Roof Construction	Metal deck, open web steel joists, beams, columns		S.F. Roof	10.38	10.38	
	B20 Exterior Enclosure						
2010	Exterior Walls	Face brick with concrete block backup	80% of wall	S.F. Wall	33.40	18.34	
2020	Exterior Windows	Outward projecting steel	20% of wall	Each	1765	12.12	24.7%
2030	Exterior Doors	Single aluminum glass with transom		Each	4030	1.62	
	B30 Roofing						
3010	Roof Coverings	Built-up tar and gravel with flashing; perlite/EPS composite insulation		S.F. Roof	7.40	7.40	5.7%
3020	Roof Openings	N/A		—	—	—	
C.	**INTERIORS**						
1010	Partitions	Double layer gypsum board on metal studs	15 S.F. Floor/L.F. Partition	S.F. Partition	6.48	4.32	
1020	Interior Doors	Single leaf hollow metal	150 S.F. Floor/Door	Each	1168	7.78	
1030	Fittings	Toilet partitions		S.F. Floor	1.22	1.22	
2010	Stair Construction	N/A		—	—	—	21.6%
3010	Wall Finishes	Paint		S.F. Surface	2.22	2.96	
3020	Floor Finishes	90% carpet, 10% terrazzo		S.F. Floor	6.05	6.05	
3030	Ceiling Finishes	Fiberglass board on exposed grid system, suspended		S.F. Ceiling	5.78	5.78	
D.	**SERVICES**						
	D10 Conveying						
1010	Elevators & Lifts	N/A		—	—	—	0.0 %
1020	Escalators & Moving Walks	N/A		—	—	—	
	D20 Plumbing						
2010	Plumbing Fixtures	Kitchen, toilet and service fixtures, supply and drainage	1 Fixture/715 S.F. Floor	Each	3775	5.28	
2020	Domestic Water Distribution	Gas fired water heater		S.F. Floor	2.33	2.33	7.3%
2040	Rain Water Drainage	Roof drains		S.F. Roof	1.82	1.82	
	D30 HVAC						
3010	Energy Supply	N/A		—	—	—	
3020	Heat Generating Systems	Included in D3050		—	—	—	
3030	Cooling Generating Systems	N/A		—	—	—	6.6 %
3050	Terminal & Package Units	Single zone unit, gas heating, electric cooling		S.F. Floor	8.59	8.59	
3090	Other HVAC Sys. & Equipment	N/A		—	—	—	
	D40 Fire Protection						
4010	Sprinklers	Wet pipe sprinkler system		S.F. Floor	5.50	5.50	5.9%
4020	Standpipes	Standpipe		S.F. Floor	2.21	2.21	
	D50 Electrical						
5010	Electrical Service/Distribution	200 ampere service, panel board and feeders		S.F. Floor	1.99	1.99	
5020	Lighting & Branch Wiring	High efficiency fluorescent fixtures, receptacles, switches, A.C. and misc. power		S.F. Floor	6.50	6.50	9.0%
5030	Communications & Security	Addressable alarm systems and emergency lighting		S.F. Floor	3.01	3.01	
5090	Other Electrical Systems	Emergency generator, 15 kW		S.F. Floor	.19	.19	
E.	**EQUIPMENT & FURNISHINGS**						
1010	Commercial Equipment	N/A		—	—	—	
1020	Institutional Equipment	N/A		—	—	—	0.0 %
1030	Vehicular Equipment	N/A		—	—	—	
1090	Other Equipment	N/A		—	—	—	
F.	**SPECIAL CONSTRUCTION**						
1020	Integrated Construction	N/A		—	—	—	0.0 %
1040	Special Facilities	N/A		—	—	—	
G.	**BUILDING SITEWORK**	**N/A**					

		Sub-Total	130.05	100%
CONTRACTOR FEES (General Requirements: 10%, Overhead: 5%, Profit: 10%)		25%	32.52	
ARCHITECT FEES		11%	17.88	

Total Building Cost	**180.45**

For customer support on your Square Foot Cost Data, call 877.756.2789.

221

Costs per square foot of floor area

Exterior Wall	S.F. Area	5000	6500	8000	9500	11000	14000	17500	21000	24000
	L.F. Perimeter	300	360	386	396	435	510	550	620	680
Face Brick with Concrete Block Back-up	Steel Joists	170.10	163.65	156.10	149.50	146.90	143.20	138.25	136.15	134.85
	Wood Joists	168.45	162.25	155.10	148.85	146.45	142.90	138.30	136.25	135.05
Stone with Concrete Block Back-up	Steel Joists	178.20	171.15	162.65	155.15	152.30	148.15	142.45	140.10	138.65
	Wood Joists	176.55	169.75	161.65	154.50	151.85	147.85	142.55	140.30	138.85
Brick Veneer	Wood Frame	156.75	150.95	144.45	138.85	136.60	133.40	129.20	127.40	126.25
E.I.F.S.	Wood Frame	148.55	143.20	137.30	132.25	130.25	127.25	123.50	121.85	120.80
Perimeter Adj., Add or Deduct	Per 100 L.F.	16.65	12.80	10.35	8.70	7.55	6.00	4.80	3.95	3.45
Story Hgt. Adj., Add or Deduct	Per 1 Ft.	3.00	2.80	2.45	2.05	1.95	1.80	1.55	1.50	1.35
For Basement, add $32.50 per square foot of basement area										

The above costs were calculated using the basic specifications shown on the facing page. These costs should be adjusted where necessary for design alternatives and owner's requirements. Reported completed project costs, for this type of structure, range from $99.60 to $261.90 per S.F.

Common additives

Description	Unit	$ Cost
Directory Boards, Plastic, glass covered		
30" x 20"	Ea.	610
36" x 48"	Ea.	1525
Aluminum, 24" x 18"	Ea.	640
36" x 24"	Ea.	765
48" x 32"	Ea.	1075
48" x 60"	Ea.	2275
Emergency Lighting, 25 watt, battery operated		
Lead battery	Ea.	345
Nickel cadmium	Ea.	685
Flagpoles, Complete		
Aluminum, 20' high	Ea.	1900
40' high	Ea.	4425
70' high	Ea.	11,500
Fiberglass, 23' high	Ea.	1350
39'-5" high	Ea.	3400
59' high	Ea.	6650
Safe, Office type, 1 hour rating		
30" x 18" x 18"	Ea.	2500
60" x 36" x 18", double door	Ea.	9350

Description	Unit	$ Cost
Smoke Detectors		
Ceiling type	Ea.	227
Duct type	Ea.	565
Vault Front, Door & frame		
1 Hour test, 32" x 78"	Opng.	8250
2 Hour test, 32" door	Opng.	10,100
40" door	Opng.	11,700
4 Hour test, 32" door	Opng.	10,700
40" door	Opng.	12,600
Time lock movement; two movement	Ea.	2300

Important: See the Reference Section for Location Factors.

Model costs calculated for a 1 story building with 12' story height and 11,000 square feet of floor area

Town Hall, 1 Story

				Unit	Unit Cost	Cost Per S.F.	% Of Sub-Total
A. SUBSTRUCTURE							
1010	Standard Foundations	Poured concrete; strip and spread footings		S.F. Ground	1.79	1.79	
1020	Special Foundations	N/A		—	—	—	
1030	Slab on Grade	4" reinforced concrete with vapor barrier and granular base		S.F. Slab	5.56	5.56	10.3%
2010	Basement Excavation	Site preparation for slab and trench for foundation wall and footing		S.F. Ground	.32	.35	
2020	Basement Walls	4' foundation wall		L.F. Wall	86	3.38	
B. SHELL							
B10 Superstructure							
1010	Floor Construction	Steel column fireproofing		S.F. Floor	.30	.30	7.6%
1020	Roof Construction	Metal deck, open web steel joists, beams, interior columns		S.F. Roof	7.87	7.87	
B20 Exterior Enclosure							
2010	Exterior Walls	Face brick with concrete block backup	70% of wall	S.F. Wall	33.39	11.09	
2020	Exterior Windows	Metal outward projecting	30% of wall	Each	746	4.62	15.6%
2030	Exterior Doors	Metal and glass with transom		Each	2920	1.06	
B30 Roofing							
3010	Roof Coverings	Built-up tar and gravel with flashing; perlite/EPS composite insulation		S.F. Roof	6.92	6.92	6.4%
3020	Roof Openings	N/A		—	—	—	
C. INTERIORS							
1010	Partitions	Gypsum board on metal studs	20 S.F. Floor/L.F. Partition	S.F. Partition	9.82	4.91	
1020	Interior Doors	Wood solid core	200 S.F. Floor/Door	Each	626	3.14	
1030	Fittings	Toilet partitions		S.F. Floor	.48	.48	
2010	Stair Construction	N/A		—	—	—	23.7%
3010	Wall Finishes	90% paint, 10% ceramic tile		S.F. Surface	1.53	1.53	
3020	Floor Finishes	70% carpet tile, 15% terrazzo, 15% vinyl composition tile		S.F. Floor	8.28	8.28	
3030	Ceiling Finishes	Mineral fiber tile on concealed zee bars		S.F. Ceiling	7.22	7.22	
D. SERVICES							
D10 Conveying							
1010	Elevators & Lifts	N/A		—	—	—	0.0%
1020	Escalators & Moving Walks	N/A		—	—	—	
D20 Plumbing							
2010	Plumbing Fixtures	Kitchen, toilet and service fixtures, supply and drainage	1 Fixture/500 S.F. Floor	Each	2470	4.94	
2020	Domestic Water Distribution	Gas fired water heater		S.F. Floor	1.72	1.72	8.4%
2040	Rain Water Drainage	Roof drains		S.F. Roof	2.36	2.36	
D30 HVAC							
3010	Energy Supply	N/A		—	—	—	
3020	Heat Generating Systems	Included in D3050		—	—	—	
3030	Cooling Generating Systems	N/A		—	—	—	8.9%
3050	Terminal & Package Units	Multizone unit, gas heating, electric cooling		S.F. Floor	9.57	9.57	
3090	Other HVAC Sys. & Equipment	N/A		—	—	—	
D40 Fire Protection							
4010	Sprinklers	Wet pipe sprinkler system		S.F. Floor	3.65	3.65	4.2%
4020	Standpipes	Standpipe, wet, Class III		S.F. Floor	.84	.84	
D50 Electrical							
5010	Electrical Service/Distribution	400 ampere service, panel board and feeders		S.F. Floor	2.05	2.05	
5020	Lighting & Branch Wiring	High efficiency fluorescent fixtures, receptacles, switches, A.C. and misc. power		S.F. Floor	10.76	10.76	15.1%
5030	Communications & Security	Addressable alarm systems, internet wiring, and emergency lighting		S.F. Floor	3.27	3.27	
5090	Other Electrical Systems	Emergency generator, 15 kW		S.F. Floor	.17	.17	
E. EQUIPMENT & FURNISHINGS							
1010	Commercial Equipment	N/A		—	—	—	
1020	Institutional Equipment	N/A		—	—	—	0.0%
1030	Vehicular Equipment	N/A		—	—	—	
1090	Other Equipment	N/A		—	—	—	
F. SPECIAL CONSTRUCTION							
1020	Integrated Construction	N/A		—	—	—	0.0%
1040	Special Facilities	N/A		—	—	—	
G. BUILDING SITEWORK	**N/A**						

		Sub-Total	107.83	**100%**
CONTRACTOR FEES (General Requirements: 10%, Overhead: 5%, Profit: 10%)			25%	26.94
ARCHITECT FEES			9%	12.13
		Total Building Cost	**146.90**	

For customer support on your Square Foot Cost Data, call 877.756.2789.

223

Costs per square foot of floor area

Exterior Wall	S.F. Area	8000	10000	12000	15000	18000	24000	28000	35000	40000
	L.F. Perimeter	206	233	260	300	320	360	393	451	493
Face Brick with Concrete Block Back-up	Steel Frame	224.50	212.85	205.00	197.15	189.60	180.30	176.70	172.50	170.40
	R/Conc. Frame	228.40	216.75	208.95	201.05	193.50	184.15	180.60	176.40	174.20
Stone with Concrete Block Back-up	Steel Frame	236.05	223.95	215.80	207.55	199.05	188.45	184.55	179.85	177.50
	R/Conc. Frame	238.85	226.20	217.70	209.15	200.70	190.25	186.25	181.55	179.25
Limestone with Concrete Block Back-up	Steel Frame	242.05	227.70	218.15	208.40	198.45	186.05	181.45	175.95	173.20
	R/Conc. Frame	245.95	231.65	222.05	212.30	202.35	189.95	185.35	179.85	177.05
Perimeter Adj., Add or Deduct	Per 100 L.F.	25.40	20.40	17.00	13.55	11.35	8.45	7.25	5.80	5.05
Story Hgt. Adj., Add or Deduct	Per 1 Ft.	3.85	3.55	3.25	2.95	2.70	2.20	2.10	1.85	1.80

For Basement, add $31.65 per square foot of basement area

The above costs were calculated using the basic specifications shown on the facing page. These costs should be adjusted where necessary for design alternatives and owner's requirements. Reported completed project costs, for this type of structure, range from $89.25 to $261.90 per S.F.

Common additives

Description	Unit	$ Cost
Directory Boards, Plastic, glass covered		
30" x 20"	Ea.	610
36" x 48"	Ea.	1525
Aluminum, 24" x 18"	Ea.	640
36" x 24"	Ea.	765
48" x 32"	Ea.	1075
48" x 60"	Ea.	2275
Elevators, Hydraulic passenger, 2 stops		
1500# capacity	Ea.	67,400
2500# capacity	Ea.	70,400
3500# capacity	Ea.	75,400
Additional stop, add	Ea.	8075
Emergency Lighting, 25 watt, battery operated		
Lead battery	Ea.	345
Nickel cadmium	Ea.	685

Description	Unit	$ Cost
Flagpoles, Complete		
Aluminum, 20' high	Ea.	1900
40' high	Ea.	4425
70' high	Ea.	11,500
Fiberglass, 23' high	Ea.	1350
39'-5" high	Ea.	3400
59' high	Ea.	6650
Safe, Office type, 1 hour rating		
30" x 18" x 18"	Ea.	2500
60" x 36" x 18", double door	Ea.	9350
Smoke Detectors		
Ceiling type	Ea.	227
Duct type	Ea.	565
Vault Front, Door & frame		
1 Hour test, 32" x 78"	Opng.	8250
2 Hour test, 32" door	Opng.	10,100
40" door	Opng.	11,700
4 Hour test, 32" door	Opng.	10,700
40" door	Opng.	12,600
Time lock movement; two movement	Ea.	2300

Important: See the Reference Section for Location Factors.

Model costs calculated for a 3 story building with 12' story height and 18,000 square feet of floor area

				Unit	Unit Cost	Cost Per S.F.	% Of Sub-Total
A. SUBSTRUCTURE							
1010	Standard Foundations	Poured concrete; strip and spread footings		S.F. Ground	4.56	1.52	
1020	Special Foundations	N/A		—	—	—	
1030	Slab on Grade	4" reinforced concrete with vapor barrier and granular base		S.F. Slab	5.56	1.85	3.5%
2010	Basement Excavation	Site preparation for slab and trench for foundation wall and footing		S.F. Ground	.57	.29	
2020	Basement Walls	4' foundation wall		L.F. Wall	42.75	1.52	
B. SHELL							
	B10 Superstructure						
1010	Floor Construction	Open web steel joists, slab form, concrete, wide flange steel columns		S.F. Floor	28.49	18.99	15.0%
1020	Roof Construction	Metal deck, open web steel joists, beams, interior columns		S.F. Roof	8.61	2.87	
	B20 Exterior Enclosure						
2010	Exterior Walls	Stone with concrete block backup	70% of wall	S.F. Wall	45.20	20.25	
2020	Exterior Windows	Metal outward projecting	10% of wall	Each	1170	9.77	21.1%
2030	Exterior Doors	Metal and glass with transoms		Each	2723	.76	
	B30 Roofing						
3010	Roof Coverings	Built-up tar and gravel with flashing; perlite/EPS composite insulation		S.F. Roof	7.53	2.51	1.7%
3020	Roof Openings	N/A		—	—	—	
C. INTERIORS							
1010	Partitions	Gypsum board on metal studs	20 S.F. Floor/L.F. Partition	S.F. Partition	10.96	5.48	
1020	Interior Doors	Wood solid core	200 S.F. Floor/Door	Each	626	3.14	
1030	Fittings	Toilet partitions		S.F. Floor	.29	.29	
2010	Stair Construction	Concrete filled metal pan		Flight	17,275	5.76	21.7%
3010	Wall Finishes	90% paint, 10% ceramic tile		S.F. Surface	1.53	1.53	
3020	Floor Finishes	70% carpet tile, 15% terrazzo, 15% vinyl composition tile		S.F. Floor	8.28	8.28	
3030	Ceiling Finishes	Mineral fiber tile on concealed zee bars		S.F. Ceiling	7.22	7.22	
D. SERVICES							
	D10 Conveying						
1010	Elevators & Lifts	Two hydraulic elevators		Each	120,060	13.34	9.1%
1020	Escalators & Moving Walks	N/A		—	—	—	
	D20 Plumbing						
2010	Plumbing Fixtures	Toilet and service fixtures, supply and drainage	1 Fixture/1385 S.F. Floor	Each	6523	4.71	
2020	Domestic Water Distribution	Gas fired water heater		S.F. Floor	.62	.62	4.2%
2040	Rain Water Drainage	Roof drains		S.F. Roof	2.52	.84	
	D30 HVAC						
3010	Energy Supply	N/A		—	—	—	
3020	Heat Generating Systems	Included in D3050		—	—	—	
3030	Cooling Generating Systems	N/A		—	—	—	6.6%
3050	Terminal & Package Units	Multizone unit, gas heating, electric cooling		S.F. Floor	9.57	9.57	
3090	Other HVAC Sys. & Equipment	N/A		—	—	—	
	D40 Fire Protection						
4010	Sprinklers	Sprinklers, light hazard		S.F. Floor	3.33	3.33	4.7%
4020	Standpipes	Standpipe, wet, Class III		S.F. Floor	3.54	3.54	
	D50 Electrical						
5010	Electrical Service/Distribution	400 ampere service, panel board and feeders		S.F. Floor	1.73	1.73	
5020	Lighting & Branch Wiring	High efficiency fluorescent fixtures, receptacles, switches, A.C. and misc. power		S.F. Floor	12.79	12.79	12.4%
5030	Communications & Security	Addressable alarm systems, internet wiring, and emergency lighting		S.F. Floor	3.39	3.39	
5090	Other Electrical Systems	Emergency generator, 15 kW		S.F. Floor	.20	.20	
E. EQUIPMENT & FURNISHINGS							
1010	Commercial Equipment	N/A		—	—	—	
1020	Institutional Equipment	N/A		—	—	—	
1030	Vehicular Equipment	N/A		—	—	—	0.0%
1090	Other Equipment	N/A		—	—	—	
F. SPECIAL CONSTRUCTION							
1020	Integrated Construction	N/A		—	—	—	0.0%
1040	Special Facilities	N/A		—	—	—	
G. BUILDING SITEWORK	**N/A**						

	Sub-Total	146.09	**100%**
CONTRACTOR FEES (General Requirements: 10%, Overhead: 5%, Profit: 10%)	25%	36.53	
ARCHITECT FEES	9%	16.43	
	Total Building Cost	199.05	

For customer support on your Square Foot Cost Data, call 877.756.2789.

225

Costs per square foot of floor area

Exterior Wall	S.F. Area	3500	5000	7500	9000	11000	12500	15000	20000	25000
	L.F. Perimeter	240	290	350	390	430	460	510	580	650
E.I.F.S.	Wood Truss	177.05	167.55	158.70	156.00	153.00	151.35	149.30	146.10	144.20
	Steel Joists	188.25	178.35	169.10	166.20	163.05	161.30	159.20	155.80	153.75
Wood Siding	Wood Truss	179.35	170.10	161.45	158.75	155.85	154.20	152.25	149.15	147.30
Brick Veneer	Wood Truss	190.40	179.40	168.95	165.75	162.15	160.15	157.75	153.85	151.50
Brick on Block	Wood Truss	194.05	181.90	170.30	166.70	162.70	160.40	157.70	153.30	150.60
Tilt Up Concrete	Steel Joists	179.75	170.90	162.75	160.15	157.40	155.90	154.00	151.10	149.35
Perimeter Adj., Add or Deduct	Per 100 L.F.	13.30	9.35	6.25	5.20	4.20	3.70	3.15	2.35	1.90
Story Hgt. Adj., Add or Deduct	Per 1 Ft.	1.65	1.35	1.20	1.05	0.90	0.85	0.85	0.75	0.60
For Basement, add $34.15 per square foot of basement area										

The above costs were calculated using the basic specifications shown on the facing page. These costs should be adjusted where necessary for design alternatives and owner's requirements. Reported completed project costs, for this type of structure, range from $73.10 to $282.60 per S.F.

Common additives

Description	Unit	$ Cost
Closed circuit surveillance, one station		
Camera and monitor	Ea.	2025
For additional camera stations, add	Ea.	1100
Directory boards, plastic, glass covered		
30" x 20"	Ea.	610
36" x 48"	Ea.	1525
Aluminum, 24" x 18"	Ea.	640
36" x 24"	Ea.	765
48" x 32"	Ea.	1075
48" x 60"	Ea.	2275
Electronic, wall mounted	S.F.	2900
Free standing	S.F.	4125
Kennel fencing		
Kennel fencing, 1-1/2" mesh, 6' long, 3'-6" wide, 6'-2" high	Ea.	790
12' long	Ea.	1025
Top covers, 1-1/2" mesh, 6' long	Ea.	210
12' long	Ea.	286

Description	Unit	$ Cost
Kennel doors		
2 way, swinging type, 13" x 19" opening	Opng.	207
17" x 29" opening	Opng.	229
9" x 9" opening, electronic with accessories	Opng.	266

Veterinary Hospital

Model costs calculated for a 1 story building with 12' story height and 7,500 square feet of floor area

					Unit	Unit Cost	Cost Per S.F.	% Of Sub-Total
A. SUBSTRUCTURE								
1010	Standard Foundations	Poured concrete; strip and spread footings			S.F. Ground	2.12	2.12	
1020	Special Foundations	N/A			—	—	—	
1030	Slab on Grade	4" and 5" reinforced concrete with vapor barrier and granular base			S.F. Slab	5.56	5.56	10.8%
2010	Basement Excavation	Site preparation for slab and trench for foundation wall and footing			S.F. Ground	.57	1.07	
2020	Basement Walls	4' foundation wall			L.F. Wall	86	3.99	
B. SHELL								
	B10 Superstructure							
1010	Floor Construction	6 x 6 wood columns			S.F. Floor	.24	.24	6.8%
1020	Roof Construction	Wood roof truss with plywood sheathing			S.F. Roof	7.85	7.85	
	B20 Exterior Enclosure							
2010	Exterior Walls	Cedar bevel siding on wood studs, insulated	85% of wall		S.F. Wall	13.74	6.54	
2020	Exterior Windows	Wood double hung and awning	15% of wall		Each	1186	4.06	10.8%
2030	Exterior Doors	Aluminum and glass			Each	2745	2.20	
	B30 Roofing							
3010	Roof Coverings	Ashphalt shingles with flashing (pitched); al. gutters and downspouts			S.F. Roof	5.19	5.19	4.4%
3020	Roof Openings	N/A			—	—	—	
C. INTERIORS								
1010	Partitions	Gypsum board on wood studs	8 S.F. Floor/L.F. Partition		S.F. Partition	5.27	6.59	
1020	Interior Doors	Solid core wood	180 S.F. Floor/Door		Each	691	3.84	
1030	Fittings	Lockers			Each	.45	.45	
2010	Stair Construction	N/A			—	—	—	20.1%
3010	Wall Finishes	Paint on drywall			S.F. Surface	1.01	2.53	
3020	Floor Finishes	90% vinyl composition tile, 5% ceramic tile, 5% sealed concrete			S.F. Floor	3.39	3.39	
3030	Ceiling Finishes	90% acoustic tile, 10% gypsum board			S.F. Ceiling	7.07	7.06	
D. SERVICES								
	D10 Conveying							
1010	Elevators & Lifts	N/A			—	—	—	0.0 %
1020	Escalators & Moving Walks	N/A			—	—	—	
	D20 Plumbing							
2010	Plumbing Fixtures	Toilet and service fixtures, supply and drainage	1 Fixture/1075 S.F. Floor		Each	1803	3.22	
2020	Domestic Water Distribution	Electric water heater			S.F. Floor	1.52	1.52	4.0%
2040	Rain Water Drainage	N/A			—	—	—	
	D30 HVAC							
3010	Energy Supply	N/A			—	—	—	
3020	Heat Generating Systems	N/A			—	—	—	
3030	Cooling Generating Systems	N/A			—	—	—	5.4 %
3050	Terminal & Package Units	Split system with air cooled condensing units			S.F. Floor	6.38	6.38	
3090	Other HVAC Sys. & Equipment	N/A			—	—	—	
	D40 Fire Protection							
4010	Sprinklers	Wet pipe sprinkler system			S.F. Floor	3.65	3.65	3.1%
4020	Standpipes	N/A			—	—	—	
	D50 Electrical							
5010	Electrical Service/Distribution	800 ampere service, panel board and feeders			S.F. Floor	6.74	6.74	
5020	Lighting & Branch Wiring	High efficiency fluorescent fixtures, receptacles, switches, A.C. and misc. power			S.F. Floor	12.69	12.69	20.5%
5030	Communications & Security	Addressable alarm systems, internet wiring, and emergency lighting			S.F. Floor	4.86	4.86	
5090	Other Electrical Systems	Emergency generator, 100 kW			S.F. Floor	.05	.05	
E. EQUIPMENT & FURNISHINGS								
1010	Commercial Equipment	N/A			—	—	—	
1020	Institutional Equipment	Eye wash station			S.F. Floor	14.75	14.75	14.1 %
1030	Vehicular Equipment	N/A			—	—	—	
1090	Other Equipment	Countertop, plastic laminate			S.F. Floor	1.96	1.96	
F. SPECIAL CONSTRUCTION & DEMOLITION								
1020	Integrated Construction	N/A			—	—	—	0.0 %
1040	Special Facilities	N/A			—	—	—	
G. BUILDING SITEWORK	**N/A**							

			Sub-Total	118.50	100%
CONTRACTOR FEES (General Requirements: 10%, Overhead: 5%, Profit: 10%)			25%	29.62	
ARCHITECT FEES			9%	13.33	
		Total Building Cost		**161.45**	

For customer support on your Square Foot Cost Data, call 877.756.2789.

227

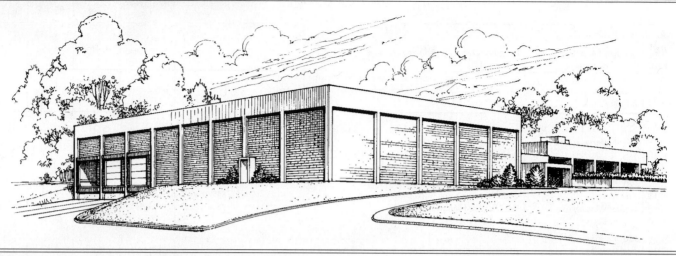

Costs per square foot of floor area

Exterior Wall	S.F. Area	10000	15000	20000	25000	30000	35000	40000	50000	60000
	L.F. Perimeter	410	500	600	640	700	766	833	966	1000
Metal Panel	Steel Frame	111.20	101.50	97.20	92.40	89.70	88.05	86.80	85.00	82.30
Pre-Engineered Metal Building	Steel Frame	119.95	109.10	104.30	98.85	95.95	94.00	92.55	90.60	87.55
E.I.F.S.	Steel Frame	128.50	117.15	112.05	106.25	103.15	101.10	99.60	97.55	94.20
Brick Veneer	Reinforced Concrete	177.90	159.55	151.45	141.60	136.35	133.05	130.55	127.10	121.35
Precast Concrete Panel	Reinforced Concrete	187.65	167.45	158.45	147.60	141.85	138.15	135.45	131.55	125.15
Tilt-up Concrete Panel	R/Conc. Frame	139.70	127.10	121.50	115.00	111.45	109.20	107.55	105.20	101.45
Perimeter Adj., Add or Deduct	Per 100 L.F.	9.30	6.25	4.65	3.65	3.15	2.60	2.30	1.90	1.50
Story Hgt. Adj., Add or Deduct	Per 1 Ft.	1.10	0.95	0.85	0.65	0.65	0.60	0.60	0.55	0.45

For Basement, add $30.40 per square foot of basement area

The above costs were calculated using the basic specifications shown on the facing page. These costs should be adjusted where necessary for design alternatives and owner's requirements. Reported completed project costs, for this type of structure, range from $39.65 to $158.30 per S.F.

Common additives

Description	Unit	$ Cost
Dock Leveler, 10 ton cap.		
6' x 8'	Ea.	6150
7' x 8'	Ea.	8025
Emergency Lighting, 25 watt, battery operated		
Lead battery	Ea.	345
Nickel cadmium	Ea.	685
Fence, Chain link, 6' high		
9 ga. wire	L.F.	28.50
6 ga. wire	L.F.	32
Gate	Ea.	400
Flagpoles, Complete		
Aluminum, 20' high	Ea.	1900
40' high	Ea.	4425
70' high	Ea.	11,500
Fiberglass, 23' high	Ea.	1350
39'-5" high	Ea.	3400
59' high	Ea.	6650
Paving, Bituminous		
Wearing course plus base course	S.Y.	6.70
Sidewalks, Concrete 4" thick	S.F.	4.63

Description	Unit	$ Cost
Sound System		
Amplifier, 250 watts	Ea.	2200
Speaker, ceiling or wall	Ea.	217
Trumpet	Ea.	415
Yard Lighting, 20' aluminum pole with 400 watt high pressure sodium fixture.	Ea.	3225

Important: See the Reference Section for Location Factors.

Model costs calculated for a 1 story building with 24' story height and 30,000 square feet of floor area

				Unit	Unit Cost	Cost Per S.F.	% Of Sub-Total
A.	**SUBSTRUCTURE**						
1010	Standard Foundations	Poured concrete; strip and spread footings		S.F. Ground	1.45	1.45	
1020	Special Foundations	N/A		—	—	—	
1030	Slab on Grade	5" reinforced concrete		S.F. Slab	6.01	6.01	9.4%
2010	Basement Excavation	Site preparation for slab and trench for foundation wall and footing		S.F. Ground	.18	.18	
2020	Basement Walls	4' Foundation wall		L.F. Wall	86	1.99	
B.	**SHELL**						
	B10 Superstructure						
1010	Floor Construction	Cast in place concrete columns, beams, and slab, fireproofed		S.F. Floor	74	7.36	22.2%
1020	Roof Construction	Precast double T with 2" topping		S.F. Roof	15.25	15.25	
	B20 Exterior Enclosure						
2010	Exterior Walls	Face brick with concrete block back up	98% of wall	S.F. Wall	38.79	21.29	
2020	Exterior Windows	Aluminum horizontal sliding	2% of wall	Each	525	3.92	25.9%
2030	Exterior Doors	Aluminum and glass entry doors, hollow steel doors, overhead doors		Each	3365	1.24	
	B30 Roofing						
3010	Roof Coverings	Single ply membrane, stone ballast, rigid insulation		S.F. Roof	5.34	5.34	5.6%
3020	Roof Openings	Roof and smoke hatches		S.F. Roof	.34	.34	
C.	**INTERIORS**						
1010	Partitions	Concrete block partitions, gypsum board on metal stud	100 S.F. Floor/L.F. Partition	S.F. Partition	17.25	1.38	
1020	Interior Doors	Single leaf wood hollow core doors, metal doors	2500 S.F. Floor/Door	Each	699	.27	
1030	Fittings	N/A		—	—	—	
2010	Stair Construction	Steel grate type with rails		Flight	14,300	.96	6.2%
3010	Wall Finishes	Paint		S.F. Surface	4.75	.76	
3020	Floor Finishes	90% hardener, 10% vinyl composition tile		S.F. Floor	2.18	2.18	
3030	Ceiling Finishes	Acoustic ceiling tiles on suspended channel grid		S.F. Ceiling	7.22	.73	
D.	**SERVICES**						
	D10 Conveying						
1010	Elevators & Lifts	N/A		—	—	—	0.0%
1020	Escalators & Moving Walks	N/A		—	—	—	
	D20 Plumbing						
2010	Plumbing Fixtures	Restroom and service fixtures, supply and drainage	1 Fixture/2500 S.F. Floor	Each	1375	.55	
2020	Domestic Water Distribution	Gas fired water heater		S.F. Floor	.23	.23	1.3%
2040	Rain Water Drainage	Roof drains		S.F. Roof	.58	.58	
	D30 HVAC						
3010	Energy Supply	N/A		—	—	—	
3020	Heat Generating Systems	Ventilation with heat system		Each	146,900	5.28	
3030	Cooling Generating Systems	N/A		—	—	—	6.1%
3050	Terminal & Package Units	Single zone unit gas heating, electric cooling		S.F. Floor	.90	.90	
3090	Other HVAC Sys. & Equipment	N/A		—	—	—	
	D40 Fire Protection						
4010	Sprinklers	Wet pipe sprinkler system, ordinary hazard		S.F. Floor	4.19	4.19	6.5%
4020	Standpipes	Standpipe		S.F. Floor	2.48	2.48	
	D50 Electrical						
5010	Electrical Service/Distribution	200 ampere service, panel board and feeders		S.F. Floor	.59	.59	
5020	Lighting & Branch Wiring	Fluorescent fixtures, receptacles, switches, A.C. and misc. power		S.F. Floor	6.36	6.36	9.6%
5030	Communications & Security	Addressable alarm system		S.F. Floor	2.80	2.80	
5090	Other Electrical Systems	N/A		—	—	—	
E.	**EQUIPMENT & FURNISHINGS**						
1010	Commercial Equipment	N/A		—	—	—	
1020	Institutional Equipment	N/A		—	—	—	1.8%
1030	Vehicular Equipment	Dock boards and levelers		S.F. Floor	1.87	1.87	
1090	Other Equipment	N/A		—	—	—	
F.	**SPECIAL CONSTRUCTION**						
1020	Integrated Construction	Shipping and receiving air curtain		S.F. Floor	5.48	5.48	5.4%
1040	Special Facilities	N/A		—	—	—	
G.	**BUILDING SITEWORK**	**N/A**					

			Sub-Total	101.96	100%
CONTRACTOR FEES (General Requirements: 10%, Overhead: 5%, Profit: 10%)			25%	25.47	
ARCHITECT FEES			7%	8.92	
			Total Building Cost	**136.35**	

For customer support on your Square Foot Cost Data, call 877.756.2789.

229

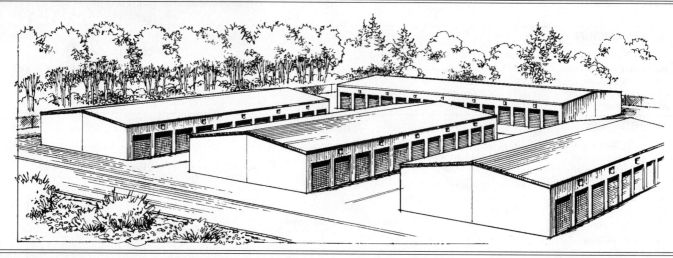

Costs per square foot of floor area

Exterior Wall	S.F. Area	2000	3000	5000	8000	12000	20000	30000	50000	100000
	L.F. Perimeter	180	220	300	420	580	900	1300	2100	4100
Concrete Block	Steel Frame	105.15	92.35	82.15	76.40	73.20	70.70	69.40	68.30	67.55
	R/Conc. Frame	107.30	98.50	91.55	87.60	85.35	83.65	82.75	82.00	81.50
Metal Sandwich Panel	Steel Frame	95.60	86.00	78.35	74.05	71.60	69.70	68.70	67.90	67.40
Tilt-up Concrete Panel	Steel and Concrete	97.65	88.75	81.75	77.80	75.55	73.80	72.90	72.20	71.65
Precast Concrete Panel	Steel Frame	116.05	102.25	91.15	84.95	81.45	78.75	77.30	76.20	75.35
	R/Conc. Frame	137.65	124.55	114.05	108.10	104.80	102.15	100.85	99.75	99.00
Perimeter Adj., Add or Deduct	Per 100 L.F.	21.05	14.05	8.45	5.30	3.50	2.10	1.40	0.90	0.45
Story Hgt. Adj., Add or Deduct	Per 1 Ft.	1.35	1.15	0.95	0.80	0.75	0.65	0.65	0.75	0.65
Basement—Not Applicable										

The above costs were calculated using the basic specifications shown on the facing page. These costs should be adjusted where necessary for design alternatives and owner's requirements. Reported completed project costs, for this type of structure, range from $38.00 to $217.80 per S.F.

Common additives

Description	Unit	$ Cost
Dock Leveler, 10 ton cap.		
6' x 8'	Ea.	6150
7' x 8'	Ea.	8025
Emergency Lighting, 25 watt, battery operated		
Lead battery	Ea.	345
Nickel cadmium	Ea.	685
Fence, Chain link, 6' high		
9 ga. wire	L.F.	28.50
6 ga. wire	L.F.	32
Gate	Ea.	400
Flagpoles, Complete		
Aluminum, 20' high	Ea.	1900
40' high	Ea.	4425
70' high	Ea.	11,500
Fiberglass, 23' high	Ea.	1350
39'-5" high	Ea.	3400
59' high	Ea.	6650
Paving, Bituminous		
Wearing course plus base course	S.Y.	6.70
Sidewalks, Concrete 4" thick	S.F.	4.63

Description	Unit	$ Cost
Sound System		
Amplifier, 250 watts	Ea.	2200
Speaker, ceiling or wall	Ea.	217
Trumpet	Ea.	415
Yard Lighting, 20' aluminum pole with 400 watt high pressure sodium fixture	Ea.	3225

Important: See the Reference Section for Location Factors.

Model costs calculated for a 1 story building with 12' story height and 20,000 square feet of floor area

Warehouse Self Storage

				Unit	Unit Cost	Cost Per S.F.	% Of Sub-Total
A. SUBSTRUCTURE							
1010	Standard Foundations	Poured concrete; strip and spread footings		S.F. Ground	2.35	2.35	
1020	Special Foundations	N/A		—	—	—	
1030	Slab on Grade	4" reinforced concrete with vapor barrier and granular base		S.F. Slab	6.90	6.90	26.0%
2010	Basement Excavation	Site preparation for slab and trench for foundation wall and footing		S.F. Ground	.32	.64	
2020	Basement Walls	4' foundation wall		L.F. Wall	86	3.85	
B. SHELL							
B10 Superstructure							
1010	Floor Construction	Steel column fireproofing		S.F. Floor	.75	.75	14.2%
1020	Roof Construction	Metal deck, open web steel joists, beams, columns		S.F. Roof	6.73	6.73	
B20 Exterior Enclosure							
2010	Exterior Walls	Concrete block	53% of wall	S.F. Wall	22.12	6.33	
2020	Exterior Windows	Aluminum projecting	1% of wall	Each	1170	.12	24.4%
2030	Exterior Doors	Steel overhead, hollow metal	46% of wall	Each	1589	6.44	
B30 Roofing							
3010	Roof Coverings	Metal panel roof; perlite/EPS composite insulation		S.F. Roof	7.09	7.09	13.4%
3020	Roof Openings	N/A		—	—	—	
C. INTERIORS							
1010	Partitions	Metal panels on metal studs	11 S.F. Floor/L.F. Partition	S.F. Partition	1.20	1.31	
1020	Interior Doors	Single leaf hollow metal	20,000 S.F. Floor/Door	Each	600	.06	
1030	Fittings	N/A		—	—	—	
2010	Stair Construction	N/A		—	—	—	2.6%
3010	Wall Finishes	N/A		—	—	—	
3020	Floor Finishes	N/A		—	—	—	
3030	Ceiling Finishes	N/A		—	—	—	
D. SERVICES							
D10 Conveying							
1010	Elevators & Lifts	N/A		—	—	—	0.0%
1020	Escalators & Moving Walks	N/A		—	—	—	
D20 Plumbing							
2010	Plumbing Fixtures	Toilet and service fixtures, supply and drainage	1 Fixture/10,000 S.F. Floor	Each	2800	.28	
2020	Domestic Water Distribution	Gas fired water heater		S.F. Floor	.34	.34	2.4%
2040	Rain Water Drainage	Roof drain		S.F. Roof	.64	.64	
D30 HVAC							
3010	Energy Supply	N/A		—	—	—	
3020	Heat Generating Systems	N/A		—	—	—	
3030	Cooling Generating Systems	N/A		—	—	—	2.4%
3050	Terminal & Package Units	Single zone rooftop unit		S.F. Floor	1.27	1.27	
3090	Other HVAC Sys. & Equipment	N/A		—	—	—	
D40 Fire Protection							
4010	Sprinklers	N/A		—	—	—	0.0%
4020	Standpipes	N/A		—	—	—	
D50 Electrical							
5010	Electrical Service/Distribution	60 ampere service, panel board and feeders		S.F. Floor	.38	.38	
5020	Lighting & Branch Wiring	High bay fixtures, receptacles, switches and misc. power		S.F. Floor	6	6	14.7%
5030	Communications & Security	Addressable alarm systems		S.F. Floor	1.37	1.37	
5090	Other Electrical Systems	N/A		—	—	—	
E. EQUIPMENT & FURNISHINGS							
1010	Commercial Equipment	N/A		—	—	—	
1020	Institutional Equipment	N/A		—	—	—	0.0%
1030	Vehicular Equipment	N/A		—	—	—	
1090	Other Equipment	N/A		—	—	—	
F. SPECIAL CONSTRUCTION							
1020	Integrated Construction	N/A		—	—	—	0.0%
1040	Special Facilities	N/A		—	—	—	
G. BUILDING SITEWORK	**N/A**						

	Sub-Total	52.85	100%
CONTRACTOR FEES (General Requirements: 10%, Overhead: 5%, Profit: 10%)	25%	13.23	
ARCHITECT FEES	7%	4.62	
Total Building Cost		**70.70**	

For customer support on your Square Foot Cost Data, call 877.756.2789.

231

Deterioration occurs within the structure itself and is determined by the observation of both materials and equipment.

Curable deterioration can be remedied either by maintenance, repair or replacement, within prudent economic limits.

Incurable deterioration that has progressed to the point of actually affecting the structural integrity of the structure, making repair or replacement not economically feasible.

Actual Versus Observed Age

The observed age of a structure refers to the age that the structure appears to be. Periodic maintenance, remodeling, and renovation all tend to reduce the amount of deterioration that has taken place, thereby decreasing the observed age. Actual age, on the other hand, relates solely to the year that the structure was built.

The Depreciation Table shown here relates to the observed age.

Obsolescence arises from conditions either occurring within the structure (functional) or caused by factors outside the limits of the structure (economic).

Functional obsolescence is any inadequacy caused by outmoded design, dated construction materials, or oversized or undersized areas, all of which cause excessive operation costs.

Incurable is so costly as not to economically justify the capital expenditure required to correct the deficiency.

Economic obsolescence is caused by factors outside the limits of the structure. The prime causes of economic obsolescence are:

- zoning and environmental laws
- government legislation
- negative neighborhood influences
- business climate
- proximity to transportation facilities

Depreciation Table
Commercial/Industrial/Institutional

Building Material			
Observed Age (Years)	Frame	Masonry On Wood	Masonry On Masonry Or Steel
1	1%	0%	0%
2	2	1	0
3	3	2	1
4	4	3	2
5	6	5	3
10	20	15	8
15	25	20	15
20	30	25	20
25	35	30	25
30	40	35	30
35	45	40	35
40	50	45	40
45	55	50	45
50	60	55	50
55	65	60	55
60	70	65	60

Table of Contents

Costs per square foot of floor area

Exterior Wall	S.F. Area	2000	2700	3400	4100	4800	5500	6200	6900	7600
	L.F. Perimeter	180	208	236	256	280	303	317	337	357
Face Brick with Concrete Block Back-up	Steel Frame	299.30	279.75	268.35	258.55	252.55	247.95	242.70	239.45	236.90
	R/Conc. Frame	315.50	295.95	284.55	274.75	268.75	264.15	258.90	255.65	253.10
Precast Concrete Panel	Steel Frame	298.60	279.10	267.70	257.90	251.90	247.30	242.10	238.90	236.25
	R/Conc. Frame	318.50	297.80	285.70	275.45	269.15	264.25	258.85	255.50	252.75
Limestone with Concrete Block Back-up	Steel Frame	325.45	302.15	288.50	276.65	269.50	264.00	257.55	253.70	250.55
	R/Conc. Frame	342.40	319.10	305.45	293.60	286.45	280.95	274.50	270.65	267.50
Perimeter Adj., Add or Deduct	Per 100 L.F.	55.90	41.45	32.85	27.25	23.35	20.35	18.00	16.20	14.75
Story Hgt. Adj., Add or Deduct	Per 1 Ft.	5.35	4.60	4.10	3.70	3.50	3.30	3.05	2.90	2.80

For Basement, add $38.25 per square foot of basement area

The above costs were calculated using the basic specifications shown on the facing page. These costs should be adjusted where necessary for design alternatives and owner's requirements. Reported completed project costs, for this type of structure, range from $158.85 to $391.05 per S.F.

Common additives

Description	Unit	$ Cost
Bulletproof Teller Window, 44" x 60"	Ea.	5600
60" x 48"	Ea.	7425
Closed Circuit Surveillance, one station, camera & monitor	Ea.	2025
For additional camera stations, add	Ea.	1100
Counters, complete, dr & frame, 3' x 6'-8", bullet resist stl with vision panel	Ea.	7425 - 9600
Drive-up Window, drawer & micr., not incl. glass	Ea.	9175 - 13,000
Night Depository	Ea.	9750 - 14,700
Package Receiver, painted	Ea.	1975
Stainless steel	Ea.	3025
Partitions, bullet resistant to 8' high	L.F.	340 - 540
Pneumatic Tube Systems, 2 station	Ea.	33,900
With TV viewer	Ea.	62,000
Service Windows, pass thru, steel, 24" x 36"	Ea.	3975
48" x 48"	Ea.	4175
Twenty-four Hour Teller, automatic deposit cash & memo	Ea.	55,000
Vault Front, door & frame, 2 hour test, 32" door	Opng.	10,100
4 hour test, 40" door	Opng.	12,600
Time lock, two movement, add	Ea.	2300

Description	Unit	$ Cost
Commissioning Fees, sustainable commercial construction	S.F.	0.24 - 3.04
Energy Modelling Fees, banks to 10,000 SF	Ea.	11,000
Green Bldg Cert Fees for comm construction project reg	Project	900
Photovoltaic Pwr Sys, grid connected, 20 kW (~2400 SF), roof	Ea.	245,600
Green Roofs, 6" soil depth, w/treated wd edging & sedum mats	S.F.	11.76
10" Soil depth, with treated wood edging & sedum mats	S.F.	13.45
Greywater Recovery Systems, prepackaged comm, 1530 gal	Ea.	53,625
Rainwater Harvest Sys, prepckged comm, 10,000 gal, sys contrller	Ea.	31,525
20,000 gal. w/system controller	Ea.	50,575
30,000 gal. w/system controller	Ea.	80,825
Solar Domestic HW, closed loop, add-on sys, ext heat exchanger	Ea.	10,850
Draindown, hot water system, 120 gal tank	Ea.	13,850

Important: See the Reference Section for Location Factors.

Model costs calculated for a 1 story building with 14' story height and 4,100 square feet of floor area

				Unit	Unit Cost	Cost Per S.F.	% Of Sub-Total
A. SUBSTRUCTURE							
1010	Standard Foundations	Poured concrete; strip and spread footings		S.F. Ground	4.41	4.41	
1020	Special Foundations	N/A		—	—	—	
1030	Slab on Grade	4" reinforced concrete with recycled vapor barrier and granular base		S.F. Slab	5.58	5.58	8.7%
2010	Basement Excavation	Site preparation for slab and trench for foundation wall and footing		S.F. Ground	.32	.32	
2020	Basement Walls	4' Foundation wall		L.F. Wall	93	6.97	
B. SHELL							
B10 Superstructure							
1010	Floor Construction	Cast-in-place columns		L.F. Column	6.14	6.14	11.4%
1020	Roof Construction	Cast-in-place concrete flat plate		S.F. Roof	16.45	16.45	
B20 Exterior Enclosure							
2010	Exterior Walls	Face brick with concrete block backup	80% of wall	S.F. Wall	38.79	27.13	
2020	Exterior Windows	Horizontal aluminum sliding	20% of wall	Each	585	6.81	18.0%
2030	Exterior Doors	Double aluminum and glass and hollow metal, low VOC paint		Each	3368	1.65	
B30 Roofing							
3010	Roof Coverings	Single-ply TPO membrane, 60 mils, heat welded seams w/4" thk. R20 insul.		S.F. Roof	7.27	7.27	3.7%
3020	Roof Openings	N/A		—	—	—	
C. INTERIORS							
1010	Partitions	Gypsum board on metal studs w/sound attenuation	20 SF of Flr./LF Part.	S.F. Partition	12.68	6.34	
1020	Interior Doors	Single leaf hollow metal, low VOC paint	200 S.F. Floor/Door	Each	1168	5.85	
1030	Fittings	N/A		—	—	—	
2010	Stair Construction	N/A		—	—	—	13.5%
3010	Wall Finishes	50% vinyl wall covering, 50% paint, low VOC		S.F. Surface	1.50	1.50	
3020	Floor Finishes	50% carpet tile, 40% vinyl composition tile, recycled, 10% quarry tile		S.F. Floor	5.74	5.74	
3030	Ceiling Finishes	Mineral fiber tile on concealed zee bars		S.F. Ceiling	7.22	7.22	
D. SERVICES							
D10 Conveying							
1010	Elevators & Lifts	N/A		—	—	—	0.0 %
1020	Escalators & Moving Walks	N/A		—	—	—	
D20 Plumbing							
2010	Plumbing Fixtures	Toilet low flow, auto sensor and service fixt., supply and drainage	1 Fixt./580 SF Flr.	Each	5870	10.12	
2020	Domestic Water Distribution	Tankless, on demand water heaters, natural gas/propane		S.F. Floor	.98	.98	6.4%
2040	Rain Water Drainage	Roof drains		S.F. Roof	1.61	1.61	
D30 HVAC							
3010	Energy Supply	N/A		—	—	—	
3020	Heat Generating Systems	Included in D3050		—	—	—	
3040	Distribution Systems	Enthalpy heat recovery packages		Each	14,850	3.63	6.1 %
3050	Terminal & Package Units	Single zone rooftop air conditioner		S.F. Floor	8.37	8.37	
3090	Other HVAC Sys. & Equipment	N/A		—	—	—	
D40 Fire Protection							
4010	Sprinklers	Wet pipe sprinkler system		S.F. Floor	5.02	5.02	4.1%
4020	Standpipes	Standpipe		S.F. Floor	3.14	3.14	
D50 Electrical							
5010	Electrical Service/Distribution	200 ampere service, panel board and feeders		S.F. Floor	1.52	1.52	
5020	Lighting & Branch Wiring	LED fixtures, daylt. dimming & ltg. on/off control, receptacles, switches, A.C.		S.F. Floor	13.16	13.16	15.2%
5030	Communications & Security	Alarm systems, internet/phone wiring, and security television		S.F. Floor	11	11	
5090	Other Electrical Systems	Emergency generator, 15 kW, UPS, energy monitoring systems		S.F. Floor	4.50	4.50	
E. EQUIPMENT & FURNISHINGS							
1010	Commercial Equipment	Automatic teller, drive up window, night depository		S.F. Floor	8.10	8.10	
1020	Institutional Equipment	Closed circuit TV monitoring system		S.F. Floor	2.93	2.93	6.3%
1090	Other Equipment	Waste handling recycling tilt truck		Each	1.39	1.39	
2020	Moveable Furnishings	No smoking signage		Each	.02	.02	
F. SPECIAL CONSTRUCTION							
1020	Integrated Construction	N/A		—	—	—	6.6 %
1040	Special Facilities	Security vault door		S.F. Floor	13.14	13.14	
G. BUILDING SITEWORK	**N/A**						

			Sub-Total	198.01	100%
CONTRACTOR FEES (General Requirements: 10%, Overhead: 5%, Profit: 10%)			25%	49.51	
ARCHITECT FEES			11%	27.23	
		Total Building Cost		**274.75**	

Costs per square foot of floor area

Exterior Wall	S.F. Area	15000	20000	28000	38000	50000	65000	85000	100000	150000
	L.F. Perimeter	350	400	480	550	630	660	750	825	1035
Face Brick with Concrete Block Back-up	Steel Frame	239.65	229.40	220.60	213.55	208.60	203.15	199.90	198.45	195.25
	Bearing Walls	244.70	232.20	221.50	212.60	206.40	199.25	195.10	193.25	189.15
Decorative Concrete Block	Steel Frame	230.90	221.90	214.25	208.10	203.95	199.35	196.60	195.40	192.70
	Bearing Walls	235.50	224.20	214.55	206.70	201.20	194.95	191.30	189.70	186.05
Stucco on Concrete Block	Steel Frame	244.70	233.70	224.30	216.65	211.30	205.35	201.75	200.20	196.75
	Bearing Walls	249.75	236.50	225.15	215.65	209.10	201.40	196.95	195.05	190.60
Perimeter Adj., Add or Deduct	Per 100 L.F.	11.35	8.50	6.10	4.45	3.35	2.65	2.00	1.70	1.10
Story Hgt. Adj., Add or Deduct	Per 1 Ft.	2.80	2.40	2.10	1.75	1.55	1.20	1.10	1.00	0.80

For Basement, add $41.05 per square foot of basement area

The above costs were calculated using the basic specifications shown on the facing page. These costs should be adjusted where necessary for design alternatives and owner's requirements. Reported completed project costs, for this type of structure, range from $142.30 to $334.45 per S.F.

Common additives

Description	Unit	$ Cost
Clock System, 20 room	Ea.	20,600
50 room	Ea.	48,500
Elevators, hydraulic passenger, 2 stops, 2000# capacity	Ea.	68,400
3500# capacity	Ea.	75,400
Additional stop, add	Ea.	8075
Seating, auditorium chair, all veneer	Ea.	315
Veneer back, padded seat	Ea.	325
Upholstered, spring seat	Ea.	320
Classroom, movable chair & desk	Set	81 - 171
Lecture hall, pedestal type	Ea.	275 - 620
Sound System, amplifier, 250 watts	Ea.	2200
Speaker, ceiling or wall	Ea.	217
Trumpet	Ea.	415
TV Antenna, Master system, 30 outlet	Outlet	225
100 outlet	Outlet	213

Description	Unit	$ Cost
Commissioning Fees, sustainable institutional construction	S.F.	0.58 - 2.47
Energy Modelling Fees, academic buildings to 10,000 SF	Ea.	9000
Greater than 10,000 SF add	S.F.	0.20
Green Bldg Cert Fees for school construction project reg	Project	900
Photovoltaic Pwr Sys, grid connected, 20 kW (~2400 SF), roof	Ea.	245,600
Green Roofs, 6" soil depth, w/treated wd edging & sedum mats	S.F.	11.76
10" Soil depth, with treated wood edging & sedum mats	S.F.	13.45
Greywater Recovery Systems, prepackaged comm, 3060 gal.	Ea.	58,265
4590 gal.	Ea.	73,075
Rainwater Harvest Sys, prepckged comm, 10,000 gal, sys contrller	Ea.	31,525
20,000 gal. w/system controller	Ea.	50,575
30,000 gal. w/system controller	Ea.	80,825
Solar Domestic HW, closed loop, add-on sys, ext heat exchanger	Ea.	10,850
Drainback, hot water system, 120 gal tank	Ea.	13,525
Draindown, hot water system, 120 gal tank	Ea.	13,850

Important: See the Reference Section for Location Factors.

Model costs calculated for a 2 story building with 12' story height and 50,000 square feet of floor area

					Unit	Unit Cost	Cost Per S.F.	% Of Sub-Total
A. SUBSTRUCTURE								
1010	Standard Foundations	Poured concrete; strip and spread footings			S.F. Ground	1.32	.66	
1020	Special Foundations	N/A			—	—	—	
1030	Slab on Grade	4" reinforced concrete with vapor barrier and granular base			S.F. Slab	5.58	2.79	3.6%
2010	Basement Excavation	Site preparation for slab and trench for foundation wall and footing			S.F. Ground	.32	.16	
2020	Basement Walls	4' Foundation wall			L.F. Wall	177	1.82	
B. SHELL								
	B10 Superstructure							
1010	Floor Construction	Open web steel joists, slab form, concrete			S.F. Floor	17.40	8.70	9.1%
1020	Roof Construction	Metal deck on open web steel joists, columns			S.F. Roof	9.96	4.98	
	B20 Exterior Enclosure							
2010	Exterior Walls	Decorative concrete block	65% of wall		S.F. Wall	20.81	4.09	
2020	Exterior Windows	Window wall	35% of wall		Each	54	5.35	6.7%
2030	Exterior Doors	Double glass and aluminum with transom			Each	5875	.71	
	B30 Roofing							
3010	Roof Coverings	Built-up tar and gravel with flashing; perlite/EPS composite insulation			S.F. Roof	5.62	2.81	1.9%
3020	Roof Openings	N/A			—	—	—	
C. INTERIORS								
1010	Partitions	Concrete block	20 SF Flr.LF Part.		S.F. Partition	22.84	11.42	
1020	Interior Doors	Single leaf hollow metal	200 S.F. Floor/Door		Each	1168	5.85	
1030	Fittings	Chalkboards, counters, cabinets			S.F. Floor	6.04	6.04	
2010	Stair Construction	Concrete filled metal pan			Flight	17,275	3.46	28.3%
3010	Wall Finishes	95% paint, 5% ceramic tile			S.F. Surface	4.40	4.40	
3020	Floor Finishes	70% vinyl composition tile, 25% carpet, 5% ceramic tile			S.F. Floor	4.13	4.13	
3030	Ceiling Finishes	Mineral fiber tile on concealed zee bars			S.F. Ceiling	7.22	7.22	
D. SERVICES								
	D10 Conveying							
1010	Elevators & Lifts	Two hydraulic passenger elevators			Each	85,250	3.41	2.3%
1020	Escalators & Moving Walks	N/A			—	—	—	
	D20 Plumbing							
2010	Plumbing Fixtures	Toilet, low flow, auto sensor & service fixt., supply & drain.	1 Fixt./455 SF Flr.		Each	8549	18.79	
2020	Domestic Water Distribution	Oil fired hot water heater			S.F. Floor	.81	.81	13.5%
2040	Rain Water Drainage	Roof drains			S.F. Roof	1.44	.72	
	D30 HVAC							
3010	Energy Supply	N/A			—	—	—	
3020	Heat Generating Systems	Included in D3050			—	—	—	
3040	Distribution Systems	Enthalpy heat recovery packages			Each	37,000	2.22	14.7 %
3050	Terminal & Package Units	Multizone unit, gas heating, electric cooling, SEER 14			S.F. Floor	19.85	19.85	
3090	Other HVAC Sys. & Equipment	N/A			—	—	—	
	D40 Fire Protection							
4010	Sprinklers	Sprinklers, light hazard			S.F. Floor	3.15	3.15	2.4%
4020	Standpipes	Standpipe, dry, Class III			S.F. Floor	.39	.39	
	D50 Electrical							
5010	Electrical Service/Distribution	2000 ampere service, panel board and feeders			S.F. Floor	3.05	3.05	
5020	Lighting & Branch Wiring	LED fixtures, receptacles, switches, A.C. & misc. power			S.F. Floor	14.38	14.38	17.5%
5030	Communications & Security	Addressable alarm sys., internet wiring, comm. sys. & emerg. ltg.			S.F. Floor	7.29	7.29	
5090	Other Electrical Systems	Emergency generator, 100 kW			S.F. Floor	1.65	1.65	
E. EQUIPMENT & FURNISHINGS								
1010	Commercial Equipment	N/A			—	—	—	
1020	Institutional Equipment	N/A			—	—	—	
1090	Other Equipment	Waste handling recycling tilt truck			Each	.12	.12	0.1 %
2020	Moveable Furnishings	No smoking signage			Each			
F. SPECIAL CONSTRUCTION								
1020	Integrated Construction	N/A			—	—	—	0.0 %
1040	Special Facilities	N/A			—	—	—	
G. BUILDING SITEWORK	**N/A**							

		Sub-Total	150.42	100%
CONTRACTOR FEES (General Requirements: 10%, Overhead: 5%, Profit: 10%)			25%	37.62
ARCHITECT FEES			7%	13.16
		Total Building Cost	**201.20**	

For customer support on your Square Foot Cost Data, call 877.756.2789.

237

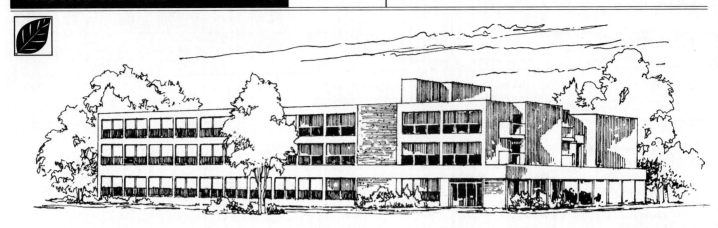

Costs per square foot of floor area

Exterior Wall	S.F. Area	10000	15000	25000	40000	55000	70000	80000	90000	100000
	L.F. Perimeter	260	320	400	476	575	628	684	721	772
Face Brick with Concrete Block Back-up	R/Conc. Frame	247.70	232.80	217.45	206.40	202.30	198.50	197.30	195.95	195.15
	Steel Frame	271.15	256.25	240.85	229.90	225.70	221.90	220.75	219.40	218.55
Decorative Concrete Block	R/Conc. Frame	240.50	228.90	217.20	209.00	205.90	203.15	202.25	201.25	200.65
	Steel Frame	253.10	241.45	229.80	221.60	218.50	215.70	214.85	213.90	213.25
Precast Concrete Panels	R/Conc. Frame	268.10	251.55	234.15	221.65	217.00	212.65	211.35	209.75	208.85
	Steel Frame	280.65	264.10	246.75	234.25	229.60	225.25	223.90	222.35	221.45
Perimeter Adj., Add or Deduct	Per 100 L.F.	21.80	14.55	8.75	5.50	3.95	3.10	2.75	2.45	2.20
Story Hgt. Adj., Add or Deduct	Per 1 Ft.	4.20	3.50	2.60	1.95	1.70	1.45	1.45	1.30	1.25

For Basement, add $39.95 per square foot of basement area

The above costs were calculated using the basic specifications shown on the facing page. These costs should be adjusted where necessary for design alternatives and owner's requirements. Reported completed project costs, for this type of structure, range from $91.90 to $270.90 per S.F.

Common additives

Description	Unit	$ Cost
Closed Circuit Surveillance, one station, camera & monitor	Ea.	2025
For additional camera stations, add	Ea.	1100
Elevators, hydraulic passenger, 2 stops, 3500# capacity	Ea.	75,400
Additional stop, add	Ea.	8075
Furniture	Student	2800 - 5350
Intercom System, 25 station capacity		
Master station	Ea.	3175
Intercom outlets	Ea.	192
Kitchen Equipment		
Broiler	Ea.	4075
Coffee urn, twin 6 gallon	Ea.	3025
Cooler, 6 ft. long	Ea.	3925
Dishwasher, 10-12 racks per hr.	Ea.	4275
Food warmer	Ea.	770
Freezer, 44 C.F., reach-in	Ea.	6175
Ice cube maker, 50 lb. per day	Ea.	2000
Range with 1 oven	Ea.	3300
Laundry Equipment, dryer, 30 lb. capacity	Ea.	3950
Washer, commercial	Ea.	1625
TV Antenna, Master system, 30 outlet	Outlet	225
100 outlet	Outlet	213

Description	Unit	$ Cost
Commissioning Fees, sustainable institutional construction	S.F.	0.58 - 2.47
Energy Modelling Fees, academic buildings to 10,000 SF	Ea.	9000
Greater than 10,000 SF add	S.F.	0.20
Green Bldg Cert Fees for school construction project reg	Project	900
Photovoltaic Pwr Sys, grid connected, 20 kW (~2400 SF), roof	Ea.	245,600
Green Roofs, 6" soil depth, w/treated wd edging & sedum mats	S.F.	11.76
10" Soil depth, with treated wood edging & sedum mats	S.F.	13.45
Greywater Recovery Systems, prepackaged comm, 3060 gal.	Ea.	58,265
4590 gal.	Ea.	73,075
Rainwater Harvest Sys, prepckged comm, 30,000 gal, sys contrller	Ea.	80,825
Solar Domestic HW, closed loop, add-on sys, ext heat exchanger	Ea.	10,850
Drainback, hot water system, 120 gal tank	Ea.	13,525
Draindown, hot water system, 120 gal tank	Ea.	13,850

Important: See the Reference Section for Location Factors.

Model costs calculated for a 3 story building with 12' story height and 25,000 square feet of floor area

G College, Dormitory, 2-3 Story

				Unit	Unit Cost	Cost Per S.F.	% Of Sub-Total
A. SUBSTRUCTURE							
1010	Standard Foundations	Poured concrete; strip and spread footings		S.F. Ground	6.45	2.15	
1020	Special Foundations	N/A					
1030	Slab on Grade	4" reinforced concrete with recycled vapor barrier and granular base		S.F. Slab	5.58	1.86	3.4%
2010	Basement Excavation	Site preparation for slab and trench for foundation wall and footing		S.F. Ground	.18	.06	
2020	Basement Walls	4' Foundation wall		L.F. Wall	93	1.49	
B. SHELL							
B10 Superstructure							
1010	Floor Construction	Concrete flat plate		S.F. Floor	26.10	17.40	13.6%
1020	Roof Construction	Concrete flat plate		S.F. Roof	14.25	4.75	
B20 Exterior Enclosure							
2010	Exterior Walls	Face brick with concrete block backup	80% of wall	S.F. Wall	38.80	17.88	
2020	Exterior Windows	Aluminum horizontal sliding	20% of wall	Each	746	3.73	14.2%
2030	Exterior Doors	Double glass & aluminum doors		Each	6125	1.47	
B30 Roofing							
3010	Roof Coverings	Single-ply TPO membrane, heat welded w/R-20 insul.		S.F. Roof	6.66	2.22	1.4%
3020	Roof Openings	N/A		—	—	—	
C. INTERIORS							
1010	Partitions	Gypsum board on metal studs, CMU w/foamed-in insul.	9 SF Flr./LF Part.	S.F. Partition	7.83	8.70	
1020	Interior Doors	Single leaf wood, low VOC paint	90 S.F. Floor/Door	Each	626	6.96	
1030	Fittings	Closet shelving, mirrors, bathroom accessories		S.F. Floor	1.77	1.77	
2010	Stair Construction	Cast in place concrete		Flight	5475	2.85	19.2%
3010	Wall Finishes	95% paint, low VOC paint, 5% ceramic tile		S.F. Surface	2.08	4.63	
3020	Floor Finishes	80% carpet tile, 10% vinyl comp. tile, recycled content, 10% ceramic tile		S.F. Floor	5.06	5.06	
3030	Ceiling Finishes	90% paint, low VOC, 10% suspended fiberglass board		S.F. Ceiling	1.21	1.21	
D. SERVICES							
D10 Conveying							
1010	Elevators & Lifts	One hydraulic passenger elevator		Each	117,000	4.68	2.9%
1020	Escalators & Moving Walks	N/A		—	—	—	
D20 Plumbing							
2010	Plumbing Fixtures	Toilet, low flow, auto sensor, & service fixt., supply & drain.	1 Fixt./455 SF Flr.	Each	11,507	25.29	
2020	Domestic Water Distribution	Electric, point-of-use water heater		S.F. Floor	1.26	1.26	16.6%
2040	Rain Water Drainage	Roof drains		S.F. Roof	1.50	.50	
D30 HVAC							
3010	Energy Supply	N/A		—	—	—	
3020	Heat Generating Systems	Included in D3050		—	—	—	
3040	Distribution Systems	Enthalpy heat recovery packages		Each	30,450	2.44	9.6%
3050	Terminal & Package Units	Rooftop multizone unit system, SEER 14		S.F. Floor	13.14	13.14	
3090	Other HVAC Sys. & Equipment	N/A		—	—	—	
D40 Fire Protection							
4010	Sprinklers	Wet pipe sprinkler system		S.F. Floor	2.96	2.96	2.4%
4020	Standpipes	Standpipe, dry, Class III		S.F. Floor	.96	.96	
D50 Electrical							
5010	Electrical Service/Distribution	800 ampere service, panel board and feeders		S.F. Floor	1.66	1.66	
5020	Lighting & Branch Wiring	LED fixtures, daylt. dim., ltg. on/off, recept., switches, and A.C. power		S.F. Floor	13.14	13.14	14.4%
5030	Communications & Security	Alarm sys., internet & phone wiring, comm. sys. and emergency ltg.		S.F. Floor	7.47	7.47	
5090	Other Electrical Systems	Emergency generator, 7.5 kW, energy monitoring systems		S.F. Floor	1.13	1.13	
E. EQUIPMENT & FURNISHINGS							
1010	Commercial Equipment	N/A		—	—	—	
1020	Institutional Equipment	N/A		—	—	—	2.3%
1090	Other Equipment	Waste handling recycling tilt truck		Each	.23	.23	
2020	Moveable Furnishings	Dormitory furniture, no smoking signage		S.F. Floor	3.52	3.52	
F. SPECIAL CONSTRUCTION							
1020	Integrated Construction	N/A		—	—	—	0.0%
1040	Special Facilities	N/A		—	—	—	
G. BUILDING SITEWORK	**N/A**						

		Sub-Total	162.57	100%
CONTRACTOR FEES (General Requirements: 10%, Overhead: 5%, Profit: 10%)		25%	40.66	
ARCHITECT FEES		7%	14.22	
	Total Building Cost		**217.45**	

For customer support on your Square Foot Cost Data, call 877.756.2789.

239

Costs per square foot of floor area

Exterior Wall	S.F. Area	20000	35000	45000	65000	85000	110000	135000	160000	200000
	L.F. Perimeter	260	340	400	440	500	540	560	590	640
Face Brick with Concrete Block Back-up	R/Conc. Frame	256.75	238.00	233.05	222.60	218.05	213.30	209.70	207.50	205.15
	Steel Frame	271.00	252.25	247.25	236.80	232.20	227.50	223.90	221.70	219.35
Decorative Concrete Block	R/Conc. Frame	240.50	225.90	221.95	214.10	210.65	207.15	204.50	202.90	201.15
	Steel Frame	247.15	232.50	228.55	220.80	217.35	213.80	211.20	209.55	207.80
Precast Concrete Panels With Exposed Aggregate	R/Conc. Frame	257.10	238.25	233.25	222.75	218.20	213.40	209.80	207.55	205.25
	Steel Frame	271.30	252.45	247.45	236.95	232.35	227.65	224.00	221.75	219.40
Perimeter Adj., Add or Deduct	Per 100 L.F.	21.20	12.10	9.45	6.50	4.95	3.85	3.05	2.65	2.15
Story Hgt. Adj., Add or Deduct	Per 1 Ft.	4.30	3.20	2.90	2.20	1.90	1.60	1.35	1.20	1.05

For Basement, add $39.95 per square foot of basement area

The above costs were calculated using the basic specifications shown on the facing page. These costs should be adjusted where necessary for design alternatives and owner's requirements. Reported completed project costs, for this type of structure, range from $130.65 to $315.05 per S.F.

Common additives

Description	Unit	$ Cost
Closed Circuit Surveillance, one station, camera & monitor	Ea.	2025
For additional camera stations, add	Ea.	1100
Elevators, electric passenger, 5 stops, 3500# capacity	Ea.	177,600
Additional stop, add	Ea.	10,500
Furniture	Student	2800 - 5350
Intercom System, 25 station capacity		
Master station	Ea.	3175
Intercom outlets	Ea.	192
Kitchen Equipment		
Broiler	Ea.	4075
Coffee urn, twin, 6 gallon	Ea.	3025
Cooler, 6 ft. long	Ea.	3925
Dishwasher, 10-12 racks per hr.	Ea.	4275
Food warmer	Ea.	770
Freezer, 44 C.F., reach-in	Ea.	6175
Ice cube maker, 50 lb. per day	Ea.	2000
Range with 1 oven	Ea.	3300
Laundry Equipment, dryer, 30 lb. capacity	Ea.	3950
Washer, commercial	Ea.	1625
TV Antenna, Master system, 30 outlet	Outlet	225
100 outlet	Outlet	213

Description	Unit	$ Cost
Commissioning Fees, sustainable institutional construction	S.F.	0.58 - 2.47
Energy Modelling Fees, academic buildings to 10,000 SF	Ea.	9000
Greater than 10,000 SF add	S.F.	0.20
Green Bldg Cert Fees for school construction project reg	Project	900
Photovoltaic Pwr Sys, grid connected, 20 kW (~2400 SF), roof	Ea.	245,600
Green Roofs, 6" soil depth, w/treated wd edging & sedum mats	S.F.	11.76
10" Soil depth, with treated wood edging & sedum mats	S.F.	13.45
Greywater Recovery Systems, prepackaged comm, 3060 gal.	Ea.	58,265
4590 gal.	Ea.	73,075
Rainwater Harvest Sys, prepckged comm, 30,000 gal, sys contrller	Ea.	80,825
Solar Domestic HW, closed loop, add-on sys, ext heat exchanger	Ea.	10,850
Drainback, hot water system, 120 gal tank	Ea.	13,525
Draindown, hot water system, 120 gal tank	Ea.	13,850

Important: See the Reference Section for Location Factors.

				Unit	Unit Cost	Cost Per S.F.	% Of Sub-Total
A. SUBSTRUCTURE							
1010	Standard Foundations	Poured concrete; strip and spread footings		S.F. Ground	9.54	1.59	
1020	Special Foundations	N/A		—	—	—	
1030	Slab on Grade	4″ reinforced concrete with recycled vapor barrier and granular base		S.F. Slab	5.58	.93	2.0%
2010	Basement Excavation	Site preparation for slab and trench for foundation wall and footing		S.F. Ground	.32	.05	
2020	Basement Walls	4′ foundation wall		L.F. Wall	177	.76	
B. SHELL							
	B10 Superstructure						
1010	Floor Construction	Concrete slab with metal deck and beams		S.F. Floor	30.35	25.29	17.9%
1020	Roof Construction	Concrete slab with metal deck and beams		S.F. Roof	22.80	3.80	
	B20 Exterior Enclosure						
2010	Exterior Walls	Decorative concrete block	80% of wall	S.F. Wall	22.55	7.64	
2020	Exterior Windows	Aluminum horizontal sliding	20% of wall	Each	525	2.96	6.7%
2030	Exterior Doors	Double glass & aluminum doors		Each	3725	.34	
	B30 Roofing						
3010	Roof Coverings	Single-ply TPO membrane, 60 mils, heat welded seams w/R-20 insul.		S.F. Roof	6	1	0.6%
3020	Roof Openings	N/A		—	—	—	
C. INTERIORS							
1010	Partitions	Concrete block w/foamed-in insulation	9 SF Floor/LF Part.	S.F. Partition	12.47	13.85	
1020	Interior Doors	Single leaf wood, low VOC paint	90 S.F. Floor/Door	Each	626	6.96	
1030	Fittings	Closet shelving, mirrors, bathroom accessories		S.F. Floor	1.44	1.44	
2010	Stair Construction	Concrete filled metal pan		Flight	17,275	3.66	23.7%
3010	Wall Finishes	95% paint, low VOC, 5% ceramic tile		S.F. Surface	2.82	6.27	
3020	Floor Finishes	80% carpet tile, 10% vinyl comp. tile, recycled, 10% ceramic tile		S.F. Floor	5.06	5.06	
3030	Ceiling Finishes	Mineral fiber tile on concealed zee bars, paint		S.F. Ceiling	1.21	1.21	
D. SERVICES							
	D10 Conveying						
1010	Elevators & Lifts	Four geared passenger elevators		Each	249,050	11.72	7.2%
1020	Escalators & Moving Walks	N/A		—	—	—	
	D20 Plumbing						
2010	Plumbing Fixtures	Toilet, low flow, auto sensor, & service fixt., supply & drain.	1 Fixt./390 SF Flr.	Each	9824	25.19	
2020	Domestic Water Distribution	Electric water heater, point-of-use, energy saver		S.F. Floor	.70	.70	16.2%
2040	Rain Water Drainage	Roof drains		S.F. Roof	2.22	.37	
	D30 HVAC						
3010	Energy Supply	N/A		—	—	—	
3020	Heat Generating Systems	N/A		—	—	—	
3040	Distribution Systems	Enthalpy heat recovery packages		Each	30,450	2.15	9.4%
3050	Terminal & Package Units	Multizone rooftop air conditioner, SEER 14		S.F. Floor	13.14	13.14	
3090	Other HVAC Sys. & Equipment	N/A		—	—	—	
	D40 Fire Protection						
4010	Sprinklers	Sprinklers, light hazard		S.F. Floor	2.89	2.89	2.2%
4020	Standpipes	Standpipe, dry, Class III		S.F. Floor	.71	.71	
	D50 Electrical						
5010	Electrical Service/Distribution	1200 ampere service, panel board and feeders		S.F. Floor	.85	.85	
5020	Lighting & Branch Wiring	LED fixtures, daylt. dim., ltg. on/off, recept. switches, and A.C. power		S.F. Floor	11.70	11.70	12.2%
5030	Communications & Security	Alarm systems, internet & phone wiring, comm. systems and emerg. ltg.		S.F. Floor	6.42	6.42	
5090	Other Electrical Systems	Emergency generator, 30 kW, energy monitoring systems		S.F. Floor	.89	.89	
E. EQUIPMENT & FURNISHINGS							
1010	Commercial Equipment	N/A		—	—	—	
1020	Institutional Equipment	N/A		—	—	—	
1090	Other Equipment	Waste handling recycling tilt truck		Each	.08	.08	1.8%
2020	Moveable Furnishings	Built-in dormitory furnishings, no smoking signage		S.F. Floor	2.86	2.86	
F. SPECIAL CONSTRUCTION							
1020	Integrated Construction	N/A		—	—	—	0.0%
1040	Special Facilities	N/A		—	—	—	
G. BUILDING SITEWORK	**N/A**						

			Sub-Total	162.48	**100%**
CONTRACTOR FEES (General Requirements: 10%, Overhead: 5%, Profit: 10%)			25%	40.65	
ARCHITECT FEES			7%	14.22	

Total Building Cost 217.35

For customer support on your Square Foot Cost Data, call 877.756.2789.

241

Costs per square foot of floor area

Exterior Wall	S.F. Area	12000	20000	28000	37000	45000	57000	68000	80000	92000
	L.F. Perimeter	470	600	698	793	900	1000	1075	1180	1275
Face Brick with Concrete Brick Back-up	Steel Frame	341.55	286.85	261.70	245.80	237.70	228.25	222.15	217.85	214.50
	Bearing Walls	321.30	266.55	241.40	225.50	217.40	207.95	201.85	197.55	194.20
Decorative Concrete Block	Steel Frame	333.65	280.80	256.70	241.45	233.60	224.70	219.00	214.90	211.75
	Bearing Walls	328.65	275.80	251.65	236.45	228.65	219.70	213.95	209.90	206.70
Stucco on Concrete Block	Steel Frame	346.55	290.75	264.90	248.50	240.25	230.50	224.20	219.80	216.30
	Bearing Walls	341.55	285.70	259.90	243.50	235.20	225.50	219.15	214.80	211.30
Perimeter Adj., Add or Deduct	Per 100 L.F.	12.70	7.60	5.35	4.05	3.40	2.70	2.20	1.90	1.65
Story Hgt. Adj., Add or Deduct	Per 1 Ft.	2.40	1.85	1.50	1.20	1.20	1.10	0.95	0.90	0.80

For Basement, add $25.80 per square foot of basement area

The above costs were calculated using the basic specifications shown on the facing page. These costs should be adjusted where necessary for design alternatives and owner's requirements. Reported completed project costs, for this type of structure, range from $188.15 to $351.40 per S.F.

Common additives

Description	Unit	$ Cost
Cabinets, Base, door units, metal	L.F.	325
Drawer units	L.F.	645
Tall storage cabinets, open	L.F.	620
With doors	L.F.	900
Wall, metal 12-1/2" deep, open	L.F.	263
With doors	L.F.	460
Countertops, not incl. base cabinets, acid proof	S.F.	61.50 - 74.50
Stainless steel	S.F.	180
Fume Hood, Not incl. ductwork	L.F.	780 - 1900
Ductwork	Hood	5950 - 9775
Glassware Washer, Distilled water rinse	Ea.	7625 - 16,700
Seating, auditorium chair, veneer back, padded seat	Ea.	325
Upholstered, spring seat	Ea.	320
Classroom, movable chair & desk	Set	81 - 171
Lecture hall, pedestal type	Ea.	275 - 620
Safety Equipment, Eye wash, hand held	Ea.	445
Deluge shower	Ea.	850
Tables, acid resist. top, drawers	L.F.	221
Titration Unit, Four 2000 ml reservoirs	Ea.	6100

Description	Unit	$ Cost
Commissioning Fees, sustainable institutional construction	S.F.	0.58 - 2.47
Energy Modelling Fees, academic buildings to 10,000 SF	Ea.	9000
Greater than 10,000 SF add	S.F.	0.20
Green Bldg Cert Fees for school construction project reg	Project	900
Photovoltaic Pwr Sys, grid connected, 20 kW (~2400 SF), roof	Ea.	245,600
Green Roofs, 6" soil depth, w/treated wd edging & sedum mats	S.F.	11.76
10" Soil depth, with treated wood edging & sedum mats	S.F.	13.45
Greywater Recovery Systems, prepackaged comm, 3060 gal.	Ea.	58,265
4590 gal.	Ea.	73,075
Rainwater Harvest Sys, prepckged comm, 30,000 gal, sys contrller	Ea.	80,825
Solar Domestic HW, closed loop, add-on sys, ext heat exchanger	Ea.	10,850
Drainback, hot water system, 120 gal tank	Ea.	13,525
Draindown, hot water system, 120 gal tank	Ea.	13,850

Model costs calculated for a 1 story building with 12' story height and 45,000 square feet of floor area

G College, Laboratory

				Unit	Unit Cost	Cost Per S.F.	% Of Sub-Total
A. SUBSTRUCTURE							
1010	Standard Foundations	Poured concrete; strip and spread footings		S.F. Ground	3.29	3.29	
1020	Special Foundations	N/A		—	—	—	
1030	Slab on Grade	4" reinforced concrete with recycled vapor barrier and granular base		S.F. Slab	5.58	5.58	10.3%
2010	Basement Excavation	Site preparation for slab and trench for foundation wall and footing		S.F. Ground	.18	.18	
2020	Basement Walls	4' foundation wall		L.F. Wall	177	7.26	
B. SHELL							
B10 Superstructure							
1010	Floor Construction	Metal deck on open web steel joists	(5680 S.F.)	S.F. Floor	2.27	2.27	4.7%
1020	Roof Construction	Metal deck on open web steel joists		S.F. Roof	5.14	5.14	
B20 Exterior Enclosure							
2010	Exterior Walls	Face brick with concrete block backup	75% of wall	S.F. Wall	38.78	6.98	
2020	Exterior Windows	Window wall	25% of wall	Each	54	3.03	7.7%
2030	Exterior Doors	Glass and metal doors and entrances with transom		Each	5053	2.24	
B30 Roofing							
3010	Roof Coverings	Single-ply TPO membrane, 60 mils, heat welded w/R-20 insul.		S.F. Roof	5.39	5.39	3.6%
3020	Roof Openings	Skylight		S.F. Roof	.35	.35	
C. INTERIORS							
1010	Partitions	Concrete block, foamed-in insulation	10 SF Flr./LF Part.	S.F. Partition	14.32	14.32	
1020	Interior Doors	Single leaf-kalamein fire doors, low VOC paint	820 S.F. Floor/Door	Each	1417	1.73	
1030	Fittings	Lockers		S.F. Floor	.05	.05	
2010	Stair Construction	N/A		—	—	—	22.2%
3010	Wall Finishes	60% paint, low VOC, 40% epoxy coating		S.F. Surface	2.78	5.55	
3020	Floor Finishes	60% epoxy, 20% carpet, 20% vinyl composition tile, recycled content		S.F. Floor	6.20	6.20	
3030	Ceiling Finishes	Mineral fiber tile on concealed zee runners		S.F. Ceiling	7.22	7.22	
D. SERVICES							
D10 Conveying							
1010	Elevators & Lifts	N/A		—	—	—	0.0 %
1020	Escalators & Moving Walks	N/A		—	—	—	
D20 Plumbing							
2010	Plumbing Fixtures	Toilet, low flow, auto sensor & service fixt., supply & drain.	1 Fixt./260 SF Flr.	Each	8000	30.77	
2020	Domestic Water Distribution	Electric, point-of-use water heater		S.F. Floor	1.60	1.60	21.0%
2040	Rain Water Drainage	Roof drains		S.F. Roof	.79	.79	
D30 HVAC							
3010	Energy Supply	N/A		—	—	—	
3020	Heat Generating Systems	Included in D3050		—	—	—	
3040	Distribution Systems	Enthalpy heat recovery packages		Each	30,450	4.06	15.1 %
3050	Terminal & Package Units	Multizone rooftop air conditioner, SEER 14		S.F. Floor	19.85	19.85	
3090	Other HVAC Sys. & Equipment	N/A		—	—	—	
D40 Fire Protection							
4010	Sprinklers	Sprinklers, light hazard		S.F. Floor	2.84	2.84	2.0%
4020	Standpipes	Standpipe, wet, Class III		S.F. Floor	.35	.35	
D50 Electrical							
5010	Electrical Service/Distribution	1000 ampere service, panel board and feeders		S.F. Floor	1.38	1.38	
5020	Lighting & Branch Wiring	LED fixtures, daylt. dim., ltg. on/off, recept., switches, & A.C. power		S.F. Floor	13.65	13.65	12.3%
5030	Communications & Security	Addressable alarm systems, internet wiring, & emergency lighting		S.F. Floor	3.50	3.50	
5090	Other Electrical Systems	Emergency generator, 11.5 kW, UPS, and energy monitoring systems		S.F. Floor	.91	.91	
E. EQUIPMENT & FURNISHINGS							
1010	Commercial Equipment	N/A		—	—	—	
1020	Institutional Equipment	Cabinets, fume hoods, lockers, glassware washer		S.F. Floor	1.47	1.47	1.0 %
1090	Other Equipment	Waste handling recycling tilt truck		Each	.13	.13	
2020	Moveable Furnishings	No smoking signage		Each	.02	.02	
F. SPECIAL CONSTRUCTION							
1020	Integrated Construction	N/A		—	—	—	0.0 %
1040	Special Facilities	N/A		—	—	—	
G. BUILDING SITEWORK	**N/A**						

	Sub-Total	158.10	**100%**
CONTRACTOR FEES (General Requirements: 10%, Overhead: 5%, Profit: 10%)	25%	39.54	
ARCHITECT FEES	10%	19.76	
	Total Building Cost	**217.40**	

For customer support on your Square Foot Cost Data, call 877.756.2789.

243

Costs per square foot of floor area

Exterior Wall	S.F. Area	4000	4500	5000	5500	6000	6500	7000	7500	8000
	L.F. Perimeter	260	280	300	310	320	336	353	370	386
Face Brick Concrete Block Back-up	Steel Joists	213.80	209.50	206.05	201.45	197.70	195.40	193.60	191.95	190.40
	Bearing Walls	213.60	209.25	205.85	201.20	197.50	195.10	193.30	191.65	190.15
Decorative Concrete Block	Steel Joists	204.20	200.55	197.75	194.05	191.05	189.15	187.70	186.30	185.10
	Bearing Walls	197.15	193.55	190.70	187.00	184.00	182.05	180.60	179.20	178.00
Limestone with Concrete Block Back-up	Steel Joists	238.00	232.95	229.00	223.40	218.85	216.05	213.95	212.00	210.20
	Bearing Walls	230.95	225.85	221.90	216.30	211.75	208.95	206.85	204.90	203.15
Perimeter Adj., Add or Deduct	Per 100 L.F.	24.10	21.45	19.35	17.60	16.10	14.80	13.75	12.90	12.05
Story Hgt. Adj., Add or Deduct	Per 1 Ft.	3.20	3.05	3.00	2.85	2.65	2.55	2.45	2.40	2.35

For Basement, add $43.80 per square foot of basement area

The above costs were calculated using the basic specifications shown on the facing page. These costs should be adjusted where necessary for design alternatives and owner's requirements. Reported completed project costs, for this type of structure, range from $86.20 to $255.00 per S.F.

Common additives

Description	Unit	$ Cost
Appliances, cooking range, 30" free standing		
1 oven	Ea.	610 - 2300
2 oven	Ea.	1475 - 3175
Microwave oven	Ea.	296 - 890
Compactor, residential, 4-1 compaction	Ea.	880 - 1475
Dishwasher, built-in, 2 cycles	Ea.	675 - 1175
4 cycles	Ea.	815 - 2000
Garbage disposer, sink type	Ea.	239 - 405
Hood for range, 2 speed, vented, 30" wide	Ea.	345 - 1375
Refrigerator, no frost 10-12 C.F.	Ea.	615 - 770
14-16 C.F.	Ea.	740 - 900
18-20 C.F.	Ea.	925 - 2100
Lockers, Steel, single tier, 60" or 72"	Opng.	231 - 380
Locker bench, lam. maple top only	L.F.	34.50
Pedestals, steel pipe	Ea.	76.50
Sound System, amplifier, 250 watts	Ea.	2200
Speaker, ceiling or wall	Ea.	217
Trumpet	Ea.	415

Description	Unit	$ Cost
Commissioning Fees, sustainable commercial construction	S.F.	0.24 - 3.04
Energy Modelling Fees, Fire Stations to 10,000 SF	Ea.	11,000
Green Bldg Cert Fees for comm construction project reg	Project	900
Photovoltaic Pwr Sys, grid connected, 20 kW (~2400 SF), roof	Ea.	245,600
Green Roofs, 6" soil depth, w/treated wd edging & sedum mats	S.F.	11.76
10" Soil depth, with treated wood edging & sedum mats	S.F.	13.45
Greywater Recovery Systems, prepackaged comm, 3060 gal.	Ea.	58,265
4590 gal.	Ea.	73,075
Rainwater Harvest Sys, prepckged comm, 10,000 gal, sys contrller	Ea.	31,525
20,000 gal. w/system controller	Ea.	50,575
30,000 gal. w/system controller	Ea.	80,825
Solar Domestic HW, closed loop, add-on sys, ext heat exchanger	Ea.	10,850
Drainback, hot water system, 120 gal tank	Ea.	13,525

Important: See the Reference Section for Location Factors.

G Fire Station, 1 Story

				Unit	Unit Cost	Cost Per S.F.	% Of Sub-Total
A. SUBSTRUCTURE							
1010	Standard Foundations	Poured concrete; strip and spread footings		S.F. Ground	3.51	3.51	
1020	Special Foundations	N/A		—	—	—	
1030	Slab on Grade	6" reinforced concrete with recycled vapor barrier and granular base		S.F. Slab	7.37	7.37	11.2%
2010	Basement Excavation	Site preparation for slab and trench for foundation wall and footing		S.F. Ground	.57	.57	
2020	Basement Walls	4' foundation wall		L.F. Wall	93	4.96	
B. SHELL							
B10 Superstructure							
1010	Floor Construction	N/A		—	—	—	7.1%
1020	Roof Construction	Metal deck, open web steel joists, beams on columns		S.F. Roof	10.42	10.42	
B20 Exterior Enclosure							
2010	Exterior Walls	Face brick with concrete block backup	75% of wall	S.F. Wall	38.80	21.73	
2020	Exterior Windows	Aluminum insulated glass	10% of wall	Each	860	2.01	19.5%
2030	Exterior Doors	Single aluminum and glass, overhead, hollow metal, low VOC paint	15% of wall	S.F. Door	2672	4.80	
B30 Roofing							
3010	Roof Coverings	Single-ply TPO membrane, 60 mil, with flashing, R-20 insulation and roof edges		S.F. Roof	6.87	6.87	4.8%
3020	Roof Openings	Skylights, roof hatches		S.F. Roof	.16	.16	
C. INTERIORS							
1010	Partitions	Concrete block w/foamed-in insulation	17 S.F. Floor/LF Part.	S.F. Partition	11.37	6.69	
1020	Interior Doors	Single leaf hollow metal, low VOC paint	500 S.F. Floor/Door	Each	1168	2.34	
1030	Fittings	Toilet partitions		S.F. Floor	.51	.51	
2010	Stair Construction	N/A		—	—	—	12.8%
3010	Wall Finishes	Paint, low VOC		S.F. Surface	2.45	2.88	
3020	Floor Finishes	50% vinyl tile with recycled content, 50% paint, low VOC		S.F. Floor	2.76	2.76	
3030	Ceiling Finishes	Mineral board acoustic ceiling tiles, concealed grid, suspended	50% of area	S.F. Ceiling	3.61	3.62	
D. SERVICES							
D10 Conveying							
1010	Elevators & Lifts	N/A		—	—	—	0.0%
1020	Escalators & Moving Walks	N/A		—	—	—	
D20 Plumbing							
2010	Plumbing Fixtures	Kitchen, toilet, low flow, auto sensor & service fixt., supply & drain.	1 Fixt./375 SF Flr.	Each	4170	11.12	
2020	Domestic Water Distribution	Water heater, tankless, on-demand, natural gas/propane		S.F. Floor	2.11	2.11	10.0%
2040	Rain Water Drainage	Roof drains		S.F. Roof	1.40	1.40	
D30 HVAC							
3010	Energy Supply	N/A		—	—	—	
3020	Heat Generating Systems	Included in D3050		—	—	—	
3040	Distribution Systems	Enthalpy heat recovery packages		Each	17,575	2.93	17.2%
3050	Terminal & Package Units	Multizone rooftop air conditioner, SEER 14		S.F. Floor	22.28	22.28	
3090	Other HVAC Sys. & Equipment	N/A		—	—	—	
D40 Fire Protection							
4010	Sprinklers	Wet pipe sprinkler system		S.F. Floor	5.02	5.02	4.5%
4020	Standpipes	Standpipe, wet, Class III		S.F. Floor	1.54	1.54	
D50 Electrical							
5010	Electrical Service/Distribution	200 ampere service, panel board and feeders		S.F. Floor	1.40	1.40	
5020	Lighting & Branch Wiring	LED fixt., daylt. dimming & ltg. on/off control, receptacles, switches, A.C. & misc. pwr.		S.F. Floor	10.17	10.17	12.2%
5030	Communications & Security	Addressable alarm systems		S.F. Floor	2	2	
5090	Other Electrical Systems	Energy monitoring systems		S.F. Floor	4.29	4.29	
E. EQUIPMENT & FURNISHINGS							
1010	Commercial Equipment	N/A		—	—	—	
1020	Institutional Equipment	N/A		—	—	—	
1090	Other Equipment	Waste handling recycling tilt truck		Each	.95	.95	0.7%
2020	Moveable Furnishings	No smoking signage		Each	.04	.04	
F. SPECIAL CONSTRUCTION							
1020	Integrated Construction	N/A		—	—	—	0.0%
1040	Special Facilities	N/A		—	—	—	
G. BUILDING SITEWORK	**N/A**						

		Sub-Total	146.45	100%
CONTRACTOR FEES (General Requirements: 10%, Overhead: 5%, Profit: 10%)			25%	36.61
ARCHITECT FEES			8%	14.64

Total Building Cost	**197.70**

For customer support on your Square Foot Cost Data, call 877.756.2789.

245

Costs per square foot of floor area

Exterior Wall	S.F. Area	6000	7000	8000	9000	10000	11000	12000	13000	14000
	L.F. Perimeter	220	240	260	280	286	303	320	336	353
Face Brick with Concrete Block Back-up	Steel Joists	228.95	222.45	217.50	213.60	208.25	205.45	203.15	201.05	199.45
	Precast Conc.	238.45	232.00	227.10	223.25	217.95	215.15	212.90	210.80	209.20
Decorative Concrete Block	Steel Joists	197.55	192.00	187.85	184.55	180.20	177.85	175.95	174.20	172.85
	Precast Conc.	223.15	217.70	213.55	210.30	206.00	203.65	201.80	200.00	198.70
Limestone with Concrete Block Back-up	Steel Joists	248.40	240.60	234.70	230.10	223.40	220.05	217.30	214.75	212.80
	Precast Conc.	257.70	249.95	244.10	239.50	232.90	229.55	226.85	224.30	222.40
Perimeter Adj., Add or Deduct	Per 100 L.F.	28.00	23.95	20.95	18.65	16.75	15.20	14.00	12.90	12.00
Story Hgt. Adj., Add or Deduct	Per 1 Ft.	3.65	3.40	3.25	3.15	2.80	2.75	2.65	2.55	2.50

For Basement, add $42.50 per square foot of basement area

The above costs were calculated using the basic specifications shown on the facing page. These costs should be adjusted where necessary for design alternatives and owner's requirements. Reported completed project costs, for this type of structure, range from $85.95 to $255.00 per S.F.

Common additives

Description	Unit	$ Cost
Appliances		
Cooking range, 30" free standing		
1 oven	Ea.	610 - 2300
2 oven	Ea.	1475 - 3175
Microwave oven	Ea.	296 - 890
Compactor, residential, 4-1 compaction	Ea.	880 - 1475
Dishwasher, built-in, 2 cycles	Ea.	675 - 1175
4 cycles	Ea.	815 - 2000
Garbage diposer, sink type	Ea.	239 - 405
Hood for range, 2 speed, vented, 30" wide	Ea.	345 - 1375
Refrigerator, no frost 10-12 C.F.	Ea.	615 - 770
14-16 C.F.	Ea.	740 - 900
18-20 C.F.	Ea.	925 - 2100
Elevators, hydraulic passenger, 2 stops, 2500# capacity	Ea.	70,400
3500# capacity	Ea.	75,400
Lockers, Steel, single tier, 60" or 72"	Opng.	231 - 380
Locker bench, lam. maple top only	L.F.	34.50
Pedestals, steel pipe	Ea.	76.50
Sound System, amplifier, 250 watts	Ea.	2200
Speaker, ceiling or wall	Ea.	217
Trumpet	Ea.	415

Description	Unit	$ Cost
Commissioning Fees, sustainable commercial construction	S.F.	0.24 - 3.04
Energy Modelling Fees, Fire Stations to 10,000 SF	Ea.	11,000
Greater than 10,000 SF add	S.F.	0.04
Green Bldg Cert Fees for comm construction project reg	Project	900
Photovoltaic Pwr Sys, grid connected, 20 kW (~2400 SF), roof	Ea.	245,600
Green Roofs, 6" soil depth, w/treated wd edging & sedum mats	S.F.	11.76
10" Soil depth, with treated wood edging & sedum mats	S.F.	13.45
Greywater Recovery Systems, prepackaged comm, 3060 gal.	Ea.	58,265
4590 gal.	Ea.	73,075
Rainwater Harvest Sys, prepckged comm, 10,000 gal, sys contrller	Ea.	31,525
20,000 gal. w/system controller	Ea.	50,575
30,000 gal. w/system controller	Ea.	80,825
Solar Domestic HW, closed loop, add-on sys, ext heat exchanger	Ea.	10,850
Drainback, hot water system, 120 gal tank	Ea.	13,525

Important: See the Reference Section for Location Factors.

G Fire Station, 2 Story

Model costs calculated for a 2 story building with 14' story height and 10,000 square feet of floor area

				Unit	Unit Cost	Cost Per S.F.	% Of Sub-Total
A. SUBSTRUCTURE							
1010	Standard Foundations	Poured concrete; strip and spread footings		S.F. Ground	3.30	1.65	
1020	Special Foundations	N/A		—	—	—	
1030	Slab on Grade	6" reinforced concrete with recycled vapor barrier and granular base		S.F. Slab	7.37	3.69	6.4%
2010	Basement Excavation	Site preparation for slab and trench for foundation wall and footing		S.F. Ground	.57	.29	
2020	Basement Walls	4' foundation wall with insulation		L.F. Wall	102	2.91	
B. SHELL							
B10 Superstructure							
1010	Floor Construction	Open web steel joists, slab form, concrete		S.F. Floor	14.88	7.44	7.5%
1020	Roof Construction	Metal deck on open web steel joists		S.F. Roof	5.14	2.57	
B20 Exterior Enclosure							
2010	Exterior Walls	Decorative concrete block, insul.	75% of wall	S.F. Wall	24.11	14.48	
2020	Exterior Windows	Aluminum insulated glass	10% of wall	Each	746	2.60	15.8%
2030	Exterior Doors	Single aluminum and glass, steel overhead, hollow metal, low VOC paint	15% of wall	S.F. Door	1958	4.01	
B30 Roofing							
3010	Roof Coverings	Single-ply TPO, 60 mils, heat welded seams w/ R-20 insul.		S.F. Roof	7.04	3.52	2.6%
3020	Roof Openings	N/A		—	—	—	
C. INTERIORS							
1010	Partitions	Concrete block, foamed-in insul.	10 SF Floor/LF Part.	S.F. Partition	11.37	6.69	
1020	Interior Doors	Single leaf hollow metal, low VOC paint	500 S.F. Floor/Door	Each	1417	2.83	
1030	Fittings	Toilet partitions		S.F. Floor	.46	.46	
2010	Stair Construction	Concrete filled metal pan		Flight	20,150	4.03	18.2%
3010	Wall Finishes	Paint, low VOC		S.F. Surface	3.96	4.66	
3020	Floor Finishes	50% vinyl tile, (recycled content) 50% paint (low VOC)		S.F. Floor	2.76	2.76	
3030	Ceiling Finishes	Fiberglass board on exposed grid, suspended	50% of area	S.F. Ceiling	2.89	2.89	
D. SERVICES							
D10 Conveying							
1010	Elevators & Lifts	One hydraulic passenger elevator		Each	85,300	8.53	6.4%
1020	Escalators & Moving Walks	N/A		—	—	—	
D20 Plumbing							
2010	Plumbing Fixtures	Kitchen, toilet low flow, auto sensor & service fixt., supply & drain.	1 Fixt./400 SF Flr.	Each	4300	10.75	
2020	Domestic Water Distribution	Gas fired tankless water heater		S.F. Floor	.81	.81	9.3%
2040	Rain Water Drainage	Roof drains		S.F. Roof	1.82	.91	
D30 HVAC							
3010	Energy Supply	N/A		—	—	—	
3020	Heat Generating Systems	Included in D3050		—	—	—	
3040	Distribution Systems	Enthalpy heat recovery packages		Each	14,850	2.97	18.9 %
3050	Terminal & Package Units	Multizone rooftop air conditioner, SEER 14		S.F. Floor	22.28	22.28	
3090	Other HVAC Sys. & Equipment	N/A		—	—	—	
D40 Fire Protection							
4010	Sprinklers	Wet pipe sprinkler system		S.F. Floor	3.95	3.95	4.1%
4020	Standpipes	Standpipe, wet, Class III		S.F. Floor	1.57	1.57	
D50 Electrical							
5010	Electrical Service/Distribution	400 ampere service, panel board and feeders		S.F. Floor	1.14	1.14	
5020	Lighting & Branch Wiring	LED fixt., daylt. dimming & ltg. on/off control., receptacles, switches, A.C. and misc. pwr.		S.F. Floor	8.94	8.94	10.4%
5030	Communications & Security	Addressable alarm systems and emergency lighting		S.F. Floor	1.57	1.57	
5090	Other Electrical Systems	Emergency generator, 15 kW, and energy monitoring systems		S.F. Floor	2.22	2.22	
E. EQUIPMENT & FURNISHINGS							
1010	Commercial Equipment	N/A		—	—	—	
1020	Institutional Equipment	N/A		—	—	—	0.3 %
1090	Other Equipment	Waste handling recycling tilt truck		Each	.35	.35	
2020	Moveable Furnishings	No smoking signage		Each	.02	.02	
F. SPECIAL CONSTRUCTION							
1020	Integrated Construction	N/A		—	—	—	0.0 %
1040	Special Facilities	N/A		—	—	—	
G. BUILDING SITEWORK	**N/A**						

			Sub-Total	133.49	100%
CONTRACTOR FEES (General Requirements: 10%, Overhead: 5%, Profit: 10%)		25%	33.36		
ARCHITECT FEES		8%	13.35		
Total Building Cost			**180.20**		

For customer support on your Square Foot Cost Data, call 877.756.2789.

247

Costs per square foot of floor area

Exterior Wall	S.F. Area	4000	6000	8000	10000	12000	14000	16000	18000	20000
	L.F. Perimeter	260	320	384	424	460	484	510	540	576
Vertical Redwood Siding	Wood Frame	167.80	156.20	150.65	146.05	142.75	139.95	137.90	136.50	135.40
Brick Veneer	Wood Frame	182.25	168.75	162.35	156.80	152.90	149.45	146.95	145.25	144.00
Aluminum Siding	Wood Frame	168.00	157.05	151.80	147.50	144.45	141.85	140.00	138.70	137.70
Brick on Block	Wood Truss	194.60	179.20	171.85	165.40	160.80	156.70	153.85	151.75	150.35
Limestone on Block	Wood Truss	215.05	195.85	186.90	178.60	172.70	167.40	163.65	161.00	159.15
Stucco on Block	Wood Truss	202.45	184.90	176.65	169.15	163.85	159.10	155.70	153.30	151.65
Perimeter Adj., Add or Deduct	Per 100 L.F.	13.75	9.15	6.85	5.45	4.60	3.90	3.45	3.05	2.80
Story Hgt. Adj., Add or Deduct	Per 1 Ft.	1.70	1.40	1.25	1.05	1.00	0.90	0.85	0.80	0.80

For Basement, add $36.05 per square foot of basement area

The above costs were calculated using the basic specifications shown on the facing page. These costs should be adjusted where necessary for design alternatives and owner's requirements. Reported completed project costs, for this type of structure, range from $122.40 to $376.25 per S.F.

Common additives

Description	Unit	$ Cost	Description	Unit	$ Cost
Autopsy Table, standard	Ea.	11,400	Commissioning Fees, sustainable commercial construction	S.F.	0.24 - 3.04
Deluxe	Ea.	19,100	Energy Modelling Fees, commercial buildings to 10,000 SF	Ea.	11,000
Directory Boards, plastic, glass covered, 30" x 20"	Ea.	610	Greater than 10,000 SF add	S.F.	0.04
36" x 48"	Ea.	1525	Green Bldg Cert Fees for comm construction project reg	Project	900
Aluminum, 24" x 18"	Ea.	640	Photovoltaic Pwr Sys, grid connected, 20 kW (~2400 SF), roof	Ea.	245,600
36" x 24"	Ea.	765	Green Roofs, 6" soil depth, w/treated wd edging & sedum mats	S.F.	11.76
48" x 32"	Ea.	1075	10" Soil depth, with treated wood edging & sedum mats	S.F.	13.45
Mortuary Refrigerator, end operated			Greywater Recovery Systems, prepackaged comm, 3060 gal.	Ea.	58,265
Two capacity	Ea.	10,800	4590 gal.	Ea.	73,075
Six capacity	Ea.	16,300	Rainwater Harvest Sys, prepckged comm, 10,000 gal, sys contrller	Ea.	31,525
Planters, precast concrete, 48" diam., 24" high	Ea.	705	20,000 gal. w/system controller	Ea.	50,575
7" diam., 36" high	Ea.	1825	30,000 gal. w/system controller	Ea.	80,825
Fiberglass, 36" diam., 24" high	Ea.	785	Solar Domestic HW, closed loop, add-on sys, ext heat exchanger	Ea.	10,850
60" diam., 24" high	Ea.	1325	Drainback, hot water system, 120 gal tank	Ea.	13,525
			Draindown, hot water system, 120 gal tank	Ea.	13,850

Model costs calculated for a 1 story building with 12' story height and 10,000 square feet of floor area

G Funeral Home

				Unit	Unit Cost	Cost Per S.F.	% Of Sub-Total
A. SUBSTRUCTURE							
1010	Standard Foundations	Poured concrete; strip and spread footings		S.F. Ground	1.92	1.92	
1020	Special Foundations	N/A		—	—	—	
1030	Slab on Grade	4" reinforced concrete with recycled plastic vapor barrier and granular base		S.F. Slab	5.58	5.58	11.0%
2010	Basement Excavation	Site preparation for slab and trench for foundation wall and footing		S.F. Ground	.32	.32	
2020	Basement Walls	4' foundation wall		L.F. Wall	84	3.92	
B. SHELL							
	B10 Superstructure						
1010	Floor Construction	N/A		—	—	—	5.0 %
1020	Roof Construction	Plywood on wood truss		S.F. Roof	5.39	5.39	
	B20 Exterior Enclosure						
2010	Exterior Walls	1" x 4" vertical T & G redwood siding on 2x6 wood studs with insul.	90% of wall	S.F. Wall	14.43	6.61	
2020	Exterior Windows	Double hung wood	10% of wall	Each	521	2.21	9.3%
2030	Exterior Doors	Wood swinging double doors, single leaf hollow metal, low VOC paint		Each	1995	1.20	
	B30 Roofing						
3010	Roof Coverings	Single ply membrane, TPO, 45 mil, fully adhered; polyisocyanurate sheets		S.F. Roof	6.37	6.37	5.9%
3020	Roof Openings	N/A		—	—	—	
C. INTERIORS							
1010	Partitions	Gypsum board on wood studs with sound deadening board	15 SF Flr./LF Part.	S.F. Partition	10.01	5.34	
1020	Interior Doors	Single leaf wood, low VOC paint	150 S.F. Floor/Door	Each	626	4.17	
1030	Fittings	N/A		—	—	—	
2010	Stair Construction	N/A		—	—	—	24.2%
3010	Wall Finishes	50% wallpaper, 25% wood paneling, 25% paint (low VOC)		S.F. Surface	3.26	3.48	
3020	Floor Finishes	70% carpet tile, 30% ceramic tile		S.F. Floor	7.21	7.21	
3030	Ceiling Finishes	Fiberglass board on exposed grid, suspended		S.F. Ceiling	5.78	5.78	
D. SERVICES							
	D10 Conveying						
1010	Elevators & Lifts	N/A		—	—	—	0.0 %
1020	Escalators & Moving Walks	N/A		—	—	—	
	D20 Plumbing						
2010	Plumbing Fixtures	Toilets, low flow, auto sensor, urinals & service fixt., supply & drain.	1 Fixt./770 SF Flr.	Each	3403	4.42	
2020	Domestic Water Distribution	Electric water heater, point-of-use		S.F. Floor	.47	.47	5.5%
2040	Rain Water Drainage	Roof drain		S.F. Roof	1.05	1.05	
	D30 HVAC						
3010	Energy Supply	N/A		—	—	—	
3020	Heat Generating Systems	Included in D3050		—	—	—	
3040	Distribution Systems	Enthalpy heat recovery packages		Each	37,000	3.70	19.7 %
3050	Terminal & Package Units	Multizone rooftop air conditioner, SEER 14		S.F. Floor	17.36	17.36	
3090	Other HVAC Sys. & Equipment	N/A		—	—	—	
	D40 Fire Protection						
4010	Sprinklers	Wet pipe sprinkler system		S.F. Floor	3.65	3.65	4.3%
4020	Standpipes	Standpipe, wet, Class III		S.F. Floor	.92	.92	
	D50 Electrical						
5010	Electrical Service/Distribution	400 ampere service, panel board and feeders		S.F. Floor	1.60	1.60	
5020	Lighting & Branch Wiring	LED light fixtures, daylt. dimming control & ltg. on/off sys., recept., switches, & A.C. power		S.F. Floor	10.34	10.34	14.5%
5030	Communications & Security	Addressable alarm systems		S.F. Floor	1.57	1.57	
5090	Other Electrical Systems	Emergency generator, 15 kW and energy monitoring systems		S.F. Floor	1.98	1.98	
E. EQUIPMENT & FURNISHINGS							
1010	Commercial Equipment	N/A		—	—	—	
1020	Institutional Equipment	N/A		—	—	—	0.6 %
1090	Other Equipment	Waste handling recycling tilt truck		Each	.57	.57	
2020	Moveable Furnishings	No smoking signage		Each	.04	.04	
F. SPECIAL CONSTRUCTION							
1020	Integrated Construction	N/A		—	—	—	0.0 %
1040	Special Facilities	N/A		—	—	—	
G. BUILDING SITEWORK	**N/A**						

		Sub-Total	107.17	100%
CONTRACTOR FEES (General Requirements: 10%, Overhead: 5%, Profit: 10%)		25%	26.82	
ARCHITECT FEES		9%	12.06	
	Total Building Cost		**146.05**	

For customer support on your Square Foot Cost Data, call 877.756.2789.

249

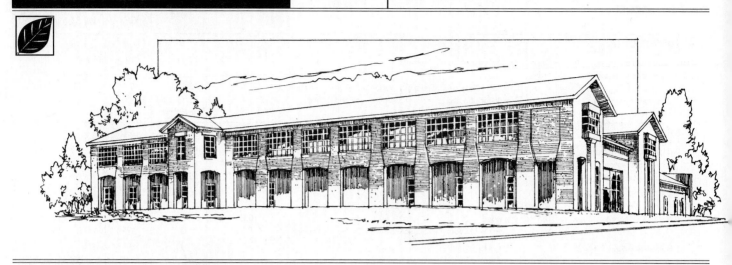

Costs per square foot of floor area

Exterior Wall	S.F. Area	7000	10000	13000	16000	19000	22000	25000	28000	31000
	L.F. Perimeter	240	300	336	386	411	435	472	510	524
Face Brick with Concrete Block Back-up	R/Conc. Frame	212.45	201.45	192.10	187.95	182.60	178.65	176.60	175.05	172.45
	Steel Frame	213.45	202.45	193.10	188.90	183.60	179.60	177.55	176.00	173.40
Limestone with Concrete Block	R/Conc. Frame	240.70	225.75	212.85	207.10	199.80	194.30	191.50	189.35	185.70
	Steel Frame	238.80	224.65	212.20	206.80	199.60	194.25	191.55	189.50	185.90
Precast Concrete Panels	R/Conc. Frame	228.75	215.30	203.80	198.70	192.25	187.35	184.90	183.00	179.80
	Steel Frame	228.70	215.25	203.80	198.65	192.20	187.35	184.85	182.95	179.75
Perimeter Adj., Add or Deduct	Per 100 L.F.	26.50	18.60	14.25	11.55	9.75	8.40	7.45	6.65	5.95
Story Hgt. Adj., Add or Deduct	Per 1 Ft.	4.05	3.55	3.05	2.80	2.55	2.35	2.20	2.10	1.95

For Basement, add $53.80 per square foot of basement area

The above costs were calculated using the basic specifications shown on the facing page. These costs should be adjusted where necessary for design alternatives and owner's requirements. Reported completed project costs, for this type of structure, range from $117.10 to $300.50 per S.F.

Common additives

Description	Unit	$ Cost
Carrels Hardwood	Ea.	945 - 2400
Closed Circuit Surveillance, one station, camera and monitor	Ea.	2025
For additional camera stations, add	Ea.	1100
Elevators, hydraulic passenger, 2 stops, 1500# capacity	Ea.	67,400
2500# capacity	Ea.	70,400
3500# capacity	Ea.	75,400
Library Furnishings, bookshelf, 90" high, 10" shelf double face	L.F.	288
Single face	L.F.	184
Charging desk, built-in with counter		
Plastic laminated top	L.F.	420
Reading table, laminated		
Top 60" x 36"	Ea.	775

Description	Unit	$ Cost
Commissioning Fees, sustainable institutional construction	S.F.	0.58 - 2.47
Energy Modelling Fees, commercial buildings to 10,000 SF	Ea.	11,000
Greater than 10,000 SF add	S.F.	0.04
Green Bldg Cert Fees for comm construction project reg	Project	900
Photovoltaic Pwr Sys, grid connected, 20 kW (~2400 SF), roof	Ea.	245,600
Green Roofs, 6" soil depth, w/treated wd edging & sedum mats	S.F.	11.76
10" Soil depth, with treated wood edging & sedum mats	S.F.	13.45
Greywater Recovery Systems, prepackaged comm, 3060 gal.	Ea.	58,265
4590 gal.	Ea.	73,075
Rainwater Harvest Sys, prepckged comm, 10,000 gal, sys contrller	Ea.	31,525
20,000 gal. w/system controller	Ea.	50,575
30,000 gal. w/system controller	Ea.	80,825
Solar Domestic HW, closed loop, add-on sys, ext heat exchanger	Ea.	10,850
Drainback, hot water system, 120 gal tank	Ea.	13,525
Draindown, hot water system, 120 gal tank	Ea.	13,850

Important: See the Reference Section for Location Factors.

Model costs calculated for a 2 story building with 14' story height and 22,000 square feet of floor area

				Unit	Unit Cost	Cost Per S.F.	% Of Sub-Total
A. SUBSTRUCTURE							
1010	Standard Foundations	Poured concrete; strip and spread footings		S.F. Ground	4.06	2.03	
1020	Special Foundations	N/A		—	—	—	
1030	Slab on Grade	4" reinforced concrete with recycled vapor barrier and granular base		S.F. Slab	5.58	2.79	5.2%
2010	Basement Excavation	Site preparation for slab and trench for foundation wall and footing		S.F. Ground	.32	.16	
2020	Basement Walls	4' foundation wall		L.F. Wall	93	1.84	
B. SHELL							
B10 Superstructure							
1010	Floor Construction	Concrete waffle slab		S.F. Floor	28.04	14.02	18.2%
1020	Roof Construction	Concrete waffle slab		S.F. Roof	20.26	10.13	
B20 Exterior Enclosure							
2010	Exterior Walls	Face brick with concrete block backup	90% of wall	S.F. Wall	36.85	18.36	
2020	Exterior Windows	Window wall	10% of wall	Each	56	3.08	16.6%
2030	Exterior Doors	Double aluminum and glass, single leaf hollow metal, low VOC paint		Each	5875	.53	
B30 Roofing							
3010	Roof Coverings	Single-ply TPO membrane, 60 mil, heat welded w/R-20 insulation		S.F. Roof	6.04	3.02	2.3%
3020	Roof Openings	Roof hatches		S.F. Roof	.08	.04	
C. INTERIORS							
1010	Partitions	Gypsum board on metal studs w/sound attenuation insul.	30 SF Flr./LF Part.	S.F. Partition	14	5.60	
1020	Interior Doors	Single leaf wood, low VOC paint	300 S.F. Floor/Door	Each	626	2.09	
1030	Fittings	N/A		—	—	—	
2010	Stair Construction	Concrete filled metal pan		Flight	9850	.89	16.4%
3010	Wall Finishes	Paint, low VOC		S.F. Surface	2.01	1.61	
3020	Floor Finishes	50% carpet tile, 50% vinyl composition tile, recycled content		S.F. Floor	4.29	4.29	
3030	Ceiling Finishes	Mineral fiber on concealed zee bars		S.F. Ceiling	7.22	7.22	
D. SERVICES							
D10 Conveying							
1010	Elevators & Lifts	One hydraulic passenger elevator		Each	88,880	4.04	3.1%
1020	Escalators & Moving Walks	N/A		—	—	—	
D20 Plumbing							
2010	Plumbing Fixtures	Toilet, low flow, auto sensor & service fixt., supply & drain.	1 Fixt./1835 SF.Flr.	Each	8129	4.43	
2020	Domestic Water Distribution	Tankless, on demand water heater, gas/propane		S.F. Floor	.57	.57	4.3%
2040	Rain Water Drainage	Roof drains		S.F. Roof	1.28	.64	
D30 HVAC							
3010	Energy Supply	N/A		—	—	—	
3020	Heat Generating Systems	Included in D3050		—	—	—	
3040	Distribution Systems	Enthalpy heat recovery packages		Each	30,450	2.77	17.5 %
3050	Terminal & Package Units	Multizone rooftop air conditioner, SEER 14		S.F. Floor	20.45	20.45	
3090	Other HVAC Sys. & Equipment	N/A		—	—	—	
D40 Fire Protection							
4010	Sprinklers	Wet pipe sprinkler system		S.F. Floor	3.15	3.15	3.2%
4020	Standpipes	Standpipe		S.F. Floor	1.05	1.05	
D50 Electrical							
5010	Electrical Service/Distribution	400 ampere service, panel board and feeders		S.F. Floor	.55	.55	
5020	Lighting & Branch Wiring	LED fixtures, daylt. dim., ltg. on/off, recept., switches, and A.C. power		S.F. Floor	13.21	13.21	13.1%
5030	Communications & Security	Addressable alarm systems, internet wiring, and emergency lighting		S.F. Floor	2.11	2.11	
5090	Other Electrical Systems	Emergency generator, 7.5 kW, UPS, energy monitoring systems		S.F. Floor	1.40	1.40	
E. EQUIPMENT & FURNISHINGS							
1010	Commercial Equipment	N/A		—	—	—	
1020	Institutional Equipment	N/A		—	—	—	
1090	Other Equipment	Waste handling recycling tilt truck		Each	.26	.26	0.2 %
2020	Moveable Furnishings	No smoking signage		Each			
F. SPECIAL CONSTRUCTION							
1020	Integrated Construction	N/A		—	—	—	0.0 %
1040	Special Facilities	N/A		—	—	—	
G. BUILDING SITEWORK	**N/A**						

	Sub-Total	132.33	**100%**
CONTRACTOR FEES (General Requirements: 10%, Overhead: 5%, Profit: 10%)	25%	33.09	
ARCHITECT FEES	8%	13.23	
Total Building Cost		178.65	

For customer support on your Square Foot Cost Data, call 877.756.2789.

251

Costs per square foot of floor area

Exterior Wall	S.F. Area	4000	5500	7000	8500	10000	11500	13000	14500	16000
	L.F. Perimeter	280	320	380	440	453	503	510	522	560
Face Brick with Concrete Block Back-up	Steel Joists	251.35	240.00	235.65	232.85	227.35	225.70	221.90	219.15	218.20
	Wood Truss	254.45	243.10	238.75	236.00	230.45	228.80	225.00	222.30	221.35
Stucco on Concrete Block	Steel Joists	259.55	246.80	242.00	238.90	232.65	230.80	226.45	223.40	222.30
	Wood Truss	267.10	254.35	249.55	246.45	240.20	238.35	233.95	230.95	229.85
Brick Veneer	Wood Truss	248.85	238.75	234.85	232.35	227.50	226.05	222.70	220.35	219.50
Wood Siding	Wood Frame	241.15	232.85	229.45	227.35	223.45	222.20	219.50	217.60	216.90
Perimeter Adj., Add or Deduct	Per 100 L.F.	19.15	13.85	10.95	9.00	7.65	6.70	5.90	5.35	4.75
Story Hgt. Adj., Add or Deduct	Per 1 Ft.	3.95	3.20	3.00	2.90	2.50	2.40	2.15	2.00	1.95

For Basement, add $35.00 per square foot of basement area

The above costs were calculated using the basic specifications shown on the facing page. These costs should be adjusted where necessary for design alternatives and owner's requirements. Reported completed project costs, for this type of structure, range from $98.65 to $253.80 per S.F.

Common additives

Description	Unit	$ Cost
Cabinets, hospital, base, laminated plastic	L.F.	420
Counter top, laminated plastic	L.F.	81.50
Nurses station, door type , laminated plastic	L.F.	470
Wall cabinets, laminated plastic	L.F.	305
Directory Boards, plastic, glass covered, 30" x 20"	Ea.	610
36" x 48"	Ea.	1525
Aluminum, 36" x 24"	Ea.	765
48" x 32"	Ea.	1075
Heat Therapy Unit, humidified, 26" x 78" x 28"	Ea.	3450
Tables, examining, vinyl top, with base cabinets	Ea.	1650 - 6775
Utensil Washer, Sanitizer	Ea.	9650
X-Ray, Mobile	Ea.	18,500 - 85,500

Description	Unit	$ Cost
Commissioning Fees, sustainable commercial construction	S.F.	0.24 - 3.04
Energy Modelling Fees, commercial buildings to 10,000 SF	Ea.	11,000
Greater than 10,000 SF add	S.F.	0.04
Green Bldg Cert Fees for comm construction project reg	Project	900
Photovoltaic Pwr Sys, grid connected, 20 kW (~2400 SF), roof	Ea.	245,600
Green Roofs, 6" soil depth, w/treated wd edging & sedum mats	S.F.	11.76
10" Soil depth, with treated wood edging & sedum mats	S.F.	13.45
Greywater Recovery Systems, prepackaged comm, 3060 gal.	Ea.	58,265
4590 gal.	Ea.	73,075
Rainwater Harvest Sys, prepckged comm, 10,000 gal, sys contrller	Ea.	31,525
20,000 gal. w/system controller	Ea.	50,575
30,000 gal. w/system controller	Ea.	80,825
Solar Domestic HW, closed loop, add-on sys, ext heat exchanger	Ea.	10,850
Drainback, hot water system, 120 gal tank	Ea.	13,525
Draindown, hot water system, 120 gal tank	Ea.	13,850

Important: See the Reference Section for Location Factors.

Model costs calculated for a 1 story building with 10' story height and 7,000 square feet of floor area

				Unit	Unit Cost	Cost Per S.F.	% Of Sub-Total
A. SUBSTRUCTURE							
1010	Standard Foundations	Poured concrete; strip and spread footings		S.F. Ground	2.23	2.23	
1020	Special Foundations	N/A		—	—	—	
1030	Slab on Grade	4" reinforced concrete with recycled vapor barrier and granular base		S.F. Slab	5.58	5.58	7.5%
2010	Basement Excavation	Site preparation for slab and trench for foundation wall and footing		S.F. Ground	.32	.32	
2020	Basement Walls	4' foundation wall		L.F. Wall	93	5.05	
B. SHELL							
B10 Superstructure							
1010	Floor Construction	N/A		—	—	—	4.2 %
1020	Roof Construction	Plywood on wood trusses		S.F. Roof	7.42	7.42	
B20 Exterior Enclosure							
2010	Exterior Walls	Face brick with concrete block backup	70% of wall	S.F. Wall	36.84	14	
2020	Exterior Windows	Wood double hung	30% of wall	Each	646	6.19	13.0%
2030	Exterior Doors	Aluminum and glass doors and entrance with transoms		Each	1475	2.53	
B30 Roofing							
3010	Roof Coverings	Asphalt shingles with flashing (Pitched); rigid fiber glass insulation, gutters		S.F. Roof	8.68	8.68	5.0%
3020	Roof Openings	N/A		—	—	—	
C. INTERIORS							
1010	Partitions	Gypsum bd. & sound deadening bd. on wood studs w/insul.	6 SF Flr./LF Part.	S.F. Partition	9.74	12.98	
1020	Interior Doors	Single leaf wood, low VOC paint	60 S.F. Floor/Door	Each	626	10.43	
1030	Fittings	N/A		—	—	—	
2010	Stair Construction	N/A		—	—	—	21.5%
3010	Wall Finishes	50% paint, low VOC, 50% vinyl wall covering		S.F. Surface	1.43	3.81	
3020	Floor Finishes	50% carpet tile, 50% vinyl composition tile, recycled content		S.F. Floor	3.98	3.98	
3030	Ceiling Finishes	Mineral fiber tile on concealed zee bars		S.F. Ceiling	6.56	6.56	
D. SERVICES							
D10 Conveying							
1010	Elevators & Lifts	N/A		—	—	—	0.0 %
1020	Escalators & Moving Walks	N/A		—	—	—	
D20 Plumbing							
2010	Plumbing Fixtures	Toilet, low flow, auto sensor, exam room & service fixt., supply & drain.	1 Fixt./195 SF Flr.	Each	4861	24.93	
2020	Domestic Water Distribution	Gas fired tankless water heater		S.F. Floor	2.81	2.81	16.5%
2040	Rain Water Drainage	Roof drains		S.F. Roof	1.11	1.11	
D30 HVAC							
3010	Energy Supply	N/A		—	—	—	
3020	Heat Generating Systems	Included in D3050		—	—	—	
3040	Distribution Systems	Enthalpy heat recovery packages		Each	17,575	2.51	9.6 %
3050	Terminal & Package Units	Multizone rooftop air conditioner, SEER 14		S.F. Floor	14.31	14.31	
3090	Other HVAC Sys. & Equipment	N/A		—	—	—	
D40 Fire Protection							
4010	Sprinklers	Wet pipe sprinkler system		S.F. Floor	5.02	5.02	3.6%
4020	Standpipes	Standpipe		S.F. Floor	1.31	1.31	
D50 Electrical							
5010	Electrical Service/Distribution	200 ampere service, panel board and feeders		S.F. Floor	1.33	1.33	
5020	Lighting & Branch Wiring	LED fixtures, daylt. dim., ltg. on/off, recept., switches, and A.C. power		S.F. Floor	10.12	10.12	12.9%
5030	Communications & Security	Alarm systems, internet & phone wiring, intercom system, & emerg. ltg.		S.F. Floor	8.04	8.04	
5090	Other Electrical Systems	Emergency generator, 7.5 kW, and energy monitoring systems		S.F. Floor	3.18	3.18	
E. EQUIPMENT & FURNISHINGS							
1010	Commercial Equipment	N/A		—	—	—	
1020	Institutional Equipment	Exam room casework and countertops		S.F. Floor	9.89	9.89	6.2 %
1090	Other Equipment	Waste handling recycling tilt truck		Each	.81	.81	
2020	Moveable Furnishings	No smoking signage		Each	.10	.10	
F. SPECIAL CONSTRUCTION							
1020	Integrated Construction	N/A		—	—	—	0.0 %
1040	Special Facilities	N/A		—	—	—	
G. BUILDING SITEWORK	**N/A**						

			Sub-Total	175.23	100%
CONTRACTOR FEES (General Requirements: 10%, Overhead: 5%, Profit: 10%)			25%	43.81	
ARCHITECT FEES			9%	19.71	
		Total Building Cost		**238.75**	

For customer support on your Square Foot Cost Data, call 877.756.2789.

253

Costs per square foot of floor area

Exterior Wall		S.F. Area	4000	5500	7000	8500	10000	11500	13000	14500	16000
		L.F. Perimeter	180	210	240	270	286	311	336	361	386
Face Brick with Concrete Block Back-up	Steel Joists		287.95	275.85	269.05	264.55	259.55	256.85	254.75	253.25	251.85
	Wood Joists		293.40	281.30	274.50	270.05	265.00	262.35	260.25	258.70	257.35
Stucco on Concrete Block	Steel Joists		271.15	261.55	256.20	252.60	248.75	246.65	245.00	243.80	242.70
	Wood Joists		281.95	271.60	265.80	261.95	257.70	255.40	253.65	252.40	251.25
Brick Veneer	Wood Frame		282.00	271.65	265.80	262.00	257.75	255.50	253.65	252.40	251.25
Wood Siding	Wood Frame		273.70	264.65	259.50	256.10	252.45	250.50	248.95	247.80	246.80
Perimeter Adj., Add or Deduct	Per 100 L.F.		34.20	24.95	19.50	16.10	13.70	11.90	10.65	9.45	8.55
Story Hgt. Adj., Add or Deduct	Per 1 Ft.		5.05	4.30	3.85	3.60	3.20	3.05	2.90	2.80	2.75
For Basement, add $38.85 per square foot of basement area											

The above costs were calculated using the basic specifications shown on the facing page. These costs should be adjusted where necessary for design alternatives and owner's requirements. Reported completed project costs, for this type of structure, range from $101.30 to $346.00 per S.F.

Common additives

Description	Unit	$ Cost	Description	Unit	$ Cost
Cabinets, hospital, base, laminated plastic	L.F.	420	Commissioning Fees, sustainable commercial construction	S.F.	0.24 - 3.04
Counter top, laminated plastic	L.F.	81.50	Energy Modelling Fees, commercial buildings to 10,000 SF	Ea.	11,000
Nurses station, door type , laminated plastic	L.F.	470	Greater than 10,000 SF add	S.F.	0.04
Wall cabinets, laminated plastic	L.F.	305	Green Bldg Cert Fees for comm construction project reg	Project	900
Elevators, hydraulic passenger, 2 stops, 2500# capacity	Ea.	70,400	Photovoltaic Pwr Sys, grid connected, 20 kW (~2400 SF), roof	Ea.	245,600
3500# capacity	Ea.	75,400	Green Roofs, 6" soil depth, w/treated wd edging & sedum mats	S.F.	11.76
Directory Boards, plastic, glass covered, 30" x 20"	Ea.	610	10" Soil depth, with treated wood edging & sedum mats	S.F.	13.45
36" x 48"	Ea.	1525	Greywater Recovery Systems, prepackaged comm, 3060 gal.	Ea.	58,265
Aluminum, 36" x 24"	Ea.	765	4590 gal.	Ea.	73,075
48" x 32"	Ea.	1075	Rainwater Harvest Sys, prepckged comm, 10,000 gal, sys contrller	Ea.	31,525
Heat Therapy Unit, humidified, 26" x 78" x 28"	Ea.	3450	20,000 gal. w/system controller	Ea.	50,575
Tables, examining, vinyl top, with base cabinets	Ea.	1650 - 6775	30,000 gal. w/system controller	Ea.	80,825
Utensil Washer, Sanitizer	Ea.	9650	Solar Domestic HW, closed loop, add-on sys, ext heat exchanger	Ea.	10,850
X-Ray, Mobile	Ea.	18,500 - 85,500	Drainback, hot water system, 120 gal tank	Ea.	13,525
			Draindown, hot water system, 120 gal tank	Ea.	13,850

Important: See the Reference Section for Location Factors.

Model costs calculated for a 2 story building with 10′ story height and 7,000 square feet of floor area

G Medical Office, 2 Story

				Unit	Unit Cost	Cost Per S.F.	% Of Sub-Total
A. SUBSTRUCTURE							
1010	Standard Foundations	Poured concrete; strip and spread footings		S.F. Ground	3.58	1.79	
1020	Special Foundations	N/A		—	—	—	
1030	Slab on Grade	4″ reinforced concrete with vapor barrier and granular base		S.F. Slab	5.58	2.79	4.2%
2010	Basement Excavation	Site preparation for slab and trench for foundation wall and footing		S.F. Ground	.18	.09	
2020	Basement Walls	4′ foundation wall		L.F. Wall	93	3.19	
B. SHELL							
B10 Superstructure							
1010	Floor Construction	Open web steel joists, slab form, concrete, columns		S.F. Floor	13.80	6.90	5.3%
1020	Roof Construction	Metal deck, open web steel joists, beams, columns		S.F. Roof	6.26	3.13	
B20 Exterior Enclosure							
2010	Exterior Walls	Stucco on concrete block		S.F. Wall	19.60	9.41	
2020	Exterior Windows	Outward projecting metal	30% of wall	Each	525	7.20	9.7%
2030	Exterior Doors	Aluminum and glass doors with transoms		Each	5875	1.68	
B30 Roofing							
3010	Roof Coverings	Built-up tar and gravel with flashing; perlite/EPS composite insulation		S.F. Roof	7.18	3.59	2.0%
3020	Roof Openings	Roof hatches		S.F. Roof	.26	.13	
C. INTERIORS							
1010	Partitions	Gypsum bd. & acous. insul. on metal studs	6 SF Floor/LF Part.	S.F. Partition	8.12	10.83	
1020	Interior Doors	Single leaf wood	60 S.F. Floor/Door	Each	626	10.43	
1030	Fittings	N/A		—	—	—	
2010	Stair Construction	Concrete filled metal pan		Flight	14,450	4.13	21.1%
3010	Wall Finishes	45% paint, 50% vinyl wall coating, 5% ceramic tile		S.F. Surface	1.43	3.81	
3020	Floor Finishes	50% carpet, 50% vinyl composition tile		S.F. Floor	3.98	3.98	
3030	Ceiling Finishes	Mineral fiber tile on concealed zee bars		S.F. Ceiling	6.56	6.56	
D. SERVICES							
D10 Conveying							
1010	Elevators & Lifts	One hydraulic hospital elevator		Each	110,320	15.76	8.4%
1020	Escalators & Moving Walks	N/A		—	—	—	
D20 Plumbing							
2010	Plumbing Fixtures	Toilet, exam room & service fixt., supply & drain.	1 Fixt./160 SF Flr.	Each	4526	28.29	
2020	Domestic Water Distribution	Gas fired water heater		S.F. Floor	2.81	2.81	17.2%
2040	Rain Water Drainage	Roof drains		S.F. Roof	2.38	1.19	
D30 HVAC							
3010	Energy Supply	N/A		—	—	—	
3020	Heat Generating Systems	Included in D3050		—	—	—	
3040	Distribution Systems	Enthalpy heat recovery packages		Each	17,575	2.51	8.9 %
3050	Terminal & Package Units	Multizone rooftop air conditioner, SEER 14		S.F. Floor	14.31	14.31	
3090	Other HVAC Sys. & Equipment	N/A		—	—	—	
D40 Fire Protection							
4010	Sprinklers	Wet pipe sprinkler system		S.F. Floor	3.95	3.95	3.2%
4020	Standpipes	Standpipe		S.F. Floor	1.98	1.98	
D50 Electrical							
5010	Electrical Service/Distribution	400 ampere service, panel board and feeders		S.F. Floor	1.65	1.65	
5020	Lighting & Branch Wiring	LED fixtures, receptacles, switches, A.C. and misc. power		S.F. Floor	14.19	14.19	14.4%
5030	Communications & Security	Alarm system, internet & phone wiring, intercom system, & emergency ltg.		S.F. Floor	8.04	8.04	
5090	Other Electrical Systems	Emergency generator, 7.5 kW		S.F. Floor	3.28	3.28	
E. EQUIPMENT & FURNISHINGS							
1010	Commercial Equipment	N/A		—	—	—	
1020	Institutional Equipment	Exam room casework and contertops		S.F. Floor	9.89	9.89	5.5 %
1090	Other Equipment	Waste handling recycling tilt truck		Each	.50	.50	
2020	Moveable Furnishings	No smoking signage		Each	.02	.02	
F. SPECIAL CONSTRUCTION							
1020	Integrated Construction	N/A		—	—	—	0.0 %
1040	Special Facilities	N/A		—	—	—	
G. BUILDING SITEWORK	**N/A**						

			Sub-Total	188.01	100%
CONTRACTOR FEES (General Requirements: 10%, Overhead: 5%, Profit: 10%)		25%		47.04	
ARCHITECT FEES		9%		21.15	
		Total Building Cost		**256.20**	

For customer support on your Square Foot Cost Data, call 877.756.2789.

255

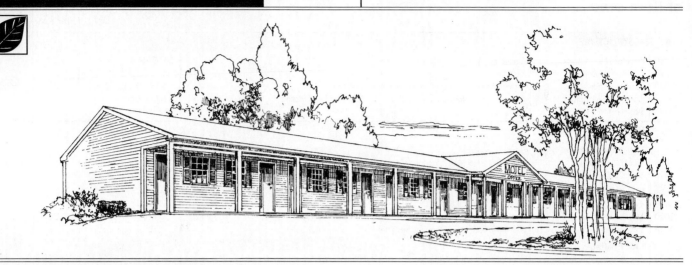

Costs per square foot of floor area

Exterior Wall	S.F. Area	2000	3000	4000	6000	8000	10000	12000	14000	16000
	L.F. Perimeter	240	260	280	380	480	560	580	660	740
Brick Veneer	Wood Frame	228.55	200.55	186.60	179.00	175.25	171.65	166.10	164.85	163.90
Aluminum Siding	Wood Frame	211.10	187.90	176.30	169.65	166.35	163.35	158.90	157.85	157.00
Wood Siding	Wood Frame	210.75	187.65	176.05	169.45	166.20	163.20	158.75	157.65	156.90
Wood Shingles	Wood Frame	214.20	190.10	178.05	171.25	167.90	164.70	160.10	158.95	158.15
Precast Concrete Block	Wood Truss	213.95	189.90	177.90	171.15	167.80	164.70	160.00	158.90	158.15
Brick on Concrete Block	Wood Truss	243.65	211.40	195.30	186.85	182.65	178.55	172.00	170.55	169.55
Perimeter Adj., Add or Deduct	Per 100 L.F.	32.10	21.45	16.00	10.70	8.00	6.40	5.30	4.55	4.05
Story Hgt. Adj., Add or Deduct	Per 1 Ft.	5.05	3.65	2.90	2.65	2.50	2.35	2.00	2.00	1.95

For Basement, add $ 26.75 per square foot of basement area

The above costs were calculated using the basic specifications shown on the facing page. These costs should be adjusted where necessary for design alternatives and owner's requirements. Reported completed project costs, for this type of structure, range from $ 81.75 to $ 387.00 per S.F.

Common additives

Description	Unit	$ Cost
Closed Circuit Surveillance, one station, camera & monitor	Ea.	2025
For additional camera stations, add	Ea.	1100
Laundry Equipment, dryer, gas, 30 lb. capacity	Ea.	3950
Washer, commercial	Ea.	1625
Sauna, prefabricated, complete, 6' x 9'	Ea.	10,500
8' x 8'	Ea.	12,000
Swimming Pools, Complete, gunite	S.F.	104 - 129
TV Antenna, Master system, 12 outlet	Outlet	263
30 outlet	Outlet	225
100 outlet	Outlet	213

Description	Unit	$ Cost
Commissioning Fees, sustainable commercial construction	S.F.	0.24 - 3.04
Energy Modelling Fees, commercial buildings to 10,000 SF	Ea.	11,000
Greater than 10,000 SF add	S.F.	0.04
Green Bldg Cert Fees for comm construction project reg	Project	900
Photovoltaic Pwr Sys, grid connected, 20 kW (~2400 SF), roof	Ea.	245,600
Green Roofs, 6" soil depth, w/treated wd edging & sedum mats	S.F.	11.76
10" Soil depth, with treated wood edging & sedum mats	S.F.	13.45
Greywater Recovery Systems, prepackaged comm, 3060 gal.	Ea.	58,265
4590 gal.	Ea.	73,075
Rainwater Harvest Sys, prepckged comm, 10,000 gal, sys contrller	Ea.	31,525
20,000 gal. w/system controller	Ea.	50,575
30,000 gal. w/system controller	Ea.	80,825
Solar Domestic HW, closed loop, add-on sys, ext heat exchanger	Ea.	10,850
Drainback, hot water system, 120 gal tank	Ea.	13,525
Draindown, hot water system, 120 gal tank	Ea.	13,850

Important: See the Reference Section for Location Factors.

Model costs calculated for a 1 story building with 9' story height and 8,000 square feet of floor area

G Motel, 1 Story

				Unit	Unit Cost	Cost Per S.F.	% Of Sub-Total
A. SUBSTRUCTURE							
1010	Standard Foundations	Poured concrete; strip and spread footings		S.F. Ground	3.38	3.38	
1020	Special Foundations	N/A		—	—	—	
1030	Slab on Grade	4" reinforced concrete with recycled vapor barrier and granular base		S.F. Slab	5.58	5.58	12.9%
2010	Basement Excavation	Site preparation for slab and trench for foundation wall and footing		S.F. Ground	.32	.32	
2020	Basement Walls	4' foundation wall		L.F. Wall	93	7.64	
B. SHELL							
B10 Superstructure							
1010	Floor Construction	N/A		—	—	—	5.7 %
1020	Roof Construction	Plywood on wood trusses		S.F. Roof	7.42	7.42	
B20 Exterior Enclosure							
2010	Exterior Walls	Face brick on wood studs with sheathing, insulation and paper	80% of wall	S.F. Wall	25.49	11.01	
2020	Exterior Windows	Wood double hung	20% of wall	Each	521	4.02	14.6%
2030	Exterior Doors	Wood solid core		Each	1475	4.05	
B30 Roofing							
3010	Roof Coverings	Asphalt strip shingles, 210-235 lbs/Sq., R-10 insul. & weather barrier-recycled		S.F. Roof	8.49	8.49	6.5%
3020	Roof Openings	N/A		—	—	—	
C. INTERIORS							
1010	Partitions	Gypsum bd. and sound deadening bd. on wood studs	9 SF Flr./LF Part.	S.F. Partition	10.46	9.30	
1020	Interior Doors	Single leaf hollow core wood, low VOC paint	300 S.F. Floor/Door	Each	542	1.81	
1030	Fittings	N/A		—	—	—	
2010	Stair Construction	N/A		—	—	—	18.5%
3010	Wall Finishes	90% paint, low VOC, 10% ceramic tile		S.F. Surface	1.65	2.93	
3020	Floor Finishes	85% carpet tile, 15% ceramic tile		S.F. Floor	5.50	5.50	
3030	Ceiling Finishes	Painted gypsum board on furring, low VOC paint		S.F. Ceiling	4.72	4.72	
D. SERVICES							
D10 Conveying							
1010	Elevators & Lifts	N/A		—	—	—	0.0 %
1020	Escalators & Moving Walks	N/A		—	—	—	
D20 Plumbing							
2010	Plumbing Fixtures	Toilet, low flow, auto sensor, & service fixt., supply & drainage	1 Fixt./90 SF Flr.	Each	2182	24.24	
2020	Domestic Water Distribution	Gas fired tankless water heater		S.F. Floor	.42	.42	20.1%
2040	Rain Water Drainage	Roof Drains		S.F. Roof	1.63	1.63	
D30 HVAC							
3010	Energy Supply	N/A		—	—	—	
3020	Heat Generating Systems	Included in D3050		—	—	—	
3030	Cooling Generating Systems	N/A		—	—	—	2.9 %
3050	Terminal & Package Units	Through the wall electric heating and cooling units		S.F. Floor	3.84	3.84	
3090	Other HVAC Sys. & Equipment	N/A		—	—	—	
D40 Fire Protection							
4010	Sprinklers	Wet pipe sprinkler system		S.F. Floor	5.02	5.02	4.7%
4020	Standpipes	Standpipe, wet, Class III		S.F. Floor	1.15	1.15	
D50 Electrical							
5010	Electrical Service/Distribution	200 ampere service, panel board and feeders		S.F. Floor	1.32	1.32	
5020	Lighting & Branch Wiring	LED fixt., daylit. dim., ltg. on/off control, recept., switches and misc. pwr.		S.F. Floor	10.41	10.41	12.7%
5030	Communications & Security	Addressable alarm systems		S.F. Floor	2.78	2.78	
5090	Other Electrical Systems	Emergency generator, 7.5 kW, and energy monitoring systems		S.F. Floor	2.17	2.17	
E. EQUIPMENT & FURNISHINGS							
1010	Commercial Equipment	Laundry equipment		S.F. Floor	.99	.99	
1020	Institutional Equipment	N/A		—	—	—	1.4%
1090	Other Equipment	Waste handling recycling tilt truck		Each	.71	.71	
2020	Moveable Furnishings	No smoking signage		Each	.16	.16	
F. SPECIAL CONSTRUCTION							
1020	Integrated Construction	N/A		—	—	—	0.0 %
1040	Special Facilities	N/A		—	—	—	
G. BUILDING SITEWORK	**N/A**						

	Sub-Total	131.01	**100%**
CONTRACTOR FEES (General Requirements:10%, Overhead: 5%, Profit:10%)	25%	32.78	
ARCHITECT FEES	7%	11.46	
	Total Building Cost	**175.25**	

For customer support on your Square Foot Cost Data, call 877.756.2789.

257

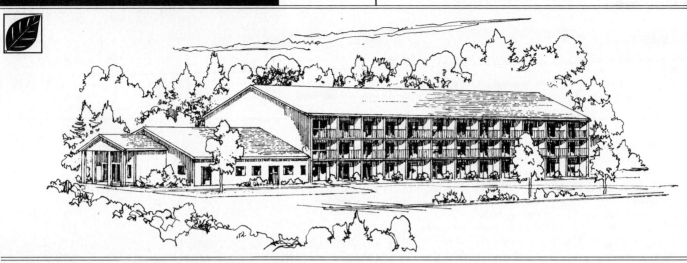

Costs per square foot of floor area

Exterior Wall	S.F. Area	25000	37000	49000	61000	73000	81000	88000	96000	104000
	L.F. Perimeter	433	593	606	720	835	911	978	1054	1074
Decorative Concrete Block	Wood Joists	182.05	178.25	172.00	170.60	169.65	169.20	168.85	168.50	167.45
	Precast Conc.	195.00	190.90	184.60	183.10	182.10	181.65	181.20	180.85	179.80
Stucco on Concrete Block	Wood Joists	183.75	179.20	172.35	170.65	169.55	169.00	168.55	168.10	167.05
	Precast Conc.	197.25	192.60	185.75	184.05	183.00	182.40	181.95	181.55	180.45
Wood Siding	Wood Frame	176.35	172.90	167.75	166.45	165.65	165.20	164.85	164.60	163.75
Brick Veneer	Wood Frame	183.85	179.20	172.35	170.65	169.55	169.00	168.55	168.15	167.05
Perimeter Adj., Add or Deduct	Per 100 L.F.	5.70	3.85	2.95	2.35	2.00	1.75	1.65	1.45	1.35
Story Hgt. Adj., Add or Deduct	Per 1 Ft.	2.05	1.85	1.50	1.40	1.35	1.35	1.25	1.25	1.20

For Basement, add $35.10 per square foot of basement area

The above costs were calculated using the basic specifications shown on the facing page. These costs should be adjusted where necessary for design alternatives and owner's requirements. Reported completed project costs, for this type of structure, range from $75.30 to $387.00 per S.F.

Common additives

Description	Unit	$ Cost
Closed Circuit Surveillance, one station, camera & monitor	Ea.	2025
For additional camera station, add	Ea.	1100
Elevators, hydraulic passenger, 2 stops, 2500# capacity	Ea.	70,400
3500# capacity	Ea.	75,400
Additional stop, add	Ea.	8075
Laundry Equipment, dryer, gas, 30 lb. capacity	Ea.	3950
Washer, commercial	Ea.	1625
Sauna, prefabricated, complete, 6' x 9'	Ea.	10,500
8' x 8'	Ea.	12,000
Swimming Pools, Complete, gunite	S.F.	104 - 129
TV Antenna, Master system, 12 outlet	Outlet	263
30 outlet	Outlet	225
100 outlet	Outlet	213

Description	Unit	$ Cost
Commissioning Fees, sustainable commercial construction	S.F.	0.24 - 3.04
Energy Modelling Fees, commercial buildings to 10,000 SF	Ea.	11,000
Greater than 10,000 SF add	S.F.	0.04
Green Bldg Cert Fees for comm construction project reg	Project	900
Photovoltaic Pwr Sys, grid connected, 20 kW (~2400 SF), roof	Ea.	245,600
Green Roofs, 6" soil depth, w/treated wd edging & sedum mats	S.F.	11.76
10" Soil depth, with treated wood edging & sedum mats	S.F.	13.45
Greywater Recovery Systems, prepackaged comm, 3060 gal.	Ea.	58,265
4590 gal.	Ea.	73,075
Rainwater Harvest Sys, prepckged comm, 10,000 gal, sys contrller	Ea.	31,525
20,000 gal. w/system controller	Ea.	50,575
30,000 gal. w/system controller	Ea.	80,825
Solar Domestic HW, closed loop, add-on sys, ext heat exchanger	Ea.	10,850
Drainback, hot water system, 120 gal tank	Ea.	13,525
Draindown, hot water system, 120 gal tank	Ea.	13,850

Model costs calculated for a 3 story building with 9' story height and 49,000 square feet of floor area

				Unit	Unit Cost	Cost Per S.F.	% Of Sub-Total
A.	**SUBSTRUCTURE**						
1010	Standard Foundations	Poured concrete; strip and spread footings		S.F. Ground	1.56	.52	
1020	Special Foundations	N/A		—	—	—	
1030	Slab on Grade	4" reinforced concrete with recycled vapor barrier and granular base		S.F. Slab	5.58	1.86	3.4%
2010	Basement Excavation	Site preparation for slab and trench for foundation wall and footing		S.F. Ground	.18	.06	
2020	Basement Walls	4' foundation wall		L.F. Wall	93	2.30	
B.	**SHELL**						
	B10 Superstructure						
1010	Floor Construction	Precast concrete plank		S.F. Floor	11.04	7.36	7.8%
1020	Roof Construction	Precast concrete plank		S.F. Roof	10.56	3.52	
	B20 Exterior Enclosure						
2010	Exterior Walls	Decorative concrete block	85% of wall	S.F. Wall	22.55	6.40	
2020	Exterior Windows	Aluminum sliding	15% of wall	Each	585	1.95	10.7%
2030	Exterior Doors	Aluminum and glass doors and entrance with transom		Each	1984	6.56	
	B30 Roofing						
3010	Roof Coverings	Single-ply TPO membrane, 60 mils w/R-20 insul.		S.F. Roof	6.51	2.17	1.6%
3020	Roof Openings	Roof hatches		S.F. Roof	.12	.04	
C.	**INTERIORS**						
1010	Partitions	Concrete block w/foamed-in insul.	7 SF Flr./LF Part.	S.F. Partition	22.04	25.19	
1020	Interior Doors	Wood hollow core, low VOC paint	70 S.F. Floor/Door	Each	542	7.74	
1030	Fittings	N/A		—	—	—	
2010	Stair Construction	Concrete filled metal pan		Flight	14,450	3.54	36.1%
3010	Wall Finishes	90% paint, low VOC, 10% ceramic tile		S.F. Surface	1.65	3.78	
3020	Floor Finishes	85% carpet tile, 5% vinyl composition tile, recycled content, 10% ceramic tile		S.F. Floor	5.50	5.50	
3030	Ceiling Finishes	Textured finish		S.F. Ceiling	4.49	4.49	
D.	**SERVICES**						
	D10 Conveying						
1010	Elevators & Lifts	Two hydraulic passenger elevators		Each	110,985	4.53	3.3%
1020	Escalators & Moving Walks	N/A		—	—	—	
	D20 Plumbing						
2010	Plumbing Fixtures	Toilet, low flow, auto sensor & service fixtures, supply & drain.	1 Fixt./180 SF Flr.	Each	5132	28.51	
2020	Domestic Water Distribution	Gas fired, tankless water heater		S.F. Floor	.29	.29	21.0%
2040	Rain Water Drainage	Roof drains		S.F. Roof	1.50	.50	
	D30 HVAC						
3010	Energy Supply	N/A		—	—	—	
3020	Heat Generating Systems	Included in D3050		—	—	—	
3030	Cooling Generating Systems	N/A		—	—	—	2.5 %
3050	Terminal & Package Units	Through the wall electric heating and cooling units		S.F. Floor	3.46	3.46	
3090	Other HVAC Sys. & Equipment	N/A		—	—	—	
	D40 Fire Protection						
4010	Sprinklers	Sprinklers, wet, light hazard		S.F. Floor	2.97	2.97	2.3%
4020	Standpipes	Standpipe, wet, Class III		S.F. Floor	.26	.26	
	D50 Electrical						
5010	Electrical Service/Distribution	800 ampere service, panel board and feeders		S.F. Floor	1.29	1.29	
5020	Lighting & Branch Wiring	LED fixtures, daylt. dim., ltg. on/off, recept., switches and misc. power		S.F. Floor	11.13	11.13	11.1%
5030	Communications & Security	Addressable alarm systems and emergency lighting		S.F. Floor	1.92	1.92	
5090	Other Electrical Systems	Emergency generator, 7.5 kW, energy monitoring systems		S.F. Floor	1.07	1.07	
E.	**EQUIPMENT & FURNISHINGS**						
1010	Commercial Equipment	Commercial laundry equipment		S.F. Floor	.32	.32	
1020	Institutional Equipment	N/A		—	—	—	
1090	Other Equipment	Waste handling recycling tilt truck		Each	.07	.07	0.3%
2020	Moveable Furnishings	No smoking signage		Each	.02	.02	
F.	**SPECIAL CONSTRUCTION**						
1020	Integrated Construction	N/A		—	—	—	0.0 %
1040	Special Facilities	N/A		—	—	—	
G.	**BUILDING SITEWORK**	**N/A**					

			Sub-Total	139.32	100%
CONTRACTOR FEES (General Requirements: 10%, Overhead: 5%, Profit: 10%)			25%	34.83	
ARCHITECT FEES			6%	10.45	
		Total Building Cost		**184.60**	

For customer support on your Square Foot Cost Data, call 877.756.2789.

259

Costs per square foot of floor area

Exterior Wall	S.F. Area	2000	3000	5000	7000	9000	12000	15000	20000	25000
	L.F. Perimeter	220	260	320	360	420	480	520	640	700
Wood Siding	Wood Truss	221.70	201.15	182.50	172.85	168.70	163.75	160.00	157.80	155.10
Brick Veneer	Wood Truss	265.75	240.85	218.15	206.40	201.45	195.30	190.75	188.05	184.70
Brick on Block	Wood Truss	213.25	185.30	159.60	146.20	140.60	133.60	128.30	125.25	121.40
Brick on Block	Steel Roof Deck	268.95	239.40	212.35	198.15	192.25	184.90	179.30	176.10	172.05
EIFS on Metal Studs	Steel Roof Deck	245.10	221.75	200.50	189.55	184.80	179.15	174.85	172.35	169.25
Tiltup Concrete Panel	Steel Roof Deck	252.05	226.80	203.90	192.05	187.00	180.90	176.30	173.50	170.20
Perimeter Adj., Add or Deduct	Per 100 L.F.	28.05	18.70	11.20	8.05	6.30	4.70	3.80	2.80	2.25
Story Hgt. Adj., Add or Deduct	Per 1 Ft.	3.40	2.70	1.95	1.65	1.50	1.30	1.10	1.00	0.85

For Basement, add $36.90 per square foot of basement area

The above costs were calculated using the basic specifications shown on the facing page. These costs should be adjusted where necessary for design alternatives and owner's requirements. Reported completed project costs, for this type of structure, range from $78.95 to $305.20 per S.F.

Common additives

Description	Unit	$ Cost
Closed Circuit Surveillance, one station, camera & monitor	Ea.	2025
For additional camera stations, add	Ea.	1100
Directory boards, plastic, glass covered, 30" x 20"	Ea.	610
36" x 48"	Ea.	1525
Aluminum, 36" x 24"	Ea.	765
48" x 32"	Ea.	1075
Electronic, wall mounted	S.F.	2900
Pedestal access flr sys w/PLAM cover, comp rm,		
Less than 6000 SF	S.F.	21
Greater than 6000 SF	S.F.	20.50
Office, greater than 6000 S.F.	S.F.	16.95
Uninterruptible power supply, 15 kVA/12.75 kW	kW	1890

Description	Unit	$ Cost
Commissioning Fees, sustainable commercial construction	S.F.	0.24 - 3.04
Energy Modelling Fees, commercial buildings to 10,000 SF	Ea.	11,000
Greater than 10,000 SF add	S.F.	0.04
Green Bldg Cert Fees for comm construction project reg	Project	900
Photovoltaic Pwr Sys, grid connected, 20 kW (~2400 SF), roof	Ea.	245,600
Green Roofs, 6" soil depth, w/treated wd edging & sedum mats	S.F.	11.76
10" Soil depth, with treated wood edging & sedum mats	S.F.	13.45
Greywater Recovery Systems, prepackaged comm, 3060 gal.	Ea.	58,265
4590 gal.	Ea.	73,075
Rainwater Harvest Sys, prepckged comm, 10,000 gal, sys contrller	Ea.	31,525
20,000 gal. w/system controller	Ea.	50,575
30,000 gal. w/system controller	Ea.	80,825
Solar Domestic HW, closed loop, add-on sys, ext heat exchanger	Ea.	10,850
Drainback, hot water system, 120 gal tank	Ea.	13,525
Draindown, hot water system, 120 gal tank	Ea.	13,850

Important: See the Reference Section for Location Factors.

Model costs calculated for a 1 story building with 12' story height and 7,000 square feet of floor area

				Unit	Unit Cost	Cost Per S.F.	% Of Sub-Total
A.	**SUBSTRUCTURE**						
1010	Standard Foundations	Poured concrete; strip and spread footings		S.F. Ground	2.93	2.93	
1020	Special Foundations	N/A		—	—	—	
1030	Slab on Grade	4" reinforced concrete with recycled vapor barrier and granular base		S.F. Slab	5.58	5.58	9.8%
2010	Basement Excavation	Site preparation for slab and trench for foundation wall and footing		S.F. Ground	.32	.32	
2020	Basement Walls	4' foundation wall		L.F. Wall	177	5.03	
B.	**SHELL**						
	B10 Superstructure						
1010	Floor Construction	N/A		—	—	—	7.4 %
1020	Roof Construction	Steel joists, girders & deck on columns		S.F. Roof	10.42	10.42	
	B20 Exterior Enclosure						
2010	Exterior Walls	E.I.F.S. on metal studs		S.F. Wall	18.35	9.06	
2020	Exterior Windows	Aluminum outward projecting	20% of wall	Each	746	4	11.4%
2030	Exterior Doors	Aluminum and glass, hollow metal		Each	3553	3.05	
	B30 Roofing						
3010	Roof Coverings	Single-ply TPO membrane, 45 mils, heat welded seams w/R-20 insul.		S.F. Roof	10.13	10.13	7.3%
3020	Roof Openings	Roof hatch		S.F. Roof	.24	.24	
C.	**INTERIORS**						
1010	Partitions	Gyp. bd. on mtl. studs w/sound attenuation	20 SF Flr/LF Part.	S.F. Partition	10.60	5.30	
1020	Interior Doors	Single leaf hollow metal, low VOC paint	200 S.F. Floor/Door	Each	1168	5.85	
1030	Fittings	Toilet partitions		S.F. Floor	.87	.87	
2010	Stair Construction	N/A		—	—	—	18.1%
3010	Wall Finishes	60% vinyl wall covering, 40% paint, low VOC		S.F. Surface	1.60	1.60	
3020	Floor Finishes	60% carpet tile, 30% vinyl composition tile, recycled content, 10% ceramic tile		S.F. Floor	4.87	4.87	
3030	Ceiling Finishes	Mineral fiber tile on concealed zee bars		S.F. Ceiling	7.22	7.22	
D.	**SERVICES**						
	D10 Conveying						
1010	Elevators & Lifts	N/A		—	—	—	0.0 %
1020	Escalators & Moving Walks	N/A		—	—	—	
	D20 Plumbing						
2010	Plumbing Fixtures	Toilet, low flow, auto sensor, & service fixt., supply & drain.	1 Fixt./1320 SF Flr.	Each	9412	7.13	
2020	Domestic Water Distribution	Tankless, on-demand water heater, gas/propane		S.F. Floor	.91	.91	6.3%
2040	Rain Water Drainage	Roof drains		S.F. Roof	.87	.87	
	D30 HVAC						
3010	Energy Supply	N/A		—	—	—	
3020	Heat Generating Systems	Included in D3050		—	—	—	
3040	Distribution Systems	Enthalpy heat recovery packages		Each	17,575	2.51	16.1 %
3050	Terminal & Package Units	Multizone rooftop air conditioner, SEER 14		S.F. Floor	20.25	20.25	
3090	Other HVAC Sys. & Equipment	N/A		—	—	—	
	D40 Fire Protection						
4010	Sprinklers	Sprinkler system, light hazard		S.F. Floor	3.65	3.65	3.7%
4020	Standpipes	Standpipes and hose systems		S.F. Floor	1.57	1.57	
	D50 Electrical						
5010	Electrical Service/Distribution	400 ampere service, panel board and feeders		S.F. Floor	2.04	2.04	
5020	Lighting & Branch Wiring	LED fixtures, daylt. dim., ltg. on/off, recept., switches, and A.C. power		S.F. Floor	15.66	15.66	19.4%
5030	Communications & Security	Addressable alarm systems, internet and phone wiring, and emergency lighting		S.F. Floor	7.30	7.30	
5090	Other Electrical Systems	Emergency generator, 7.5 kW, energy monitoring systems		S.F. Floor	2.47	2.47	
E.	**EQUIPMENT & FURNISHINGS**						
1010	Commercial Equipment	N/A		—	—	—	
1020	Institutional Equipment	N/A		—	—	—	0.6 %
1090	Other Equipment	Waste handling recycling tilt truck		Each	.81	.81	
2020	Moveable Furnishings	No smoking signage		Each	.06	.06	
F.	**SPECIAL CONSTRUCTION**						
1020	Integrated Construction	N/A		—	—	—	0.0 %
1040	Special Facilities	N/A		—	—	—	
G.	**BUILDING SITEWORK**	**N/A**					

			Sub-Total	141.70	100%
CONTRACTOR FEES (General Requirements: 10%, Overhead: 5%, Profit: 10%)			25%	35.45	
ARCHITECT FEES			7%	12.40	
		Total Building Cost		**189.55**	

For customer support on your Square Foot Cost Data, call 877.756.2789.

261

Costs per square foot of floor area

Exterior Wall	S.F. Area	5000	8000	12000	16000	20000	35000	50000	65000	80000
	L.F. Perimeter	220	260	310	330	360	440	550	600	675
Face Brick with Concrete Block Back-up	Wood Joists	272.85	238.20	218.30	204.15	196.85	182.10	177.65	173.05	171.00
	Steel Joists	273.30	238.70	218.75	204.60	197.30	182.50	178.05	173.50	171.40
Glass and Metal Curtain Wall	Steel Frame	326.05	279.90	253.30	233.95	224.00	203.80	197.70	191.40	188.45
	R/Conc. Frame	322.55	276.90	250.65	231.35	221.50	201.45	195.45	189.10	186.20
Wood Siding	Wood Frame	218.50	194.05	180.05	170.50	165.45	155.55	152.40	149.45	148.00
Brick Veneer	Wood Frame	234.85	206.15	189.65	178.15	172.20	160.20	156.45	152.85	151.15
Perimeter Adj., Add or Deduct	Per 100 L.F.	44.90	28.10	18.75	14.05	11.25	6.40	4.45	3.50	2.80
Story Hgt. Adj., Add or Deduct	Per 1 Ft.	7.35	5.45	4.35	3.50	3.00	2.10	1.75	1.55	1.35

For Basement, add $41.30 per square foot of basement area

The above costs were calculated using the basic specifications shown on the facing page. These costs should be adjusted where necessary for design alternatives and owner's requirements. Reported completed project costs, for this type of structure, range from $82.60 to $320.05 per S.F.

Common additives

Description	Unit	$ Cost
Closed Circuit Surveillance, one station, camera and monitor	Ea.	2025
For additional camera stations, add	Ea.	1100
Directory Boards, plastic, glass covered, 30" x 20"	Ea.	610
36" x 48"	Ea.	1525
Aluminum, 36" x 24"	Ea.	765
48" x 32"	Ea.	1075
Electronic, wall mounted	S.F.	2900
Escalators, 10' rise, 32" wide, glass balustrade	Ea.	147,500
48" wide, glass balustrade	Ea.	155,000
Pedestal access flr sys w/PLAM cover, comp rm,		
Less than 6000 SF	S.F.	21
Greater than 6000 SF	S.F.	20.50
Office, greater than 6000 S.F.	S.F.	16.95
Uninterruptible power supply, 15 kVA/12.75 kW	kW	1890

Description	Unit	$ Cost
Commissioning Fees, sustainable commercial construction	S.F.	0.24 - 3.04
Energy Modelling Fees, commercial buildings to 10,000 SF	Ea.	11,000
Greater than 10,000 SF add	S.F.	0.04
Green Bldg Cert Fees for comm construction project reg	Project	900
Photovoltaic Pwr Sys, grid connected, 20 kW (~2400 SF), roof	Ea.	245,600
Green Roofs, 6" soil depth, w/treated wd edging & sedum mats	S.F.	11.76
10" Soil depth, with treated wood edging & sedum mats	S.F.	13.45
Greywater Recovery Systems, prepackaged comm, 3060 gal.	Ea.	58,265
4590 gal.	Ea.	73,075
Rainwater Harvest Sys, prepckged comm, 10,000 gal, sys contrller	Ea.	31,525
20,000 gal. w/system controller	Ea.	50,575
30,000 gal. w/system controller	Ea.	80,825
Solar Domestic HW, closed loop, add-on sys, ext heat exchanger	Ea.	10,850
Drainback, hot water system, 120 gal tank	Ea.	13,525
Draindown, hot water system, 120 gal tank	Ea.	13,850

Important: See the Reference Section for Location Factors.

Model costs calculated for a 3 story building with 12' story height and 20,000 square feet of floor area

			Unit	Unit Cost	Cost Per S.F.	% Of Sub-Total
A.	**SUBSTRUCTURE**					
1010	Standard Foundations	Poured concrete; strip and spread footings	S.F. Ground	7.86	2.62	
1020	Special Foundations	N/A	—	—	—	
1030	Slab on Grade	4" reinforced concrete with recycled vapor barrier and granular base	S.F. Slab	5.58	1.86	4.4%
2010	Basement Excavation	Site preparation for slab and trench for foundation wall and footing	S.F. Ground	.18	.06	
2020	Basement Walls	4' foundation wall	L.F. Wall	177	1.96	
B.	**SHELL**					
	B10 Superstructure					
1010	Floor Construction	Open web steel joists, slab form, concrete, columns	S.F. Floor	19.56	13.04	10.5%
1020	Roof Construction	Metal deck, open web steel joists, columns	S.F. Roof	7.41	2.47	
	B20 Exterior Enclosure					
2010	Exterior Walls	Face brick with concrete block backup 80% of wall	S.F. Wall	38.81	20.12	
2020	Exterior Windows	Aluminum outward projecting 20% of wall	Each	746	4.21	17.2%
2030	Exterior Doors	Aluminum and glass, hollow metal	Each	3553	1.07	
	B30 Roofing					
3010	Roof Coverings	Single-ply, TPO membrane, 60 mils, heat welded seams w/R-20 insul.	S.F. Roof	7.41	2.47	1.7%
3020	Roof Openings	N/A	—	—	—	
C.	**INTERIORS**					
1010	Partitions	Gypsum board on metal studs 20 SF Flr/LF Part.	S.F. Partition	10.95	4.38	
1020	Interior Doors	Single leaf hollow metal, low VOC paint 200 S.F. Floor/Door	Each	1168	5.85	
1030	Fittings	Toilet partitions	S.F. Floor	1.83	1.83	
2010	Stair Construction	Concrete filled metal pan	Flight	14,450	5.06	20.7%
3010	Wall Finishes	60% vinyl wall covering, 40% paint, low VOC	S.F. Surface	1.60	1.28	
3020	Floor Finishes	60% carpet tile, 30% vinyl composition tile, recycled content, 10% ceramic tile	S.F. Floor	4.87	4.87	
3030	Ceiling Finishes	Mineral fiber tile on concealed zee bars	S.F. Ceiling	7.22	7.22	
D.	**SERVICES**					
	D10 Conveying					
1010	Elevators & Lifts	Two hydraulic passenger elevators	Each	120,000	12	8.1%
1020	Escalators & Moving Walks	N/A	—	—	—	
	D20 Plumbing					
2010	Plumbing Fixtures	Toilet, low flow, auto sensor, & service fixt., supply & drain. 1 Fixt./1320 SF Flr.	Each	5518	4.18	
2020	Domestic Water Distribution	Gas fired, tankless water heater	S.F. Floor	.20	.20	3.5%
2040	Rain Water Drainage	Roof drains	S.F. Roof	2.19	.73	
	D30 HVAC					
3010	Energy Supply	N/A	—	—	—	
3020	Heat Generating Systems	Included in D3050	—	—	—	
3040	Distribution Systems	Enthalpy heat recovery packages	Each	17,575	2.64	12.9 %
3050	Terminal & Package Units	Multizone rooftop air conditioner. SEER 14	S.F. Floor	16.35	16.35	
3090	Other HVAC Sys. & Equipment	N/A	—	—	—	
	D40 Fire Protection					
4010	Sprinklers	Wet pipe sprinkler system	S.F. Floor	3.86	3.86	3.3%
4020	Standpipes	Standpipes and hose systems	S.F. Floor	1.02	1.02	
	D50 Electrical					
5010	Electrical Service/Distribution	1000 ampere service, panel board and feeders	S.F. Floor	2.87	2.87	
5020	Lighting & Branch Wiring	LED fixtures, daylt. dim., ltg. on/off, recept., switches, and A.C. power	S.F. Floor	15.37	15.37	17.5%
5030	Communications & Security	Addressable alarm systems, internet and phone wiring, & emergency ltg.	S.F. Floor	6.09	6.09	
5090	Other Electrical Systems	Emergency generator, 7.5 kW, UPS, energy monitoring systems	S.F. Floor	1.52	1.52	
E.	**EQUIPMENT & FURNISHINGS**					
1010	Commercial Equipment	N/A	—	—	—	
1020	Institutional Equipment	N/A	—	—	—	
1090	Other Equipment	Waste handling recycling tilt truck	Each	.29	.29	0.2 %
2020	Moveable Furnishings	No smoking signage	Each	.02	.02	
F.	**SPECIAL CONSTRUCTION**					
1020	Integrated Construction	N/A	—	—	—	0.0%
1040	Special Facilities	N/A	—	—	—	
G.	**BUILDING SITEWORK**	**N/A**				

		Sub-Total	147.51	**100%**
CONTRACTOR FEES (General Requirements: 10%, Overhead: 5%, Profit: 10%)		25%	36.88	
ARCHITECT FEES		7%	12.91	
		Total Building Cost	197.30	

Costs per square foot of floor area

Exterior Wall	S.F. Area	20000	40000	60000	80000	100000	150000	200000	250000	300000
	L.F. Perimeter	260	360	400	420	460	520	650	720	800
Precast Concrete Panel	Steel Frame	245.95	212.40	194.45	183.80	178.70	170.20	168.25	165.45	163.80
	R/Conc. Frame	244.60	210.85	192.90	182.20	177.05	168.50	166.60	163.75	162.15
Face Brick with Concrete Block Back-up	Steel Frame	232.90	203.50	187.95	178.80	174.40	167.10	165.35	163.00	161.55
	R/Conc. Frame	230.60	201.15	185.70	176.50	172.10	164.80	163.05	160.65	159.25
Limestone Panel Concrete Block Back-up	Steel Frame	264.25	225.15	204.05	191.50	185.45	175.40	173.20	169.90	167.95
	R/Conc. Frame	261.95	222.90	201.80	189.15	183.20	173.15	170.95	167.65	165.70
Perimeter Adj., Add or Deduct	Per 100 L.F.	34.05	17.05	11.35	8.45	6.85	4.50	3.45	2.70	2.30
Story Hgt. Adj., Add or Deduct	Per 1 Ft.	7.10	4.95	3.60	2.85	2.50	1.85	1.80	1.55	1.45

For Basement, add $44.50 per square foot of basement area

The above costs were calculated using the basic specifications shown on the facing page. These costs should be adjusted where necessary for design alternatives and owner's requirements. Reported completed project costs, for this type of structure, range from $92.95 to $273.35 per S.F.

Common additives

Description	Unit	$ Cost
Closed Circuit Surveillance, one station, camera and monitor	Ea.	2025
For additional camera stations, add	Ea.	1100
Directory Boards, plastic, glass covered, 30" x 20"	Ea.	610
36" x 48"	Ea.	1525
Aluminum, 36" x 24"	Ea.	765
48" x 32"	Ea.	1075
Electronic, wall mounted	S.F.	2900
Escalators, 10' rise, 32" wide, glass balustrade	Ea.	147,500
48" wide, glass balustrade	Ea.	155,000
Pedestal access flr sys w/PLAM cover, comp rm,		
Less than 6000 SF	S.F.	21
Greater than 6000 SF	S.F.	20.50
Office, greater than 6000 S.F.	S.F.	16.95
Uninterruptible power supply, 15 kVA/12.75 kW	kW	1890

Description	Unit	$ Cost
Commissioning Fees, sustainable commercial construction	S.F.	0.24 - 3.04
Energy Modelling Fees, commercial buildings to 10,000 SF	Ea.	11,000
Greater than 10,000 SF add	S.F.	0.04
Green Bldg Cert Fees for comm construction project reg	Project	900
Photovoltaic Pwr Sys, grid connected, 20 kW (~2400 SF), roof	Ea.	245,600
Green Roofs, 6" soil depth, w/treated wd edging & sedum mats	S.F.	11.76
10" Soil depth, with treated wood edging & sedum mats	S.F.	13.45
Greywater Recovery Systems, prepackaged comm, 3060 gal.	Ea.	58,265
4590 gal.	Ea.	73,075
Rainwater Harvest Sys, prepckged comm, 10,000 gal, sys contrller	Ea.	31,525
20,000 gal. w/system controller	Ea.	50,575
30,000 gal. w/system controller	Ea.	80,825
Solar Domestic HW, closed loop, add-on sys, ext heat exchanger	Ea.	10,850
Drainback, hot water system, 120 gal tank	Ea.	13,525
Draindown, hot water system, 120 gal tank	Ea.	13,850

Model costs calculated for a 8 story building with 12' story height and 80,000 square feet of floor area

				Unit	Unit Cost	Cost Per S.F.	% Of Sub-Total
A. SUBSTRUCTURE							
1010	Standard Foundations	Poured concrete; strip and spread footings		S.F. Ground	12.72	1.59	
1020	Special Foundations	N/A		—	—	—	
1030	Slab on Grade	4" reinforced concrete with recycled vapor barrier and granular base		S.F. Slab	5.58	.70	2.1%
2010	Basement Excavation	Site preparation for slab and trench for foundation wall and footing		S.F. Ground	.32	.04	
2020	Basement Walls	4' foundation wall		L.F. Wall	93	.62	
B. SHELL							
	B10 Superstructure						
1010	Floor Construction	Concrete slab with metal deck and beams		S.F. Floor	25.38	22.21	16.7%
1020	Roof Construction	Metal deck, open web steel joists, interior columns		S.F. Roof	7.20	.90	
	B20 Exterior Enclosure						
2010	Exterior Walls	Precast concrete panels	80% of wall	S.F. Wall	49.40	19.92	
2020	Exterior Windows	Vertical pivoted steel	20% of wall	Each	585	3.94	17.4%
2030	Exterior Doors	Double aluminum and glass doors and entrance with transoms		Each	4279	.27	
	B30 Roofing						
3010	Roof Coverings	Single-ply TPO membrane, 60 mils, welded seams w/R-20 insul.		S.F. Roof	5.92	.74	0.5%
3020	Roof Openings	N/A		—	—	—	
C. INTERIORS							
1010	Partitions	Gypsum board on metal studs w/sound attenuation	30 SF Flr./LF Part.	S.F. Partition	11.70	3.90	
1020	Interior Doors	Single leaf hollow metal, low VOC paint	400 S.F. Floor/Door	Each	1168	2.92	
1030	Fittings	Toilet Partitions		S.F. Floor	1.22	1.22	
2010	Stair Construction	Concrete filled metal pan		Flight	14,450	3.07	17.5%
3010	Wall Finishes	60% vinyl wall covering, 40% paint, low VOC paint		S.F. Surface	1.61	1.07	
3020	Floor Finishes	60% carpet tile, 30% vinyl composition tile, recycled content, 10% ceramic tile		S.F. Floor	4.87	4.87	
3030	Ceiling Finishes	Mineral fiber tile on concealed zee bars		S.F. Ceiling	7.22	7.22	
D. SERVICES							
	D10 Conveying						
1010	Elevators & Lifts	Four geared passenger elevators		Each	303,200	15.16	10.9%
1020	Escalators & Moving Walks	N/A		—	—	—	
	D20 Plumbing						
2010	Plumbing Fixtures	Toilet, low flow, auto sensor, & service fixt., supply & drain.	1 Fixt./1370 SF Flr.	Each	4042	2.95	
2020	Domestic Water Distribution	Gas fired tankless water heater		S.F. Floor	.39	.39	2.6%
2040	Rain Water Drainage	Roof drains		S.F. Roof	2.48	.31	
	D30 HVAC						
3010	Energy Supply	N/A		—	—	—	
3020	Heat Generating Systems	Included in D3050		—	—	—	
3040	Distribution Systems	Enthalpy heat recovery packages		Each	37,000	1.85	13.1%
3050	Terminal & Package Units	Multizone rooftop air conditioner, SEER 14		S.F. Floor	16.35	16.35	
3090	Other HVAC Sys. & Equipment	N/A		—	—	—	
	D40 Fire Protection						
4010	Sprinklers	Wet pipe sprinkler system		S.F. Floor	2.88	2.88	2.9%
4020	Standpipes	Standpipes and hose systems		S.F. Floor	1.15	1.15	
	D50 Electrical						
5010	Electrical Service/Distribution	1600 ampere service, panel board and feeders		S.F. Floor	1.06	1.06	
5020	Lighting & Branch Wiring	LED fixtures, daylt. dim., ltg. on/off, recept., switches, and A.C. power		S.F. Floor	15.08	15.08	16.2%
5030	Communications & Security	Addressable alarm systems, internet and phone wiring, emergency lighting		S.F. Floor	4.52	4.52	
5090	Other Electrical Systems	Emergency generator, 100 kW, UPS, energy monitoring systems		S.F. Floor	1.77	1.77	
E. EQUIPMENT & FURNISHINGS							
1010	Commercial Equipment	N/A		—	—	—	
1020	Institutional Equipment	N/A		—	—	—	0.0%
1090	Other Equipment	Waste handling recycling tilt truck		Each	.04	.04	
2020	Moveable Furnishings	No smoking signage		Each			
F. SPECIAL CONSTRUCTION							
1020	Integrated Construction	N/A		—	—	—	0.0%
1040	Special Facilities	N/A		—	—	—	
G. BUILDING SITEWORK	**N/A**						

				Sub-Total	138.71	**100%**
CONTRACTOR FEES (General Requirements: 10%, Overhead: 5%, Profit: 10%)				25%	34.69	
ARCHITECT FEES				6%	10.40	

Total Building Cost	**183.80**

For customer support on your Square Foot Cost Data, call 877.756.2789.

265

Costs per square foot of floor area

Exterior Wall	S.F. Area	120000	145000	170000	200000	230000	260000	400000	600000	800000
	L.F. Perimeter	420	450	470	490	510	530	650	800	900
Double Glazed Heat Absorbing Tinted Plate Glass Panels	Steel Frame	220.45	210.95	203.05	195.85	190.55	186.45	176.70	169.90	165.40
	R/Conc. Frame	210.60	201.30	193.55	186.45	181.30	177.15	167.60	161.00	156.50
Face Brick with Concrete Block Back-up	Steel Frame	187.90	182.05	177.30	173.00	169.90	167.40	161.45	157.45	154.75
	R/Conc. Frame	215.90	210.20	205.65	201.45	198.40	195.95	190.20	186.30	183.70
Precast Concrete Panel With Exposed Aggregate	Steel Frame	192.40	186.10	180.95	176.25	172.75	170.05	163.60	159.15	156.25
	R/Conc. Frame	182.75	176.60	171.55	166.95	163.60	160.95	154.70	150.35	147.50
Perimeter Adj., Add or Deduct	Per 100 L.F.	16.25	13.45	11.50	9.75	8.50	7.45	4.85	3.30	2.40
Story Hgt. Adj., Add or Deduct	Per 1 Ft.	6.65	5.90	5.25	4.70	4.20	3.80	3.05	2.55	2.10
For Basement, add $ 44.50 per square foot of basement area										

The above costs were calculated using the basic specifications shown on the facing page. These costs should be adjusted where necessary for design alternatives and owner's requirements. Reported completed project costs, for this type of structure, range from $116.60 to $284.55 per S.F.

Common additives

Description	Unit	$ Cost
Closed Circuit Surveillance, one station, camera and monitor	Ea.	2025
For additional camera stations, add	Ea.	1100
Directory Boards, plastic, glass covered, 30" x 20"	Ea.	610
36" x 48"	Ea.	1525
Aluminum, 36" x 24"	Ea.	765
48" x 32"	Ea.	1075
Electronic, wall mounted	S.F.	2900
Escalators, 10' rise, 32" wide, glass balustrade	Ea.	147,500
48" wide, glass balustrade	Ea.	155,000
Pedestal access flr sys w/PLAM cover, comp rm,		
Less than 6000 SF	S.F.	21
Greater than 6000 SF	S.F.	20.50
Office, greater than 6000 S.F.	S.F.	16.95
Uninterruptible power supply, 15 kVA/12.75 kW	kW	1890

Description	Unit	$ Cost
Commissioning Fees, sustainable commercial construction	S.F.	0.24 - 3.04
Energy Modelling Fees, commercial buildings to 10,000 SF	Ea.	11,000
Greater than 10,000 SF add	S.F.	0.04
Green Bldg Cert Fees for comm construction project reg	Project	900
Photovoltaic Pwr Sys, grid connected, 20 kW (~2400 SF), roof	Ea.	245,600
Green Roofs, 6" soil depth, w/treated wd edging & sedum mats	S.F.	11.76
10" Soil depth, with treated wood edging & sedum mats	S.F.	13.45
Greywater Recovery Systems, prepackaged comm, 3060 gal.	Ea.	58,265
4590 gal.	Ea.	73,075
Rainwater Harvest Sys, prepckged comm, 10,000 gal, sys contrller	Ea.	31,525
20,000 gal. w/system controller	Ea.	50,575
30,000 gal. w/system controller	Ea.	80,825
Solar Domestic HW, closed loop, add-on sys, ext heat exchanger	Ea.	10,850
Drainback, hot water system, 120 gal tank	Ea.	13,525
Draindown, hot water system, 120 gal tank	Ea.	13,850

Important: See the Reference Section for Location Factors.

Model costs calculated for a 16 story building with 10' story height and 260,000 square feet of floor area

				Unit	Unit Cost	Cost Per S.F.	% Of Sub-Total
A. SUBSTRUCTURE							
1010	Standard Foundations	CIP concrete pile caps		S.F. Ground	10.40	.65	
1020	Special Foundations	Steel H-piles, concrete grade beams		S.F. Ground	63	3.92	
1030	Slab on Grade	4" reinforced concrete with recycled vapor barrier and granular base		S.F. Slab	5.58	.35	3.8%
2010	Basement Excavation	Site preparation for slab, piles and grade beams		S.F. Ground	.32	.02	
2020	Basement Walls	4' foundation wall		L.F. Wall	93	.45	
B. SHELL							
B10 Superstructure							
1010	Floor Construction	Concrete slab, metal deck, beams		S.F. Floor	28.96	27.15	19.7%
1020	Roof Construction	Metal deck, open web steel joists, beams, columns		S.F. Roof	8.64	.54	
B20 Exterior Enclosure							
2010	Exterior Walls	N/A		—	—	—	
2020	Exterior Windows	Double glazed heat absorbing, tinted plate glass wall panels	100% of wall	Each	85	27.69	20.2 %
2030	Exterior Doors	Double aluminum & glass doors		Each	6521	.71	
B30 Roofing							
3010	Roof Coverings	Single ply membrane, fully adhered; perlite/EPS R-20 insulation		S.F. Roof	6.72	.42	0.3%
3020	Roof Openings	N/A		—	—	—	
C. INTERIORS							
1010	Partitions	Gypsum board on metal studs, sound attenuation	30 SF Flr./LF Part.	S.F. Partition	13.28	3.54	
1020	Interior Doors	Single leaf hollow metal, low VOC paint	400 S.F. Floor/Door	Each	1168	2.92	
1030	Fittings	Toilet partitions		S.F. Floor	.71	.71	
2010	Stair Construction	Concrete filled metal pan		Flight	17,275	2.33	16.2%
3010	Wall Finishes	60% vinyl wall covering, 40% paint, low VOC		S.F. Surface	1.61	.86	
3020	Floor Finishes	60% carpet tile, 30% vinyl composition tile, recycled content, 10% ceramic tile		S.F. Floor	5.24	5.24	
3030	Ceiling Finishes	Mineral fiber tile on concealed zee bars		S.F. Ceiling	7.22	7.22	
D. SERVICES							
D10 Conveying							
1010	Elevators & Lifts	Four geared passenger elevators		Each	475,800	7.32	5.2%
1020	Escalators & Moving Walks	N/A		—	—	—	
D20 Plumbing							
2010	Plumbing Fixtures	Toilet, low flow, auto sensor, & service fixt., supply & drain.	1 Fixt./1345 SF Flr.	Each	6577	4.89	
2020	Domestic Water Distribution	Gas fired, tankless water heater		S.F. Floor	.18	.18	3.7%
2040	Rain Water Drainage	Roof drains		S.F. Roof	3.20	.20	
D30 HVAC							
3010	Energy Supply	N/A		—	—	—	
3020	Heat Generating Systems	N/A		—	—	—	
3040	Distribution Systems	Enthalpy heat recovery packages		Each	37,000	1.57	12.7 %
3050	Terminal & Package Units	Multizone rooftop air conditioner		S.F. Floor	16.35	16.35	
3090	Other HVAC Sys. & Equipment	N/A		—	—	—	
D40 Fire Protection							
4010	Sprinklers	Sprinkler system, light hazard		S.F. Floor	2.78	2.78	2.4%
4020	Standpipes	Standpipes and hose systems		S.F. Floor	.59	.59	
D50 Electrical							
5010	Electrical Service/Distribution	2400 ampere service, panel board and feeders		S.F. Floor	.70	.70	
5020	Lighting & Branch Wiring	LED fixtures, daylt. dim., ltg. on/off, recept., switches, and A.C. power		S.F. Floor	14.96	14.96	15.7%
5030	Communications & Security	Addressable alarm systems, internet & phone wiring, emergency ltg.		S.F. Floor	5.43	5.43	
5090	Other Electrical Systems	Emergency generator, 200 kW, UPS, energy monitoring systems		S.F. Floor	.99	.99	
E. EQUIPMENT & FURNISHINGS							
1010	Commercial Equipment	N/A		—	—	—	
1020	Institutional Equipment	N/A		—	—	—	
1090	Other Equipment	Waste handling recycling tilt truck		Each	.01	.01	0.0 %
2020	Moveable Furnishings	No smoking signage		Each			
F. SPECIAL CONSTRUCTION							
1020	Integrated Construction	N/A		—	—	—	0.0 %
1040	Special Facilities	N/A		—	—	—	
G. BUILDING SITEWORK	**N/A**						

		Sub-Total	140.69	100%
CONTRACTOR FEES (General Requirements: 10%, Overhead: 5%, Profit: 10%)		25%	35.21	
ARCHITECT FEES		6%	10.55	

Total Building Cost   186.45

For customer support on your Square Foot Cost Data, call 877.756.2789.

267

Costs per square foot of floor area

Exterior Wall	S.F. Area	7000	9000	11000	13000	15000	17000	19000	21000	23000
	L.F. Perimeter	240	280	303	325	354	372	397	422	447
Limestone with Concrete Block Back-up	Bearing Walls	292.65	276.30	262.10	252.00	245.80	239.50	235.40	232.10	229.35
	R/Conc. Frame	293.35	277.65	264.40	254.90	249.10	243.15	239.30	236.25	233.60
Face Brick with Concrete Block Back-up	Bearing Walls	249.70	235.95	224.85	216.95	212.00	207.20	203.90	201.35	199.15
	R/Conc. Frame	272.10	258.40	247.30	239.40	234.45	229.60	226.35	223.75	221.55
Decorative Concrete Block	Bearing Walls	236.50	224.00	214.25	207.30	202.90	198.70	195.85	193.55	191.65
	R/Conc. Frame	260.15	247.65	237.90	231.00	226.60	222.40	219.50	217.20	215.30
Perimeter Adj., Add or Deduct	Per 100 L.F.	35.40	27.45	22.50	19.00	16.55	14.55	13.00	11.80	10.75
Story Hgt. Adj., Add or Deduct	Per 1 Ft.	6.25	5.65	5.00	4.55	4.35	3.95	3.80	3.65	3.55

For Basement, add $35.00 per square foot of basement area

The above costs were calculated using the basic specifications shown on the facing page. These costs should be adjusted where necessary for design alternatives and owner's requirements. Reported completed project costs, for this type of structure, range from $134.00 to $349.50 per S.F.

Common additives

Description	Unit	$ Cost
Cells Prefabricated, 5'-6" x 7'-8", 7'-8' high	Ea.	12,900
Elevators, hydraulic passenger, 2 stops, 2500# capacity	Ea.	70,400
3500# capacity	Ea.	75,400
Lockers, Steel, Single tier, 60" to 72"	Opng.	231 - 380
Locker bench, lam. maple top only	L.F.	34.50
Pedestals, steel pipe	Ea.	76.50
Safe, Office type, 1 hr rating, 60"x36"x18", double door	Ea.	9350
Shooting Range, incl. bullet traps,		
Target provisions & controls, excl. struct shell	Ea.	52,500
Sound System, amplifier, 250 watts	Ea.	2200
Speaker, ceiling or wall	Ea.	217
Trumpet	Ea.	415

Description	Unit	$ Cost
Commissioning Fees, sustainable commercial construction	S.F.	0.24 - 3.04
Energy Modelling Fees, commercial buildings to 10,000 SF	Ea.	11,000
Greater than 10,000 SF add	S.F.	0.04
Green Bldg Cert Fees for comm construction project reg	Project	900
Photovoltaic Pwr Sys, grid connected, 20 kW (~2400 SF), roof	Ea.	245,600
Green Roofs, 6" soil depth, w/treated wd edging & sedum mats	S.F.	11.76
10" Soil depth, with treated wood edging & sedum mats	S.F.	13.45
Greywater Recovery Systems, prepackaged comm, 3060 gal.	Ea.	58,265
4590 gal.	Ea.	73,075
Rainwater Harvest Sys, prepckged comm, 10,000 gal, sys contrller	Ea.	31,525
20,000 gal. w/system controller	Ea.	50,575
30,000 gal. w/system controller	Ea.	80,825
Solar Domestic HW, closed loop, add-on sys, ext heat exchanger	Ea.	10,850
Drainback, hot water system, 120 gal tank	Ea.	13,525
Draindown, hot water system, 120 gal tank	Ea.	13,850

Important: See the Reference Section for Location Factors.

Model costs calculated for a 2 story building with 12' story height and 11,000 square feet of floor area

				Unit	Unit Cost	Cost Per S.F.	% Of Sub-Total
A. SUBSTRUCTURE							
1010	Standard Foundations	Poured concrete; strip and spread footings		S.F. Ground	3.02	1.51	
1020	Special Foundations	N/A		—	—	—	
1030	Slab on Grade	4" reinforced concrete with recycled vapor barrier and granular base		S.F. Slab	5.58	2.79	4.1%
2010	Basement Excavation	Site preparation for slab and trench for foundation wall and footing		S.F. Ground	.32	.16	
2020	Basement Walls	4' foundation wall		L.F. Wall	93	3.41	
B. SHELL							
B10 Superstructure							
1010	Floor Construction	Open web steel joists, slab form, concrete		S.F. Floor	12.56	6.28	4.6%
1020	Roof Construction	Metal deck on open web steel joists		S.F. Roof	4.96	2.48	
B20 Exterior Enclosure							
2010	Exterior Walls	Limestone with concrete block backup	80% of wall	S.F. Wall	63	33.06	
2020	Exterior Windows	Metal horizontal sliding	20% of wall	Each	1167	10.28	23.6%
2030	Exterior Doors	Hollow metal		Each	2826	2.04	
B30 Roofing							
3010	Roof Coverings	Single-ply TPO membrane, 60 mils, heat welded w/R-20 insul.		S.F. Roof	6.94	3.47	1.8%
3020	Roof Openings	N/A		—	—	—	
C. INTERIORS							
1010	Partitions	Concrete block w/foamed-in insul.	20 SF Flr./LF Part.	S.F. Partition	11.38	5.69	
1020	Interior Doors	Single leaf kalamein fire door, low VOC paint	200 S.F. Floor/Door	Each	1168	5.85	
1030	Fittings	Toilet partitions		S.F. Floor	.98	.98	
2010	Stair Construction	Concrete filled metal pan		Flight	17,275	3.14	15.7%
3010	Wall Finishes	90% paint, low VOC, 10% ceramic tile		S.F. Surface	2.79	2.79	
3020	Floor Finishes	70% vinyl composition tile, recycled content, 20% carpet tile, 10% ceramic tile		S.F. Floor	4.61	4.61	
3030	Ceiling Finishes	Mineral fiber tile on concealed zee bars		S.F. Ceiling	7.22	7.22	
D. SERVICES							
D10 Conveying							
1010	Elevators & Lifts	One hydraulic passenger elevator		Each	85,360	7.76	4.0%
1020	Escalators & Moving Walks	N/A		—	—	—	
D20 Plumbing							
2010	Plumbing Fixtures	Toilet, low flow, auto sensor, & service fixt., supply & drainage	1 Fixt./580 SF Flr.	Each	5261	9.07	
2020	Domestic Water Distribution	Electric water heater, point-of-use, energy saver		S.F. Floor	1.23	1.23	6.2%
2040	Rain Water Drainage	Roof drains		S.F. Roof	3.30	1.65	
D30 HVAC							
3010	Energy Supply	N/A		—	—	—	
3020	Heat Generating Systems	N/A		—	—	—	
3040	Distribution Systems	Enthalpy heat recovery packages		Each	30,450	2.77	12.0 %
3050	Terminal & Package Units	Multizone rooftop air conditioner, SEER 14		S.F. Floor	20.25	20.25	
3090	Other HVAC Sys. & Equipment	N/A		—	—	—	
D40 Fire Protection							
4010	Sprinklers	Wet pipe sprinkler system		S.F. Floor	3.95	3.95	2.6%
4020	Standpipes	Standpipe		S.F. Floor	1.11	1.11	
D50 Electrical							
5010	Electrical Service/Distribution	400 ampere service, panel board and feeders		S.F. Floor	1.03	1.03	
5020	Lighting & Branch Wiring	LED fixtures, daylt. dim., ltg. on/off cntrl., recept., switches, & A.C. pwr.		S.F. Floor	21.90	21.90	16.7%
5030	Communications & Security	Addressable alarm systems, internet wiring, intercom &emergency ltg.		S.F. Floor	7.46	7.46	
5090	Other Electrical Systems	Emergency generator, 15 kW, energy monitoring systems		S.F. Floor	1.73	1.73	
E. EQUIPMENT & FURNISHINGS							
1020	Institutional Equipment	Lockers, detention rooms, cells, gasoline dispensers		S.F. Floor	13.68	13.68	
1030	Vehicular Equipment	Gasoline dispenser system		S.F. Floor	2.47	2.47	8.7%
1090	Other Equipment	Waste handling recycling tilt truck		Each	.51	.51	
2020	Moveable Furnishings	No smoking signage		Each	.04	.04	
F. SPECIAL CONSTRUCTION							
1020	Integrated Construction	N/A		—	—	—	0.0 %
1040	Special Facilities	N/A		—	—	—	
G. BUILDING SITEWORK	**N/A**						

	Sub-Total	192.37	**100%**
CONTRACTOR FEES (General Requirements: 10%, Overhead: 5%, Profit: 10%)		25%	48.09
ARCHITECT FEES		9%	21.64
	Total Building Cost	262.10	

For customer support on your Square Foot Cost Data, call 877.756.2789.

Costs per square foot of floor area

Exterior Wall	S.F. Area	2000	2800	3500	4200	5000	5800	6500	7200	8000
	L.F. Perimeter	180	212	240	268	300	314	336	344	368
Wood Siding	Wood Frame	280.80	266.10	258.65	253.65	249.75	244.85	242.55	239.40	237.70
Brick Veneer	Wood Frame	288.85	272.20	263.85	258.20	253.80	248.10	245.45	241.70	239.80
Face Brick with Concrete Block Back-up	Wood Joists	302.35	283.55	274.05	267.75	262.70	256.15	253.15	248.90	246.70
	Steel Joists	300.95	279.80	269.10	262.00	256.40	248.85	245.45	240.60	238.05
Stucco on Concrete Block	Wood Joists	288.85	272.20	263.80	258.15	253.75	248.05	245.35	241.70	239.80
	Steel Joists	274.90	258.20	249.85	244.20	239.75	234.05	231.45	227.75	225.85
Perimeter Adj., Add or Deduct	Per 100 L.F.	33.00	23.55	18.90	15.75	13.25	11.40	10.15	9.20	8.25
Story Hgt. Adj., Add or Deduct	Per 1 Ft.	3.45	2.85	2.65	2.50	2.35	2.05	2.00	1.85	1.80

For Basement, add $39.95 per square foot of basement area

The above costs were calculated using the basic specifications shown on the facing page. These costs should be adjusted where necessary for design alternatives and owner's requirements. Reported completed project costs, for this type of structure, range from $142.55 to $332.20 per S.F.

Common additives

Description	Unit	$ Cost
Bar, Front Bar	L.F.	435
Back bar	L.F.	345
Booth, Upholstered, custom straight	L.F.	252 - 465
"L" or "U" shaped	L.F.	261 - 440
Fireplace, brick, excl. chimney or foundation, 30"x29" opening	Ea.	3200
Chimney, standard brick, single flue, 16" x 20"	V.L.F.	108
2 Flue, 20" x 24"	V.L.F.	157
Kitchen Equipment		
Broiler	Ea.	4075
Coffee urn, twin 6 gallon	Ea.	3025
Cooler, 6 ft. long	Ea.	3925
Dishwasher, 10-12 racks per hr.	Ea.	4275
Food warmer, counter, 1.2 KW	Ea.	770
Freezer, 44 C.F., reach-in	Ea.	6175
Ice cube maker, 50 lb. per day	Ea.	2000
Range with 1 oven	Ea.	3300
Refrigerators, prefabricated, walk-in, 7'-6" high, 6' x 6'	S.F.	203
10' x 10'	S.F.	159
12' x 14'	S.F.	142
12' x 20'	S.F.	125

Description	Unit	$ Cost
Commissioning Fees, sustainable commercial construction	S.F.	0.24 - 3.04
Energy Modelling Fees, commercial buildings to 10,000 SF	Ea.	11,000
Greater than 10,000 SF add	S.F.	0.04
Green Bldg Cert Fees for comm construction project reg	Project	900
Photovoltaic Pwr Sys, grid connected, 20 kW (~2400 SF), roof	Ea.	245,600
Green Roofs, 6" soil depth, w/treated wd edging & sedum mats	S.F.	11.76
10" Soil depth, with treated wood edging & sedum mats	S.F.	13.45
Greywater Recovery Systems, prepackaged comm, 3060 gal.	Ea.	58,265
4590 gal.	Ea.	73,075
Rainwater Harvest Sys, prepckged comm, 10,000 gal, sys contrller	Ea.	31,525
20,000 gal. w/system controller	Ea.	50,575
30,000 gal. w/system controller	Ea.	80,825
Solar Domestic HW, closed loop, add-on sys, ext heat exchanger	Ea.	10,850
Drainback, hot water system, 120 gal tank	Ea.	13,525
Draindown, hot water system, 120 gal tank	Ea.	13,850

Model costs calculated for a 1 story building with 12' story height and 5,000 square feet of floor area

G Restaurant

					Unit	Unit Cost	Cost Per S.F.	% Of Sub-Total

A. SUBSTRUCTURE

					Unit	Unit Cost	Cost Per S.F.	% Of Sub-Total
1010	Standard Foundations	Poured concrete; strip and spread footings			S.F. Ground	2.77	2.77	
1020	Special Foundations	N/A			—	—	—	
1030	Slab on Grade	4" reinforced concrete with recycled vapor barrier and granular base			S.F. Slab	5.58	5.58	7.8%
2010	Basement Excavation	Site preparation for slab and trench for foundation wall and footing			S.F. Ground	.57	.57	
2020	Basement Walls	4' foundation wall			L.F. Wall	93	5.58	

B. SHELL

B10 Superstructure

					Unit	Unit Cost	Cost Per S.F.	% Of Sub-Total
1010	Floor Construction	Wood columns			S.F. Floor	.83	.83	4.9%
1020	Roof Construction	Plywood on wood truss (pitched)			S.F. Roof	8.29	8.29	

B20 Exterior Enclosure

					Unit	Unit Cost	Cost Per S.F.	% Of Sub-Total
2010	Exterior Walls	Cedar siding on wood studs with insulation	70% of wall		S.F. Wall	13.29	6.70	
2020	Exterior Windows	Storefront windows	30% of wall		Each	56	11.52	12.7%
2030	Exterior Doors	Aluminum and glass doors and entrance with transom			Each	5462	5.48	

B30 Roofing

					Unit	Unit Cost	Cost Per S.F.	% Of Sub-Total
3010	Roof Coverings	Cedar shingles with flashing (pitched); rigid fiberglass insulation			S.F. Roof	12.70	12.70	6.9%
3020	Roof Openings	Skylights			S.F. Roof	.10	.10	

C. INTERIORS

					Unit	Unit Cost	Cost Per S.F.	% Of Sub-Total
1010	Partitions	Gypsum board on wood studs	25 SF Floor/LF Part.		S.F. Partition	12.48	4.99	
1020	Interior Doors	Hollow core wood, low VOC paint	250 S.F. Floor/Door		Each	542	2.17	
1030	Fittings	Toilet partitions			S.F. Floor	1.22	1.22	
2010	Stair Construction	N/A			—	—	—	14.4%
3010	Wall Finishes	75% paint, low VOC, 25% ceramic tile			S.F. Surface	2.65	2.12	
3020	Floor Finishes	65% carpet, 35% quarry tile			S.F. Floor	9.14	9.14	
3030	Ceiling Finishes	Mineral fiber tile on concealed zee bars			S.F. Ceiling	7.22	7.22	

D. SERVICES

D10 Conveying

					Unit	Unit Cost	Cost Per S.F.	% Of Sub-Total
1010	Elevators & Lifts	N/A			—	—	—	0.0%
1020	Escalators & Moving Walks	N/A			—	—	—	

D20 Plumbing

					Unit	Unit Cost	Cost Per S.F.	% Of Sub-Total
2010	Plumbing Fixtures	Kitchen, toilet, low flow, auto sensor, & service fixt., supply & drain.	1 Fixt./335 SF Flr.		Each	3774	10.63	
2020	Domestic Water Distribution	Tankless, on-demand water heater gas/propane			S.F. Floor	2.64	2.64	8.2%
2040	Rain Water Drainage	Roof drains			S.F. Roof	1.99	1.99	

D30 HVAC

					Unit	Unit Cost	Cost Per S.F.	% Of Sub-Total
3010	Energy Supply	N/A			—	—	—	
3020	Heat Generating Systems	Included in D3050			—	—	—	
3040	Distribution Systems	Enthalpy heat recovery packages and kitchen exhaust/make-up air system			Each	44,575	8.92	26.3 %
3050	Terminal & Package Units	Multizone rooftop air conditioner, SEER 14			S.F. Floor	40.15	40.15	
3090	Other HVAC Sys. & Equipment	N/A			—	—	—	

D40 Fire Protection

					Unit	Unit Cost	Cost Per S.F.	% Of Sub-Total
4010	Sprinklers	Sprinklers, light hazard and ordinary hazard kitchen			S.F. Floor	9.19	9.19	6.1%
4020	Standpipes	Standpipe			S.F. Floor	2.21	2.21	

D50 Electrical

					Unit	Unit Cost	Cost Per S.F.	% Of Sub-Total
5010	Electrical Service/Distribution	400 ampere service, panel board and feeders			S.F. Floor	2.35	2.35	
5020	Lighting & Branch Wiring	LED fixtures, daylt. dim., ltg. on/off, receptacles, switches, and A.C. power			S.F. Floor	13.28	13.28	12.2%
5030	Communications & Security	Addressable alarm systems and emergency lighting			S.F. Floor	3.67	3.67	
5090	Other Electrical Systems	Emergency generator, 15 kW and energy monitoring systems			S.F. Floor	3.53	3.53	

E. EQUIPMENT & FURNISHINGS

					Unit	Unit Cost	Cost Per S.F.	% Of Sub-Total
1010	Commercial Equipment	N/A			—	—	—	
1020	Institutional Equipment	N/A			—	—	—	0.6 %
1090	Other Equipment	Waste handling recycling tilt truck			Each	1.14	1.14	
2020	Moveable Furnishings	No smoking signage			Each	.06	.06	

F. SPECIAL CONSTRUCTION

					Unit	Unit Cost	Cost Per S.F.	% Of Sub-Total
1020	Integrated Construction	N/A			—	—	—	0.0 %
1040	Special Facilities	N/A			—	—	—	

G. BUILDING SITEWORK   N/A

	Sub-Total	186.74	100%
CONTRACTOR FEES (General Requirements: 10%, Overhead: 5%, Profit: 10%)	25%	46.67	
ARCHITECT FEES	7%	16.34	

Total Building Cost	**249.75**

For customer support on your Square Foot Cost Data, call 877.756.2789.

271

Costs per square foot of floor area

Exterior Wall	S.F. Area	25000	30000	35000	40000	45000	50000	55000	60000	65000
	L.F. Perimeter	900	1050	1200	1350	1510	1650	1800	1970	2100
Face Brick with Concrete Block Back-up	Steel Frame	201.55	199.20	197.60	196.40	195.65	194.65	194.00	193.90	193.05
	Bearing Walls	193.50	191.50	190.05	188.95	188.35	187.40	186.90	186.80	186.05
Stucco on Concrete Block	Steel Frame	194.80	192.65	191.20	190.05	189.40	188.45	187.90	187.75	187.00
	Bearing Walls	188.35	186.20	184.70	183.55	182.95	182.00	181.45	181.30	180.50
Decorative Concrete Block	Steel Frame	192.65	190.60	189.10	187.95	187.30	186.40	185.85	185.70	184.95
	Bearing Walls	189.35	187.25	185.75	184.60	183.95	183.10	182.50	182.35	181.60
Perimeter Adj., Add or Deduct	Per 100 L.F.	5.05	4.30	3.60	3.10	2.85	2.55	2.30	2.15	1.95
Story Hgt. Adj., Add or Deduct	Per 1 Ft.	1.90	1.90	1.80	1.70	1.80	1.75	1.80	1.75	1.70

For Basement, add $29.60 per square foot of basement area

The above costs were calculated using the basic specifications shown on the facing page. These costs should be adjusted where necessary for design alternatives and owner's requirements. Reported completed project costs, for this type of structure, range from $98.30 to $250.05 per S.F.

Common additives

Description	Unit	$ Cost
Bleachers, Telescoping, manual, 16-20 tier	Seat	294 - 360
21-30 tier	Seat	310 - 410
For power operation, add	Seat	57.50 - 90
Carrels Hardwood	Ea.	945 - 2400
Clock System, 20 room	Ea.	20,600
50 room	Ea.	48,500
Kitchen Equipment		
Broiler	Ea.	4075
Cooler, 6 ft. long, reach-in	Ea.	3925
Dishwasher, 10-12 racks per hr.	Ea.	4275
Food warmer, counter, 1.2 KW	Ea.	770
Freezer, 44 C.F., reach-in	Ea.	6175
Ice cube maker, 50 lb. per day	Ea.	2000
Range with 1 oven	Ea.	3300
Lockers, Steel, single tier, 60" to 72"	Opng.	231 - 380
2 tier, 60" to 72" total	Opng.	121 - 170
Locker bench, lam. maple top only	L.F.	34.50
Pedestals, steel pipe	Ea.	76.50
Seating, auditorium chair, veneer back, padded seat	Ea.	325
Classroom, movable chair & desk	Set	81 - 171
Lecture hall, pedestal type	Ea.	275 - 620

Description	Unit	$ Cost
Sound System, amplifier, 250 watts	Ea.	2200
Speaker, ceiling or wall	Ea.	217
Commissioning Fees, sustainable institutional construction	S.F.	0.58 - 2.47
Energy Modelling Fees, academic buildings to 10,000 SF	Ea.	9000
Greater than 10,000 SF add	S.F.	0.20
Green Bldg Cert Fees for school construction project reg	Project	900
Photovoltaic Pwr Sys, grid connected, 20 kW (~2400 SF), roof	Ea.	245,600
Green Roofs, 6" soil depth, w/treated wd edging & sedum mats	S.F.	11.76
10" Soil depth, with treated wood edging & sedum mats	S.F.	13.45
Greywater Recovery Systems, prepackaged comm, 3060 gal.	Ea.	58,265
4590 gal.	Ea.	73,075
Rainwater Harvest Sys, prepckged comm, 10,000 gal, sys contrller	Ea.	31,525
20,000 gal. w/system controller	Ea.	50,575
30,000 gal. w/system controller	Ea.	80,825
Solar Domestic HW, closed loop, add-on sys, ext heat exchanger	Ea.	10,850
Drainback, hot water system, 120 gal tank	Ea.	13,525
Draindown, hot water system, 120 gal tank	Ea.	13,850

Important: See the Reference Section for Location Factors.

Model costs calculated for a 1 story building with 15' story height and 45,000 square feet of floor area

				Unit	Unit Cost	Cost Per S.F.	% Of Sub-Total
A. SUBSTRUCTURE							
1010	Standard Foundations	Poured concrete; strip and spread footings		S.F. Ground	5.25	5.25	
1020	Special Foundations	N/A		—	—	—	
1030	Slab on Grade	4" reinforced concrete with recycled vapor barrier and granular base		S.F. Slab	5.58	5.58	11.6%
2010	Basement Excavation	Site preparation for slab and trench for foundation wall and footing		S.F. Ground	.18	.18	
2020	Basement Walls	4' foundation wall		L.F. Wall	93	5.30	
B. SHELL							
B10 Superstructure							
1010	Floor Construction	N/A		—	—	—	
1020	Roof Construction	Metal deck on open web steel joists		S.F. Roof	4.53	4.53	3.2 %
B20 Exterior Enclosure							
2010	Exterior Walls	Face brick with concrete block backup	70% of wall	S.F. Wall	38.80	13.67	
2020	Exterior Windows	Steel outward projecting	25% of wall	Each	746	4.90	13.7%
2030	Exterior Doors	Metal and glass	5% of wall	Each	3880	.69	
B30 Roofing							
3010	Roof Coverings	Single-ply TPO membrane, 60 mils, w/flashing; polyiso. insulation		S.F. Roof	10.43	10.43	7.4%
3020	Roof Openings	N/A		—	—	—	
C. INTERIORS							
1010	Partitions	Concrete block w/foamed-in insul.	20 SF Flr./LF Part.	S.F. Partition	11.86	5.93	
1020	Interior Doors	Single leaf kalamein fire doors, low VOC paint	700 S.F. Floor/Door	Each	1168	1.67	
1030	Fittings	Toilet partitions		S.F. Floor	1.90	1.90	
2010	Stair Construction	N/A		—	—	—	19.8%
3010	Wall Finishes	75% paint, low VOC, 15% glazed coating, 10% ceramic tile		S.F. Surface	4.96	4.96	
3020	Floor Finishes	65% vinyl composition tile, recycled content, 25% carpet tile, 10% terrazzo		S.F. Floor	6.25	6.25	
3030	Ceiling Finishes	Mineral fiber tile on concealed zee bars		S.F. Ceiling	7.22	7.22	
D. SERVICES							
D10 Conveying							
1010	Elevators & Lifts	N/A		—	—	—	0.0 %
1020	Escalators & Moving Walks	N/A		—	—	—	
D20 Plumbing							
2010	Plumbing Fixtures	Kitchen, toilet, low flow, auto sensor, & service fixt., supply & drain.	1 Fixt./625 SF Flr.	Each	9500	15.20	
2020	Domestic Water Distribution	Gas fired, tankless water heater		S.F. Floor	.16	.16	11.9%
2040	Rain Water Drainage	Roof drains		S.F. Roof	1.37	1.37	
D30 HVAC							
3010	Energy Supply	N/A		—	—	—	
3020	Heat Generating Systems	N/A		—	—	—	
3040	Distribution Systems	Enthalpy heat recovery packages		Each	37,000	2.47	15.8 %
3050	Terminal & Package Units	Multizone rooftop air conditioner		S.F. Floor	19.85	19.85	
3090	Other HVAC Sys. & Equipment	N/A		—	—	—	
D40 Fire Protection							
4010	Sprinklers	Sprinklers, light hazard		S.F. Floor	2.84	2.84	2.3%
4020	Standpipes	Standpipe		S.F. Floor	.44	.44	
D50 Electrical							
5010	Electrical Service/Distribution	800 ampere service, panel board and feeders		S.F. Floor	.90	.90	
5020	Lighting & Branch Wiring	LED fixtures, daylt. dim., ltg. on/off, recept., switches, and A.C. power		S.F. Floor	13.94	13.94	14.0%
5030	Communications & Security	Addressable alarm systems, internet wiring, comm. systems & emerg. ltg.		S.F. Floor	3.85	3.85	
5090	Other Electrical Systems	Emergency generator, 15 kW, energy monitoring systems		S.F. Floor	1.05	1.05	
E. EQUIPMENT & FURNISHINGS							
1010	Commercial Equipment	N/A		—	—	—	
1020	Institutional Equipment	Chalkboards		S.F. Floor	.20	.20	0.2 %
1090	Other Equipment	Waste handling recycling tilt truck		Each	.08	.08	
2020	Moveable Furnishings	No smoking signage		Each	.02	.02	
F. SPECIAL CONSTRUCTION							
1020	Integrated Construction	N/A		—	—	—	0.0 %
1040	Special Facilities	N/A		—	—	—	
G. BUILDING SITEWORK	**N/A**						

	Sub-Total	140.83	**100%**
CONTRACTOR FEES (General Requirements: 10%, Overhead: 5%, Profit: 10%)		25%	35.20
ARCHITECT FEES		7%	12.32
	Total Building Cost	**188.35**	

For customer support on your Square Foot Cost Data, call 877.756.2789.

273

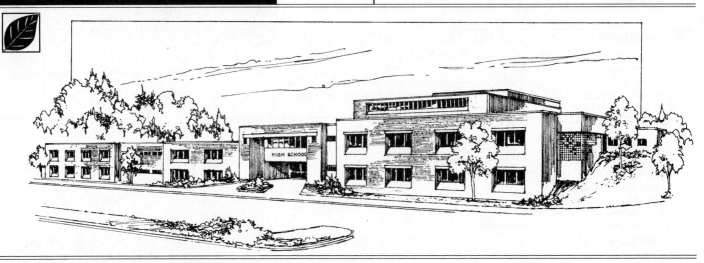

Costs per square foot of floor area

Exterior Wall	S.F. Area	50000	70000	90000	110000	130000	150000	170000	190000	210000
	L.F. Perimeter	850	1140	1420	1700	1980	2280	2560	2840	3120
Face Brick with Concrete Block Back-up	Steel Frame	188.30	185.25	183.40	182.15	181.25	180.95	180.50	180.05	179.75
	R/Conc. Frame	192.70	189.65	187.75	186.55	185.65	185.40	184.90	184.40	184.15
Decorative Concrete Block	Steel Frame	180.90	178.20	176.55	175.50	174.65	174.40	173.95	173.55	173.35
	R/Conc. Frame	182.05	179.30	177.65	176.60	175.75	175.50	175.10	174.65	174.45
Limestone with Concrete Block Back-up	Steel Frame	200.60	197.05	194.75	193.35	192.30	191.95	191.35	190.85	190.50
	R/Conc. Frame	208.35	204.80	202.60	201.15	200.05	199.80	199.20	198.65	198.30
Perimeter Adj., Add or Deduct	Per 100 L.F.	4.65	3.30	2.50	2.05	1.80	1.60	1.35	1.25	1.15
Story Hgt. Adj., Add or Deduct	Per 1 Ft.	2.35	2.25	2.15	2.15	2.15	2.15	2.05	2.10	2.10
For Basement, add $38.65 per square foot of basement area										

The above costs were calculated using the basic specifications shown on the facing page. These costs should be adjusted where necessary for design alternatives and owner's requirements. Reported completed project costs, for this type of structure, range from $114.25 to $269.10 per S.F.

Common additives

Description	Unit	$ Cost
Bleachers, Telescoping, manual, 16-20 tier	Seat	294 - 360
21-30 tier	Seat	310 - 410
For power operation, add	Seat	57.50 - 90
Carrels Hardwood	Ea.	945 - 2400
Clock System, 20 room	Ea.	20,600
50 room	Ea.	48,500
Elevators, hydraulic passenger, 2 stops, 2500# capacity	Ea.	70,400
3500# capacity	Ea.	75,400
Kitchen Equipment		
Broiler	Ea.	4075
Cooler, 6 ft. long, reach-in	Ea.	3925
Dishwasher, 10-12 racks per hr.	Ea.	4275
Food warmer, counter, 1.2 KW	Ea.	770
Freezer, 44 C.F., reach-in	Ea.	6175
Lockers, Steel, single tier, 60" or 72"	Opng.	231 - 380
2 tier, 60" or 72" total	Opng.	121 - 170
Locker bench, lam. maple top only	L.F.	34.50
Pedestals, steel pipe	Ea.	76.50
Seat, auditorium chair, veneer back, padded seat	Ea.	325
Classroom, movable chair & desk	Set	81 - 171
Lecture hall, pedestal type	Ea.	275 - 620

Description	Unit	$ Cost
Sound System, amplifier, 250 watts	Ea.	2200
Speaker, ceiling or wall	Ea.	217
Commissioning Fees, sustainable institutional construction	S.F.	0.58 - 2.47
Energy Modelling Fees, academic buildings to 10,000 SF	Ea.	9000
Greater than 10,000 SF add	S.F.	0.20
Green Bldg Cert Fees for school construction project reg	Project	900
Photovoltaic Pwr Sys, grid connected, 20 kW (~2400 SF), roof	Ea.	245,600
Green Roofs, 6" soil depth, w/treated wd edging & sedum mats	S.F.	11.76
10" Soil depth, with treated wood edging & sedum mats	S.F.	13.45
Greywater Recovery Systems, prepackaged comm, 3060 gal.	Ea.	58,265
4590 gal.	Ea.	73,075
Rainwater Harvest Sys, prepckged comm, 10,000 gal, sys contrller	Ea.	31,525
20,000 gal. w/system controller	Ea.	50,575
30,000 gal. w/system controller	Ea.	80,825
Solar Domestic HW, closed loop, add-on sys, ext heat exchanger	Ea.	10,850
Drainback, hot water system, 120 gal tank	Ea.	13,525
Draindown, hot water system, 120 gal tank	Ea.	13,850

Model costs calculated for a 2 story building with 15' story height and 130,000 square feet of floor area

			Unit	Unit Cost	Cost Per S.F.	% Of Sub-Total
A. SUBSTRUCTURE						
1010	Standard Foundations	Poured concrete; strip and spread footings	S.F. Ground	2.80	1.40	
1020	Special Foundations	N/A	—	—	—	
1030	Slab on Grade	4" reinforced concrete with recycled vapor barrier and granular base	S.F. Slab	5.58	2.79	4.1%
2010	Basement Excavation	Site preparation for slab and trench for foundation wall and footing	S.F. Ground	.18	.09	
2020	Basement Walls	4' foundation wall	L.F. Wall	93	1.41	
B. SHELL						
B10 Superstructure						
1010	Floor Construction	Concrete slab without drop panel, concrete columns	S.F. Floor	22.84	11.42	13.9%
1020	Roof Construction	Concrete slab without drop panel	S.F. Roof	15.82	7.91	
B20 Exterior Enclosure						
2010	Exterior Walls	Face brick with concrete block backup 75% of wall	S.F. Wall	38.81	13.30	
2020	Exterior Windows	Window wall 25% of wall	Each	77	8.75	16.3%
2030	Exterior Doors	Metal and glass	Each	2277	.52	
B30 Roofing						
3010	Roof Coverings	Single-ply TPO membrane and standing seam metal; polyiso. insulation	S.F. Roof	14.02	7.01	5.1%
3020	Roof Openings	Roof hatches	S.F. Roof	.10	.05	
C. INTERIORS						
1010	Partitions	Concrete block w/foamed-in insul. 25 SF Flr./LF Part.	S.F. Partition	15.02	7.21	
1020	Interior Doors	Single leaf kalamein fire doors, low VOC paint 700 S.F. Floor/Door	Each	1168	1.67	
1030	Fittings	Toilet partitions, chalkboards	S.F. Floor	1.33	1.33	
2010	Stair Construction	Concrete filled metal pan	Flight	14,450	.66	19.9%
3010	Wall Finishes	75% paint, low VOC. 15% glazed coating, 10% ceramic tile	S.F. Surface	3.52	3.38	
3020	Floor Finishes	70% vinyl comp. tile, recycled content, 20% carpet tile, 10% terrazzo	S.F. Floor	6.21	6.21	
3030	Ceiling Finishes	Mineral fiber tile on concealed zee bars	S.F. Ceiling	7.22	7.22	
D. SERVICES						
D10 Conveying						
1010	Elevators & Lifts	One hydraulic passenger elevator	Each	85,800	.66	0.5%
1020	Escalators & Moving Walks	N/A	—	—	—	
D20 Plumbing						
2010	Plumbing Fixtures	Kitchen, toilet, low flow, auto sensor & service fixt., supply & drain. 1 Fixt./860 SF Flr.	Each	4756	5.53	
2020	Domestic Water Distribution	Gas fired, tankless water heater	S.F. Floor	.30	.30	4.8%
2040	Rain Water Drainage	Roof drains	S.F. Roof	1.56	.78	
D30 HVAC						
3010	Energy Supply	N/A	—	—	—	
3020	Heat Generating Systems	N/A	—	—	—	
3040	Distribution Systems	Enthalpy heat recovery packages	Each	37,000	2.27	15.9 %
3050	Terminal & Package Units	Multizone rooftop air conditioner, SEER 14	S.F. Floor	19.85	19.85	
3090	Other HVAC Sys. & Equipment	N/A	—	—	—	
D40 Fire Protection						
4010	Sprinklers	Sprinklers, light hazard	S.F. Floor	2.49	2.49	2.1%
4020	Standpipes	Standpipe, wet, Class III	S.F. Floor	.38	.38	
D50 Electrical						
5010	Electrical Service/Distribution	2000 ampere service, panel board and feeders	S.F. Floor	.87	.87	
5020	Lighting & Branch Wiring	LED fixtures, daylt. dim., ltg. on/off, recept., switches, and A.C. power	S.F. Floor	15.23	15.23	15.2%
5030	Communications & Security	Addressable alarm systems, internet wiring, comm. systems and emerg. ltg.	S.F. Floor	4.21	4.21	
5090	Other Electrical Systems	Emergency generator, 250 kW, and energy monitoring systems	S.F. Floor	.82	.82	
E. EQUIPMENT & FURNISHINGS						
1010	Commercial Equipment	N/A	—	—	—	
1020	Institutional Equipment	Laboratory casework and counters	S.F. Floor	2.07	2.07	2.2%
1090	Other Equipment	Waste handlg. recyc. tilt truck, built-in athletic equip., bleachers & backstp.	Each	1.01	1.01	
2020	Moveable Furnishings	No smoking signage	Each	.02	.02	
F. SPECIAL CONSTRUCTION						
1020	Integrated Construction	N/A	—	—	—	0.0%
1040	Special Facilities	N/A	—	—	—	
G. BUILDING SITEWORK N/A						

		Sub-Total	138.82	100%
CONTRACTOR FEES (General Requirements: 10%, Overhead: 5%, Profit: 10%)		25%	34.68	
ARCHITECT FEES		7%	12.15	

Total Building Cost	**185.65**

Costs per square foot of floor area

Exterior Wall	S.F. Area	50000	65000	80000	95000	110000	125000	140000	155000	170000
	L.F. Perimeter	850	1060	1280	1490	1700	1920	2140	2340	2560
Face Brick with Concrete Block Back-up	Steel Frame	195.60	193.25	192.05	191.05	190.30	189.90	189.60	189.10	188.90
	Bearing Walls	186.95	184.65	183.45	182.45	181.65	181.30	181.00	180.45	180.30
Concrete Block Stucco Face	Steel Frame	201.10	198.50	197.20	196.00	195.15	194.75	194.40	193.80	193.60
	Bearing Walls	194.35	191.75	190.45	189.30	188.40	188.00	187.65	187.05	186.85
Decorative Concrete Block	Steel Frame	187.25	185.25	184.15	183.30	182.65	182.35	182.00	181.60	181.45
	Bearing Walls	178.95	176.90	175.85	174.95	174.30	173.95	173.70	173.25	173.05
Perimeter Adj., Add or Deduct	Per 100 L.F.	4.25	3.30	2.75	2.30	1.90	1.75	1.50	1.35	1.30
Story Hgt. Adj., Add or Deduct	Per 1 Ft.	2.20	2.10	2.05	2.00	1.90	1.95	1.90	1.90	1.90

For Basement, add $40.25 per square foot of basement area

The above costs were calculated using the basic specifications shown on the facing page. These costs should be adjusted where necessary for design alternatives and owner's requirements. Reported completed project costs, for this type of structure, range from $107.85 to $253.45 per S.F.

Common additives

Description	Unit	$ Cost
Bleachers, Telescoping, manual, 16-20 tier	Seat	294 - 360
21-30 tier	Seat	310 - 410
For power operation, add	Seat	57.50 - 90
Carrels Hardwood	Ea.	945 - 2400
Clock System, 20 room	Ea.	20,600
50 room	Ea.	48,500
Elevators, hydraulic passenger, 2 stops, 2500# capacity	Ea.	70,400
3500# capacity	Ea.	75,400
Kitchen Equipment		
Broiler	Ea.	4075
Cooler, 6 ft. long, reach-in	Ea.	3925
Dishwasher, 10-12 racks per hr.	Ea.	4275
Food warmer, counter, 1.2 KW	Ea.	770
Freezer, 44 C.F., reach-in	Ea.	6175
Lockers, Steel, single tier, 60" to 72"	Opng.	231 - 380
2 tier, 60" to 72" total	Opng.	121 - 170
Locker bench, lam. maple top only	L.F.	34.50
Pedestals, steel pipe	Ea.	76.50

Description	Unit	$ Cost
Seating, auditorium chair, veneer back, padded seat	Ea.	325
Upholstered, spring seat	Ea.	320
Classroom, movable chair & desk	Set	81 - 171
Lecture hall, pedestal type	Ea.	275 - 620
Sound System, amplifier, 250 watts	Ea.	2200
Speaker, ceiling or wall	Ea.	217
Commissioning Fees, sustainable institutional construction	S.F.	0.58 - 2.47
Energy Modelling Fees, academic buildings to 10,000 SF	Ea.	9000
Greater than 10,000 SF add	S.F.	0.20
Green Bldg Cert Fees for school construction project reg	Project	900
Photovoltaic Pwr Sys, grid connected, 20 kW (~2400 SF), roof	Ea.	245,600
Green Roofs, 6" soil depth, w/treated wd edging & sedum mats	S.F.	11.76
10" Soil depth, with treated wood edging & sedum mats	S.F.	13.45
Greywater Recovery Systems, prepackaged comm, 3060 gal.	Ea.	58,265
4590 gal.	Ea.	73,075
Rainwater Harvest Sys, prepckged comm, 10,000 gal, sys contrller	Ea.	31,525
20,000 gal. w/system controller	Ea.	50,575
30,000 gal. w/system controller	Ea.	80,825
Solar Domestic HW, closed loop, add-on sys, ext heat exchanger	Ea.	10,850
Drainback, hot water system, 120 gal tank	Ea.	13,525
Draindown, hot water system, 120 gal tank	Ea.	13,850

Important: See the Reference Section for Location Factors.

Model costs calculated for a 2 story building with 15' story height and 110,000 square feet of floor area

G School, Jr High, 2-3 Story

				Unit	Unit Cost	Cost Per S.F.	% Of Sub-Total
A. SUBSTRUCTURE							
1010	Standard Foundations	Poured concrete; strip and spread footings		S.F. Ground	2.58	1.29	
1020	Special Foundations	N/A		—	—	—	
1030	Slab on Grade	4" reinforced concrete with recycled vapor barrier and granular base		S.F. Slab	5.58	2.79	3.9%
2010	Basement Excavation	Site preparation for slab and trench for foundation wall and footing		S.F. Ground	.18	.09	
2020	Basement Walls	4' foundation wall		L.F. Wall	93	1.44	
B. SHELL							
	B10 Superstructure						
1010	Floor Construction	Open web steel joists, slab form, concrete, columns		S.F. Floor	28.94	14.47	14.4%
1020	Roof Construction	Metal deck, open web steel joists, columns		S.F. Roof	12.02	6.01	
	B20 Exterior Enclosure						
2010	Exterior Walls	Face brick with concrete block backup	75% of wall	S.F. Wall	38.79	13.49	
2020	Exterior Windows	Window wall	25% of wall	Each	65	7.14	15.0%
2030	Exterior Doors	Double aluminum & glass		Each	2410	.64	
	B30 Roofing						
3010	Roof Coverings	Single-ply TPO membrane & standing seam metal; polyiso. insulation		S.F. Roof	14.22	7.11	5.0%
3020	Roof Openings	Roof hatches		S.F. Roof	.04	.02	
C. INTERIORS							
1010	Partitions	Concrete block w/foamed-in insul.	20 SF Flr./LF Part.	S.F. Partition	12.47	7.48	
1020	Interior Doors	Single leaf kalamein fire doors, low VOC paint	750 S.F. Floor/Door	Each	1168	1.56	
1030	Fittings	Toilet partitions, chalkboards		S.F. Floor	1.44	1.44	
2010	Stair Construction	Concrete filled metal pan		Flight	14,450	.79	22.5%
3010	Wall Finishes	50% paint, low VOC, 40% glazed coatings, 10% ceramic tile		S.F. Surface	3.88	4.66	
3020	Floor Finishes	50% vinyl comp. tile, recycled content, 30% carpet tile, 20% terrrazzo		S.F. Floor	8.85	8.85	
3030	Ceiling Finishes	Mineral fiberboard on concealed zee bars		S.F. Ceiling	7.22	7.22	
D. SERVICES							
	D10 Conveying						
1010	Elevators & Lifts	One hydraulic passenger elevator		Each	84,700	.77	0.5%
1020	Escalators & Moving Walks	N/A		—	—	—	
	D20 Plumbing						
2010	Plumbing Fixtures	Kitchen, toilet, low flow, auto sensor, & service fixt., supply & drain.	1 Fixt./1170 SF Flr.	Each	6880	5.88	
2020	Domestic Water Distribution	Gas fired, tankless water heater		S.F. Floor	.36	.36	4.9%
2040	Rain Water Drainage	Roof drains		S.F. Roof	1.56	.78	
	D30 HVAC						
3010	Energy Supply	N/A		—	—	—	
3020	Heat Generating Systems	Included in D3050		—	—	—	
3040	Distribution Systems	Enthalpy heat recovery packages		Each	37,000	2.36	15.6 %
3050	Terminal & Package Units	Multizone rooftop air conditioner, SEER 14		S.F. Floor	19.85	19.85	
3090	Other HVAC Sys. & Equipment	N/A		—	—	—	
	D40 Fire Protection						
4010	Sprinklers	Sprinklers, light hazard	10% of area	S.F. Floor	2.49	2.49	2.1%
4020	Standpipes	Standpipe, wet, Class III		S.F. Floor	.43	.43	
	D50 Electrical						
5010	Electrical Service/Distribution	1600 ampere service, panel board and feeders		S.F. Floor	.76	.76	
5020	Lighting & Branch Wiring	LED fixtures, daylt. dim., ltg. on/off, recept., switches, and A.C. power		S.F. Floor	13.15	13.15	13.5%
5030	Communications & Security	Addressable alarm systems, internet wiring, comm. systems and emerg. ltg.		S.F. Floor	4.50	4.50	
5090	Other Electrical Systems	Emergency generator, 100 kW, and energy monitoring systems		S.F. Floor	.76	.76	
E. EQUIPMENT & FURNISHINGS							
1010	Commercial Equipment	N/A		—	—	—	
1020	Institutional Equipment	Laboratory casework and counters		S.F. Floor	2.45	2.45	2.6 %
1090	Other Equipment	Waste handlg. recyc. tilt truck, built-in athletic equip., bleachers & backstps.		Each	1.22	1.22	
2020	Moveable Furnishings	No smoking signage		Each	.02	.02	
F. SPECIAL CONSTRUCTION							
1020	Integrated Construction	N/A		—	—	—	0.0 %
1040	Special Facilities	N/A		—	—	—	
G. BUILDING SITEWORK **N/A**							

		Sub-Total	142.27	100%
CONTRACTOR FEES (General Requirements: 10%, Overhead: 5%, Profit: 10%)		25%	35.58	
ARCHITECT FEES		7%	12.45	
	Total Building Cost		**190.30**	

For customer support on your Square Foot Cost Data, call 877.756.2789.

277

Costs per square foot of floor area

Exterior Wall	S.F. Area	20000	30000	40000	50000	60000	70000	80000	90000	100000
	L.F. Perimeter	440	590	740	900	1050	1200	1360	1510	1660
Face Brick with Concrete Block Back-up	Steel Frame	199.25	192.25	188.75	187.10	185.60	184.55	184.10	183.45	182.95
	Bearing Walls	195.25	188.35	184.90	183.20	181.80	180.70	180.25	179.60	179.10
Decorative Concrete Block	Steel Frame	184.20	178.80	176.15	174.75	173.65	172.80	172.45	171.95	171.55
	Bearing Walls	176.95	171.55	168.85	167.45	166.35	165.55	165.15	164.65	164.25
Steel Siding on Steel Studs	Steel Frame	174.45	170.05	167.85	166.75	165.85	165.20	164.90	164.45	164.15
Metal Sandwich Panel	Steel Frame	202.60	195.30	191.55	189.80	188.25	187.15	186.65	185.95	185.45
Perimeter Adj., Add or Deduct	Per 100 L.F.	10.90	7.25	5.45	4.40	3.65	3.10	2.70	2.40	2.15
Story Hgt. Adj., Add or Deduct	Per 1 Ft.	2.65	2.40	2.25	2.20	2.15	2.10	2.05	2.00	2.00

For Basement, add $39.05 per square foot of basement area

The above costs were calculated using the basic specifications shown on the facing page. These costs should be adjusted where necessary for design alternatives and owner's requirements. Reported completed project costs, for this type of structure, range from $96.80 to $276.60 per S.F.

Common additives

Description	Unit	$ Cost
Carrels Hardwood	Ea.	945 - 2400
Clock System, 20 room	Ea.	20,600
50 room	Ea.	48,500
Directory Boards, plastic, glass covered, 36" x 48"	Ea.	1525
Aluminum, 36" x 24"	Ea.	765
48" x 60"	Ea.	2275
Elevators, hydraulic passenger, 2 stops, 2500# capacity	Ea.	70,400
3500# capacity	Ea.	75,400
Seating, auditorium chair, veneer back, padded seat	Ea.	325
Classroom, movable chair & desk	Set	81 - 171
Lecture hall, pedestal type	Ea.	275 - 620
Shops & Workroom, benches, metal	Ea.	870
Parts bins 6'-3" high, 3' wide, 12" deep, 72 bins	Ea.	720
Shelving, metal 1' x 3'	S.F.	12
Wide span 6' wide x 24" deep	S.F.	10.30

Description	Unit	$ Cost
Sound System, amplifier, 250 watts	Ea.	2200
Speaker, ceiling or wall	Ea.	217
Commissioning Fees, sustainable institutional construction	S.F.	0.58 - 2.47
Energy Modelling Fees, academic buildings to 10,000 SF	Ea.	9000
Greater than 10,000 SF add	S.F.	0.20
Green Bldg Cert Fees for school construction project reg	Project	900
Photovoltaic Pwr Sys, grid connected, 20 kW (~2400 SF), roof	Ea.	245,600
Green Roofs, 6" soil depth, w/treated wd edging & sedum mats	S.F.	11.76
10" Soil depth, with treated wood edging & sedum mats	S.F.	13.45
Greywater Recovery Systems, prepackaged comm, 3060 gal.	Ea.	58,265
4590 gal.	Ea.	73,075
Rainwater Harvest Sys, prepckged comm, 10,000 gal, sys contrller	Ea.	31,525
20,000 gal. w/system controller	Ea.	50,575
30,000 gal. w/system controller	Ea.	80,825
Solar Domestic HW, closed loop, add-on sys, ext heat exchanger	Ea.	10,850
Drainback, hot water system, 120 gal tank	Ea.	13,525
Draindown, hot water system, 120 gal tank	Ea.	13,850

Model costs calculated for a 2 story building with 16' story height and 40,000 square feet of floor area

G School, Vocational

				Unit	Unit Cost	Cost Per S.F.	% Of Sub-Total
A.	**SUBSTRUCTURE**						
1010	Standard Foundations	Poured concrete; strip and spread footings		S.F. Ground	2.90	1.45	
1020	Special Foundations	N/A		—	—	—	
1030	Slab on Grade	5" reinforced concrete with recycled vapor barrier and granular base		S.F. Slab	13.82	6.91	7.4%
2010	Basement Excavation	Site preparation for slab and trench for foundation wall and footing		S.F. Ground	.18	.09	
2020	Basement Walls	4' foundation wall		L.F. Wall	93	1.72	
B.	**SHELL**						
	B10 Superstructure						
1010	Floor Construction	Open web steel joists, slab form, concrete, beams, columns		S.F. Floor	19	9.50	9.3%
1020	Roof Construction	Metal deck, open web steel joists, beams, columns		S.F. Roof	6.60	3.30	
	B20 Exterior Enclosure						
2010	Exterior Walls	Face brick with concrete block backup	85% of wall	S.F. Wall	38.79	19.52	
2020	Exterior Windows	Tubular aluminum framing with insulated glass	15% of wall	Each	65	5.48	18.5%
2030	Exterior Doors	Metal and glass doors, low VOC paint		Each	3035	.61	
	B30 Roofing						
3010	Roof Coverings	Single-ply TPO membrane, 60 mils, heat welded seams w/polyiso. insul.		S.F. Roof	9	4.50	3.4%
3020	Roof Openings	Roof hatches		S.F. Roof	.32	.16	
C.	**INTERIORS**						
1010	Partitions	Concrete block w/foamed-in insul.	20 SF Flr./LF Part.	S.F. Partition	11.37	6.82	
1020	Interior Doors	Single leaf kalamein fire doors, low VOC paint	600 S.F. Floor/Door	Each	1168	1.95	
1030	Fittings	Toilet partitions, chalkboards		S.F. Floor	1.57	1.57	
2010	Stair Construction	Concrete filled metal pan		Flight	14,450	1.45	20.2%
3010	Wall Finishes	50% paint, low VOC, 40% glazed coating, 10% ceramic tile		S.F. Surface	4.70	5.64	
3020	Floor Finishes	70% vinyl comp. tile, recycled content, 20% carpet tile, 10% terrazzo		S.F. Floor	6.21	6.21	
3030	Ceiling Finishes	Mineral fiber tile on concealed zee bars	60% of area	S.F. Ceiling	4.33	4.33	
D.	**SERVICES**						
	D10 Conveying						
1010	Elevators & Lifts	One hydraulic passenger elevator		Each	85,600	2.14	1.5%
1020	Escalators & Moving Walks	N/A		—	—	—	
	D20 Plumbing						
2010	Plumbing Fixtures	Toilet, low flow, auto sensor, & service fixt., supply & drain.	1 Fixt./700 SF.Flr.	Each	3535	5.05	
2020	Domestic Water Distribution	Gas fired, tankless water heater		S.F. Floor	.30	.30	4.6%
2040	Rain Water Drainage	Roof drains		S.F. Roof	2.02	1.01	
	D30 HVAC						
3010	Energy Supply	N/A		—	—	—	
3020	Heat Generating Systems	N/A		—	—	—	
3040	Distribution Systems	Enthalpy heat recovery packages		Each	30,450	5.33	18.2 %
3050	Terminal & Package Units	Multizone rooftop air conditioner, SEER 14		S.F. Floor	19.85	19.85	
3090	Other HVAC Sys. & Equipment	N/A		—	—	—	
	D40 Fire Protection						
4010	Sprinklers	Sprinklers, light hazard		S.F. Floor	3.15	3.15	3.0%
4020	Standpipes	Standpipe, wet, Class III		S.F. Floor	1.03	1.03	
	D50 Electrical						
5010	Electrical Service/Distribution	800 ampere service, panel board and feeders		S.F. Floor	1.02	1.02	
5020	Lighting & Branch Wiring	LED fixtures, daylt. dim., ltg. on/off, recept., switches, and A.C. power		S.F. Floor	13.32	13.32	13.6%
5030	Communications & Security	Addressable alarm systems, internet wiring, comm. systems & emerg. ltg.		S.F. Floor	3.26	3.26	
5090	Other Electrical Systems	Emergency generator, 11.5 kW, and energy monitoring systems		S.F. Floor	1.22	1.22	
E.	**EQUIPMENT & FURNISHINGS**						
1010	Commercial Equipment	N/A		—	—	—	
1020	Institutional Equipment	Stainless steel countertops		S.F. Floor	.18	.18	0.2%
1090	Other Equipment	Waste handling recycling tilt truck		Each	.14	.14	
2020	Moveable Furnishings	No smoking signage		Each	.02	.02	
F.	**SPECIAL CONSTRUCTION**						
1020	Integrated Construction	N/A		—	—	—	0.0 %
1040	Special Facilities	N/A		—	—	—	
G.	**BUILDING SITEWORK**	**N/A**					

		Sub-Total	138.23	100%
CONTRACTOR FEES (General Requirements: 10%, Overhead: 5%, Profit: 10%)		25%	34.57	
ARCHITECT FEES		7%	12.10	
	Total Building Cost		**184.90**	

For customer support on your Square Foot Cost Data, call 877.756.2789.

Costs per square foot of floor area

Exterior Wall	S.F. Area	12000	16000	20000	26000	32000	38000	44000	52000	60000
	L.F. Perimeter	450	510	570	650	716	780	840	920	1020
Face Brick with Concrete Block Back-up	Steel Frame	181.30	170.10	163.45	156.70	151.90	148.55	146.00	143.45	142.15
	Bearing Walls	164.70	153.55	146.85	140.10	135.35	131.95	129.40	126.90	125.55
Stucco on Concrete Block	Steel Frame	171.10	161.50	155.70	149.90	145.90	142.95	140.80	138.70	137.55
	Bearing Walls	168.60	158.90	153.10	147.35	143.30	140.45	138.30	136.10	135.00
Precast Concrete Panels	Steel Frame	186.75	174.70	167.50	160.30	155.15	151.45	148.75	146.00	144.55
Metal Sandwich Panels	Steel Frame	181.65	170.45	163.65	156.90	152.15	148.70	146.20	143.60	142.30
Perimeter Adj., Add or Deduct	Per 100 L.F.	11.15	8.35	6.60	5.10	4.20	3.50	3.05	2.60	2.20
Story Hgt. Adj., Add or Deduct	Per 1 Ft.	2.35	2.00	1.75	1.50	1.45	1.25	1.15	1.10	1.05
For Basement, add $29.65 per square foot of basement area										

The above costs were calculated using the basic specifications shown on the facing page. These costs should be adjusted where necessary for design alternatives and owner's requirements. Reported completed project costs, for this type of structure, range from $81.50 to $233.40 per S.F.

Common additives

Description	Unit	$ Cost
Check Out Counter, single belt	Ea.	3600
Scanner, registers, guns & memory 10 lanes	Ea.	173,500
Power take away	Ea.	6025
Refrigerators, prefabricated, walk-in, 7'-6" High, 10' x 10'	S.F.	159
12' x 14'	S.F.	142
12' x 20'	S.F.	125
Refrigerated Food Cases, dairy, multi deck, 12' long	Ea.	12,800
Delicatessen case, multi deck, 18 S.F. shelf display	Ea.	8875
Freezer, glass door upright, 78 C.F.	Ea.	13,600
Frozen food, chest type, 12' long	Ea.	9450
Glass door reach in, 5 door	Ea.	15,900
Island case 12' long, multi deck	Ea.	9800
Meat case, 12' long, multi deck	Ea.	11,800
Produce, 12' long multi deck	Ea.	10,000
Safe, Office type, 1 hr rating, 60"x36"x18", double door	Ea.	9350

Description	Unit	$ Cost
Sound System, amplifier, 250 watts	Ea.	2200
Speaker, ceiling or wall	Ea.	217
Commissioning Fees, sustainable commercial construction	S.F.	0.24 - 3.04
Energy Modelling Fees, commercial buildings to 10,000 SF	Ea.	11,000
Greater than 10,000 SF add	S.F.	0.04
Green Bldg Cert Fees for comm construction project reg	Project	900
Photovoltaic Pwr Sys, grid connected, 20 kW (~2400 SF), roof	Ea.	245,600
Green Roofs, 6" soil depth, w/treated wd edging & sedum mats	S.F.	11.76
10" Soil depth, with treated wood edging & sedum mats	S.F.	13.45
Greywater Recovery Systems, prepackaged comm, 3060 gal.	Ea.	58,265
4590 gal.	Ea.	73,075
Rainwater Harvest Sys, prepckged comm, 10,000 gal, sys contrller	Ea.	31,525
20,000 gal. w/system controller	Ea.	50,575
30,000 gal. w/system controller	Ea.	80,825
Solar Domestic HW, closed loop, add-on sys, ext heat exchanger	Ea.	10,850
Drainback, hot water system, 120 gal tank	Ea.	13,525
Draindown, hot water system, 120 gal tank	Ea.	13,850

Important: See the Reference Section for Location Factors.

Model costs calculated for a 1 story building with 18' story height and 44,000 square feet of floor area

G Supermarket

				Unit	Unit Cost	Cost Per S.F.	% Of Sub-Total
A. SUBSTRUCTURE							
1010	Standard Foundations	Poured concrete; strip and spread footings		S.F. Ground	1.19	1.19	
1020	Special Foundations	N/A		—	—	—	
1030	Slab on Grade	4" reinforced concrete with recycled vapor barrier and granular base		S.F. Slab	5.58	5.58	9.0%
2010	Basement Excavation	Site preparation for slab and trench for foundation wall and footing		S.F. Ground	.18	.18	
2020	Basement Walls	4' foundation wall		L.F. Wall	93	1.78	
B. SHELL							
	B10 Superstructure						
1010	Floor Construction	N/A		—	—	—	8.5%
1020	Roof Construction	Metal deck, open web steel joists, beams, interior columns		S.F. Roof	8.19	8.19	
	B20 Exterior Enclosure						
2010	Exterior Walls	Face brick with concrete block backup	85% of wall	S.F. Wall	38.79	11.33	
2020	Exterior Windows	Storefront windows	15% of wall	Each	65	3.34	16.6%
2030	Exterior Doors	Sliding entrance doors with electrical operator, hollow metal		Each	5594	1.40	
	B30 Roofing						
3010	Roof Coverings	Single-ply TPO membrane, 60 mils, heat welded seams w/R-20 insul.		S.F. Roof	5.23	5.23	5.5%
3020	Roof Openings	Roof hatches		S.F. Roof	.14	.14	
C. INTERIORS							
1010	Partitions	50% CMU w/foamed-in insul., 50% gyp. bd. on mtl. studs	50 SF Flr./LF Part.	S.F. Partition	9.50	2.28	
1020	Interior Doors	Single leaf hollow metal, low VOC paint	2000 S.F. Floor/Door	Each	1168	.59	
1030	Fittings	N/A		—	—	—	
2010	Stair Construction	N/A		—	—	—	16.8%
3010	Wall Finishes	Paint, low VOC		S.F. Surface	5.63	2.70	
3020	Floor Finishes	Vinyl composition tile, recycled content		S.F. Floor	3.48	3.48	
3030	Ceiling Finishes	Mineral fiber tile on concealed zee bars		S.F. Ceiling	7.22	7.22	
D. SERVICES							
	D10 Conveying						
1010	Elevators & Lifts	N/A		—	—	—	0.0%
1020	Escalators & Moving Walks	N/A		—	—	—	
	D20 Plumbing						
2010	Plumbing Fixtures	Toilet, low flow, auto sensor, & service fixt., supply & drain.	1 Fixt./1820 SF Flr.	Each	2402	1.32	
2020	Domestic Water Distribution	Point of use water heater, electric, energy saver		S.F. Floor	.20	.20	2.7%
2040	Rain Water Drainage	Roof drains		S.F. Roof	1.07	1.07	
	D30 HVAC						
3010	Energy Supply	N/A		—	—	—	
3020	Heat Generating Systems	Included in D3050		—	—	—	
3040	Distribution Systems	Enthalpy heat recovery packages		Each	30,450	2.77	7.3%
3050	Terminal & Package Units	Single zone rooftop air conditioner		S.F. Floor	4.28	4.28	
3090	Other HVAC Sys. & Equipment	N/A		—	—	—	
	D40 Fire Protection						
4010	Sprinklers	Sprinklers, light hazard		S.F. Floor	3.65	3.65	4.4%
4020	Standpipes	Standpipe		S.F. Floor	.59	.59	
	D50 Electrical						
5010	Electrical Service/Distribution	1600 ampere service, panel board and feeders		S.F. Floor	1.91	1.91	
5020	Lighting & Branch Wiring	LED fixtures, daylt. dim., ltg. on/off, receptacles, switches, and A.C.power		S.F. Floor	22.86	22.86	29.0%
5030	Communications & Security	Addressable alarm systems, partial internet wiring and emergency lighting		S.F. Floor	2.28	2.28	
5090	Other Electrical Systems	Emergency generator, 15 kW and energy monitoring systems		S.F. Floor	1.05	1.05	
E. EQUIPMENT & FURNISHINGS							
1010	Commercial Equipment	N/A		—	—	—	
1020	Institutional Equipment	N/A		—	—	—	0.2%
1090	Other Equipment	Waste handling recycling tilt truck		Each	.13	.13	
2020	Moveable Furnishings	No smoking signage		Each	.02	.02	
F. SPECIAL CONSTRUCTION							
1020	Integrated Construction	N/A		—	—	—	0.0%
1040	Special Facilities	N/A		—	—	—	
G. BUILDING SITEWORK	**N/A**						

					Sub-Total	96.76	100%
	CONTRACTOR FEES (General Requirements: 10%, Overhead: 5%, Profit: 10%)				25%	24.17	
	ARCHITECT FEES				7%	8.47	

		Total Building Cost	**129.40**

For customer support on your Square Foot Cost Data, call 877.756.2789.

281

Costs per square foot of floor area

Exterior Wall	S.F. Area	10000	15000	20000	25000	30000	35000	40000	50000	60000
	L.F. Perimeter	410	500	600	640	700	766	833	966	1000
Tiltup Concrete Panels	Steel Frame	153.90	141.65	136.00	130.35	127.20	125.05	123.55	121.45	118.45
Brick with Block Back-up	Bearing Walls	176.05	158.80	151.00	142.55	137.90	134.85	132.70	129.55	124.85
Concrete Block	Steel Frame	159.70	148.85	143.85	138.65	135.80	133.90	132.50	130.55	127.70
	Bearing Walls	143.95	133.05	128.10	122.90	120.05	118.10	116.80	114.85	111.95
Galvanized Steel Siding	Steel Frame	166.70	155.50	150.40	145.00	142.05	140.10	138.70	136.70	133.70
Metal Sandwich Panels	Steel Frame	168.45	155.95	150.30	144.15	140.80	138.55	136.95	134.75	131.30
Perimeter Adj., Add or Deduct	Per 100 L.F.	9.25	6.15	4.60	3.65	3.10	2.65	2.40	1.80	1.55
Story Hgt. Adj., Add or Deduct	Per 1 Ft.	1.10	0.90	0.80	0.65	0.65	0.65	0.60	0.45	0.45

For Basement, add $32.85 per square foot of basement area

The above costs were calculated using the basic specifications shown on the facing page. These costs should be adjusted where necessary for design alternatives and owner's requirements. Reported completed project costs, for this type of structure, range from $42.80 to $170.95 per S.F.

Common additives

Description	Unit	$ Cost
Dock Leveler, 10 ton cap., 6' x 8'	Ea.	6150
7' x 8'	Ea.	8025
Fence, Chain link, 6' high, 9 ga. wire	L.F.	28.50
6 ga. wire	L.F.	32
Gate	Ea.	400
Paving, Bituminous, wearing course plus base course	S.Y.	6.70
Sidewalks, Concrete 4" thick	S.F.	4.63
Sound System, amplifier, 250 watts	Ea.	2200
Speaker, ceiling or wall	Ea.	217
Yard Lighting, 20' aluminum pole, 400W, HP sodium fixture	Ea.	3225

Description	Unit	$ Cost
Commissioning Fees, sustainable industrial construction	S.F.	0.20 - 1.06
Energy Modelling Fees, commercial buildings to 10,000 SF	Ea.	11,000
Greater than 10,000 SF add	S.F.	0.04
Green Bldg Cert Fees for comm construction project reg	Project	900
Photovoltaic Pwr Sys, grid connected, 20 kW (~2400 SF), roof	Ea.	245,600
Green Roofs, 6" soil depth, w/treated wd edging & sedum mats	S.F.	11.76
10" Soil depth, with treated wood edging & sedum mats	S.F.	13.45
Greywater Recovery Systems, prepackaged comm, 3060 gal.	Ea.	58,265
4590 gal.	Ea.	73,075
Rainwater Harvest Sys, prepckged comm, 10,000 gal, sys contrller	Ea.	31,525
20,000 gal. w/system controller	Ea.	50,575
30,000 gal. w/system controller	Ea.	80,825
Solar Domestic HW, closed loop, add-on sys, ext heat exchanger	Ea.	10,850
Drainback, hot water system, 120 gal tank	Ea.	13,525
Draindown, hot water system, 120 gal tank	Ea.	13,850

Important: See the Reference Section for Location Factors.

Model costs calculated for a 1 story building with 24' story height and 30,000 square feet of floor area

G Warehouse

				Unit	Unit Cost	Cost Per S.F.	% Of Sub-Total
A. SUBSTRUCTURE							
1010	Standard Foundations	Poured concrete; strip and spread footings		S.F. Ground	1.52	1.52	
1020	Special Foundations	N/A		—	—	—	
1030	Slab on Grade	5" reinforced concrete with recycled vapor barrier and granular base		S.F. Slab	13.82	13.82	20.4%
2010	Basement Excavation	Site preparation for slab and trench for foundation wall and footing		S.F. Ground	.18	.18	
2020	Basement Walls	4' foundation wall		L.F. Wall	93	2.82	
B. SHELL							
B10 Superstructure							
1010	Floor Construction	Mezzanine: open web steel joists, slab form, concrete beams, columns	10% of area	S.F. Floor	2.20	2.20	10.7%
1020	Roof Construction	Metal deck, open web steel joists, beams, columns		S.F. Roof	7.42	7.42	
B20 Exterior Enclosure							
2010	Exterior Walls	Concrete block	95% of wall	S.F. Wall	17.56	9.34	
2020	Exterior Windows	N/A		—	—	—	11.8%
2030	Exterior Doors	Steel overhead, hollow metal	5% of wall	Each	3347	1.23	
B30 Roofing							
3010	Roof Coverings	Single-ply TPO membrane, 60 mils, heat welded w/R-20 insul.		S.F. Roof	5.40	5.40	6.4%
3020	Roof Openings	Roof hatches and skylight		S.F. Roof	.34	.34	
C. INTERIORS							
1010	Partitions	Concrete block w/foamed-in insul. (office and washrooms)	100 SF Flr./LF Part.	S.F. Partition	11.38	.91	
1020	Interior Doors	Single leaf hollow metal, low VOC paint	5000 S.F. Floor/Door	Each	1168	.24	
1030	Fittings	N/A		—	—	—	
2010	Stair Construction	Steel gate with rails		Flight	14,300	.96	8.3%
3010	Wall Finishes	Paint, low VOC		S.F. Surface	14.63	2.34	
3020	Floor Finishes	90% hardener, 10% vinyl composition tile, recycled content		S.F. Floor	2.25	2.25	
3030	Ceiling Finishes	Suspended mineral tile on zee channels in office area	10% of area	S.F. Ceiling	.72	.73	
D. SERVICES							
D10 Conveying							
1010	Elevators & Lifts	N/A		—	—	—	0.0 %
1020	Escalators & Moving Walks	N/A		—	—	—	
D20 Plumbing							
2010	Plumbing Fixtures	Toilet, low flow, auto sensor & service fixt., supply & drainage	1 Fixt./2500 SF Flr.	Each	4275	1.71	
2020	Domestic Water Distribution	Tankless, on-demand water heater, gas/propane		S.F. Floor	.43	.43	3.9%
2040	Rain Water Drainage	Roof drains		S.F. Roof	1.34	1.34	
D30 HVAC							
3010	Energy Supply	N/A		—	—	—	
3020	Heat Generating Systems	Ventilation with heat		Each	146,900	5.28	
3030	Cooling Generating Systems	N/A		—	—	—	6.7 %
3050	Terminal & Package Units	Single zone rooftop air conditioner, SEER 14		S.F. Floor	.74	.74	
3090	Other HVAC Sys. & Equipment	N/A		—	—	—	
D40 Fire Protection							
4010	Sprinklers	Sprinklers, ordinary hazard		S.F. Floor	4.19	4.19	7.4%
4020	Standpipes	Standpipe		S.F. Floor	2.48	2.48	
D50 Electrical							
5010	Electrical Service/Distribution	200 ampere service, panel board and feeders		S.F. Floor	.31	.31	
5020	Lighting & Branch Wiring	LED fixtures, daylt. dim., ltg. on/off, receptacles, switches, and A.C. power		S.F. Floor	9.97	9.97	16.0%
5030	Communications & Security	Addressable alarm systems		S.F. Floor	2.80	2.80	
5090	Other Electrical Systems	Energy monitoring systems		S.F. Foot	1.25	1.25	
E. EQUIPMENT & FURNISHINGS							
1010	Commercial Equipment	N/A		—	—	—	
1030	Vehicular Equipment	Dock boards, dock levelers		S.F. Floor	1.87	1.87	2.3 %
1090	Other Equipment	Waste handling recycling tilt truck		Each	.19	.19	
2020	Moveable Furnishings	No smoking signage		Each	.02	.02	
F. SPECIAL CONSTRUCTION							
1020	Integrated Construction	Shipping & receiving air curtain		S.F. Floor	5.48	5.48	6.1%
1040	Special Facilities	N/A		—	—	—	
G. BUILDING SITEWORK	**N/A**						

	Sub-Total	89.76	100%
CONTRACTOR FEES (General Requirements: 10%, Overhead: 5%, Profit: 10%)	25%	22.44	
ARCHITECT FEES	7%	7.85	
Total Building Cost		**120.05**	

For customer support on your Square Foot Cost Data, call 877.756.2789.

283

Assemblies Section

Table of Contents

Table of Contents

Table of Contents

Table of Contents

Introduction to the Assemblies Section

This section contains the technical description and installed cost of all components used in the commercial/industrial/ institutional section. In most cases, there is a graphic of the component to aid in identification.

All Costs Include:

Materials purchased in lot sizes typical of normal construction projects with discounts appropriate to established contractors.

Installation work performed by a union labor working under standard conditions at a normal pace.

Installing Contractor's Overhead & Profit

Standard installing contractor's overhead & profit are included in the assemblies costs.

These assemblies costs contain no provisions for the general contractor's overhead and profit or the architect's fees, nor do they include premiums for labor or materials, or savings which may be realized under certain economic situations.

The assemblies tables are arranged by physical size (dimensions) of the component wherever possible.

How RSMeans Assemblies Data Works

Assemblies estimating provides a fast and reasonably accurate way to develop construction costs. An assembly is the grouping of individual work items, with appropriate quantities, to provide a cost for a major construction component in a convenient unit of measure.

An assemblies estimate is often used during early stages of design development to compare the cost impact of various design alternatives on total building cost.

Assemblies estimates are also used as an efficient tool to verify construction estimates.

Assemblies estimates do not require a completed design or detailed drawings. Instead, they are based on the general size of the structure and other known parameters of the project. The degree of accuracy of an assemblies estimate is generally within +/- 15%.

Most RSMeans assemblies consist of three major elements: a graphic, the system components, and the cost data itself. The **Graphic** is a visual representation showing

❶ Unique 12-character Identifier

RSMeans assemblies are identified by a **unique 12-character identifier**. The assemblies are numbered using UNIFORMAT II, ASTM Standard E1557. The first 5 characters represent this system to Level 3. The last 7 characters represent further breakdown by RSMeans in order to arrange items in understandable groups of similar tasks. Line numbers are consistent across all RSMeans publications, so a line number in any RSMeans assemblies data set will always refer to the same work.

❷ Narrative Descriptions

RSMeans assemblies descriptions appear in two formats: narrative and table. **Narrative descriptions** are shown in a hierarchical structure to make them readable. In order to read a complete description, read up through the indents to the top of the section. Include everything that is above and to the left that is not contradicted by information below.

Narrative Format

C20 Stairs

C2010 Stair Construction

The table below lists the cost per flight for 4'-0" wide stairs. Side walls are not included. Railings are included in the prices.

C2010 110	Stairs	COST PER FLIGHT		
		MAT.	INST.	TOTAL
0470	Stairs, C.I.P. concrete, w/o landing, 12 risers, w/o nosing	1,125	2,475	3,600
0480	With nosing	2,100	2,675	4,775
0550	W/landing, 12 risers, w/o nosing	1,300	3,025	4,325
0560	With nosing	2,250	3,225	5,475
0570	16 risers, w/o nosing	1,575	3,775	5,350
0580	With nosing	2,875	4,075	6,950
0590	20 risers, w/o nosing	1,875	4,575	6,450
0600	With nosing	3,475	4,925	8,400
0610	24 risers, w/o nosing	2,150	5,325	7,475
0620	With nosing	4,100	5,750	9,850
0630	Steel, grate type w/nosing & rails, 12 risers, w/o landing	5,700	1,250	6,950
0640	With landing	7,975	1,700	9,675
0660	16 risers, with landing	9,875	2,100	11,975
0680	20 risers, with landing	11,800	2,500	14,300
0700	24 risers, with landing	13,700	2,925	16,625
0701				
0710	Concrete fill metal pan & picket rail, 12 risers, w/o landing	7,375	1,250	8,625
0720	With landing	9,725	1,850	11,575
0740	16 risers, with landing	12,200	2,250	14,450
0760	20 risers, with landing	14,600	2,675	17,275
0780	24 risers, with landing	17,100	3,050	20,150
0790	Cast iron tread & pipe rail, 12 risers, w/o landing	7,375	1,250	8,625
0800	With landing	9,725	1,850	11,575
1120	Wood, prefab box type, oak treads, wood rails 3'-6" wide, 14 risers	2,250	495	2,745
1150	Prefab basement type, oak treads, wood rails 3'-0" wide, 14 risers	1,100	122	1,222

For supplemental customizable square foot estimating forms, visit: **www.RSMeans.com/2016extras**

the typical appearance of the assembly in question, frequently accompanied by additional explanatory technical information describing the class of items. The **Assemblies Data** below lists prices for other similar systems with dimensional and/or size variations.

All RSMeans assemblies costs represent the cost for the installing contractor. An allowance for profit has been added to all material, labor, and equipment rental costs. A markup for labor burdens, including workers' compensation, fixed overhead, and business overhead, is included with installation costs.

The information in RSMeans cost data represents a "national average" cost. This data should be modified to the project location using the **Location Factors** tables found in the Reference Section.

Table Format

A20 Basement Construction

A2020 Basement Walls

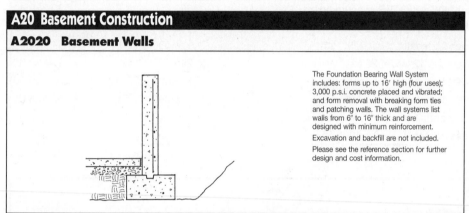

The Foundation Bearing Wall System includes: forms up to 16' high (four uses); 3,000 p.s.i. concrete placed and vibrated; and form removal with breaking form ties and patching walls. The wall systems list walls from 6" to 16" thick and are designed with minimum reinforcement.

Excavation and backfill are not included.

Please see the reference section for further design and cost information.

③ Unit of Measure

All RSMeans assemblies data includes a typical **Unit of Measure** used for estimating that item. For instance, for continuous footings or foundation walls the unit is linear feet (L.F.). For spread footings the unit is each (Ea.). The estimator needs to take special care that the unit in the data matches the unit in the takeoff. Abbreviations and unit conversions can be found in the Reference Section.

④ Table Descriptions

Table descriptions work in a similar fashion, except that if there is a blank in the column at a particular line number, read up to the description above in the same column.

A2020 110		Walls, Cast in Place						③
	WALL HEIGHT (FT.)	PLACING METHOD	CONCRETE (C.Y. per L.F.)	REINFORCING (LBS. per L.F.)	WALL THICKNESS (IN.)	MAT.	INST.	TOTAL
1500	4'	direct chute	.074	3.3	6	17	54	71
1520			.099	4.8	8	20.50	55.50	76
1540			.123	6.0	10	24	56.50	80.50
1560			.148	7.2	12	28	57.50	85.50
1580			.173	8.1	14	31	58.50	89.50
1600			.197	9.44	16	34.50	60	94.50
3000	6'	direct chute	.111	4.95	6	25.50	81	106.50
3020			.149	7.20	8	31	83	114
3040			.184	9.00	10	36	84.50	120.50
3060			.222	10.8	12	41.50	86.50	128
5000	8'	direct chute	.148	6.6	6	34	108	142
5020			.199	9.6	8	41.50	114	155.50
5040			.250	12	10	49	113	162
5061			.296	14.39	12	55.50	115	170.50
6020	10'	direct chute	.248	12	8	52	138	190
6040			.307	14.99	10	60.50	141	201.50
6061			.370	17.99	12	69.50	144	213.50
7220	12'	pumped	.298	14.39	8	62.50	172	234.50
7240			.369	17.99	10	72.50	175	247.50
7260			.444	21.59	12	83.50	180	263.50
9220	16'	pumped	.397	19.19	8	83	229	312
9240			.492	23.99	10	96.50	233	329.50
9260			.593	28.79	12	111	240	351

How RSMeans Assemblies Data Works (Continued)

Sample Estimate

This sample demonstrates the elements of an estimate, including a tally of the RSMeans data lines. Published assemblies costs include all markups for labor burden and profit for the installing contractor. This estimate adds a summary of the markups applied by a general contractor on the installing contractor's work. These figures represent the total cost to the owner. The RSMeans location factor is applied at the bottom of the estimate to adjust the cost of the work to a specific location.

Project Name:	Interior Fit-out, ABC Office			
Location:	**Anywhere, USA**		**Date: 1/1/2016**	**STD**
Assembly Number	**Description**	**Qty.**	**Unit**	**Subtotal**
C1010 ❶ 4 1200	Wood partition, 2 x 4 @ 16" OC w/5/8" FR gypsum board	560.000	S.F.	$2,738.40
C1020 114 1800	Metal door & frame, flush hollow core, 3'-0" x 7'-0"	2.000	Ea.	$2,470.00
C3010 230 0080	Painting, brushwork, primer & 2 coats	1,120.000	S.F.	$1,344.00
C3020 410 0140	Carpet, tufted, nylon, roll goods, 12' wide, 26 oz	240.000	S.F.	$813.60
C3030 210 6000	Acoustic ceilings, 24" x 48" tile, tee grid suspension	200.000	S.F.	$1,156.00
D5020 125 0560	Receptacles incl plate, box, conduit, wire, 20 A duplex	8.000	Ea.	$2,168.00
D5020 125 0720	Light switch incl plate, box, conduit, wire, 20 A single pole	2.000	Ea.	$532.00
D5020 210 0560	Fluorescent fixtures, recess mounted, 20 per 1000 SF	200.000	S.F.	$2,186.00
	Assembly Subtotal			**$13,408.00**
	Sales Tax @ ❷	5 %		$ 335.20
	General Requirements @ ❸	7 %		$ 938.56
	Subtotal A			**$14,681.76**
	GC Overhead @ ❹	5 %		$ 734.09
	Subtotal B			**$15,415.85**
	GC Profit @ ❺	5 %		$ 770.79
	Subtotal C			**$16,186.64**
Adjusted by Location Factor ❻		118.1		$ 19,116.42
	Architects Fee @ ❼	8 %		$ 1,529.31
	Contingency @ ❽	15 %		$ 2,867.46
	Project Total Cost			**$ 23,513.20**

This estimate is based on an interactive spreadsheet.
A copy of this spreadsheet is located on the RSMeans website at
www.RSMeans.com/2016extras.
You are free to download it and adjust it to your methodology.

1. Work Performed

The body of the estimate shows the RSMeans data selected, including line numbers, a brief description of each item, its takeoff quantity and unit, and the total installed cost, including the installing contractor's overhead and profit.

2. Sales Tax

If the work is subject to state or local sales taxes, the amount must be added to the estimate. In a conceptual estimate it can be assumed that one half of the total represents material costs. Therefore, apply the sales tax rate to 50% of the assembly subtotal.

3. General Requirements

This item covers project-wide needs provided by the general contractor. These items vary by project but may include temporary facilities and utilities, security, testing, project cleanup, etc. In assemblies estimates a percentage is used, typically between 5% and 15% of project cost.

4. General Contractor Overhead

This entry represents the general contractor's markup on all work to cover project administration costs.

5. General Contractor Profit

This entry represents the GC's profit on all work performed. The value included here can vary widely by project, and is influenced by the GC's perception of the project's financial risk and market conditions.

6. Location Factor

RSMeans published data is based on national average costs. If necessary, adjust the total cost of the project using a location factor from the "Location Factor" table or the "City Cost Indexes" table found in the Reference Section. Use location factors if the work is general, covering the work of multiple trades. If the work is by a single trade (e.g., masonry) use the more specific data found in the City Cost Indexes.

To adjust costs by location factors, multiply the base cost by the factor and divide by 100.

7. Architect's Fee

If appropriate, add the design cost to the project estimate. These fees vary based on project complexity and size. Typical design and engineering fees can be found in the reference section.

8. Contingency

A factor for contingency may be added to any estimate to represent the cost of unknowns that may occur between the time that the estimate is performed and the time the project is constructed. The amount of the allowance will depend on the stage of design at which the estimate is done, and the contractor's assessment of the risk involved.

Example

Assemblies costs can be used to calculate the cost for each component of a building and, accumulated with appropriate markups, produce a cost for the complete structure. Components from a model building with similar characteristics in conjunction with Assemblies components may be used to calculate costs for a complete structure.

Example:

Outline Specifications

General

Building size—60' x 100', 8 suspended floors, 12' floor to floor, 4' high parapet above roof, full basement 11'-8" floor to floor, bay size 25' x 30', ceiling heights -9' in office area and 8' in core area.

A Substructure

Concrete, spread and strip footings, concrete walls waterproofed, 4" slab on grade.

B Shell

B10 Superstructure

Columns, wide flange; 3 hr. fire rated; Floors, composite steel frame & deck with concrete slab; Roof, steel beams, open web joists and deck.

B20 Exterior Enclosure

Walls; North, East & West, brick & lightweight concrete block with 2" cavity insulation, 25% window; South, 8" lightweight concrete block insulated, 10% window. Doors, aluminum & glass. Windows, aluminum, 3'-0" x 5'-4", insulating glass.

B30 Roofing

Tar & gravel, 2" rigid insulation, R 12.5.

C Interiors

Core—6" lightweight concrete block, full height with plaster on exposed faces to ceiling height.

Corridors—1st & 2nd floor; 3-5/8" steel studs with F.R. gypsum board, full height.

Exterior Wall—plaster, ceiling height.

Fittings—Toilet accessories, directory boards.

Doors—hollow metal.

Wall Finishes—lobby, mahogany paneling on furring, remainder plaster & gypsum board, paint.

Floor Finishes—1st floor lobby, corridors & toilet rooms, terrazzo; remainder concrete, tenant developed. 2nd through 8th, toilet rooms, ceramic tile; office & corridor, carpet.

Ceiling Finishes—24" x 48" fiberglass board on Tee grid.

D Services

D10 Conveying

2-2500 lb. capacity, 200 F.P.M. geared elevators, 9 stops.

D20 Plumbing

Fixtures, see sketch.

Roof Drains—2-4" C.I. pipe.

D30 HVAC

Heating—fin tube radiation, forced hot water.

Air Conditioning—chilled water with water cooled condenser.

D40 Fire Protection

Fire Protection—4" standpipe, 9 hose cabinets.

D50 Electrical

Lighting, 1st through 8th, 15 fluorescent fixtures/1000 S.F., 3 Watts/S.F.

Basement, 10 fluorescent fixtures/1000 S.F., 2 Watts/S.F.

Receptacles, 1st through 8th, 16.5/1000 S.F., 2 Watts/S.F.

Basement, 10 receptacles/1000 S.F., 1.2 Watts/S.F.

Air Conditioning, 4 Watts/S.F.

Miscellaneous Connections, 1.2 Watts/S.F.

Elevator Power, 2-10 H.P., 230 volt motors.

Wall Switches, 2/1000 S.F.

Service, panel board & feeder, 2000 Amp.

Fire Detection System, pull stations, signals, smoke and heat detectors.

Emergency Lighting Generator, 30 KW.

E Equipment & Furnishings — NA.

F Special Construction — NA.

G Building Sitework — NA.

Front Elevation

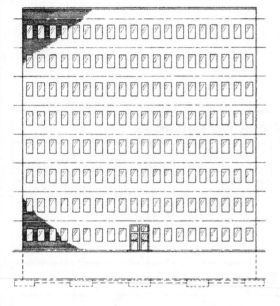

Basement Plan

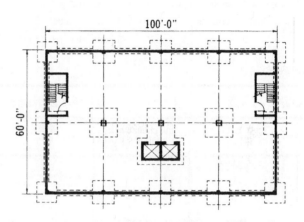

Ground Floor Plan

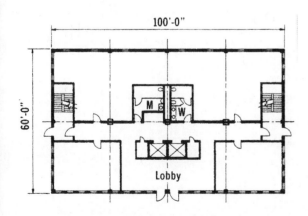

Typical Floor Plan

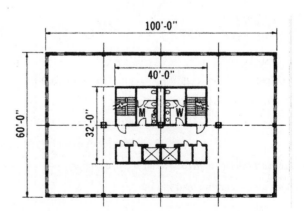

Example

If spread footing & column sizes are unknown, develop approximate loads as follows. Enter tables with these loads to determine costs.

Approximate loads/S.F. for roof & floors.

Roof. Assume 40 psf superimposed load.

Steel joists, beams & deck.

Table B1020 112—Line 3900

Superimposed Load Ranges

Apartments & Residential Structures	65	to	75 psf
Assembly Areas & Retail Stores	110	to	125 psf
Commercial & Manufacturing	150	to	250 psf
Offices	75	to	100 psf

B1020 112	**Steel Joists, Beams, & Deck on Columns**							
	BAY SIZE (FT.)	SUPERIMPOSED LOAD (P.S.F.)	DEPTH (IN.)	TOTAL LOAD (P.S.F.)	COLUMN ADD	COST PER S.F.		
						MAT.	INST.	TOTAL
3500	25x30	20	22	40		5.55	1.63	7.18
3600					columns	1.21	.28	1.49
3900		40	25	60		6.70	1.93	8.63
4000					columns	1.45	.34	1.79

B1010 256	**Composite Beams, Deck & Slab**							
	BAY SIZE (FT.)	SUPERIMPOSED LOAD (P.S.F.)	SLAB THICKNESS (IN.)	TOTAL DEPTH (FT.-IN.)	TOTAL LOAD (P.S.F.)	COST PER S.F.		
						MAT.	INST.	TOTAL
3400	25x30	40	5-1/2	1 - 11-1/2	83	12.90	5.95	18.85
3600		75	5-1/2	1 - 11-1/2	119	13.85	6.05	19.90
3900		125	5-1/2	1 - 11-1/2	170	16.10	6.80	22.90
4000		200	6-1/4	2 - 6-1/4	252	19.90	7.70	27.60

Floors—Total load, 119 psf.

Interior foundation load.

Roof

[(25' x 30' x 60 psf) + 8 floors x (25' x 30' x 119 psf)] x 1/1000 lb./Kip	=	759 Kips
Approximate Footing Loads, Interior footing	=	759 Kips
Exterior footing (1/2 bay) 759 k x .6	=	455 Kips
Corner footing (1/4 bay) 759 k x .45	=	342 Kips
[Factors to convert Interior load to Exterior & Corner loads]		
Approximate average Column load 759 k/2	=	379 Kips

PRELIMINARY ESTIMATE

PROJECT	**Office Building**	TOTAL SITE AREA	
BUILDING TYPE		OWNER	
LOCATION		ARCHITECT	
DATE OF CONSTRUCTION		ESTIMATED CONSTRUCTION PERIOD	

BRIEF DESCRIPTION **Building size: 60' x 100' - 8 structural floors - 12' floor to floor**

4' high parapet above roof, full basement 11' - 8" floor to floor, bay size

25' x 30', ceiling heights - 9' in office area + 8' in core area

TYPE OF PLAN		TYPE OF CONSTRUCTION	
QUALITY		BUILDING CAPACITY	

Floor				**Wall Area**				
Below Grade Levels				Foundation Walls	L.F.		Ht.	S.F.
Area			S.F.	Frost Walls	L.F.		Ht.	S.F.
Area			S.F.	Exterior Closure		Total		S.F.
Total Area			S.F.	Comment				
Ground Floor				Fenestration		%		S.F.
Area			S.F.			%		S.F.
Area			S.F.	Exterior Wall		%		S.F.
Total Area			S.F.			%		S.F.
Supported Levels				**Site Work**				
Area				Parking		S.F. (For		Cars)
Area			S.F.	Access Roads		L.F. (X		Ft. Wide)
Area			S.F.	Sidewalk		L.F. (X		Ft. Wide)
Area			S.F.	Landscaping		S.F. (		% Unbuilt Site)
Area			S.F.	**Building Codes**				
Total Area			S.F.	City		Country		
Miscellaneous				National		Other		
Area			S.F.	**Loading**				
Area			S.F.	Roof	psf	Ground Floor		psf
Area			S.F.	Supported Floors	psf	Corridor		psf
Area			S.F.	Balcony	psf	Partition, allow		psf
Total Area			S.F.	Miscellaneous				psf
Net Finished Area			S.F.	Live Load Reduction				
Net Floor Area			S.F.	Wind				
Gross Floor Area		**54,000**	S.F.	Earthquake		Zone		
Roof				Comment				
Total Area			S.F.	Soil Type				
Comments				Bearing Capacity				K.S.F.
				Frost Depth				Ft.
Volume				**Frame**				
Depth of Floor System				Type		Bay Spacing		
Minimum			In.	Foundation				
Maximum			In.	Special				
Foundation Wall Height			Ft.	Substructure				
Floor to Floor Height			Ft.	Comment				
Floor to Ceiling Height			Ft.	Superstructure, Vertical				
Subgrade Volume			C.F.	Fireproofing		☐ Columns		Hrs.
Above Grade Volume			C.F.	☐ Girders	Hrs.	☐ Beams		Hrs.
Total Building Volume		**648,000**	C.F.	☐ Floor	Hrs.	☐ None		

NUMBER		QTY.	UNIT	TOTAL COST UNIT	TOTAL COST TOTAL	COST PER S.F.
A	**SUBSTRUCTURE**					
A1010 210 7900	Corner Footings 8'-6" SQ. x 27"	4	Ea.	1,970.00	7,880	
8010	Exterior 9' -6" SQ. x 30"	8		2,650.00	21,200	
8300	Interior 12' SQ.	3		4,850.00	14,550	
A1010 110 2700	Strip 2' Wide x 1' Thick					
	320 L.F.- [(4 x 8.5) + 8 x 9.5)] =	210	L.F.	41.15	8,642	
A2020 110 7260	Foundation Wall 12' High, 1' Thick	320		263.50	84,320	
A1010 320 2800	Foundation Waterproofing	320		24.25	7,760	
A1030 120 2240	4", non-industrial, reinf. slab on grade	6000	S.F.	5.56	33,360	
A2010 110 3440	Building Excavation + Backfill	6000	S.F.	7.43	44,580	
3500	(Interpolated ; 12' Between					
4620	8' and 16' ; 6,000 Between					
4680	4,000 and 10,000 S.F.					
	Total				222,292	4.12
B	**SHELL**					
B10	**Superstructure**					
B1010 208 5800	Columns: Load 379K Use 400K					
	Exterior 12 x 96' + Interior 3 x 108'	1476	V.L.F.	150.85	222,655	
B1010 720 3650	Column Fireproofing - Interior 4 sides					
3700	Exterior 1/2 (Interpolated for 12")	900	V.L.F	37.05	33,345	
B1010 256 3600	Floors: Composite steel + lt. wt. conc.	48000	S.F.	19.90	955,200	
B1020 112 3900	Roof: Open web joists, beams + decks	6000	S.F.	8.63	51,780	
	Total				1,262,980	23.39
B20	**Exterior Enclosure**					
B2010 134 1200	4" Brick + 6" Block - Insulated					
	75% NE + W Walls 220' x 100' x .75	16500	S.F.	31.00	511,500	
B2010 110 3410	8" Block - Insulated 90%					
	100' x 100' x .9	9000	S.F.	12.30	110,700	
B2030 110 6950	Double Aluminum + Glass Door	1	Pr.	5,875.00	5,875	
6300	Single	2	Ea.	2,975.00	5,950	
B2020 106 8800	Aluminum windows - Insul. glass					
	(5,500 S.F. + 1,000 S.F.)/(3 x 5.33)	406	Ea.	909.00	369,054	
B2010 105 8480	Precast concrete coping	320	L.F.	42.65	13,648	
	Total				1,016,727	18.83
B30	**Roofing**					
B3010 320 4020	Insulation: Perlite/Polyisocyanurate	6000	S.F.	1.98	11,880	
B3010 105 1400	Roof: 4 Ply T & G composite	6000		3.07	18,420	
B3010 430 0300	Flashing: Aluminum - fabric backed	640		3.48	2,227	
B3020 210 0500	Roof hatch	1	Ea.	1,659.00	1,659	
	Total				34,186	0.63

ASSEMBLY NUMBER		QTY.	UNIT	TOTAL COST		COST PER S.F.
				UNIT	TOTAL	
C	**INTERIORS**					
C1010 104 5500	Core partitions: 6" Lt. Wt. concrete					
	block 288 L.F. x 11.5' x 8 Floors	26,449	S.F.	9.73	257,349	
C1010 144 0920	Plaster: [(196 L.F. x 8') + (144 L.F. x 9')] 8					
	+ Ext. Wall [(320 x 9 x 8 Floors) - 6500]	39,452	S.F.	3.27	129,008	
C1010 126 5400	Corridor partitions 1st + 2nd floors					
	steel studs + F.R. gypsum board	2,530	S.F.	4.34	10,980	
C1020 102 2600	Interior doors	66	Ea.	1,168.00	77,088	
	Wall finishes - Lobby: Mahogany w/furring					
C1010 128 0652	Furring: 150 L.F. x 9' + Ht.	1,350	S.F.	1.66	2,241	
C1030 710 0120	Towel Dispenser	16	Ea.	89.00	1,424	
C1030 710 0140	Grab Bar	32	Ea.	58.50	1,872	
C1030 710 0160	Mirror	31	Ea.	247.50	7,673	
C1030 710 0180	Toilet Tissue Dispenser	31	Ea.	40.80	1,265	
C1030 510 0140	Directory Board	8	Ea.	353.00	2,824	
C2010 110 0780	Stairs: Steel w/conc. fill	18	Flt.	20,150.00	362,700	
C3010 230 1662	Mahogany paneling	1,350	S.F.	6.05	8,168	
C3010 230 0080	Paint, primer + 2 coats finish, on plaster or gyp. board					
	39,452 + [(220' x 9' x 2) - 1350] + 72	42,134	S.F.	1.20	50,561	
	Floor Finishes					
C3020 410 1100	1st floor lobby + terrazzo	2,175	S.F.	21.19	46,088	
	Remainder: Concrete - Tenant finished					
C3020 410 0060	2nd - 8th: Carpet - 4725 S.F. x 7	33,075	S.F.	4.45	147,184	
C3020 410 1720	Ceramic Tile - 300 S.F. x 7	2,100	S.F.	11.45	24,045	
C3030 240 2780	Suspended Ceiling Tiles- 5200 S.F. x 8	41,600	S.F.	2.13	88,608	
C3030 240 3260	Suspended Ceiling Grid- 5200 S.F. x 8	41,600	S.F.	1.58	65,728	
C1030 110 0700	Toilet partitions	31	Ea.	755.00	23,405	
0760	Handicapped addition	16	Ea.	370.00	5,920	
	Total				1,314,130	24.34
D	**SERVICES**					
D10	**Conveying**					
	2 Elevators - 2500# 200'/min. Geared					
D1010 140 1600	5 Floors $175,600					
1800	15 Floors $402,500					
	Diff. $226,900/10 = 22,690/Flr.					
	$175,600 + (4 x22,690) = $266,360/Ea.	2	Ea.	266,360.00	532,720	
	Total				532,720	9.87
D20	**Plumbing**					
D2010 310 1600	Plumbing - Lavatories	31	Ea.	1,155.00	35,805	
D2010 440 4340	-Service Sink	8	Ea.	3,450.00	27,600	
D2010 210 2000	-Urinals	8	Ea.	1,405.00	11,240	

ASSEMBLY NUMBER		QTY.	UNIT	TOTAL COST UNIT	TOTAL	COST PER S.F.
D2010 110 2080	-Water Closets	31	Ea.	3,085.00	95,635	
	Water Control, Waste Vent Piping	45%			76,626	
D2040 210 4200	- Roof Drains, 4" C.I.	2	Ea.	1,980.00	3,960	
4240	- Pipe, 9 Flr. x 12' x 12' Ea.	216	L.F.	45.00	9,720	
	Total				260,586	4.83
D30	HVAC					
D3010 520 2000	10,000 S.F. @ $10.65/S.F. Interpolate					
2040	100,000 S.F. @ $4.57/S.F.	48,000	S.F.	7.41	355,680	
	Cooling - Chilled Water, Air Cool, Cond.					
D3030 115 3840	4,000 S.F. @17.75/S.F. Interpolate	48,000	S.F.	15.83	759,840	
4040	60,000 S.F. @15.30/S.F.					
	Total				1,115,520	20.66
D40	Fire Protection					
D4020 310 0560	Wet Stand Pipe: 4" x 10' 1st floor	12/10	Ea.	9,075.00	10,890	
0580	Additional Floors: (12'/10 x 8)	9.6	Ea.	2,425.00	23,280	
D4020 410 8400	Cabinet Assembly	9	Ea.	1,305.00	11,745	
	Total				45,915	0.85
D50	Electrical					
D5020 210 0280	Office Lighting, 15/1000 S.F. - 3 Watts/S.F.	48,000	S.F.	8.05	386,400	
0240	Basement Lighting, 10/1000 S.F. - 2 Watts/S.F.	6,000	S.F.	5.35	32,100	
D5020 110 0640	Office Receptacles 16.5/1000 S.F. - 2 Watts/S.F.	48,000	S.F.	4.85	232,800	
0560	Basement Receptacles 10/1000 S.F. - 1.2 Watts/S.F.	6,000	S.F.	3.71	22,260	
D5020 140 0280	Central A.C. - 4 Watts/S.F.	48,000	S.F.	0.65	31,200	
D5020 135 0320	Misc. Connections - 1.2 Watts/S.F.	48,000	S.F.	0.35	16,800	
D5020 145 0680	Elevator Motor Power - 10 H.P.	2	Ea.	2,850.00	5,700	
D5020 130 0280	Wall Switches - 2/1000 S.F.	54,000	S.F.	0.50	27,000	
D5010 120 0560	2000 Amp Service	1	Ea.	30,300.00	30,300	
D5010 240 0400	Switchgear	1	Ea.	44,500.00	44,500	
D5010 230 0560	Feeder	50	L.F.	498.00	24,900	
D5030 910 0400	Fire Detection System - 50 Detectors	1	Ea.	35,400.00	35,400	
D5090 210 0320	Emergency Generator - 30 kW	30	kW	628.00	18,840	
	Total				908,200	16.82
E	EQUIPMENT & FURNISHINGS					
G	BUILDING SITEWORK					
	Miscellaneous					

Preliminary Estimate Cost Summary

PROJECT: Office Building	TOTAL AREA	54,000 S.F.	SHEET NO.
LOCATION	TOTAL VOLUME	648,000 C.F.	ESTIMATE NO.
ARCHITECT	COST PER S.F.		DATE
OWNER	COST PER C.F.		NO OF STORIES
QUANTITIES BY	EXTENSIONS BY		CHECKED BY

DIV	DESCRIPTION	SUBTOTAL COST	COST/S.F.		PERCENTAGE
A	SUBSTRUCTURE	222,292	$	4.12	
B10	SHELL: SUPERSTRUCTURE	1,262,980	$	23.39	
B20	SHELL: EXTERIOR ENCLOSURE	1,016,727	$	18.83	
B30	SHELL: ROOFING	34,186	$	0.63	
C	INTERIORS	1,314,130	$	24.34	
D10	SERVICES: CONVEYING	532,720	$	9.87	
D20	SERVICES: PLUMBING	260,586	$	4.83	
D30	SERVICES: HVAC	1,115,520	$	20.66	
D40	SERVICES: FIRE PROTECTION	45,915	$	0.85	
D50	SERVICES: ELECTRICAL	908,200	$	16.82	
E	EQUIPMENT & FURNISHINGS				
F	SPECIAL CONSTRUCTION & DEMO.				
G	BUILDING SITEWORK				
	BUILDING SUBTOTAL	$ 6,713,255			$ 6,713,255

Sales Tax % x Subtotal /2	N/A		$ -
General Conditions 10 % x Subtotal			$ 671,326
		Subtotal "A"	$ 7,384,581
Overhead 5 % x Subtotal "A"			$ 369,229
		Subtotal "B"	$ 7,753,810
Profit 10% x Subtotal "B"			$ 775,381
		Subtotal "C"	$ 8,529,191
Location Factor % x Subtotal "C"	N/A	Localized Cost	
Architects Fee 6.5% x Localized Cost =			$ 554,397
Contingency x Localized Cost =	N/A		$ -
		Project Total Cost	$ 9,083,588
Square Foot Cost $9,081,534/ 54,000 S.F. =		S.F. Cost	$ 168.21
Cubic Foot Cost $9,081,534/ 648,000 C.F. =		C.F. Cost	$ 14.02

Did you know?

RSMeans Online gives you the same access
to RSMeans' data with 24/7 access:

- Quickly locate costs in the searchable database.
- Build cost lists, estimates, and reports in minutes.
- Adjust costs to any location in the U.S. and Canada with the click of a button.

Start your free trial today at **www.RSMeansOnline.com**

RSMeans Online
FROM THE GORDIAN GROUP

A1010 Standard Foundations

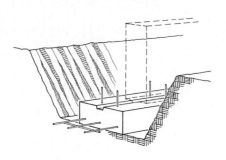

The Strip Footing System includes: excavation; hand trim; all forms needed for footing placement; forms for 2″ x 6″ keyway (four uses); dowels; and 3,000 p.s.i. concrete.

The footing size required varies for different soils. Soil bearing capacities are listed for 3 KSF and 6 KSF. Depths of the system range from 8″ and deeper. Widths range from 16″ and wider. Smaller strip footings may not require reinforcement.

Please see the reference section for further design and cost information.

A1010 110	Strip Footings	COST PER L.F.		
		MAT.	INST.	TOTAL
2100	Strip footing, load 2.6 KLF, soil capacity 3 KSF, 16″ wide x 8″ deep, plain	7.50	11.15	18.65
2300	Load 3.9 KLF, soil capacity 3 KSF, 24″ wide x 8″ deep, plain	9.50	12.55	22.05
2500	Load 5.1 KLF, soil capacity 3 KSF, 24″ wide x 12″ deep, reinf.	17.15	24	41.15
2700	Load 11.1 KLF, soil capacity 6 KSF, 24″ wide x 12″ deep, reinf.	17.15	24	41.15
2900	Load 6.8 KLF, soil capacity 3 KSF, 32″ wide x 12″ deep, reinf.	20.50	26	46.50
3100	Load 14.8 KLF, soil capacity 6 KSF, 32″ wide x 12″ deep, reinf.	20.50	26	46.50
3300	Load 9.3 KLF, soil capacity 3 KSF, 40″ wide x 12″ deep, reinf.	24	28.50	52.50
3500	Load 18.4 KLF, soil capacity 6 KSF, 40″ wide x 12″ deep, reinf.	24.50	28.50	53
4500	Load 10 KLF, soil capacity 3 KSF, 48″ wide x 16″ deep, reinf.	34.50	36	70.50
4700	Load 22 KLF, soil capacity 6 KSF, 48″ wide, 16″ deep, reinf.	35	37	72
5700	Load 15 KLF, soil capacity 3 KSF, 72″ wide x 20″ deep, reinf.	59	51	110
5900	Load 33 KLF, soil capacity 6 KSF, 72″ wide x 20″ deep, reinf.	62.50	54.50	117

For customer support on your Square Foot Cost Data, call 877.756.2789.

A1010 Standard Foundations

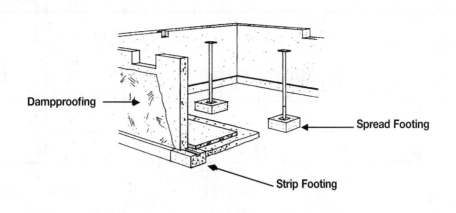

Dampproofing

Spread Footing

Strip Footing

A1010 210	Spread Footings	COST EACH		
		MAT.	INST.	TOTAL
7090	Spread footings, 3000 psi concrete, chute delivered			
7100	Load 25K, soil capacity 3 KSF, 3'-0" sq. x 12" deep	60.50	127	187.50
7150	Load 50K, soil capacity 3 KSF, 4'-6" sq. x 12" deep	129	218	347
7200	Load 50K, soil capacity 6 KSF, 3'-0" sq. x 12" deep	60.50	127	187.50
7250	Load 75K, soil capacity 3 KSF, 5'-6" sq. x 13" deep	205	305	510
7300	Load 75K, soil capacity 6 KSF, 4'-0" sq. x 12" deep	104	187	291
7350	Load 100K, soil capacity 3 KSF, 6'-0" sq. x 14" deep	259	370	629
7410	Load 100K, soil capacity 6 KSF, 4'-6" sq. x 15" deep	159	256	415
7450	Load 125K, soil capacity 3 KSF, 7'-0" sq. x 17" deep	410	525	935
7500	Load 125K, soil capacity 6 KSF, 5'-0" sq. x 16" deep	205	305	510
7550	Load 150K, soil capacity 3 KSF 7'-6" sq. x 18" deep	500	610	1,110
7610	Load 150K, soil capacity 6 KSF, 5'-6" sq. x 18" deep	273	385	658
7650	Load 200K, soil capacity 3 KSF, 8'-6" sq. x 20" deep	710	810	1,520
7700	Load 200K, soil capacity 6 KSF, 6'-0" sq. x 20" deep	360	475	835
7750	Load 300K, soil capacity 3 KSF, 10'-6" sq. x 25" deep	1,300	1,325	2,625
7810	Load 300K, soil capacity 6 KSF, 7'-6" sq. x 25" deep	680	795	1,475
7850	Load 400K, soil capacity 3 KSF, 12'-6" sq. x 28" deep	2,075	1,950	4,025
7900	Load 400K, soil capacity 6 KSF, 8'-6" sq. x 27" deep	945	1,025	1,970
8010	Load 500K, soil capacity 6 KSF, 9'-6" sq. x 30" deep	1,300	1,350	2,650
8100	Load 600K, soil capacity 6 KSF, 10'-6" sq. x 33" deep	1,750	1,725	3,475
8200	Load 700K, soil capacity 6 KSF, 11'-6" sq. x 36" deep	2,250	2,100	4,350
8300	Load 800K, soil capacity 6 KSF, 12'-0" sq. x 37" deep	2,525	2,325	4,850
8400	Load 900K, soil capacity 6 KSF, 13'-0" sq. x 39" deep	3,125	2,775	5,900
8500	Load 1000K, soil capacity 6 KSF, 13'-6" sq. x 41" deep	3,550	3,075	6,625

A1010 320	Foundation Dampproofing	COST PER L.F.		
		MAT.	INST.	TOTAL
1000	Foundation dampproofing, bituminous, 1 coat, 4' high	1	4.80	5.80
1400	8' high	2	9.60	11.60
1800	12' high	3	14.85	17.85
2000	2 coats, 4' high	1.96	5.95	7.91
2400	8' high	3.92	11.90	15.82
2800	12' high	5.90	18.35	24.25
3000	Asphalt with fibers, 1/16" thick, 4' high	1.56	5.95	7.51
3400	8' high	3.12	11.90	15.02
3800	12' high	4.68	18.35	23.03
4000	1/8" thick, 4' high	2.80	7.10	9.90
4400	8' high	5.60	14.15	19.75
4800	12' high	8.40	21.50	29.90
5000	Asphalt coated board and mastic, 1/4" thick, 4' high	5.10	6.45	11.55
5400	8' high	10.25	12.85	23.10

A1010 Standard Foundations

A1010 320	Foundation Dampproofing	COST PER L.F.		
		MAT.	INST.	TOTAL
5800	12' high	15.35	19.75	35.10
6000	1/2" thick, 4' high	7.70	8.95	16.65
6400	8' high	15.35	17.90	33.25
6800	12' high	23	27.50	50.50
7000	Cementitious coating, on walls, 1/8" thick coating, 4' high	3.24	8.95	12.19
7400	8' high	6.50	17.95	24.45
7800	12' high	9.70	27	36.70
8000	Cementitious/metallic slurry, 4 coat, 1/2"thick, 2' high	.77	9	9.77
8400	4' high	1.54	18	19.54
8800	6' high	2.31	27	29.31

A1030 Slab on Grade

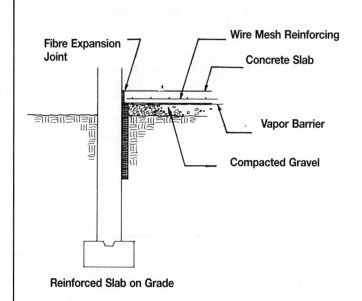

Fibre Expansion Joint

Wire Mesh Reinforcing

Concrete Slab

Vapor Barrier

Compacted Gravel

Reinforced Slab on Grade

A Slab on Grade system includes fine grading; 6″ of compacted gravel; vapor barrier; 3500 p.s.i. concrete; bituminous fiber expansion joint; all necessary edge forms 4 uses; steel trowel finish; and sprayed on membrane curing compound. Wire mesh reinforcing used in all reinforced slabs.

Non-industrial slabs are for foot traffic only with negligible abrasion. Light industrial slabs are for pneumatic wheels and light abrasion. Industrial slabs are for solid rubber wheels and moderate abrasion. Heavy industrial slabs are for steel wheels and severe abrasion.

A1030 120	Plain & Reinforced	COST PER S.F.		
		MAT.	INST.	TOTAL
2220	Slab on grade, 4″ thick, non industrial, non reinforced	2.22	2.81	5.03
2240	Reinforced	2.37	3.19	5.56
2260	Light industrial, non reinforced	2.92	3.45	6.37
2280	Reinforced	3.07	3.83	6.90
2300	Industrial, non reinforced	3.59	7.15	10.74
2320	Reinforced	3.74	7.55	11.29
3340	5″ thick, non industrial, non reinforced	2.59	2.89	5.48
3360	Reinforced	2.74	3.27	6.01
3380	Light industrial, non reinforced	3.29	3.53	6.82
3400	Reinforced	3.44	3.91	7.35
3420	Heavy industrial, non reinforced	4.71	8.55	13.26
3440	Reinforced	4.81	9	13.81
4460	6″ thick, non industrial, non reinforced	3.07	2.82	5.89
4480	Reinforced	3.30	3.33	6.63
4500	Light industrial, non reinforced	3.79	3.46	7.25
4520	Reinforced	4.16	4.15	8.31
4540	Heavy industrial, non reinforced	5.20	8.65	13.85
4560	Reinforced	5.45	9.15	14.60
5580	7″ thick, non industrial, non reinforced	3.45	2.92	6.37
5600	Reinforced	3.75	3.47	7.22
5620	Light industrial, non reinforced	4.18	3.56	7.74
5640	Reinforced	4.48	4.11	8.59
5660	Heavy industrial, non reinforced	5.60	8.55	14.15
5680	Reinforced	5.80	9.05	14.85
6700	8″ thick, non industrial, non reinforced	3.82	2.97	6.79
6720	Reinforced	4.07	3.43	7.50
6740	Light industrial, non reinforced	4.55	3.61	8.16
6760	Reinforced	4.80	4.07	8.87
6780	Heavy industrial, non reinforced	6	8.65	14.65
6800	Reinforced	6.35	9.10	15.45

A2010 Basement Excavation

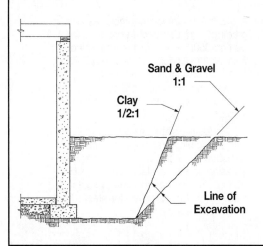

Sand & Gravel
1:1

Clay
1/2:1

Line of
Excavation

In general, the following items are accounted for in the table below.

Costs:
1) Excavation for building or other structure to depth and extent indicated.
2) Backfill compacted in place
3) Haul of excavated waste
4) Replacement of unsuitable backfill material with bank run gravel.

A2010 110	Building Excavation & Backfill	COST PER S.F.		
		MAT.	INST.	TOTAL
2220	Excav & fill, 1000 S.F., 4' sand, gravel, or common earth, on site storage		1.08	1.08
2240	Off site storage		1.64	1.64
2260	Clay excavation, bank run gravel borrow for backfill	2.25	2.53	4.78
2280	8' deep, sand, gravel, or common earth, on site storage		6.55	6.55
2300	Off site storage		14.55	14.55
2320	Clay excavation, bank run gravel borrow for backfill	9.15	11.80	20.95
2340	16' deep, sand, gravel, or common earth, on site storage		17.85	17.85
2350	Off site storage		36	36
2360	Clay excavation, bank run gravel borrow for backfill	25.50	30.50	56
3380	4000 S.F., 4' deep, sand, gravel, or common earth, on site storage		.57	.57
3400	Off site storage		1.14	1.14
3420	Clay excavation, bank run gravel borrow for backfill	1.15	1.27	2.42
3440	8' deep, sand, gravel, or common earth, on site storage		4.63	4.63
3460	Off site storage		8.15	8.15
3480	Clay excavation, bank run gravel borrow for backfill	4.16	7.10	11.26
3500	16' deep, sand, gravel, or common earth, on site storage		11.10	11.10
3520	Off site storage		22	22
3540	Clay, excavation, bank run gravel borrow for backfill	11.15	16.80	27.95
4560	10,000 S.F., 4' deep, sand gravel, or common earth, on site storage		.32	.32
4580	Off site storage		.67	.67
4600	Clay excavation, bank run gravel borrow for backfill	.70	.78	1.48
4620	8' deep, sand, gravel, or common earth, on site storage		4.03	4.03
4640	Off site storage		6.15	6.15
4660	Clay excavation, bank run gravel borrow for backfill	2.52	5.50	8.02
4680	16' deep, sand, gravel, or common earth, on site storage		9.05	9.05
4700	Off site storage		15.55	15.55
4720	Clay excavation, bank run gravel borrow for backfill	6.70	12.60	19.30
5740	30,000 S.F., 4' deep, sand, gravel, or common earth, on site storage		.18	.18
5760	Off site storage		.40	.40
5780	Clay excavation, bank run gravel borrow for backfill	.41	.46	.87
5860	16' deep, sand, gravel, or common earth, on site storage		7.80	7.80
5880	Off site storage		11.30	11.30
5900	Clay excavation, bank run gravel borrow for backfill	3.71	9.80	13.51
6910	100,000 S.F., 4' deep, sand, gravel, or common earth, on site storage		.11	.11
6940	8' deep, sand, gravel, or common earth, on site storage		3.38	3.38
6970	16' deep, sand, gravel, or common earth, on site storage		7.10	7.10
6980	Off site storage		9	9
6990	Clay excavation, bank run gravel borrow for backfill	2	8.15	10.15

A2020 Basement Walls

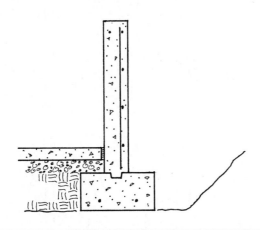

The Foundation Bearing Wall System includes: forms up to 16' high (four uses); 3,000 p.s.i. concrete placed and vibrated; and form removal with breaking form ties and patching walls. The wall systems list walls from 6" to 16" thick and are designed with minimum reinforcement.

Excavation and backfill are not included.

Please see the reference section for further design and cost information.

A2020 110		Walls, Cast in Place						
	WALL HEIGHT (FT.)	PLACING METHOD	CONCRETE (C.Y. per L.F.)	REINFORCING (LBS. per L.F.)	WALL THICKNESS (IN.)	COST PER L.F.		
						MAT.	INST.	TOTAL
1500	4'	direct chute	.074	3.3	6	17	54	71
1520			.099	4.8	8	20.50	55.50	76
1540			.123	6.0	10	24	56.50	80.50
1560			.148	7.2	12	28	57.50	85.50
1580			.173	8.1	14	31	58.50	89.50
1600			.197	9.44	16	34.50	60	94.50
3000	6'	direct chute	.111	4.95	6	25.50	81	106.50
3020			.149	7.20	8	31	83	114
3040			.184	9.00	10	36	84.50	120.50
3060			.222	10.8	12	41.50	86.50	128
5000	8'	direct chute	.148	6.6	6	34	108	142
5020			.199	9.6	8	41.50	114	155.50
5040			.250	12	10	49	113	162
5061			.296	14.39	12	55.50	115	170.50
6020	10'	direct chute	.248	12	8	52	138	190
6040			.307	14.99	10	60.50	141	201.50
6061			.370	17.99	12	69.50	144	213.50
7220	12'	pumped	.298	14.39	8	62.50	172	234.50
7240			.369	17.99	10	72.50	175	247.50
7260			.444	21.59	12	83.50	180	263.50
9220	16'	pumped	.397	19.19	8	83	229	312
9240			.492	23.99	10	96.50	233	329.50
9260			.593	28.79	12	111	240	351

309

For customer support on your Square Foot Cost Data, call 877.756.2789.

B1010 Floor Construction

General: It is desirable for purposes of consistency and simplicity to maintain constant column sizes throughout building height. To do this, concrete strength may be varied (higher strength concrete at lower stories and lower strength concrete at upper stories), as well as varying the amount of reinforcing.

The table provides probably minimum column sizes with related costs and weight per lineal foot of story height.

| B1010 203 | C.I.P. Column, Square Tied | | | | | | |

	LOAD (KIPS)	STORY HEIGHT (FT.)	COLUMN SIZE (IN.)	COLUMN WEIGHT (P.L.F.)	CONCRETE STRENGTH (PSI)	COST PER V.L.F.		
						MAT.	INST.	TOTAL
0640	100	10	10	96	4000	10.95	50	60.95
0680		12	10	97	4000	11.15	50.50	61.65
0700		14	12	142	4000	14.60	61	75.60
0840	200	10	12	140	4000	15.95	63	78.95
0860		12	12	142	4000	16.25	63.50	79.75
0900		14	14	196	4000	18.95	70.50	89.45
0920	300	10	14	192	4000	20	72	92
0960		12	14	194	4000	20.50	73	93.50
0980		14	16	253	4000	23.50	80.50	104
1020	400	10	16	248	4000	25	82.50	107.50
1060		12	16	251	4000	25.50	83.50	109
1080		14	16	253	4000	26	84.50	110.50
1200	500	10	18	315	4000	29.50	92.50	122
1250		12	20	394	4000	35.50	105	140.50
1300		14	20	397	4000	36.50	106	142.50
1350	600	10	20	388	4000	37	108	145
1400		12	20	394	4000	38	109	147
1600		14	20	397	4000	38.50	110	148.50
3400	900	10	24	560	4000	50.50	133	183.50
3800		12	24	567	4000	51.50	135	186.50
4000		14	24	571	4000	52.50	137	189.50
7300	300	10	14	192	6000	20	71.50	91.50
7500		12	14	194	6000	20.50	72	92.50
7600		14	14	196	6000	21	73	94
8000	500	10	16	248	6000	26	82.50	108.50
8050		12	16	251	6000	26.50	83.50	110
8100		14	16	253	6000	27	84.50	111.50
8200	600	10	18	315	6000	30	93	123
8300		12	18	319	6000	30.50	94	124.50
8400		14	18	321	6000	31	95	126
8800	800	10	20	388	6000	37	105	142
8900		12	20	394	6000	37.50	106	143.50
9000		14	20	397	6000	38.50	108	146.50

B1010 Floor Construction

| B1010 203 | C.I.P. Column, Square Tied | | | | | | | |

	LOAD (KIPS)	STORY HEIGHT (FT.)	COLUMN SIZE (IN.)	COLUMN WEIGHT (P.L.F.)	CONCRETE STRENGTH (PSI)	COST PER V.L.F.		
						MAT.	INST.	TOTAL
9100	900	10	20	388	6000	40	110	150
9300		12	20	394	6000	40.50	111	151.50
9600		14	20	397	6000	41.50	112	153.50

| B1010 204 | C.I.P. Column, Square Tied-Minimum Reinforcing | | | | | | | |

	LOAD (KIPS)	STORY HEIGHT (FT.)	COLUMN SIZE (IN.)	COLUMN WEIGHT (P.L.F.)	CONCRETE STRENGTH (PSI)	COST PER V.L.F.		
						MAT.	INST.	TOTAL
9913	150	10-14	12	135	4000	13.25	58.50	71.75
9918	300	10-14	16	240	4000	21.50	76.50	98
9924	500	10-14	20	375	4000	32	99	131
9930	700	10-14	24	540	4000	44	123	167
9936	1000	10-14	28	740	4000	58	150	208
9942	1400	10-14	32	965	4000	72	169	241
9948	1800	10-14	36	1220	4000	89.50	197	286.50

313

For customer support on your Square Foot Cost Data, call 877.756.2789.

B1010 Floor Construction

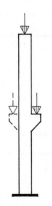

Concentric Load

Eccentric Load

General: Data presented here is for plant produced members transported 50 miles to 100 miles to the site and erected.

Design and pricing assumptions:
Normal wt. concrete, f'c = 5 KSI

Main reinforcement, fy = 60 KSI
Ties, fy = 40 KSI

Minimum design eccentricity, 0.1t.

Concrete encased structural steel haunches are assumed where practical; otherwise galvanized rebar haunches are assumed.

Base plates are integral with columns.

Foundation anchor bolts, nuts and washers are included in price.

B1010 206			**Tied, Concentric Loaded Precast Concrete Columns**					
	LOAD (KIPS)	STORY HEIGHT (FT.)	COLUMN SIZE (IN.)	COLUMN WEIGHT (P.L.F.)	LOAD LEVELS	COST PER V.L.F.		
						MAT.	INST.	TOTAL
0560	100	10	12x12	164	2	173	14.25	187.25
0570		12	12x12	162	2	168	11.85	179.85
0580		14	12x12	161	2	168	11.65	179.65
0590	150	10	12x12	166	3	168	11.55	179.55
0600		12	12x12	169	3	168	10.35	178.35
0610		14	12x12	162	3	168	10.20	178.20
0620	200	10	12x12	168	4	167	12.35	179.35
0630		12	12x12	170	4	168	11.15	179.15
0640		14	12x12	220	4	167	11.05	178.05

B1010 207			**Tied, Eccentric Loaded Precast Concrete Columns**					
	LOAD (KIPS)	STORY HEIGHT (FT.)	COLUMN SIZE (IN.)	COLUMN WEIGHT (P.L.F.)	LOAD LEVELS	COST PER V.L.F.		
						MAT.	INST.	TOTAL
1130	100	10	12x12	161	2	165	14.25	179.25
1140		12	12x12	159	2	165	11.85	176.85
1150		14	12x12	159	2	165	11.65	176.65
1390	600	10	18x18	385	4	242	12.35	254.35
1400		12	18x18	380	4	243	11.15	254.15
1410		14	18x18	375	4	242	11.05	253.05
1480	800	10	20x20	490	4	220	12.35	232.35
1490		12	20x20	480	4	220	11.15	231.15
1500		14	20x20	475	4	220	11.05	231.05

B10 Superstructure

B1010 Floor Construction

(A) Wide Flange

(B) Pipe

(C) Pipe, Concrete Filled

(G) Rectangular Tube, Concrete Filled

B1010 208	Steel Columns							
	LOAD (KIPS)	UNSUPPORTED HEIGHT (FT.)	WEIGHT (P.L.F.)	SIZE (IN.)	TYPE	COST PER V.L.F.		
						MAT.	INST.	TOTAL
1000	25	10	13	4	A	23	10.85	33.85
1020			7.58	3	B	13.40	10.85	24.25
1040			15	3-1/2	C	16.35	10.85	27.20
1120			20	4x3	G	18.85	10.85	29.70
1200		16	16	5	A	26	8.15	34.15
1220			10.79	4	B	17.65	8.15	25.80
1240			36	5-1/2	C	24.50	8.15	32.65
1320			64	8x6	G	38	8.15	46.15
1600	50	10	16	5	A	28.50	10.85	39.35
1620			14.62	5	B	26	10.85	36.85
1640			24	4-1/2	C	19.50	10.85	30.35
1720			28	6x3	G	25	10.85	35.85
1800		16	24	8	A	39	8.15	47.15
1840			36	5-1/2	C	24.50	8.15	32.65
1920			64	8x6	G	38	8.15	46.15
2000		20	28	8	A	43.50	8.15	51.65
2040			49	6-5/8	C	30.50	8.15	38.65
2120			64	8x6	G	36	8.15	44.15
2200	75	10	20	6	A	35.50	10.85	46.35
2240			36	4-1/2	C	49	10.85	59.85
2320			35	6x4	G	28	10.85	38.85
2400		16	31	8	A	50.50	8.15	58.65
2440			49	6-5/8	C	32	8.15	40.15
2520			64	8x6	G	38	8.15	46.15
2600		20	31	8	A	48	8.15	56.15
2640			81	8-5/8	C	46	8.15	54.15
2720			64	8x6	G	36	8.15	44.15
2800	100	10	24	8	A	42.50	10.85	53.35
2840			35	4-1/2	C	49	10.85	59.85
2920			46	8x4	G	34.50	10.85	45.35

	LOAD (KIPS)	UNSUPPORTED HEIGHT (FT.)	WEIGHT (P.L.F.)	SIZE (IN.)	TYPE	COST PER V.L.F.		
						MAT.	INST.	TOTAL
3000	100	16	31	8	A	50.50	8.15	58.65
3040			56	6-5/8	C	47.50	8.15	55.65
3120			64	8x6	G	38	8.15	46.15
3200		20	40	8	A	62	8.15	70.15
3240			81	8-5/8	C	46	8.15	54.15
3320			70	8x6	G	51.50	8.15	59.65
3400	125	10	31	8	A	55	10.85	65.85
3440			81	8	C	52	10.85	62.85
3520			64	8x6	G	41	10.85	51.85
3600		16	40	8	A	65.50	8.15	73.65
3640			81	8	C	48.50	8.15	56.65
3720			64	8x6	G	38	8.15	46.15
3800		20	48	8	A	74.50	8.15	82.65
3840			81	8	C	46	8.15	54.15
3920			60	8x6	G	51.50	8.15	59.65
4000	150	10	35	8	A	62	10.85	72.85
4040			81	8-5/8	C	52	10.85	62.85
4120			64	8x6	G	41	10.85	51.85
4200		16	45	10	A	73.50	8.15	81.65
4240			81	8-5/8	C	48.50	8.15	56.65
4320			70	8x6	G	54.50	8.15	62.65
4400		20	49	10	A	76	8.15	84.15
4440			123	10-3/4	C	65.50	8.15	73.65
4520			86	10x6	G	51	8.15	59.15
4600	200	10	45	10	A	79.50	10.85	90.35
4640			81	8-5/8	C	52	10.85	62.85
4720			70	8x6	G	59	10.85	69.85
4800		16	49	10	A	80	8.15	88.15
4840			123	10-3/4	C	69	8.15	77.15
4920			85	10x6	G	63.50	8.15	71.65
5200	300	10	61	14	A	108	10.85	118.85
5240			169	12-3/4	C	91.50	10.85	102.35
5320			86	10x6	G	88	10.85	98.85
5400		16	72	12	A	118	8.15	126.15
5440			169	12-3/4	C	85	8.15	93.15
5600		20	79	12	A	122	8.15	130.15
5640			169	12-3/4	C	80.50	8.15	88.65
5800	400	10	79	12	A	140	10.85	150.85
5840			178	12-3/4	C	119	10.85	129.85
6000		16	87	12	A	142	8.15	150.15
6040			178	12-3/4	C	111	8.15	119.15
6400	500	10	99	14	A	175	10.85	185.85
6600		16	109	14	A	178	8.15	186.15
6800		20	120	12	A	186	8.15	194.15
7000	600	10	120	12	A	212	10.85	222.85
7200		16	132	14	A	216	8.15	224.15
7400		20	132	14	A	204	8.15	212.15
7600	700	10	136	12	A	240	10.85	250.85
7800		16	145	14	A	237	8.15	245.15
8000		20	145	14	A	224	8.15	232.15
8200	800	10	145	14	A	256	10.85	266.85
8300		16	159	14	A	260	8.15	268.15
8400		20	176	14	A	272	8.15	280.15

B1010 Floor Construction

B1010 208		Steel Columns						
	LOAD (KIPS)	UNSUPPORTED HEIGHT (FT.)	WEIGHT (P.L.F.)	SIZE (IN.)	TYPE	COST PER V.L.F.		
						MAT.	INST.	TOTAL
8800	900	10	159	14	A	281	10.85	291.85
8900		16	176	14	A	288	8.15	296.15
9000		20	193	14	A	299	8.15	307.15
9100	1000	10	176	14	A	310	10.85	320.85
9200		16	193	14	A	315	8.15	323.15
9300		20	211	14	A	325	8.15	333.15

317

For customer support on your Square Foot Cost Data, call 877.756.2789.

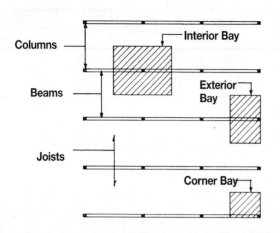

Description: Table below lists costs per S.F. of bay size for wood columns of various sizes and unsupported heights and the maximum allowable total load per S.F. per bay size.

B1010 210	Wood Columns							
	NOMINAL COLUMN SIZE (IN.)	**BAY SIZE (FT.)**	**UNSUPPORTED HEIGHT (FT.)**	**MATERIAL (BF per M.S.F.)**	**TOTAL LOAD (P.S.F.)**	**COST PER S.F.**		
						MAT.	**INST.**	**TOTAL**
1000	4 x 4	10 x 8	8	133	100	.18	.26	.44
1050			10	167	60	.23	.33	.56
1200		10 x 10	8	106	80	.15	.21	.36
1250			10	133	50	.18	.26	.44
1400		10 x 15	8	71	50	.10	.14	.24
1450			10	88	30	.12	.17	.29
1600		15 x 15	8	47	30	.06	.09	.15
1650			10	59	15	.08	.12	.20
2000	6 x 6	10 x 15	8	160	230	.25	.29	.54
2050			10	200	210	.32	.37	.69
2200		15 x 15	8	107	150	.17	.20	.37
2250			10	133	140	.21	.24	.45
2400		15 x 20	8	80	110	.13	.15	.28
2450			10	100	100	.16	.18	.34
2500			12	120	70	.19	.22	.41
2600		20 x 20	8	60	80	.09	.11	.20
2650			10	75	70	.12	.14	.26
2800		20 x 25	8	48	60	.08	.09	.17
2850			10	60	50	.09	.11	.20
3400	8 x 8	20 x 20	8	107	160	.15	.18	.33
3450			10	133	160	.19	.23	.42
3600		20 x 25	8	85	130	.12	.17	.29
3650			10	107	130	.15	.21	.36
3800		25 x 25	8	68	100	.11	.11	.22
3850			10	85	100	.14	.13	.27
4200	10 x 10	20 x 25	8	133	210	.22	.21	.43
4250			10	167	210	.28	.26	.54
4400		25 x 25	8	107	160	.18	.17	.35
4450			10	133	160	.22	.21	.43
4700	12 x 12	20 x 25	8	192	310	.31	.28	.59
4750			10	240	310	.38	.35	.73
4900		25 x 25	8	154	240	.25	.23	.48
4950			10	192	240	.31	.28	.59

B1010 Floor Construction

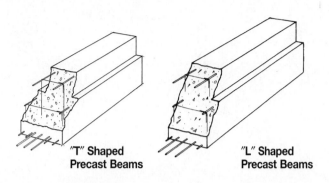

"T" Shaped Precast Beams   "L" Shaped Precast Beams

B1010 214	"T" Shaped Precast Beams							
	SPAN (FT.)	SUPERIMPOSED LOAD (K.L.F.)	SIZE W X D (IN.)	BEAM WEIGHT (P.L.F.)	TOTAL LOAD (K.L.F.)	COST PER L.F.		
						MAT.	INST.	TOTAL
2300	15	2.8	12x16	260	3.06	185	20.50	205.50
2500		8.37	12x28	515	8.89	220	22	242
8900	45	3.34	12x60	1165	4.51	355	13.20	368.20
9900		6.5	24x60	1915	8.42	380	18.45	398.45

B1010 215	"L" Shaped Precast Beams							
	SPAN (FT.)	SUPERIMPOSED LOAD (K.L.F.)	SIZE W X D (IN.)	BEAM WEIGHT (P.L.F.)	TOTAL LOAD (K.L.F.)	COST PER L.F.		
						MAT.	INST.	TOTAL
2250	15	2.58	12x16	230	2.81	174	20.50	194.50
2400		5.92	12x24	370	6.29	191	20.50	211.50
4000	25	2.64	12x28	435	3.08	192	13.30	205.30
4450		6.44	18x36	790	7.23	238	15.85	253.85
5300	30	2.80	12x36	565	3.37	214	12.10	226.10
6400		8.66	24x44	1245	9.90	284	17.35	301.35

B1010 Floor Construction

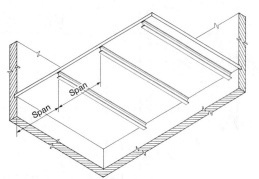

Cast in Place Floor Slab, One Way
General: Solid concrete slabs of uniform depth reinforced for flexure in one direction and for temperature and shrinkage in the other direction.

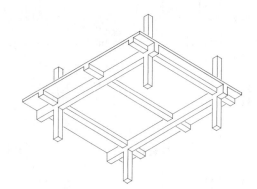

Cast in Place Beam & Slab, One Way
General: Solid concrete one way slab cast monolithically with reinforced concrete beams and girders.

B1010 217 — Cast in Place Slabs, One Way

	SLAB DESIGN & SPAN (FT.)	SUPERIMPOSED LOAD (P.S.F.)	THICKNESS (IN.)	TOTAL LOAD (P.S.F.)	COST PER S.F.		
					MAT.	INST.	TOTAL
2500	Single 8	40	4	90	4.31	8.55	12.86
2600		75	4	125	4.34	9.35	13.69
2700		125	4-1/2	181	4.52	9.40	13.92
2800		200	5	262	4.99	9.70	14.69
3000	Single 10	40	4	90	4.48	9.25	13.73
3100		75	4	125	4.48	9.25	13.73
3200		125	5	188	4.96	9.45	14.41
3300		200	7-1/2	293	6.50	10.60	17.10
3500	Single 15	40	5-1/2	90	5.30	9.35	14.65
3600		75	6-1/2	156	5.90	9.85	15.75
3700		125	7-1/2	219	6.40	10.10	16.50
3800		200	8-1/2	306	6.95	10.30	17.25
4000	Single 20	40	7-1/2	115	6.40	9.85	16.25
4100		75	9	200	6.95	9.80	16.75
4200		125	10	250	7.65	10.20	17.85
4300		200	10	324	8.20	10.65	18.85

B1010 219 — Cast in Place Beam & Slab, One Way

	BAY SIZE (FT.)	SUPERIMPOSED LOAD (P.S.F.)	MINIMUM COL. SIZE (IN.)	SLAB THICKNESS (IN.)	TOTAL LOAD (P.S.F.)	COST PER S.F.		
						MAT.	INST.	TOTAL
5000	20x25	40	12	5-1/2	121	5.30	11.75	17.05
5200		125	16	5-1/2	215	6.25	13.45	19.70
5500	25x25	40	12	6	129	5.55	11.55	17.10
5700		125	18	6	227	7.10	14.15	21.25
7500	30x35	40	16	8	158	7	13.05	20.05
7600		75	18	8	196	7.45	13.45	20.90
7700		125	22	8	254	8.35	14.95	23.30
7800		200	26	8	332	9.05	15.40	24.45
8000	35x35	40	16	9	169	7.80	13.40	21.20
8200		75	20	9	213	8.45	14.65	23.10
8400		125	24	9	272	9.30	15.20	24.50
8600		200	26	9	355	10.20	16.20	26.40

B10 Superstructure

B1010   Floor Construction

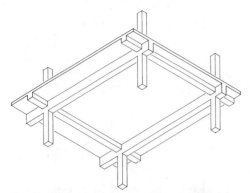

General: Solid concrete two way slab cast monolithically with reinforced concrete support beams and girders.

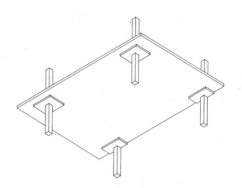

General: Flat Slab: Solid uniform depth concrete two way slabs with drop panels at columns and no column capitals.

B1010 220 — Cast in Place Beam & Slab, Two Way

	BAY SIZE (FT.)	SUPERIMPOSED LOAD (P.S.F.)	MINIMUM COL. SIZE (IN.)	SLAB THICKNESS (IN.)	TOTAL LOAD (P.S.F.)	COST PER S.F. MAT.	COST PER S.F. INST.	COST PER S.F. TOTAL
4000	20 x 25	40	12	7	141	6.15	11.85	18
4300		75	14	7	181	7	12.95	19.95
4500		125	16	7	236	7.15	13.30	20.45
5100	25 x 25	40	12	7-1/2	149	6.40	12	18.40
5200		75	16	7-1/2	185	7	12.85	19.85
5300		125	18	7-1/2	250	7.60	13.90	21.50
7600	30 x 35	40	16	10	188	8.55	13.55	22.10
7700		75	18	10	225	9.05	14.05	23.10
8000		125	22	10	282	10.05	15.10	25.15
8500	35 x 35	40	16	10-1/2	193	9.15	13.80	22.95
8600		75	20	10-1/2	233	9.55	14.30	23.85
9000		125	24	10-1/2	287	10.70	15.35	26.05

B1010 222 — Cast in Place Flat Slab with Drop Panels

	BAY SIZE (FT.)	SUPERIMPOSED LOAD (P.S.F.)	MINIMUM COL. SIZE (IN.)	SLAB & DROP (IN.)	TOTAL LOAD (P.S.F.)	COST PER S.F. MAT.	COST PER S.F. INST.	COST PER S.F. TOTAL
1960	20 x 20	40	12	7 - 3	132	5.45	9.40	14.85
1980		75	16	7 - 4	168	5.80	9.65	15.45
2000		125	18	7 - 6	221	6.45	9.95	16.40
3200	25 x 25	40	12	8-1/2 - 5-1/2	154	6.45	9.85	16.30
4000		125	20	8-1/2 - 8-1/2	243	7.30	10.55	17.85
4400		200	24	9 - 8-1/2	329	7.65	10.80	18.45
5000	25 x 30	40	14	9-1/2 - 7	168	7	10.20	17.20
5200		75	18	9-1/2 - 7	203	7.45	10.60	18.05
5600		125	22	9-1/2 - 8	256	7.85	10.85	18.70
6400	30 x 30	40	14	10-1/2 - 7-1/2	182	7.60	10.50	18.10
6600		75	18	10-1/2 - 7-1/2	217	8.10	10.90	19
6800		125	22	10-1/2 - 9	269	8.45	11.15	19.60
7400	30 x 35	40	16	11-1/2 - 9	196	8.30	10.90	19.20
7900		75	20	11-1/2 - 9	231	8.85	11.30	20.15
8000		125	24	11-1/2 - 11	284	9.20	11.60	20.80
9000	35 x 35	40	16	12 - 9	202	8.55	10.95	19.50
9400		75	20	12 - 11	240	9.20	11.50	20.70
9600		125	24	12 - 11	290	9.45	11.70	21.15

321

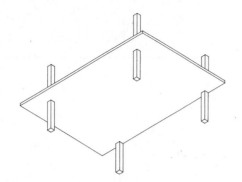

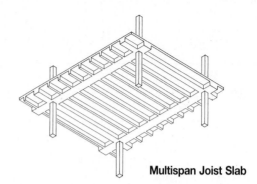

Multispan Joist Slab

General: Flat Plates: Solid uniform depth concrete two way slab without drops or interior beams. Primary design limit is shear at columns.

General: Combination of thin concrete slab and monolithic ribs at uniform spacing to reduce dead weight and increase rigidity.

B1010 223	**Cast in Place Flat Plate**							
	BAY SIZE (FT.)	SUPERIMPOSED LOAD (P.S.F.)	MINIMUM COL. SIZE (IN.)	SLAB THICKNESS (IN.)	TOTAL LOAD (P.S.F.)	COST PER S.F.		
						MAT.	INST.	TOTAL
3000	15 x 20	40	14	7	127	5.15	9.10	14.25
3400		75	16	7-1/2	169	5.50	9.30	14.80
3600		125	22	8-1/2	231	6.05	9.55	15.60
3800		175	24	8-1/2	281	6.10	9.55	15.65
4200	20 x 20	40	16	7	127	5.15	9.10	14.25
4400		75	20	7-1/2	175	5.55	9.30	14.85
4600		125	24	8-1/2	231	6.05	9.50	15.55
5000		175	24	8-1/2	281	6.10	9.55	15.65
5600	20 x 25	40	18	8-1/2	146	6	9.50	15.50
6000		75	20	9	188	6.20	9.60	15.80
6400		125	26	9-1/2	244	6.75	9.90	16.65
6600		175	30	10	300	7	10	17
7000	25 x 25	40	20	9	152	6.25	9.55	15.80
7400		75	24	9-1/2	194	6.60	9.85	16.45
7600		125	30	10	250	7	10.05	17.05

B1010 226	**Cast in Place Multispan Joist Slab**							
	BAY SIZE (FT.)	SUPERIMPOSED LOAD (P.S.F.)	MINIMUM COL. SIZE (IN.)	RIB DEPTH (IN.)	TOTAL LOAD (P.S.F.)	COST PER S.F.		
						MAT.	INST.	TOTAL
2000	15 x 15	40	12	8	115	6.85	11.15	18
2100		75	12	8	150	6.90	11.15	18.05
2200		125	12	8	200	7.05	11.30	18.35
2300		200	14	8	275	7.20	11.70	18.90
2600	15 x 20	40	12	8	115	7	11.15	18.15
2800		75	12	8	150	7.10	11.80	18.90
3000		125	14	8	200	7.35	11.95	19.30
3300		200	16	8	275	7.70	12.15	19.85
3600	20 x 20	40	12	10	120	7.15	10.95	18.10
3900		75	14	10	155	7.45	11.65	19.10
4000		125	16	10	205	7.50	11.85	19.35
4100		200	18	10	280	7.80	12.35	20.15
6200	30 x 30	40	14	14	131	7.90	11.50	19.40
6400		75	18	14	166	8.10	11.85	19.95
6600		125	20	14	216	8.55	12.45	21
6700		200	24	16	297	9.15	12.95	22.10

B10 Superstructure

B1010 Floor Construction

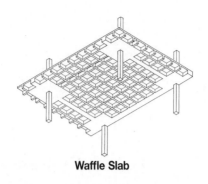

Waffle Slab

B1010 227		Cast in Place Waffle Slab						
	BAY SIZE (FT.)	SUPERIMPOSED LOAD (P.S.F.)	MINIMUM COL. SIZE (IN.)	RIB DEPTH (IN.)	TOTAL LOAD (P.S.F.)	COST PER S.F.		
						MAT.	INST.	TOTAL
3900	20 x 20	40	12	8	144	8.65	11	19.65
4000		75	12	8	179	8.80	11.15	19.95
4100		125	16	8	229	8.95	11.30	20.25
4200		200	18	8	304	9.40	11.65	21.05
4400	20 x 25	40	12	8	146	8.80	11.10	19.90
4500		75	14	8	181	9	11.25	20.25
4600		125	16	8	231	9.20	11.40	20.60
4700		200	18	8	306	9.60	11.70	21.30
4900	25 x 25	40	12	10	150	9.05	11.20	20.25
5000		75	16	10	185	9.30	11.40	20.70
5300		125	18	10	235	9.55	11.60	21.15
5500		200	20	10	310	9.80	11.85	21.65
5700	25 x 30	40	14	10	154	9.20	11.30	20.50
5800		75	16	10	189	9.45	11.50	20.95
5900		125	18	10	239	9.70	11.70	21.40
6000		200	20	12	329	10.65	12.20	22.85
6400	30 x 30	40	14	12	169	9.85	11.55	21.40
6500		75	18	12	204	10.05	11.70	21.75
6600		125	20	12	254	10.20	11.85	22.05
6700		200	24	12	329	11.05	12.50	23.55
6900	30 x 35	40	16	12	169	10.05	11.65	21.70
7000		75	18	12	204	10.05	11.65	21.70
7100		125	22	12	254	10.50	12.05	22.55
7200		200	26	14	334	11.60	12.80	24.40
7400	35 x 35	40	16	14	174	10.60	11.95	22.55
7500		75	20	14	209	10.80	12.15	22.95
7600		125	24	14	259	11.05	12.30	23.35
7700		200	26	16	346	11.80	12.95	24.75
8000	35 x 40	40	18	14	176	10.85	12.15	23
8300		75	22	14	211	11.20	12.45	23.65
8500		125	26	16	271	11.75	12.70	24.45
8750		200	30	20	372	12.80	13.40	26.20
9200	40 x 40	40	18	14	176	11.20	12.45	23.65
9400		75	24	14	211	11.60	12.80	24.40
9500		125	26	16	271	11.95	12.85	24.80
9700	40 x 45	40	20	16	186	11.65	12.60	24.25
9800		75	24	16	221	12.05	12.95	25
9900		125	28	16	271	12.25	13.10	25.35

For customer support on your Square Foot Cost Data, call 877.756.2789.

323

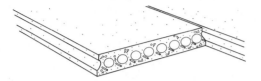

Precast Plank with No Topping

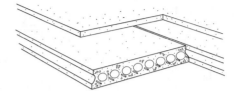

Precast Plank with 2″ Concrete Topping

B1010 229		**Precast Plank with No Topping**						
	SPAN (FT.)	SUPERIMPOSED LOAD (P.S.F.)	TOTAL DEPTH (IN.)	DEAD LOAD (P.S.F.)	TOTAL LOAD (P.S.F.)	COST PER S.F.		
						MAT.	INST.	TOTAL
0720	10	40	4	50	90	6.75	3.47	10.22
0750		75	6	50	125	7.60	2.97	10.57
0770		100	6	50	150	7.60	2.97	10.57
0800	15	40	6	50	90	7.60	2.97	10.57
0820		75	6	50	125	7.60	2.97	10.57
0850		100	6	50	150	7.60	2.97	10.57
0950	25	40	6	50	90	7.60	2.97	10.57
0970		75	8	55	130	8.45	2.60	11.05
1000		100	8	55	155	8.45	2.60	11.05
1200	30	40	8	55	95	8.45	2.60	11.05
1300		75	8	55	130	8.45	2.60	11.05
1400		100	10	70	170	8.80	2.31	11.11
1500	40	40	10	70	110	8.80	2.31	11.11
1600		75	12	70	145	9.10	2.08	11.18
1700	45	40	12	70	110	9.10	2.08	11.18

B1010 230		**Precast Plank with 2″ Concrete Topping**						
	SPAN (FT.)	SUPERIMPOSED LOAD (P.S.F.)	TOTAL DEPTH (IN.)	DEAD LOAD (P.S.F.)	TOTAL LOAD (P.S.F.)	COST PER S.F.		
						MAT.	INST.	TOTAL
2000	10	40	6	75	115	7.80	5.90	13.70
2100		75	8	75	150	8.65	5.40	14.05
2200		100	8	75	175	8.65	5.40	14.05
2500	15	40	8	75	115	8.65	5.40	14.05
2600		75	8	75	150	8.65	5.40	14.05
2700		100	8	75	175	8.65	5.40	14.05
3100	25	40	8	75	115	8.65	5.40	14.05
3200		75	8	75	150	8.65	5.40	14.05
3300		100	10	80	180	9.50	5.05	14.55
3400	30	40	10	80	120	9.50	5.05	14.55
3500		75	10	80	155	9.50	5.05	14.55
3600		100	10	80	180	9.50	5.05	14.55
4000	40	40	12	95	135	9.85	4.74	14.59
4500		75	14	95	170	10.15	4.51	14.66
5000	45	40	14	95	135	10.15	4.51	14.66

324

For customer support on your Square Foot Cost Data, call 877.756.2789.

B1010 Floor Construction

Precast Double "T" Beams with No Topping

Precast Double "T" Beams with 2″ Topping

Most widely used for moderate span floors and moderate and long span roofs. At shorter spans, they tend to be competitive with hollow core slabs. They are also used as wall panels.

B1010 234					Precast Double "T" Beams with No Topping			
	SPAN (FT.)	SUPERIMPOSED LOAD (P.S.F.)	DBL. "T" SIZE D (IN.) W (FT.)	CONCRETE "T" TYPE	TOTAL LOAD (P.S.F.)	COST PER S.F.		
						MAT.	INST.	TOTAL
4300	50	30	20x8	Lt. Wt.	66	9.80	1.59	11.39
4400		40	20x8	Lt. Wt.	76	9.80	1.46	11.26
4500		50	20x8	Lt. Wt.	86	9.90	1.75	11.65
4600		75	20x8	Lt. Wt.	111	9.95	2.01	11.96
5600	70	30	32x10	Lt. Wt.	78	9.80	1.09	10.89
5750		40	32x10	Lt. Wt.	88	9.85	1.32	11.17
5900		50	32x10	Lt. Wt.	98	9.90	1.52	11.42
6000		75	32x10	Lt. Wt.	123	10.05	1.85	11.90
6100		100	32x10	Lt. Wt.	148	10.25	2.51	12.76
6200	80	30	32x10	Lt. Wt.	78	9.90	1.52	11.42
6300		40	32x10	Lt. Wt.	88	10.05	1.85	11.90
6400		50	32x10	Lt. Wt.	98	10.15	2.18	12.33

B1010 235					Precast Double "T" Beams With 2″ Topping			
	SPAN (FT.)	SUPERIMPOSED LOAD (P.S.F.)	DBL. "T" SIZE D (IN.) W (FT.)	CONCRETE "T" TYPE	TOTAL LOAD (P.S.F.)	COST PER S.F.		
						MAT.	INST.	TOTAL
7100	40	30	18x8	Reg. Wt.	120	8.60	3.26	11.86
7200		40	20x8	Reg. Wt.	130	10.05	3.12	13.17
7300		50	20x8	Reg. Wt.	140	10.15	3.41	13.56
7400		75	20x8	Reg. Wt.	165	10.20	3.55	13.75
7500		100	20x8	Reg. Wt.	190	10.35	3.96	14.31
7550	50	30	24x8	Reg. Wt.	120	10.15	3.26	13.41
7600		40	24x8	Reg. Wt.	130	10.20	3.37	13.57
7750		50	24x8	Reg. Wt.	140	10.20	3.39	13.59
7800		75	24x8	Reg. Wt.	165	10.30	3.81	14.11
7900		100	32x10	Reg. Wt.	189	10.20	3.13	13.33

325

For customer support on your Square Foot Cost Data, call 877.756.2789.

B1010 Floor Construction

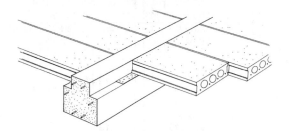

Precast Beam and Plank with No Topping

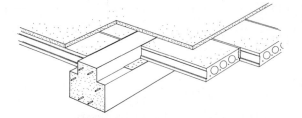

Precast Beam and Plank with 2″ Topping

B1010 236		Precast Beam & Plank with No Topping						
	BAY SIZE (FT.)	SUPERIMPOSED LOAD (P.S.F.)	PLANK THICKNESS (IN.)	TOTAL DEPTH (IN.)	TOTAL LOAD (P.S.F.)	COST PER S.F.		
						MAT.	INST.	TOTAL
6200	25x25	40	8	28	118	22	3.47	25.47
6400		75	8	36	158	22	3.46	25.46
6500		100	8	36	183	22	3.46	25.46
7000	25x30	40	8	36	110	20	3.35	23.35
7200		75	10	36	159	21	3.07	24.07
7400		100	12	36	188	21	2.80	23.80
7600	30x30	40	8	36	121	20.50	3.24	23.74
8000		75	10	44	140	20.50	3.25	23.75
8250		100	12	52	206	22.50	2.75	25.25
8500	30x35	40	12	44	135	19.60	2.65	22.25
8750		75	12	52	176	20.50	2.66	23.16
9000	35x35	40	12	52	141	20.50	2.60	23.10
9250		75	12	60	181	21.50	2.59	24.09
9500	35x40	40	12	52	137	20.50	2.60	23.10

B1010 238		Precast Beam & Plank with 2″ Topping						
	BAY SIZE (FT.)	SUPERIMPOSED LOAD (P.S.F.)	PLANK THICKNESS (IN.)	TOTAL DEPTH (IN.)	TOTAL LOAD (P.S.F.)	COST PER S.F.		
						MAT.	INST.	TOTAL
4300	20x20	40	6	22	135	23.50	6.20	29.70
4400		75	6	24	173	24	6.30	30.30
4500		100	6	28	200	24.50	6.25	30.75
4600	20x25	40	6	26	134	22	6.05	28.05
5000		75	8	30	177	22.50	5.60	28.10
5200		100	8	30	202	22.50	5.60	28.10
5400	25x25	40	6	38	143	22.50	5.80	28.30
5600		75	8	38	183	17.90	5.50	23.40
6000		100	8	46	216	18.20	5.05	23.25
6200	25x30	40	8	38	144	21	5.25	26.25
6400		75	10	46	200	21.50	4.95	26.45
6600		100	10	46	225	22	4.96	26.96
7000	30x30	40	8	46	150	21.50	5.15	26.65
7200		75	10	54	181	23	5.15	28.15
7600		100	10	54	231	23	5.15	28.15
7800	30x35	40	10	54	166	21.50	4.76	26.26
8000		75	12	54	200	21.50	4.52	26.02

B1010 Floor Construction

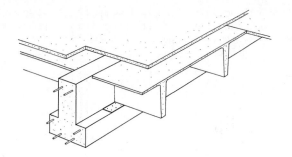

General: Beams and double tees priced here are for plant produced prestressed members transported to the site and erected.

The 2″ structural topping is applied after the beams and double tees are in place and is reinforced with W.W.F.

With Topping

B1010 239		Precast Double "T" & 2″ Topping on Precast Beams						
	BAY SIZE (FT.)	SUPERIMPOSED LOAD (P.S.F.)	DEPTH (IN.)		TOTAL LOAD (P.S.F.)	COST PER S.F.		
						MAT.	INST.	TOTAL
3000	25x30	40	38		130	22	4.40	26.40
3100		75	38		168	22	4.40	26.40
3300		100	46		196	21.50	4.37	25.87
3600	30x30	40	46		150	22	4.26	26.26
3750		75	46		174	22	4.26	26.26
4000		100	54		203	23	4.25	27.25
4100	30x40	40	46		136	19.60	3.85	23.45
4300		75	54		173	20	3.84	23.84
4400		100	62		204	21	3.85	24.85
4600	30x50	40	54		138	18.25	3.55	21.80
4800		75	54		181	18.10	3.58	21.68
5000		100	54		219	18.15	3.27	21.42
5200	30x60	40	62		151	17.85	3.22	21.07
5400		75	62		192	17.40	3.26	20.66
5600		100	62		215	17.50	3.26	20.76
5800	35x40	40	54		139	20.50	3.78	24.28
6000		75	62		179	20.50	3.62	24.12
6250		100	62		212	20.50	3.75	24.25
6500	35x50	40	62		142	18.95	3.52	22.47
6750		75	62		186	18.95	3.65	22.60
7300		100	62		231	19.05	3.34	22.39
7600	35x60	40	54		154	18.30	3.08	21.38
7750		75	54		179	18.45	3.09	21.54
8000		100	62		224	18.85	3.14	21.99
8250	40x40	40	62		145	22.50	3.79	26.29
8400		75	62		187	21.50	3.81	25.31
8750		100	62		223	22	3.89	25.89
9000	40x50	40	62		151	19.70	3.55	23.25
9300		75	62		193	19.15	2.62	21.77
9800	40x60	40	62		164	22	2.52	24.52

B1010 Floor Construction

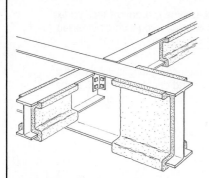

General: The following table is based upon structural wide flange (WF) beam and girder framing. Non-composite action is assumed between WF framing and decking. Deck costs not included.

The deck spans the short direction. The steel beams and girders are fireproofed with sprayed fiber fireproofing.

No columns included in price.

B1010 241	W Shape Beams & Girders							
	BAY SIZE (FT.) BEAM X GIRD	SUPERIMPOSED LOAD (P.S.F.)	STEEL FRAMING DEPTH (IN.)	FIREPROOFING (S.F. PER S.F.)	TOTAL LOAD (P.S.F.)	COST PER S.F.		
						MAT.	INST.	TOTAL
1350	15x20	40	12	.535	50	5.10	1.98	7.08
1400		40	16	.65	90	6.65	2.55	9.20
1450		75	18	.694	125	8.75	3.20	11.95
1500		125	24	.796	175	12.10	4.44	16.54
2000	20x20	40	12	.55	50	5.70	2.17	7.87
2050		40	14	.579	90	7.80	2.82	10.62
2100		75	16	.672	125	9.35	3.36	12.71
2150		125	16	.714	175	11.15	4.07	15.22
2900	20x25	40	16	.53	50	6.25	2.32	8.57
2950		40	18	.621	96	9.90	3.48	13.38
3000		75	18	.651	131	11.40	3.96	15.36
3050		125	24	.77	200	15.05	5.35	20.40
5200	25x25	40	18	.486	50	6.85	2.46	9.31
5300		40	18	.592	96	10.20	3.53	13.73
5400		75	21	.668	131	12.30	4.24	16.54
5450		125	24	.738	191	16.25	5.65	21.90
6300	25x30	40	21	.568	50	7.80	2.81	10.61
6350		40	21	.694	90	10.55	3.73	14.28
6400		75	24	.776	125	13.55	4.70	18.25
6450		125	30	.904	175	16.30	5.85	22.15
7700	30x30	40	21	.52	50	8.35	2.93	11.28
7750		40	24	.629	103	12.90	4.37	17.27
7800		75	30	.715	138	15.30	5.15	20.45
7850		125	36	.822	206	20	6.95	26.95
8250	30x35	40	21	.508	50	9.55	3.27	12.82
8300		40	24	.651	109	14.10	4.75	18.85
8350		75	33	.732	150	17.10	5.70	22.80
8400		125	36	.802	225	21.50	7.30	28.80
9550	35x35	40	27	.560	50	9.85	3.42	13.27
9600		40	36	.706	109	18	5.95	23.95
9650		75	36	.750	150	18.60	6.20	24.80
9820		125	36	.797	225	25	8.40	33.40

B1010 Floor Construction

Description: Table below lists costs for light gage CEE or PUNCHED DOUBLE joists to suit the span and loading with the minimum thickness subfloor required by the joist spacing.

	SPAN (FT.)	SUPERIMPOSED LOAD (P.S.F.)	FRAMING DEPTH (IN.)	FRAMING SPAC. (IN.)	TOTAL LOAD (P.S.F.)	COST PER S.F.		
B1010 244						**MAT.**	**INST.**	**TOTAL**
1500	15	40	8	16	54	2.86	2.28	5.14
1550			8	24	54	3	2.34	5.34
1600		65	10	16	80	3.73	2.74	6.47
1650			10	24	80	3.69	2.69	6.38
1700		75	10	16	90	3.73	2.74	6.47
1750			10	24	90	3.69	2.69	6.38
1800		100	10	16	116	4.86	3.47	8.33
1850			10	24	116	3.69	2.69	6.38
1900		125	10	16	141	4.86	3.47	8.33
1950			10	24	141	3.69	2.69	6.38
2500	20	40	8	16	55	3.29	2.57	5.86
2550			8	24	55	3	2.34	5.34
2600		65	8	16	80	3.81	2.94	6.75
2650			10	24	80	3.70	2.71	6.41
2700		75	10	16	90	3.52	2.61	6.13
2750			12	24	90	4.12	2.37	6.49
2800		100	10	16	115	3.52	2.61	6.13
2850			10	24	116	4.75	3.04	7.79
2900		125	12	16	142	5.50	2.97	8.47
2950			12	24	141	5.10	3.22	8.32
3500	25	40	10	16	55	3.73	2.74	6.47
3550			10	24	55	3.69	2.69	6.38
3600		65	10	16	81	4.86	3.47	8.33
3650			12	24	81	4.57	2.56	7.13
3700		75	12	16	92	5.50	2.96	8.46
3750			12	24	91	5.10	3.22	8.32
3800		100	12	16	117	6.20	3.27	9.47
3850		125	12	16	143	7.05	4.29	11.34
4500	30	40	12	16	57	5.50	2.96	8.46
4550			12	24	56	4.55	2.55	7.10
4600		65	12	16	82	6.15	3.26	9.41
4650		75	12	16	92	6.15	3.26	9.41

Light Gauge Steel Floor Systems

329

B1010 Floor Construction

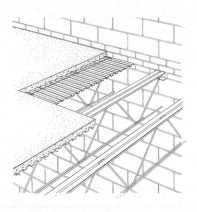

Description: The table below lists costs per S.F. for a floor system on bearing walls using open web steel joists, galvanized steel slab form and 2-1/2″ concrete slab reinforced with welded wire fabric. Costs of the bearing walls are not included.

B1010 246		Deck & Joists on Bearing Walls						
	SPAN (FT.)	SUPERIMPOSED LOAD (P.S.F.)	JOIST SPACING FT. - IN.	DEPTH (IN.)	TOTAL LOAD (P.S.F.)	COST PER S.F.		
						MAT.	INST.	TOTAL
1050	20	40	2-0	14-1/2	83	6.05	3.94	9.99
1070		65	2-0	16-1/2	109	6.50	4.17	10.67
1100		75	2-0	16-1/2	119	6.50	4.17	10.67
1120		100	2-0	18-1/2	145	6.55	4.18	10.73
1150		125	1-9	18-1/2	170	7.15	4.43	11.58
1170	25	40	2-0	18-1/2	84	6.55	4.18	10.73
1200		65	2-0	20-1/2	109	6.75	4.27	11.02
1220		75	2-0	20-1/2	119	7.20	4.45	11.65
1250		100	2-0	22-1/2	145	7.30	4.50	11.80
1270		125	1-9	22-1/2	170	8.30	4.90	13.20
1300	30	40	2-0	22-1/2	84	7.30	4.10	11.40
1320		65	2-0	24-1/2	110	7.90	4.28	12.18
1350		75	2-0	26-1/2	121	8.20	4.36	12.56
1370		100	2-0	26-1/2	146	8.70	4.52	13.22
1400		125	2-0	24-1/2	172	10.20	4.97	15.17
1420	35	40	2-0	26-1/2	85	9.65	4.80	14.45
1450		65	2-0	28-1/2	111	9.95	4.91	14.86
1470		75	2-0	28-1/2	121	9.95	4.91	14.86
1500		100	1-11	28-1/2	147	10.85	5.15	16
1520		125	1-8	28-1/2	172	11.90	5.50	17.40
1550	5/8″ gyp. fireproof.							
1560	On metal furring, add					.86	3.27	4.13

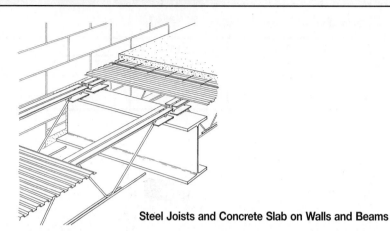

The table below lists costs for a floor system on exterior bearing walls and interior columns and beams using open web steel joists, galvanized steel slab form, 2-1/2″ concrete slab reinforced with welded wire fabric.

Columns are sized to accommodate one floor plus roof loading but costs for columns are from suspended floor to slab on grade only. Costs of the bearing walls are not included.

Steel Joists and Concrete Slab on Walls and Beams

B1010 248						**Steel Joists on Beam & Wall**		
	BAY SIZE (FT.)	SUPERIMPOSED LOAD (P.S.F.)	DEPTH (IN.)	TOTAL LOAD (P.S.F.)	COLUMN ADD	COST PER S.F.		
						MAT.	INST.	TOTAL
1720	25x25	40	23	84		8.20	4.49	12.69
1730					columns	.42	.13	.55
1750	25x25	65	29	110		8.60	4.64	13.24
1760					columns	.42	.13	.55
1770	25x25	75	26	120		9.55	4.98	14.53
1780					columns	.48	.14	.62
1800	25x25	100	29	145		10.75	5.45	16.20
1810					columns	.48	.14	.62
1820	25x25	125	29	170		11.30	5.60	16.90
1830					columns	.54	.16	.70
2020	30x30	40	29	84		9.35	4.54	13.89
2030					columns	.37	.11	.48
2050	30x30	65	29	110		10.60	4.90	15.50
2060					columns	.37	.11	.48
2070	30x30	75	32	120		10.90	4.99	15.89
2080					columns	.43	.13	.56
2100	30x30	100	35	145		12.05	5.35	17.40
2110					columns	.50	.15	.65
2120	30x30	125	35	172		13.90	5.90	19.80
2130					columns	.60	.18	.78
2170	30x35	40	29	85		10.55	4.90	15.45
2180					columns	.36	.10	.46
2200	30x35	65	29	111		12.55	5.50	18.05
2210					columns	.41	.13	.54
2220	30x35	75	32	121		12.55	5.50	18.05
2230					columns	.42	.13	.55
2250	30x35	100	35	148		12.75	5.55	18.30
2260					columns	.52	.15	.67
2270	30x35	125	38	173		14.30	6.05	20.35
2280					columns	.53	.15	.68
2320	35x35	40	32	85		10.85	4.99	15.84
2330					columns	.37	.11	.48
2350	35x35	65	35	111		13.25	5.70	18.95
2360					columns	.44	.13	.57
2370	35x35	75	35	121		13.55	5.80	19.35
2380					columns	.44	.13	.57
2400	35x35	100	38	148		14.10	5.95	20.05
2410					columns	.55	.16	.71

B1010 Floor Construction

B1010 248		Steel Joists on Beam & Wall						
	BAY SIZE (FT.)	SUPERIMPOSED LOAD (P.S.F.)	DEPTH (IN.)	TOTAL LOAD (P.S.F.)	COLUMN ADD	COST PER S.F.		
						MAT.	INST.	TOTAL
2460	5/8 gyp. fireproof.							
2475	On metal furring, add					.86	3.27	4.13

B1010 Floor Construction

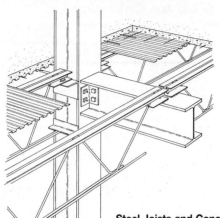

The table below lists costs per S.F. for a floor system on steel columns and beams using open web steel joists, galvanized steel slab form, 2-1/2″ concrete slab reinforced with welded wire fabric.

Columns are sized to accommodate one floor plus roof loading but costs for columns are from floor to grade only.

Steel Joists and Concrete Slab on Steel Columns and Beams

B1010 250		Steel Joists, Beams & Slab on Columns						
	BAY SIZE (FT.)	SUPERIMPOSED LOAD (P.S.F.)	DEPTH (IN.)	TOTAL LOAD (P.S.F.)	COLUMN ADD	COST PER S.F.		
						MAT.	INST.	TOTAL
2350	15x20	40	17	83		8.50	4.54	13.04
2400					column	1.38	.41	1.79
2450	15x20	65	19	108		9.45	4.81	14.26
2500					column	1.38	.41	1.79
2550	15x20	75	19	119		9.80	4.96	14.76
2600					column	1.51	.44	1.95
2650	15x20	100	19	144		10.40	5.15	15.55
2700					column	1.51	.44	1.95
2750	15x20	125	19	170		10.95	5.30	16.25
2800					column	2.01	.59	2.60
2850	20x20	40	19	83		9.25	4.75	14
2900					column	1.13	.33	1.46
2950	20x20	65	23	109		10.20	5.05	15.25
3000					column	1.51	.44	1.95
3100	20x20	75	26	119		10.75	5.25	16
3200					column	1.51	.44	1.95
3400	20x20	100	23	144		11.20	5.35	16.55
3450					column	1.51	.44	1.95
3500	20x20	125	23	170		12.40	5.80	18.20
3600					column	1.81	.53	2.34
3700	20x25	40	44	83		9.90	4.98	14.88
3800					column	1.21	.35	1.56
3900	20x25	65	26	110		10.85	5.25	16.10
4000					column	1.21	.35	1.56
4100	20x25	75	26	120		11.30	5.45	16.75
4200					column	1.45	.43	1.88
4300	20x25	100	26	145		12	5.70	17.70
4400					column	1.45	.43	1.88
4500	20x25	125	29	170		13.30	6.15	19.45
4600					column	1.69	.49	2.18
4700	25x25	40	23	84		10.75	5.20	15.95
4800					column	1.16	.34	1.50
4900	25x25	65	29	110		11.35	5.40	16.75
5000					column	1.16	.34	1.50
5100	25x25	75	26	120		12.55	5.85	18.40
5200					column	1.35	.39	1.74
5300	25x25	100	29	145		13.90	6.35	20.25
5400					column	1.35	.39	1.74

B1010 250 — Steel Joists, Beams & Slab on Columns

	BAY SIZE (FT.)	SUPERIMPOSED LOAD (P.S.F.)	DEPTH (IN.)	TOTAL LOAD (P.S.F.)	COLUMN ADD	MAT.	INST.	TOTAL
5500	25x25	125	32	170		14.70	6.60	21.30
5600					column	1.50	.44	1.94
5700	25x30	40	29	84		12.25	5.35	17.60
5800					column	1.12	.33	1.45
5900	25x30	65	29	110		12.75	5.50	18.25
6000					column	1.12	.33	1.45
6050	25x30	75	29	120		13.10	5.65	18.75
6100					column	1.24	.37	1.61
6150	25x30	100	29	145		14.20	5.95	20.15
6200					column	1.24	.37	1.61
6250	25x30	125	32	170		16	6.50	22.50
6300					column	1.43	.42	1.85
6350	30x30	40	29	84		12.25	5.35	17.60
6400					column	1.04	.30	1.34
6500	30x30	65	29	110		13.95	5.85	19.80
6600					column	1.04	.30	1.34
6700	30x30	75	32	120		14.20	5.95	20.15
6800					column	1.19	.35	1.54
6900	30x30	100	35	145		15.75	6.40	22.15
7000					column	1.39	.41	1.80
7100	30x30	125	35	172		18	7.05	25.05
7200					column	1.55	.46	2.01
7300	30x35	40	29	85		13.80	5.85	19.65
7400					column	.89	.26	1.15
7500	30x35	65	29	111		16.05	6.50	22.55
7600					column	1.15	.34	1.49
7700	30x35	75	32	121		16.05	6.50	22.55
7800					column	1.17	.34	1.51
7900	30x35	100	35	148		16.55	6.65	23.20
8000					column	1.43	.42	1.85
8100	30x35	125	38	173		18.35	7.20	25.55
8200					column	1.46	.43	1.89
8300	35x35	40	32	85		14.20	5.90	20.10
8400					column	1.02	.30	1.32
8500	35x35	65	35	111		16.90	6.75	23.65
8600					column	1.23	.37	1.60
9300	35x35	75	38	121		17.30	6.85	24.15
9400					column	1.23	.37	1.60
9500	35x35	100	38	148		18.65	7.30	25.95
9600					column	1.52	.44	1.96
9750	35x35	125	41	173		19.25	7.45	26.70
9800					column	1.54	.46	2
9810	5/8 gyp. fireproof.							
9815	On metal furring, add					.86	3.27	4.13

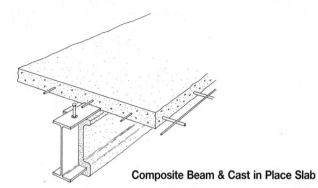

General: Composite construction of wide flange beams and concrete slabs is most efficiently used when loads are heavy and spans are moderately long. It is stiffer with less deflection than non-composite construction of similar depth and spans.

Composite Beam & Cast in Place Slab

B1010 252		Composite Beam & Cast in Place Slab						
	BAY SIZE (FT.)	SUPERIMPOSED LOAD (P.S.F.)	SLAB THICKNESS (IN.)	TOTAL DEPTH (FT.-IN.)	TOTAL LOAD (P.S.F.)	COST PER S.F.		
						MAT.	INST.	TOTAL
3800	20x25	40	4	1 - 8	94	10.45	11.30	21.75
3900		75	4	1 - 8	130	12.15	12	24.15
4000		125	4	1 - 10	181	14.15	12.80	26.95
4100		200	5	2 - 2	272	18.40	14.40	32.80
4200	25x25	40	4-1/2	1 - 8-1/2	99	10.90	11.45	22.35
4300		75	4-1/2	1 - 10-1/2	136	13.15	12.30	25.45
4400		125	5-1/2	2 - 0-1/2	200	15.55	13.30	28.85
4500		200	5-1/2	2 - 2-1/2	278	19.35	14.70	34.05
4600	25x30	40	4-1/2	1 - 8-1/2	100	12.20	11.90	24.10
4700		75	4-1/2	1 - 10-1/2	136	14.30	12.70	27
4800		125	5-1/2	2 - 2-1/2	202	17.55	13.95	31.50
4900		200	5-1/2	2 - 5-1/2	279	21	15.20	36.20
5000	30x30	40	4	1 - 8	95	12.20	11.90	24.10
5200		75	4	2 - 1	131	14.20	12.55	26.75
5400		125	4	2 - 4	183	17.55	13.85	31.40
5600		200	5	2 - 10	274	22.50	15.65	38.15
5800	30x35	40	4	2 - 1	95	12.95	12.10	25.05
6000		75	4	2 - 4	139	15.65	13.05	28.70
6250		125	4	2 - 4	185	20.50	14.75	35.25
6500		200	5	2 - 11	276	25	16.40	41.40
8000	35x35	40	4	2 - 1	96	13.65	12.35	26
8250		75	4	2 - 4	133	16.25	13.35	29.60
8500		125	4	2 - 10	185	19.55	14.50	34.05
8750		200	5	3 - 5	276	25.50	16.60	42.10
9000	35x40	40	4	2 - 4	97	15.20	12.90	28.10
9250		75	4	2 - 4	134	17.80	13.70	31.50
9500		125	4	2 - 7	186	21.50	15.15	36.65
9750		200	5	3 - 5	278	28	17.35	45.35

For customer support on your Square Foot Cost Data, call 877.756.2789.

B1010 Floor Construction

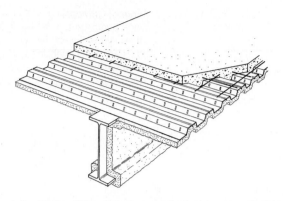

The table below lists costs per S.F. for floors using steel beams and girders, composite steel deck, concrete slab reinforced with W.W.F. and sprayed fiber fireproofing (non- asbestos) on the steel beams and girders and on the steel deck.

Wide Flange, Composite Deck & Slab

B1010 254						W Shape, Composite Deck, & Slab			
	BAY SIZE (FT.) BEAM X GIRD	SUPERIMPOSED LOAD (P.S.F.)	SLAB THICKNESS (IN.)	TOTAL DEPTH (FT.-IN.)	TOTAL LOAD (P.S.F.)	COST PER S.F.			
						MAT.	INST.	TOTAL	
0540	15x20	40	5	1-7	89	11.95	6.80	18.75	
0560		75	5	1-9	125	13.15	7.20	20.35	
0580		125	5	1-11	176	15.40	7.85	23.25	
0600		200	5	2-2	254	18.95	8.95	27.90	
0700	20x20	40	5	1-9	90	13	7.05	20.05	
0720		75	5	1-9	126	14.70	7.65	22.35	
0740		125	5	1-11	177	16.70	8.30	25	
0760		200	5	2-5	255	20	9.25	29.25	
0780	20x25	40	5	1-9	90	13.25	7	20.25	
0800		75	5	1-9	127	16.55	8.10	24.65	
0820		125	5	1-11	180	19.80	8.80	28.60	
0840		200	5	2-5	256	22	9.95	31.95	
0940	25x25	40	5	1-11	91	15.05	7.55	22.60	
0960		75	5	2-5	178	17.35	8.30	25.65	
0980		125	5	2-5	181	21	9.20	30.20	
1000		200	5-1/2	2-8-1/2	263	24	10.65	34.65	
1400	25x30	40	5	2-5	91	14.90	7.70	22.60	
1500		75	5	2-5	128	18.20	8.60	26.80	
1600		125	5	2-8	180	20.50	9.50	30	
1700		200	5	2-11	259	26	11	37	
2200	30x30	40	5	2-2	92	16.55	8.20	24.75	
2300		75	5	2-5	129	19.40	9.15	28.55	
2400		125	5	2-11	182	23	10.25	33.25	
2500		200	5	3-2	263	31.50	12.80	44.30	
2600	30x35	40	5	2-5	94	18.05	8.70	26.75	
2700		75	5	2-11	131	21	9.60	30.60	
2800		125	5	3-2	183	24.50	10.75	35.25	
2900		200	5-1/2	3-5-1/2	268	30.50	12.35	42.85	
3800	35x35	40	5	2-8	94	18.60	8.65	27.25	
3900		75	5	2-11	131	21.50	9.65	31.15	
4000		125	5	3-5	184	26	11	37	
4100		200	5-1/2	3-5-1/2	270	33.50	13	46.50	

B1010 Floor Construction

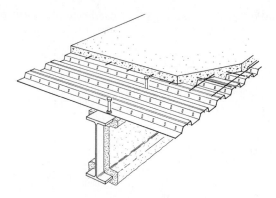

The table below lists costs per S.F. for floors using composite steel beams with welded shear studs, composite steel deck and light weight concrete slab reinforced with W.W.F.

Composite Beam, Deck & Slab

B1010 256	Composite Beams, Deck & Slab							
	BAY SIZE (FT.)	SUPERIMPOSED LOAD (P.S.F.)	SLAB THICKNESS (IN.)	TOTAL DEPTH (FT.-IN.)	TOTAL LOAD (P.S.F.)	COST PER S.F.		
						MAT.	INST.	TOTAL
2400	20x25	40	5-1/2	1 - 5-1/2	80	12.85	6.30	19.15
2500		75	5-1/2	1 - 9-1/2	115	13.35	6.30	19.65
2750		125	5-1/2	1 - 9-1/2	167	16.25	7.35	23.60
2900		200	6-1/4	1 - 11-1/2	251	18.30	7.90	26.20
3000	25x25	40	5-1/2	1 - 9-1/2	82	12.65	6	18.65
3100		75	5-1/2	1 - 11-1/2	118	14.05	6.05	20.10
3200		125	5-1/2	2 - 2-1/2	169	14.65	6.55	21.20
3300		200	6-1/4	2 - 6-1/4	252	19.80	7.65	27.45
3400	25x30	40	5-1/2	1 - 11-1/2	83	12.90	5.95	18.85
3600		75	5-1/2	1 - 11-1/2	119	13.85	6.05	19.90
3900		125	5-1/2	1 - 11-1/2	170	16.10	6.80	22.90
4000		200	6-1/4	2 - 6-1/4	252	19.90	7.70	27.60
4200	30x30	40	5-1/2	1 - 11-1/2	81	13	6.15	19.15
4400		75	5-1/2	2 - 2-1/2	116	14	6.40	20.40
4500		125	5-1/2	2 - 5-1/2	168	16.90	7.15	24.05
4700		200	6-1/4	2 - 9-1/4	252	20	8.30	28.30
4900	30x35	40	5-1/2	2 - 2-1/2	82	13.55	6.35	19.90
5100		75	5-1/2	2 - 5-1/2	117	14.80	6.50	21.30
5300		125	5-1/2	2 - 5-1/2	169	17.40	7.30	24.70
5500		200	6-1/4	2 - 9-1/4	254	20.50	8.30	28.80
5750	35x35	40	5-1/2	2 - 5-1/2	84	14.45	6.35	20.80
6000		75	5-1/2	2 - 5-1/2	121	16.45	6.80	23.25
7000		125	5-1/2	2 - 8-1/2	170	19.30	7.80	27.10
7200		200	5-1/2	2 - 11-1/2	254	22	8.60	30.60
7400	35x40	40	5-1/2	2 - 5-1/2	85	15.95	6.85	22.80
7600		75	5-1/2	2 - 5-1/2	121	17.30	7.05	24.35
8000		125	5-1/2	2 - 5-1/2	171	19.80	7.90	27.70
9000		200	5-1/2	2 - 11-1/2	255	24	8.95	32.95

B1010 Floor Construction

Description: Table B1010 258 lists S.F. costs for steel deck and concrete slabs for various spans.

Description: Table B1010 261 lists the S.F. costs for wood joists and a minimum thickness plywood subfloor.

Description: Table B1010 264 lists the S.F. costs, total load, and member sizes, for various bay sizes and loading conditions.

Metal Deck/Concrete Fill

Wood Joist

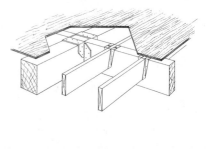

Wood Beam and Joist

B1010 258 — Metal Deck/Concrete Fill

	SUPERIMPOSED LOAD (P.S.F.)	DECK SPAN (FT.)	DECK GAGE DEPTH	SLAB THICKNESS (IN.)	TOTAL LOAD (P.S.F.)	COST PER S.F.		
						MAT.	INST.	TOTAL
0900	125	6	22 1-1/2	4	164	3.08	2.59	5.67
0920		7	20 1-1/2	4	164	3.39	2.74	6.13
0950		8	20 1-1/2	4	165	3.39	2.74	6.13
0970		9	18 1-1/2	4	165	3.96	2.74	6.70
1000		10	18 2	4	165	3.86	2.89	6.75
1020		11	18 3	5	169	4.75	3.12	7.87

B1010 261 — Wood Joist

		COST PER S.F.		
		MAT.	INST.	TOTAL
2900	2"x8", 12" O.C.	1.81	2.02	3.83
2950	16" O.C.	1.56	1.73	3.29
3000	24" O.C.	1.57	1.60	3.17
3300	2"x10", 12" O.C.	2.41	2.30	4.71
3350	16" O.C.	2	1.93	3.93
3400	24" O.C.	1.85	1.73	3.58
3700	2"x12", 12" O.C.	2.85	2.33	5.18
3750	16" O.C.	2.33	1.95	4.28
3800	24" O.C.	2.08	1.75	3.83
4100	2"x14", 12" O.C.	3.41	2.53	5.94
4150	16" O.C.	2.77	2.11	4.88
4200	24" O.C.	2.37	1.86	4.23
7101	Note: Subfloor cost is included in these prices.			

B1010 264 — Wood Beam & Joist

	BAY SIZE (FT.)	SUPERIMPOSED LOAD (P.S.F.)	GIRDER BEAM (IN.)	JOISTS (IN.)	TOTAL LOAD (P.S.F.)	COST PER S.F.		
						MAT.	INST.	TOTAL
2000	15x15	40	8 x 12 4 x 12	2 x 6 @ 16	53	9.05	4.69	13.74
2050		75	8 x 16 4 x 16	2 x 8 @ 16	90	12.60	5.10	17.70
2100		125	12 x 16 6 x 16	2 x 8 @ 12	144	19.50	6.55	26.05
2150		200	14 x 22 12 x 16	2 x 10 @ 12	227	39	10.05	49.05
3000	20x20	40	10 x 14 10 x 12	2 x 8 @ 16	63	11.90	4.71	16.61
3050		75	12 x 16 8 x 16	2 x 10 @ 16	102	16.85	5.25	22.10
3100		125	14 x 22 12 x 16	2 x 10 @ 12	163	33	8.40	41.40

B1010 Floor Construction

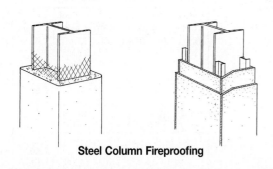

Steel Column Fireproofing

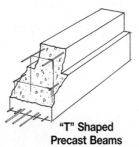

"T" Shaped
Precast Beams

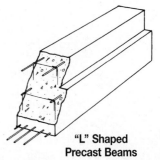

"L" Shaped
Precast Beams

B1010 720		Steel Column Fireproofing						
	ENCASEMENT SYSTEM	COLUMN SIZE (IN.)	THICKNESS (IN.)	FIRE RATING (HRS.)	WEIGHT (P.L.F.)	COST PER V.L.F.		
						MAT.	INST.	TOTAL
3000	Concrete	8	1	1	110	6.90	37.50	44.40
3300		14	1	1	258	11.40	54	65.40
3400			2	3	325	13.65	61.50	75.15
3450	Gypsum board	8	1/2	2	8	3.66	23	26.66
3550	1 layer	14	1/2	2	18	3.96	24.50	28.46
3600	Gypsum board	8	1	3	14	5.10	29.50	34.60
3650	1/2" fire rated	10	1	3	17	5.45	31	36.45
3700	2 layers	14	1	3	22	5.65	32	37.65
3750	Gypsum board	8	1-1/2	3	23	6.80	37	43.80
3800	1/2" fire rated	10	1-1/2	3	27	7.65	41	48.65
3850	3 layers	14	1-1/2	3	35	8.45	45	53.45
3900	Sprayed fiber	8	1-1/2	2	6.3	4	7.30	11.30
3950	Direct application		2	3	8.3	5.50	10	15.50
4050		10	1-1/2	2	7.9	4.83	8.75	13.58
4200	Sprayed fiber	14	1-1/2	2	10.8	6	10.85	16.85

B1020 Roof Construction

The table below lists prices per S.F. for roof rafters and sheathing by nominal size and spacing. Sheathing is 5/16″ CDX for 12″ and 16″ spacing and 3/8″ CDX for 24″ spacing.

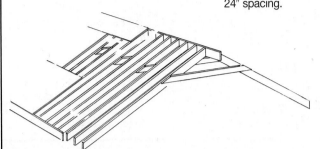

Factors for Converting Inclined to Horizontal

Roof Slope	Approx. Angle	Factor	Roof Slope	Approx. Angle	Factor
Flat	0°	1.000	12 in 12	45.0°	1.414
1 in 12	4.8°	1.003	13 in 12	47.3°	1.474
2 in 12	9.5°	1.014	14 in 12	49.4°	1.537
3 in 12	14.0°	1.031	15 in 12	51.3°	1.601
4 in 12	18.4°	1.054	16 in 12	53.1°	1.667
5 in 12	22.6°	1.083	17 in 12	54.8°	1.734
6 in 12	26.6°	1.118	18 in 12	56.3°	1.803
7 in 12	30.3°	1.158	19 in 12	57.7°	1.873
8 in 12	33.7°	1.202	20 in 12	59.0°	1.943
9 in 12	36.9°	1.250	21 in 12	60.3°	2.015
10 in 12	39.8°	1.302	22 in 12	61.4°	2.088
11 in 12	42.5°	1.357	23 in 12	62.4°	2.162

B1020 102	Wood/Flat or Pitched	MAT.	INST.	TOTAL
2500	Flat rafter, 2″x4″, 12″ O.C.	1.20	1.82	3.02
2550	16″ O.C.	1.08	1.56	2.64
2600	24″ O.C.	.95	1.35	2.30
2900	2″x6″, 12″ O.C.	1.46	1.83	3.29
2950	16″ O.C.	1.28	1.57	2.85
3000	24″ O.C.	1.08	1.34	2.42
3300	2″x8″, 12″ O.C.	1.74	1.97	3.71
3350	16″ O.C.	1.49	1.68	3.17
3400	24″ O.C.	1.23	1.42	2.65
3700	2″x10″, 12″ O.C.	2.34	2.25	4.59
3750	16″ O.C.	1.93	1.88	3.81
3800	24″ O.C.	1.51	1.55	3.06
4100	2″x12″, 12″ O.C.	2.78	2.28	5.06
4150	16″ O.C.	2.42	2.08	4.50
4200	24″ O.C.	1.74	1.57	3.31
4500	2″x14″, 12″ O.C.	3.34	3.31	6.65
4550	16″ O.C.	2.70	2.69	5.39
4600	24″ O.C.	2.03	2.10	4.13
4900	3″x6″, 12″ O.C.	2.99	1.94	4.93
4950	16″ O.C.	2.43	1.65	4.08
5000	24″ O.C.	1.85	1.40	3.25
5300	3″x8″, 12″ O.C.	3.97	2.16	6.13
5350	16″ O.C.	3.15	1.81	4.96
5400	24″ O.C.	2.33	1.51	3.84
5700	3″x10″, 12″ O.C.	4.85	2.46	7.31
5750	16″ O.C.	3.83	2.04	5.87
5800	24″ O.C.	2.78	1.66	4.44
6100	3″x12″, 12″ O.C.	5.65	2.96	8.61
6150	16″ O.C.	4.44	2.42	6.86
6200	24″ O.C.	3.19	1.91	5.10
7001	Wood truss, 4 in 12 slope, 24″ O.C., 24′ to 29′ span	3.79	2.72	6.51
7100	30′ to 43′ span	4.44	2.98	7.42
7200	44′ to 60′ span	4.39	2.72	7.11

B1020 Roof Construction

The table below lists the cost per S.F. for a roof system with steel columns, beams and deck using open web steel joists and 1-1/2″ galvanized metal deck. Perimeter of system is supported on bearing walls.

Fireproofing is not included. Costs/S.F. are based on a building 4 bays long and 4 bays wide.

Column costs are additive. Costs for the bearing walls are not included.

Steel Joists, Beams and Deck on Bearing Walls

B1020 108	Steel Joists, Beams & Deck on Columns & Walls							
	BAY SIZE (FT.)	SUPERIMPOSED LOAD (P.S.F.)	DEPTH (IN.)	TOTAL LOAD (P.S.F.)	COLUMN ADD	COST PER S.F.		
						MAT.	INST.	TOTAL
3000	25x25	20	18	40		4.31	1.43	5.74
3100					columns	.39	.09	.48
3200		30	22	50		4.62	1.52	6.14
3300					columns	.52	.12	.64
3400		40	20	60		5.10	1.68	6.78
3500					columns	.52	.12	.64
3600	25x30	20	22	40		4.54	1.40	5.94
3700					columns	.44	.10	.54
3800		30	20	50		5.15	1.56	6.71
3900					columns	.44	.10	.54
4000		40	25	60		5.35	1.60	6.95
4100					columns	.52	.12	.64
4200	30x30	20	25	42		5	1.50	6.50
4300					columns	.36	.09	.45
4400		30	22	52		5.45	1.63	7.08
4500					columns	.44	.10	.54
4600		40	28	62		5.65	1.68	7.33
4700					columns	.44	.10	.54
4800	30x35	20	22	42		5.20	1.56	6.76
4900					columns	.37	.09	.46
5000		30	28	52		5.45	1.63	7.08
5100					columns	.37	.09	.46
5200		40	25	62		5.90	1.75	7.65
5300					columns	.44	.10	.54
5400	35x35	20	28	42		5.20	1.57	6.77
5500					columns	.32	.08	.40

341

B1020 Roof Construction

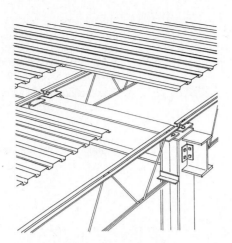

Description: The table below lists the cost per S.F. for a roof system with steel columns, beams, and deck, using open web steel joists and 1-1/2″ galvanized metal deck.

Column costs are additive. Fireproofing is not included.

Steel Joists, Beams and Deck on Columns

B1020 112	Steel Joists, Beams, & Deck on Columns							
	BAY SIZE (FT.)	SUPERIMPOSED LOAD (P.S.F.)	DEPTH (IN.)	TOTAL LOAD (P.S.F.)	COLUMN ADD	COST PER S.F.		
						MAT.	INST.	TOTAL
1100	15x20	20	16	40		4.26	1.43	5.69
1200					columns	2.26	.53	2.79
1500		40	18	60		4.75	1.58	6.33
1600					columns	2.26	.53	2.79
2300	20x25	20	18	40		4.84	1.59	6.43
2400					columns	1.35	.31	1.66
2700		40	20	60		5.45	1.77	7.22
2800					columns	1.81	.42	2.23
2900	25x25	20	18	40		5.70	1.80	7.50
3000					columns	1.08	.25	1.33
3100		30	22	50		5.90	1.84	7.74
3200					columns	1.45	.34	1.79
3300		40	20	60		6.55	2.04	8.59
3400					columns	1.45	.34	1.79
3500	25x30	20	22	40		5.55	1.63	7.18
3600					columns	1.21	.28	1.49
3900		40	25	60		6.70	1.93	8.63
4000					columns	1.45	.34	1.79
4100	30x30	20	25	42		6.30	1.81	8.11
4200					columns	1	.24	1.24
4300		30	22	52		6.95	1.98	8.93
4400					columns	1.21	.28	1.49
4500		40	28	62		7.30	2.07	9.37
4600					columns	1.21	.28	1.49
5300	35x35	20	28	42		6.85	1.96	8.81
5400					columns	.88	.20	1.08
5500		30	25	52		7.70	2.17	9.87
5600					columns	1.03	.24	1.27
5700		40	28	62		8.30	2.32	10.62
5800					columns	1.14	.26	1.40

B1020 Roof Construction

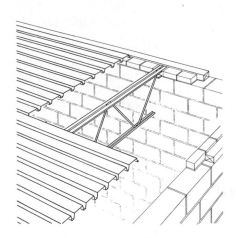

Description: The table below lists the cost per S.F. for a roof system using open web steel joists and 1-1/2″ galvanized metal deck. The system is assumed supported on bearing walls or other suitable support. Costs for the supports are not included.

B1020 116				Steel Joists & Deck on Bearing Walls				
	BAY SIZE (FT.)	SUPERIMPOSED LOAD (P.S.F.)	DEPTH (IN.)	TOTAL LOAD (P.S.F.)		COST PER S.F.		
						MAT.	INST.	TOTAL
1100	20	20	13-1/2	40		2.73	1.04	3.77
1200		30	15-1/2	50		2.79	1.06	3.85
1300		40	15-1/2	60		2.99	1.14	4.13
1400	25	20	17-1/2	40		2.99	1.15	4.14
1500		30	17-1/2	50		3.22	1.24	4.46
1600		40	19-1/2	60		3.27	1.26	4.53
1700	30	20	19-1/2	40		3.26	1.26	4.52
1800		30	21-1/2	50		3.33	1.28	4.61
1900		40	23-1/2	60		3.58	1.38	4.96
2000	35	20	23-1/2	40		3.54	1.18	4.72
2100		30	25-1/2	50		3.64	1.21	4.85
2200		40	25-1/2	60		3.86	1.28	5.14
2300	40	20	25-1/2	41		3.92	1.30	5.22
2400		30	25-1/2	51		4.16	1.37	5.53
2500		40	25-1/2	61		4.30	1.41	5.71
2600	45	20	27-1/2	41		4.71	1.54	6.25
2700		30	31-1/2	51		4.98	1.62	6.60
2800		40	31-1/2	61		5.25	1.70	6.95
2900	50	20	29-1/2	42		5.25	1.71	6.96
3000		30	31-1/2	52		5.75	1.85	7.60
3100		40	31-1/2	62		6.10	1.97	8.07
3200	60	20	37-1/2	42		6.25	2.20	8.45
3300		30	37-1/2	52		6.90	2.43	9.33
3400		40	37-1/2	62		6.90	2.43	9.33
3500	70	20	41-1/2	42		6.90	2.43	9.33
3600		30	41-1/2	52		7.35	2.59	9.94
3700		40	41-1/2	64		8.95	3.13	12.08
3800	80	20	45-1/2	44		8.70	3.04	11.74
3900		30	45-1/2	54		8.70	3.04	11.74
4000		40	45-1/2	64		9.60	3.36	12.96
4400	100	20	57-1/2	44		8.55	2.89	11.44
4500		30	57-1/2	54		10.20	3.43	13.63
4600		40	57-1/2	65		11.25	3.78	15.03
4700	125	20	69-1/2	44		10.20	3.42	13.62
4800		30	69-1/2	56		11.85	3.97	15.82
4900		40	69-1/2	67		13.45	4.51	17.96

343

B1020   Roof Construction

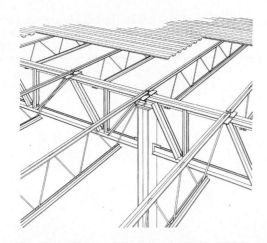

Description: The table below lists costs for a roof system supported on exterior bearing walls and interior columns. Costs include bracing, joist girders, open web steel joists and 1-1/2″ galvanized metal deck.

Column costs are additive. Fireproofing is not included.

Costs/S.F. are based on a building 4 bays long and 4 bays wide. Costs for bearing walls are not included.

B1020 120		Steel Joists, Joist Girders & Deck on Columns & Walls						
	BAY SIZE (FT.) GIRD X JOISTS	SUPERIMPOSED LOAD (P.S.F.)	DEPTH (IN.)	TOTAL LOAD (P.S.F.)	COLUMN ADD	COST PER S.F.		
						MAT.	INST.	TOTAL
2350	30x35	20	32-1/2	40		3.98	1.45	5.43
2400					columns	.37	.09	.46
2550		40	36-1/2	60		4.46	1.60	6.06
2600					columns	.43	.10	.53
3000	35x35	20	36-1/2	40		4.34	1.79	6.13
3050					columns	.32	.08	.40
3200		40	36-1/2	60		4.86	1.97	6.83
3250					columns	.41	.10	.51
3300	35x40	20	36-1/2	40		4.40	1.63	6.03
3350					columns	.33	.08	.41
3500		40	36-1/2	60		4.95	1.81	6.76
3550					columns	.36	.09	.45
3900	40x40	20	40-1/2	41		5.10	1.82	6.92
3950					columns	.32	.08	.40
4100		40	40-1/2	61		5.70	1.99	7.69
4150					columns	.32	.08	.40
5100	45x50	20	52-1/2	41		5.75	2.07	7.82
5150					columns	.23	.05	.28
5300		40	52-1/2	61		7	2.45	9.45
5350					columns	.33	.08	.41
5400	50x45	20	56-1/2	41		5.55	2.02	7.57
5450					columns	.23	.05	.28
5600		40	56-1/2	61		7.15	2.51	9.66
5650					columns	.31	.07	.38
5700	50x50	20	56-1/2	42		6.25	2.25	8.50
5750					columns	.23	.05	.28
5900		40	59	64		7.20	2.75	9.95
5950					columns	.32	.08	.40
6300	60x50	20	62-1/2	43		6.30	2.29	8.59
6350					columns	.36	.09	.45
6500		40	71	65		7.35	2.85	10.20
6550					columns	.48	.11	.59

B1020 Roof Construction

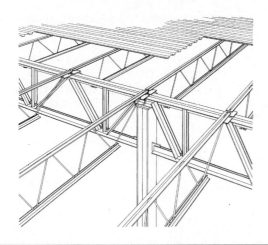

Description: The table below lists the cost per S.F. for a roof system supported on columns. Costs include joist girders, open web steel joists and 1-1/2" galvanized metal deck. Column costs are additive.

Fireproofing is not included. Costs/S.F. are based on a building 4 bays long and 4 bays wide.

Costs for wind columns are not included.

B1020 124	Steel Joists & Joist Girders on Columns							
	BAY SIZE (FT.) GIRD X JOISTS	SUPERIMPOSED LOAD (P.S.F.)	DEPTH (IN.)	TOTAL LOAD (P.S.F.)	COLUMN ADD	COST PER S.F.		
						MAT.	INST.	TOTAL
2000	30x30	20	17-1/2	40		3.99	1.73	5.72
2050					columns	1.21	.28	1.49
2100		30	17-1/2	50		4.32	1.88	6.20
2150					columns	1.21	.28	1.49
2200		40	21-1/2	60		4.44	1.91	6.35
2250					columns	1.21	.28	1.49
2300	30x35	20	32-1/2	40		4.27	1.62	5.89
2350					columns	1.03	.24	1.27
2500		40	36-1/2	60		4.80	1.80	6.60
2550					columns	1.21	.28	1.49
3200	35x40	20	36-1/2	40		5.55	2.19	7.74
3250					columns	.90	.21	1.11
3400		40	36-1/2	60		6.25	2.42	8.67
3450					columns	1	.24	1.24
3800	40x40	20	40-1/2	41		5.60	2.19	7.79
3850					columns	.88	.20	1.08
4000		40	40-1/2	61		6.30	2.41	8.71
4050					columns	.88	.20	1.08
4100	40x45	20	40-1/2	41		6.05	2.50	8.55
4150					columns	.78	.19	.97
4200		30	40-1/2	51		6.55	2.65	9.20
4250					columns	.78	.19	.97
4300		40	40-1/2	61		7.30	2.88	10.18
4350					columns	.88	.20	1.08
5000	45x50	20	52-1/2	41		6.35	2.51	8.86
5050					columns	.62	.15	.77
5200		40	52-1/2	61		7.80	2.96	10.76
5250					columns	.80	.19	.99
5300	50x45	20	56-1/2	41		6.15	2.47	8.62
5350					columns	.62	.15	.77
5500		40	56-1/2	61		8	3.03	11.03
5550					columns	.80	.19	.99
5600	50x50	20	56-1/2	42		6.90	2.71	9.61
5650					columns	.63	.15	.78
5800		40	59	64		7.90	3.23	11.13
5850					columns	.89	.21	1.10

B2010 Exterior Walls

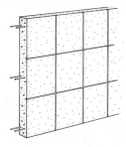 

Cast in Place Concrete Wall

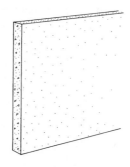

Flat Precast Concrete Wall

Table B2010 101 describes a concrete wall system for exterior closure. There are several types of wall finishes priced from plain finish to a finish with 3/4″ rustication strip.

In Table B2010 103, precast concrete wall panels are either solid or insulated. Prices below are based on delivery within 50 miles of a plant. Small jobs can double the prices below. For large, highly repetitive jobs, deduct up to 15%.

B2010 101	Cast In Place Concrete	COST PER S.F.		
		MAT.	INST.	TOTAL
2100	Conc wall reinforced, 8' high, 6″ thick, plain finish, 3000 PSI	4.83	16.90	21.73
2700	Aged wood liner, 3000 PSI	6.05	19.20	25.25
3000	Sand blast light 1 side, 3000 PSI	5.40	19.15	24.55
3700	3/4″ bevel rustication strip, 3000 PSI	4.95	17.95	22.90
4000	8″ thick, plain finish, 3000 PSI	5.70	17.35	23.05
4100	4000 PSI	5.85	17.35	23.20
4300	Rub concrete 1 side, 3000 PSI	5.70	20	25.70
4550	8″ thick, aged wood liner, 3000 PSI	6.95	19.65	26.60
4750	Sand blast light 1 side, 3000 PSI	6.30	19.60	25.90
5300	3/4″ bevel rustication strip, 3000 PSI	5.85	18.35	24.20
5600	10″ thick, plain finish, 3000 PSI	6.55	17.75	24.30
5900	Rub concrete 1 side, 3000 PSI	6.55	21	27.55
6200	Aged wood liner, 3000 PSI	7.80	20	27.80
6500	Sand blast light 1 side, 3000 PSI	7.15	20	27.15
7100	3/4″ bevel rustication strip, 3000 PSI	6.70	18.75	25.45
7400	12″ thick, plain finish, 3000 PSI	7.60	18.20	25.80
7700	Rub concrete 1 side, 3000 PSI	7.60	21.50	29.10
8000	Aged wood liner, 3000 PSI	8.80	20.50	29.30
8300	Sand blast light 1 side, 3000 PSI	8.15	20.50	28.65
8900	3/4″ bevel rustication strip, 3000 PSI	7.70	19.25	26.95

B2010 102			Flat Precast Concrete			COST PER S.F.		
	THICKNESS (IN.)	PANEL SIZE (FT.)	FINISHES	RIGID INSULATION (IN)	TYPE	MAT.	INST.	TOTAL
3000	4	5x18	smooth gray	none	low rise	12.65	7.90	20.55
3050		6x18				10.60	6.65	17.25
3100		8x20				21	3.20	24.20
3150		12x20				19.90	3.03	22.93
3200	6	5x18	smooth gray	2	low rise	13.70	8.55	22.25
3250		6x18				11.65	7.20	18.85
3300		8x20				22.50	3.91	26.41
3350		12x20				20.50	3.58	24.08
3400	8	5x18	smooth gray	2	low rise	29	4.92	33.92
3450		6x18				27.50	4.68	32.18
3500		8x20				25	4.31	29.31
3550		12x20				23	3.97	26.97

B2010 Exterior Walls

B2010 102	Flat Precast Concrete

	THICKNESS (IN.)	PANEL SIZE (FT.)	FINISHES	RIGID INSULATION (IN)	TYPE	COST PER S.F.		
						MAT.	INST.	TOTAL
3600	4	4x8	white face	none	low rise	48.50	4.60	53.10
3650		8x8				36.50	5.15	41.65
3700		10x10				32	3.02	35.02
3750		20x10				29	2.74	31.74
3800	5	4x8	white face	none	low rise	49.50	4.69	54.19
3850		8x8				37.50	3.54	41.04
3900		10x10				33.50	3.14	36.64
3950		20x20				30.50	2.87	33.37
4000	6	4x8	white face	none	low rise	51.50	4.86	56.36
4050		8x8				39	3.69	42.69
4100		10x10				34.50	3.25	37.75
4150		20x10				31.50	2.99	34.49
4200	6	4x8	white face	2	low rise	52.50	5.55	58.05
4250		8x8				40.50	4.36	44.86
4300		10x10				35.50	3.92	39.42
4350		20x10				31.50	2.99	34.49
4400	7	4x8	white face	none	low rise	53	4.97	57.97
4450		8x8				40.50	3.83	44.33
4500		10x10				36.50	3.43	39.93
4550		20x10				33.50	3.14	36.64
4600	7	4x8	white face	2	low rise	54	5.65	59.65
4650		8x8				41.50	4.50	46
4700		10x10				37.50	4.10	41.60
4750		20x10				34.50	3.81	38.31
4800	8	4x8	white face	none	low rise	54	5.10	59.10
4850		8x8				41.50	3.92	45.42
4900		10x10				37.50	3.52	41.02
4950		20x10				34.50	3.24	37.74
5000	8	4x8	white face	2	low rise	55	5.75	60.75
5050		8x8				42.50	4.59	47.09
5100		10x10				43.50	4.84	48.34
5150		20x10				43.50	4.84	48.34

B2010 103	Fluted Window or Mullion Precast Concrete

	THICKNESS (IN.)	PANEL SIZE (FT.)	FINISHES	RIGID INSULATION (IN)	TYPE	COST PER S.F.		
						MAT.	INST.	TOTAL
5200	4	4x8	smooth gray	none	high rise	29	18.15	47.15
5250		8x8				20.50	12.95	33.45
5300		10x10				40	6.15	46.15
5350		20x10				35.50	5.35	40.85
5400	5	4x8	smooth gray	none	high rise	29.50	18.45	47.95
5450		8x8				21.50	13.45	34.95
5500		10x10				42	6.40	48.40
5550		20x10				37	5.65	42.65
5600	6	4x8	smooth gray	none	high rise	30.50	18.95	49.45
5650		8x8				22	13.90	35.90
5700		10x10				43	6.55	49.55
5750		20x10				38.50	5.85	44.35
5800	6	4x8	smooth gray	2	high rise	31.50	19.65	51.15
5850		8x8				23.50	14.55	38.05
5900		10x10				44.50	7.25	51.75
5950		20x10				39.50	6.50	46

B20 Exterior Enclosure

B2010 Exterior Walls

B2010 103 — Fluted Window or Mullion Precast Concrete

	THICKNESS (IN.)	PANEL SIZE (FT.)	FINISHES	RIGID INSULATION (IN)	TYPE	COST PER S.F.		
						MAT.	INST.	TOTAL
6000	7	4x8	smooth gray	none	high rise	31	19.35	50.35
6050		8x8				23	14.30	37.30
6100		10x10				45	6.85	51.85
6150		20x10				39.50	6	45.50
6200	7	4x8	smooth gray	2	high rise	32	20	52
6250		8x8				24	14.95	38.95
6300		10x10				46	7.55	53.55
6350		20x10				40.50	6.70	47.20
6400	8	4x8	smooth gray	none	high rise	31.50	19.65	51.15
6450		8x8				23.50	14.60	38.10
6500		10x10				46.50	7.05	53.55
6550		20x10				41.50	6.30	47.80
6600	8	4x8	smooth gray	2	high rise	32.50	20.50	53
6650		8x8				24.50	15.30	39.80
6700		10x10				47.50	7.75	55.25
6750		20x10				42.50	6.95	49.45

B2010 104 — Ribbed Precast Concrete

	THICKNESS (IN.)	PANEL SIZE (FT.)	FINISHES	RIGID INSULATION(IN.)	TYPE	COST PER S.F.		
						MAT.	INST.	TOTAL
6800	4	4x8	aggregate	none	high rise	34.50	18.15	52.65
6850		8x8				25.50	13.25	38.75
6900		10x10				42.50	4.81	47.31
6950		20x10				37.50	4.25	41.75
7000	5	4x8	aggregate	none	high rise	35	18.45	53.45
7050		8x8				26	13.65	39.65
7100		10x10				44.50	4.99	49.49
7150		20x10				39.50	4.45	43.95
7200	6	4x8	aggregate	none	high rise	36	18.80	54.80
7250		8x8				26.50	14	40.50
7300		10x10				45.50	5.15	50.65
7350		20x10				41	4.62	45.62
7400	6	4x8	aggregate	2	high rise	37	19.50	56.50
7450		8x8				28	14.65	42.65
7500		10x10				47	5.80	52.80
7550		20x10				42	5.30	47.30
7600	7	4x8	aggregate	none	high rise	36.50	19.25	55.75
7650		8x8				27.50	14.35	41.85
7700		10x10				47	5.30	52.30
7750		20x10				42	4.75	46.75
7800	7	4x8	aggregate	2	high rise	38	19.90	57.90
7850		8x8				28.50	15	43.50
7900		10x10				48.50	6	54.50
7950		20x10				43.50	5.40	48.90
8000	8	4x8	aggregate	none	high rise	37.50	19.60	57.10
8050		8x8				28	14.75	42.75
8100		10x10				49	5.50	54.50
8150		20x10				44	4.97	48.97
8200	8	4x8	aggregate	2	high rise	38.50	20.50	59
8250		8x8				29.50	15.45	44.95
8300		10x10				50	7.90	57.90
8350		20x10				45	5.65	50.65

B2010 Exterior Walls

B2010 105	Precast Concrete Specialties							

	TYPE	SIZE				COST PER L.F.		
						MAT.	INST.	TOTAL
8400	Coping, precast	6" wide				19.15	13.70	32.85
8450	Stock units	10" wide				20.50	14.70	35.20
8460		12" wide				23.50	15.85	39.35
8480		14" wide				25.50	17.15	42.65
8500	Window sills	6" wide				16.10	14.70	30.80
8550	Precast	10" wide				26	17.15	43.15
8600		14" wide				26	20.50	46.50
8610								

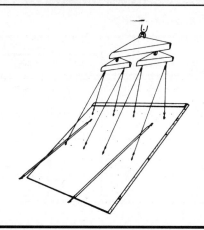

The advantage of tilt up construction is in the low cost of forms and placing of concrete and reinforcing. Tilt up has been used for several types of buildings, including warehouses, stores, offices, and schools. The panels are cast in forms on the ground, or floor slab. Most jobs use 5-1/2" thick solid reinforced concrete panels.

Design Assumptions:
Conc. f'c = 3000 psi
Reinf. fy = 60,000

B2010 106	Tilt-Up Concrete Panel	COST PER S.F.		
		MAT.	INST.	TOTAL
3200	Tilt-up conc panels, broom finish, 5-1/2" thick, 3000 PSI	4.37	6.35	10.72
3250	5000 PSI	4.48	6.25	10.73
3300	6" thick, 3000 PSI	4.79	6.50	11.29
3350	5000 PSI	4.94	6.40	11.34
3400	7-1/2" thick, 3000 PSI	6.15	6.75	12.90
3450	5000 PSI	6.30	6.65	12.95
3500	8" thick, 3000 PSI	6.55	6.90	13.45
3550	5000 PSI	6.80	6.80	13.60
3700	Steel trowel finish, 5-1/2" thick, 3000 PSI	4.37	6.45	10.82
3750	5000 PSI	4.48	6.30	10.78
3800	6" thick, 3000 PSI	4.79	6.60	11.39
3850	5000 PSI	4.94	6.45	11.39
3900	7-1/2" thick, 3000 PSI	6.15	6.80	12.95
3950	5000 PSI	6.30	6.70	13
4000	8" thick, 3000 PSI	6.55	7.05	13.60
4050	5000 PSI	6.80	6.90	13.70
4200	Exp. aggregate finish, 5-1/2" thick, 3000 PSI	4.69	6.50	11.19
4250	5000 PSI	4.81	6.40	11.21
4300	6" thick, 3000 PSI	5.10	6.65	11.75
4350	5000 PSI	5.25	6.55	11.80
4400	7-1/2" thick, 3000 PSI	6.45	6.90	13.35
4450	5000 PSI	6.65	6.80	13.45
4500	8" thick, 3000 PSI	6.90	7.05	13.95
4550	5000 PSI	7.10	6.95	14.05
4600	Exposed aggregate & vert. rustication 5-1/2" thick, 3000 PSI	7.05	8.05	15.10
4650	5000 PSI	7.15	7.95	15.10
4700	6" thick, 3000 PSI	7.50	8.20	15.70
4750	5000 PSI	7.60	8.10	15.70
4800	7-1/2" thick, 3000 PSI	8.80	8.45	17.25
4850	5000 PSI	9	8.35	17.35
4900	8" thick, 3000 PSI	9.25	8.60	17.85
4950	5000 PSI	9.45	8.50	17.95
5000	Vertical rib & light sandblast, 5-1/2" thick, 3000 PSI	6.85	10.50	17.35
5100	6" thick, 3000 PSI	7.25	10.65	17.90
5200	7-1/2" thick, 3000 PSI	8.60	10.90	19.50
5300	8" thick, 3000 PSI	9.05	11.10	20.15
6000	Broom finish w/2" polystyrene insulation, 6" thick, 3000 PSI	4.21	7.85	12.06
6100	Broom finish 2" fiberplank insulation, 6" thick, 3000 PSI	4.75	7.75	12.50
6200	Exposed aggregate w/2" polystyrene insulation, 6" thick, 3000 PSI	4.43	7.85	12.28
6300	Exposed aggregate 2" fiberplank insulation, 6" thick, 3000 PSI	4.97	7.75	12.72

B2010 Exterior Walls

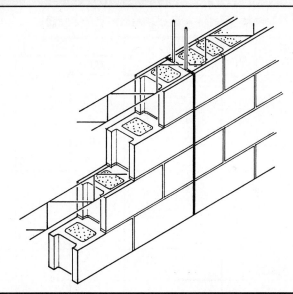

Exterior concrete block walls are defined in the following terms; structural reinforcement, weight, percent solid, size, strength and insulation. Within each of these categories, two to four variations are shown. No costs are included for brick shelf or relieving angles.

B2010 109					Concrete Block Wall - Regular Weight			
	TYPE	**SIZE (IN.)**	**STRENGTH (P.S.I.)**	**CORE FILL**		**COST PER S.F.**		
						MAT.	**INST.**	**TOTAL**
1200	Hollow	4x8x16	2,000	none		2.18	6.30	8.48
1250			4,500	none		2.66	6.30	8.96
1400		8x8x16	2,000	perlite		4.46	7.70	12.16
1410				styrofoam		4.46	7.25	11.71
1440				none		2.97	7.25	10.22
1450			4,500	perlite		5.45	7.70	13.15
1460				styrofoam		5.45	7.25	12.70
1490				none		3.95	7.25	11.20
2000	75% solid	4x8x16	2,000	none		2.62	6.35	8.97
2100		6x8x16	2,000	perlite		3.86	7.05	10.91
2500	Solid	4x8x16	2,000	none		2.45	6.50	8.95
2700		8x8x16	2,000	none		4.13	7.55	11.68

B2010 110					Concrete Block Wall - Lightweight			
	TYPE	**SIZE (IN.)**	**WEIGHT (P.C.F.)**	**CORE FILL**		**COST PER S.F.**		
						MAT.	**INST.**	**TOTAL**
3100	Hollow	8x4x16	105	perlite		4.40	7.25	11.65
3110				styrofoam		4.40	6.80	11.20
3200		4x8x16	105	none		2.30	6.15	8.45
3250			85	none		2.92	6.05	8.97
3300		6x8x16	105	perlite		4.14	6.95	11.09
3310				styrofoam		4.45	6.60	11.05
3340				none		3.13	6.60	9.73
3400		8x8x16	105	perlite		5.25	7.50	12.75
3410				styrofoam		5.25	7.05	12.30
3440				none		3.78	7.05	10.83
3450			85	perlite		5.45	7.35	12.80
4000	75% solid	4x8x16	105	none		3.31	6.25	9.56
4050			85	none		3.59	6.10	9.69
4100		6x8x16	105	perlite		5.05	6.90	11.95
4500	Solid	4x8x16	105	none		2.08	6.45	8.53
4700		8x8x16	105	none		4.12	7.45	11.57

B2010 Exterior Walls

B2010 111 — Reinforced Concrete Block Wall - Regular Weight

	TYPE	SIZE (IN.)	STRENGTH (P.S.I.)	VERT. REINF & GROUT SPACING		COST PER S.F. MAT.	INST.	TOTAL
5200	Hollow	4x8x16	2,000	#4 @ 48"		2.34	7	9.34
5300		6x8x16	2,000	#4 @ 48"		3.03	7.50	10.53
5330				#5 @ 32"		3.25	7.75	11
5340				#5 @ 16"		3.71	8.75	12.46
5350			4,500	#4 @ 28"		3.51	7.50	11.01
5390				#5 @ 16"		4.19	8.75	12.94
5400		8x8x16	2,000	#4 @ 48"		3.36	8	11.36
5430				#5 @ 32"		3.53	8.45	11.98
5440				#5 @ 16"		4.08	9.65	13.73
5450		8x8x16	4,500	#4 @ 48"		4.27	8.10	12.37
5490				#5 @ 16"		5.05	9.65	14.70
5500		12x8x16	2,000	#4 @ 48"		5.20	10.25	15.45
5540				#5 @ 16"		6.30	11.85	18.15
6100	75% solid	6x8x16	2,000	#4 @ 48"		3.49	7.45	10.94
6140				#5 @ 16"		3.93	8.50	12.43
6150			4,500	#4 @ 48"		4.13	7.45	11.58
6190				#5 @ 16"		4.57	8.50	13.07
6200		8x8x16	2,000	#4 @ 48"		3.72	8.05	11.77
6230				#5 @ 32"		3.88	8.30	12.18
6240				#5 @ 16"		4.18	9.30	13.48
6250			4,500	#4 @ 48"		5	8.05	13.05
6280				#5 @ 32"		5.15	8.30	13.45
6290				#5 @ 16"		5.45	9.30	14.75
6500	Solid-double	2-4x8x16	2,000	#4 @ 48" E.W.		5.75	14.75	20.50
6530	Wythe			#5 @ 16" E.W.		6.45	15.65	22.10
6550			4,500	#4 @ 48" E.W.		8.15	14.65	22.80
6580				#5 @ 16" E.W.		8.85	15.55	24.40

B2010 112 — Reinforced Concrete Block Wall - Lightweight

	TYPE	SIZE (IN.)	WEIGHT (P.C.F.)	VERT REINF. & GROUT SPACING		COST PER S.F. MAT.	INST.	TOTAL
7100	Hollow	8x4x16	105	#4 @ 48"		3.23	7.65	10.88
7140				#5 @ 16"		4.02	9.20	13.22
7150			85	#4 @ 48"		4.77	7.50	12.27
7190				#5 @ 16"		5.55	9.05	14.60
7400		8x8x16	105	#4 @ 48"		4.10	7.90	12
7440				#5 @ 16"		4.89	9.45	14.34
7450		8x8x16	85	#4 @ 48"		4.27	7.75	12.02
7490				#5 @ 16"		5.05	9.30	14.35
7800		8x8x24	105	#4 @ 48"		3.16	8.55	11.71
7840				#5 @ 16"		3.95	10.10	14.05
7850			85	#4 @ 48"		7.15	7.25	14.40
7890				#5 @ 16"		7.90	8.80	16.70
8100	75% solid	6x8x16	105	#4 @ 48"		4.70	7.30	12
8130				#5 @ 32"		4.85	7.55	12.40
8150			85	#4 @ 48"		4.76	7.10	11.86
8180				#5 @ 32"		4.91	7.35	12.26
8200		8x8x16	105	#4 @ 48"		5.65	7.85	13.50
8230				#5 @ 32"		5.85	8.10	13.95
8250			85	#4 @ 48"		5	7.70	12.70
8280				#5 @ 32"		5.15	7.95	13.10

B2010 Exterior Walls

| B2010 112 | | Reinforced Concrete Block Wall - Lightweight | | | | | | | |

	TYPE	SIZE (IN.)	WEIGHT (P.C.F.)	VERT REINF. & GROUT SPACING		COST PER S.F.		
						MAT.	INST.	TOTAL
8500	Solid-double	2-4x8x16	105	#4 @ 48"		5	14.65	19.65
8530	Wythe		105	#5 @ 16"		5.70	15.55	21.25
8600		2-6x8x16	105	#4 @ 48"		8.15	15.70	23.85
8630				#5 @ 16"		8.85	16.60	25.45
8650			85	#4 @ 48"		11.50	15	26.50

B2010 Exterior Walls

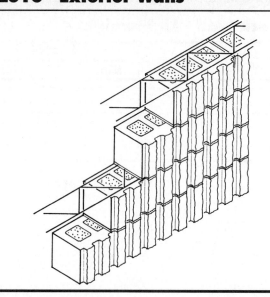

Exterior split ribbed block walls are defined in the following terms; structural reinforcement, weight, percent solid, size, number of ribs and insulation. Within each of these categories two to four variations are shown. No costs are included for brick shelf or relieving angles. Costs include control joints every 20′ and horizontal reinforcing.

B2010 113 — Split Ribbed Block Wall - Regular Weight

	TYPE	SIZE (IN.)	RIBS	CORE FILL		MAT.	INST.	TOTAL
1220	Hollow	4x8x16	4	none		4.41	7.80	12.21
1250			8	none		4.78	7.80	12.58
1280			16	none		5.15	7.90	13.05
1430		8x8x16	8	perlite		7.60	9.30	16.90
1440				styrofoam		7.45	8.85	16.30
1450				none		6.10	8.85	14.95
1530		12x8x16	8	perlite		9.60	12.40	22
1540				styrofoam		8.90	11.55	20.45
1550				none		7.20	11.55	18.75
2120	75% solid	4x8x16	4	none		5.50	7.90	13.40
2150			8	none		6	7.90	13.90
2180			16	none		6.50	8.05	14.55
2520	Solid	4x8x16	4	none		6.25	8.05	14.30
2550			8	none		7.40	8.05	15.45
2580			16	none		7.40	8.15	15.55

B2010 115 — Reinforced Split Ribbed Block Wall - Regular Weight

	TYPE	SIZE (IN.)	RIBS	VERT. REINF. & GROUT SPACING		MAT.	INST.	TOTAL
5200	Hollow	4x8x16	4	#4 @ 48″		4.57	8.50	13.07
5230			8	#4 @ 48″		4.94	8.50	13.44
5260			16	#4 @ 48″		5.30	8.60	13.90
5430		8x8x16	8	#4 @ 48″		6.45	9.70	16.15
5440				#5 @ 32″		6.65	10.05	16.70
5450				#5 @ 16″		7.20	11.25	18.45
5530		12x8x16	8	#4 @ 48″		7.65	12.40	20.05
5540				#5 @ 32″		7.95	12.80	20.75
5550				#5 @ 16″		8.70	14	22.70
6230	75% solid	6x8x16	8	#4 @ 48″		7	9	16
6240				#5 @ 32″		7.15	9.25	16.40
6250				#5 @ 16″		7.40	10.05	17.45
6330		8x8x16	8	#4 @ 48″		7.80	9.70	17.50
6340				#5 @ 32″		8	9.95	17.95
6350				#5 @ 16″		8.30	10.95	19.25

B2010 Exterior Walls

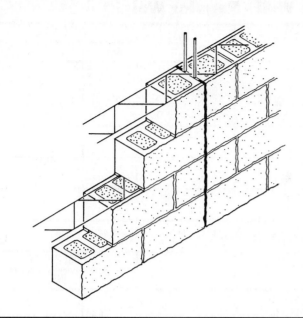

Exterior split face block walls are defined in the following terms; structural reinforcement, weight, percent solid, size, scores and insulation. Within each of these categories two to four variations are shown. No costs are included for brick shelf or relieving angles. Costs include control joints every 20′ and horizontal reinforcing.

B2010 117				Split Face Block Wall - Regular Weight				
	TYPE	SIZE (IN.)	SCORES	CORE FILL		COST PER S.F.		
						MAT.	INST.	TOTAL
1200	Hollow	8x4x16	0	perlite		7.35	10.25	17.60
1210				styrofoam		7.35	9.80	17.15
1240				none		5.85	9.80	15.65
1250			1	perlite		7.75	10.25	18
1260				styrofoam		7.75	9.80	17.55
1290				none		6.25	9.80	16.05
1300		12x4x16	0	perlite		9.65	11.65	21.30
1310				styrofoam		8.95	10.80	19.75
1340				none		7.25	10.80	18.05
1350			1	perlite		10.10	11.65	21.75
1360				styrofoam		9.40	10.80	20.20
1390				none		7.70	10.80	18.50
1400		4x8x16	0	none		4.03	7.70	11.73
1450			1	none		4.36	7.80	12.16
1500		6x8x16	0	perlite		5.60	8.90	14.50
1510				styrofoam		5.90	8.55	14.45
1540				none		4.57	8.55	13.12
1550			1	perlite		5.95	9	14.95
1560				styrofoam		6.25	8.65	14.90
1590				none		4.94	8.65	13.59
1600		8x8x16	0	perlite		6.65	9.60	16.25
1610				styrofoam		6.65	9.15	15.80
1640				none		5.15	9.15	14.30
1650		8x8x16	1	perlite		7.05	9.75	16.80
1660				styrofoam		7.05	9.30	16.35
1690				none		5.55	9.30	14.85
1700		12x8x16	0	perlite		8.60	12.60	21.20
1710				styrofoam		7.90	11.75	19.65
1740				none		6.20	11.75	17.95
1750			1	perlite		9.10	12.80	21.90
1760				styrofoam		8.40	11.95	20.35
1790				none		6.70	11.95	18.65

B20 Exterior Enclosure

B2010 Exterior Walls

B2010 117 — Split Face Block Wall - Regular Weight

	TYPE	SIZE (IN.)	SCORES	CORE FILL		MAT.	INST.	TOTAL
1800	75% solid	8x4x16	0	perlite		7.85	10.15	18
1840				none		7.10	9.95	17.05
1850			1	perlite		8.40	10.15	18.55
1890				none		7.65	9.95	17.60
2000		4x8x16	0	none		5.05	7.80	12.85
2050			1	none		5.45	7.90	13.35
2400	Solid	8x4x16	0	none		7.95	10.15	18.10
2450			1	none		8.55	10.15	18.70
2900		12x8x16	0	none		8.65	12.20	20.85
2950			1	none		9.35	12.40	21.75

B2010 119 — Reinforced Split Face Block Wall - Regular Weight

	TYPE	SIZE (IN.)	SCORES	VERT. REINF. & GROUT SPACING		MAT.	INST.	TOTAL
5200	Hollow	8x4x16	0	#4 @ 48"		6.20	10.65	16.85
5210				#5 @ 32"		6.40	11	17.40
5240				#5 @ 16"		6.95	12.20	19.15
5250			1	#4 @ 48"		6.60	10.65	17.25
5260				#5 @ 32"		6.80	11	17.80
5290				#5 @ 16"		7.35	12.20	19.55
5700		12x8x16	0	#4 @ 48"		6.65	12.60	19.25
5710				#5 @ 32"		6.95	13	19.95
5740				#5 @ 16"		7.70	14.20	21.90
5750			1	#4 @ 48"		7.15	12.80	19.95
5760				#5 @ 32"		7.45	13.20	20.65
5790				#5 @ 16"		8.20	14.40	22.60
6000	75% solid	8x4x16	0	#4 @ 48"		7.25	10.65	17.90
6010				#5 @ 32"		7.45	10.90	18.35
6040				#5 @ 16"		7.75	11.90	19.65
6050			1	#4 @ 48"		7.80	10.65	18.45
6060				#5 @ 32"		8	10.90	18.90
6090				#5 @ 16"		8.30	11.90	20.20
6100		12x4x16	0	#4 @ 48"		8.95	11.70	20.65
6110				#5 @ 32"		9.15	12	21.15
6140				#5 @ 16"		9.55	12.80	22.35
6150			1	#4 @ 48"		9.60	11.70	21.30
6160				#5 @ 32"		9.80	12	21.80
6190				#5 @ 16"		10.25	13	23.25
6700	Solid-double Wythe	2-4x8x16	0	#4 @ 48" E.W.		12.25	17.55	29.80
6710				#5 @ 32" E.W.		12.50	17.70	30.20
6740				#5 @ 16" E.W.		12.95	18.45	31.40
6750			1	#4 @ 48" E.W.		13.25	17.85	31.10
6760				#5 @ 32" E.W.		13.50	18	31.50
6790				#5 @ 16" E.W.		13.95	18.75	32.70
6800		2-6x8x16	0	#4 @ 48" E.W.		13.70	19.40	33.10
6810				#5 @ 32" E.W.		14	19.55	33.55
6840				#5 @ 16" E.W.		14.45	20	34.45
6850			1	#4 @ 48" E.W.		14.80	19.70	34.50
6860				#5 @ 32" E.W.		15.10	19.85	34.95
6890				#5 @ 16" E.W.		15.55	20.50	36.05

B2010 Exterior Walls

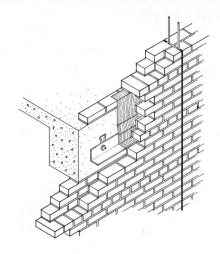

Exterior solid brick walls are defined in the following terms; structural reinforcement, type of brick, thickness, and bond. Nine different types of face bricks are presented, with single wythes shown in four different bonds. Shelf angles are included in system components. These walls do not include ties and as such are not tied to a backup wall.

B2010 125			Solid Brick Walls - Single Wythe			COST PER S.F.		
	TYPE	THICKNESS (IN.)	BOND			MAT.	INST.	TOTAL
1000	Common	4	running			5.75	14.20	19.95
1010			common			6.95	16.25	23.20
1050			Flemish			7.75	19.70	27.45
1100			English			8.70	21	29.70
1150	Standard	4	running			5.60	14.70	20.30
1160			common			6.60	17	23.60
1200			Flemish			7.35	20.50	27.85
1250			English			8.25	21.50	29.75
1300	Glazed	4	running			21	15.30	36.30
1310			common			25.50	17.80	43.30
1350			Flemish			28	21.50	49.50
1400			English			31.50	23	54.50
1600	Economy	4	running			8.35	11.25	19.60
1610			common			9.85	12.90	22.75
1650			Flemish			10.90	15.30	26.20
1700			English			23.50	16.25	39.75
1900	Fire	4-1/2	running			13.55	12.90	26.45
1910			common			16.15	14.70	30.85
1950			Flemish			17.90	17.80	35.70
2000			English			20	18.70	38.70
2050	King	3-1/2	running			4.58	11.55	16.13
2060			common			5.35	13.30	18.65
2100			Flemish			6.50	15.90	22.40
2150			English			7.20	16.60	23.80
2200	Roman	4	running			8.90	13.30	22.20
2210			common			10.65	15.30	25.95
2250			Flemish			11.80	18.25	30.05
2300			English			13.35	19.70	33.05
2350	Norman	4	running			6.90	11	17.90
2360			common			8.15	12.50	20.65
2400			Flemish			19.65	15	34.65
2450			English			10.15	15.90	26.05

357

B2010 Exterior Walls

B2010 125 — Solid Brick Walls - Single Wythe

	TYPE	THICKNESS (IN.)	BOND		COST PER S.F. MAT.	INST.	TOTAL
2500	Norwegian	4	running		6.10	9.80	15.90
2510			common		7.25	11.15	18.40
2550			Flemish		8	13.30	21.30
2600			English		8.95	13.95	22.90

B2010 126 — Solid Brick Walls - Double Wythe

	TYPE	THICKNESS (IN.)	COLLAR JOINT THICKNESS (IN.)		COST PER S.F. MAT.	INST.	TOTAL
4100	Common	8	3/4		11.15	23.50	34.65
4150	Standard	8	1/2		10.60	24.50	35.10
4200	Glazed	8	1/2		41.50	25.50	67
4250	Engineer	8	1/2		10.55	22	32.55
4300	Economy	8	3/4		16.10	18.70	34.80
4350	Double	8	3/4		14.85	15.10	29.95
4400	Fire	8	3/4		26.50	22	48.50
4450	King	7	3/4		8.60	19.15	27.75
4500	Roman	8	1		17.25	22.50	39.75
4550	Norman	8	3/4		13.15	18.25	31.40
4600	Norwegian	8	3/4		11.65	16.20	27.85
4650	Utility	8	3/4		12.45	14	26.45
4700	Triple	8	3/4		9.60	12.60	22.20
4750	SCR	12	3/4		15.90	18.75	34.65
4800	Norwegian	12	3/4		13.70	16.55	30.25

B2010 127 — Reinforced Brick Walls

	TYPE	THICKNESS (IN.)	WYTHE	REINF. & SPACING		COST PER S.F. MAT.	INST.	TOTAL
7200	Common	4"	1	#4 @ 48"vert		5.95	14.75	20.70
7220				#5 @ 32"vert		6.10	14.95	21.05
7230				#5 @ 16"vert		6.40	15.60	22
7700	King	9"	2	#4 @ 48" E.W.		8.95	20	28.95
7720				#5 @ 32" E.W.		9.25	20	29.25
7730				#5 @ 16" E.W.		9.65	21	30.65
7800	Common	10"	2	#4 @ 48" E.W.		11.50	24.50	36
7820				#5 @ 32" E.W.		11.80	24.50	36.30
7830				#5 @ 16" E.W.		12.20	25.50	37.70
7900	Standard	10"	2	#4 @ 48" E.W.		10.95	25.50	36.45
7920				#5 @ 32" E.W.		11.25	25.50	36.75
7930				#5 @ 16" E.W.		11.65	26.50	38.15
8100	Economy	10"	2	#4 @ 48" E.W.		16.45	19.65	36.10
8120				#5 @ 32" E.W.		16.75	19.80	36.55
8130				#5 @ 16" E.W.		17.15	20.50	37.65
8200	Double	10"	2	#4 @ 48" E.W.		15.20	16.05	31.25
8220				#5 @ 32" E.W.		15.50	16.20	31.70
8230				#5 @ 16" E.W.		15.90	16.80	32.70
8300	Fire	10"	2	#4 @ 48" E.W.		27	22.50	49.50
8320				#5 @ 32" E.W.		27	23	50
8330				#5 @ 16" E.W.		27.50	23.50	51
8400	Roman	10"	2	#4 @ 48" E.W.		17.60	23.50	41.10
8420				#5 @ 32" E.W.		17.90	23.50	41.40
8430				#5 @ 16" E.W.		18.30	24	42.30

B2010 Exterior Walls

| B2010 127 | | | | Reinforced Brick Walls | | | | |

	TYPE	THICKNESS (IN.)	WYTHE	REINF. & SPACING		COST PER S.F.		
						MAT.	INST.	TOTAL
8500	Norman	10"	2	#4 @ 48" E.W.		13.50	19.20	32.70
8520				#5 @ 32" E.W.		13.80	19.35	33.15
8530				#5 @ 16" E.W.		14.20	19.95	34.15
8600	Norwegian	10"	2	#4 @ 48" E.W.		12	17.15	29.15
8620				#5 @ 32" E.W.		12.30	17.30	29.60
8630				#5 @ 16" E.W.		12.70	17.90	30.60
8800	Triple	10"	2	#4 @ 48" E.W.		9.95	13.55	23.50
8820				#5 @ 32" E.W.		10.25	13.70	23.95
8830				#5 @ 16" E.W.		10.65	14.30	24.95

B2010 Exterior Walls

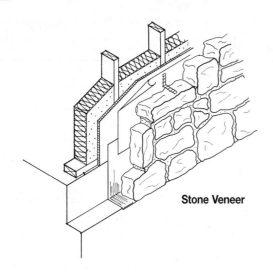

The table below lists costs per S.F. for stone veneer walls on various backup using different stone.

Stone Veneer

B2010 128	Stone Veneer	MAT.	INST.	TOTAL
2000	Ashlar veneer, 4", 2" x 4" stud backup, 16" O.C., 8' high, low priced stone	15.40	22.50	37.90
2050	2" x 6" stud backup, 16" O.C.	17.15	26.50	43.65
2100	Metal stud backup, 8' high, 16" O.C.	16.25	23	39.25
2150	24" O.C.	15.90	22.50	38.40
2200	Conc. block backup, 4" thick	16.90	26.50	43.40
2300	6" thick	17.55	27	44.55
2350	8" thick	17.70	27.50	45.20
2400	10" thick	18.25	29	47.25
2500	12" thick	19.50	31	50.50
3100	High priced stone, wood stud backup, 10' high, 16" O.C.	22	26	48
3200	Metal stud backup, 10' high, 16" O.C.	23	26	49
3250	24" O.C.	22.50	25.50	48
3300	Conc. block backup, 10' high, 4" thick	23.50	30	53.50
3350	6" thick	24	30	54
3400	8" thick	25	32	57
3450	10" thick	25	32	57
3500	12" thick	26	34	60
4000	Indiana limestone 2" thk., sawn finish, wood stud backup, 10' high, 16" O.C	34	13.85	47.85
4100	Metal stud backup, 10' high, 16" O.C.	36.50	17.55	54.05
4150	24" O.C.	36.50	17.55	54.05
4200	Conc. block backup, 4" thick	36	18.05	54.05
4250	6" thick	36.50	18.30	54.80
4300	8" thick	36.50	18.90	55.40
4350	10" thick	37	20.50	57.50
4400	12" thick	38.50	22.50	61
4450	2" thick, smooth finish, wood stud backup, 8' high, 16" O.C.	34	13.85	47.85
4550	Metal stud backup, 8' high, 16" O.C.	35	13.95	48.95
4600	24" O.C.	34.50	13.45	47.95
4650	Conc. block backup, 4" thick	36	17.85	53.85
4700	6" thick	36.50	18.10	54.60
4750	8" thick	36.50	18.70	55.20
4800	10" thick	37	20.50	57.50
4850	12" thick	38.50	22.50	61
5350	4" thick, smooth finish, wood stud backup, 8' high, 16" O.C.	36.50	13.85	50.35
5450	Metal stud backup, 8' high, 16" O.C.	37.50	14.15	51.65
5500	24" O.C.	37	13.65	50.65
5550	Conc. block backup, 4" thick	38.50	18.05	56.55
5600	6" thick	39	18.30	57.30

B2010 Exterior Walls

B2010 128	Stone Veneer	COST PER S.F.		
		MAT.	INST.	TOTAL
5650	8" thick	43	23	66
5700	10" thick	39.50	20.50	60
5750	12" thick	41	22.50	63.50
6000	Granite, gray or pink, 2" thick, wood stud backup, 8' high, 16" O.C.	33.50	25	58.50
6100	Metal studs, 8' high, 16" O.C.	34.50	25.50	60
6150	24" O.C.	34	25	59
6200	Conc. block backup, 4" thick	35.50	29.50	65
6250	6" thick	36	29.50	65.50
6300	8" thick	36	30	66
6350	10" thick	36.50	31.50	68
6400	12" thick	38	33.50	71.50
6900	4" thick, wood stud backup, 8' high, 16" O.C.	44.50	29.50	74
7000	Metal studs, 8' high, 16" O.C.	45.50	29.50	75
7050	24" O.C.	45	29	74
7100	Conc. block backup, 4" thick	46.50	33.50	80
7150	6" thick	47	33.50	80.50
7200	8" thick	47	34.50	81.50
7250	10" thick	47.50	35.50	83
7300	12" thick	49	37.50	86.50

361

For customer support on your Square Foot Cost Data, call 877.756.2789.

B2010 Exterior Walls

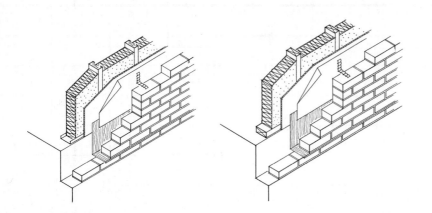

Exterior brick veneer/stud backup walls are defined in the following terms; type of brick and studs, stud spacing and bond. All systems include a brick shelf, ties to the backup and all necessary dampproofing and insulation.

B2010 129				Brick Veneer/Wood Stud Backup					
	FACE BRICK	STUD BACKUP	STUD SPACING (IN.)	BOND	FACE	COST PER S.F.			
							MAT.	INST.	TOTAL
1100	Standard	2x4-wood	16	running			7.40	17.50	24.90
1120				common			8.45	19.80	28.25
1140				Flemish			9.20	23	32.20
1160				English			10.10	24.50	34.60
1400		2x6-wood	16	running			7.90	17.45	25.35
1420				common			8.70	19.95	28.65
1440				Flemish			9.45	23	32.45
1460				English			10.35	24.50	34.85
1700	Glazed	2x4-wood	16	running			23	18.10	41.10
1720				common			27	20.50	47.50
1740				Flemish			29.50	24.50	54
1760				English			33	25.50	58.50
2300	Engineer	2x4-wood	16	running			7.40	15.70	23.10
2320				common			8.40	17.50	25.90
2340				Flemish			9.10	20.50	29.60
2360				English			10	21.50	31.50
2900	Roman	2x4-wood	16	running			10.75	16.10	26.85
2920				common			12.50	18.10	30.60
2940				Flemish			13.65	21	34.65
2960				English			15.20	22.50	37.70
4100	Norwegian	2x4-wood	16	running			7.95	12.60	20.55
4120				common			9.10	13.95	23.05
4140				Flemish			9.85	16.10	25.95
4160				English			10.80	16.75	27.55

B2010 130				Brick Veneer/Metal Stud Backup					
	FACE BRICK	STUD BACKUP	STUD SPACING (IN.)	BOND	FACE	COST PER S.F.			
							MAT.	INST.	TOTAL
5050	Standard	16 ga x 6"LB	16	running			9	18.35	27.35
5100		25ga.x6"NLB	24	running			6.90	17.10	24
5120				common			7.90	19.40	27.30
5140				Flemish			8.30	22	30.30
5160				English			9.55	24	33.55

B2010 130				Brick Veneer/Metal Stud Backup					

	FACE BRICK	STUD BACKUP	STUD SPACING (IN.)	BOND	FACE	COST PER S.F.		
						MAT.	INST.	TOTAL
5200	Standard	20ga.x3-5/8"NLB	16	running		7	17.85	24.85
5220				common		8.05	20	28.05
5240				Flemish		8.80	23.50	32.30
5260				English		9.70	24.50	34.20
5400		16ga.x3-5/8"LB	16	running		7.85	18.10	25.95
5420				common		8.85	20.50	29.35
5440				Flemish		9.60	23.50	33.10
5460				English		10.50	25	35.50
5500			24	running		7.50	17.60	25.10
5520				common		8.50	19.90	28.40
5540				Flemish		9.25	23	32.25
5560				English		10.15	24.50	34.65
5700	Glazed	25ga.x6"NLB	24	running		22.50	17.70	40.20
5720				common		26.50	20	46.50
5740				Flemish		29	24	53
5760				English		32.50	25	57.50
5800		20ga.x3-5/8"NLB	24	running		22.50	17.90	40.40
5820				common		26.50	20.50	47
5840				Flemish		29	24	53
5860				English		32.50	25.50	58
6000		16ga.x3-5/8"LB	16	running		23.50	18.70	42.20
6020				common		27.50	21	48.50
6040				Flemish		30	25	55
6060				English		33.50	26	59.50
6300	Engineer	25ga.x6"NLB	24	running		6.90	15.30	22.20
6320				common		7.85	17.10	24.95
6340				Flemish		8.55	20	28.55
6360				English		9.45	21	30.45
6400		20ga.x3-5/8"NLB	16	running		7	16.05	23.05
6420				common		8	17.85	25.85
6440				Flemish		8.70	21	29.70
6460				English		9.60	22	31.60
6900	Roman	25ga.x6"NLB	24	running		10.20	15.70	25.90
6920				common		11.95	17.65	29.60
6940				Flemish		13.10	20.50	33.60
6960				English		14.65	22	36.65
7000		20ga.x3-5/8"NLB	16	running		10.35	16.45	26.80
7020				common		12.10	18.45	30.55
7040				Flemish		13.25	21.50	34.75
7060				English		14.80	23	37.80
7500	Norman	25ga.x6"NLB	24	running		8.20	13.40	21.60
7520				common		9.45	14.90	24.35
7540				Flemish		21	17.40	38.40
7560				English		11.45	18.30	29.75
7600		20ga.x3-5/8"NLB	24	running		8.20	13.60	21.80
7620				common		9.45	15.10	24.55
7640				Flcmish		21	17.60	38.60
7660				English		11.45	18.50	29.95
8100	Norwegian	25ga.x6"NLB	24	running		7.45	12.20	19.65
8120				common		8.55	13.55	22.10
8140				Flemish		9.30	15.70	25
8160				English		10.25	16.35	26.60

B2010 Exterior Walls

B2010 130		Brick Veneer/Metal Stud Backup						

	FACE BRICK	STUD BACKUP	STUD SPACING (IN.)	BOND	FACE	COST PER S.F.		
						MAT.	INST.	TOTAL
8400	Norwegian	16ga.x3-5/8"LB	16	running		8.40	13.20	21.60
8420				common		9.50	14.55	24.05
8440				Flemish		10.25	16.70	26.95
8460				English		11.20	17.35	28.55

B2010 Exterior Walls

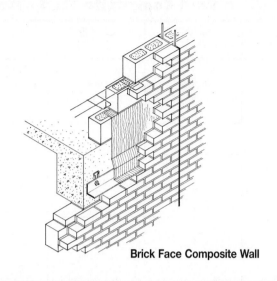

Brick Face Composite Wall

B2010 132	Brick Face Composite Wall - Double Wythe							

	FACE BRICK	BACKUP MASONRY	BACKUP THICKNESS (IN.)	BACKUP CORE FILL		COST PER S.F.		
						MAT.	INST.	TOTAL
1000	Standard	common brick	4	none		10.60	27	37.60
1040		SCR brick	6	none		13.10	24	37.10
1080		conc. block	4	none		8.10	21.50	29.60
1120			6	perlite		9.70	22.50	32.20
1160				styrofoam		10.05	22	32.05
1200			8	perlite		10.40	23	33.40
1240				styrofoam		10.40	22.50	32.90
1520		glazed block	4	none		17.35	23	40.35
1560			6	perlite		19.25	24	43.25
1600				styrofoam		19.55	23.50	43.05
1640			8	perlite		21	24.50	45.50
1680				styrofoam		21	24	45
1720		clay tile	4	none		13	21	34
1760			6	none		15.35	21.50	36.85
1800			8	none		17.60	22	39.60
1840		glazed tile	4	none		19	27.50	46.50
1880								
2000	Glazed	common brick	4	none		26	27.50	53.50
2040		SCR brick	6	none		28.50	24.50	53
2080		conc. block	4	none		23.50	22.50	46
2200			8	perlite		26	23.50	49.50
2240				styrofoam		26	23	49
2280		L.W. block	4	none		23.50	22	45.50
2400			8	perlite		26.50	23.50	50
2520		glazed block	4	none		33	24	57
2560			6	perlite		35	24.50	59.50
2600				styrofoam		35	24	59
2640			8	perlite		36.50	25	61.50
2680				styrofoam		36.50	24.50	61

365

For customer support on your Square Foot Cost Data, call 877.756.2789.

B2010 132		Brick Face Composite Wall - Double Wythe						

	FACE BRICK	BACKUP MASONRY	BACKUP THICKNESS (IN.)	BACKUP CORE FILL		COST PER S.F.		
						MAT.	INST.	TOTAL
2720	Glazed	clay tile	4	none		28.50	21.50	50
2760			6	none		31	22	53
2800			8	none		33	22.50	55.50
2840		glazed tile	4	none		34.50	28	62.50
3000	Engineer	common brick	4	none		10.60	25	35.60
3040		SCR brick	6	none		13.05	22	35.05
3080		conc. block	4	none		8.10	19.90	28
3200			8	perlite		10.35	21	31.35
3280		L.W. block	4	none		6.25	14.10	20.35
3320			6	perlite		10.05	20.50	30.55
3520		glazed block	4	none		17.30	21.50	38.80
3560			6	perlite		19.25	22	41.25
3600				styrofoam		19.55	21.50	41.05
3640			8	perlite		21	22.50	43.50
3680				styrofoam		21	22	43
3720		clay tile	4	none		12.95	19.05	32
3800			8	none		17.55	20.50	38.05
3840		glazed tile	4	none		18.95	25.50	44.45
4000	Roman	common brick	4	none		13.90	25.50	39.40
4040		SCR brick	6	none		16.40	22.50	38.90
4080		conc. block	4	none		11.40	20.50	31.90
4120			6	perlite		13.05	21	34.05
4200			8	perlite		13.70	21.50	35.20
5000	Norman	common brick	4	none		11.90	23.50	35.40
5040		SCR brick	6	none		14.40	20.50	34.90
5280		L.W. block	4	none		9.55	17.85	27.40
5320			6	perlite		11.40	18.75	30.15
5720		clay tile	4	none		14.30	17.15	31.45
5760			6	none		16.65	17.75	34.40
5840		glazed tile	4	none		20.50	24	44.50
6000	Norwegian	common brick	4	none		11.15	22	33.15
6040		SCR brick	6	none		13.60	19.15	32.75
6080		conc. block	4	none		8.65	16.80	25.45
6120			6	perlite		10.25	17.60	27.85
6160				styrofoam		10.55	17.25	27.80
6200			8	perlite		10.90	18.15	29.05
6520		glazed block	4	none		17.85	18.30	36.15
6560			6	perlite		19.80	19	38.80
7000	Utility	common brick	4	none		11.50	21	32.50
7040		SCR brick	6	none		14	18	32
7080		conc. block	4	none		9	15.65	24.65
7120			6	perlite		10.65	16.45	27.10
7160				styrofoam		10.95	16.10	27.05
7200			8	perlite		11.30	17	28.30
7240				styrofoam		11.30	16.55	27.85
7280		L.W. block	4	none		9.15	15.50	24.65
7320			6	perlite		11	16.30	27.30
7520		glazed block	4	none		18.25	17.15	35.40
7560			6	perlite		20	17.85	37.85
7720		clay tile	4	none		13.90	14.80	28.70
7760			6	none		16.25	15.40	31.65
7840		glazed tile	4	none		19.90	21.50	41.40

B2010 Exterior Walls

| B2010 133 | Brick Face Composite Wall - Triple Wythe | | | | | | | |

	FACE BRICK	MIDDLE WYTHE	INSIDE MASONRY	TOTAL THICKNESS (IN.)		COST PER S.F.		
						MAT.	INST.	TOTAL
8000	Standard	common brick	standard brick	12		15.40	38.50	53.90
8100		4" conc. brick	standard brick	12		14.65	39.50	54.15
8120		4" conc. brick	common brick	12		14.85	39	53.85
8200	Glazed	common brick	standard brick	12		31	39	70
8300		4" conc. brick	standard brick	12		30	40.50	70.50
8320		4" conc. brick	glazed brick	12		30.50	40	70.50

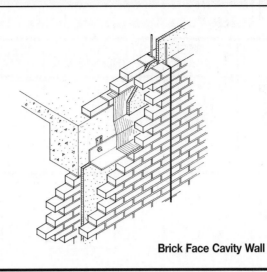

Brick Face Cavity Wall

B2010 134					Brick Face Cavity Wall				
	FACE BRICK	**BACKUP MASONRY**	**TOTAL THICKNESS (IN.)**	**CAVITY INSULATION**		**COST PER S.F.**			
						MAT.	**INST.**	**TOTAL**	
1000	Standard	4" common brick	10	polystyrene		10.50	27	37.50	
1020				none		10.20	26.50	36.70	
1040		6" SCR brick	12	polystyrene		13	24	37	
1060				none		12.70	23.50	36.20	
1080		4" conc. block	10	polystyrene		8	22	30	
1100				none		7.70	21	28.70	
1120		6" conc. block	12	polystyrene		8.65	22	30.65	
1140				none		8.35	21.50	29.85	
1160		4" L.W. block	10	polystyrene		8.10	21.50	29.60	
1180				none		7.85	21	28.85	
1200		6" L.W. block	12	polystyrene		9	22	31	
1220				none		8.70	21.50	30.20	
1240		4" glazed block	10	polystyrene		17.25	23.50	40.75	
1260				none		16.95	22.50	39.45	
1280		6" glazed block	12	polystyrene		17.25	23.50	40.75	
1300				none		16.95	22.50	39.45	
1320		4" clay tile	10	polystyrene		12.90	21	33.90	
1340				none		12.60	20	32.60	
1360		4" glazed tile	10	polystyrene		18.90	27.50	46.40	
1380				none		18.60	27	45.60	
1500	Glazed	4" common brick	10	polystyrene		26	27.50	53.50	
1520				none		25.50	27	52.50	
1580		4" conc. block	10	polystyrene		23.50	22.50	46	
1600				none		23	21.50	44.50	
1660		4" L.W. block	10	polystyrene		23.50	22	45.50	
1680				none		23.50	21.50	45	
1740		4" glazed block	10	polystyrene		32.50	24	56.50	
1760				none		32.50	23	55.50	
1820		4" clay tile	10	polystyrene		28.50	21.50	50	
1840				none		28	21	49	
1860		4" glazed tile	10	polystyrene		30.50	22	52.50	
1880				none		30.50	21.50	52	
2000	Engineer	4" common brick	10	polystyrene		10.50	25.50	36	
2020				none		10.20	24.50	34.70	
2080		4" conc. block	10	polystyrene		8	19.95	27.95	
2100				none		7.70	19.25	26.95	

For customer support on your Square Foot Cost Data, call 877.756.2789.

B2010 Exterior Walls

B2010 134	Brick Face Cavity Wall

	FACE BRICK	BACKUP MASONRY	TOTAL THICKNESS (IN.)	CAVITY INSULATION		COST PER S.F.		
						MAT.	INST.	TOTAL
2160	Engineer	4" L.W. block	10	polystyrene		8.10	19.80	27.90
2180				none		7.80	19.10	26.90
2240		4" glazed block	10	polystyrene		17.20	21.50	38.70
2260				none		16.95	20.50	37.45
2320		4" clay tile	10	polystyrene		12.85	19.10	31.95
2340				none		12.60	18.40	31
2360		4" glazed tile	10	polystyrene		18.85	26	44.85
2380				none		18.60	25	43.60
2500	Roman	4" common brick	10	polystyrene		13.80	25.50	39.30
2520				none		13.50	25	38.50
2580		4" conc. block	10	polystyrene		11.30	20.50	31.80
2600				none		11	19.65	30.65
2660		4" L.W. block	10	polystyrene		11.45	20	31.45
2680				none		11.15	19.50	30.65
2740		4" glazed block	10	polystyrene		20.50	22	42.50
2760				none		20.50	21	41.50
2820		4" clay tile	10	polystyrene		16.20	19.50	35.70
2840				none		15.90	18.80	34.70
2860		4" glazed tile	10	polystyrene		22	26	48
2880				none		22	25.50	47.50
3000	Norman	4" common brick	10	polystyrene		11.80	23.50	35.30
3020						11.50	22.50	34
3080		4" conc. block	10	polystyrene		9.30	18.05	27.35
3100				none		9	17.35	26.35
3160		4" L.W. block	10	polystyrene		9.45	17.90	27.35
3180				none		9.15	17.20	26.35
3240		4" glazed block	10	polystyrene		18.55	19.55	38.10
3260				none		18.25	18.85	37.10
3320		4" clay tile	10	polystyrene		14.20	17.20	31.40
3340				none		13.90	16.50	30.40
3360		4" glazed tile	10	polystyrene		20	24	44
3380				none		19.90	23	42.90
3500	Norwegian	4" common brick	10	polystyrene		11.05	22	33.05
3520				none		10.75	21.50	32.25
3580		4" conc. block	10	polystyrene		8.55	16.85	25.40
3600				none		8.25	16.15	24.40
3660		4" L.W. block	10	polystyrene		8.65	16.70	25.35
3680				none		8.35	16	24.35
3740		4" glazed block	10	polystyrene		17.75	18.35	36.10
3760				none		17.50	17.65	35.15
3820		4" clay tile	10	polystyrene		13.40	16	29.40
3840				none		13.15	15.30	28.45
3860		4" glazed tile	10	polystyrene		19.40	22.50	41.90
3880				none		19.15	22	41.15
4000	Utility	4" common brick	10	polystyrene		11.40	21	32.40
4020				none		11.10	20.50	31.60
4080		4" conc. block	10	polystyrene		8.90	15.70	24.60
4100				none		8.60	15	23.60
4160		4" L.W. block	10	polystyrene		9.05	15.55	24.60
4180				none		8.75	14.85	23.60
4240		4" glazed block	10	polystyrene		18.15	17.20	35.35
4260				none		17.85	16.50	34.35

B2010 Exterior Walls

B2010 134 — Brick Face Cavity Wall

	FACE BRICK	BACKUP MASONRY	TOTAL THICKNESS (IN.)	CAVITY INSULATION		COST PER S.F. MAT.	INST.	TOTAL
4320	Utility	4" clay tile	10	polystyrene		13.80	14.85	28.65
4340				none		13.50	14.15	27.65
4360		4" glazed tile	10	polystyrene		19.80	21.50	41.30
4380				none		19.50	21	40.50

B2010 135 — Brick Face Cavity Wall - Insulated Backup

	FACE BRICK	BACKUP MASONRY	TOTAL THICKNESS (IN.)	BACKUP CORE FILL		COST PER S.F. MAT.	INST.	TOTAL
5100	Standard	6" conc. block	10	perlite		9.35	22	31.35
5120				styrofoam		9.65	21.50	31.15
5180		6" L.W. block	10	perlite		9.70	21.50	31.20
5200				styrofoam		10	21.50	31.50
5260		6" glazed block	10	perlite		18.85	23	41.85
5280				styrofoam		19.15	23	42.15
5340		6" clay tile	10	none		14.95	21	35.95
5360		8" clay tile	12	none		17.20	21.50	38.70
5600	Glazed	6" conc. block	10	perlite		25	22.50	47.50
5620				styrofoam		25	22	47
5680		6" L.W. block	10	perlite		25	22.50	47.50
5700				styrofoam		25.50	22	47.50
5760		6" glazed block	10	perlite		34.50	24	58.50
5780				styrofoam		34.50	23.50	58
5840		6" clay tile	10	none		30.50	21.50	52
5860		8" clay tile	8	none		32.50	22	54.50
6100	Engineer	6" conc. block	10	perlite		9.30	20	29.30
6120				styrofoam		9.65	19.70	29.35
6180		6" L.W. block	10	perlite		9.70	19.85	29.55
6200				styrofoam		10	19.55	29.55
6260		6" glazed block	10	perlite		18.85	21.50	40.35
6280				styrofoam		19.15	21	40.15
6340		6" clay tile	10	none		14.95	19	33.95
6360		8" clay tile	12	none		17.20	19.70	36.90
6600	Roman	6" conc. block	10	perlite		12.65	20.50	33.15
6620				styrofoam		12.95	20	32.95
6680		6" L.W. block	10	perlite		13	20.50	33.50
6700				styrofoam		13.30	19.95	33.25
6760		6" glazed block	10	perlite		22	22	44
6780				styrofoam		22.50	21.50	44
6840		6" clay tile	10	none		18.25	19.40	37.65
6860		8" clay tile	12	none		20.50	20	40.50
7100	Norman	6" conc. block	10	perlite		10.65	18.10	28.75
7120				styrofoam		10.95	17.80	28.75
7180		6" L.W. block	10	perlite		11	17.95	28.95
7200				styrofoam		11.30	17.65	28.95
7260		6" glazed block	10	perlite		20	19.50	39.50
7280				styrofoam		20.50	19.20	39.70
7340		6" clay tile	10	none		16.25	17.10	33.35
7360		8" clay tile	12	none		18.55	17.80	36.35
7600	Norwegian	6" conc. block	10	perlite		9.85	16.90	26.75
7620				styrofoam		10.20	16.60	26.80
7680		6" L.W. block	10	perlite		10.25	16.75	27
7700				styrofoam		10.55	16.45	27

B2010 Exterior Walls

B2010 135	Brick Face Cavity Wall - Insulated Backup

	FACE BRICK	BACKUP MASONRY	TOTAL THICKNESS (IN.)	BACKUP CORE FILL		COST PER S.F.		
						MAT.	INST.	TOTAL
7760	Norwegian	6" glazed block	10	perlite		19.40	18.30	37.70
7780				styrofoam		19.70	18	37.70
7840		6" clay tile	10	none		15.50	15.90	31.40
7860		8" clay tile	12	none		17.75	16.60	34.35
8100	Utility	6" conc. block	10	perlite		10.25	15.75	26
8120				styrofoam		10.55	15.45	26
8180		6" L.W. block	10	perlite		10.60	15.60	26.20
8200				styrofoam		10.90	15.30	26.20
8220		8" L.W. block	12	perlite		11.75	16.10	27.85
8240				styrofoam		11.75	15.70	27.45
8260		6" glazed block	10	perlite		19.75	17.15	36.90
8280				styrofoam		20	16.85	36.85
8300		8" glazed block	12	perlite		21.50	17.75	39.25
8340		6" clay tile	10	none		15.85	14.75	30.60
8360		8" clay tile	12	none		18.15	15.45	33.60

B2010 Exterior Walls

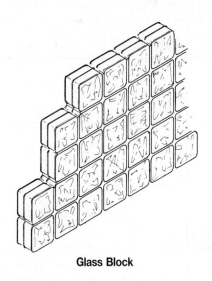

Glass Block

B2010 140	Glass Block	COST PER S.F.		
		MAT.	INST.	TOTAL
2300	Glass block 4″ thick, 6″x6″ plain, under 1,000 S.F.	30.50	25.50	56
2400	1,000 to 5,000 S.F.	30	22	52
2500	Over 5,000 S.F.	29	20.50	49.50
2600	Solar reflective, under 1,000 S.F.	43	35	78
2700	1,000 to 5,000 S.F.	42	30	72
2800	Over 5,000 S.F.	41	28	69
3500	8″x8″ plain, under 1,000 S.F.	17.75	19	36.75
3600	1,000 to 5,000 S.F.	17.40	16.40	33.80
3700	Over 5,000 S.F.	16.90	14.80	31.70
3800	Solar reflective, under 1,000 S.F.	25	25.50	50.50
3900	1,000 to 5,000 S.F.	24.50	22	46.50
4000	Over 5,000 S.F.	23.50	19.70	43.20
5000	12″x12″ plain, under 1,000 S.F.	26.50	17.60	44.10
5100	1,000 to 5,000 S.F.	26	14.80	40.80
5200	Over 5,000 S.F.	25	13.55	38.55
5300	Solar reflective, under 1,000 S.F.	37.50	23.50	61
5400	1,000 to 5,000 S.F.	36.50	19.70	56.20
5600	Over 5,000 S.F.	35	17.95	52.95
5800	3″ thinline, 6″x6″ plain, under 1,000 S.F.	24.50	25.50	50
5900	Over 5,000 S.F.	23.50	20.50	44
6000	Solar reflective, under 1,000 S.F.	34.50	35	69.50
6100	Over 5,000 S.F.	33	28	61
6200	8″x8″ plain, under 1,000 S.F.	14	19	33
6300	Over 5,000 S.F.	13.40	14.80	28.20
6400	Solar reflective, under 1,000 S.F.	19.50	25.50	45
6500	Over 5,000 S.F.	18.65	19.70	38.35

B2010 Exterior Walls

Metal Siding

The table below lists costs for metal siding of various descriptions, not including the steel frame, or the structural steel, of a building. Costs are per S.F. including all accessories and insulation.

B2010 146	Metal Siding Panel	COST PER S.F.		
		MAT.	INST.	TOTAL
1400	Metal siding aluminum panel, corrugated, .024" thick, natural	3.26	4.29	7.55
1450	Painted	3.45	4.29	7.74
1500	.032" thick, natural	3.48	4.29	7.77
1550	Painted	4.03	4.29	8.32
1600	Ribbed 4" pitch, .032" thick, natural	3.55	4.29	7.84
1650	Painted	4.09	4.29	8.38
1700	.040" thick, natural	4.07	4.29	8.36
1750	Painted	4.64	4.29	8.93
1800	.050" thick, natural	4.59	4.29	8.88
1850	Painted	5.20	4.29	9.49
1900	Ribbed 8" pitch, .032" thick, natural	3.39	4.11	7.50
1950	Painted	3.92	4.13	8.05
2000	.040" thick, natural	3.91	4.14	8.05
2050	Painted	4.46	4.18	8.64
2100	.050" thick, natural	4.42	4.16	8.58
2150	Painted	5.05	4.19	9.24
3000	Steel, corrugated or ribbed, 29 Ga., .0135" thick, galvanized	2.14	3.90	6.04
3050	Colored	3.07	3.94	7.01
3100	26 Ga., .0179" thick, galvanized	2.25	3.92	6.17
3150	Colored	3.25	3.96	7.21
3200	24 Ga., .0239" thick, galvanized	2.80	3.94	6.74
3250	Colored	3.54	3.98	7.52
3300	22 Ga., .0299" thick, galvanized	3.02	3.96	6.98
3350	Colored	4.24	4	8.24
3400	20 Ga., .0359" thick, galvanized	3.02	3.96	6.98
3450	Colored	4.50	4.23	8.73
4100	Sandwich panels, factory fab., 1" polystyrene, steel core, 26 Ga., galv.	5.90	6.15	12.05
4200	Colored, 1 side	7.25	6.15	13.40
4300	2 sides	9.10	6.15	15.25
4400	2" polystyrene core, 26 Ga., galvanized	10.05	6.15	16.20
4500	Colored, 1 side	8.35	6.15	14.50
4600	2 sides	10.20	6.15	16.35
4700	22 Ga., baked enamel exterior	13.35	6.45	19.80
4800	Polyvinyl chloride exterior	14.10	6.45	20.55
5100	Textured aluminum, 4' x 8' x 5/16" plywood backing, single face	4.50	3.73	8.23
5200	Double face	5.85	3.73	9.58

373

For customer support on your Square Foot Cost Data, call 877.756.2789.

B2010   Exterior Walls

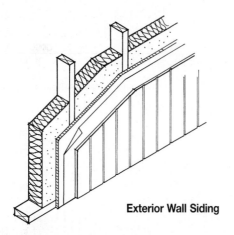

The table below lists costs per S.F. for exterior walls with wood siding. A variety of systems are presented using both wood and metal studs at 16″ and 24″ O.C.

Exterior Wall Siding

B2010 148	Panel, Shingle & Lap Siding	COST PER S.F.		
		MAT.	INST.	TOTAL
1400	Wood siding w/2″ x 4″ studs, 16″ O.C., insul. wall, 5/8″ text 1-11 fir ply.	3.68	6.05	9.73
1450	5/8″ text 1-11 cedar plywood	5.05	5.85	10.90
1500	1″ x 4″ vert T.&G. redwood	5.65	7	12.65
1600	1″ x 8″ vert T.&G. redwood	7.75	6.10	13.85
1650	1″ x 5″ rabbetted cedar bev. siding	6.75	6.30	13.05
1700	1″ x 6″ cedar drop siding	6.85	6.35	13.20
1750	1″ x 12″ rough sawn cedar	3.49	6.10	9.59
1800	1″ x 12″ sawn cedar, 1″ x 4″ battens	7	5.70	12.70
1850	1″ x 10″ redwood shiplap siding	7	5.85	12.85
1900	18″ no. 1 red cedar shingles, 5-1/2″ exposed	4.88	8	12.88
1950	6″ exposed	4.68	7.70	12.38
2000	6-1/2″ exposed	4.48	7.45	11.93
2100	7″ exposed	4.28	7.20	11.48
2150	7-1/2″ exposed	4.08	6.95	11.03
3000	8″ wide aluminum siding	4.54	5.45	9.99
3150	8″ plain vinyl siding	3.31	5.55	8.86
3250	8″ insulated vinyl siding	3.67	6.15	9.82
3300				
3400	2″ x 6″ studs, 16″ O.C., insul. wall, w/ 5/8″ text 1-11 fir plywood	4.11	6.20	10.31
3500	5/8″ text 1-11 cedar plywood	5.45	6.20	11.65
3600	1″ x 4″ vert T.&G. redwood	6.10	7.15	13.25
3700	1″ x 8″ vert T.&G. redwood	8.20	6.25	14.45
3800	1″ x 5″ rabbetted cedar bev siding	7.20	6.45	13.65
3900	1″ x 6″ cedar drop siding	7.25	6.50	13.75
4000	1″ x 12″ rough sawn cedar	3.92	6.25	10.17
4200	1″ x 12″ sawn cedar, 1″ x 4″ battens	7.45	5.85	13.30
4500	1″ x 10″ redwood shiplap siding	7.45	6.05	13.50
4550	18″ no. 1 red cedar shingles, 5-1/2″ exposed	5.30	8.15	13.45
4600	6″ exposed	5.10	7.90	13
4650	6-1/2″ exposed	4.91	7.60	12.51
4700	7″ exposed	4.71	7.35	12.06
4750	7-1/2″ exposed	4.51	7.10	11.61
4800	8″ wide aluminum siding	4.97	5.60	10.57
4850	8″ plain vinyl siding	4.82	9.75	14.57
4900	8″ insulated vinyl siding	5.20	10.35	15.55
4910				
4950	8″ fiber cement siding	4.70	10.55	15.25
5000	2″ x 6″ studs, 24″ O.C., insul. wall, 5/8″ text 1-11, fir plywood	3.94	5.85	9.79
5050	5/8″ text 1-11 cedar plywood	5.30	5.85	11.15
5100	1″ x 4″ vert T.&G. redwood	7	10.85	17.85

B2010 Exterior Walls

B2010 148	Panel, Shingle & Lap Siding	COST PER S.F.		
		MAT.	INST.	TOTAL
5150	1" x 8" vert T.&G. redwood	8.05	5.90	13.95
5200	1" x 5" rabbetted cedar bev siding	7	6.10	13.10
5250	1" x 6" cedar drop siding	7.10	6.15	13.25
5300	1" x 12" rough sawn cedar	3.75	5.90	9.65
5400	1" x 12" sawn cedar, 1" x 4" battens	7.25	5.50	12.75
5450	1" x 10" redwood shiplap siding	7.30	5.65	12.95
5500	18" no. 1 red cedar shingles, 5-1/2" exposed	5.15	7.80	12.95
5550	6" exposed	4.94	7.50	12.44
5650	7" exposed	4.54	7	11.54
5700	7-1/2" exposed	4.34	6.75	11.09
5750	8" wide aluminum siding	4.80	5.25	10.05
5800	8" plain vinyl siding	3.57	5.35	8.92
5850	8" insulated vinyl siding	3.93	5.95	9.88
5875	8" fiber cement siding	4.53	10.15	14.68
5900	3-5/8" metal studs, 16 Ga. 16" O.C. insul. wall, 5/8" text 1-11 fir plywood	4.54	6.35	10.89
5950	5/8" text 1-11 cedar plywood	5.90	6.35	12.25
6000	1" x 4" vert T.&G. redwood	6.50	7.30	13.80
6050	1" x 8" vert T.&G. redwood	8.65	6.40	15.05
6100	1" x 5" rabbetted cedar bev siding	7.60	6.60	14.20
6150	1" x 6" cedar drop siding	7.70	6.65	14.35
6200	1" x 12" rough sawn cedar	4.35	6.40	10.75
6250	1" x 12" sawn cedar, 1" x 4" battens	7.85	6	13.85
6300	1" x 10" redwood shiplap siding	7.90	6.20	14.10
6350	18" no. 1 red cedar shingles, 5-1/2" exposed	5.75	8.30	14.05
6500	6" exposed	5.55	8.05	13.60
6550	6-1/2" exposed	5.35	7.75	13.10
6600	7" exposed	5.15	7.50	12.65
6650	7-1/2" exposed	5.15	7.30	12.45
6700	8" wide aluminum siding	5.40	5.75	11.15
6750	8" plain vinyl siding	4.17	5.65	9.82
6800	8" insulated vinyl siding	4.53	6.25	10.78
7000	3-5/8" metal studs, 16 Ga. 24" O.C. insul wall, 5/8" text 1-11 fir plywood	4.17	5.65	9.82
7050	5/8" text 1-11 cedar plywood	5.55	5.65	11.20
7100	1" x 4" vert T.&G. redwood	6.15	6.80	12.95
7150	1" x 8" vert T.&G. redwood	8.25	5.90	14.15
7200	1" x 5" rabbetted cedar bev siding	7.25	6.10	13.35
7250	1" x 6" cedar drop siding	7.35	6.15	13.50
7300	1" x 12" rough sawn cedar	3.98	5.90	9.88
7350	1" x 12" sawn cedar 1" x 4" battens	7.50	5.50	13
7400	1" x 10" redwood shiplap siding	7.50	5.65	13.15
7450	18" no. 1 red cedar shingles, 5-1/2" exposed	5.60	8.05	13.65
7500	6" exposed	5.35	7.80	13.15
7550	6-1/2" exposed	5.15	7.50	12.65
7600	7" exposed	4.97	7.25	12.22
7650	7-1/2" exposed	4.57	6.75	11.32
7700	8" wide aluminum siding	5.05	5.25	10.30
7750	8" plain vinyl siding	3.80	5.35	9.15
7800	8" insul. vinyl siding	4.16	5.95	10.11

B2010 Exterior Walls

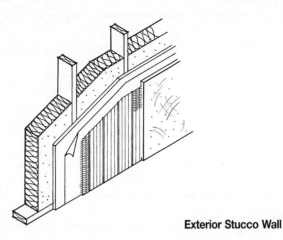

The table below lists costs for some typical stucco walls including all the components as demonstrated in the component block below. Prices are presented for backup walls using wood studs, metal studs and CMU.

Exterior Stucco Wall

B2010 151	Stucco Wall	COST PER S.F.		
		MAT.	INST.	TOTAL
2100	Cement stucco, 7/8" th., plywood sheathing, stud wall, 2" x 4", 16" O.C.	2.93	9.90	12.83
2200	24" O.C.	2.82	9.60	12.42
2300	2" x 6", 16" O.C.	3.36	10.05	13.41
2400	24" O.C.	3.19	9.70	12.89
2500	No sheathing, metal lath on stud wall, 2" x 4", 16" O.C.	2.03	8.60	10.63
2600	24" O.C.	1.92	8.30	10.22
2700	2" x 6", 16" O.C.	2.46	8.75	11.21
2800	24" O.C.	2.29	8.40	10.69
2900	1/2" gypsum sheathing, 3-5/8" metal studs, 16" O.C.	3.16	9	12.16
2950	24" O.C.	2.79	8.50	11.29
3000	Cement stucco, 5/8" th., 2 coats on std. CMU block, 8"x 16", 8" thick	3.43	11.80	15.23
3100	10" thick	5	12.10	17.10
3200	12" thick	5.15	13.90	19.05
3300	Std. light Wt. block 8" x 16", 8" thick	4.19	11.60	15.79
3400	10" thick	4.95	11.90	16.85
3500	12" thick	5.20	13.65	18.85
3600	3 coat stucco, self furring metal lath 3.4 Lb/SY, on 8" x 16", 8" thick	4.30	13.90	18.20
3700	10" thick	5.05	12.80	17.85
3800	12" thick	8.25	17.25	25.50
3900	Lt. Wt. block, 8" thick	4.24	12.30	16.54
4000	10" thick	5	12.60	17.60
4100	12" thick	5.25	14.35	19.60

B2010 152	E.I.F.S.	COST PER S.F.		
		MAT.	INST.	TOTAL
5100	E.I.F.S., plywood sheathing, stud wall, 2" x 4", 16" O.C., 1" EPS	3.97	9.90	13.87
5110	2" EPS	4.26	9.90	14.16
5120	3" EPS	4.54	9.90	14.44
5130	4" EPS	4.83	9.90	14.73
5140	2" x 6", 16" O.C., 1" EPS	4.40	10.05	14.45
5150	2" EPS	4.87	10.20	15.07
5160	3" EPS	4.97	10.05	15.02
5170	4" EPS	5.25	10.05	15.30
5180	Cement board sheathing, 3-5/8" metal studs, 16" O.C., 1" EPS	4.85	12.25	17.10
5190	2" EPS	5.15	12.25	17.40
5200	3" EPS	5.45	12.25	17.70
5210	4" EPS	6.60	12.25	18.85
5220	6" metal studs, 16" O.C., 1" EPS	5.45	12.30	17.75
5230	2" EPS	5.90	12.45	18.35

B2010 Exterior Walls

B2010 152	E.I.F.S.	COST PER S.F.		
		MAT.	INST.	TOTAL
5240	3" EPS	6	12.30	18.30
5250	4" EPS	7.15	12.30	19.45
5260	CMU block, 8" x 8" x 16", 1" EPS	5.05	13.95	19
5270	2" EPS	6.80	18.35	25.15
5280	3" EPS	5.60	13.95	19.55
5290	4" EPS	6.75	15.40	22.15
5300	8" x 10" x 16", 1" EPS	6.65	14.25	20.90
5310	2" EPS	6.90	14.25	21.15
5320	3" EPS	7.20	14.25	21.45
5330	4" EPS	7.50	14.25	21.75
5340	8" x 12" x 16", 1" EPS	6.80	16.05	22.85
5350	2" EPS	7.10	16.05	23.15
5360	3" EPS	7.35	16.05	23.40
5370	4" EPS	7.65	16.05	23.70

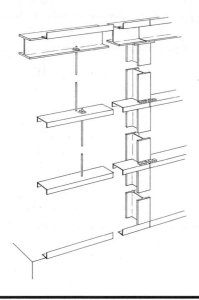

Description: The table below lists costs, $/S.F., for channel girts with sag rods and connector angles top and bottom for various column spacings, building heights and wind loads. Additive costs are shown for wind columns.

How to Use this Table: Add the cost of girts, sag rods, and angles to the framing costs of steel buildings clad in metal or composition siding. If the column spacing is in excess of the column spacing shown, use intermediate wind columns. Additive costs are shown under "Wind Columns".

B2010 154	Metal Siding Support							
	BLDG. HEIGHT (FT.)	WIND LOAD (P.S.F.)	COL. SPACING (FT.)		INTERMEDIATE COLUMNS	COST PER S.F.		
						MAT.	INST.	TOTAL
3000	18	20	20			2.42	4.02	6.44
3100					wind cols.	1.25	.29	1.54
3200		20	25			2.65	4.07	6.72
3300					wind cols.	1	.23	1.23
3400		20	30			2.92	4.14	7.06
3500					wind cols.	.83	.19	1.02
3600		20	35			3.21	4.21	7.42
3700					wind cols.	.71	.16	.87
3800		30	20			2.68	4.08	6.76
3900					wind cols.	1.25	.29	1.54
4000		30	25			2.92	4.14	7.06
4100					wind cols.	1	.23	1.23
4200		30	30			3.23	4.21	7.44
4300					wind cols.	1.13	.26	1.39
4600	30	20	20			2.35	2.83	5.18
4700					wind cols.	1.77	.41	2.18
4800		20	25			2.64	2.90	5.54
4900					wind cols.	1.41	.33	1.74
5000		20	30			2.97	2.98	5.95
5100					wind cols.	1.39	.33	1.72
5200		20	35			3.32	3.06	6.38
5300					wind cols.	1.38	.32	1.70
5400		30	20			2.67	2.91	5.58
5500					wind cols.	2.09	.49	2.58
5600		30	25			2.96	2.97	5.93
5700					wind cols.	1.99	.47	2.46
5800		30	30			3.34	3.06	6.40
5900					wind cols.	1.87	.44	2.31

B2010 Exterior Walls

B2010 160	Pole Barn Exterior Wall	COST PER S.F.		
		MAT.	INST.	TOTAL
2000	Pole barn exterior wall, pressure treated pole in concrete, 8' O.C.			
3000	Steel siding, 8' eave	4.42	10.05	14.47
3050	10' eave	4.23	9.25	13.48
3100	12' eave	4.01	8.45	12.46
3150	14' eave	3.90	8	11.90
3200	16' eave	3.82	7.65	11.47
4000	Aluminum siding, 8' eave	4.21	10.05	14.26
4050	10' eave	3.96	9.10	13.06
4100	12' eave	3.80	8.45	12.25
4150	14' eave	3.69	8	11.69
4200	16' eave	3.61	7.65	11.26
5000	Wood siding, 8' eave	4.68	8.85	13.53
5050	10' eave	4.43	7.85	12.28
5100	12' eave	4.27	7.20	11.47
5150	14' eave	4.16	6.80	10.96
5200	16' eave	4.08	6.40	10.48
6000	Plywood siding, 8' eave	3.88	8.80	12.68
6050	10' eave	3.63	7.85	11.48
6100	12' eave	3.47	7.20	10.67
6150	14' eave	3.36	6.75	10.11
6200	16' eave	3.28	6.40	9.68

The table below lists window systems by material, type and size. Prices between sizes listed can be interpolated with reasonable accuracy. Prices include frame, hardware, and casing as illustrated in the component block below.

B2020 102				Wood Windows				
	MATERIAL	TYPE	GLAZING	SIZE	DETAIL	COST PER UNIT		
						MAT.	INST.	TOTAL
3000	Wood	double hung	std. glass	2'-8" x 4'-6"		247	258	505
3050				3'-0" x 5'-6"		325	296	621
3100			insul. glass	2'-8" x 4'-6"		263	258	521
3150				3'-0" x 5'-6"		350	296	646
3200		sliding	std. glass	3'-4" x 2'-7"		330	215	545
3250				4'-4" x 3'-3"		345	235	580
3300				5'-4" x 6'-0"		435	283	718
3350			insul. glass	3'-4" x 2'-7"		405	251	656
3400				4'-4" x 3'-3"		415	273	688
3450				5'-4" x 6'-0"		530	320	850
3500		awning	std. glass	2'-10" x 1'-9"		230	129	359
3600				4'-4" x 2'-8"		375	125	500
3700			insul. glass	2'-10" x 1'-9"		281	149	430
3800				4'-4" x 2'-8"		460	137	597
3900		casement	std. glass	1'-10" x 3'-2"	1 lite	340	170	510
3950				4'-2" x 4'-2"	2 lite	600	225	825
4000				5'-11" x 5'-2"	3 lite	900	290	1,190
4050				7'-11" x 6'-3"	4 lite	1,275	350	1,625
4100				9'-11" x 6'-3"	5 lite	1,675	405	2,080
4150			insul. glass	1'-10" x 3'-2"	1 lite	340	170	510
4200				4'-2" x 4'-2"	2 lite	615	225	840
4250				5'-11" x 5'-2"	3 lite	950	290	1,240
4300				7'-11" x 6'-3"	4 lite	1,350	350	1,700
4350				9'-11" x 6'-3"	5 lite	1,675	405	2,080
4400		picture	std. glass	4'-6" x 4'-6"		475	291	766
4450				5'-8" x 4'-6"		535	325	860
4500		picture	insul. glass	4'-6" x 4'-6"		580	340	920
4550				5'-8" x 4'-6"		650	375	1,025
4600		fixed bay	std. glass	8' x 5'		1,650	535	2,185
4650				9'-9" x 5'-4"		1,175	750	1,925
4700			insul. glass	8' x 5'		2,175	535	2,710
4750				9'-9" x 5'-4"		1,275	750	2,025
4800		casement bay	std. glass	8' x 5'		1,725	615	2,340
4850			insul. glass	8' x 5'		1,875	615	2,490
4900		vert. bay	std. glass	8' x 5'		1,875	615	2,490
4950			insul. glass	8' x 5'		1,950	615	2,565

For customer support on your Square Foot Cost Data, call 877.756.2789.

B2020 Exterior Windows

B2020 104 — Steel Windows

	MATERIAL	TYPE	GLAZING	SIZE	DETAIL	COST PER UNIT MAT.	INST.	TOTAL
5000	Steel	double hung	1/4" tempered	2'-8" x 4'-6"		910	200	1,110
5050				3'-4" x 5'-6"		1,400	305	1,705
5100			insul. glass	2'-8" x 4'-6"		930	230	1,160
5150				3'-4" x 5'-6"		1,425	350	1,775
5200		horz. pivoted	std. glass	2' x 2'		276	66.50	342.50
5250				3' x 3'		620	150	770
5300				4' x 4'		1,100	266	1,366
5350				6' x 4'		1,650	400	2,050
5400			insul. glass	2' x 2'		281	76.50	357.50
5450				3' x 3'		635	172	807
5500				4' x 4'		1,125	305	1,430
5550				6' x 4'		1,675	460	2,135
5600		picture window	std. glass	3' x 3'		390	150	540
5650				6' x 4'		1,050	400	1,450
5700			insul. glass	3' x 3'		405	172	577
5750				6' x 4'		1,075	460	1,535
5800		industrial security	std. glass	2'-9" x 4'-1"		830	187	1,017
5850				4'-1" x 5'-5"		1,625	370	1,995
5900			insul. glass	2'-9" x 4'-1"		845	215	1,060
5950				4'-1" x 5'-5"		1,675	425	2,100
6000		comm. projected	std. glass	3'-9" x 5'-5"		1,350	340	1,690
6050				6'-9" x 4'-1"		1,825	460	2,285
6100			insul. glass	3'-9" x 5'-5"		1,375	390	1,765
6150				6'-9" x 4'-1"		1,875	525	2,400
6200		casement	std. glass	4'-2" x 4'-2"	2 lite	1,225	289	1,514
6250			insul. glass	4'-2" x 4'-2"		1,250	330	1,580
6300			std. glass	5'-11" x 5'-2"	3 lite	2,550	510	3,060
6350			insul. glass	5'-11" x 5'-2"		2,600	585	3,185

B2020 106 — Aluminum Windows

	MATERIAL	TYPE	GLAZING	SIZE	DETAIL	COST PER UNIT MAT.	INST.	TOTAL
6400	Aluminum	double hung	insul. glass	3'-0" x 4'-0"		530	144	674
6450				4'-5" x 5'-3"		440	180	620
6500			insul. glass	3'-1" x 3'-2"		470	173	643
6550				4'-5" x 5'-3"		530	216	746
6600		sliding	std. glass	3' x 2'		239	144	383
6650				5' x 3'		365	160	525
6700				8' x 4'		385	240	625
6750				9' x 5'		585	360	945
6800			insul. glass	3' x 2'		255	144	399
6850				5' x 3'		425	160	585
6900				8' x 4'		620	240	860
6950				9' x 5'		935	360	1,295
7000		single hung	std. glass	2' x 3'		229	144	373
7050				2'-8" x 6'-8"		400	180	580
7100				3'-4" x 5'-0"		330	160	490
7150			insul. glass	2' x 3'		277	144	421
7200				2'-8" x 6'-8"		520	180	700
7250				3'-4" x 5'		365	160	525
7300		double hung	std. glass	2' x 3'		320	100	420
7350				2'-8" x 6'-8"		950	296	1,246

B2020 Exterior Windows

B2020 106			Aluminum Windows			COST PER UNIT		
	MATERIAL	TYPE	GLAZING	SIZE	DETAIL	MAT.	INST.	TOTAL
7400	Aluminum	double hung	std. glass	3'-4" x 5'		890	277	1,167
7450			insul. glass	2' x 3'		330	115	445
7500				2'-8" x 6'-8"		975	340	1,315
7550				3'-4" x 5'-0"		910	320	1,230
7600		casement	std. glass	3'-1" x 3'-2"		283	162	445
7650				4'-5" x 5'-3"		685	395	1,080
7700			insul. glass	3'-1" x 3'-2"		296	187	483
7750				4'-5" x 5'-3"		720	450	1,170
7800		hinged swing	std. glass	3' x 4'		585	200	785
7850				4' x 5'		980	335	1,315
7900			insul. glass	3' x 4'		605	230	835
7950				4' x 5'		1,000	385	1,385
8200		picture unit	std. glass	2'-0" x 3'-0"		182	100	282
8250				2'-8" x 6'-8"		540	296	836
8300				3'-4" x 5'-0"		505	277	782
8350			insul. glass	2'-0" x 3'-0"		191	115	306
8400				2'-8" x 6'-8"		565	340	905
8450				3'-4" x 5'-0"		530	320	850
8500		awning type	std. glass	3'-0" x 3'-0"	2 lite	470	103	573
8550				3'-0" x 4'-0"	3 lite	540	144	684
8600				3'-0" x 5'-4"	4 lite	650	144	794
8650				4'-0" x 5'-4"	4 lite	715	160	875
8700			insul. glass	3'-0" x 3'-0"	2 lite	495	103	598
8750				3'-0" x 4'-0"	3 lite	620	144	764
8800				3'-0" x 5'-4"	4 lite	765	144	909
8850				4'-0" x 5'-4"	4 lite	855	160	1,015
9051								
9052								

B2020 210	Tubular Aluminum Framing	COST/S.F. OPNG.		
		MAT.	INST.	TOTAL
1100	Alum flush tube frame, for 1/4" glass, 1-3/4"x4", 5'x6'opng, no inter horiz	14.95	11.85	26.80
1150	One intermediate horizontal	20.50	14.15	34.65
1200	Two intermediate horizontals	26	16.40	42.40
1250	5' x 20' opening, three intermediate horizontals	13.75	10.10	23.85
1400	1-3/4" x 4-1/2", 5' x 6' opening, no intermediate horizontals	17.20	11.85	29.05
1450	One intermediate horizontal	23	14.15	37.15
1500	Two intermediate horizontals	28.50	16.40	44.90
1550	5' x 20' opening, three intermediate horizontals	15.45	10.10	25.55
1700	For insulating glass, 2"x4-1/2", 5'x6' opening, no intermediate horizontals	17.50	12.50	30
1750	One intermediate horizontal	22.50	14.95	37.45
1800	Two intermediate horizontals	27	17.35	44.35
1850	5' x 20' opening, three intermediate horizontals	14.95	10.70	25.65
2000	Thermal break frame, 2-1/4"x4-1/2", 5'x6'opng, no intermediate horizontals	18.30	12.70	31
2050	One intermediate horizontal	24.50	15.60	40.10
2100	Two intermediate horizontals	30.50	18.45	48.95
2150	5' x 20' opening, three intermediate horizontals	16.55	11.25	27.80

B2020 Exterior Windows

The table below lists costs of curtain wall and spandrel panels per S.F. Costs do not include structural framing used to hang the panels.

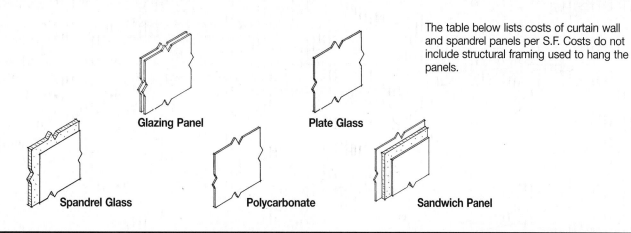

Glazing Panel

Plate Glass

Spandrel Glass

Polycarbonate

Sandwich Panel

B2020 220	Curtain Wall Panels	COST PER S.F.		
		MAT.	INST.	TOTAL
1000	Glazing panel, insulating, 1/2" thick, 2 lites 1/8" float, clear	11.30	11.95	23.25
1100	Tinted	15.65	11.95	27.60
1200	5/8" thick units, 2 lites 3/16" float, clear	15.05	12.65	27.70
1300	Tinted	15.15	12.65	27.80
1400	1" thick units, 2 lites, 1/4" float, clear	18.85	15.15	34
1500	Tinted	26	15.15	41.15
1600	Heat reflective film inside	30.50	13.35	43.85
1700	Light and heat reflective glass, tinted	34.50	13.35	47.85
2000	Plate glass, 1/4" thick, clear	7.05	9.45	16.50
2050	Tempered	9.90	9.45	19.35
2100	Tinted	10	9.45	19.45
2200	3/8" thick, clear	11.45	15.15	26.60
2250	Tempered	18.45	15.15	33.60
2300	Tinted	17.25	15.15	32.40
2400	1/2" thick, clear	19.40	20.50	39.90
2450	Tempered	27.50	20.50	48
2500	Tinted	30	20.50	50.50
2600	3/4" thick, clear	39	32.50	71.50
2650	Tempered	45.50	32.50	78
3000	Spandrel glass, panels, 1/4" plate glass insul w/fiberglass, 1" thick	18.55	9.45	28
3100	2" thick	22.50	9.45	31.95
3200	Galvanized steel backing, add	6.45		6.45
3300	3/8" plate glass, 1" thick	31.50	9.45	40.95
3400	2" thick	35.50	9.45	44.95
4000	Polycarbonate, masked, clear or colored, 1/8" thick	15.75	6.70	22.45
4100	3/16" thick	19.10	6.90	26
4200	1/4" thick	19.10	7.35	26.45
4300	3/8" thick	29.50	7.60	37.10
5000	Facing panel, textured alum., 4' x 8' x 5/16" plywood backing, sgl face	4.50	3.73	8.23
5100	Double face	5.85	3.73	9.58
5200	4' x 10' x 5/16" plywood backing, single face	4.73	3.73	8.46
5300	Double face	6.40	3.73	10.13
5400	4' x 12' x 5/16" plywood backing, single face	4.73	3.73	8.46
5500	Sandwich panel, 22 Ga. galv., both sides 2" insulation, enamel exterior	13.35	6.45	19.80
5600	Polyvinylidene fluoride exterior finish	14.10	6.45	20.55
5700	26 Ga., galv. both sides, 1" insulation, colored 1 side	7.25	6.15	13.40
5800	Colored 2 sides	9.10	6.15	15.25

383

B2030 Exterior Doors

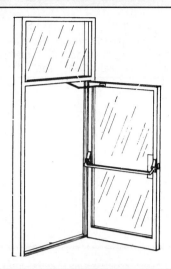

Costs are listed for exterior door systems by material, type and size. Prices between sizes listed can be interpolated with reasonable accuracy. Prices are per opening for a complete door system including frame.

B2030 110				Glazed Doors, Steel or Aluminum				
	MATERIAL	TYPE	DOORS	SPECIFICATION	OPENING	COST PER OPNG.		
						MAT.	INST.	TOTAL
5600	St. Stl. & glass	revolving	stock unit	manual oper.	6'-0" x 7'-0"	52,000	9,550	61,550
5650				auto Cntrls.	6'-10" x 7'-0"	68,500	10,200	78,700
5700	Bronze	revolving	stock unit	manual oper.	6'-10" x 7'-0"	53,000	19,200	72,200
5750				auto Cntrls.	6'-10" x 7'-0"	69,500	19,900	89,400
5800	St. Stl. & glass	balanced	standard	economy	3'-0" x 7'-0"	10,300	1,600	11,900
5850				premium	3'-0" x 7'-0"	17,500	2,100	19,600
6300	Alum. & glass	w/o transom	narrow stile	w/panic Hrdwre.	3'-0" x 7'-0"	1,900	1,075	2,975
6350				dbl. door, Hrdwre.	6'-0" x 7'-0"	3,100	1,750	4,850
6400			wide stile	hdwre.	3'-0" x 7'-0"	2,100	1,050	3,150
6450				dbl. door, Hdwre.	6'-0" x 7'-0"	4,075	2,100	6,175
6500			full vision	hdwre.	3'-0" x 7'-0"	2,800	1,675	4,475
6550				dbl. door, Hdwre.	6'-0" x 7'-0"	3,750	2,375	6,125
6600			non-standard	hdwre.	3'-0" x 7'-0"	2,300	1,050	3,350
6650				dbl. door, Hdwre.	6'-0" x 7'-0"	4,600	2,100	6,700
6700			bronze fin.	hdwre.	3'-0" x 7'-0"	1,675	1,050	2,725
6750				dbl. door, Hrdwre.	6'-0" x 7'-0"	3,350	2,100	5,450
6800			black fin.	hdwre.	3'-0" x 7'-0"	2,350	1,050	3,400
6850				dbl. door, Hdwre.	6'-0" x 7'-0"	4,700	2,100	6,800
6900		w/transom	narrow stile	hdwre.	3'-0" x 10'-0"	2,400	1,225	3,625
6950				dbl. door, Hdwre.	6'-0" x 10'-0"	3,750	2,125	5,875
7000			wide stile	hdwre.	3'-0" x 10'-0"	2,725	1,450	4,175
7050				dbl. door, Hdwre.	6'-0" x 10'-0"	4,150	2,525	6,675
7100			full vision	hdwre.	3'-0" x 10'-0"	3,025	1,625	4,650
7150				dbl. door, Hdwre.	6'-0" x 10'-0"	4,525	2,775	7,300
7200			non-standard	hdwre.	3'-0" x 10'-0"	2,400	1,125	3,525
7250				dbl. door, Hdwre.	6'-0" x 10'-0"	4,775	2,275	7,050
7300			bronze fin.	hdwre.	3'-0" x 10'-0"	1,775	1,125	2,900
7350				dbl. door, Hdwre.	6'-0" x 10'-0"	3,525	2,275	5,800
7400			black fin.	hdwre.	3'-0" x 10'-0"	2,450	1,125	3,575
7450				dbl. door, Hdwre.	6'-0" x 10'-0"	4,875	2,275	7,150
7500		revolving	stock design	minimum	6'-10" x 7'-0"	24,900	3,850	28,750
7550				average	6'-0" x 7'-0"	29,300	4,800	34,100

B20 Exterior Enclosure

B2030 Exterior Doors

B2030 110		Glazed Doors, Steel or Aluminum					

	MATERIAL	TYPE	DOORS	SPECIFICATION	OPENING	COST PER OPNG.		
						MAT.	INST.	TOTAL
7600	Alum. & glass	revolving	stock design	maximum	6'-10" x 7'-0"	34,600	6,400	41,000
7650				min., automatic	6'-10" x 7'-0"	41,500	4,500	46,000
7700				avg., automatic	6'-10" x 7'-0"	45,900	5,450	51,350
7750				max., automatic	6'-10" x 7'-0"	51,000	7,050	58,050
7800		balanced	standard	economy	3'-0" x 7'-0"	7,375	1,600	8,975
7850				premium	3'-0" x 7'-0"	8,825	2,050	10,875
7900		mall front	sliding panels	alum. fin.	16'-0" x 9'-0"	3,900	875	4,775
7950					24'-0" x 9'-0"	5,525	1,625	7,150
8000				bronze fin.	16'-0" x 9'-0"	4,550	1,025	5,575
8050					24'-0" x 9'-0"	6,425	1,900	8,325
8100			fixed panels	alum. fin.	48'-0" x 9'-0"	10,200	1,275	11,475
8150				bronze fin.	48'-0" x 9'-0"	11,900	1,475	13,375
8200		sliding entrance	5' x 7' door	electric oper.	12'-0" x 7'-6"	9,575	1,625	11,200
8250		sliding patio	temp. glass	economy	6'-0" x 7'-0"	1,825	297	2,122
8300			temp. glass	economy	12'-0" x 7'-0"	4,275	395	4,670
8350				premium	6'-0" x 7'-0"	2,750	445	3,195
8400					12'-0" x 7'-0"	6,425	595	7,020

B2030 Exterior Doors

Exterior Door System

The table below lists exterior door systems by material, type and size. Prices between sizes listed can be interpolated with reasonable accuracy. Prices are per opening for a complete door system including frame and required hardware.

Wood doors in this table are designed with wood frames and metal doors with hollow metal frames. Depending upon quality the total material and installation cost of a wood door frame is about the same as a hollow metal frame.

B2030 210 — Wood Doors

	MATERIAL	TYPE	DOORS	SPECIFICATION	OPENING	MAT.	INST.	TOTAL
2350	Birch	solid core	single door	hinged	2'-6" x 6'-8"	1,150	325	1,475
2400					2'-6" x 7'-0"	1,150	325	1,475
2450					2'-8" x 7'-0"	1,150	325	1,475
2500					3'-0" x 7'-0"	1,150	325	1,475
2550			double door		2'-6" x 6'-8"	2,150	590	2,740
2600					2'-6" x 7'-0"	2,150	590	2,740
2650					2'-8" x 7'-0"	2,125	590	2,715
2700					3'-0" x 7'-0"	2,000	515	2,515
2750	Wood	combination	storm & screen		3'-0" x 6'-8"	335	79.50	414.50
2800					3'-0" x 7'-0"	370	87.50	457.50
2850		overhead	panels, H.D.	manual oper.	8'-0" x 8'-0"	1,250	595	1,845
2900					10'-0" x 10'-0"	1,775	660	2,435
2950					12'-0" x 12'-0"	2,450	795	3,245
3000					14'-0" x 14'-0"	3,700	915	4,615
3050					20'-0" x 16'-0"	6,200	1,825	8,025
3100				electric oper.	8'-0" x 8'-0"	2,425	890	3,315
3150					10'-0" x 10'-0"	2,950	955	3,905
3200					12'-0" x 12'-0"	3,625	1,100	4,725
3250					14'-0" x 14'-0"	4,875	1,200	6,075
3300					20'-0" x 16'-0"	7,450	2,425	9,875

B2030 220 — Steel Doors

	MATERIAL	TYPE	DOORS	SPECIFICATION	OPENING	MAT.	INST.	TOTAL
3350	Steel 18 Ga.	hollow metal	1 door w/frame	no label	2'-6" x 7'-0"	1,550	335	1,885
3400					2'-8" x 7'-0"	1,550	335	1,885
3450					3'-0" x 7'-0"	1,575	360	1,935
3500					3'-6" x 7'-0"	1,675	350	2,025
3550		hollow metal	1 door w/frame	no label	4'-0" x 8'-0"	1,975	355	2,330
3600			2 doors w/frame	no label	5'-0" x 7'-0"	3,000	620	3,620
3650					5'-4" x 7'-0"	3,000	620	3,620
3700					6'-0" x 7'-0"	3,000	620	3,620
3750					7'-0" x 7'-0"	3,250	655	3,905
3800					8'-0" x 8'-0"	3,825	655	4,480

For customer support on your Square Foot Cost Data, call 877.756.2789.

B20 Exterior Enclosure

B2030 Exterior Doors

B2030 220 — Steel Doors

	MATERIAL	TYPE	DOORS	SPECIFICATION	OPENING	COST PER OPNG. MAT.	INST.	TOTAL
3850	Steel 18 Ga.	hollow metal	1 door w/frame	"A" label	2'-6" x 7'-0"	1,950	405	2,355
3900					2'-8" x 7'-0"	1,950	410	2,360
3950					3'-0" x 7'-0"	1,950	410	2,360
4000					3'-6" x 7'-0"	2,100	415	2,515
4050					4'-0" x 8'-0"	2,325	435	2,760
4100			2 doors w/frame	"A" label	5'-0" x 7'-0"	3,750	745	4,495
4150					5'-4" x 7'-0"	3,750	755	4,505
4200					6'-0" x 7'-0"	3,750	755	4,505
4250					7'-0" x 7'-0"	4,075	775	4,850
4300					8'-0" x 8'-0"	4,075	775	4,850
4350	Steel 24 Ga.	overhead	sectional	manual oper.	8'-0" x 8'-0"	990	595	1,585
4400					10'-0" x 10'-0"	1,350	660	2,010
4450					12'-0" x 12'-0"	1,625	795	2,420
4500					20'-0" x 14'-0"	3,775	1,700	5,475
4550				electric oper.	8'-0" x 8'-0"	2,175	890	3,065
4600					10'-0" x 10'-0"	2,525	955	3,480
4650					12'-0" x 12'-0"	2,800	1,100	3,900
4700					20'-0" x 14'-0"	5,025	2,300	7,325
4750	Steel	overhead	rolling	manual oper.	8'-0" x 8'-0"	1,300	900	2,200
4800					10'-0" x 10'-0"	2,150	1,025	3,175
4850					12'-0" x 12'-0"	2,175	1,200	3,375
4900					14'-0" x 14'-0"	3,475	1,800	5,275
4950					20'-0" x 12'-0"	2,425	1,600	4,025
5000					20'-0" x 16'-0"	4,175	2,400	6,575
5050				electric oper.	8'-0" x 8'-0"	2,600	1,200	3,800
5100					10'-0" x 10'-0"	3,450	1,325	4,775
5150					12'-0" x 12'-0"	3,475	1,500	4,975
5200					14'-0" x 14'-0"	4,775	2,100	6,875
5250					20'-0" x 12'-0"	3,625	1,900	5,525
5300					20'-0" x 16'-0"	5,375	2,700	8,075
5350				fire rated	10'-0" x 10'-0"	2,450	1,300	3,750
5400			rolling grille	manual oper.	10'-0" x 10'-0"	2,850	1,450	4,300
5450					15'-0" x 8'-0"	3,200	1,800	5,000
5500		vertical lift	1 door w/frame	motor operator	16"-0" x 16'-0"	23,700	5,650	29,350
5550					32'-0" x 24'-0"	53,500	3,775	57,275

B2030 230 — Aluminum Doors

	MATERIAL	TYPE	DOORS	SPECIFICATION	OPENING	COST PER OPNG. MAT.	INST.	TOTAL
6000	Aluminum	combination	storm & screen	hinged	3'-0" x 6'-8"	340	85	425
6050					3'-0" x 7'-0"	375	93.50	468.50
6100		overhead	rolling grille	manual oper.	12'-0" x 12'-0"	4,525	2,525	7,050
6150				motor oper.	12'-0" x 12'-0"	5,900	2,825	8,725
6200	Alum. & Fbrgls.	overhead	heavy duty	manual oper.	12'-0" x 12'-0"	3,350	795	4,145
6250				electric oper.	12'-0" x 12'-0"	4,525	1,100	5,625

387

B3010 Roof Coverings

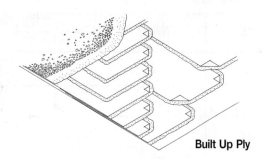

Built Up Ply

Multiple ply roofing is the most popular covering for minimum pitch roofs.

B3010 105	Built-Up	COST PER S.F.		
		MAT.	INST.	TOTAL
1200	Asphalt flood coat w/gravel; not incl. insul, flash., nailers			
1300				
1400	Asphalt base sheets & 3 plies #15 asphalt felt, mopped	1.09	1.98	3.07
1500	On nailable deck	1.13	2.08	3.21
1600	4 plies #15 asphalt felt, mopped	1.50	2.18	3.68
1700	On nailable deck	1.33	2.30	3.63
1800	Coated glass base sheet, 2 plies glass (type IV), mopped	1.21	1.98	3.19
1900	For 3 plies	1.47	2.18	3.65
2000	On nailable deck	1.39	2.30	3.69
2300	4 plies glass fiber felt (type IV), mopped	1.82	2.18	4
2400	On nailable deck	1.65	2.30	3.95
2500	Organic base sheet & 3 plies #15 organic felt, mopped	1.16	2.19	3.35
2600	On nailable deck	1.14	2.30	3.44
2700	4 plies #15 organic felt, mopped	1.43	1.98	3.41
2750				
2800	Asphalt flood coat, smooth surface, not incl. insul, flash., nailers			
2900	Asphalt base sheet & 3 plies #15 asphalt felt, mopped	1.16	1.82	2.98
3000	On nailable deck	1.08	1.90	2.98
3100	Coated glass fiber base sheet & 2 plies glass fiber felt, mopped	1.16	1.75	2.91
3200	On nailable deck	1.10	1.82	2.92
3300	For 3 plies, mopped	1.42	1.90	3.32
3400	On nailable deck	1.34	1.98	3.32
3700	4 plies glass fiber felt (type IV), mopped	1.68	1.90	3.58
3800	On nailable deck	1.60	1.98	3.58
3900	Organic base sheet & 3 plies #15 organic felt, mopped	1.18	1.82	3
4000	On nailable decks	1.09	1.90	2.99
4100	4 plies #15 organic felt, mopped	1.38	1.98	3.36
4200	Coal tar pitch with gravel surfacing			
4300	4 plies #15 tarred felt, mopped	2.20	2.08	4.28
4400	3 plies glass fiber felt (type IV), mopped	1.81	2.30	4.11
4500	Coated glass fiber base sheets 2 plies glass fiber felt, mopped	1.86	2.30	4.16
4600	On nailable decks	1.62	2.43	4.05
4800	3 plies glass fiber felt (type IV), mopped	2.52	2.08	4.60
4900	On nailable decks	2.28	2.18	4.46

B3010 Roof Coverings

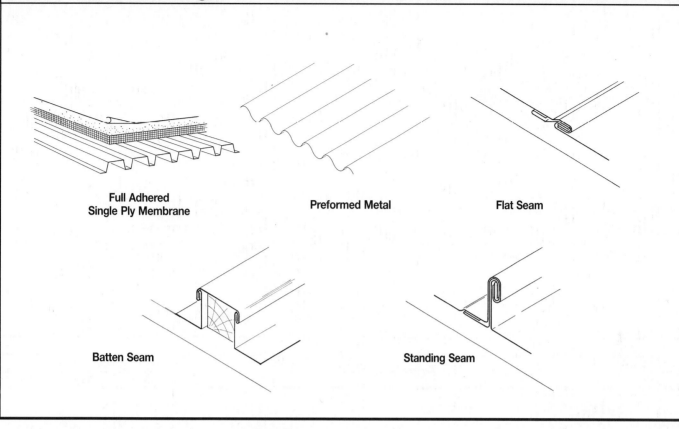

Full Adhered Single Ply Membrane

Preformed Metal

Flat Seam

Batten Seam

Standing Seam

B3010 120	Single Ply Membrane	COST PER S.F.		
		MAT.	INST.	TOTAL
1000	CSPE (Chlorosulfonated polyethylene), 45 mils, plate attached	2.76	.80	3.56
2000	EPDM (Ethylene propylene diene monomer), 45 mils, fully adhered	1.22	1.08	2.30
4000	Modified bit., SBS modified, granule surface cap sheet, mopped, 150 mils	1.40	2.18	3.58
4500	APP modified, granule surface cap sheet, torched, 180 mils	1.02	1.40	2.42
6000	Reinforced PVC, 48 mils, loose laid and ballasted with stone	1.29	.55	1.84
6200	Fully adhered with adhesive	1.70	1.08	2.78

B3010 130	Preformed Metal Roofing	COST PER S.F.		
		MAT.	INST.	TOTAL
0200	Corrugated roofing, aluminum, mill finish, .0175" thick, .272 P.S.F.	1.10	1.94	3.04
0250	.0215" thick, .334 P.S.F.	1.38	1.94	3.32

B3010 135	Formed Metal	COST PER S.F.		
		MAT.	INST.	TOTAL
1000	Batten seam, formed copper roofing, 3" min slope, 16 oz., 1.2 P.S.F.	13.85	6.45	20.30
1100	18 oz., 1.35 P.S.F.	15.35	7.10	22.45
2000	Zinc copper alloy, 3" min slope, .020" thick, .88 P.S.F.	14.10	5.95	20.05
3000	Flat seam, copper, 1/4" min. slope, 16 oz. 1.2 P.S.F.	10.10	5.95	16.05
3100	18 oz., 1.35 P.S.F.	11.35	6.20	17.55
5000	Standing seam, copper, 2-1/2" min. slope, 16 oz., 1.25 P.S.F.	10.85	5.50	16.35
5100	18 oz., 1.40 P.S.F.	12.10	5.95	18.05
6000	Zinc copper alloy, 2-1/2" min. slope, .020" thick, .87 P.S.F.	13.60	5.95	19.55
6100	.032" thick, 1.39 P.S.F.	19.60	6.45	26.05

389

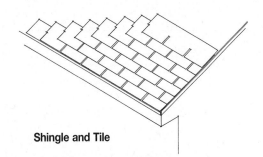

Shingle and Tile

Shingles and tiles are practical in applications where the roof slope is more than 3-1/2" per foot of rise. Lines 1100 through 6000 list the various materials and the weight per square foot.

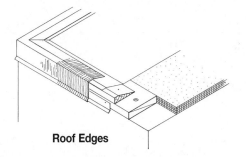

Roof Edges

The table below lists the costs for various types of roof perimeter edge treatments.

Roof edge systems include the cost per L.F. for a 2" x 8" treated wood nailer fastened at 4'-0" O.C. and a diagonally cut 4" x 6" treated wood cant.

Roof edge and base flashing are assumed to be made from the same material.

B3010 140	Shingle & Tile	COST PER S.F.		
		MAT.	INST.	TOTAL
1095	Asphalt roofing			
1100	Strip shingles, 4" slope, inorganic class A 210-235 lb./sq.	.95	1.19	2.14
1150	Organic, class C, 235-240 lb./sq.	1.12	1.30	2.42
1200	Premium laminated multi-layered, class A, 260-300 lb./sq.	1.80	1.78	3.58
1545	Metal roofing			
1550	Alum., shingles, colors, 3" min slope, .019" thick, 0.4 PSF	2.77	1.29	4.06
1850	Steel, colors, 3" min slope, 26 gauge, 1.0 PSF	4.53	2.69	7.22
2795	Slate roofing			
2800	Slate roofing, 4" min. slope, shingles, 3/16" thick, 8.0 PSF	6.05	3.35	9.40
3495	Wood roofing			
3500	4" min slope, cedar shingles, 16" x 5", 5" exposure 1.6 PSF	3.22	2.57	5.79
4000	Shakes, 18", 8-1/2" exposure, 2.8 PSF	2.09	3.16	5.25
5095	Tile roofing			
5100	Aluminum, mission, 3" min slope, .019" thick, 0.65 PSF	9.20	2.48	11.68
6000	Clay tile, flat shingle, interlocking, 15", 166 pcs/sq, fireflashed blend	5.20	2.95	8.15

B3010 Roof Coverings

B3010 320	Roof Deck Rigid Insulation	COST PER S.F.		
		MAT.	INST.	TOTAL
0100	Fiberboard low density			
0150	1" thick R2.78	.73	.73	1.46
0300	1 1/2" thick R4.17	1.04	.73	1.77
0350	2" thick R5.56	1.33	.73	2.06
0370	Fiberboard high density, 1/2" thick R1.3	.45	.62	1.07
0380	1" thick R2.5	.74	.73	1.47
0390	1 1/2" thick R3.8	1.07	.73	1.80
0410	Fiberglass, 3/4" thick R2.78	.79	.62	1.41
0450	15/16" thick R3.70	1.01	.62	1.63
0550	1-5/16" thick R5.26	1.63	.62	2.25
0650	2 7/16" thick R10	1.97	.73	2.70
1510	Polyisocyanurate 2#/CF density, 1" thick	.64	.49	1.13
1550	1 1/2" thick	.82	.53	1.35
1600	2" thick	1.02	.58	1.60
1650	2 1/2" thick	1.33	.60	1.93
1700	3" thick	1.46	.62	2.08
1750	3 1/2" thick	2.21	.62	2.83
1800	Tapered for drainage	.77	.49	1.26
1810	Expanded polystyrene, 1#/CF density, 3/4" thick R2.89	.33	.47	.80
1820	2" thick R7.69	.69	.53	1.22
1830	Extruded polystyrene, 15 PSI compressive strength, 1" thick, R5	.78	.47	1.25
1835	2" thick R10	.98	.53	1.51
1840	3" thick R15	1.84	.62	2.46
2550	40 PSI compressive strength, 1" thick R5	.70	.47	1.17
2600	2" thick R10	1.23	.53	1.76
2650	3" thick R15	1.72	.62	2.34
2700	4" thick R20	2.22	.62	2.84
2750	Tapered for drainage	.96	.49	1.45
2810	60 PSI compressive strength, 1" thick R5	.93	.48	1.41
2850	2" thick R10	1.67	.54	2.21
2900	Tapered for drainage	1.18	.49	1.67
3050	Composites with 2" EPS			
3060	1" Fiberboard	1.66	.61	2.27
3070	7/16" oriented strand board	1.36	.73	2.09
3080	1/2" plywood	1.65	.73	2.38
3090	1" perlite	1.37	.73	2.10
4000	Composites with 1-1/2" polyisocyanurate			
4010	1" fiberboard	1.43	.73	2.16
4020	1" perlite	1.29	.69	1.98
4030	7/16" oriented strand board	1.14	.73	1.87

B3010 420			Roof Edges					
	EDGE TYPE	DESCRIPTION	SPECIFICATION	FACE HEIGHT		COST PER L.F.		
						MAT.	INST.	TOTAL
1000	Aluminum	mill finish	.050" thick	4"		14.45	10.85	25.30
1100				6"		14.65	11.20	25.85
1300		duranodic	.050" thick	4"		15.50	10.85	26.35
1400				6"		15.95	11.20	27.15
1600		painted	.050" thick	4"		15.60	10.85	26.45
1700				6"		16.95	11.20	28.15
2000	Copper	plain	16 oz.	4"		34	10.85	44.85
2100				6"		44	11.20	55.20
2300			20 oz.	4"		42.50	12.45	54.95
2400				6"		49.50	12	61.50
2700	Sheet Metal	galvanized	20 Ga.	4"		17.25	12.95	30.20
2800				6"		17.65	12.95	30.60
3000			24 Ga.	4"		14.30	10.85	25.15

B3010 Roof Coverings

B3010 420 — Roof Edges

	EDGE TYPE	DESCRIPTION	SPECIFICATION	FACE HEIGHT		COST PER L.F.		
						MAT.	INST.	TOTAL
3100	Sheet Metal	galvanized	20 Ga.	6"		14.55	10.85	25.40

B3010 430 — Flashing

	MATERIAL	BACKING	SIDES	SPECIFICATION	QUANTITY	COST PER S.F.		
						MAT.	INST.	TOTAL
0040	Aluminum	none		.019"		1.38	3.92	5.30
0050				.032"		1.49	3.92	5.41
0300		fabric	2	.004"		1.76	1.72	3.48
0400		mastic		.004"		1.76	1.72	3.48
0700	Copper	none		16 oz.	<500 lbs.	8.65	4.94	13.59
0800				24 oz.	<500 lbs.	16.05	5.40	21.45
3500	PVC black	none		.010"		.23	1.99	2.22
3700				.030"		.37	1.99	2.36
4200	Neoprene			1/16"		2.81	1.99	4.80
4500	Stainless steel	none		.015"	<500 lbs.	7.05	4.94	11.99
4600	Copper clad				>2000 lbs.	7.35	3.67	11.02
5000	Plain			32 ga.		3.58	3.67	7.25

B3010 Roof Coverings

B3010 610 — Gutters

	SECTION	MATERIAL	THICKNESS	SIZE	FINISH	COST PER L.F. MAT.	COST PER L.F. INST.	COST PER L.F. TOTAL
0050	Box	aluminum	.027"	5"	enameled	3.17	5.80	8.97
0100					mill	3.19	5.80	8.99
0200			.032"	5"	enameled	3.95	5.60	9.55
0500		copper	16 Oz.	5"	lead coated	20	5.60	25.60
0600					mill	8.85	5.60	14.45
1000		steel galv.	28 Ga.	5"	enameled	2.40	5.60	8
1200			26 Ga.	5"	mill	2.64	5.60	8.24
1800		vinyl		4"	colors	1.43	5.15	6.58
1900				5"	colors	1.76	5.15	6.91
2300		hemlock or fir		4"x5"	treated	18.45	5.95	24.40
3000	Half round	copper	16 Oz.	4"	lead coated	16.15	5.60	21.75
3100					mill	10.45	5.60	16.05
3600		steel galv.	28 Ga.	5"	enameled	2.40	5.60	8
4102		stainless steel		5"	mill	10.40	5.60	16
5000		vinyl		4"	white	1.49	5.15	6.64

B3010 620 — Downspouts

	MATERIALS	SECTION	SIZE	FINISH	THICKNESS	COST PER V.L.F. MAT.	COST PER V.L.F. INST.	COST PER V.L.F. TOTAL
0100	Aluminum	rectangular	2"x3"	embossed mill	.020"	1.16	3.68	4.84
0150				enameled	.020"	1.53	3.68	5.21
0250			3"x4"	enameled	.024"	2.61	5	7.61
0300		round corrugated	3"	enameled	.020"	2.20	3.68	5.88
0350			4"	enameled	.025"	3.25	5	8.25
0500	Copper	rectangular corr.	2"x3"	mill	16 Oz.	9.60	3.68	13.28
0600		smooth		mill	16 Oz.	12.15	3.68	15.83
0700		rectangular corr.	3"x4"	mill	16 Oz.	11.45	4.83	16.28
1300	Steel	rectangular corr.	2"x3"	galvanized	28 Ga.	2.13	3.68	5.81
1350				epoxy coated	24 Ga.	2.56	3.68	6.24
1400		smooth		galvanized	28 Ga.	4.02	3.68	7.70
1450		rectangular corr.	3"x4"	galvanized	28 Ga.	2	4.83	6.83
1500				epoxy coated	24 Ga.	3.11	4.83	7.94
1550		smooth		galvanized	28 Ga.	4.47	4.83	9.30
1600		round corrugated	2"	galvanized	28 Ga.	2.31	3.68	5.99
1650			3"	galvanized	28 Ga.	2.31	3.68	5.99
1700			4"	galvanized	28 Ga.	2.77	4.83	7.60
1750			5"	galvanized	28 Ga.	4.10	5.40	9.50
2552	S.S. tubing sch.5	rectangular	3"x4"	mill		54	4.83	58.83

B3010 630 — Gravel Stop

	MATERIALS	SECTION	SIZE	FINISH	THICKNESS	COST PER L.F. MAT.	COST PER L.F. INST.	COST PER L.F. TOTAL
5100	Aluminum	extruded	4"	mill	.050"	7.05	4.83	11.88
5200			4"	duranodic	.050"	8.10	4.83	12.93
5300			8"	mill	.050"	8.30	5.60	13.90
5400			8"	duranodic	.050"	10	5.60	15.60
6000			12"-2 pc.	duranodic	.050"	11.90	7	18.90
6100	Stainless	formed	6"	mill	24 Ga.	16.85	5.20	22.05

B3020 Roof Openings

Roof Hatch

Smoke Hatch

Skylight

B3020 110	Skylights	COST PER S.F.		
		MAT.	INST.	TOTAL
5100	Skylights, plastic domes, insul curbs, nom. size to 10 S.F., single glaze	32	14.55	46.55
5200	Double glazing	36.50	17.95	54.45
5300	10 S.F. to 20 S.F., single glazing	32.50	5.90	38.40
5400	Double glazing	31	7.40	38.40
5500	20 S.F. to 30 S.F., single glazing	24.50	5	29.50
5600	Double glazing	29.50	5.90	35.40
5700	30 S.F. to 65 S.F., single glazing	24	3.82	27.82
5800	Double glazing	30	5	35
6000	Sandwich panels fiberglass, 1-9/16" thick, 2 S.F. to 10 S.F.	20.50	11.65	32.15
6100	10 S.F. to 18 S.F.	18.25	8.80	27.05
6200	2-3/4" thick, 25 S.F. to 40 S.F.	29.50	7.90	37.40
6300	40 S.F. to 70 S.F.	24	7.05	31.05

B3020 210	Hatches	COST PER OPNG.		
		MAT.	INST.	TOTAL
0200	Roof hatches with curb and 1" fiberglass insulation, 2'-6"x3'-0", aluminum	1,075	233	1,308
0300	Galvanized steel 165 lbs.	700	233	933
0400	Primed steel 164 lbs.	775	233	1,008
0500	2'-6"x4'-6" aluminum curb and cover, 150 lbs.	1,400	259	1,659
0600	Galvanized steel 220 lbs.	975	259	1,234
0650	Primed steel 218 lbs.	1,050	259	1,309
0800	2'x6"x8'-0" aluminum curb and cover, 260 lbs.	2,075	355	2,430
0900	Galvanized steel, 360 lbs.	1,975	355	2,330
0950	Primed steel 358 lbs.	1,575	355	1,930
1200	For plexiglass panels, add to the above	495		495
2100	Smoke hatches, unlabeled not incl. hand winch operator, 2'-6"x3', galv	845	282	1,127
2200	Plain steel, 160 lbs.	940	282	1,222
2400	2'-6"x8'-0",galvanized steel, 360 lbs.	2,150	390	2,540
2500	Plain steel, 350 lbs.	1,725	390	2,115
3000	4'-0"x8'-0", double leaf low profile, aluminum cover, 359 lb.	2,925	291	3,216
3100	Galvanized steel 475 lbs.	2,425	291	2,716
3200	High profile, aluminum cover, galvanized curb, 361 lbs.	2,850	291	3,141

C1010 Partitions

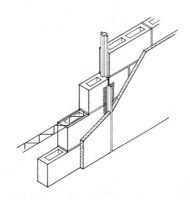

The Concrete Block Partition Systems are defined by weight and type of block, thickness, type of finish and number of sides finished. System components include joint reinforcing on alternate courses and vertical control joints.

C1010 102		Concrete Block Partitions - Regular Weight						
	TYPE	THICKNESS (IN.)	TYPE FINISH	SIDES FINISHED		COST PER S.F.		
						MAT.	INST.	TOTAL
1000	Hollow	4	none	0		2.18	6.30	8.48
1010			gyp. plaster 2 coat	1		2.57	9.25	11.82
1020				2		2.95	12.20	15.15
1200			portland - 3 coat	1		2.45	9.75	12.20
1400			5/8" drywall	1		2.82	8.50	11.32
1500		6	none	0		2.77	6.75	9.52
1510			gyp. plaster 2 coat	1		3.16	9.70	12.86
1520				2		3.54	12.65	16.19
1700			portland - 3 coat	1		3.04	10.20	13.24
1900			5/8" drywall	1		3.41	8.95	12.36
1910				2		4.05	11.15	15.20
2000		8	none	0		2.97	7.25	10.22
2010			gyp. plaster 2 coat	1		3.36	10.20	13.56
2020			gyp. plaster 2 coat	2		3.74	13.15	16.89
2200			portland - 3 coat	1		3.24	10.65	13.89
2400			5/8" drywall	1		3.61	9.45	13.06
2410				2		4.25	11.65	15.90
2500		10	none	0		3.51	7.60	11.11
2510			gyp. plaster 2 coat	1		3.90	10.50	14.40
2520				2		4.28	13.50	17.78
2700			portland - 3 coat	1		3.78	11	14.78
2900			5/8" drywall	1		4.15	9.80	13.95
2910				2		4.79	11.95	16.74
3000	Solid	2	none	0		1.88	6.25	8.13
3010			gyp. plaster	1		2.27	9.20	11.47
3020				2		2.65	12.15	14.80
3200			portland - 3 coat	1		2.15	9.70	11.85
3400			5/8" drywall	1		2.52	8.45	10.97
3410				2		3.16	10.65	13.81
3500		4	none	0		2.45	6.50	8.95
3510			gyp. plaster	1		2.95	9.50	12.45
3520				2		3.22	12.40	15.62
3700			portland - 3 coat	1		2.72	9.95	12.67
3900			5/8" drywall	1		3.09	8.70	11.79
3910				2		3.73	10.90	14.63

C1010 Partitions

C1010 102	Concrete Block Partitions - Regular Weight

	TYPE	THICKNESS (IN.)	TYPE FINISH	SIDES FINISHED		COST PER S.F.		
						MAT.	INST.	TOTAL
4000	Solid	6	none	0		2.83	7	9.83
4010			gyp. plaster	1		3.22	9.95	13.17
4020				2		3.60	12.90	16.50
4200			portland - 3 coat	1		3.10	10.45	13.55
4400			5/8" drywall	1		3.47	9.20	12.67
4410				2		4.11	11.40	15.51

C1010 104	Concrete Block Partitions - Lightweight

	TYPE	THICKNESS (IN.)	TYPE FINISH	SIDES FINISHED		COST PER S.F.		
						MAT.	INST.	TOTAL
5000	Hollow	4	none	0		2.30	6.15	8.45
5010			gyp. plaster	1		2.69	9.10	11.79
5020				2		3.07	12.05	15.12
5200			portland - 3 coat	1		2.57	9.60	12.17
5400			5/8" drywall	1		2.94	8.35	11.29
5410				2		3.58	10.55	14.13
5500		6	none	0		3.13	6.60	9.73
5520			gyp. plaster	2		3.90	12.50	16.40
5700			portland - 3 coat	1		3.40	10.05	13.45
5900			5/8" drywall	1		3.77	8.80	12.57
5910				2		4.41	11	15.41
6000		8	none	0		3.78	7.05	10.83
6010			gyp. plaster	1		4.17	10	14.17
6020				2		4.55	12.95	17.50
6200			portland - 3 coat	1		4.05	10.45	14.50
6400			5/8" drywall	1		4.42	9.25	13.67
6410				2		5.05	11.45	16.50
6500		10	none	0		4.55	7.40	11.95
6510			gyp. plaster	1		4.94	10.30	15.24
6520				2		5.30	13.30	18.60
6700			portland - 3 coat	1		4.82	10.80	15.62
6900			5/8" drywall	1		5.20	9.60	14.80
6910				2		5.85	11.75	17.60
7000	Solid	4	none	0		2.08	6.45	8.53
7010			gyp. plaster	1		2.47	9.40	11.87
7020				2		2.85	12.35	15.20
7200			portland - 3 coat	1		2.35	9.90	12.25
7400			5/8" drywall	1		2.72	8.65	11.37
7410				2		3.36	10.85	14.21
7500		6	none	0		3.65	6.95	10.60
7510			gyp. plaster	1		4.18	9.95	14.13
7520				2		4.42	12.85	17.27
7700			portland - 3 coat	1		3.92	10.40	14.32
7900			5/8" drywall	1		4.29	9.15	13.44
7910				2		4.93	11.35	16.28
8000		8	none	0		4.12	7.45	11.57
8010			gyp. plaster	1		4.51	10.40	14.91
8020				2		4.89	13.35	18.24
8200			portland - 3 coat	1		4.39	10.85	15.24
8400			5/8" drywall	1		4.76	9.65	14.41
8410				2		5.40	11.85	17.25

397

For customer support on your Square Foot Cost Data, call 877.756.2789.

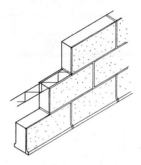

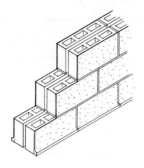

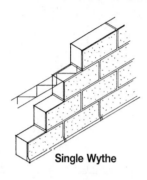

Single Wythe

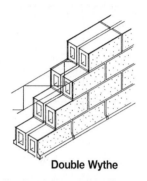

Double Wythe

C1010 120	Tile Partitions	COST PER S.F.		
		MAT.	INST.	TOTAL
1000	8W series 8"x16", 4" thick wall, reinf every 2 courses, glazed 1 side	15.05	7.60	22.65
1100	Glazed 2 sides	17.90	8.05	25.95
1200	Glazed 2 sides, using 2 wythes of 2" thick tile	22	14.60	36.60
1300	6" thick wall, horizontal reinf every 2 courses, glazed 1 side	26	7.95	33.95
1400	Glazed 2 sides, each face different color, 2" and 4" tile	26	14.90	40.90
1500	8" thick wall, glazed 2 sides using 2 wythes of 4" thick tile	30	15.20	45.20
1600	10" thick wall, glazed 2 sides using 1 wythe of 4" & 1 wythe of 6" tile	41	15.55	56.55
1700	Glazed 2 sides cavity wall, using 2 wythes of 4" thick tile	30	15.20	45.20
1800	12" thick wall, glazed 2 sides using 2 wythes of 6" thick tile	52	15.90	67.90
1900	Glazed 2 sides cavity wall, using 2 wythes of 4" thick tile	30	15.20	45.20
2100	6T series 5-1/3"x12" tile, 4" thick, non load bearing glazed one side	12.80	11.90	24.70
2200	Glazed two sides	17.05	13.45	30.50
2300	Glazed two sides, using two wythes of 2" thick tile	18.80	23.50	42.30
2400	6" thick, glazed one side	19.20	12.50	31.70
2500	Glazed two sides	22.50	14.20	36.70
2600	Glazed two sides using 2" thick tile and 4" thick tile	22	23.50	45.50
2700	8" thick, glazed one side	25.50	14.55	40.05
2800	Glazed two sides using two wythes of 4" thick tile	25.50	24	49.50
2900	Glazed two sides using 6" thick tile and 2" thick tile	28.50	24	52.50
3000	10" thick cavity wall, glazed two sides using two wythes of 4" tile	25.50	24	49.50
3100	12" thick, glazed two sides using 4" thick tile and 8" thick tile	38.50	26.50	65
3200	2" thick facing tile, glazed one side, on 6" concrete block	13.60	13.85	27.45
3300	On 8" concrete block	13.75	14.30	28.05
3400	On 10" concrete block	14.30	14.60	28.90

C1010 Partitions

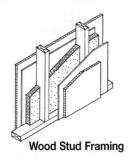

Wood Stud Framing

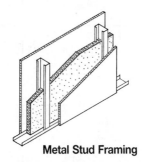

Metal Stud Framing

The Drywall Partitions/Stud Framing Systems are defined by type of drywall and number of layers, type and spacing of stud framing, and treatment on the opposite face. Components include taping and finishing.

Cost differences between regular and fire resistant drywall are negligible, and terminology is interchangeable. In some cases fiberglass insulation is included for additional sound deadening.

C1010 124 — Drywall Partitions/Wood Stud Framing

	FACE LAYER	BASE LAYER	FRAMING	OPPOSITE FACE	INSULATION	MAT.	INST.	TOTAL
						COST PER S.F.		
1200	5/8" FR drywall	none	2 x 4, @ 16" O.C.	same	0	1.34	3.55	4.89
1250				5/8" reg. drywall	0	1.32	3.55	4.87
1300				nothing	0	.90	2.37	3.27
1400		1/4" SD gypsum	2 x 4 @ 16" O.C.	same	1-1/2" fiberglass	2.64	5.45	8.09
1450				5/8" FR drywall	1-1/2" fiberglass	2.20	4.80	7
1500				nothing	1-1/2" fiberglass	1.76	3.62	5.38
1600		resil. channels	2 x 4 @ 16", O.C.	same	1-1/2" fiberglass	2.09	6.95	9.04
1650				5/8" FR drywall	1-1/2" fiberglass	1.93	5.55	7.48
1700				nothing	1-1/2" fiberglass	1.49	4.36	5.85
1800		5/8" FR drywall	2 x 4 @ 24" O.C.	same	0	2.01	4.49	6.50
1850				5/8" FR drywall	0	1.62	3.90	5.52
1900				nothing	0	1.18	2.72	3.90
2200		5/8" FR drywall	2 rows-2 x 4	same	2" fiberglass	3.12	6.50	9.62
2250			16"O.C.	5/8" FR drywall	2" fiberglass	2.73	5.90	8.63
2300				nothing	2" fiberglass	2.29	4.74	7.03
2400	5/8" WR drywall	none	2 x 4, @ 16" O.C.	same	0	1.50	3.55	5.05
2450				5/8" FR drywall	0	1.42	3.55	4.97
2500				nothing	0	.98	2.37	3.35
2600		5/8" FR drywall	2 x 4, @ 24" O.C.	same	0	2.17	4.49	6.66
2650				5/8" FR drywall	0	1.70	3.90	5.60
2700				nothing	0	1.26	2.72	3.98
2800	5/8" VF drywall	none	2 x 4, @ 16" O.C.	same	0	2.14	3.83	5.97
2850				5/8" FR drywall	0	1.74	3.69	5.43
2900				nothing	0	1.30	2.51	3.81
3000		5/8" FR drywall	2 x 4, 24" O.C.	same	0	2.81	4.77	7.58
3050				5/8" FR drywall	0	2.02	4.04	6.06
3100				nothing	0	1.58	2.86	4.44
3200	1/2" reg drywall	3/8" reg drywall	2 x 4, @ 16" O.C.	same	0	2.06	4.73	6.79
3250				5/8" FR drywall	0	1.70	4.14	5.84
3300				nothing	0	1.26	2.96	4.22

C1010 126 — Drywall Partitions/Metal Stud Framing

	FACE LAYER	BASE LAYER	FRAMING	OPPOSITE FACE	INSULATION	MAT.	INST.	TOTAL
						COST PER S.F.		
5000	5/8" FR drywall	none	1-5/8" @ 16" O.C.	same	0	1.15	3.56	4.71
5010				5/8" reg. drywall	0	1.13	3.56	4.69
5020				nothing	0	.71	2.38	3.09
5030			2-1/2"" @ 16" O.C.	same	0	1.22	3.57	4.79

C1010 Partitions

C1010 126	Drywall Partitions/Metal Stud Framing

	FACE LAYER	BASE LAYER	FRAMING	OPPOSITE FACE	INSULATION	COST PER S.F.		
						MAT.	INST.	TOTAL
5040	5/8" FR drywall	none	2-1/2" @ 16" O.C.	5/8" reg. drywall	0	1.20	3.57	4.77
5050				nothing	0	.78	2.39	3.17
5060			3-5/8" @ 16" O.C.	same	0	1.29	3.60	4.89
5070				5/8" reg. drywall	0	1.27	3.60	4.87
5080				nothing	0	.85	2.42	3.27
5200			1-5/8" @ 24" O.C.	same	0	1.08	3.14	4.22
5250				5/8" reg. drywall	0	1.06	3.14	4.20
5300				nothing	0	.64	1.96	2.60
5310			2-1/2" @ 24" O.C.	same	0	1.13	3.15	4.28
5320				5/8" reg. drywall	0	1.11	3.15	4.26
5330				nothing	0	.69	1.97	2.66
5400			3-5/8" @ 24" O.C.	same	0	1.18	3.16	4.34
5450				5/8" reg. drywall	0	1.16	3.16	4.32
5500				nothing	0	.74	1.98	2.72
5530		1/4" SD gypsum	1-5/8" @ 16" O.C.	same	0	2.03	4.88	6.91
5535				5/8" FR drywall	0	1.15	3.56	4.71
5540				nothing	0	1.15	3.04	4.19
5545			2-1/2" @ 16" O.C.	same	0	2.10	4.89	6.99
5550				5/8" FR drywall	0	1.22	3.57	4.79
5555				nothing	0	1.22	3.05	4.27
5560			3-5/8" @ 16" O.C.	same	0	2.17	4.92	7.09
5565				5/8" FR drywall	0	1.29	3.60	4.89
5570				nothing	0	1.29	3.08	4.37
5600			1-5/8" @ 24" O.C.	same	0	1.96	4.46	6.42
5650				5/8" FR drywall	0	1.52	3.80	5.32
5700				nothing	0	1.08	2.62	3.70
5800			2-1/2" @ 24" O.C.	same	0	2.01	4.47	6.48
5850				5/8" FR drywall	0	1.57	3.81	5.38
5900				nothing	0	1.13	2.63	3.76
5910			3-5/8" @ 24" O.C.	same	0	2.06	4.48	6.54
5920				5/8" FR drywall	0	1.62	3.82	5.44
5930				nothing	0	1.18	2.64	3.82
6000		5/8" FR drywall	2-1/2" @ 16" O.C.	same	0	2.10	5.05	7.15
6050				5/8" FR drywall	0	1.71	4.47	6.18
6100				nothing	0	1.27	3.29	4.56
6110			3-5/8" @ 16" O.C.	same	0	2.07	4.78	6.85
6120				5/8" FR drywall	0	1.68	4.19	5.87
6130				nothing	0	1.24	3.01	4.25
6170			2-1/2" @ 24" O.C.	same	0	1.91	4.33	6.24
6180				5/8" FR drywall	0	1.52	3.74	5.26
6190				nothing	0	1.08	2.56	3.64
6200			3-5/8" @ 24" O.C.	same	0	1.96	4.34	6.30
6250				5/8"FR drywall	3-1/2" fiberglass	2.09	4.19	6.28
6300				nothing	0	1.13	2.57	3.70
6310	5/8" WR drywall	none	1-5/8" @ 16" O.C.	same	0	1.31	3.56	4.87
6320				5/8" WR drywall	0	1.78	4.15	5.93
6330				nothing	0	1.26	2.97	4.23
6340			2-1/2" @ 16" O.C.	same	0	2.32	4.75	7.07
6350				5/8" WR drywall	0	1.85	4.16	6.01
6360				nothing	0	1.33	2.98	4.31
6370			3-5/8" @ 16" O.C.	same	0	2.39	4.78	7.17
6380				5/8" WR drywall	0	1.92	4.19	6.11
6390				nothing	0	1.40	3.01	4.41

400

For customer support on your Square Foot Cost Data, call 877.756.2789.

C1010 Partitions

C1010 126	Drywall Partitions/Metal Stud Framing

	FACE LAYER	BASE LAYER	FRAMING	OPPOSITE FACE	INSULATION	COST PER S.F.		
						MAT.	INST.	TOTAL
6400	5/8" WR drywall	none	1-5/8" @ 24" O.C.	same	0	1.24	3.14	4.38
6450				5/8" WR drywall	0	1.16	3.14	4.30
6500				nothing	0	.72	1.96	2.68
6510			2-1/2" @ 24" O.C.	same	0	2.23	4.33	6.56
6520				5/8" WR drywall	0	1.21	3.15	4.36
6530				nothing	0	.77	1.97	2.74
6600			3-5/8" @ 24" O.C.	same	0	1.34	3.16	4.50
6650				5/8" WR drywall	0	1.26	3.16	4.42
6700				nothing	0	.82	1.98	2.80
6800		5/8" FR drywall	2-1/2" @ 16" O.C.	same	0	2.26	5.05	7.31
6850				5/8" FR drywall	0	1.79	4.47	6.26
6900				nothing	0	1.35	3.29	4.64
6910			3-5/8" @ 16" O.C.	same	0	2.23	4.78	7.01
6920				5/8" FR drywall	0	1.76	4.19	5.95
6930				nothing	0	1.32	3.01	4.33
6940			2-1/2" @ 24" O.C.	same	0	1.91	4.33	6.24
6950				5/8" FR drywall	0	1.60	3.74	5.34
6960				nothing	0	1.16	2.56	3.72
7000			3-5/8" @ 24" O.C.	same	0	2.12	4.34	6.46
7050				5/8"FR drywall	3-1/2" fiberglass	2.17	4.19	6.36
7100				nothing	0	1.21	2.57	3.78
7110	5/8" VF drywall	none	1-5/8" @ 16" O.C.	same	0	1.95	3.84	5.79
7120				5/8" FR drywall	0	1.62	4.15	5.77
7130				nothing	0	1.18	2.97	4.15
7140			3-5/8" @ 16" O.C.	same	0	2.23	4.78	7.01
7150				5/8" FR drywall	0	1.76	4.19	5.95
7160				nothing	0	1.32	3.01	4.33
7200		none	1-5/8" @ 24" O.C.	same	0	1.88	3.42	5.30
7250				5/8" FR drywall	0	1.48	3.28	4.76
7300				nothing	0	1.04	2.10	3.14
7400			3-5/8" @ 24" O.C.	same	0	1.98	3.44	5.42
7450				5/8" FR drywall	0	1.58	3.30	4.88
7500				nothing	0	1.14	2.12	3.26
7600		5/8" FR drywall	2-1/2" @ 16" O.C.	same	0	2.90	5.35	8.25
7650				5/8" FR drywall	0	2.11	4.61	6.72
7700				nothing	0	1.67	3.43	5.10
7710			3-5/8" @ 16" O.C.	same	0	2.87	5.05	7.92
7720				5/8" FR drywall	0	2.08	4.33	6.41
7730				nothing	0	1.64	3.15	4.79
7740			2-1/2" @ 24" O.C.	same	0	2.71	4.61	7.32
7750				5/8" FR drywall	0	1.92	3.88	5.80
7760				nothing	0	1.48	2.70	4.18
7800			3-5/8" @ 24" O.C.	same	0	2.76	4.62	7.38
7850				5/8"FR drywall	3-1/2" fiberglass	2.49	4.33	6.82
7900				nothing	0	1.53	2.71	4.24

C1010 Partitions

C1010 128	Drywall Components	COST PER S.F.		
		MAT.	INST.	TOTAL
0060	Metal studs, 24" O.C. including track, load bearing, 20 gage, 2-1/2"	.70	1.11	1.81
0080	3-5/8"	.84	1.13	1.97
0100	4"	.75	1.16	1.91
0120	6"	1.12	1.18	2.30
0140	Metal studs, 24" O.C. including track, load bearing, 18 gage, 2-1/2"	.70	1.11	1.81
0160	3-5/8"	.84	1.13	1.97
0180	4"	.75	1.16	1.91
0200	6"	1.12	1.18	2.30
0220	16 gage, 2-1/2"	.82	1.27	2.09
0240	3-5/8"	.98	1.29	2.27
0260	4"	1.03	1.32	2.35
0280	6"	1.30	1.35	2.65
0300	Non load bearing, 25 gage, 1-5/8"	.20	.78	.98
0340	3-5/8"	.30	.80	1.10
0360	4"	.33	.80	1.13
0380	6"	.40	.82	1.22
0400	20 gage, 2-1/2"	.33	.99	1.32
0420	3-5/8"	.37	1.01	1.38
0440	4"	.44	1.01	1.45
0460	6"	.52	1.02	1.54
0540	Wood studs including blocking, shoe and double top plate, 2"x4", 12" O.C.	.57	1.49	2.06
0560	16" O.C.	.46	1.19	1.65
0580	24" O.C.	.35	.95	1.30
0600	2"x6", 12" O.C.	.88	1.70	2.58
0620	16" O.C.	.71	1.32	2.03
0640	24" O.C.	.54	1.04	1.58
0642	Furring one side only, steel channels, 3/4", 12" O.C.	.42	2.38	2.80
0644	16" O.C.	.38	2.11	2.49
0646	24" O.C.	.25	1.60	1.85
0647	1-1/2" , 12" O.C.	.57	2.66	3.23
0648	16" O.C.	.51	2.33	2.84
0649	24" O.C.	.34	1.83	2.17
0650	Wood strips, 1" x 3", on wood, 12" O.C.	.43	1.08	1.51
0651	16" O.C.	.32	.81	1.13
0652	On masonry, 12" O.C.	.46	1.20	1.66
0653	16" O.C.	.35	.90	1.25
0654	On concrete, 12" O.C.	.46	2.29	2.75
0655	16" O.C.	.35	1.72	2.07
0665	Gypsum board, one face only, exterior sheathing, 1/2"	.51	1.06	1.57
0680	Interior, fire resistant, 1/2"	.40	.59	.99
0700	5/8"	.39	.59	.98
0720	Sound deadening board 1/4"	.44	.66	1.10
0740	Standard drywall 3/8"	.39	.59	.98
0760	1/2"	.36	.59	.95
0780	5/8"	.37	.59	.96
0800	Tongue & groove coreboard 1"	.83	2.48	3.31
0820	Water resistant, 1/2"	.45	.59	1.04
0840	5/8"	.47	.59	1.06
0860	Add for the following:, foil backing	.17		.17
0880	Fiberglass insulation, 3-1/2"	.52	.44	.96
0900	6"	.68	.44	1.12
0920	Rigid insulation 1"	.57	.59	1.16
0940	Resilient furring @ 16" O.C.	.22	1.86	2.08
0960	Taping and finishing	.05	.59	.64
0980	Texture spray	.04	.70	.74
1000	Thin coat plaster	.12	.74	.86
1050	Sound wall framing, 2x6 plates, 2x4 staggered studs, 12" O.C.	.69	1.46	2.15

C1010 Partitions

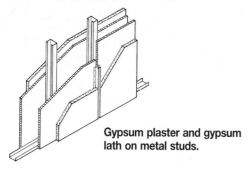

Gypsum plaster and gypsum lath on metal studs.

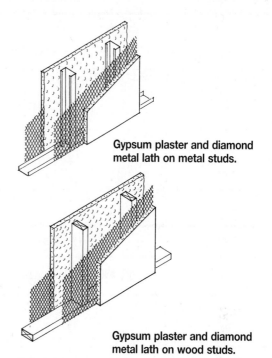

Gypsum plaster and diamond metal lath on metal studs.

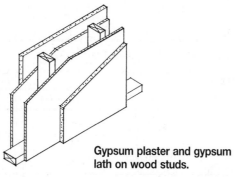

Gypsum plaster and gypsum lath on wood studs.

Gypsum plaster and diamond metal lath on wood studs.

C1010 140	Plaster Partitions/Metal Stud Framing							
	TYPE	FRAMING	LATH	OPPOSITE FACE		COST PER S.F.		
						MAT.	INST.	TOTAL
1000	2 coat gypsum	2-1/2" @ 16"O.C.	3/8" gypsum	same		2.36	8.75	11.11
1010				nothing		1.38	5.05	6.43
1100		3-1/4" @ 24"O.C.	1/2" gypsum	same		2.16	8.45	10.61
1110				nothing		1.27	4.72	5.99
1500	2 coat vermiculite	2-1/2" @ 16"O.C.	3/8" gypsum	same		2.60	9.55	12.15
1510				nothing		1.51	5.50	7.01
1600		3-1/4" @ 24"O.C.	1/2" gypsum	same		2.40	9.25	11.65
1610				nothing		1.40	5.15	6.55
2000	3 coat gypsum	2-1/2" @ 16"O.C.	3/8" gypsum	same		2.40	9.90	12.30
2010				nothing		1.41	5.65	7.06
2020			3.4lb. diamond	same		2.70	9.90	12.60
2030				nothing		1.57	5.65	7.22
2040			2.75lb. ribbed	same		2.50	9.90	12.40
2050				nothing		1.47	5.65	7.12
2100		3-1/4" @ 24"O.C.	1/2" gypsum	same		2.20	9.65	11.85
2110				nothing		1.30	5.30	6.60
2120			3.4lb. ribbed	same		2.71	9.65	12.36
2130				nothing		1.55	5.30	6.85
3500	3 coat gypsum	2-1/2" @ 16"O.C.	3/8" gypsum	same		2.88	12.65	15.53
3510	W/med. Keenes			nothing		1.65	7.05	8.70
3520			3.4lb. diamond	same		3.18	12.65	15.83
3530				nothing		1.81	7.05	8.86
3540			2.75lb. ribbed	same		2.98	12.65	15.63
3550				nothing		1.71	7.05	8.76
3600		3-1/4" @ 24"O.C.	1/2" gypsum	same		2.68	12.40	15.08
3610				nothing		1.54	6.70	8.24
3620			3.4lb. ribbed	same		3.19	12.40	15.59

For customer support on your Square Foot Cost Data, call 877.756.2789.

403

C1010 Partitions

C1010 140	Plaster Partitions/Metal Stud Framing

	TYPE	FRAMING	LATH	OPPOSITE FACE		COST PER S.F.		
						MAT.	INST.	TOTAL
3630	3 coat gypsum	3-1/4" @ 24"O.C.	3.4lb. ribbed	nothing		1.79	6.70	8.49
4000	3 coat gypsum	2-1/2" @ 16"O.C.	3/8" gypsum	same		2.88	14.05	16.93
4010	W/hard Keenes			nothing		1.65	7.70	9.35
4020			3.4lb. diamond	same		3.18	14.05	17.23
4030				nothing		1.81	7.70	9.51
4040			2.75lb. ribbed	same		2.98	14.05	17.03
4050				nothing		1.71	7.70	9.41
4100		3-1/4" @ 24"O.C.	1/2" gypsum	same		2.68	13.75	16.43
4110				nothing		1.54	7.35	8.89
4120			3.4lb. ribbed	same		3.19	13.75	16.94
4130				nothing		1.79	7.35	9.14

C1010 142	Plaster Partitions/Wood Stud Framing

	TYPE	FRAMING	LATH	OPPOSITE FACE		COST PER S.F.		
						MAT.	INST.	TOTAL
5000	2 coat gypsum	2"x4" @ 16"O.C.	3/8" gypsum	same		2.48	8.55	11.03
5010				nothing		1.50	4.95	6.45
5100		2"x4" @ 24"O.C.	1/2" gypsum	same		2.21	8.40	10.61
5110				nothing		1.32	4.76	6.08
5500	2 coat vermiculite	2"x4" @ 16"O.C.	3/8" gypsum	same		2.72	9.30	12.02
5510				nothing		1.63	5.40	7.03
5600		2"x4" @ 24"O.C.	1/2" gypsum	same		2.45	9.15	11.60
5610				nothing		1.45	5.20	6.65
6000	3 coat gypsum	2"x4" @ 16"O.C.	3/8" gypsum	same		2.52	9.70	12.22
6010				nothing		1.53	5.55	7.08
6020			3.4lb. diamond	same		2.82	9.80	12.62
6030				nothing		1.69	5.60	7.29
6040			2.75lb. ribbed	same		2.59	9.85	12.44
6050				nothing		1.57	5.60	7.17
6100		2"x4" @ 24"O.C.	1/2" gypsum	same		2.25	9.55	11.80
6110				nothing		1.35	5.35	6.70
6120			3.4lb. ribbed	same		2.51	9.65	12.16
6130				nothing		1.48	5.40	6.88
7500	3 coat gypsum	2"x4" @ 16"O.C.	3/8" gypsum	same		3	12.45	15.45
7510	W/med Keenes			nothing		1.77	6.95	8.72
7520			3.4lb. diamond	same		3.30	12.55	15.85
7530				nothing		1.93	7	8.93
7540			2.75lb. ribbed	same		3.07	12.60	15.67
7550				nothing		1.81	7	8.81
7600		2"x4" @ 24"O.C.	1/2" gypsum	same		2.73	12.30	15.03
7610				nothing		1.59	6.75	8.34
7620			3.4lb. ribbed	same		3.24	12.55	15.79
7630				nothing		1.84	6.85	8.69
8000	3 coat gypsum	2"x4" @ 16"O.C.	3/8" gypsum	same		3	13.85	16.85
8010	W/hard Keenes			nothing		1.77	7.60	9.37
8020			3.4lb. diamond	same		3.30	13.95	17.25
8030				nothing		1.93	7.65	9.58
8040			2.75lb. ribbed	same		3.07	14	17.07
8050				nothing		1.81	7.65	9.46
8100		2"x4" @ 24"O.C.	1/2" gypsum	same		2.73	13.70	16.43
8110				nothing		1.59	7.40	8.99
8120			3.4lb. ribbed	same		3.24	13.90	17.14
8130				nothing		1.84	7.50	9.34

C1010 Partitions

C1010 144	Plaster Partition Components	COST PER S.F.		
		MAT.	INST.	TOTAL
0060	Metal studs, 16" O.C., including track, non load bearing, 25 gage, 1-5/8"	.34	1.21	1.55
0080	2-1/2"	.34	1.21	1.55
0100	3-1/4"	.41	1.24	1.65
0120	3-5/8"	.41	1.24	1.65
0140	4"	.45	1.25	1.70
0160	6"	.54	1.26	1.80
0180	Load bearing, 20 gage, 2-1/2"	.95	1.55	2.50
0200	3-5/8"	1.14	1.57	2.71
0220	4"	1.18	1.61	2.79
0240	6"	1.51	1.63	3.14
0260	16 gage 2-1/2"	1.13	1.75	2.88
0280	3-5/8"	1.35	1.80	3.15
0300	4"	1.41	1.83	3.24
0320	6"	1.78	1.86	3.64
0340	Wood studs, including blocking, shoe and double plate, 2"x4" , 12" O.C.	.57	1.49	2.06
0360	16" O.C.	.46	1.19	1.65
0380	24" O.C.	.35	.95	1.30
0400	2"x6" , 12" O.C.	.88	1.70	2.58
0420	16" O.C.	.71	1.32	2.03
0440	24" O.C.	.54	1.04	1.58
0460	Furring one face only, steel channels, 3/4", 12" O.C.	.42	2.38	2.80
0480	16" O.C.	.38	2.11	2.49
0500	24" O.C.	.25	1.60	1.85
0520	1-1/2" , 12" O.C.	.57	2.66	3.23
0540	16" O.C.	.51	2.33	2.84
0560	24"O.C.	.34	1.83	2.17
0580	Wood strips 1"x3", on wood., 12" O.C.	.43	1.08	1.51
0600	16"O.C.	.32	.81	1.13
0620	On masonry, 12" O.C.	.46	1.20	1.66
0640	16" O.C.	.35	.90	1.25
0660	On concrete, 12" O.C.	.46	2.29	2.75
0680	16" O.C.	.35	1.72	2.07
0700	Gypsum lath. plain or perforated, nailed to studs, 3/8" thick	.37	.73	1.10
0720	1/2" thick	.30	.78	1.08
0740	Clipped to studs, 3/8" thick	.37	.83	1.20
0760	1/2" thick	.30	.89	1.19
0780	Metal lath, diamond painted, nailed to wood studs, 2.5 lb.	.50	.73	1.23
0800	3.4 lb.	.53	.78	1.31
0820	Screwed to steel studs, 2.5 lb.	.50	.78	1.28
0840	3.4 lb.	.51	.83	1.34
0860	Rib painted, wired to steel, 2.75 lb	.43	.83	1.26
0880	3.4 lb	.55	.89	1.44
0900	4.0 lb	.69	.95	1.64
0910				
0920	Gypsum plaster, 2 coats	.45	2.82	3.27
0940	3 coats	.65	3.42	4.07
0960	Perlite or vermiculite plaster, 2 coats	.72	3.24	3.96
0980	3 coats	.78	4	4.78
1000	Stucco, 3 coats, 1" thick, on wood framing	.78	5.70	6.48
1020	On masonry	.31	4.42	4.73
1100	Metal base galvanized and painted 2-1/2" high	.72	2.33	3.05

C1010 Partitions

Folding Accordion

Folding Leaf

Movable and Borrow Lites

C1010 205	Partitions	COST PER S.F.		
		MAT.	INST.	TOTAL
0360	Folding accordion, vinyl covered, acoustical, 3 lb. S.F., 17 ft max. hgt	35	11.90	46.90
0380	5 lb. per S.F. 27 ft max height	48.50	12.50	61
0400	5.5 lb. per S.F., 17 ft. max height	56.50	13.20	69.70
0420	Commercial, 1.75 lb per S.F., 8 ft. max height	31.50	5.30	36.80
0440	2.0 Lb per S.F., 17 ft. max height	32.50	7.95	40.45
0460	Industrial, 4.0 lb. per S.F. 27 ft max height	49	15.85	64.85
0480	Vinyl clad wood or steel, electric operation 6 psf	69	7.45	76.45
0500	Wood, non acoustic, birch or mahogany	37	3.96	40.96
0560	Folding leaf, alum framed acoustical 12' high., 5.5 lb/S.F. standard trim	45	19.80	64.80
0580	Premium trim	54	39.50	93.50
0600	6.5 lb. per S.F., standard trim	47	19.80	66.80
0620	Premium trim	58.50	39.50	98
0640	Steel acoustical, 7.5 per S.F., vinyl faced, standard trim	66.50	19.80	86.30
0660	Premium trim	81.50	39.50	121
0680	Wood acoustic type, vinyl faced to 18' high 6 psf, economy trim	62.50	19.80	82.30
0700	Standard trim	74.50	26.50	101
0720	Premium trim	96.50	39.50	136
0740	Plastic lam. or hardwood faced, standard trim	64.50	19.80	84.30
0760	Premium trim	68.50	39.50	108
0780	Wood, low acoustical type to 12 ft. high 4.5 psf	47	24	71
0840	Demountable, trackless wall, cork finish, semi acous, 1-5/8" th, unsealed	42.50	3.66	46.16
0860	Sealed	47	6.25	53.25
0880	Acoustic, 2" thick, unsealed	40	3.90	43.90
0900	Sealed	61.50	5.30	66.80
0920	In-plant modular office system, w/prehung steel door			
0940	3" thick honeycomb core panels			
0960	12' x 12', 2 wall	13.60	.74	14.34
0970	4 wall	13.80	.98	14.78
0980	16' x 16', 2 wall	13.75	.52	14.27
0990	4 wall	9.35	.52	9.87
1000	Gypsum, demountable, 3" to 3-3/4" thick x 9' high, vinyl clad	7.35	2.78	10.13
1020	Fabric clad	18.30	3	21.30
1040	1.75 system, vinyl clad hardboard, paper honeycomb core panel			
1060	1-3/4" to 2-1/2" thick x 9' high	12.30	2.78	15.08
1080	Unitized gypsum panel system, 2" to 2-1/2" thick x 9' high			
1100	Vinyl clad gypsum	15.85	2.78	18.63
1120	Fabric clad gypsum	26	3	29
1140	Movable steel walls, modular system			
1160	Unitized panels, 48" wide x 9' high			
1180	Baked enamel, pre-finished	17.85	2.20	20.05
1200	Fabric clad	26	2.33	28.33

For customer support on your Square Foot Cost Data, call 877.756.2789.

C1020 Interior Doors

Single Leaf

Sliding Entrance

Rolling Overhead

C1020 102	Special Doors	COST PER OPNG.		
		MAT.	INST.	TOTAL
2500	Single leaf, wood, 3'-0"x7'-0"x1 3/8", birch, solid core	405	221	626
2510	Hollow core	330	212	542
2530	Hollow core, lauan	320	212	532
2540	Louvered pine	465	212	677
2550	Paneled pine	495	212	707
2600	Hollow metal, comm. quality, flush, 3'-0"x7'-0"x1-3/8"	935	233	1,168
2650	3'-0"x10'-0" openings with panel	1,475	300	1,775
2700	Metal fire, comm. quality, 3'-0"x7'-0"x1-3/8"	1,175	242	1,417
3200	Double leaf, wood, hollow core, 2 - 3'-0"x7'-0"x1-3/8"	590	405	995
3300	Hollow metal, comm. quality, B label, 2'-3'-0"x7'-0"x1-3/8"	2,225	510	2,735
3400	6'-0"x10'-0" opening, with panel	2,850	630	3,480
3500	Double swing door system, 12'-0"x7'-0", mill finish	9,450	3,225	12,675
3700	Black finish	9,600	3,325	12,925
3800	Sliding entrance door and system mill finish	11,000	2,875	13,875
3900	Bronze finish	12,000	3,100	15,100
4000	Black finish	12,500	3,225	15,725
4100	Sliding panel mall front, 16'x9' opening, mill finish	3,900	875	4,775
4200	Bronze finish	5,075	1,150	6,225
4300	Black finish	6,250	1,400	7,650
4400	24'x9' opening mill finish	5,525	1,625	7,150
4500	Bronze finish	7,175	2,125	9,300
4600	Black finish	8,850	2,600	11,450
4700	48'x9' opening mill finish	10,200	1,275	11,475
4800	Bronze finish	13,300	1,650	14,950
4900	Black finish	16,300	2,050	18,350
5000	Rolling overhead steel door, manual, 8' x 8' high	1,300	900	2,200
5100	10' x 10' high	2,150	1,025	3,175
5200	20' x 10' high	3,500	1,450	4,950
5300	12' x 12' high	2,175	1,200	3,375
5400	Motor operated, 8' x 8' high	2,600	1,200	3,800
5500	10' x 10' high	3,450	1,325	4,775
5600	20' x 10' high	4,800	1,750	6,550
5700	12' x 12' high	3,475	1,500	4,975
5800	Roll up grille, aluminum, manual, 10' x 10' high, mill finish	3,150	1,750	4,900
5900	Bronze anodized	4,950	1,750	6,700
6000	Motor operated, 10' x 10' high, mill finish	4,525	2,050	6,575
6100	Bronze anodized	6,325	2,050	8,375
6200	Steel, manual, 10' x 10' high	2,850	1,450	4,300
6300	15' x 8' high	3,200	1,800	5,000
6400	Motor operated, 10' x 10' high	4,225	1,750	5,975
6500	15' x 8' high	4,575	2,100	6,675
8970	Counter door, rolling, 6' high, 14' wide, aluminum	3,025	850	3,875

C1030 Fittings

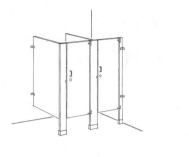

Toilet Units

Entrance Screens

Urinal Screens

C1030 110	Toilet Partitions	COST PER UNIT		
		MAT.	INST.	TOTAL
0380	Toilet partitions, cubicles, ceiling hung, marble	1,975	560	2,535
0400	Painted metal	575	297	872
0420	Plastic laminate	580	297	877
0460	Stainless steel	1,225	297	1,522
0480	Handicap addition	495		495
0520	Floor and ceiling anchored, marble	2,175	450	2,625
0540	Painted metal	630	238	868
0560	Plastic laminate	905	238	1,143
0600	Stainless steel	1,350	238	1,588
0620	Handicap addition	405		405
0660	Floor mounted marble	1,350	375	1,725
0680	Painted metal	660	170	830
0700	Plastic laminate	585	170	755
0740	Stainless steel	1,575	170	1,745
0760	Handicap addition	370		370
0780	Juvenile deduction	45.50		45.50
0820	Floor mounted with headrail marble	1,275	375	1,650
0840	Painted metal	420	198	618
0860	Plastic laminate	845	198	1,043
0900	Stainless steel	1,125	198	1,323
0920	Handicap addition	410		410
0960	Wall hung, painted metal	700	170	870
1020	Stainless steel	1,825	170	1,995
1040	Handicap addition	410		410
1080	Entrance screens, floor mounted, 54" high, marble	870	125	995
1100	Painted metal	266	79.50	345.50
1140	Stainless steel	1,000	79.50	1,079.50
1300	Urinal screens, floor mounted, 24" wide, laminated plastic	224	149	373
1320	Marble	710	172	882
1340	Painted metal	253	149	402
1380	Stainless steel	640	149	789
1428	Wall mounted wedge type, painted metal	148	119	267
1460	Stainless steel	635	119	754
1500	Partitions, shower stall, single wall, painted steel, 2'-8" x 2'-8"	1,050	280	1,330
1510	Fiberglass, 2'-8" x 2'-8"	875	310	1,185
1520	Double wall, enameled steel, 2'-8" x 2'-8"	1,125	280	1,405
1530	Stainless steel, 2'-8" x 2'-8"	2,525	280	2,805
1560	Tub enclosure, sliding panels, tempered glass, aluminum frame	450	350	800
1570	Chrome/brass frame, clear glass	1,325	465	1,790

For customer support on your Square Foot Cost Data, call 877.756.2789.

C10 Interior Construction

C1030 Fittings

C1030 310	Storage Specialties, EACH	COST EACH		
		MAT.	INST.	TOTAL
0200	Lockers, steel, single tier, 5' to 6' high, per opening, 1 wide	181	50	231
0210	3 wide	320	58.50	378.50
0600	Shelving, metal industrial, braced, 3' wide, 1' deep	23.50	12.35	35.85
0610	2' deep	34	13.10	47.10

C1030 510	Identifying/Visual Aid Specialties, EACH	COST EACH		
		MAT.	INST.	TOTAL
0100	Control boards, magnetic, porcelain finish, framed, 24" x 18"	213	149	362
0110	96" x 48"	1,175	238	1,413
0120	Directory boards, outdoor, black plastic, 36" x 24"	840	595	1,435
0130	36" x 36"	970	795	1,765
0140	Indoor, economy, open faced, 18" x 24"	183	170	353
0500	Signs, interior electric exit sign, wall mounted, 6"	102	82.50	184.50
0510	Street, reflective alum., dbl. face, 4 way, w/bracket	269	39.50	308.50
0520	Letters, cast aluminum, 1/2" deep, 4" high	27.50	33	60.50
0530	1" deep, 10" high	58	33	91
0540	Plaques, cast aluminum, 20" x 30"	2,025	297	2,322
0550	Cast bronze, 36" x 48"	5,125	595	5,720

C1030 520	Identifying/Visual Aid Specialties, S.F.	COST PER S.F.		
		MAT.	INST.	TOTAL
0100	Bulletin board, cork sheets, no frame, 1/4" thick	1.52	4.10	5.62
0120	Aluminum frame, 1/4" thick, 3' x 5'	9.90	4.99	14.89
0200	Chalkboards, wall hung, alum, frame & chalktrough	14.90	2.64	17.54
0210	Wood frame & chalktrough	10.85	2.84	13.69
0220	Sliding board, one board with back panel	57	2.54	59.54
0230	Two boards with back panel	89	2.54	91.54
0240	Liquid chalk type, alum. frame & chalktrough	13.65	2.64	16.29
0250	Wood frame & chalktrough	33.50	2.64	36.14

C1030 710	Bath and Toilet Accessories, EACH	COST EACH		
		MAT.	INST.	TOTAL
0100	Specialties, bathroom accessories, st. steel, curtain rod, 5' long, 1" diam	31	45.50	76.50
0120	Dispenser, towel, surface mounted	52	37	89
0140	Grab bar, 1-1/4" diam., 12" long	33.50	25	58.50
0160	Mirror, framed with shelf, 18" x 24"	218	29.50	247.50
0170	72" x 24"	285	99	384
0180	Toilet tissue dispenser, surface mounted, single roll	21	19.80	40.80
0200	Towel bar, 18" long	45.50	26	71.50
0300	Medicine cabinets, sliding mirror doors, 20" x 16" x 4-3/4", unlighted	146	85	231
0310	24" x 19" x 8-1/2", lighted	202	119	321

C1030 730	Bath and Toilet Accessories, L.F.	COST PER L.F.		
		MAT.	INST.	TOTAL
0100	Partitions, hospital curtain, ceiling hung, polyester oxford cloth	24.50	5.80	30.30
0110	Designer oxford cloth	17.20	7.35	24.55

C1030 830	Fabricated Cabinets, EACH	COST EACH		
		MAT.	INST.	TOTAL
0110	Household, base, hardwood, one top drawer & one door below x 12" wide	298	48	346
0115	24" wide	410	53.50	465
0120	Four drawer x 24" wide	385	53.50	438.50
0130	Wall, hardwood, 30" high with one door x 12" wide	258	54	312
0140	Two doors x 48" wide	565	64.50	629.50

C1030 Fittings

C1030 830	Fabricated Counters, L.F.	COST PER L.F.		
		MAT.	INST.	TOTAL
0150	Counter top-laminated plastic, stock, economy	19	19.80	38.80
0160	Custom-square edge, 7/8" thick	16.85	44.50	61.35
0170	School, counter, wood, 32" high	264	59.50	323.50
0180	Metal, 84" high	470	79.50	549.50

C1030 910	Other Fittings, EACH	COST EACH		
		MAT.	INST.	TOTAL
0500	Mail boxes, horizontal, rear loaded, aluminum, 5" x 6" x 15" deep	47	17.50	64.50
0510	Front loaded, aluminum, 10" x 12" x 15" deep	102	29.50	131.50
0520	Vertical, front loaded, aluminum, 15" x 5" x 6" deep	47	17.50	64.50
0530	Bronze, duranodic finish	53	17.50	70.50
0540	Letter slot, post office	131	74.50	205.50
0550	Mail counter, window, post office, with grille	605	297	902
0700	Turnstiles, one way, 4' arm, 46" diam., manual	2,300	238	2,538
0710	Electric	2,525	990	3,515
0720	3 arm, 5'-5" diam. & 7' high, manual	8,425	1,200	9,625
0730	Electric	10,200	1,975	12,175

C2010 Stair Construction

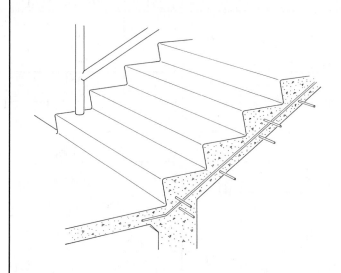

The table below lists the cost per flight for 4'-0" wide stairs. Side walls are not included. Railings are included in the prices.

C2010 110	Stairs	COST PER FLIGHT		
		MAT.	INST.	TOTAL
0470	Stairs, C.I.P. concrete, w/o landing, 12 risers, w/o nosing	1,125	2,475	3,600
0480	With nosing	2,100	2,675	4,775
0550	W/landing, 12 risers, w/o nosing	1,300	3,025	4,325
0560	With nosing	2,250	3,225	5,475
0570	16 risers, w/o nosing	1,575	3,775	5,350
0580	With nosing	2,875	4,075	6,950
0590	20 risers, w/o nosing	1,875	4,575	6,450
0600	With nosing	3,475	4,925	8,400
0610	24 risers, w/o nosing	2,150	5,325	7,475
0620	With nosing	4,100	5,750	9,850
0630	Steel, grate type w/nosing & rails, 12 risers, w/o landing	5,700	1,250	6,950
0640	With landing	7,975	1,700	9,675
0660	16 risers, with landing	9,875	2,100	11,975
0680	20 risers, with landing	11,800	2,500	14,300
0700	24 risers, with landing	13,700	2,925	16,625
0701				
0710	Concrete fill metal pan & picket rail, 12 risers, w/o landing	7,375	1,250	8,625
0720	With landing	9,725	1,850	11,575
0740	16 risers, with landing	12,200	2,250	14,450
0760	20 risers, with landing	14,600	2,675	17,275
0780	24 risers, with landing	17,100	3,050	20,150
0790	Cast iron tread & pipe rail, 12 risers, w/o landing	7,375	1,250	8,625
0800	With landing	9,725	1,850	11,575
1120	Wood, prefab box type, oak treads, wood rails 3'-6" wide, 14 risers	2,250	495	2,745
1150	Prefab basement type, oak treads, wood rails 3'-0" wide, 14 risers	1,100	122	1,222

C3010 Wall Finishes

C3010 230	Paint & Covering	COST PER S.F.		
		MAT.	INST.	TOTAL
0060	Painting, interior on plaster and drywall, brushwork, primer & 1 coat	.14	.75	.89
0080	Primer & 2 coats	.21	.99	1.20
0100	Primer & 3 coats	.29	1.22	1.51
0120	Walls & ceilings, roller work, primer & 1 coat	.14	.50	.64
0140	Primer & 2 coats	.21	.64	.85
0160	Woodwork incl. puttying, brushwork, primer & 1 coat	.14	1.08	1.22
0180	Primer & 2 coats	.21	1.43	1.64
0200	Primer & 3 coats	.29	1.95	2.24
0260	Cabinets and casework, enamel, primer & 1 coat	.14	1.22	1.36
0280	Primer & 2 coats	.22	1.50	1.72
0300	Masonry or concrete, latex, brushwork, primer & 1 coat	.31	1.01	1.32
0320	Primer & 2 coats	.40	1.45	1.85
0340	Addition for block filler	.23	1.25	1.48
0380	Fireproof paints, intumescent, 1/8" thick 3/4 hour	2.43	.99	3.42
0400	3/16" thick 1 hour	5.40	1.50	6.90
0420	7/16" thick 2 hour	7	3.47	10.47
0440	1-1/16" thick 3 hour	11.40	6.95	18.35
0500	Gratings, primer & 1 coat	.33	1.52	1.85
0600	Pipes over 12" diameter	.84	4.86	5.70
0700	Structural steel, brushwork, light framing 300-500 S.F./Ton	.11	1.77	1.88
0720	Heavy framing 50-100 S.F./Ton	.11	.89	1
0740	Spraywork, light framing 300-500 S.F./Ton	.12	.39	.51
0760	Heavy framing 50-100 S.F./Ton	.12	.44	.56
0800	Varnish, interior wood trim, no sanding sealer & 1 coat	.08	1.22	1.30
0820	Hardwood floor, no sanding 2 coats	.17	.26	.43
0840	Wall coatings, acrylic glazed coatings, minimum	.34	.93	1.27
0860	Maximum	.72	1.59	2.31
0880	Epoxy coatings, solvent based	.44	.93	1.37
0900	Water based	.30	2.86	3.16
0940	Exposed epoxy aggregate, troweled on, 1/16" to 1/4" aggregate, topping mix	.68	2.07	2.75
0960	Integral mix	1.46	3.74	5.20
0980	1/2" to 5/8" aggregate, topping mix	1.32	3.74	5.06
1000	Integral mix	2.29	6.10	8.39
1020	1" aggregate, topping mix	2.33	5.40	7.73
1040	Integral mix	3.55	8.85	12.40
1060	Sprayed on, topping mix	.63	1.65	2.28
1080	Water based	1.16	3.35	4.51
1100	High build epoxy 50 mil, solvent based	.75	1.25	2
1120	Water based	1.27	5.10	6.37
1140	Laminated epoxy with fiberglass solvent based	.80	1.65	2.45
1160	Water based	1.45	3.35	4.80
1180	Sprayed perlite or vermiculite 1/16" thick, solvent based	.30	.17	.47
1200	Water based	.81	.76	1.57
1260	Wall coatings, vinyl plastic, solvent based	.36	.66	1.02
1280	Water based	.90	2.03	2.93
1300	Urethane on smooth surface, 2 coats, solvent based	.34	.43	.77
1320	Water based	.65	.73	1.38
1340	3 coats, solvent based	.48	.58	1.06
1360	Water based	.87	1.03	1.90
1380	Ceramic-like glazed coating, cementitious, solvent based	.53	1.11	1.64
1400	Water based	.89	1.41	2.30
1420	Resin base, solvent based	.42	.76	1.18
1440	Water based	.64	1.47	2.11
1460	Wall coverings, aluminum foil	1.14	1.78	2.92
1500	Vinyl backing	6.10	2.04	8.14
1520	Cork tiles, 12"x12", light or dark, 3/16" thick	5.15	2.04	7.19

C3010 Wall Finishes

C3010 230	Paint & Covering	COST PER S.F.		
		MAT.	INST.	TOTAL
1540	5/16" thick	3.44	2.08	5.52
1560	Basketweave, 1/4" thick	3.85	2.04	5.89
1580	Natural, non-directional, 1/2" thick	7.20	2.04	9.24
1600	12"x36", granular, 3/16" thick	1.44	1.27	2.71
1620	1" thick	1.85	1.32	3.17
1640	12"x12", polyurethane coated, 3/16" thick	4.49	2.04	6.53
1660	5/16" thick	7.25	2.08	9.33
1661	Paneling, prefinished plywood, birch	1.32	2.83	4.15
1662	Mahogany, African	3.08	2.97	6.05
1663	Philippine (lauan)	.72	2.38	3.10
1664	Oak or cherry	2.31	2.97	5.28
1665	Rosewood	3.25	3.72	6.97
1666	Teak	3.25	2.97	6.22
1667	Chestnut	5.35	3.17	8.52
1668	Pecan	2.81	2.97	5.78
1669	Walnut	5.80	2.97	8.77
1670	Wood board, knotty pine, finished	2.49	4.89	7.38
1671	Rough sawn cedar	3.86	4.89	8.75
1672	Redwood	5.80	4.89	10.69
1673	Aromatic cedar	2.98	5.25	8.23
1680	Cork wallpaper, paper backed, natural	2.20	1.02	3.22
1700	Color	3.16	1.02	4.18
1720	Gypsum based, fabric backed, minimum	.85	.61	1.46
1740	Average	1.40	.68	2.08
1760	Small quantities	.77	.77	1.54
1780	Vinyl wall covering, fabric back, light weight	1.07	.77	1.84
1800	Medium weight	.99	1.02	2.01
1820	Heavy weight	1.51	1.13	2.64
1840	Wall paper, double roll, solid pattern, avg. workmanship	.65	.77	1.42
1860	Basic pattern, avg. workmanship	1.24	.91	2.15
1880	Basic pattern, quality workmanship	1.89	1.13	3.02
1900	Grass cloths with lining paper, minimum	1.06	1.22	2.28
1920	Maximum	3.14	1.40	4.54
1940	Ceramic tile, thin set, 4-1/4" x 4-1/4"	2.74	4.93	7.67
1942				
1960	12" x 12"	5	5.85	10.85

C3010 235	Paint Trim	COST PER L.F.		
		MAT.	INST.	TOTAL
2040	Painting, wood trim, to 6" wide, enamel, primer & 1 coat	.14	.61	.75
2060	Primer & 2 coats	.22	.77	.99
2080	Misc. metal brushwork, ladders	.66	6.10	6.76
2100	Pipes, to 4" dia.	.11	1.28	1.39
2120	6" to 8" dia.	.22	2.56	2.78
2140	10" to 12" dia.	.66	3.83	4.49
2160	Railings, 2" pipe	.23	3.04	3.27
2180	Handrail, single	.18	1.22	1.40
2185	Caulking & Sealants, Polyurethane, In place, 1 or 2 component, 1/2" X 1/4"	.38	1.96	2.34

For customer support on your Square Foot Cost Data, call 877.756.2789.

413

C3020 Floor Finishes

C3020 410	Tile & Covering	COST PER S.F.		
		MAT.	INST.	TOTAL
0060	Carpet tile, nylon, fusion bonded, 18″ x 18″ or 24″ x 24″, 24 oz.	3.72	.73	4.45
0080	35 oz.	4.33	.73	5.06
0100	42 oz.	5.65	.73	6.38
0140	Carpet, tufted, nylon, roll goods, 12′ wide, 26 oz.	2.61	.78	3.39
0160	36 oz.	4.27	.78	5.05
0180	Woven, wool, 36 oz.	11.10	.83	11.93
0200	42 oz.	12.20	.83	13.03
0220	Padding, add to above, 2.7 density	.73	.39	1.12
0240	13.0 density	.98	.39	1.37
0260	Composition flooring, acrylic, 1/4″ thick	1.83	5.70	7.53
0280	3/8″ thick	2.31	6.60	8.91
0300	Epoxy, 3/8″ thick	3.09	4.38	7.47
0320	1/2″ thick	4.46	6.05	10.51
0340	Epoxy terrazzo, granite chips	6.60	6.55	13.15
0360	Recycled porcelain	10.05	8.70	18.75
0380	Mastic, hot laid, 1-1/2″ thick, minimum	4.57	4.29	8.86
0400	Maximum	5.85	5.70	11.55
0420	Neoprene 1/4″ thick, minimum	4.50	5.45	9.95
0440	Maximum	6.15	6.90	13.05
0460	Polyacrylate with ground granite 1/4″, granite chips	3.87	4.02	7.89
0480	Recycled porcelain	7.10	6.20	13.30
0500	Polyester with colored quartz chips 1/16″, minimum	3.50	2.79	6.29
0520	Maximum	5.25	4.38	9.63
0540	Polyurethane with vinyl chips, clear	8.35	2.79	11.14
0560	Pigmented	12.15	3.44	15.59
0600	Concrete topping, granolithic concrete, 1/2″ thick	.36	4.71	5.07
0620	1″ thick	.71	4.83	5.54
0640	2″ thick	1.43	5.55	6.98
0660	Heavy duty 3/4″ thick, minimum	.48	7.30	7.78
0680	Maximum	.89	8.70	9.59
0700	For colors, add to above, minimum	.52	1.68	2.20
0720	Maximum	.86	1.85	2.71
0740	Exposed aggregate finish, minimum	.22	.87	1.09
0760	Maximum	.37	1.16	1.53
0780	Abrasives, .25 P.S.F. add to above, minimum	.58	.64	1.22
0800	Maximum	.87	.64	1.51
0820	Dust on coloring, add, minimum	.52	.42	.94
0840	Maximum	.86	.87	1.73
0860	Floor coloring using 0.6 psf powdered color, 1/2″ integral, minimum	5.75	4.71	10.46
0880	Maximum	6.10	4.71	10.81
0900	Dustproofing, add, minimum	.21	.29	.50
0920	Maximum	.74	.42	1.16
0930	Paint	.40	1.45	1.85
0940	Hardeners, metallic add, minimum	.66	.64	1.30
0960	Maximum	1.98	.94	2.92
0980	Non-metallic, minimum	.17	.64	.81
1000	Maximum	.51	.94	1.45
1020	Integral topping and finish, 1:1:2 mix, 3/16″ thick	.12	2.78	2.90
1040	1/2″ thick	.32	2.93	3.25
1060	3/4″ thick	.48	3.27	3.75
1080	1″ thick	.64	3.70	4.34
1100	Terrazzo, minimum	3.69	17.50	21.19
1120	Maximum	6.80	22	28.80
1340	Cork tile, minimum	7.65	1.67	9.32
1360	Maximum	13.35	1.67	15.02
1380	Polyethylene, in rolls, minimum	4.58	1.91	6.49
1400	Maximum	7.50	1.91	9.41
1420	Polyurethane, thermoset, minimum	5.95	5.25	11.20

C3020 Floor Finishes

C3020 410	Tile & Covering	COST PER S.F.		
		MAT.	INST.	TOTAL
1440	Maximum	7.45	10.50	17.95
1460	Rubber, sheet goods, minimum	7.80	4.38	12.18
1480	Maximum	13.55	5.85	19.40
1500	Tile, minimum	6.30	1.32	7.62
1520	Maximum	11.20	1.91	13.11
1580	Vinyl, composition tile, minimum	1.31	1.05	2.36
1600	Maximum	1.71	1.05	2.76
1620	Vinyl tile, 3/32"	4.08	1.05	5.13
1640	Maximum	3.58	1.05	4.63
1660	Sheet goods, plain pattern/colors	4.06	2.10	6.16
1680	Intricate pattern/colors	8.25	2.63	10.88
1720	Tile, ceramic natural clay	6.35	5.10	11.45
1730	Marble, synthetic 12"x12"x5/8"	11.35	15.65	27
1740	Porcelain type, minimum	5.70	5.10	10.80
1760	Maximum	7.60	6.05	13.65
1800	Quarry tile, mud set, minimum	8.75	6.70	15.45
1820	Maximum	9.55	8.50	18.05
1840	Thin set, deduct		1.34	1.34
1850	Tile, natural stone, marble, in mortar bed, 12" x 12" x 3/8" thick	17.60	25	42.60
1860	Terrazzo precast, minimum	5.30	7	12.30
1880	Maximum	10.25	7	17.25
1900	Non-slip, minimum	22.50	17.15	39.65
1920	Maximum	22.50	23	45.50
1960	Stone flooring, polished marble in mortar bed	19.25	25	44.25
2020	Wood block, end grain factory type, natural finish, 2" thick	4.94	4.76	9.70
2040	Fir, vertical grain, 1"x4", no finish, minimum	2.89	2.33	5.22
2060	Maximum	3.07	2.33	5.40
2080	Prefinished white oak, prime grade, 2-1/4" wide	5.30	3.50	8.80
2100	3-1/4" wide	5.75	3.21	8.96
2120	Maple strip, sanded and finished, minimum	5.85	5.10	10.95
2140	Maximum	6	5.10	11.10
2160	Oak strip, sanded and finished, minimum	4.65	5.10	9.75
2180	Maximum	5.50	5.10	10.60
2200	Parquetry, sanded and finished, plain pattern	6.95	5.30	12.25
2220	Intricate pattern	11.65	7.55	19.20
2260	Add for sleepers on concrete, treated, 24" O.C., 1"x2"	1.80	3.05	4.85
2280	1"x3"	1.95	2.38	4.33
2300	2"x4"	1	1.20	2.20
2340	Underlayment, plywood, 3/8" thick	1.09	.79	1.88
2350	1/2" thick	1.29	.82	2.11
2360	5/8" thick	1.42	.85	2.27
2370	3/4" thick	1.63	.91	2.54
2380	Particle board, 3/8" thick	.45	.79	1.24
2390	1/2" thick	.50	.82	1.32
2400	5/8" thick	.59	.85	1.44
2410	3/4" thick	.75	.91	1.66
2420	Hardboard, 4' x 4', .215" thick	.69	.79	1.48
9200	Vinyl, composition tile, 12" x 12" x 1/8" thick, recycled content	2.43	1.05	3.48

C3030 Ceiling Finishes

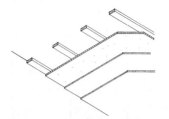

2 Coats of Plaster on Gypsum
Lath on Wood Furring

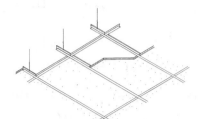

Fiberglass Board on
Exposed Suspended Grid System

Plaster and Metal Lath
on Metal Furring

C3030 105 — Plaster Ceilings

	TYPE	LATH	FURRING	SUPPORT		MAT.	INST.	TOTAL
2400	2 coat gypsum	3/8″ gypsum	1″x3″ wood, 16″ O.C.	wood		1.28	6.05	7.33
2500	Painted			masonry		1.31	6.15	7.46
2600				concrete		1.31	6.90	8.21
2700	3 coat gypsum	3.4# metal	1″x3″ wood, 16″ O.C.	wood		1.64	6.50	8.14
2800	Painted			masonry		1.67	6.60	8.27
2900				concrete		1.67	7.30	8.97
3000	2 coat perlite	3/8″ gypsum	1″x3″ wood, 16″ O.C.	wood		1.55	6.25	7.80
3100	Painted			masonry		1.58	6.40	7.98
3200				concrete		1.58	7.10	8.68
3300	3 coat perlite	3.4# metal	1″x3″ wood, 16″ O.C.	wood		1.71	6.40	8.11
3400	Painted			masonry		1.74	6.55	8.29
3500				concrete		1.74	7.25	8.99
3600	2 coat gypsum	3/8″ gypsum	3/4″ CRC, 12″ O.C.	1-1/2″ CRC, 48″O.C.		1.38	7.45	8.83
3700	Painted		3/4″ CRC, 16″ O.C.	1-1/2″ CRC, 48″O.C.		1.34	6.70	8.04
3800			3/4″ CRC, 24″ O.C.	1-1/2″ CRC, 48″O.C.		1.21	6.10	7.31
3900	2 coat perlite	3/8″ gypsum	3/4″ CRC, 12″ O.C.	1-1/2″ CRC, 48″O.C		1.65	7.95	9.60
4000	Painted		3/4″ CRC, 16″ O.C.	1-1/2″ CRC, 48″O.C.		1.61	7.20	8.81
4100			3/4″ CRC, 24″ O.C.	1-1/2″ CRC, 48″O.C.		1.48	6.60	8.08
4200	3 coat gypsum	3.4# metal	3/4″ CRC, 12″ O.C.	1-1/2″ CRC, 36″ O.C.		1.91	9.65	11.56
4300	Painted		3/4″ CRC, 16″ O.C.	1-1/2″ CRC, 36″ O.C.		1.87	8.90	10.77
4400			3/4″ CRC, 24″ O.C.	1-1/2″ CRC, 36″ O.C.		1.74	8.30	10.04
4500	3 coat perlite	3.4# metal	3/4″ CRC, 12″ O.C.	1-1/2″ CRC,36″ O.C.		2.04	10.50	12.54
4600	Painted		3/4″ CRC, 16″ O.C.	1-1/2″ CRC, 36″ O.C.		2	9.80	11.80
4700			3/4″ CRC, 24″ O.C.	1-1/2″ CRC, 36″ O.C.		1.87	9.20	11.07

C3030 110 — Drywall Ceilings

	TYPE	FINISH	FURRING	SUPPORT		MAT.	INST.	TOTAL
4800	1/2″ F.R. drywall	painted and textured	1″x3″ wood, 16″ O.C.	wood		.95	3.71	4.66
4900				masonry		.98	3.83	4.81
5000				concrete		.98	4.55	5.53
5100	5/8″ F.R. drywall	painted and textured	1″x3″ wood, 16″ O.C.	wood		.94	3.71	4.65
5200				masonry		.97	3.83	4.80
5300				concrete		.97	4.55	5.52
5400	1/2″ F.R. drywall	painted and textured	7/8″ resil. channels	24″ O.C.		.77	3.60	4.37

For customer support on your Square Foot Cost Data, call 877.756.2789.

C3030 Ceiling Finishes

C3030 110 — Drywall Ceilings

	TYPE	FINISH	FURRING	SUPPORT		COST PER S.F.		
						MAT.	INST.	TOTAL
5500	1/2" F.R. drywall	painted and textured	1"x2" wood	stud clips		.95	3.48	4.43
5602		painted	1-5/8" metal studs	24" O.C.		.83	3.21	4.04
5700	5/8" F.R. drywall	painted and textured	1-5/8"metal studs	24" O.C.		.82	3.21	4.03
5702								

C3030 140 — Plaster Ceiling Components

		COST PER S.F.		
		MAT.	INST.	TOTAL
0060	Plaster, gypsum incl. finish			
0080	3 coats	.65	3.83	4.48
0100	Perlite, incl. finish, 2 coats	.72	3.77	4.49
0120	3 coats	.78	4.71	5.49
0140	Thin coat on drywall	.12	.74	.86
0200	Lath, gypsum, 3/8" thick	.37	1.02	1.39
0220	1/2" thick	.30	1.07	1.37
0240	5/8" thick	.36	1.24	1.60
0260	Metal, diamond, 2.5 lb.	.50	.83	1.33
0280	3.4 lb.	.53	.89	1.42
0300	Flat rib, 2.75 lb.	.43	.83	1.26
0320	3.4 lb.	.55	.89	1.44
0440	Furring, steel channels, 3/4" galvanized , 12" O.C.	.42	2.66	3.08
0460	16" O.C.	.38	1.93	2.31
0480	24" O.C.	.25	1.33	1.58
0500	1-1/2" galvanized , 12" O.C.	.57	2.94	3.51
0520	16" O.C.	.51	2.15	2.66
0540	24" O.C.	.34	1.43	1.77
0560	Wood strips, 1"x3", on wood, 12" O.C.	.43	1.70	2.13
0580	16" O.C.	.32	1.28	1.60
0600	24" O.C.	.22	.85	1.07
0620	On masonry, 12" O.C.	.46	1.86	2.32
0640	16" O.C.	.35	1.40	1.75
0660	24" O.C.	.23	.93	1.16
0680	On concrete, 12" O.C.	.46	2.83	3.29
0700	16" O.C.	.35	2.12	2.47
0720	24" O.C.	.23	1.42	1.65
0940	Paint on plaster or drywall, roller work, primer + 1 coat	.14	.50	.64
0960	Primer + 2 coats	.21	.64	.85

C3030 210 — Acoustical Ceilings

	TYPE	TILE	GRID	SUPPORT		COST PER S.F.		
						MAT.	INST.	TOTAL
5800	5/8" fiberglass board	24" x 48"	tee	suspended		2.45	1.78	4.23
5900		24" x 24"	tee	suspended		2.70	1.95	4.65
6000	3/4" fiberglass board	24" x 48"	tee	suspended		3.96	1.82	5.78
6100		24" x 24"	tee	suspended		4.21	1.99	6.20
6500	5/8" mineral fiber	12" x 12"	1"x3" wood, 12" O.C.	wood		2.88	3.68	6.56
6600				masonry		2.91	3.84	6.75
6700				concrete		2.91	4.81	7.72
6800	3/4" mineral fiber	12" x 12"	1"x3" wood, 12" O.C.	wood		3.65	3.68	7.33
6900				masonry		3.65	3.68	7.33
7000				concrete		3.65	3.68	7.33
7100	3/4"mineral fiber on	12" x 12"	25 ga. channels	runners		4.10	4.59	8.69
7102	5/8" F.R. drywall							

C30 Interior Finishes

C3030 Ceiling Finishes

C3030 210 — Acoustical Ceilings

	TYPE	TILE	GRID	SUPPORT		MAT.	INST.	TOTAL
7200 7201 7202	5/8" plastic coated Mineral fiber	12" x 12"	25 ga. channels	adhesive backed		2.89	1.98	4.87
7300 7301 7302	3/4" plastic coated Mineral fiber	12" x 12"		adhesive backed		3.66	1.98	5.64
7400 7401 7402	3/4" mineral fiber	12" x 12"	conceal 2" bar & channels	suspended		2.75	4.47	7.22

COST PER S.F. columns: MAT., INST., TOTAL

C3030 240 — Acoustical Ceiling Components

		MAT.	INST.	TOTAL
2480	Ceiling boards, eggcrate, acrylic, 1/2" x 1/2" x 1/2" cubes	1.97	1.19	3.16
2500	Polystyrene, 3/8" x 3/8" x 1/2" cubes	1.73	1.17	2.90
2520	1/2" x 1/2" x 1/2" cubes	2	1.19	3.19
2540	Fiberglass boards, plain, 5/8" thick	1.38	.95	2.33
2560	3/4" thick	2.89	.99	3.88
2580	Grass cloth faced, 3/4" thick	2.99	1.19	4.18
2600	1" thick	3.64	1.23	4.87
2620	Luminous panels, prismatic, acrylic	2.90	1.49	4.39
2640	Polystyrene	1.78	1.49	3.27
2660	Flat or ribbed, acrylic	4.37	1.49	5.86
2680	Polystyrene	2.52	1.49	4.01
2700	Drop pan, white, acrylic	6.25	1.49	7.74
2720	Polystyrene	4.90	1.49	6.39
2740	Mineral fiber boards, 5/8" thick, standard	.75	.88	1.63
2760	Plastic faced	2.76	1.49	4.25
2780	2 hour rating	1.25	.88	2.13
2800	Perforated aluminum sheets, .024 thick, corrugated painted	2.71	1.21	3.92
2820	Plain	5.05	1.19	6.24
3080	Mineral fiber, plastic coated, 12" x 12" or 12" x 24", 5/8" thick	2.45	1.98	4.43
3100	3/4" thick	3.22	1.98	5.20
3120	Fire rated, 3/4" thick, plain faced	1.44	1.98	3.42
3140	Mylar faced	2.12	1.98	4.10
3160	Add for ceiling primer	.14		.14
3180	Add for ceiling cement	.44		.44
3240	Suspension system, furring, 1" x 3" wood 12" O.C.	.43	1.70	2.13
3260	T bar suspension system, 2' x 4' grid	.84	.74	1.58
3280	2' x 2' grid	1.09	.91	2
3300	Concealed Z bar suspension system 12" module	.95	1.14	2.09
3320	Add to above for 1-1/2" carrier channels 4' O.C.	.13	1.26	1.39
3340	Add to above for carrier channels for recessed lighting	.24	1.29	1.53

COST PER S.F. columns: MAT., INST., TOTAL

D1010 Elevators and Lifts

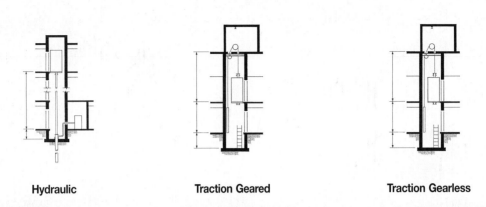

Hydraulic   Traction Geared   Traction Gearless

D1010 110	Hydraulic	COST EACH		
		MAT.	INST.	TOTAL
1300	Pass. elev., 1500 lb., 2 Floors, 100 FPM	48,000	19,400	67,400
1400	5 Floors, 100 FPM	84,500	51,500	136,000
1600	2000 lb., 2 Floors, 100 FPM	49,000	19,400	68,400
1700	5 floors, 100 FPM	85,500	51,500	137,000
1900	2500 lb., 2 Floors, 100 FPM	51,000	19,400	70,400
2000	5 floors, 100 FPM	88,000	51,500	139,500
2200	3000 lb., 2 Floors, 100 FPM	52,500	19,400	71,900
2300	5 floors, 100 FPM	89,000	51,500	140,500
2500	3500 lb., 2 Floors, 100 FPM	56,000	19,400	75,400
2600	5 floors, 100 FPM	92,500	51,500	144,000
2800	4000 lb., 2 Floors, 100 FPM	57,500	19,400	76,900
2900	5 floors, 100 FPM	94,000	51,500	145,500
3100	4500 lb., 2 Floors, 100 FPM	60,500	19,400	79,900
3200	5 floors, 100 FPM	97,000	51,500	148,500
4000	Hospital elevators, 3500 lb., 2 Floors, 100 FPM	82,000	19,400	101,400
4100	5 floors, 100 FPM	131,500	51,500	183,000
4300	4000 lb., 2 Floors, 100 FPM	82,000	19,400	101,400
4400	5 floors, 100 FPM	131,500	51,500	183,000
4600	4500 lb., 2 Floors, 100 FPM	90,000	19,400	109,400
4800	5 floors, 100 FPM	139,500	51,500	191,000
4900	5000 lb., 2 Floors, 100 FPM	93,500	19,400	112,900
5000	5 floors, 100 FPM	143,000	51,500	194,500
6700	Freight elevators (Class "B"), 3000 lb., 2 Floors, 50 FPM	120,500	24,600	145,100
6800	5 floors, 100 FPM	193,000	64,500	257,500
7000	4000 lb., 2 Floors, 50 FPM	125,500	24,600	150,100
7100	5 floors, 100 FPM	198,000	64,500	262,500
7500	10,000 lb., 2 Floors, 50 FPM	157,000	24,600	181,600
7600	5 floors, 100 FPM	229,500	64,500	294,000
8100	20,000 lb., 2 Floors, 50 FPM	204,500	24,600	229,100
8200	5 Floors, 100 FPM	277,000	64,500	341,500

D1010 140	Traction Geared Elevators	COST EACH		
		MAT.	INST.	TOTAL
1300	Passenger, 2000 Lb., 5 floors, 200 FPM	122,500	48,600	171,100
1500	15 floors, 350 FPM	251,000	147,000	398,000
1600	2500 Lb., 5 floors, 200 FPM	127,000	48,600	175,600
1800	15 floors, 350 FPM	255,500	147,000	402,500
2200	3500 Lb., 5 floors, 200 FPM	129,000	48,600	177,600
2400	15 floors, 350 FPM	257,500	147,000	404,500
2500	4000 Lb., 5 floors, 200 FPM	130,000	48,600	178,600
2700	15 floors, 350 FPM	258,500	147,000	405,500

D10 Conveying

D1010 Elevators and Lifts

D1010 140	Traction Geared Elevators	COST EACH		
		MAT.	INST.	TOTAL
2800	4500 Lb., 5 floors, 200 FPM	133,000	48,600	181,600
3000	15 floors, 350 FPM	261,500	147,000	408,500
3100	5000 Lb., 5 floors, 200 FPM	135,500	48,600	184,100
3300	15 floors, 350 FPM	264,000	147,000	411,000
4000	Hospital, 3500 Lb., 5 floors, 200 FPM	130,500	48,600	179,100
4200	15 floors, 350 FPM	306,500	147,000	453,500
4300	4000 Lb., 5 floors, 200 FPM	130,500	48,600	179,100
4500	15 floors, 350 FPM	306,500	147,000	453,500
4600	4500 Lb., 5 floors, 200 FPM	137,000	48,600	185,600
4800	15 floors, 350 FPM	313,000	147,000	460,000
4900	5000 Lb., 5 floors, 200 FPM	139,000	48,600	187,600
5100	15 floors, 350 FPM	315,000	147,000	462,000
6000	Freight, 4000 Lb., 5 floors, 50 FPM class 'B'	156,000	50,500	206,500
6200	15 floors, 200 FPM class 'B'	360,000	175,000	535,000
6300	8000 Lb., 5 floors, 50 FPM class 'B'	181,500	50,500	232,000
6500	15 floors, 200 FPM class 'B'	385,500	175,000	560,500
7000	10,000 Lb., 5 floors, 50 FPM class 'B'	209,000	50,500	259,500
7200	15 floors, 200 FPM class 'B'	597,000	175,000	772,000
8000	20,000 Lb., 5 floors, 50 FPM class 'B'	231,000	50,500	281,500
8200	15 floors, 200 FPM class 'B'	619,000	175,000	794,000

D1010 150	Traction Gearless Elevators	COST EACH		
		MAT.	INST.	TOTAL
1700	Passenger, 2500 Lb., 10 floors, 200 FPM	284,500	126,500	411,000
1900	30 floors, 600 FPM	571,000	323,500	894,500
2000	3000 Lb., 10 floors, 200 FPM	285,000	126,500	411,500
2200	30 floors, 600 FPM	571,500	323,500	895,000
2300	3500 Lb., 10 floors, 200 FPM	286,500	126,500	413,000
2500	30 floors, 600 FPM	573,000	323,500	896,500
2700	50 floors, 800 FPM	825,500	520,000	1,345,500
2800	4000 Lb., 10 floors, 200 FPM	287,500	126,500	414,000
3000	30 floors, 600 FPM	574,500	323,500	898,000
3200	50 floors, 800 FPM	826,500	520,000	1,346,500
3300	4500 Lb., 10 floors, 200 FPM	290,500	126,500	417,000
3500	30 floors, 600 FPM	577,000	323,500	900,500
3700	50 floors, 800 FPM	829,500	520,000	1,349,500
3800	5000 Lb., 10 floors, 200 FPM	293,000	126,500	419,500
4000	30 floors, 600 FPM	579,500	323,500	903,000
4200	50 floors, 800 FPM	831,500	520,000	1,351,500
6000	Hospital, 3500 Lb., 10 floors, 200 FPM	311,500	126,500	438,000
6200	30 floors, 600 FPM	698,500	323,500	1,022,000
6400	4000 Lb., 10 floors, 200 FPM	311,500	126,500	438,000
6600	30 floors, 600 FPM	698,500	323,500	1,022,000
6800	4500 Lb., 10 floors, 200 FPM	318,000	126,500	444,500
7000	30 floors, 600 FPM	705,000	323,500	1,028,500
7200	5000 Lb., 10 floors, 200 FPM	320,000	126,500	446,500
7400	30 floors, 600 FPM	707,000	323,500	1,030,500

D20 Plumbing

D2010 Plumbing Fixtures

Minimum Plumbing Fixture Requirements

Classification	Occupancy	Description	Water		Lavatories		Bathtubs/Showers	Drinking Fountains	Other
			Male	Female	Male	Female			
Assembly	A-1	Theaters and other buildings for the performing arts and motion pictures	1:125	1:65	1:200			1:500	1 Service Sink
	A-2	Nightclubs, bars, taverns dance halls	1:40		1:75			1:500	1 Service Sink
		Restaurants, banquet halls, food courts	1:75		1:200			1:500	1 Service Sink
	A-3	Auditorium w/o permanent seating, art galleries, exhibition halls, museums, lecture halls, libraries, arcades & gymnasiums	1:125	1:65	1:200			1:500	1 Service Sink
		Passenger terminals and transportation facilities	1:500		1:750			1:1000	1 Service Sink
		Places of worship and other religious services	1:150	1:75	1:200			1:1000	1 Service Sink
	A-4	Indoor sporting events and activities, coliseums, arenas, skating rinks, pools, and tennis courts	1:75 for the first 1500, then 1:120 for the remainder	1:40 for the first 1520, then 1:60 for the remainder	1:200	1:150		1:1000	1 Service Sink
	A-5	Outdoor sporting events and activities, stadiums, amusement parks, bleachers, grandstands	1:75 for the first 1500, then 1:120 for the remainder	1:40 for the first 1520, then 1:60 for the remainder	1:200	1:150		1:1000	1 Service Sink
Business	B	Buildings for the transaction of business, professional services, other services involving merchandise, office buildings, banks, light industrial	1:25 for the first 50, then 1:50 for the remainder		1:40 for the first 80, then 1:80 for the remainder			1:100	1 Service Sink
Educational	E	Educational Facilities	1:50		1:50			1:100	1 Service Sink
Factory and industrial	F-1 and F-2	Structures in which occupants are engaged in work fabricating, assembly or processing of products or materials	1:100		1:100		See International Plumbing Code	1:400	1 Service Sink
Institutional	I-1	Residential Care	1:10		1:10		1:8	1:100	1 Service Sink
	I-2	Hospitals, ambulatory nursing home care recipient	1 per room		1 per room		1:15	1:100	1 Service Sink
		Employees, other than residential care	1:25		1:35			1:100	
		Visitors, other than residential care	1:75		1:100			1:500	
	I-3	Prisons	1 per cell		1 per cell		1:15	1:100	1 Service Sink
		Reformatories, detention and correction centers	1:15		1:15		1:15	1:100	1 Service Sink
		Employees	1:25		1:35			1:100	
	I-4	Adult and child day care	1:15		1:15		1	1:100	1 Service Sink
Mercantile	M	Retail stores, service stations, shops, salesrooms, markets and shopping centers	1:500		1:750			1:1000	1 Service Sink
Residential	R-1	Hotels, Motels, boarding houses (transient)	1 per sleeping unit		1 per sleeping unit		1 per sleeping unit		1 Service Sink
	R-2	Dormitories, fraternities, sororities and boarding houses (not transient)	1:10		1:10		1:8	1:100	1 Service Sink
		Apartment House	1 per dwelling unit		1 per dwelling unit		1 per dwelling unit		1 Kitchen sink per dwelling; 1 clothes washer connection per 20 dwellings
	R-3	1 and 2 Family Dwellings	1 per dwelling unit		1:10		1 per dwelling unit		1 Kitchen sink per dwelling; 1 clothes washer connection per dwelling
	R-3	Congregate living facilities w/<16 people	1:10		1:10		1:8	1:100	1 Service Sink
	R-4	Congregate living facilities w/<16 people	1:10		1:10		1:8	1:100	1 Service Sink
Storage	S-1 and S-2	Structures for the storage of good, warehouses, storehouses and freight depots, low and moderate hazard	1:100		1:100		See International Plumbing Code	1;1000	1 Service Sink

Table 2902.1

D20 Plumbing

D2010 Plumbing Fixtures

One Piece Wall Hung Water Closet

Wall Hung Urinal

Side By Side Water Closet Group

Floor Mount Water Closet

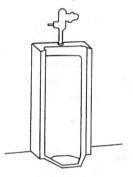

Stall Type Urinal

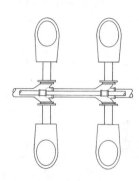

Back to Back Water Closet Group

D2010 110	Water Closet Systems	COST EACH		
		MAT.	INST.	TOTAL
1800	Water closet, vitreous china			
1840	Tank type, wall hung			
1880	Close coupled two piece	1,625	760	2,385
1920	Floor mount, one piece	1,225	805	2,030
1960	One piece low profile	1,200	805	2,005
2000	Two piece close coupled	660	805	1,465
2040	Bowl only with flush valve			
2080	Wall hung	2,225	860	3,085
2120	Floor mount	790	820	1,610
2122				
2160	Floor mount, ADA compliant with 18" high bowl	785	840	1,625

D2010 120	Water Closets, Group	COST EACH		
		MAT.	INST.	TOTAL
1760	Water closets, battery mount, wall hung, side by side, first closet	2,325	885	3,210
1800	Each additional water closet, add	2,225	835	3,060
3000	Back to back, first pair of closets	4,350	1,175	5,525
3100	Each additional pair of closets, back to back	4,275	1,150	5,425
9000	Back to back, first pair of closets, auto sensor flush valve, 1.28 gpf	4,875	1,275	6,150
9100	Ea additional pair of cls, back to back, auto sensor flush valve, 1.28 gpf	4,725	1,200	5,925

D2010 210	Urinal Systems	COST EACH		
		MAT.	INST.	TOTAL
2000	Urinal, vitreous china, wall hung	555	850	1,405
2040	Stall type	1,225	1,025	2,250

For customer support on your Square Foot Cost Data, call 877.756.2789.

423

D2010 Plumbing Fixtures

Systems are complete with trim and rough-in (supply, waste and vent) to connect to supply branches and waste mains.

Vanity Top

Wall Hung

Counter Top Single Bowl

Counter Top Double Bowl

D2010 310	Lavatory Systems	COST EACH		
		MAT.	INST.	TOTAL
1560	Lavatory w/trim, vanity top, PE on CI, 20" x 18", Vanity top by others.	585	760	1,345
1600	19" x 16" oval	395	760	1,155
1640	18" round	700	760	1,460
1680	Cultured marble, 19" x 17"	405	760	1,165
1720	25" x 19"	435	760	1,195
1760	Stainless, self-rimming, 25" x 22"	615	760	1,375
1800	17" x 22"	605	760	1,365
1840	Steel enameled, 20" x 17"	405	780	1,185
1880	19" round	405	780	1,185
1920	Vitreous china, 20" x 16"	500	795	1,295
1960	19" x 16"	510	795	1,305
2000	22" x 13"	510	795	1,305
2040	Wall hung, PE on CI, 18" x 15"	905	840	1,745
2080	19" x 17"	935	840	1,775
2120	20" x 18"	770	840	1,610
2160	Vitreous china, 18" x 15"	685	860	1,545
2200	19" x 17"	640	860	1,500
2240	24" x 20"	770	860	1,630
2300	20" x 27", handicap	1,425	930	2,355

D2010 410	Kitchen Sink Systems	COST EACH		
		MAT.	INST.	TOTAL
1720	Kitchen sink w/trim, countertop, PE on CI, 24"x21", single bowl	600	835	1,435
1760	30" x 21" single bowl	875	835	1,710
1800	32" x 21" double bowl	705	900	1,605
1880	Stainless steel, 19" x 18" single bowl	940	835	1,775
1920	25" x 22" single bowl	1,025	835	1,860
1960	33" x 22" double bowl	1,375	900	2,275
2000	43" x 22" double bowl	1,550	910	2,460
2040	44" x 22" triple bowl	1,575	950	2,525
2080	44" x 24" corner double bowl	1,100	910	2,010
2120	Steel, enameled, 24" x 21" single bowl	855	835	1,690
2160	32" x 21" double bowl	870	900	1,770
2240	Raised deck, PE on CI, 32" x 21", dual level, double bowl	795	1,150	1,945
2280	42" x 21" dual level, triple bowl	1,650	1,250	2,900

D2010 Plumbing Fixtures

Laboratory Sink

Service Sink

Systems are complete with trim and rough-in (supply, waste and vent) to connect to supply branches and waste mains.

Single Compartment Sink

Double Compartment Sink

D2010 420	Laundry Sink Systems	COST EACH		
		MAT.	INST.	TOTAL
1740	Laundry sink w/trim, PE on CI, black iron frame			
1760	24" x 20", single compartment	920	815	1,735
1840	48" x 21" double compartment	1,175	885	2,060
1920	Molded stone, on wall, 22" x 21" single compartment	470	815	1,285
1960	45"x 21" double compartment	675	885	1,560
2040	Plastic, on wall or legs, 18" x 23" single compartment	445	800	1,245
2080	20" x 24" single compartment	460	800	1,260
2120	36" x 23" double compartment	520	865	1,385
2160	40" x 24" double compartment	625	865	1,490

D2010 430	Laboratory Sink Systems	COST EACH		
		MAT.	INST.	TOTAL
1580	Laboratory sink w/trim, stainless steel, single bowl,			
1590	Stainless steel, single bowl,			
1600	Double drainboard, 54" x 24" O.D.	1,450	1,075	2,525
1640	Single drainboard, 47" x 24"O.D.	1,175	1,075	2,250
1670	Stainless steel, double bowl,			
1680	70" x 24" O.D.	1,575	1,075	2,650
1750	Polyethylene, single bowl,			
1760	Flanged, 14-1/2" x 14-1/2" O.D.	520	955	1,475
1800	18-1/2" x 18-1/2" O.D.	630	955	1,585
1840	23-1/2" x 20-1/2" O.D.	650	955	1,605
1920	Polypropylene, cup sink, oval, 7" x 4" O.D.	410	845	1,255
1960	10" x 4-1/2" O.D.	435	845	1,280
1961				

D2010 Plumbing Fixtures

D2010 440	Service Sink Systems	COST EACH		
		MAT.	INST.	TOTAL
4260	Service sink w/trim, PE on Cl, corner floor, 28″ x 28″, w/rim guard	1,900	1,075	2,975
4300	Wall hung w/rim guard, 22″ x 18″	2,125	1,250	3,375
4340	24″ x 20″	2,200	1,250	3,450
4380	Vitreous china, wall hung 22″ x 20″	1,925	1,250	3,175
4383				

D2010 Plumbing Fixtures

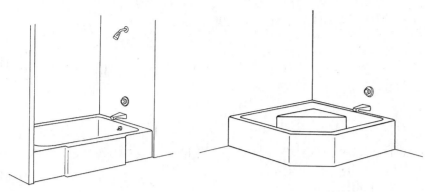

Recessed Bathtub

Corner Bathtub

Systems are complete with trim and rough-in (supply, waste and vent) to connect to supply branches and waste mains.

Circular Wash Fountain

Semi-Circular Wash Fountain

D2010 510	Bathtub Systems	COST EACH		
		MAT.	INST.	TOTAL
2000	Bathtub, recessed, P.E. on Cl., 48" x 42"	3,325	940	4,265
2040	72" x 36"	3,425	1,050	4,475
2080	Mat bottom, 5' long	1,650	915	2,565
2120	5'-6" long	2,425	940	3,365
2160	Corner, 48" x 42"	3,325	915	4,240
2200	Formed steel, enameled, 4'-6" long	945	845	1,790

D2010 610	Group Wash Fountain Systems	COST EACH		
		MAT.	INST.	TOTAL
1740	Group wash fountain, precast terrazzo			
1760	Circular, 36" diameter	8,500	1,375	9,875
1800	54" diameter	10,500	1,500	12,000
1840	Semi-circular, 36" diameter	7,500	1,375	8,875
1880	54" diameter	9,950	1,500	11,450
1960	Stainless steel, circular, 36" diameter	7,700	1,275	8,975
2000	54" diameter	9,375	1,425	10,800
2040	Semi-circular, 36" diameter	6,025	1,275	7,300
2080	54" diameter	8,225	1,425	9,650
2160	Thermoplastic, circular, 36" diameter	5,375	1,050	6,425
2200	54" diameter	6,175	1,200	7,375
2240	Semi-circular, 36" diameter	4,975	1,050	6,025
2280	54" diameter	5,975	1,200	7,175

427

For customer support on your Square Foot Cost Data, call 877.756.2789.

D2010 Plumbing Fixtures

Square Shower Stall

Corner Angle Shower Stall

Systems are complete with trim and rough-in (supply, waste and vent) to connect to supply branches and waste mains.

Wall Mounted, Low Back

Wall Mounted, No Back

D2010 710	Shower Systems	COST EACH		
		MAT.	INST.	TOTAL
1560	Shower, stall, baked enamel, molded stone receptor, 30" square	1,575	875	2,450
1600	32" square	1,575	885	2,460
1640	Terrazzo receptor, 32" square	1,775	885	2,660
1680	36" square	1,925	895	2,820
1720	36" corner angle	2,200	895	3,095
1800	Fiberglass one piece, three walls, 32" square	665	865	1,530
1840	36" square	715	865	1,580
1880	Polypropylene, molded stone receptor, 30" square	995	1,275	2,270
1920	32" square	1,000	1,275	2,275
1960	Built-in head, arm, bypass, stops and handles	113	330	443
2050	Shower, stainless steel panels, handicap			
2100	w/fixed and handheld head, control valves, grab bar, and seat	4,000	3,950	7,950
2500	Shower, group with six heads, thermostatic mix valves & balancing valve	12,500	965	13,465
2520	Five heads	8,950	880	9,830

D2010 810	Drinking Fountain Systems	COST EACH		
		MAT.	INST.	TOTAL
1740	Drinking fountain, one bubbler, wall mounted			
1760	Non recessed			
1800	Bronze, no back	1,300	500	1,800
1840	Cast iron, enameled, low back	1,250	500	1,750
1880	Fiberglass, 12" back	2,325	500	2,825
1920	Stainless steel, no back	1,375	500	1,875
1960	Semi-recessed, poly marble	1,300	500	1,800
2040	Stainless steel	1,675	500	2,175
2080	Vitreous china	1,225	500	1,725
2120	Full recessed, poly marble	2,025	500	2,525
2200	Stainless steel	1,875	500	2,375
2240	Floor mounted, pedestal type, aluminum	2,675	680	3,355
2320	Bronze	2,325	680	3,005
2360	Stainless steel	2,150	680	2,830

D2010 Plumbing Fixtures

Wall Hung Water Cooler

Floor Mounted Water Cooler

Systems are complete with trim and rough-in (supply, waste and vent) to connect to supply branches and waste mains.

D2010 820	Water Cooler Systems	COST EACH		
		MAT.	INST.	TOTAL
1840	Water cooler, electric, wall hung, 8.2 G.P.H.	1,075	645	1,720
1880	Dual height, 14.3 G.P.H.	1,375	665	2,040
1920	Wheelchair type, 7.5 G.P.H.	1,225	645	1,870
1960	Semi recessed, 8.1 G.P.H.	1,050	645	1,695
2000	Full recessed, 8 G.P.H.	2,550	695	3,245
2040	Floor mounted, 14.3 G.P.H.	1,225	560	1,785
2080	Dual height, 14.3 G.P.H.	1,575	680	2,255
2120	Refrigerated compartment type, 1.5 G.P.H.	1,875	560	2,435

D2010 Plumbing Fixtures

Two Fixture   Three Fixture

*Common wall is with adjacent bathroom.

D2010 920	Two Fixture Bathroom, Two Wall Plumbing	COST EACH		
		MAT.	INST.	TOTAL
1180	Bathroom, lavatory & water closet, 2 wall plumbing, stand alone	1,675	2,100	3,775
1200	Share common plumbing wall*	1,550	1,800	3,350

D2010 922	Two Fixture Bathroom, One Wall Plumbing	COST EACH		
		MAT.	INST.	TOTAL
2220	Bathroom, lavatory & water closet, one wall plumbing, stand alone	1,600	1,875	3,475
2240	Share common plumbing wall*	1,375	1,600	2,975

D2010 924	Three Fixture Bathroom, One Wall Plumbing	COST EACH		
		MAT.	INST.	TOTAL
1150	Bathroom, three fixture, one wall plumbing			
1160	Lavatory, water closet & bathtub			
1170	Stand alone	2,975	2,450	5,425
1180	Share common plumbing wall *	2,575	1,775	4,350

D2010 926	Three Fixture Bathroom, Two Wall Plumbing	COST EACH		
		MAT.	INST.	TOTAL
2130	Bathroom, three fixture, two wall plumbing			
2140	Lavatory, water closet & bathtub			
2160	Stand alone	3,000	2,475	5,475
2180	Long plumbing wall common *	2,700	1,975	4,675
3610	Lavatory, bathtub & water closet			
3620	Stand alone	3,200	2,825	6,025
3640	Long plumbing wall common *	3,050	2,550	5,600
4660	Water closet, corner bathtub & lavatory			
4680	Stand alone	4,700	2,500	7,200
4700	Long plumbing wall common *	4,300	1,900	6,200
6100	Water closet, stall shower & lavatory			
6120	Stand alone	3,275	2,825	6,100
6140	Long plumbing wall common *	3,150	2,600	5,750
7060	Lavatory, corner stall shower & water closet			
7080	Stand alone	3,675	2,500	6,175
7100	Short plumbing wall common *	3,175	1,675	4,850

D20 Plumbing

D2010 Plumbing Fixtures

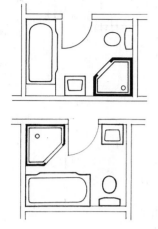

Four Fixture Bathroom Systems consisting of a lavatory, water closet, bathtub, shower and rough-in service piping.

- Prices for plumbing and fixtures only.

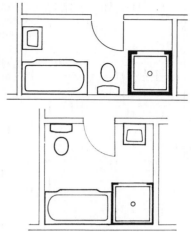

*Common wall is with an adjacent bathroom.

D2010 928	Four Fixture Bathroom, Two Wall Plumbing	COST EACH		
		MAT.	INST.	TOTAL
1140	Bathroom, four fixture, two wall plumbing			
1150	Bathtub, water closet, stall shower & lavatory			
1160	Stand alone	4,250	2,700	6,950
1180	Long plumbing wall common *	3,850	2,075	5,925
2260	Bathtub, lavatory, corner stall shower & water closet			
2280	Stand alone	4,875	2,700	7,575
2320	Long plumbing wall common *	4,475	2,100	6,575
3620	Bathtub, stall shower, lavatory & water closet			
3640	Stand alone	4,675	3,400	8,075
3660	Long plumbing wall (opp. door) common *	4,300	2,775	7,075

D2010 930	Four Fixture Bathroom, Three Wall Plumbing	COST EACH		
		MAT.	INST.	TOTAL
4680	Bathroom, four fixture, three wall plumbing			
4700	Bathtub, stall shower, lavatory & water closet			
4720	Stand alone	5,450	3,750	9,200
4760	Long plumbing wall (opposite door) common *	5,325	3,450	8,775

D2010 Plumbing Fixtures

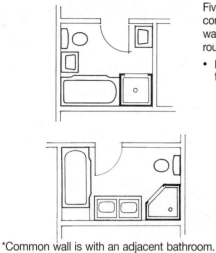

Five Fixture Bathroom Systems consisting of two lavatories, a water closet, bathtub, shower and rough-in service piping.

- Prices for plumbing and fixtures only.

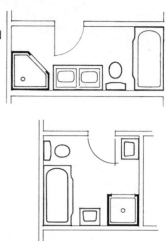

*Common wall is with an adjacent bathroom.

D2010 932	Five Fixture Bathroom, Two Wall Plumbing	COST EACH		
		MAT.	INST.	TOTAL
1320	Bathroom, five fixture, two wall plumbing			
1340	Bathtub, water closet, stall shower & two lavatories			
1360	Stand alone	5,450	4,325	9,775
1400	One short plumbing wall common *	5,050	3,725	8,775

D2010 934	Five Fixture Bathroom, Three Wall Plumbing	COST EACH		
		MAT.	INST.	TOTAL
2360	Bathroom, five fixture, three wall plumbing			
2380	Water closet, bathtub, two lavatories & stall shower			
2400	Stand alone	6,075	4,350	10,425
2440	One short plumbing wall common *	5,675	3,725	9,400

D2010 936	Five Fixture Bathroom, One Wall Plumbing	COST EACH		
		MAT.	INST.	TOTAL
4080	Bathroom, five fixture, one wall plumbing			
4100	Bathtub, two lavatories, corner stall shower & water closet			
4120	Stand alone	5,775	3,875	9,650
4160	Share common wall *	4,875	2,525	7,400

432

For customer support on your Square Foot Cost Data, call 877.756.2789.

D2020 Domestic Water Distribution

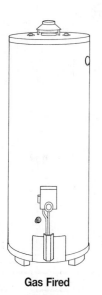

Installation includes piping and fittings within 10' of heater. Gas and oil fired heaters require vent piping (not included with these units).

Gas Fired

Oil Fired

D2020 220	Gas Fired Water Heaters - Residential Systems	COST EACH		
		MAT.	INST.	TOTAL
2200	Gas fired water heater, residential, 100°F rise			
2260	30 gallon tank, 32 GPH	2,475	1,475	3,950
2300	40 gallon tank, 32 GPH	2,650	1,650	4,300
2340	50 gallon tank, 63 GPH	2,775	1,650	4,425
2380	75 gallon tank, 63 GPH	3,850	1,850	5,700
2420	100 gallon tank, 63 GPH	3,975	1,950	5,925
2422				

D2020 230	Oil Fired Water Heaters - Residential Systems	COST EACH		
		MAT.	INST.	TOTAL
2200	Oil fired water heater, residential, 100°F rise			
2220	30 gallon tank, 103 GPH	2,000	1,375	3,375
2260	50 gallon tank, 145 GPH	2,400	1,525	3,925
2300	70 gallon tank, 164 GPH	3,350	1,725	5,075
2340	85 gallon tank, 181 GPH	10,800	1,775	12,575

433

For customer support on your Square Foot Cost Data, call 877.756.2789.

D2020 Domestic Water Distribution

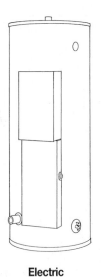

Systems below include piping and fittings within 10' of heater. Electric water heaters do not require venting. Gas fired heaters require vent piping (not included in these prices).

Electric

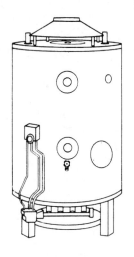

Gas Fired

D2020 240	Electric Water Heaters - Commercial Systems	COST EACH		
		MAT.	INST.	TOTAL
1800	Electric water heater, commercial, 100°F rise			
1820	50 gallon tank, 9 KW 37 GPH	7,800	1,250	9,050
1860	80 gal, 12 KW 49 GPH	9,900	1,525	11,425
1900	36 KW 147 GPH	13,500	1,650	15,150
1940	120 gal, 36 KW 147 GPH	14,600	1,800	16,400
1980	150 gal, 120 KW 490 GPH	45,100	1,925	47,025
2020	200 gal, 120 KW 490 GPH	47,500	1,975	49,475
2060	250 gal, 150 KW 615 GPH	53,000	2,300	55,300
2100	300 gal, 180 KW 738 GPH	57,000	2,425	59,425
2140	350 gal, 30 KW 123 GPH	42,500	2,625	45,125
2180	180 KW 738 GPH	59,000	2,625	61,625
2220	500 gal, 30 KW 123 GPH	55,500	3,075	58,575
2260	240 KW 984 GPH	89,500	3,075	92,575
2300	700 gal, 30 KW 123 GPH	67,000	3,500	70,500
2340	300 KW 1230 GPH	99,500	3,500	103,000
2380	1000 gal, 60 KW 245 GPH	81,000	4,900	85,900
2420	480 KW 1970 GPH	131,500	4,900	136,400
2460	1500 gal, 60 KW 245 GPH	118,500	6,050	124,550
2500	480 KW 1970 GPH	162,500	6,050	168,550

D2020 250	Gas Fired Water Heaters - Commercial Systems	COST EACH		
		MAT.	INST.	TOTAL
1760	Gas fired water heater, commercial, 100°F rise			
1780	75.5 MBH input, 63 GPH	4,950	1,900	6,850
1820	95 MBH input, 86 GPH	8,550	1,900	10,450
1860	100 MBH input, 91 GPH	8,800	2,000	10,800
1900	115 MBH input, 110 GPH	8,850	2,050	10,900
1980	155 MBH input, 150 GPH	11,100	2,325	13,425
2020	175 MBH input, 168 GPH	11,400	2,475	13,875
2060	200 MBH input, 192 GPH	11,700	2,800	14,500
2100	240 MBH input, 230 GPH	12,400	3,025	15,425
2140	300 MBH input, 278 GPH	13,500	3,475	16,975
2180	390 MBH input, 374 GPH	16,000	3,500	19,500
2220	500 MBH input, 480 GPH	21,600	3,800	25,400
2260	600 MBH input, 576 GPH	24,900	4,100	29,000

D2020 Domestic Water Distribution

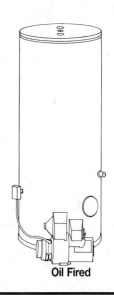

Oil Fired

Oil fired water heater systems include piping and fittings within 10′ of heater. Oil fired heaters require vent piping (not included in these systems).

D2020 260	Oil Fired Water Heaters - Commercial Systems	COST EACH		
		MAT.	INST.	TOTAL
1800	Oil fired water heater, commercial, 100°F rise			
1820	140 gal., 140 MBH input, 134 GPH	26,400	1,625	28,025
1900	140 gal., 255 MBH input, 247 GPH	28,400	2,050	30,450
1940	140 gal., 270 MBH input, 259 GPH	35,000	2,325	37,325
1980	140 gal., 400 MBH input, 384 GPH	36,100	2,675	38,775
2060	140 gal., 720 MBH input, 691 GPH	38,400	2,800	41,200
2100	221 gal., 300 MBH input, 288 GPH	51,000	3,075	54,075
2140	221 gal., 600 MBH input, 576 GPH	56,500	3,125	59,625
2180	221 gal., 800 MBH input, 768 GPH	57,500	3,250	60,750
2220	201 gal., 1000 MBH input, 960 GPH	58,500	3,275	61,775
2260	201 gal., 1250 MBH input, 1200 GPH	59,500	3,350	62,850
2300	201 gal., 1500 MBH input, 1441 GPH	64,500	3,425	67,925
2340	411 gal., 600 MBH input, 576 GPH	65,000	3,500	68,500
2380	411 gal., 800 MBH input, 768 GPH	68,000	3,600	71,600
2420	411 gal., 1000 MBH input, 960 GPH	71,000	4,150	75,150
2460	411 gal., 1250 MBH input, 1200 GPH	72,500	4,275	76,775
2500	397 gal., 1500 MBH input, 1441 GPH	77,000	4,400	81,400
2540	397 gal., 1750 MBH input, 1681 GPH	78,500	4,550	83,050

D2040 Rain Water Drainage

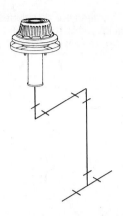

Design Assumptions: Vertical conductor size is based on a maximum rate of rainfall of 4″ per hour. To convert roof area to other rates multiply "Max. S.F. Roof Area" shown by four and divide the result by desired local rate. The answer is the local roof area that may be handled by the indicated pipe diameter.

Basic cost is for roof drain, 10′ of vertical leader and 10′ of horizontal, plus connection to the main.

Pipe Dia.	Max. S.F. Roof Area	Gallons per Min.
2″	544	23
3″	1610	67
4″	3460	144
5″	6280	261
6″	10,200	424
8″	22,000	913

D2040 210	Roof Drain Systems	MAT.	INST.	TOTAL
1880	Roof drain, DWV PVC, 2″ diam., piping, 10′ high	305	720	1,025
1920	For each additional foot add	6.30	22	28.30
1960	3″ diam., 10′ high	385	840	1,225
2000	For each additional foot add	9	24.50	33.50
2040	4″ diam., 10′ high	460	945	1,405
2080	For each additional foot add	11.25	27	38.25
2120	5″ diam., 10′ high	1,475	1,075	2,550
2160	For each additional foot add	30	30	60
2200	6″ diam., 10′ high	1,475	1,200	2,675
2240	For each additional foot add	18.60	33	51.60
2280	8″ diam., 10′ high	3,175	2,050	5,225
2320	For each additional foot add	39	41.50	80.50
3940	C.I., soil, single hub, service wt., 2″ diam. piping, 10′ high	715	785	1,500
3980	For each additional foot add	12.85	20.50	33.35
4120	3″ diam., 10′ high	920	850	1,770
4160	For each additional foot add	17	21.50	38.50
4200	4″ diam., 10′ high	1,050	930	1,980
4240	For each additional foot add	21.50	23.50	45
4280	5″ diam., 10′ high	1,475	1,025	2,500
4320	For each additional foot add	29	26.50	55.50
4360	6″ diam., 10′ high	1,850	1,100	2,950
4400	For each additional foot add	35.50	27.50	63
4440	8″ diam., 10′ high	3,625	2,225	5,850
4480	For each additional foot add	53	46	99
6040	Steel galv. sch 40 threaded, 2″ diam. piping, 10′ high	750	760	1,510
6080	For each additional foot add	9.30	20	29.30
6120	3″ diam., 10′ high	1,375	1,100	2,475
6160	For each additional foot add	18.55	30	48.55
6200	4″ diam., 10′ high	1,950	1,400	3,350
6240	For each additional foot add	27	35.50	62.50
6280	5″ diam., 10′ high	1,875	1,175	3,050
6320	For each additional foot add	30	35	65
6360	6″ diam, 10′ high	2,375	1,500	3,875
6400	For each additional foot add	37.50	47.50	85
6440	8″ diam., 10′ high	4,400	2,175	6,575
6480	For each additional foot add	56.50	54	110.50

D2090 Other Plumbing Systems

D2090 810	Piping - Installed - Unit Costs	COST PER L.F.		
		MAT.	INST.	TOTAL
0840	Cast iron, soil, B & S, service weight, 2" diameter	12.85	20.50	33.35
0860	3" diameter	17	21.50	38.50
0880	4" diameter	21.50	23.50	45
0900	5" diameter	29	26.50	55.50
0920	6" diameter	35.50	27.50	63
0940	8" diameter	53	46	99
0960	10" diameter	85	50.50	135.50
0980	12" diameter	120	56.50	176.50
1040	No hub, 1-1/2" diameter	12.65	18.10	30.75
1060	2" diameter	13.25	19.20	32.45
1080	3" diameter	17	20	37
1100	4" diameter	21.50	22	43.50
1120	5" diameter	29.50	24	53.50
1140	6" diameter	36	25.50	61.50
1160	8" diameter	60.50	39.50	100
1180	10" diameter	99.50	44.50	144
1220	Copper tubing, hard temper, solder, type K, 1/2" diameter	5.50	9.15	14.65
1260	3/4" diameter	8.95	9.65	18.60
1280	1" diameter	11.65	10.85	22.50
1300	1-1/4" diameter	14.30	12.75	27.05
1320	1-1/2" diameter	17.95	14.30	32.25
1340	2" diameter	27	17.85	44.85
1360	2-1/2" diameter	40	21.50	61.50
1380	3" diameter	54.50	24	78.50
1400	4" diameter	85	34	119
1480	5" diameter	161	40	201
1500	6" diameter	235	52.50	287.50
1520	8" diameter	410	59	469
1560	Type L, 1/2" diameter	4.05	8.85	12.90
1600	3/4" diameter	5.75	9.40	15.15
1620	1" diameter	8.15	10.50	18.65
1640	1-1/4" diameter	11.05	12.30	23.35
1660	1-1/2" diameter	13.85	13.75	27.60
1680	2" diameter	20	17	37
1700	2-1/2" diameter	30.50	21	51.50
1720	3" diameter	40.50	23	63.50
1740	4" diameter	63.50	33	96.50
1760	5" diameter	117	38	155
1780	6" diameter	160	50	210
1800	8" diameter	273	55.50	328.50
1840	Type M, 1/2" diameter	3.53	8.50	12.03
1880	3/4" diameter	4.82	9.15	13.97
1900	1" diameter	7.25	10.20	17.45
1920	1-1/4" diameter	10.10	11.90	22
1940	1-1/2" diameter	13.30	13.25	26.55
1960	2" diameter	19.95	16.25	36.20
1980	2-1/2" diameter	29.50	20	49.50
2000	3" diameter	39	22	61
2020	4" diameter	63.50	32	95.50
2060	6" diameter	175	47.50	222.50
2080	8" diameter	300	52.50	352.50
2120	Type DWV, 1-1/4" diameter	10.10	11.90	22
2160	1-1/2" diameter	12.40	13.25	25.65
2180	2" diameter	16.25	16.25	32.50
2200	3" diameter	29	22	51
2220	4" diameter	46.50	32	78.50
2240	5" diameter	114	35.50	149.50
2260	6" diameter	162	47.50	209.50

437

For customer support on your Square Foot Cost Data, call 877.756.2789.

D2090 Other Plumbing Systems

D2090 810	Piping - Installed - Unit Costs	COST PER L.F.		
		MAT.	INST.	TOTAL
2280	8" diameter	340	52.50	392.50
2800	Plastic, PVC, DWV, schedule 40, 1-1/4" diameter	6.05	17	23.05
2820	1-1/2" diameter	5.85	19.85	25.70
2830	2" diameter	6.30	22	28.30
2840	3" diameter	9	24.50	33.50
2850	4" diameter	11.25	27	38.25
2890	6" diameter	18.60	33	51.60
3010	Pressure pipe 200 PSI, 1/2" diameter	5.30	13.25	18.55
3030	3/4" diameter	5.70	14	19.70
3040	1" diameter	6.30	15.55	21.85
3050	1-1/4" diameter	7.20	17	24.20
3060	1-1/2" diameter	7.65	19.85	27.50
3070	2" diameter	8.65	22	30.65
3080	2-1/2" diameter	13.50	23	36.50
3090	3" diameter	14.95	24.50	39.45
3100	4" diameter	23	27	50
3110	6" diameter	39	33	72
3120	8" diameter	57.50	41.50	99
4000	Steel, schedule 40, black, threaded, 1/2" diameter	3.19	11.35	14.54
4020	3/4" diameter	3.68	11.70	15.38
4030	1" diameter	4.73	13.50	18.23
4040	1-1/4" diameter	5.70	14.45	20.15
4050	1-1/2" diameter	6.45	16.10	22.55
4060	2" diameter	8.15	20	28.15
4070	2-1/2" diameter	12.70	25.50	38.20
4080	3" diameter	16.05	30	46.05
4090	4" diameter	24.50	35.50	60
4100	Grooved, 5" diameter	44.50	35	79.50
4110	6" diameter	57	47.50	104.50
4120	8" diameter	92.50	54	146.50
4130	10" diameter	125	64.50	189.50
4140	12" diameter	139	74	213
4200	Galvanized, threaded, 1/2" diameter	3.49	11.35	14.84
4220	3/4" diameter	3.96	11.70	15.66
4230	1" diameter	5.30	13.50	18.80
4240	1-1/4" diameter	6.40	14.45	20.85
4250	1-1/2" diameter	7.25	16.10	23.35
4260	2" diameter	9.30	20	29.30
4270	2-1/2" diameter	14.70	25.50	40.20
4280	3" diameter	18.55	30	48.55
4290	4" diameter	27	35.50	62.50
4300	Grooved, 5" diameter	30	35	65
4310	6" diameter	37.50	47.50	85
4320	8" diameter	56.50	54	110.50
4330	10" diameter	118	64.50	182.50
4340	12" diameter	144	74	218
5010	Flanged, black, 1" diameter	10.25	19.30	29.55
5020	1-1/4" diameter	11.40	21	32.40
5030	1-1/2" diameter	11.80	23	34.80
5040	2" diameter	13.90	30	43.90
5050	2-1/2" diameter	19.05	37.50	56.55
5060	3" diameter	22	42	64
5070	4" diameter	29.50	52	81.50
5080	5" diameter	57	64.50	121.50
5090	6" diameter	68	82.50	150.50
5100	8" diameter	111	108	219
5110	10" diameter	157	129	286
5120	12" diameter	189	148	337

438

For customer support on your Square Foot Cost Data, call 877.756.2789.

D2090   Other Plumbing Systems

D2090 810	Piping - Installed - Unit Costs	COST PER L.F.		
		MAT.	INST.	TOTAL
5720	Grooved joints, black, 3/4" diameter	5.50	10.05	15.55
5730	1" diameter	5.30	11.35	16.65
5740	1-1/4" diameter	6.55	12.30	18.85
5750	1-1/2" diameter	7.25	14	21.25
5760	2" diameter	8.40	17.85	26.25
5770	2-1/2" diameter	12.80	22.50	35.30
5900	3" diameter	15.40	25.50	40.90
5910	4" diameter	21.50	28.50	50
5920	5" diameter	44.50	35	79.50
5930	6" diameter	57	47.50	104.50
5940	8" diameter	92.50	54	146.50
5950	10" diameter	125	64.50	189.50
5960	12" diameter	139	74	213

For customer support on your Square Foot Cost Data, call 877.756.2789.

439

D3010 Energy Supply

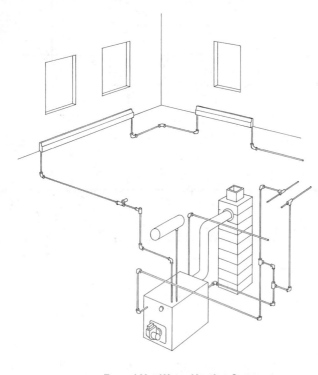

Forced Hot Water Heating System

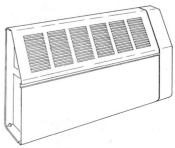

Fin Tube Radiation

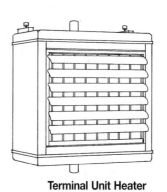

Terminal Unit Heater

D3010 510	Apartment Building Heating - Fin Tube Radiation	COST PER S.F.		
		MAT.	INST.	TOTAL
1740	Heating systems, fin tube radiation, forced hot water			
1760	1,000 S.F. area, 10,000 C.F. volume	7.50	6.95	14.45
1800	10,000 S.F. area, 100,000 C.F. volume	3.14	4.18	7.32
1840	20,000 S.F. area, 200,000 C.F. volume	3.43	4.69	8.12
1880	30,000 S.F. area, 300,000 C.F. volume	3.32	4.55	7.87

D3010 520	Commercial Building Heating - Fin Tube Radiation	COST PER S.F.		
		MAT.	INST.	TOTAL
1940	Heating systems, fin tube radiation, forced hot water			
1960	1,000 S.F. bldg, one floor	18.20	15.70	33.90
2000	10,000 S.F., 100,000 C.F., total two floors	4.70	5.95	10.65
2040	100,000 S.F., 1,000,000 C.F., total three floors	1.90	2.67	4.57
2080	1,000,000 S.F., 10,000,000 C.F., total five floors	1.01	1.40	2.41

D3010 530	Commercial Bldg. Heating - Terminal Unit Heaters	COST PER S.F.		
		MAT.	INST.	TOTAL
1860	Heating systems, terminal unit heaters, forced hot water			
1880	1,000 S.F. bldg., one floor	18	14.40	32.40
1920	10,000 S.F. bldg., 100,000 C.F. total two floors	4.25	4.91	9.16
1960	100,000 S.F. bldg., 1,000,000 C.F. total three floors	1.95	2.43	4.38
2000	1,000,000 S.F. bldg., 10,000,000 C.F. total five floors	1.32	1.57	2.89

D3020 Heat Generating Systems

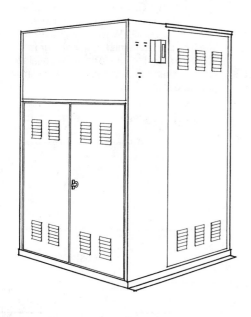

Small Electric Boiler Systems Considerations:

1. Terminal units are fin tube baseboard radiation rated at 720 BTU/hr with 200° water temperature or 820 BTU/hr steam.
2. Primary use being for residential or smaller supplementary areas, the floor levels are based on 7-1/2" ceiling heights.
3. All distribution piping is copper for boilers through 205 MBH. All piping for larger systems is steel pipe.

Large Electric Boiler System Considerations:

1. Terminal units are all unit heaters of the same size. Quantities are varied to accommodate total requirements.
2. All air is circulated through the heaters a minimum of three times per hour.
3. As the capacities are adequate for commercial use, gross output rating by floor levels are based on 10' ceiling height.
4. All distribution piping is black steel pipe.

D3020 102	Small Heating Systems, Hydronic, Electric Boilers	COST PER S.F.		
		MAT.	INST.	TOTAL
1100	Small heating systems, hydronic, electric boilers			
1120	Steam, 1 floor, 1480 S.F., 61 M.B.H.	7.50	8.92	16.42
1160	3,000 S.F., 123 M.B.H.	5.75	7.85	13.60
1200	5,000 S.F., 205 M.B.H.	4.98	7.25	12.23
1240	2 floors, 12,400 S.F., 512 M.B.H.	3.88	7.20	11.08
1280	3 floors, 24,800 S.F., 1023 M.B.H.	4.32	7.10	11.42
1320	34,750 S.F., 1,433 M.B.H.	3.97	6.90	10.87
1360	Hot water, 1 floor, 1,000 S.F., 41 M.B.H.	12.65	4.95	17.60
1400	2,500 S.F., 103 M.B.H.	8.85	8.95	17.80
1440	2 floors, 4,850 S.F., 205 M.B.H.	8.45	10.75	19.20
1480	3 floors, 9,700 S.F., 410 M.B.H.	8.70	11.10	19.80

D3020 104	Large Heating Systems, Hydronic, Electric Boilers	COST PER S.F.		
		MAT.	INST.	TOTAL
1230	Large heating systems, hydronic, electric boilers			
1240	9,280 S.F., 135 K.W., 461 M.B.H., 1 floor	4.13	3.34	7.47
1280	14,900 S.F., 240 K.W., 820 M.B.H., 2 floors	5.35	5.40	10.75
1320	18,600 S.F., 296 K.W., 1,010 M.B.H., 3 floors	5.25	5.85	11.10
1360	26,100 S.F., 420 K.W., 1,432 M.B.H., 4 floors	5.15	5.75	10.90
1400	39,100 S.F., 666 K.W., 2,273 M.B.H., 4 floors	4.43	4.81	9.24
1440	57,700 S.F., 900 K.W., 3,071 M.B.H., 5 floors	4.24	4.74	8.98
1480	111,700 S.F., 1,800 K.W., 6,148 M.B.H., 6 floors	3.77	4.07	7.84
1520	149,000 S.F., 2,400 K.W., 8,191 M.B.H., 8 floors	3.71	4.05	7.76
1560	223,300 S.F., 3,600 K.W., 12,283 M.B.H., 14 floors	3.91	4.60	8.51

D3020 Heat Generating Systems

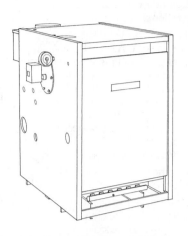

Boiler Selection: The maximum allowable working pressures are limited by ASME "Code for Heating Boilers" to 15 PSI for steam and 160 PSI for hot water heating boilers, with a maximum temperature limitation of 250° F. Hot water boilers are generally rated for a working pressure of 30 PSI. High pressure boilers are governed by the ASME "Code for Power Boilers" which is used almost universally for boilers operating over 15 PSIG. High pressure boilers used for a combination of heating/process loads are usually designed for 150 PSIG.

Boiler ratings are usually indicated as either Gross or Net Output. The Gross Load is equal to the Net Load plus a piping and pickup allowance. When this allowance cannot be determined, divide the gross output rating by 1.25 for a value equal to or greater than the next heat loss requirement of the building.

Table below lists installed cost per boiler and includes insulating jacket, standard controls, burner and safety controls. Costs do not include piping or boiler base pad. Outputs are Gross.

D3020 106	Boilers, Hot Water & Steam	COST EACH		
		MAT.	INST.	TOTAL
0600	Boiler, electric, steel, hot water, 12 K.W., 41 M.B.H.	5,450	1,525	6,975
0620	30 K.W., 103 M.B.H.	5,950	1,650	7,600
0640	60 K.W., 205 M.B.H.	6,600	1,800	8,400
0660	120 K.W., 410 M.B.H.	7,025	2,200	9,225
0680	210 K.W., 716 M.B.H.	8,375	3,300	11,675
0700	510 K.W., 1,739 M.B.H.	25,400	6,150	31,550
0720	720 K.W., 2,452 M.B.H.	31,400	6,950	38,350
0740	1,200 K.W., 4,095 M.B.H.	39,900	7,975	47,875
0760	2,100 K.W., 7,167 M.B.H.	70,000	10,000	80,000
0780	3,600 K.W., 12,283 M.B.H.	103,500	16,900	120,400
0820	Steam, 6 K.W., 20.5 M.B.H.	4,500	1,650	6,150
0840	24 K.W., 81.8 M.B.H.	5,500	1,800	7,300
0860	60 K.W., 205 M.B.H.	7,600	1,975	9,575
0880	150 K.W., 512 M.B.H.	11,000	3,050	14,050
0900	510 K.W., 1,740 M.B.H.	35,900	7,525	43,425
0920	1,080 K.W., 3,685 M.B.H.	44,700	10,800	55,500
0940	2,340 K.W., 7,984 M.B.H.	95,000	16,900	111,900
0980	Gas, cast iron, hot water, 80 M.B.H.	2,175	1,900	4,075
1000	100 M.B.H.	2,825	2,075	4,900
1020	163 M.B.H.	3,475	2,800	6,275
1040	280 M.B.H.	4,750	3,100	7,850
1060	544 M.B.H.	10,800	5,500	16,300
1080	1,088 M.B.H.	14,900	6,975	21,875
1100	2,000 M.B.H.	23,500	10,900	34,400
1120	2,856 M.B.H.	33,900	14,000	47,900
1140	4,720 M.B.H.	81,500	19,200	100,700
1160	6,970 M.B.H.	144,500	31,400	175,900
1180	For steam systems under 2,856 M.B.H., add 8%			
1520	Oil, cast iron, hot water, 109 M.B.H.	2,550	2,325	4,875
1540	173 M.B.H.	3,225	2,800	6,025
1560	236 M.B.H.	4,100	3,275	7,375
1580	1,084 M.B.H.	12,000	7,425	19,425
1600	1,600 M.B.H.	15,400	10,700	26,100
1620	2,480 M.B.H.	22,600	13,600	36,200
1640	3,550 M.B.H.	30,100	16,300	46,400
1660	Steam systems same price as hot water			

D3020 Heat Generating Systems

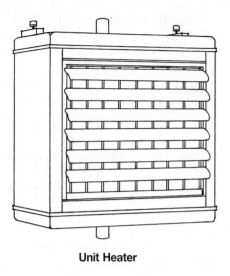

Unit Heater

Fossil Fuel Boiler System Considerations:

1. Terminal units are horizontal unit heaters. Quantities are varied to accommodate total heat loss per building.
2. Unit heater selection was determined by their capacity to circulate the building volume a minimum of three times per hour in addition to the BTU output.
3. Systems shown are forced hot water. Steam boilers cost slightly more than hot water boilers. However, this is compensated for by the smaller size or fewer terminal units required with steam.
4. Floor levels are based on 10′ story heights.
5. MBH requirements are gross boiler output.

D3020 108	Heating Systems, Unit Heaters	COST PER S.F.		
		MAT.	INST.	TOTAL
1260	Heating systems, hydronic, fossil fuel, terminal unit heaters,			
1280	Cast iron boiler, gas, 80 M.B.H., 1,070 S.F. bldg.	10.84	10.65	21.49
1320	163 M.B.H., 2,140 S.F. bldg.	7.30	7.25	14.55
1360	544 M.B.H., 7,250 S.F. bldg.	5.45	5.15	10.60
1400	1,088 M.B.H., 14,500 S.F. bldg.	4.47	4.85	9.32
1440	3,264 M.B.H., 43,500 S.F. bldg.	3.67	3.66	7.33
1480	5,032 M.B.H., 67,100 S.F. bldg.	4.21	3.75	7.96
1520	Oil, 109 M.B.H., 1,420 S.F. bldg.	12.10	9.50	21.60
1560	235 M.B.H., 3,150 S.F. bldg.	7.50	6.80	14.30
1600	940 M.B.H., 12,500 S.F. bldg.	5.35	4.43	9.78
1640	1,600 M.B.H., 21,300 S.F. bldg.	4.96	4.24	9.20
1680	2,480 M.B.H., 33,100 S.F. bldg.	4.95	3.82	8.77
1720	3,350 M.B.H., 44,500 S.F. bldg.	4.39	3.91	8.30
1760	Coal, 148 M.B.H., 1,975 S.F. bldg.	94.50	6.20	100.70
1800	300 M.B.H., 4,000 S.F. bldg.	51.50	4.86	56.36
1840	2,360 M.B.H., 31,500 S.F. bldg.	11.35	3.89	15.24

D3020 Heat Generating Systems

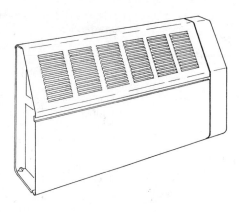

Fin Tube Radiator

Fossil Fuel Boiler System Considerations:

1. Terminal units are commercial steel fin tube radiation. Quantities are varied to accommodate total heat loss per building.
2. Systems shown are forced hot water. Steam boilers cost slightly more than hot water boilers. However, this is compensated for by the smaller size or fewer terminal units required with steam.
3. Floor levels are based on 10′ story heights.
4. MBH requirements are gross boiler output.

D3020 110	Heating System, Fin Tube Radiation	COST PER S.F.		
		MAT.	INST.	TOTAL
3230	Heating systems, hydronic, fossil fuel, fin tube radiation			
3240	Cast iron boiler, gas, 80 MBH, 1,070 S.F. bldg.	13.27	16.07	29.34
3280	169 M.B.H., 2,140 S.F. bldg.	8.40	10.20	18.60
3320	544 M.B.H., 7,250 S.F. bldg.	7.20	8.75	15.95
3360	1,088 M.B.H., 14,500 S.F. bldg.	6.40	8.55	14.95
3400	3,264 M.B.H., 43,500 S.F. bldg.	5.75	7.55	13.30
3440	5,032 M.B.H., 67,100 S.F. bldg.	6.30	7.65	13.95
3480	Oil, 109 M.B.H., 1,420 S.F. bldg.	16.45	17.55	34
3520	235 M.B.H., 3,150 S.F. bldg.	9.25	10.30	19.55
3560	940 M.B.H., 12,500 S.F. bldg.	7.30	8.15	15.45
3600	1,600 M.B.H., 21,300 S.F. bldg.	7.05	8.10	15.15
3640	2,480 M.B.H., 33,100 S.F. bldg.	7.05	7.70	14.75
3680	3,350 M.B.H., 44,500 S.F. bldg.	6.45	7.80	14.25
3720	Coal, 148 M.B.H., 1,975 S.F. bldg.	96	9.70	105.70
3760	300 M.B.H., 4,000 S.F. bldg.	53.50	8.35	61.85
3800	2,360 M.B.H., 31,500 S.F. bldg.	13.35	7.70	21.05
4080	Steel boiler, oil, 97 M.B.H., 1,300 S.F. bldg.	12.25	14.40	26.65
4120	315 M.B.H., 4,550 S.F. bldg.	6.30	7.30	13.60
4160	525 M.B.H., 7,000 S.F. bldg.	9.10	8.45	17.55
4200	1,050 M.B.H., 14,000 S.F. bldg.	8.20	8.45	16.65
4240	2,310 M.B.H., 30,800 S.F. bldg.	7.10	7.75	14.85
4280	3,150 M.B.H., 42,000 S.F. bldg.	6.60	7.85	14.45

D3030 Cooling Generating Systems

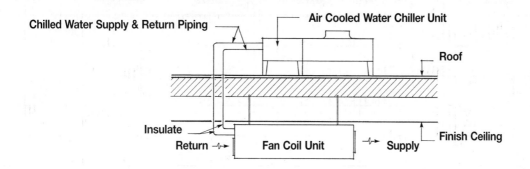

Chilled Water Supply & Return Piping

Air Cooled Water Chiller Unit

Roof

Insulate

Return

Fan Coil Unit

Supply

Finish Ceiling

*Cooling requirements would lead to a
 choice of multiple chillers.

D3030 110	Chilled Water, Air Cooled Condenser Systems	COST PER S.F.		
		MAT.	INST.	TOTAL
1180	Packaged chiller, air cooled, with fan coil unit			
1200	Apartment corridors, 3,000 S.F., 5.50 ton	6.50	8.90	15.40
1240	6,000 S.F., 11.00 ton	5.40	7.10	12.50
1280	10,000 S.F., 18.33 ton	4.65	5.35	10
1320	20,000 S.F., 36.66 ton	3.63	3.87	7.50
1360	40,000 S.F., 73.33 ton	5.10	3.89	8.99
1440	Banks and libraries, 3,000 S.F., 12.50 ton	9.95	10.35	20.30
1480	6,000 S.F., 25.00 ton	9.10	8.40	17.50
1520	10,000 S.F., 41.66 ton	7.15	5.85	13
1560	20,000 S.F., 83.33 ton	8.50	5.25	13.75
1600	40,000 S.F., 167 ton*			
1680	Bars and taverns, 3,000 S.F., 33.25 ton	17.55	12.10	29.65
1720	6,000 S.F., 66.50 ton	15.75	9.20	24.95
1760	10,000 S.F., 110.83 ton	12.70	2.76	15.46
1800	20,000 S.F., 220 ton*			
1840	40,000 S.F., 440 ton*			
1920	Bowling alleys, 3,000 S.F., 17.00 ton	12.55	11.60	24.15
1960	6,000 S.F., 34.00 ton	9.70	8.10	17.80
2000	10,000 S.F., 56.66 ton	8.40	5.80	14.20
2040	20,000 S.F., 113.33 ton	9.45	5.30	14.75
2080	40,000 S.F., 227 ton*			
2160	Department stores, 3,000 S.F., 8.75 ton	9.15	9.90	19.05
2200	6,000 S.F., 17.50 ton	6.95	7.60	14.55
2240	10,000 S.F., 29.17 ton	5.70	5.45	11.15
2280	20,000 S.F., 58.33 ton	4.79	3.99	8.78
2320	40,000 S.F., 116.66 ton	6.20	4	10.20
2400	Drug stores, 3,000 S.F., 20.00 ton	14.25	11.95	26.20
2440	6,000 S.F., 40.00 ton	11.55	8.85	20.40
2480	10,000 S.F., 66.66 ton	12.90	7.50	20.40
2520	20,000 S.F., 133.33 ton	11.10	5.55	16.65
2560	40,000 S.F., 267 ton*			
2640	Factories, 2,000 S.F., 10.00 ton	8.75	9.95	18.70
2680	6,000 S.F., 20.00 ton	7.90	8	15.90
2720	10,000 S.F., 33.33 ton	6.15	5.60	11.75
2760	20,000 S.F., 66.66 ton	7.50	5	12.50
2800	40,000 S.F., 133.33 ton	6.70	4.10	10.80
2880	Food supermarkets, 3,000 S.F., 8.50 ton	9	9.85	18.85
2920	6,000 S.F., 17.00 ton	6.80	7.55	14.35
2960	10,000 S.F., 28.33 ton	5.40	5.30	10.70
3000	20,000 S.F., 56.66 ton	4.59	3.91	8.50

445

D3030 Cooling Generating Systems

D3030 110	Chilled Water, Air Cooled Condenser Systems	COST PER S.F.		
		MAT.	INST.	TOTAL
3040	40,000 S.F., 113.33 ton	6.05	4.01	10.06
3120	Medical centers, 3,000 S.F., 7.00 ton	8.10	9.60	17.70
3160	6,000 S.F., 14.00 ton	6.10	7.30	13.40
3200	10,000 S.F., 23.33 ton	5.45	5.60	11.05
3240	20,000 S.F., 46.66 ton	4.35	4.07	8.42
3280	40,000 S.F., 93.33 ton	5.60	3.94	9.54
3360	Offices, 3,000 S.F., 9.50 ton	8.25	9.80	18.05
3400	6,000 S.F., 19.00 ton	7.65	7.95	15.60
3440	10,000 S.F., 31.66 ton	6	5.55	11.55
3480	20,000 S.F., 63.33 ton	7.50	5.10	12.60
3520	40,000 S.F., 126.66 ton	6.60	4.14	10.74
3600	Restaurants, 3,000 S.F., 15.00 ton	11.15	10.75	21.90
3640	6,000 S.F., 30.00 ton	9.25	8.15	17.40
3680	10,000 S.F., 50.00 ton	8.10	5.90	14
3720	20,000 S.F., 100.00 ton	9.65	5.55	15.20
3760	40,000 S.F., 200 ton*			
3840	Schools and colleges, 3,000 S.F., 11.50 ton	9.45	10.20	19.65
3880	6,000 S.F., 23.00 ton	8.60	8.25	16.85
3920	10,000 S.F., 38.33 ton	6.75	5.75	12.50
3960	20,000 S.F., 76.66 ton	8.35	5.30	13.65
4000	40,000 S.F., 153 ton*			

D3030 Cooling Generating Systems

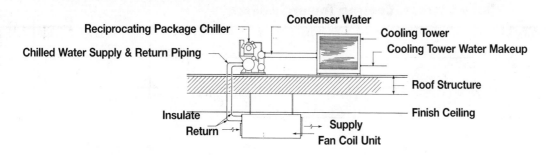

Reciprocating Package Chiller
Condenser Water
Cooling Tower
Cooling Tower Water Makeup
Chilled Water Supply & Return Piping
Roof Structure
Finish Ceiling
Insulate
Return
Supply
Fan Coil Unit

*Cooling requirements would lead to a choice of multiple chillers.

D3030 115	Chilled Water, Cooling Tower Systems	COST PER S.F.		
		MAT.	INST.	TOTAL
1300	Packaged chiller, water cooled, with fan coil unit			
1320	Apartment corridors, 4,000 S.F., 7.33 ton	6.30	8.73	15.03
1360	6,000 S.F., 11.00 ton	5.30	7.45	12.75
1400	10,000 S.F., 18.33 ton	4.96	5.50	10.46
1440	20,000 S.F., 26.66 ton	4	4.10	8.10
1480	40,000 S.F., 73.33 ton	4.99	4.20	9.19
1520	60,000 S.F., 110.00 ton	5.25	4.32	9.57
1600	Banks and libraries, 4,000 S.F., 16.66 ton	10.55	9.45	20
1640	6,000 S.F., 25.00 ton	8.95	8.35	17.30
1680	10,000 S.F., 41.66 ton	7.25	6.15	13.40
1720	20,000 S.F., 83.33 ton	8.10	5.90	14
1760	40,000 S.F., 166.66 ton	8.95	7.20	16.15
1800	60,000 S.F., 250.00 ton	9.20	7.60	16.80
1880	Bars and taverns, 4,000 S.F., 44.33 ton	17	11.75	28.75
1920	6,000 S.F., 66.50 ton	21	12.20	33.20
1960	10,000 S.F., 110.83 ton	17.70	9.60	27.30
2000	20,000 S.F., 221.66 ton	20.50	9.70	30.20
2040	40,000 S.F., 440 ton*			
2080	60,000 S.F., 660 ton*			
2160	Bowling alleys, 4,000 S.F., 22.66 ton	12.35	10.35	22.70
2200	6,000 S.F., 34.00 ton	11.05	9.10	20.15
2240	10,000 S.F., 56.66 ton	9.65	6.65	16.30
2280	20,000 S.F., 113.33 ton	10.05	6.15	16.20
2320	40,000 S.F., 226.66 ton	11.80	7.15	18.95
2360	60,000 S.F., 340 ton			
2440	Department stores, 4,000 S.F., 11.66 ton	7.20	9.40	16.60
2480	6,000 S.F., 17.50 ton	7.75	7.70	15.45
2520	10,000 S.F., 29.17 ton	6	5.75	11.75
2560	20,000 S.F., 58.33 ton	4.93	4.24	9.17
2600	40,000 S.F., 116.66 ton	6.20	4.40	10.60
2640	60,000 S.F., 175.00 ton	8.10	6.95	15.05
2720	Drug stores, 4,000 S.F., 26.66 ton	13.30	10.65	23.95
2760	6,000 S.F., 40.00 ton	11.45	9.10	20.55
2800	10,000 S.F., 66.66 ton	13.50	8.25	21.75
2840	20,000 S.F., 133.33 ton	11.50	6.40	17.90
2880	40,000 S.F., 266.67 ton	12.10	8.10	20.20
2920	60,000 S.F., 400 ton*			
3000	Factories, 4,000 S.F., 13.33 ton	9.10	9	18.10
3040	6,000 S.F., 20.00 ton	7.85	7.75	15.60
3080	10,000 S.F., 33.33 ton	6.60	5.95	12.55
3120	20,000 S.F., 66.66 ton	7.65	5.30	12.95
3160	40,000 S.F., 133.33 ton	6.85	4.54	11.39

447

For customer support on your Square Foot Cost Data, call 877.756.2789.

D3030 Cooling Generating Systems

D3030 115	Chilled Water, Cooling Tower Systems	COST PER S.F.		
		MAT.	INST.	TOTAL
3200	60,000 S.F., 200.00 ton	8.35	7.30	15.65
3280	Food supermarkets, 4,000 S.F., 11.33 ton	7.10	9.30	16.40
3320	6,000 S.F., 17.00 ton	6.95	7.55	14.50
3360	10,000 S.F., 28.33 ton	5.90	5.70	11.60
3400	20,000 S.F., 56.66 ton	5.10	4.26	9.36
3440	40,000 S.F., 113.33 ton	6.15	4.41	10.56
3480	60,000 S.F., 170.00 ton	8.05	6.95	15
3560	Medical centers, 4.000 S.F., 9.33 ton	6.15	8.55	14.70
3600	6,000 S.F., 14.00 ton	6.75	7.40	14.15
3640	10,000 S.F., 23.33 ton	5.40	5.55	10.95
3680	20,000 S.F., 46.66 ton	4.58	4.16	8.74
3720	40,000 S.F., 93.33 ton	5.60	4.32	9.92
3760	60,000 S.F., 140.00 ton	6.95	7.05	14
3840	Offices, 4,000 S.F., 12.66 ton	8.80	8.95	17.75
3880	6,000 S.F., 19.00 ton	7.80	7.95	15.75
3920	10,000 S.F., 31.66 ton	6.55	6	12.55
3960	20,000 S.F., 63.33 ton	7.60	5.30	12.90
4000	40,000 S.F., 126.66 ton	7.50	6.90	14.40
4040	60,000 S.F., 190.00 ton	8.10	7.20	15.30
4120	Restaurants, 4,000 S.F., 20.00 ton	11.05	9.60	20.65
4160	6,000 S.F., 30.00 ton	9.95	8.55	18.50
4200	10,000 S.F., 50.00 ton	8.90	6.40	15.30
4240	20,000 S.F., 100.00 ton	9.45	6.20	15.65
4280	40,000 S.F., 200.00 ton	9.45	6.95	16.40
4320	60,000 S.F., 300.00 ton	10.15	7.85	18
4400	Schools and colleges, 4,000 S.F., 15.33 ton	9.95	9.30	19.25
4440	6,000 S.F., 23.00 ton	8.50	8.15	16.65
4480	10,000 S.F., 38.33 ton	6.85	6.05	12.90
4520	20,000 S.F., 76.66 ton	7.85	5.85	13.70
4560	40,000 S.F., 153.33 ton	8.45	7.05	15.50
4600	60,000 S.F., 230.00 ton	8.55	7.30	15.85

D3050 Terminal & Package Units

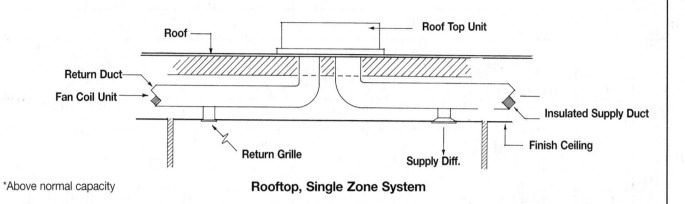

Roof Top Unit

Roof

Return Duct

Fan Coil Unit

Insulated Supply Duct

Finish Ceiling

Return Grille

Supply Diff.

*Above normal capacity

Rooftop, Single Zone System

D3050 150	Rooftop Single Zone Unit Systems	COST PER S.F.		
		MAT.	INST.	TOTAL
1260	Rooftop, single zone, air conditioner			
1280	Apartment corridors, 500 S.F., .92 ton	3.15	4.15	7.30
1320	1,000 S.F., 1.83 ton	3.15	4.14	7.29
1360	1500 S.F., 2.75 ton	1.98	3.43	5.41
1400	3,000 S.F., 5.50 ton	2.12	3.29	5.41
1440	5,000 S.F., 9.17 ton	2.17	2.99	5.16
1480	10,000 S.F., 18.33 ton	3.32	2.81	6.13
1560	Banks or libraries, 500 S.F., 2.08 ton	7.20	9.40	16.60
1600	1,000 S.F., 4.17 ton	4.51	7.80	12.31
1640	1,500 S.F., 6.25 ton	4.82	7.50	12.32
1680	3,000 S.F., 12.50 ton	4.95	6.80	11.75
1720	5,000 S.F., 20.80 ton	7.55	6.40	13.95
1760	10,000 S.F., 41.67 ton	5.80	6.35	12.15
1840	Bars and taverns, 500 S.F. 5.54 ton	11.80	12.55	24.35
1880	1,000 S.F., 11.08 ton	12.10	10.75	22.85
1920	1,500 S.F., 16.62 ton	11.70	10	21.70
1960	3,000 S.F., 33.25 ton	15.50	9.55	25.05
2000	5,000 S.F., 55.42 ton	13.25	9.55	22.80
2040	10,000 S.F., 110.83 ton*			
2080	Bowling alleys, 500 S.F., 2.83 ton	6.10	10.60	16.70
2120	1,000 S.F., 5.67 ton	6.55	10.15	16.70
2160	1,500 S.F., 8.50 ton	6.75	9.25	16
2200	3,000 S.F., 17.00 ton	6.50	8.85	15.35
2240	5,000 S.F., 28.33 ton	8.45	8.65	17.10
2280	10,000 S.F., 56.67 ton	7.30	8.65	15.95
2360	Department stores, 500 S.F., 1.46 ton	5.05	6.60	11.65
2400	1,000 S.F., 2.92 ton	3.16	5.45	8.61
2480	3,000 S.F., 8.75 ton	3.47	4.76	8.23
2520	5,000 S.F., 14.58 ton	3.35	4.56	7.91
2560	10,000 S.F., 29.17 ton	4.36	4.45	8.81
2640	Drug stores, 500 S.F., 3.33 ton	7.20	12.45	19.65
2680	1,000 S.F., 6.67 ton	7.70	11.95	19.65
2720	1,500 S.F., 10.00 ton	7.90	10.90	18.80
2760	3,000 S.F., 20.00 ton	12.10	10.20	22.30
2800	5,000 S.F., 33.33 ton	9.95	10.15	20.10
2840	10,000 S.F., 66.67 ton	8.60	10.15	18.75
2920	Factories, 500 S.F., 1.67 ton	5.75	7.55	13.30
3000	1,500 S.F., 5.00 ton	3.86	6	9.86
3040	3,000 S.F., 10.00 ton	3.96	5.45	9.41
3080	5,000 S.F., 16.67 ton	3.83	5.20	9.03
3120	10,000 S.F., 33.33 ton	4.99	5.10	10.09
3200	Food supermarkets, 500 S.F., 1.42 ton	4.88	6.40	11.28

449

D3050 Terminal & Package Units

D3050 150	Rooftop Single Zone Unit Systems	COST PER S.F.		
		MAT.	INST.	TOTAL
3240	1,000 S.F., 2.83 ton	3.03	5.25	8.28
3280	1,500 S.F., 4.25 ton	3.28	5.10	8.38
3320	3,000 S.F., 8.50 ton	3.36	4.62	7.98
3360	5,000 S.F., 14.17 ton	3.25	4.44	7.69
3400	10,000 S.F., 28.33 ton	4.23	4.31	8.54
3480	Medical centers, 500 S.F., 1.17 ton	4.03	5.30	9.33
3520	1,000 S.F., 2.33 ton	4.03	5.30	9.33
3560	1,500 S.F., 3.50 ton	2.53	4.36	6.89
3640	5,000 S.F., 11.67 ton	2.77	3.81	6.58
3680	10,000 S.F., 23.33 ton	4.23	3.58	7.81
3760	Offices, 500 S.F., 1.58 ton	5.45	7.15	12.60
3800	1,000 S.F., 3.17 ton	3.43	5.90	9.33
3840	1,500 S.F., 4.75 ton	3.67	5.70	9.37
3880	3,000 S.F., 9.50 ton	3.77	5.15	8.92
3920	5,000 S.F., 15.83 ton	3.64	4.95	8.59
3960	10,000 S.F., 31.67 ton	4.74	4.83	9.57
4000	Restaurants, 500 S.F., 2.50 ton	8.65	11.30	19.95
4040	1,000 S.F., 5.00 ton	5.80	9	14.80
4080	1,500 S.F., 7.50 ton	5.95	8.15	14.10
4120	3,000 S.F., 15.00 ton	5.75	7.85	13.60
4160	5,000 S.F., 25.00 ton	9.05	7.65	16.70
4200	10,000 S.F., 50.00 ton	6.45	7.60	14.05
4240	Schools and colleges, 500 S.F., 1.92 ton	6.60	8.70	15.30
4280	1,000 S.F., 3.83 ton	4.14	7.15	11.29
4360	3,000 S.F., 11.50 ton	4.55	6.25	10.80
4400	5,000 S.F., 19.17 ton	6.95	5.90	12.85

D3050 Terminal & Package Units

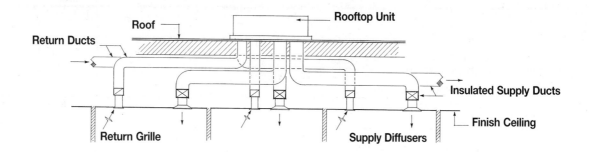

*Note A: Small single zone unit
recommended.

*Note B: A combination of multizone units
recommended.

D3050 155	Rooftop Multizone Unit Systems	COST PER S.F.		
		MAT.	INST.	TOTAL
1240	Rooftop, multizone, air conditioner			
1260	Apartment corridors, 1,500 S.F., 2.75 ton. See Note A.			
1280	3,000 S.F., 5.50 ton	11.13	5.30	16.43
1320	10,000 S.F., 18.30 ton	7.70	5.10	12.80
1360	15,000 S.F., 27.50 ton	7.50	5.10	12.60
1400	20,000 S.F., 36.70 ton	7.85	5.10	12.95
1440	25,000 S.F., 45.80 ton	7.30	5.10	12.40
1520	Banks or libraries, 1,500 S.F., 6.25 ton	25.50	12.05	37.55
1560	3,000 S.F., 12.50 ton	21.50	11.80	33.30
1600	10,000 S.F., 41.67 ton	16.65	11.65	28.30
1640	15,000 S.F., 62.50 ton	12.15	11.55	23.70
1680	20,000 S.F., 83.33 ton	12.15	11.55	23.70
1720	25,000 S.F., 104.00 ton	10.80	11.50	22.30
1800	Bars and taverns, 1,500 S.F., 16.62 ton	55	17.35	72.35
1840	3,000 S.F., 33.24 ton	46	16.75	62.75
1880	10,000 S.F., 110.83 ton	27.50	16.65	44.15
1920	15,000 S.F., 165 ton, See Note B.			
1960	20,000 S.F., 220 ton, See Note B.			
2000	25,000 S.F., 275 ton, See Note B.			
2080	Bowling alleys, 1,500 S.F., 8.50 ton	34.50	16.40	50.90
2120	3,000 S.F., 17.00 ton	29	16.05	45.05
2160	10,000 S.F., 56.70 ton	22.50	15.85	38.35
2200	15,000 S.F., 85.00 ton	16.50	15.70	32.20
2240	20,000 S.F., 113.00 ton	14.70	15.65	30.35
2280	25,000 S.F., 140.00 ton see Note B.			
2360	Department stores, 1,500 S.F., 4.37 ton, See Note A.			
2400	3,000 S.F., 8.75 ton	17.75	8.45	26.20
2440	10,000 S.F., 29.17 ton	12.45	8.10	20.55
2520	20,000 S.F., 58.33 ton	8.50	8.10	16.60
2560	25,000 S.F., 72.92 ton	8.50	8.10	16.60
2640	Drug stores, 1,500 S.F., 10.00 ton	40.50	19.25	59.75
2680	3,000 S.F., 20.00 ton	28	18.55	46.55
2720	10,000 S.F., 66.66 ton	19.45	18.50	37.95
2760	15,000 S.F., 100.00 ton	17.35	18.45	35.80
2800	20,000 S.F., 135 ton, See Note B.			
2840	25,000 S.F., 165 ton, See Note B.			
2920	Factories, 1,500 S.F., 5 ton, See Note A.			
3000	10,000 S.F., 33.33 ton	14.25	9.25	23.50

D3050 Terminal & Package Units

D3050 155	Rooftop Multizone Unit Systems	COST PER S.F.		
		MAT.	INST.	TOTAL
3040	15,000 S.F., 50.00 ton	13.30	9.30	22.60
3080	20,000 S.F., 66.66 ton	9.75	9.25	19
3120	25,000 S.F., 83.33 ton	9.70	9.25	18.95
3200	Food supermarkets, 1,500 S.F., 4.25 ton, See Note A.			
3240	3,000 S.F., 8.50 ton	17.20	8.20	25.40
3280	10,000 S.F., 28.33 ton	11.60	7.85	19.45
3320	15,000 S.F., 42.50 ton	11.30	7.90	19.20
3360	20,000 S.F., 56.67 ton	8.25	7.85	16.10
3400	25,000 S.F., 70.83 ton	8.25	7.85	16.10
3480	Medical centers, 1,500 S.F., 3.5 ton, See Note A.			
3520	3,000 S.F., 7.00 ton	14.20	6.75	20.95
3560	10,000 S.F., 23.33 ton	9.40	6.50	15.90
3600	15,000 S.F., 35.00 ton	10	6.45	16.45
3640	20,000 S.F., 46.66 ton	9.30	6.50	15.80
3680	25,000 S.F., 58.33 ton	6.80	6.45	13.25
3760	Offices, 1,500 S.F., 4.75 ton, See Note A.			
3800	3,000 S.F., 9.50 ton	19.25	9.15	28.40
3840	10,000 S.F., 31.66 ton	13.55	8.80	22.35
3920	20,000 S.F., 63.33 ton	9.25	8.80	18.05
3960	25,000 S.F., 79.16 ton	9.25	8.80	18.05
4000	Restaurants, 1,500 S.F., 7.50 ton	30.50	14.45	44.95
4040	3,000 S.F., 15.00 ton	25.50	14.15	39.65
4080	10,000 S.F., 50.00 ton	19.95	13.95	33.90
4120	15,000 S.F., 75.00 ton	14.60	13.85	28.45
4160	20,000 S.F., 100.00 ton	13	13.85	26.85
4200	25,000 S.F., 125 ton, See Note B.			
4240	Schools and colleges, 1,500 S.F., 5.75 ton	23.50	11.10	34.60
4320	10,000 S.F., 38.33 ton	15.30	10.70	26
4360	15,000 S.F., 57.50 ton	11.20	10.65	21.85
4400	20,000 S.F.,76.66 ton	11.15	10.65	21.80
4440	25,000 S.F., 95.83 ton	10.50	10.65	21.15

D3050 Terminal & Package Units

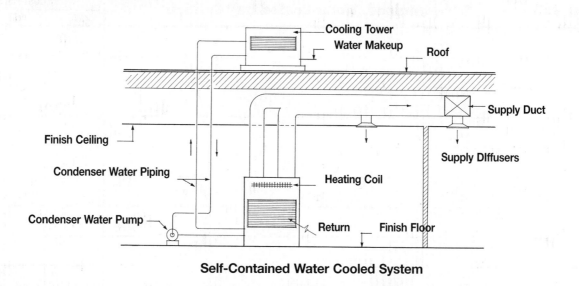

Self-Contained Water Cooled System

D3050 160	Self-contained, Water Cooled Unit Systems	COST PER S.F.		
		MAT.	INST.	TOTAL
1280	Self-contained, water cooled unit	3.42	3.36	6.78
1300	Apartment corridors, 500 S.F., .92 ton	3.40	3.35	6.75
1320	1,000 S.F., 1.83 ton	3.40	3.35	6.75
1360	3,000 S.F., 5.50 ton	2.83	2.80	5.63
1400	5,000 S.F., 9.17 ton	2.73	2.60	5.33
1440	10,000 S.F., 18.33 ton	3.94	2.32	6.26
1520	Banks or libraries, 500 S.F., 2.08 ton	7.45	3.44	10.89
1560	1,000 S.F., 4.17 ton	6.45	6.35	12.80
1600	3,000 S.F., 12.50 ton	6.20	5.90	12.10
1640	5,000 S.F., 20.80 ton	8.95	5.30	14.25
1680	10,000 S.F., 41.66 ton	7.80	5.30	13.10
1760	Bars and taverns, 500 S.F., 5.54 ton	16.30	5.80	22.10
1800	1,000 S.F., 11.08 ton	16.05	10.15	26.20
1840	3,000 S.F., 33.25 ton	21	8	29
1880	5,000 S.F., 55.42 ton	19.85	8.35	28.20
1920	10,000 S.F., 110.00 ton	19.30	8.20	27.50
2000	Bowling alleys, 500 S.F., 2.83 ton	10.15	4.70	14.85
2040	1,000 S.F., 5.66 ton	8.75	8.65	17.40
2080	3,000 S.F., 17.00 ton	12.20	7.20	19.40
2120	5,000 S.F., 28.33 ton	11.05	6.95	18
2160	10,000 S.F., 56.66 ton	10.35	7.10	17.45
2200	Department stores, 500 S.F., 1.46 ton	5.25	2.42	7.67
2240	1,000 S.F., 2.92 ton	4.49	4.47	8.96
2280	3,000 S.F., 8.75 ton	4.33	4.15	8.48
2320	5,000 S.F., 14.58 ton	6.25	3.71	9.96
2360	10,000 S.F., 29.17 ton	5.70	3.57	9.27
2440	Drug stores, 500 S.F., 3.33 ton	11.95	5.50	17.45
2480	1,000 S.F., 6.66 ton	10.30	10.20	20.50
2520	3,000 S.F., 20.00 ton	14.30	8.45	22.75
2560	5,000 S.F., 33.33 ton	15.05	8.35	23.40
2600	10,000 S.F., 66.66 ton	12.15	8.35	20.50
2680	Factories, 500 S.F., 1.66 ton	5.95	2.76	8.71
2720	1,000 S.F. 3.37 ton	5.20	5.15	10.35
2760	3,000 S.F., 10.00 ton	4.96	4.73	9.69
2800	5,000 S.F., 16.66 ton	7.15	4.23	11.38
2840	10,000 S.F., 33.33 ton	6.50	4.08	10.58

D3050 Terminal & Package Units

D3050 160	Self-contained, Water Cooled Unit Systems	COST PER S.F.		
		MAT.	INST.	TOTAL
2920	Food supermarkets, 500 S.F., 1.42 ton	5.05	2.35	7.40
2960	1,000 S.F., 2.83 ton	5.25	5.20	10.45
3000	3,000 S.F., 8.50 ton	4.37	4.33	8.70
3040	5,000 S.F., 14.17 ton	4.21	4.02	8.23
3080	10,000 S.F., 28.33 ton	5.50	3.46	8.96
3160	Medical centers, 500 S.F., 1.17 ton	4.18	1.93	6.11
3200	1,000 S.F., 2.33 ton	4.35	4.26	8.61
3240	3,000 S.F., 7.00 ton	3.61	3.56	7.17
3280	5,000 S.F., 11.66 ton	3.48	3.31	6.79
3320	10,000 S.F., 23.33 ton	5	2.96	7.96
3400	Offices, 500 S.F., 1.58 ton	5.65	2.62	8.27
3440	1,000 S.F., 3.17 ton	5.90	5.80	11.70
3480	3,000 S.F., 9.50 ton	4.71	4.49	9.20
3520	5,000 S.F., 15.83 ton	6.80	4.02	10.82
3560	10,000 S.F., 31.67 ton	6.15	3.87	10.02
3640	Restaurants, 500 S.F., 2.50 ton	8.95	4.15	13.10
3680	1,000 S.F., 5.00 ton	7.70	7.65	15.35
3720	3,000 S.F., 15.00 ton	7.45	7.10	14.55
3760	5,000 S.F., 25.00 ton	10.75	6.35	17.10
3800	10,000 S.F., 50.00 ton	6.70	5.85	12.55
3880	Schools and colleges, 500 S.F., 1.92 ton	6.85	3.18	10.03
3920	1,000 S.F., 3.83 ton	5.90	5.85	11.75
3960	3,000 S.F., 11.50 ton	5.70	5.45	11.15
4000	5,000 S.F., 19.17 ton	8.25	4.87	13.12
4040	10,000 S.F., 38.33 ton	7.20	4.87	12.07

D3050 Terminal & Package Units

D30 HVAC

D3050 Terminal & Package Units

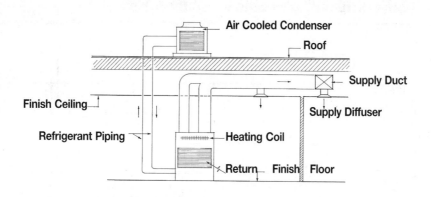

Air Cooled Condenser

Roof

Supply Duct

Finish Ceiling

Supply Diffuser

Refrigerant Piping

Heating Coil

Return **Finish** **Floor**

Self-Contained Air Cooled System

D3050 165	Self-contained, Air Cooled Unit Systems	COST PER S.F.		
		MAT.	INST.	TOTAL
1300	Self-contained, air cooled unit			
1320	Apartment corridors, 500 S.F., .92 ton	5.30	4.30	9.60
1360	1,000 S.F., 1.83 ton	5.20	4.27	9.47
1400	3,000 S.F., 5.50 ton	4.85	3.97	8.82
1440	5,000 S.F., 9.17 ton	4	3.78	7.78
1480	10,000 S.F., 18.33 ton	3.34	3.47	6.81
1560	Banks or libraries, 500 S.F., 2.08 ton	11.50	5.50	17
1600	1,000 S.F., 4.17 ton	10.95	9	19.95
1640	3,000 S.F., 12.50 ton	9.10	8.60	17.70
1680	5,000 S.F., 20.80 ton	7.70	7.90	15.60
1720	10,000 S.F., 41.66 ton	8.50	7.70	16.20
1800	Bars and taverns, 500 S.F., 5.54 ton	26.50	12	38.50
1840	1,000 S.F., 11.08 ton	24	17.30	41.30
1880	3,000 S.F., 33.25 ton	22.50	14.95	37.45
1920	5,000 S.F., 55.42 ton	22.50	15	37.50
1960	10,000 S.F., 110.00 ton	23	14.90	37.90
2040	Bowling alleys, 500 S.F., 2.83 ton	15.80	7.55	23.35
2080	1,000 S.F., 5.66 ton	14.95	12.30	27.25
2120	3,000 S.F., 17.00 ton	10.40	10.75	21.15
2160	5,000 S.F., 28.33 ton	11.75	10.50	22.25
2200	10,000 S.F., 56.66 ton	11.80	10.45	22.25
2240	Department stores, 500 S.F., 1.46 ton	8.15	3.87	12.02
2280	1,000 S.F., 2.92 ton	7.70	6.30	14
2320	3,000 S.F., 8.75 ton	6.35	6	12.35
2360	5,000 S.F., 14.58 ton	6.35	6	12.35
2400	10,000 S.F., 29.17 ton	6.05	5.40	11.45
2480	Drug stores, 500 S.F., 3.33 ton	18.60	8.85	27.45
2520	1,000 S.F., 6.66 ton	17.60	14.45	32.05
2560	3,000 S.F., 20.00 ton	12.40	12.65	25.05
2600	5,000 S.F., 33.33 ton	13.85	12.35	26.20
2640	10,000 S.F., 66.66 ton	14.05	12.35	26.40
2720	Factories, 500 S.F., 1.66 ton	9.45	4.46	13.91
2760	1,000 S.F., 3.33 ton	8.85	7.25	16.10
2800	3,000 S.F., 10.00 ton	7.30	6.85	14.15
2840	5,000 S.F., 16.66 ton	6.05	6.30	12.35
2880	10,000 S.F., 33.33 ton	6.90	6.15	13.05
2960	Food supermarkets, 500 S.F., 1.42 ton	7.90	3.77	11.67
3000	1,000 S.F., 2.83 ton	8.05	6.60	14.65
3040	3,000 S.F., 8.50 ton	7.50	6.15	13.65
3080	5,000 S.F., 14.17 ton	6.20	5.85	12.05

455

D3050 Terminal & Package Units

D3050 165	Self-contained, Air Cooled Unit Systems	COST PER S.F.		
		MAT.	INST.	TOTAL
3120	10,000 S.F., 28.33 ton	5.90	5.25	11.15
3200	Medical centers, 500 S.F., 1.17 ton	6.55	3.11	9.66
3240	1,000 S.F., 2.33 ton	6.70	5.45	12.15
3280	3,000 S.F., 7.00 ton	6.20	5.05	11.25
3320	5,000 S.F., 16.66 ton	5.10	4.81	9.91
3360	10,000 S.F., 23.33 ton	4.27	4.42	8.69
3440	Offices, 500 S.F., 1.58 ton	8.85	4.21	13.06
3480	1,000 S.F., 3.16 ton	9.10	7.40	16.50
3520	3,000 S.F., 9.50 ton	6.95	6.55	13.50
3560	5,000 S.F., 15.83 ton	5.85	6	11.85
3600	10,000 S.F., 31.66 ton	6.60	5.85	12.45
3680	Restaurants, 500 S.F., 2.50 ton	13.90	6.60	20.50
3720	1,000 S.F., 5.00 ton	13.25	10.85	24.10
3760	3,000 S.F., 15.00 ton	10.95	10.30	21.25
3800	5,000 S.F., 25.00 ton	10.35	9.25	19.60
3840	10,000 S.F., 50.00 ton	10.75	9.20	19.95
3920	Schools and colleges, 500 S.F., 1.92 ton	10.70	5.10	15.80
3960	1,000 S.F., 3.83 ton	10.15	8.30	18.45
4000	3,000 S.F., 11.50 ton	8.35	7.90	16.25
4040	5,000 S.F., 19.17 ton	7	7.25	14.25
4080	10,000 S.F., 38.33 ton	7.80	7.10	14.90

D3050 Terminal & Package Units

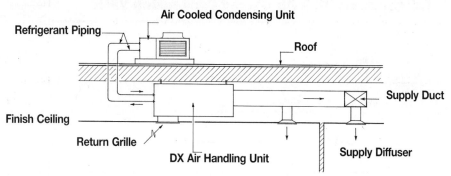

Air Cooled Condensing Unit

Refrigerant Piping

Roof

Finish Ceiling

Return Grille

DX Air Handling Unit

Supply Duct

Supply Diffuser

*Cooling requirements would lead to more than one system.

D3050 170	Split Systems With Air Cooled Condensing Units	COST PER S.F.		
		MAT.	INST.	TOTAL
1260	Split system, air cooled condensing unit			
1280	Apartment corridors, 1,000 S.F., 1.83 ton	2.53	2.70	5.23
1320	2,000 S.F., 3.66 ton	1.79	2.71	4.50
1360	5,000 S.F., 9.17 ton	1.63	3.39	5.02
1400	10,000 S.F., 18.33 ton	2.03	3.70	5.73
1440	20,000 S.F., 36.66 ton	2.15	3.76	5.91
1520	Banks and libraries, 1,000 S.F., 4.17 ton	4.07	6.20	10.27
1560	2,000 S.F., 8.33 ton	3.70	7.70	11.40
1600	5,000 S.F., 20.80 ton	4.63	8.40	13.03
1640	10,000 S.F., 41.66 ton	4.92	8.55	13.47
1680	20,000 S.F., 83.32 ton	5.95	8.90	14.85
1760	Bars and taverns, 1,000 S.F., 11.08 ton	8.45	12.15	20.60
1800	2,000 S.F., 22.16 ton	12.60	14.55	27.15
1840	5,000 S.F., 55.42 ton	10.80	13.60	24.40
1880	10,000 S.F., 110.84 ton	13.10	14.20	27.30
1920	20,000 S.F., 220 ton*			
2000	Bowling alleys, 1,000 S.F., 5.66 ton	5.60	11.50	17.10
2040	2,000 S.F., 11.33 ton	5	10.50	15.50
2080	5,000 S.F., 28.33 ton	6.30	11.45	17.75
2120	10,000 S.F., 56.66 ton	6.70	11.65	18.35
2160	20,000 S.F., 113.32 ton	8.75	12.60	21.35
2320	Department stores, 1,000 S.F., 2.92 ton	2.88	4.25	7.13
2360	2,000 S.F., 5.83 ton	2.89	5.95	8.84
2400	5,000 S.F., 14.58 ton	2.58	5.40	7.98
2440	10,000 S.F., 29.17 ton	3.24	5.90	9.14
2480	20,000 S.F., 58.33 ton	3.44	6	9.44
2560	Drug stores, 1,000 S.F., 6.66 ton	6.60	13.55	20.15
2600	2,000 S.F., 13.32 ton	5.90	12.30	18.20
2640	5,000 S.F., 33.33 ton	7.85	13.70	21.55
2680	10,000 S.F., 66.66 ton	8.10	14.30	22.40
2720	20,000 S.F., 133.32 ton*			
2800	Factories, 1,000 S.F., 3.33 ton	3.28	4.85	8.13
2840	2,000 S.F., 6.66 ton	3.31	6.75	10.06
2880	5,000 S.F., 16.66 ton	3.70	6.70	10.40
2920	10,000 S.F., 33.33 ton	3.93	6.85	10.78
2960	20,000 S.F., 66.66 ton	4.05	7.15	11.20
3040	Food supermarkets, 1,000 S.F., 2.83 ton	2.80	4.13	6.93
3080	2,000 S.F., 5.66 ton	2.82	5.75	8.57
3120	5,000 S.F., 14.66 ton	2.51	5.25	7.76
3160	10,000 S.F., 28.33 ton	3.15	5.70	8.85
3200	20,000 S.F., 56.66 ton	3.34	5.80	9.14
3280	Medical centers, 1,000 S.F., 2.33 ton	2.51	3.32	5.83

457

D30 HVAC

D3050 Terminal & Package Units

D3050 170	Split Systems With Air Cooled Condensing Units	COST PER S.F.		
		MAT.	INST.	TOTAL
3320	2,000 S.F., 4.66 ton	2.31	4.75	7.06
3360	5,000 S.F., 11.66 ton	2.07	4.31	6.38
3400	10,000 S.F., 23.33 ton	2.60	4.71	7.31
3440	20,000 S.F., 46.66 ton	2.76	4.79	7.55
3520	Offices, 1,000 S.F., 3.17 ton	3.12	4.61	7.73
3560	2,000 S.F., 6.33 ton	3.14	6.45	9.59
3600	5,000 S.F., 15.83 ton	2.81	5.85	8.66
3640	10,000 S.F., 31.66 ton	3.53	6.40	9.93
3680	20,000 S.F., 63.32 ton	3.84	6.80	10.64
3760	Restaurants, 1,000 S.F., 5.00 ton	4.97	10.15	15.12
3800	2,000 S.F., 10.00 ton	4.44	9.25	13.69
3840	5,000 S.F., 25.00 ton	5.55	10.10	15.65
3880	10,000 S.F., 50.00 ton	5.90	10.30	16.20
3920	20,000 S.F., 100.00 ton	7.75	11.15	18.90
4000	Schools and colleges, 1,000 S.F., 3.83 ton	3.74	5.70	9.44
4040	2,000 S.F., 7.66 ton	3.39	7.10	10.49
4080	5,000 S.F., 19.17 ton	4.26	7.75	12.01
4120	10,000 S.F., 38.33 ton	4.52	7.85	12.37

D3090 Other HVAC Systems/Equip

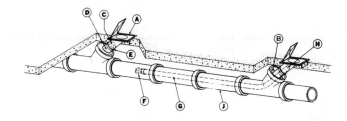

Cast Iron Garage Exhaust System

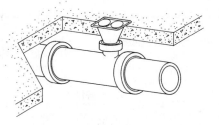

Dual Exhaust System

D3090 320	Garage Exhaust Systems	COST PER BAY		
		MAT.	INST.	TOTAL
1040	Garage, single 3" exhaust outlet, cars & light trucks, one bay	4,400	1,975	6,375
1060	Additional bays up to seven bays	1,000	530	1,530
1500	4" outlet, trucks, one bay	4,425	1,975	6,400
1520	Additional bays up to six bays	1,025	530	1,555
1600	5" outlet, diesel trucks, one bay	4,725	1,975	6,700
1650	Additional single bays up to six bays	1,425	625	2,050
1700	Two adjoining bays	4,725	1,975	6,700
2000	Dual exhaust, 3" outlets, pair of adjoining bays	5,475	2,475	7,950
2100	Additional pairs of adjoining bays	1,600	625	2,225

Reference - System Classification

System Classification
Rules for installation of sprinkler systems vary depending on the classification of occupancy falling into one of three categories as follows:

Light Hazard Occupancy
The protection area allotted per sprinkler should not exceed 225 S.F., with the maximum distance between lines and sprinklers on lines being 15'. The sprinklers do not need to be staggered. Branch lines should not exceed eight sprinklers on either side of a cross main. Each large area requiring more than 100 sprinklers and without a sub-dividing partition should be supplied by feed mains or risers sized for ordinary hazard occupancy.
Maximum system area = 52,000 S.F.

Included in this group are:
Churches	Nursing Homes
Clubs	Offices
Educational	Residential
Hospitals	Restaurants
Institutional	Theaters and Auditoriums
Libraries	(except stages and prosceniums)
(except large stack rooms)	Unused Attics
Museums	

Ordinary Hazard Occupancy
The protection area allotted per sprinkler shall not exceed 130 S.F. of noncombustible ceiling and 130 S.F. of combustible ceiling. The maximum allowable distance between sprinkler lines and sprinklers on line is 15'. Sprinklers shall be staggered if the distance between heads exceeds 12'. Branch lines should not exceed eight sprinklers on either side of a cross main.
Maximum system area = 52,000 S.F.

Included in this group are:
Group 1	Group 2
Automotive Parking and Showrooms	Cereal Mills
Bakeries	Chemical Plants—Ordinary
Beverage manufacturing	Confectionery Products
Canneries	Distilleries
Dairy Products Manufacturing/Processing	Dry Cleaners
Electronic Plans	Feed Mills
Glass and Glass Products Manufacturing	Horse Stables
Laundries	Leather Goods Manufacturing
Restaurant Service Areas	Libraries—Large Stack Room Areas
	Machine Shops
	Metal Working
	Mercantile
	Paper and Pulp Mills
	Paper Process Plants
	Piers and Wharves
	Post Offices
	Printing and Publishing
	Repair Garages
	Stages
	Textile Manufacturing
	Tire Manufacturing
	Tobacco Products Manufacturing
	Wood Machining
	Wood Product Assembly

Extra Hazard Occupancy
The protection area allotted per sprinkler shall not exceed 100 S.F. of noncombustible ceiling and 100 S.F. of combustible ceiling. The maximum allowable distance between lines and between sprinklers on lines is 12'. Sprinklers on alternate lines shall be staggered if the distance between sprinklers on lines exceeds 8'. Branch lines should not exceed six sprinklers on either side of a cross main.
Maximum system area:
   Design by pipe schedule = 25,000 S.F.
   Design by hydraulic calculation = 40,000 S.F.

Included in this group are:
Group 1	Group 2
Aircraft hangars	Asphalt Saturating
Combustible Hydraulic Fluid Use Area	Flammable Liquids Spraying
Die Casting	Flow Coating
Metal Extruding	Manufactured/Modular Home
Plywood/Particle Board Manufacturing	Building Assemblies (where
Printing (inks with flash points < 100	finished enclosure is present
degrees F	and has combustible interiors)
Rubber Reclaiming, Compounding,	Open Oil Quenching
Drying, Milling, Vulcanizing	Plastics Processing
Saw Mills	Solvent Cleaning
Textile Picking, Opening, Blending,	Varnish and Paint Dipping
Garnetting, Carding, Combing of	
Cotton, Synthetics, Wood Shoddy,	
or Burlap	
Upholstering with Plastic Foams	

460

For customer support on your Square Foot Cost Data, call 877.756.2789.

Reference - Sprinkler Systems (Automatic)

Sprinkler systems may be classified by type as follows:

1. **Wet Pipe System.** A system employing automatic sprinklers attached to a piping system containing water and connected to a water supply so that water discharges immediately from sprinklers opened by a fire.

2. **Dry Pipe System.** A system employing automatic sprinklers attached to a piping system containing air under pressure, the release of which as from the opening of sprinklers permits the water pressure to open a valve known as a "dry pipe valve". The water then flows into the piping system and out the opened sprinklers.

3. **Pre-Action System.** A system employing automatic sprinklers attached to a piping system containing air that may or may not be under pressure, with a supplemental heat responsive system of generally more sensitive characteristics than the automatic sprinklers themselves, installed in the same areas as the sprinklers; actuation of the heat responsive system, as from a fire, opens a valve which permits water to flow into the sprinkler piping system and to be discharged from any sprinklers which may be open.

4. **Deluge System.** A system employing open sprinklers attached to a piping system connected to a water supply through a valve which is opened by the operation of a heat responsive system installed in the same areas as the sprinklers. When this valve opens, water flows into the piping system and discharges from all sprinklers attached thereto.

5. **Combined Dry Pipe and Pre-Action Sprinkler System.** A system employing automatic sprinklers attached to a piping system containing air under pressure with a supplemental heat responsive system of generally more sensitive characteristics than the automatic sprinklers themselves, installed in the same areas as the sprinklers; operation of the heat responsive system, as from a fire, actuates tripping devices which open dry pipe valves simultaneously and without loss of air pressure in the system. Operation of the heat responsive system also opens approved air exhaust valves at the end of the feed main which facilitates the filling of the system with water which usually precedes the opening of sprinklers. The heat responsive system also serves as an automatic fire alarm system.

6. **Limited Water Supply System.** A system employing automatic sprinklers and conforming to these standards but supplied by a pressure tank of limited capacity.

7. **Chemical Systems.** Systems using halon, carbon dioxide, dry chemical or high expansion foam as selected for special requirements. Agent may extinguish flames by chemically inhibiting flame propagation, suffocate flames by excluding oxygen, interrupting chemical action of oxygen uniting with fuel or sealing and cooling the combustion center.

8. **Firecycle System.** Firecycle is a fixed fire protection sprinkler system utilizing water as its extinguishing agent. It is a time delayed, recycling, preaction type which automatically shuts the water off when heat is reduced below the detector operating temperature and turns the water back on when that temperature is exceeded. The system senses a fire condition through a closed circuit electrical detector system which controls water flow to the fire automatically. Batteries supply up to 90 hour emergency power supply for system operation. The piping system is dry (until water is required) and is monitored with pressurized air. Should any leak in the system piping occur, an alarm will sound, but water will not enter the system until heat is sensed by a firecycle detector.

Area coverage sprinkler systems may be laid out and fed from the supply in any one of several patterns as shown below. It is desirable, if possible, to utilize a central feed and achieve a shorter flow path from the riser to the furthest sprinkler. This permits use of the smallest sizes of pipe possible with resulting savings.

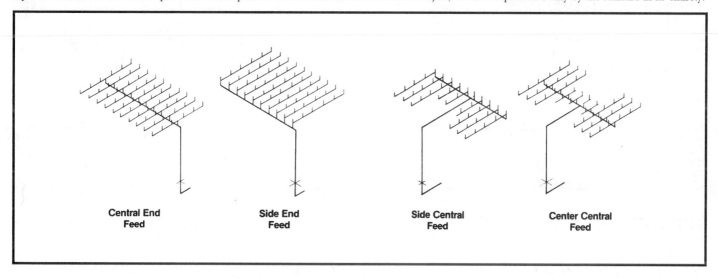

| Central End Feed | Side End Feed | Side Central Feed | Center Central Feed |

461

D4010 Sprinklers

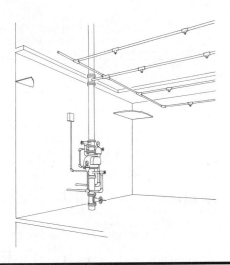

Dry Pipe System: A system employing automatic sprinklers attached to a piping system containing air under pressure, the release of which from the opening of sprinklers permits the water pressure to open a valve known as a "dry pipe valve". The water then flows into the piping system and out the opened sprinklers.

All areas are assumed to be open.

D4010 310	Dry Pipe Sprinkler Systems	COST PER S.F.		
		MAT.	INST.	TOTAL
0520	Dry pipe sprinkler systems, steel, black, sch. 40 pipe			
0530	Light hazard, one floor, 500 S.F.	10.05	6.30	16.35
0560	1000 S.F.	5.70	3.72	9.42
0580	2000 S.F.	5.50	3.74	9.24
0600	5000 S.F.	2.87	2.54	5.41
0620	10,000 S.F.	2.01	2.09	4.10
0640	50,000 S.F.	1.47	1.88	3.35
0660	Each additional floor, 500 S.F.	2.04	3.09	5.13
0680	1000 S.F.	1.85	2.57	4.42
0700	2000 S.F.	1.81	2.33	4.14
0720	5000 S.F.	1.48	2.03	3.51
0740	10,000 S.F.	1.33	1.86	3.19
0760	50,000 S.F.	1.22	1.68	2.90
1000	Ordinary hazard, one floor, 500 S.F.	10.25	6.40	16.65
1020	1000 S.F.	5.90	3.73	9.63
1040	2000 S.F.	5.70	3.89	9.59
1060	5000 S.F.	3.33	2.72	6.05
1080	10,000 S.F.	2.65	2.75	5.40
1100	50,000 S.F.	2.37	2.60	4.97
1140	Each additional floor, 500 S.F.	2.26	3.17	5.43
1160	1000 S.F.	2.14	2.85	4.99
1180	2000 S.F.	2.18	2.59	4.77
1200	5000 S.F.	2.03	2.25	4.28
1220	10,000 S.F.	1.84	2.17	4.01
1240	50,000 S.F.	1.76	1.89	3.65
1500	Extra hazard, one floor, 500 S.F.	14.40	7.95	22.35
1520	1000 S.F.	8.80	5.75	14.55
1540	2000 S.F.	6.30	5	11.30
1560	5000 S.F.	3.83	3.79	7.62
1580	10,000 S.F.	4.14	3.59	7.73
1600	50,000 S.F.	4.47	3.49	7.96
1660	Each additional floor, 500 S.F.	3.32	3.92	7.24
1680	1000 S.F.	3.28	3.72	7
1700	2000 S.F.	3	3.74	6.74
1720	5000 S.F.	2.55	3.26	5.81
1740	10,000 S.F.	3.14	2.96	6.10
1760	50,000 S.F.	3.16	2.86	6.02
2020	Grooved steel, black, sch. 40 pipe, light hazard, one floor, 2000 S.F.	5.35	3.20	8.55

D40 Fire Protection

D4010 Sprinklers

D4010 310	Dry Pipe Sprinkler Systems	COST PER S.F.		
		MAT.	INST.	TOTAL
2060	10,000 S.F.	2.05	1.79	3.84
2100	Each additional floor, 2000 S.F.	1.84	1.90	3.74
2150	10,000 S.F.	1.37	1.56	2.93
2200	Ordinary hazard, one floor, 2000 S.F.	5.65	3.39	9.04
2250	10,000 S.F.	2.55	2.32	4.87
2300	Each additional floor, 2000 S.F.	2.12	2.09	4.21
2350	10,000 S.F.	1.88	2.09	3.97
2400	Extra hazard, one floor, 2000 S.F.	6.35	4.29	10.64
2450	10,000 S.F.	3.64	2.98	6.62
2500	Each additional floor, 2000 S.F.	3.06	3.08	6.14
2550	10,000 S.F.	2.78	2.60	5.38
3050	Grooved steel, black, sch. 10 pipe, light hazard, one floor, 2000 S.F.	5.30	3.18	8.48
3100	10,000 S.F.	2.03	1.77	3.80
3150	Each additional floor, 2000 S.F.	1.79	1.88	3.67
3200	10,000 S.F.	1.35	1.54	2.89
3250	Ordinary hazard, one floor, 2000 S.F.	5.60	3.36	8.96
3300	10,000 S.F.	2.49	2.27	4.76
3350	Each additional floor, 2000 S.F.	2.08	2.06	4.14
3400	10,000 S.F.	1.82	2.04	3.86
3450	Extra hazard, one floor, 2000 S.F.	6.35	4.26	10.61
3500	10,000 S.F.	3.46	2.94	6.40
3550	Each additional floor, 2000 S.F.	3.03	3.05	6.08
3600	10,000 S.F.	2.72	2.57	5.29
4050	Copper tubing, type M, light hazard, one floor, 2000 S.F.	5.65	3.16	8.81
4100	10,000 S.F.	2.51	1.80	4.31
4150	Each additional floor, 2000 S.F.	2.18	1.90	4.08
4200	10,000 S.F.	1.84	1.58	3.42
4250	Ordinary hazard, one floor, 2000 S.F.	6.10	3.53	9.63
4300	10,000 S.F.	3.08	2.14	5.22
4350	Each additional floor, 2000 S.F.	2.84	2.19	5.03
4400	10,000 S.F.	2.37	1.87	4.24
4450	Extra hazard, one floor, 2000 S.F.	7	4.31	11.31
4500	10,000 S.F.	5.15	3.25	8.40
4550	Each additional floor, 2000 S.F.	3.69	3.10	6.79
4600	10,000 S.F.	3.83	2.84	6.67
5050	Copper tubing, type M, T-drill system, light hazard, one floor			
5060	2000 S.F.	5.70	2.95	8.65
5100	10,000 S.F.	2.40	1.50	3.90
5150	Each additional floor, 2000 S.F.	2.20	1.69	3.89
5200	10,000 S.F.	1.73	1.28	3.01
5250	Ordinary hazard, one floor, 2000 S.F.	5.90	3.02	8.92
5300	10,000 S.F.	2.99	1.91	4.90
5350	Each additional floor, 2000 S.F.	2.40	1.72	4.12
5400	10,000 S.F.	2.22	1.59	3.81
5450	Extra hazard, one floor, 2000 S.F.	6.50	3.58	10.08
5500	10,000 S.F.	4.46	2.39	6.85
5550	Each additional floor, 2000 S.F.	3.22	2.37	5.59
5600	10,000 S.F.	3.12	1.98	5.10

For customer support on your Square Foot Cost Data, call 877.756.2789.

463

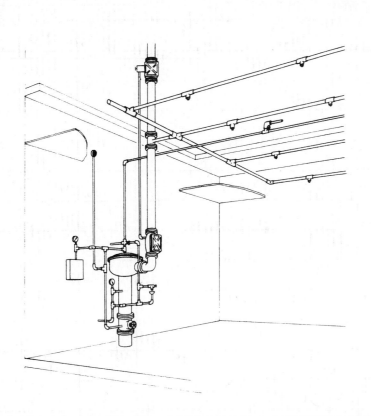

Preaction System: A system employing automatic sprinklers attached to a piping system containing air that may or may not be under pressure, with a supplemental heat responsive system of generally more sensitive characteristics than the automatic sprinklers themselves, installed in the same areas as the sprinklers. Actuation of the heat responsive system, as from a fire, opens a valve which permits water to flow into the sprinkler piping system and to be discharged from those sprinklers which were opened by heat from the fire.

All areas are assumed to be opened.

D4010 350	Preaction Sprinkler Systems	COST PER S.F.		
		MAT.	INST.	TOTAL
0520	Preaction sprinkler systems, steel, black, sch. 40 pipe			
0530	Light hazard, one floor, 500 S.F.	9.90	5.05	14.95
0560	1000 S.F.	5.75	3.80	9.55
0580	2000 S.F.	5.35	3.73	9.08
0600	5000 S.F.	2.73	2.53	5.26
0620	10,000 S.F.	1.88	2.08	3.96
0640	50,000 S.F.	1.34	1.87	3.21
0660	Each additional floor, 500 S.F.	2.40	2.76	5.16
0680	1000 S.F.	1.87	2.57	4.44
0700	2000 S.F.	1.82	2.34	4.16
0720	5000 S.F.	1.34	2.02	3.36
0740	10,000 S.F.	1.20	1.85	3.05
0760	50,000 S.F.	1.17	1.72	2.89
1000	Ordinary hazard, one floor, 500 S.F.	10.20	5.45	15.65
1020	1000 S.F.	5.70	3.72	9.42
1040	2000 S.F.	5.80	3.89	9.69
1060	5000 S.F.	3.04	2.70	5.74
1080	10,000 S.F.	2.30	2.73	5.03
1100	50,000 S.F.	2.01	2.57	4.58
1140	Each additional floor, 500 S.F.	2.67	3.18	5.85
1160	1000 S.F.	1.86	2.57	4.43
1180	2000 S.F.	1.71	2.56	4.27
1200	5000 S.F.	1.85	2.42	4.27
1220	10,000 S.F.	1.63	2.50	4.13
1240	50,000 S.F.	1.53	2.23	3.76
1500	Extra hazard, one floor, 500 S.F.	13.95	7	20.95
1520	1000 S.F.	8	5.25	13.25
1540	2000 S.F.	5.75	4.99	10.74

D4010 Sprinklers

D4010 350	Preaction Sprinkler Systems	COST PER S.F.		
		MAT.	INST.	TOTAL
1560	5000 S.F.	3.48	4.09	7.57
1580	10,000 S.F.	3.56	3.97	7.53
1600	50,000 S.F.	3.81	3.88	7.69
1660	Each additional floor, 500 S.F.	3.35	3.91	7.26
1680	1000 S.F.	2.72	3.69	6.41
1700	2000 S.F.	2.44	3.71	6.15
1720	5000 S.F.	2	3.28	5.28
1740	10,000 S.F.	2.42	2.99	5.41
1760	50,000 S.F.	2.33	2.82	5.15
2020	Grooved steel, black, sch. 40 pipe, light hazard, one floor, 2000 S.F.	5.35	3.21	8.56
2060	10,000 S.F.	1.92	1.78	3.70
2100	Each additional floor of 2000 S.F.	1.85	1.91	3.76
2150	10,000 S.F.	1.24	1.55	2.79
2200	Ordinary hazard, one floor, 2000 S.F.	5.45	3.38	8.83
2250	10,000 S.F.	2.20	2.30	4.50
2300	Each additional floor, 2000 S.F.	1.94	2.08	4.02
2350	10,000 S.F.	1.53	2.07	3.60
2400	Extra hazard, one floor, 2000 S.F.	5.80	4.26	10.06
2450	10,000 S.F.	2.93	2.95	5.88
2500	Each additional floor, 2000 S.F.	2.50	3.05	5.55
2550	10,000 S.F.	2.03	2.56	4.59
3050	Grooved steel, black, sch. 10 pipe light hazard, one floor, 2000 S.F.	5.30	3.19	8.49
3100	10,000 S.F.	1.90	1.76	3.66
3150	Each additional floor, 2000 S.F.	1.80	1.89	3.69
3200	10,000 S.F.	1.22	1.53	2.75
3250	Ordinary hazard, one floor, 2000 S.F.	5.30	3.14	8.44
3300	10,000 S.F.	1.85	2.24	4.09
3350	Each additional floor, 2000 S.F.	1.90	2.05	3.95
3400	10,000 S.F.	1.47	2.02	3.49
3450	Extra hazard, one floor, 2000 S.F.	5.75	4.23	9.98
3500	10,000 S.F.	2.71	2.90	5.61
3550	Each additional floor, 2000 S.F.	2.47	3.02	5.49
3600	10,000 S.F.	1.97	2.53	4.50
4050	Copper tubing, type M, light hazard, one floor, 2000 S.F.	5.70	3.17	8.87
4100	10,000 S.F.	2.38	1.79	4.17
4150	Each additional floor, 2000 S.F.	2.20	1.91	4.11
4200	10,000 S.F.	1.42	1.56	2.98
4250	Ordinary hazard, one floor, 2000 S.F.	5.90	3.52	9.42
4300	10,000 S.F.	2.73	2.12	4.85
4350	Each additional floor, 2000 S.F.	2.19	1.94	4.13
4400	10,000 S.F.	1.84	1.70	3.54
4450	Extra hazard, one floor, 2000 S.F.	6.45	4.28	10.73
4500	10,000 S.F.	4.38	3.20	7.58
4550	Each additional floor, 2000 S.F.	3.13	3.07	6.20
4600	10,000 S.F.	3.08	2.80	5.88
5050	Copper tubing, type M, T-drill system, light hazard, one floor			
5060	2000 S.F.	5.70	2.96	8.66
5100	10,000 S.F.	2.27	1.49	3.76
5150	Each additional floor, 2000 S.F.	2.21	1.70	3.91
5200	10,000 S.F.	1.60	1.27	2.87
5250	Ordinary hazard, one floor, 2000 S.F.	5.75	3.01	8.76
5300	10,000 S.F.	2.64	1.89	4.53
5350	Each additional floor, 2000 S.F.	2.22	1.72	3.94
5400	10,000 S.F.	1.97	1.66	3.63
5450	Extra hazard, one floor, 2000 S.F.	5.95	3.55	9.50
5500	10,000 S.F.	3.67	2.34	6.01
5550	Each additional floor, 2000 S.F.	2.66	2.34	5
5600	10,000 S.F.	2.37	1.94	4.31

D4010 Sprinklers

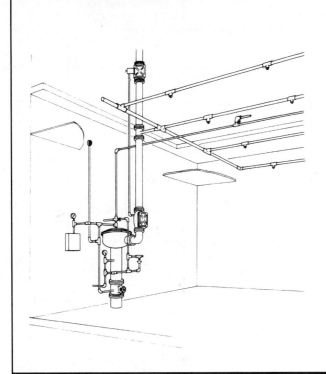

Deluge System: A system employing open sprinklers attached to a piping system connected to a water supply through a valve which is opened by the operation of a heat responsive system installed in the same areas as the sprinklers. When this valve opens, water flows into the piping system and discharges from all sprinklers attached thereto.

D4010 370	Deluge Sprinkler Systems	COST PER S.F.		
		MAT.	INST.	TOTAL
0520	Deluge sprinkler systems, steel, black, sch. 40 pipe			
0530	Light hazard, one floor, 500 S.F.	21.50	5.10	26.60
0560	1000 S.F.	11.55	3.66	15.21
0580	2000 S.F.	8.60	3.74	12.34
0600	5000 S.F.	4.03	2.54	6.57
0620	10,000 S.F.	2.53	2.08	4.61
0640	50,000 S.F.	1.47	1.87	3.34
0660	Each additional floor, 500 S.F.	2.40	2.76	5.16
0680	1000 S.F.	1.87	2.57	4.44
0700	2000 S.F.	1.82	2.34	4.16
0720	5000 S.F.	1.34	2.02	3.36
0740	10,000 S.F.	1.20	1.85	3.05
0760	50,000 S.F.	1.17	1.72	2.89
1000	Ordinary hazard, one floor, 500 S.F.	22.50	5.85	28.35
1020	1000 S.F.	11.55	3.75	15.30
1040	2000 S.F.	9.05	3.90	12.95
1060	5000 S.F.	4.34	2.71	7.05
1080	10,000 S.F.	2.95	2.73	5.68
1100	50,000 S.F.	2.20	2.61	4.81
1140	Each additional floor, 500 S.F.	2.67	3.18	5.85
1160	1000 S.F.	1.86	2.57	4.43
1180	2000 S.F.	1.71	2.56	4.27
1200	5000 S.F.	1.74	2.23	3.97
1220	10,000 S.F.	1.62	2.20	3.82
1240	50,000 S.F.	1.46	2.05	3.51
1500	Extra hazard, one floor, 500 S.F.	25.50	7.05	32.55
1520	1000 S.F.	14.30	5.45	19.75
1540	2000 S.F.	9	5	14
1560	5000 S.F.	4.55	3.77	8.32
1580	10,000 S.F.	4.09	3.63	7.72
1600	50,000 S.F.	4.22	3.56	7.78
1660	Each additional floor, 500 S.F.	3.35	3.91	7.26

D4010 Sprinklers

D4010 370	Deluge Sprinkler Systems	COST PER S.F.		
		MAT.	INST.	TOTAL
1680	1000 S.F.	2.72	3.69	6.41
1700	2000 S.F.	2.44	3.71	6.15
1720	5000 S.F.	2	3.28	5.28
1740	10,000 S.F.	2.46	3.10	5.56
1760	50,000 S.F.	2.46	3	5.46
2000	Grooved steel, black, sch. 40 pipe, light hazard, one floor			
2020	2000 S.F.	8.60	3.22	11.82
2060	10,000 S.F.	2.59	1.80	4.39
2100	Each additional floor, 2,000 S.F.	1.85	1.91	3.76
2150	10,000 S.F.	1.24	1.55	2.79
2200	Ordinary hazard, one floor, 2000 S.F.	5.45	3.38	8.83
2250	10,000 S.F.	2.85	2.30	5.15
2300	Each additional floor, 2000 S.F.	1.94	2.08	4.02
2350	10,000 S.F.	1.53	2.07	3.60
2400	Extra hazard, one floor, 2000 S.F.	9.05	4.27	13.32
2450	10,000 S.F.	3.61	2.96	6.57
2500	Each additional floor, 2000 S.F.	2.50	3.05	5.55
2550	10,000 S.F.	2.03	2.56	4.59
3000	Grooved steel, black, sch. 10 pipe, light hazard, one floor			
3050	2000 S.F.	7.90	3.06	10.96
3100	10,000 S.F.	2.55	1.76	4.31
3150	Each additional floor, 2000 S.F.	1.80	1.89	3.69
3200	10,000 S.F.	1.22	1.53	2.75
3250	Ordinary hazard, one floor, 2000 S.F.	8.65	3.36	12.01
3300	10,000 S.F.	2.50	2.24	4.74
3350	Each additional floor, 2000 S.F.	1.90	2.05	3.95
3400	10,000 S.F.	1.47	2.02	3.49
3450	Extra hazard, one floor, 2000 S.F.	9	4.24	13.24
3500	10,000 S.F.	3.36	2.90	6.26
3550	Each additional floor, 2000 S.F.	2.47	3.02	5.49
3600	10,000 S.F.	1.97	2.53	4.50
4000	Copper tubing, type M, light hazard, one floor			
4050	2000 S.F.	8.90	3.18	12.08
4100	10,000 S.F.	3.03	1.79	4.82
4150	Each additional floor, 2000 S.F.	2.19	1.91	4.10
4200	10,000 S.F.	1.42	1.56	2.98
4250	Ordinary hazard, one floor, 2000 S.F.	9.15	3.53	12.68
4300	10,000 S.F.	3.38	2.12	5.50
4350	Each additional floor, 2000 S.F.	2.19	1.94	4.13
4400	10,000 S.F.	1.84	1.70	3.54
4450	Extra hazard, one floor, 2000 S.F.	9.70	4.29	13.99
4500	10,000 S.F.	5.10	3.22	8.32
4550	Each additional floor, 2000 S.F.	3.13	3.07	6.20
4600	10,000 S.F.	3.08	2.80	5.88
5000	Copper tubing, type M, T-drill system, light hazard, one floor			
5050	2000 S.F.	8.95	2.97	11.92
5100	10,000 S.F.	2.92	1.49	4.41
5150	Each additional floor, 2000 S.F.	2.25	1.72	3.97
5200	10,000 S.F.	1.60	1.27	2.87
5250	Ordinary hazard, one floor, 2000 S.F.	9	3.02	12.02
5300	10,000 S.F.	3.29	1.89	5.18
5350	Each additional floor, 2000 S.F.	2.22	1.71	3.93
5400	10,000 S.F.	1.97	1.66	3.63
5450	Extra hazard, one floor, 2000 S.F.	9.20	3.56	12.76
5500	10,000 S.F.	4.32	2.34	6.66
5550	Each additional floor, 2000 S.F.	2.66	2.34	5
5600	10,000 S.F.	2.37	1.94	4.31

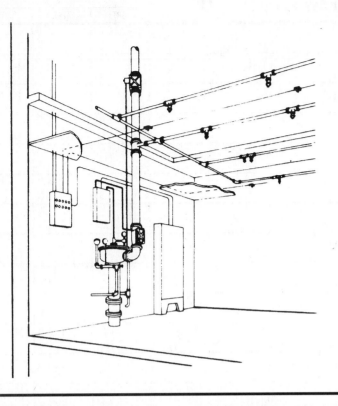

Firecycle is a fixed fire protection sprinkler system utilizing water as its extinguishing agent. It is a time delayed, recycling, preaction type which automatically shuts the water off when heat is reduced below the detector operating temperature and turns the water back on when that temperature is exceeded.

The system senses a fire condition through a closed circuit electrical detector system which controls water flow to the fire automatically. Batteries supply up to 90 hours emergency power supply for system operation. The piping system is dry (until water is required) and is monitored with pressurized air. Shouldany leak in the system piping occur, an alarm will sound, but water will not enter the system until heat is sensed by a Firecycle detector.

D4010 390	Firecycle Sprinkler Systems	COST PER S.F.		
		MAT.	INST.	TOTAL
0520	Firecycle sprinkler systems, steel black sch. 40 pipe			
0530	Light hazard, one floor, 500 S.F.	51	9.90	60.90
0560	1000 S.F.	26.50	6.20	32.70
0580	2000 S.F.	15.45	4.56	20.01
0600	5000 S.F.	6.80	2.88	9.68
0620	10,000 S.F.	3.97	2.26	6.23
0640	50,000 S.F.	1.79	1.90	3.69
0660	Each additional floor of 500 S.F.	2.39	2.78	5.17
0680	1000 S.F.	1.86	2.57	4.43
0700	2000 S.F.	1.53	2.33	3.86
0720	5000 S.F.	1.33	2.03	3.36
0740	10,000 S.F.	1.26	1.86	3.12
0760	50,000 S.F.	1.20	1.71	2.91
1000	Ordinary hazard, one floor, 500 S.F.	51.50	10.30	61.80
1020	1000 S.F.	26	6.15	32.15
1040	2000 S.F.	15.65	4.71	20.36
1060	5000 S.F.	7.10	3.05	10.15
1080	10,000 S.F.	4.39	2.91	7.30
1100	50,000 S.F.	2.73	2.91	5.64
1140	Each additional floor, 500 S.F.	2.66	3.20	5.86
1160	1000 S.F.	1.85	2.57	4.42
1180	2000 S.F.	1.92	2.36	4.28
1200	5000 S.F.	1.73	2.24	3.97
1220	10,000 S.F.	1.55	2.16	3.71
1240	50,000 S.F.	1.47	1.94	3.41
1500	Extra hazard, one floor, 500 S.F.	55	11.85	66.85
1520	1000 S.F.	28	7.65	35.65
1540	2000 S.F.	15.90	5.80	21.70
1560	5000 S.F.	7.30	4.11	11.41
1580	10,000 S.F.	5.60	4.12	9.72

468

For customer support on your Square Foot Cost Data, call 877.756.2789.

D4010 Sprinklers

D4010 390	Firecycle Sprinkler Systems	COST PER S.F.		
		MAT.	INST.	TOTAL
1600	50,000 S.F.	4.68	4.57	9.25
1660	Each additional floor, 500 S.F.	3.34	3.93	7.27
1680	1000 S.F.	2.71	3.69	6.40
1700	2000 S.F.	2.44	3.71	6.15
1720	5000 S.F.	1.99	3.29	5.28
1740	10,000 S.F.	2.48	3	5.48
1760	50,000 S.F.	2.47	2.89	5.36
2020	Grooved steel, black, sch. 40 pipe, light hazard, one floor			
2030	2000 S.F.	15.50	4.04	19.54
2060	10,000 S.F.	4.34	2.72	7.06
2100	Each additional floor, 2000 S.F.	1.85	1.91	3.76
2150	10,000 S.F.	1.30	1.56	2.86
2200	Ordinary hazard, one floor, 2000 S.F.	15.60	4.21	19.81
2250	10,000 S.F.	4.60	2.61	7.21
2300	Each additional floor, 2000 S.F.	1.94	2.08	4.02
2350	10,000 S.F.	1.59	2.08	3.67
2400	Extra hazard, one floor, 2000 S.F.	15.95	5.10	21.05
2450	10,000 S.F.	4.95	3.10	8.05
2500	Each additional floor, 2000 S.F.	2.50	3.05	5.55
2550	10,000 S.F.	2.09	2.57	4.66
3050	Grooved steel, black, sch. 10 pipe light hazard, one floor,			
3060	2000 S.F.	15.45	4.02	19.47
3100	10,000 S.F.	3.99	1.94	5.93
3150	Each additional floor, 2000 S.F.	1.80	1.89	3.69
3200	10,000 S.F.	1.28	1.54	2.82
3250	Ordinary hazard, one floor, 2000 S.F.	15.55	4.18	19.73
3300	10,000 S.F.	4.23	2.43	6.66
3350	Each additional floor, 2000 S.F.	1.90	2.05	3.95
3400	10,000 S.F.	1.53	2.03	3.56
3450	Extra hazard, one floor, 2000 S.F.	15.90	5.05	20.95
3500	10,000 S.F.	4.77	3.06	7.83
3550	Each additional floor, 2000 S.F.	2.47	3.02	5.49
3600	10,000 S.F.	2.03	2.54	4.57
4060	Copper tubing, type M, light hazard, one floor, 2000 S.F.	15.80	4	19.80
4100	10,000 S.F.	4.47	1.97	6.44
4150	Each additional floor, 2000 S.F.	2.19	1.91	4.10
4200	10,000 S.F.	1.77	1.58	3.35
4250	Ordinary hazard, one floor, 2000 S.F.	16.05	4.35	20.40
4300	10,000 S.F.	4.82	2.30	7.12
4350	Each additional floor, 2000 S.F.	2.19	1.94	4.13
4400	10,000 S.F.	1.87	1.69	3.56
4450	Extra hazard, one floor, 2000 S.F.	16.55	5.10	21.65
4500	10,000 S.F.	6.50	3.41	9.91
4550	Each additional floor, 2000 S.F.	3.13	3.07	6.20
4600	10,000 S.F.	3.14	2.81	5.95
5060	Copper tubing, type M, T-drill system, light hazard, one floor 2000 S.F.	15.80	3.79	19.59
5100	10,000 S.F.	4.36	1.67	6.03
5150	Each additional floor, 2000 S.F.	2.42	1.79	4.21
5200	10,000 S.F.	1.66	1.28	2.94
5250	Ordinary hazard, one floor, 2000 S.F.	15.85	3.84	19.69
5300	10,000 S.F.	4.73	2.07	6.80
5350	Each additional floor, 2000 S.F.	2.22	1.71	3.93
5400	10,000 S.F.	2.03	1.67	3.70
5450	Extra hazard, one floor, 2000 S.F.	16.10	4.38	20.48
5500	10,000 S.F.	5.75	2.50	8.25
5550	Each additional floor, 2000 S.F.	2.66	2.34	5
5600	10,000 S.F.	2.43	1.95	4.38

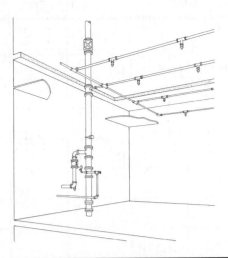

Wet Pipe System. A system employing automatic sprinklers attached to a piping system containing water and connected to a water supply so that water discharges immediately from sprinklers opened by heat from a fire.

All areas are assumed to be open.

D4010 410	Wet Pipe Sprinkler Systems	COST PER S.F.		
		MAT.	INST.	TOTAL
0520	Wet pipe sprinkler systems, steel, black, sch. 40 pipe			
0530	Light hazard, one floor, 500 S.F.	2.96	3.26	6.22
0560	1000 S.F.	6.35	3.37	9.72
0580	2000 S.F.	5.70	3.39	9.09
0600	5000 S.F.	2.64	2.38	5.02
0620	10,000 S.F.	1.65	2	3.65
0640	50,000 S.F.	1.01	1.83	2.84
0660	Each additional floor, 500 S.F.	1.26	2.77	4.03
0680	1000 S.F.	1.31	2.60	3.91
0700	2000 S.F.	1.24	2.31	3.55
0720	5000 S.F.	.87	2	2.87
0740	10,000 S.F.	.79	1.83	2.62
0760	50,000 S.F.	.66	1.45	2.11
1000	Ordinary hazard, one floor, 500 S.F.	3.26	3.49	6.75
1020	1000 S.F.	6.30	3.29	9.59
1040	2000 S.F.	5.85	3.54	9.39
1060	5000 S.F.	2.95	2.55	5.50
1080	10,000 S.F.	2.07	2.65	4.72
1100	50,000 S.F.	1.64	2.51	4.15
1140	Each additional floor, 500 S.F.	1.50	3.13	4.63
1160	1000 S.F.	1.27	2.54	3.81
1180	2000 S.F.	1.42	2.55	3.97
1200	5000 S.F.	1.40	2.44	3.84
1220	10,000 S.F.	1.22	2.48	3.70
1240	50,000 S.F.	1.16	2.21	3.37
1500	Extra hazard, one floor, 500 S.F.	12.25	5.35	17.60
1520	1000 S.F.	7.55	4.65	12.20
1540	2000 S.F.	6.15	4.78	10.93
1560	5000 S.F.	3.55	4.17	7.72
1580	10,000 S.F.	3.26	3.93	7.19
1600	50,000 S.F.	3.62	3.82	7.44
1660	Each additional floor, 500 S.F.	2.18	3.86	6.04
1680	1000 S.F.	2.13	3.66	5.79
1700	2000 S.F.	1.86	3.68	5.54
1720	5000 S.F.	1.53	3.26	4.79
1740	10,000 S.F.	2.01	2.97	4.98
1760	50,000 S.F.	1.99	2.85	4.84
2020	Grooved steel, black sch. 40 pipe, light hazard, one floor, 2000 S.F.	5.70	2.87	8.57

D4010 Sprinklers

D4010 410	Wet Pipe Sprinkler Systems	COST PER S.F.		
		MAT.	INST.	TOTAL
2060	10,000 S.F.	2.19	1.77	3.96
2100	Each additional floor, 2000 S.F.	1.27	1.88	3.15
2150	10,000 S.F.	.83	1.53	2.36
2200	Ordinary hazard, one floor, 2000 S.F.	5.80	3.04	8.84
2250	10,000 S.F.	1.97	2.22	4.19
2300	Each additional floor, 2000 S.F.	1.36	2.05	3.41
2350	10,000 S.F.	1.12	2.05	3.17
2400	Extra hazard, one floor, 2000 S.F.	6.15	3.92	10.07
2450	10,000 S.F.	2.66	2.86	5.52
2500	Each additional floor, 2000 S.F.	1.92	3.02	4.94
2550	10,000 S.F.	1.62	2.54	4.16
3050	Grooved steel, black sch. 10 pipe, light hazard, one floor, 2000 S.F.	5.65	2.85	8.50
3100	10,000 S.F.	1.67	1.68	3.35
3150	Each additional floor, 2000 S.F.	1.22	1.86	3.08
3200	10,000 S.F.	.81	1.51	2.32
3250	Ordinary hazard, one floor, 2000 S.F.	5.75	3.01	8.76
3300	10,000 S.F.	1.91	2.17	4.08
3350	Each additional floor, 2000 S.F.	1.32	2.02	3.34
3400	10,000 S.F.	1.06	2	3.06
3450	Extra hazard, one floor, 2000 S.F.	6.10	3.89	9.99
3500	10,000 S.F.	2.48	2.82	5.30
3550	Each additional floor, 2000 S.F.	1.89	2.99	4.88
3600	10,000 S.F.	1.56	2.51	4.07
4050	Copper tubing, type M, light hazard, one floor, 2000 S.F.	6	2.83	8.83
4100	10,000 S.F.	2.15	1.71	3.86
4150	Each additional floor, 2000 S.F.	1.61	1.88	3.49
4200	10,000 S.F.	1.30	1.55	2.85
4250	Ordinary hazard, one floor, 2000 S.F.	6.25	3.18	9.43
4300	10,000 S.F.	2.50	2.04	4.54
4350	Each additional floor, 2000 S.F.	1.85	2.08	3.93
4400	10,000 S.F.	1.61	1.83	3.44
4450	Extra hazard, one floor, 2000 S.F.	6.80	3.94	10.74
4500	10,000 S.F.	4.15	3.12	7.27
4550	Each additional floor, 2000 S.F.	2.55	3.04	5.59
4600	10,000 S.F.	2.67	2.78	5.45
5050	Copper tubing, type M, T-drill system, light hazard, one floor			
5060	2000 S.F.	6.05	2.62	8.67
5100	10,000 S.F.	2.04	1.41	3.45
5150	Each additional floor, 2000 S.F.	1.63	1.67	3.30
5200	10,000 S.F.	1.19	1.25	2.44
5250	Ordinary hazard, one floor, 2000 S.F.	6.10	2.67	8.77
5300	10,000 S.F.	2.41	1.81	4.22
5350	Each additional floor, 2000 S.F.	1.64	1.68	3.32
5400	10,000 S.F.	1.56	1.64	3.20
5450	Extra hazard, one floor, 2000 S.F.	6.30	3.21	9.51
5500	10,000 S.F.	3.44	2.26	5.70
5550	Each additional floor, 2000 S.F.	2.18	2.36	4.54
5600	10,000 S.F.	1.96	1.92	3.88

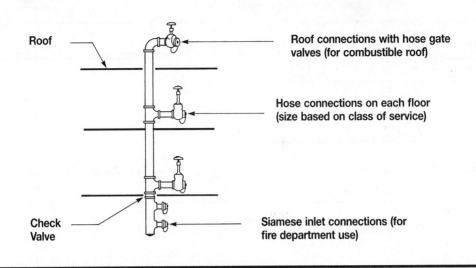

Roof — Roof connections with hose gate valves (for combustible roof)

Hose connections on each floor (size based on class of service)

Check Valve

Siamese inlet connections (for fire department use)

D4020 310	Wet Standpipe Risers, Class I	COST PER FLOOR		
		MAT.	INST.	TOTAL
0550	Wet standpipe risers, Class I, steel, black sch. 40, 10' height			
0560	4" diameter pipe, one floor	5,550	3,525	9,075
0580	Additional floors	1,325	1,100	2,425
0600	6" diameter pipe, one floor	9,375	6,200	15,575
0620	Additional floors	2,375	1,725	4,100
0640	8" diameter pipe, one floor	14,200	7,475	21,675
0660	Additional floors	3,450	2,100	5,550

D4020 310	Wet Standpipe Risers, Class II	COST PER FLOOR		
		MAT.	INST.	TOTAL
1030	Wet standpipe risers, Class II, steel, black sch. 40, 10' height			
1040	2" diameter pipe, one floor	2,475	1,275	3,750
1060	Additional floors	810	490	1,300
1080	2-1/2" diameter pipe, one floor	3,425	1,875	5,300
1100	Additional floors	890	570	1,460

D4020 310	Wet Standpipe Risers, Class III	COST PER FLOOR		
		MAT.	INST.	TOTAL
1530	Wet standpipe risers, Class III, steel, black sch. 40, 10' height			
1540	4" diameter pipe, one floor	5,675	3,525	9,200
1560	Additional floors	1,175	915	2,090
1580	6" diameter pipe, one floor	9,500	6,200	15,700
1600	Additional floors	2,450	1,725	4,175
1620	8" diameter pipe, one floor	14,400	7,475	21,875
1640	Additional floors	3,525	2,100	5,625

D4020 Standpipes

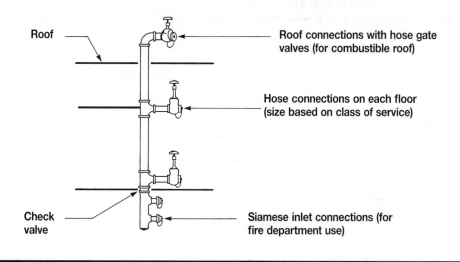

Roof

Roof connections with hose gate valves (for combustible roof)

Hose connections on each floor (size based on class of service)

Check valve

Siamese inlet connections (for fire department use)

D4020 330	Dry Standpipe Risers, Class I	COST PER FLOOR		
		MAT.	INST.	TOTAL
0530	Dry standpipe riser, Class I, steel, black sch. 40, 10' height			
0540	4" diameter pipe, one floor	3,225	2,850	6,075
0560	Additional floors	1,225	1,025	2,250
0580	6" diameter pipe, one floor	6,825	4,925	11,750
0600	Additional floors	2,275	1,675	3,950
0620	8" diameter pipe, one floor	10,600	5,975	16,575
0640	Additional floors	3,350	2,050	5,400

D4020 330	Dry Standpipe Risers, Class II	COST PER FLOOR		
		MAT.	INST.	TOTAL
1030	Dry standpipe risers, Class II, steel, black sch. 40, 10' height			
1040	2" diameter pipe, one floor	2,075	1,325	3,400
1060	Additional floors	705	435	1,140
1080	2-1/2" diameter pipe, one floor	2,750	1,550	4,300
1100	Additional floors	790	515	1,305

D4020 330	Dry Standpipe Risers, Class III	COST PER FLOOR		
		MAT.	INST.	TOTAL
1530	Dry standpipe risers, Class III, steel, black sch. 40, 10' height			
1540	4" diameter pipe, one floor	3,275	2,800	6,075
1560	Additional floors	1,075	935	2,010
1580	6" diameter pipe, one floor	6,900	4,925	11,825
1600	Additional floors	2,350	1,675	4,025
1620	8" diameter pipe, one floor	10,600	5,975	16,575
1640	Additional floors	3,425	2,050	5,475

D4020 Standpipes

D4020 410	Fire Hose Equipment	COST EACH		
		MAT.	INST.	TOTAL
0100	Adapters, reducing, 1 piece, FxM, hexagon, cast brass, 2-1/2" x 1-1/2"	64.50		64.50
0200	Pin lug, 1-1/2" x 1"	55.50		55.50
0250	3" x 2-1/2"	144		144
0300	For polished chrome, add 75% mat.			
0400	Cabinets, D.S. glass in door, recessed, steel box, not equipped			
0500	Single extinguisher, steel door & frame	134	157	291
0550	Stainless steel door & frame	237	157	394
0600	Valve, 2-1/2" angle, steel door & frame	156	105	261
0650	Aluminum door & frame	189	105	294
0700	Stainless steel door & frame	255	105	360
0750	Hose rack assy, 2-1/2" x 1-1/2" valve & 100' hose, steel door & frame	292	210	502
0800	Aluminum door & frame	430	210	640
0850	Stainless steel door & frame	575	210	785
0900	Hose rack assy & extinguisher,2-1/2"x1-1/2" valve & hose,steel door & frame	305	252	557
0950	Aluminum	550	252	802
1000	Stainless steel	605	252	857
1550	Compressor, air, dry pipe system, automatic, 200 gal., 3/4 H.P.	890	535	1,425
1600	520 gal., 1 H.P.	925	535	1,460
1650	Alarm, electric pressure switch (circuit closer)	102	27	129
2500	Couplings, hose, rocker lug, cast brass, 1-1/2"	61.50		61.50
2550	2-1/2"	81		81
3000	Escutcheon plate, for angle valves, polished brass, 1-1/2"	16.70		16.70
3050	2-1/2"	27.50		27.50
3500	Fire pump, electric, w/controller, fittings, relief valve			
3550	4" pump, 30 H.P., 500 G.P.M.	17,000	3,925	20,925
3600	5" pump, 40 H.P., 1000 G.P.M.	19,000	4,425	23,425
3650	5" pump, 100 H.P., 1000 G.P.M.	25,800	4,925	30,725
3700	For jockey pump system, add	2,975	630	3,605
5000	Hose, per linear foot, synthetic jacket, lined,			
5100	300 lb. test, 1-1/2" diameter	3.74	.48	4.22
5150	2-1/2" diameter	6.45	.57	7.02
5200	500 lb. test, 1-1/2" diameter	3.86	.48	4.34
5250	2-1/2" diameter	6.85	.57	7.42
5500	Nozzle, plain stream, polished brass, 1-1/2" x 10"	55.50		55.50
5550	2-1/2" x 15" x 13/16" or 1-1/2"	116		116
5600	Heavy duty combination adjustable fog and straight stream w/handle 1-1/2"	475		475
5650	2-1/2" direct connection	590		590
6000	Rack, for 1-1/2" diameter hose 100 ft. long, steel	71	63	134
6050	Brass	120	63	183
6500	Reel, steel, for 50 ft. long 1-1/2" diameter hose	151	90	241
6550	For 75 ft. long 2-1/2" diameter hose	245	90	335
7050	Siamese, w/plugs & chains, polished brass, sidewalk, 4" x 2-1/2" x 2-1/2"	730	505	1,235
7100	6" x 2-1/2" x 2-1/2"	945	630	1,575
7200	Wall type, flush, 4" x 2-1/2" x 2-1/2"	600	252	852
7250	6" x 2-1/2" x 2-1/2"	835	273	1,108
7300	Projecting, 4" x 2-1/2" x 2-1/2"	550	252	802
7350	6" x 2-1/2" x 2-1/2"	925	273	1,198
7400	For chrome plate, add 15% mat.			
8000	Valves, angle, wheel handle, 300 Lb., rough brass, 1-1/2"	115	58	173
8050	2-1/2"	202	100	302
8100	Combination pressure restricting, 1-1/2"	98.50	58	156.50
8150	2-1/2"	202	100	302
8200	Pressure restricting, adjustable, satin brass, 1-1/2"	350	58	408
8250	2-1/2"	410	100	510
8300	Hydrolator, vent and drain, rough brass, 1-1/2"	102	58	160
8350	2-1/2"	102	58	160
8400	Cabinet assy, incls. adapter, rack, hose, and nozzle	925	380	1,305

D4090 Other Fire Protection Systems

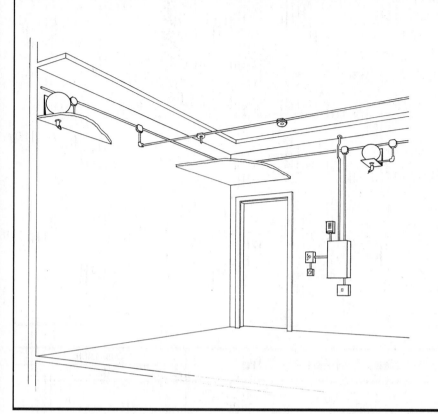

General: Automatic fire protection (suppression) systems other than water sprinklers may be desired for special environments, high risk areas, isolated locations or unusual hazards. Some typical applications would include:

1. Paint dip tanks
2. Securities vaults
3. Electronic data processing
4. Tape and data storage
5. Transformer rooms
6. Spray booths
7. Petroleum storage
8. High rack storage

Piping and wiring costs are highly variable and are not included.

D4090 910	Fire Suppression Unit Components	COST EACH		
		MAT.	INST.	TOTAL
0020	Detectors with brackets			
0040	Fixed temperature heat detector	56	82.50	138.50
0060	Rate of temperature rise detector	56	82.50	138.50
0080	Ion detector (smoke) detector	121	106	227
0200	Extinguisher agent			
0240	200 lb FM200, container	8,550	320	8,870
0280	75 lb carbon dioxide cylinder	1,400	214	1,614
0320	Dispersion nozzle			
0340	FM200 1-1/2" dispersion nozzle	73.50	51	124.50
0380	Carbon dioxide 3" x 5" dispersion nozzle	73.50	39.50	113
0420	Control station			
0440	Single zone control station with batteries	1,900	660	2,560
0470	Multizone (4) control station with batteries	3,600	1,325	4,925
0500	Electric mechanical release	184	345	529
0550	Manual pull station	66.50	119	185.50
0640	Battery standby power 10" x 10" x 17"	455	165	620
0740	Bell signalling device	106	82.50	188.50
0900	Standard low-rise sprinkler accessory package, 3 story	3,475	1,850	5,325
1000	Standard mid-rise sprinkler accessory package, 8 story	7,275	3,550	10,825
1100	Standard high-rise sprinkler accessory package, 16 story	17,300	6,925	24,225

D4090 920	FM200 Systems	COST PER C.F.		
		MAT.	INST.	TOTAL
0820	Average FM200 system, minimum			1.94
0840	Maximum			3.85

For customer support on your Square Foot Cost Data, call 877.756.2789.

475

D5010   Electrical Service/Distribution

D5010 120	Overhead Electric Service, 3 Phase - 4 Wire	COST EACH		
		MAT.	INST.	TOTAL
0200	Service installation, includes breakers, metering, 20' conduit & wire			
0220	3 phase, 4 wire, 120/208 volts, 60 A	555	1,075	1,630
0240	100 A	725	1,225	1,950
0245	100 A w/circuit breaker	1,525	1,500	3,025
0280	200 A	1,350	1,675	3,025
0285	200 A w/circuit breaker	3,025	2,100	5,125
0320	400 A	2,700	3,350	6,050
0325	400 A w/circuit breaker	5,875	4,175	10,050
0360	600 A	4,600	4,925	9,525
0365	600 A, w/switchboard	9,175	6,250	15,425
0400	800 A	6,300	5,925	12,225
0405	800 A, w/switchboard	10,900	7,425	18,325
0440	1000 A	7,850	7,300	15,150
0445	1000 A, w/switchboard	13,300	8,950	22,250
0480	1200 A	10,400	8,325	18,725
0485	1200 A, w/groundfault switchboard	34,100	10,300	44,400
0520	1600 A	12,800	10,900	23,700
0525	1600 A, w/groundfault switchboard	38,600	13,000	51,600
0560	2000 A	17,100	13,200	30,300
0565	2000 A, w/groundfault switchboard	46,200	16,300	62,500
0610	1 phase, 3 wire, 120/240 volts, 100 A (no safety switch)	182	595	777
0615	100 A w/load center	490	1,150	1,640
0620	200 A	420	815	1,235
0625	200 A w/load center	1,250	1,725	2,975

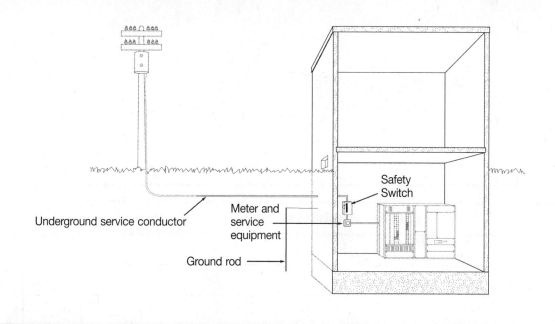

Underground service conductor

Meter and service equipment

Ground rod

Safety Switch

D5010 130	Underground Electric Service	COST EACH		
		MAT.	INST.	TOTAL
0950	Underground electric service including excavation,backfill, and compaction			
1000	3 phase, 4 wire, 277/480 volts, 2000 A	34,400	21,100	55,500
1050	2000 A w/groundfault switchboard	62,000	23,500	85,500
1100	1600 A	27,300	17,000	44,300
1150	1600 A w/groundfault switchboard	53,000	19,100	72,100
1200	1200 A	20,900	13,500	34,400
1250	1200 A w/groundfault switchboard	44,600	15,500	60,100
1400	800 A	14,700	11,200	25,900
1450	800 A w/ switchboard	19,300	12,700	32,000
1500	600 A	10,300	10,300	20,600
1550	600 A w/ switchboard	14,900	11,600	26,500
1600	1 phase, 3 wire, 120/240 volts, 200 A	3,100	3,550	6,650
1650	200 A w/load center	3,925	4,450	8,375
1700	100 A	2,150	2,900	5,050
1750	100 A w/load center	3,000	3,825	6,825

D5010 Electrical Service/Distribution

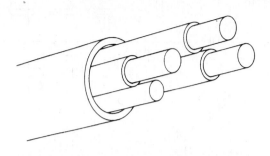

D5010 230	Feeder Installation	COST PER L.F.		
		MAT.	INST.	TOTAL
0200	Feeder installation 600 V, including RGS conduit and XHHW wire, 60 A	5.30	12.10	17.40
0240	100 A	9.50	15.80	25.30
0280	200 A	21	24	45
0320	400 A	42.50	47.50	90
0360	600 A	78.50	78	156.50
0400	800 A	102	92	194
0440	1000 A	128	121	249
0480	1200 A	157	156	313
0520	1600 A	204	184	388
0560	2000 A	256	242	498
1200	Branch installation 600 V, including EMT conduit and THW wire, 15 A	1.13	5.90	7.03
1240	20 A	1.32	6.30	7.62
1280	30 A	1.65	6.50	8.15
1320	50 A	3.27	9.05	12.32
1360	65 A	4.02	9.60	13.62
1400	85 A	6.15	11.15	17.30
1440	100 A	6.80	11.40	18.20
1480	130 A	10.20	13.30	23.50
1520	150 A	12.50	15.25	27.75
1560	200 A	17	17.40	34.40

D50 Electrical

D5010 Electrical Service/Distribution

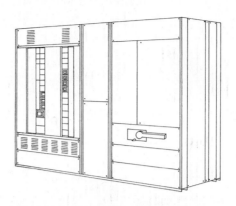

D5010 240	Switchgear	COST EACH		
		MAT.	INST.	TOTAL
0190	Switchgear installation, including switchboard, panels, & circuit breaker			
0200	120/208 V, 1 phase, 400 A	9,225	2,800	12,025
0240	600 A	13,200	3,150	16,350
0280	800 A	15,300	3,625	18,925
0300	1000 A	18,800	4,000	22,800
0320	1200 A	19,800	4,525	24,325
0360	1600 A	32,100	5,225	37,325
0400	2000 A	38,800	5,700	44,500
0500	277/480 V, 3 phase, 400 A	12,600	5,200	17,800
0520	600 A	18,400	6,000	24,400
0540	800 A	19,900	6,625	26,525
0560	1000 A	24,700	7,300	32,000
0580	1200 A	27,100	8,175	35,275
0600	1600 A	40,000	9,025	49,025
0620	2000 A	50,500	9,950	60,450

479

For customer support on your Square Foot Cost Data, call 877.756.2789.

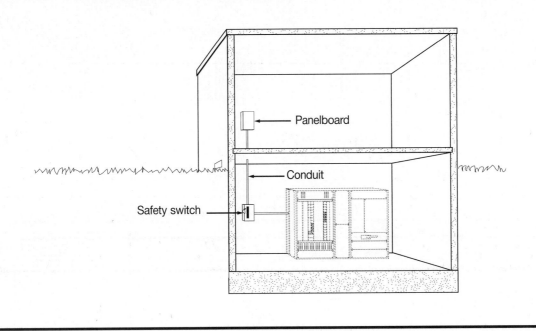

D5010 250	Panelboard	COST EACH		
		MAT.	INST.	TOTAL
0900	Panelboards, NQOD, 4 wire, 120/208 volts w/conductor & conduit			
1000	100 A, 0 stories, 0' horizontal	1,400	1,400	2,800
1020	1 stories, 25' horizontal	1,775	2,000	3,775
1040	5 stories, 50' horizontal	2,525	3,200	5,725
1060	10 stories, 75' horizontal	3,400	4,575	7,975
1080	225A, 0 stories, 0' horizontal	2,625	1,825	4,450
2000	1 stories, 25' horizontal	3,750	2,900	6,650
2020	5 stories, 50' horizontal	5,950	5,050	11,000
2040	10 stories, 75' horizontal	8,500	7,550	16,050
2060	400A, 0 stories, 0' horizontal	3,875	2,750	6,625
2080	1 stories, 25' horizontal	5,425	4,500	9,925
3000	5 stories, 50' horizontal	8,475	8,000	16,475
3020	10 stories, 75' horizontal	16,900	15,600	32,500
3040	600 A, 0 stories, 0' horizontal	5,725	3,300	9,025
3060	1 stories, 25' horizontal	8,650	6,175	14,825
3080	5 stories, 50' horizontal	14,400	11,900	26,300
4000	10 stories, 75' horizontal	30,000	23,300	53,300
4010	Panelboards, NEHB, 4 wire, 277/480 volts w/conductor, conduit, & safety switch			
4020	100 A, 0 stories, 0' horizontal, includes safety switch	3,125	1,925	5,050
4040	1 stories, 25' horizontal	3,500	2,525	6,025
4060	5 stories, 50' horizontal	4,250	3,700	7,950
4080	10 stories, 75' horizontal	5,125	5,100	10,225
5000	225 A, 0 stories, 0' horizontal	4,700	2,325	7,025
5020	1 stories, 25' horizontal	5,800	3,425	9,225
5040	5 stories, 50' horizontal	8,025	5,550	13,575
5060	10 stories, 75' horizontal	10,600	8,050	18,650
5080	400 A, 0 stories, 0' horizontal	7,450	3,600	11,050
6000	1 stories, 25' horizontal	9,000	5,375	14,375
6020	5 stories, 50' horizontal	12,100	8,850	20,950
6040	10 stories, 75' horizontal	20,400	16,500	36,900
6060	600 A, 0 stories, 0' horizontal	11,500	4,575	16,075
6080	1 stories, 25' horizontal	14,400	7,450	21,850
7000	5 stories, 50' horizontal	20,200	13,100	33,300
7020	10 stories, 75' horizontal	35,800	24,600	60,400

D5020 Lighting and Branch Wiring

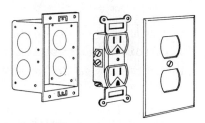

Duplex Wall Receptacle

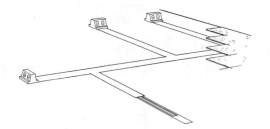

Undercarpet Receptacle System

D5020 110	Receptacle (by Wattage)	COST PER S.F.		
		MAT.	INST.	TOTAL
0190	Receptacles include plate, box, conduit, wire & transformer when required			
0200	2.5 per 1000 S.F., .3 watts per S.F.	.46	1.47	1.93
0240	With transformer	.54	1.54	2.08
0280	4 per 1000 S.F., .5 watts per S.F.	.52	1.71	2.23
0320	With transformer	.65	1.83	2.48
0360	5 per 1000 S.F., .6 watts per S.F.	.61	2.03	2.64
0400	With transformer	.78	2.18	2.96
0440	8 per 1000 S.F., .9 watts per S.F.	.64	2.24	2.88
0480	With transformer	.91	2.47	3.38
0520	10 per 1000 S.F., 1.2 watts per S.F.	.65	2.44	3.09
0560	With transformer	.98	2.73	3.71
0600	16.5 per 1000 S.F., 2.0 watts per S.F.	.78	3.04	3.82
0640	With transformer	1.33	3.52	4.85
0680	20 per 1000 S.F., 2.4 watts per S.F.	.81	3.32	4.13
0720	With transformer	1.48	3.91	5.39

D5020 115	Receptacles, Floor	COST PER S.F.		
		MAT.	INST.	TOTAL
0200	Receptacle systems, underfloor duct, 5' on center, low density	7.45	3.24	10.69
0240	High density	8	4.17	12.17
0280	7' on center, low density	5.90	2.78	8.68
0320	High density	6.45	3.71	10.16
0400	Poke thru fittings, low density	1.25	1.59	2.84
0440	High density	2.50	3.17	5.67
0520	Telepoles, using Romex, low density	1.19	.97	2.16
0560	High density	2.37	1.95	4.32
0600	Using EMT, low density	1.23	1.29	2.52
0640	High density	2.51	2.57	5.08
0720	Conduit system with floor boxes, low density	1.18	1.11	2.29
0760	High density	2.38	2.21	4.59
0840	Undercarpet power system, 3 conductor with 5 conductor feeder, low density	1.57	.39	1.96
0880	High density	3.11	.81	3.92

D5020 Lighting and Branch Wiring

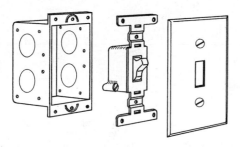

Description: Table D5020 130 includes the cost for switch, plate, box, conduit in slab or EMT exposed and copper wire. Add 20% for exposed conduit.

No power required for switches.

Federal energy guidelines recommend the maximum lighting area controlled per switch shall not exceed 1000 S.F. and that areas over 500 S.F. shall be so controlled that total illumination can be reduced by at least 50%.

D5020 130	Wall Switch by Sq. Ft.	COST PER S.F.		
		MAT.	INST.	TOTAL
0200	Wall switches, 1.0 per 1000 S.F.	.06	.25	.31
0240	1.2 per 1000 S.F.	.08	.26	.34
0280	2.0 per 1000 S.F.	.12	.38	.50
0320	2.5 per 1000 S.F.	.15	.48	.63
0360	5.0 per 1000 S.F.	.31	1.02	1.33
0400	10.0 per 1000 S.F.	.62	2.07	2.69

D5020 135	Miscellaneous Power	COST PER S.F.		
		MAT.	INST.	TOTAL
0200	Miscellaneous power, to .5 watts	.04	.12	.16
0240	.8 watts	.06	.17	.23
0280	1 watt	.08	.22	.30
0320	1.2 watts	.09	.26	.35
0360	1.5 watts	.11	.30	.41
0400	1.8 watts	.12	.35	.47
0440	2 watts	.14	.42	.56
0480	2.5 watts	.18	.52	.70
0520	3 watts	.21	.61	.82

D5020 140	Central A. C. Power (by Wattage)	COST PER S.F.		
		MAT.	INST.	TOTAL
0200	Central air conditioning power, 1 watt	.08	.26	.34
0220	2 watts	.10	.30	.40
0240	3 watts	.20	.44	.64
0280	4 watts	.20	.45	.65
0320	6 watts	.35	.59	.94
0360	8 watts	.45	.65	1.10
0400	10 watts	.63	.75	1.38

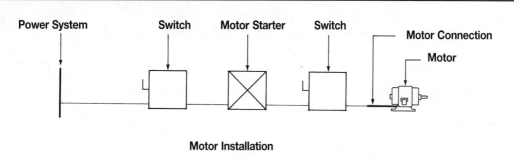

Motor Installation

D5020 145	Motor Installation	COST EACH		
		MAT.	INST.	TOTAL
0200	Motor installation, single phase, 115V, 1/3 HP motor size	530	1,000	1,530
0240	1 HP motor size	555	1,000	1,555
0280	2 HP motor size	605	1,075	1,680
0320	3 HP motor size	685	1,075	1,760
0360	230V, 1 HP motor size	535	1,025	1,560
0400	2 HP motor size	570	1,025	1,595
0440	3 HP motor size	640	1,100	1,740
0520	Three phase, 200V, 1-1/2 HP motor size	610	1,125	1,735
0560	3 HP motor size	740	1,225	1,965
0600	5 HP motor size	725	1,350	2,075
0640	7-1/2 HP motor size	755	1,375	2,130
0680	10 HP motor size	1,125	1,725	2,850
0720	15 HP motor size	1,550	1,925	3,475
0760	20 HP motor size	1,950	2,225	4,175
0800	25 HP motor size	2,000	2,225	4,225
0840	30 HP motor size	3,150	2,625	5,775
0880	40 HP motor size	3,900	3,100	7,000
0920	50 HP motor size	6,900	3,625	10,525
0960	60 HP motor size	7,200	3,825	11,025
1000	75 HP motor size	8,875	4,375	13,250
1040	100 HP motor size	26,600	5,150	31,750
1080	125 HP motor size	26,900	5,650	32,550
1120	150 HP motor size	31,300	6,650	37,950
1160	200 HP motor size	32,300	7,850	40,150
1240	230V, 1-1/2 HP motor size	585	1,100	1,685
1280	3 HP motor size	715	1,200	1,915
1320	5 HP motor size	700	1,325	2,025
1360	7-1/2 HP motor size	700	1,325	2,025
1400	10 HP motor size	1,075	1,650	2,725
1440	15 HP motor size	1,200	1,800	3,000
1480	20 HP motor size	1,850	2,150	4,000
1520	25 HP motor size	1,950	2,225	4,175
1560	30 HP motor size	2,000	2,225	4,225
1600	40 HP motor size	3,800	3,025	6,825
1640	50 HP motor size	4,075	3,200	7,275
1680	60 HP motor size	6,925	3,625	10,550
1720	75 HP motor size	8,225	4,125	12,350
1760	100 HP motor size	9,475	4,600	14,075
1800	125 HP motor size	27,100	5,325	32,425
1840	150 HP motor size	28,900	6,025	34,925
1880	200 HP motor size	30,300	6,725	37,025
1960	460V, 2 HP motor size	725	1,125	1,850
2000	5 HP motor size	855	1,225	2,080
2040	10 HP motor size	805	1,325	2,130
2080	15 HP motor size	1,075	1,525	2,600
2120	20 HP motor size	1,125	1,650	2,775

D50 Electrical

D5020 Lighting and Branch Wiring

D5020 145	Motor Installation	COST EACH		
		MAT.	INST.	TOTAL
2160	25 HP motor size	1,200	1,725	2,925
2200	30 HP motor size	1,550	1,875	3,425
2240	40 HP motor size	1,975	2,000	3,975
2280	50 HP motor size	2,200	2,225	4,425
2320	60 HP motor size	3,350	2,600	5,950
2360	75 HP motor size	3,950	2,875	6,825
2400	100 HP motor size	4,375	3,200	7,575
2440	125 HP motor size	7,225	3,650	10,875
2480	150 HP motor size	8,950	4,075	13,025
2520	200 HP motor size	10,300	4,600	14,900
2600	575V, 2 HP motor size	725	1,125	1,850
2640	5 HP motor size	855	1,225	2,080
2680	10 HP motor size	805	1,325	2,130
2720	20 HP motor size	1,075	1,525	2,600
2760	25 HP motor size	1,125	1,650	2,775
2800	30 HP motor size	1,550	1,875	3,425
2840	50 HP motor size	1,675	1,925	3,600
2880	60 HP motor size	3,300	2,575	5,875
2920	75 HP motor size	3,350	2,600	5,950
2960	100 HP motor size	3,950	2,875	6,825
3000	125 HP motor size	7,050	3,575	10,625
3040	150 HP motor size	7,225	3,650	10,875
3080	200 HP motor size	9,050	4,150	13,200

D5020 155	Motor Feeder	COST PER L.F.		
		MAT.	INST.	TOTAL
0200	Motor feeder systems, single phase, feed up to 115V 1HP or 230V 2 HP	2.66	7.80	10.46
0240	115V 2HP, 230V 3HP	2.79	7.90	10.69
0280	115V 3HP	3.09	8.25	11.34
0360	Three phase, feed to 200V 3HP, 230V 5HP, 460V 10HP, 575V 10HP	2.77	8.40	11.17
0440	200V 5HP, 230V 7.5HP, 460V 15HP, 575V 20HP	2.97	8.60	11.57
0520	200V 10HP, 230V 10HP, 460V 30HP, 575V 30HP	3.43	9.10	12.53
0600	200V 15HP, 230V 15HP, 460V 40HP, 575V 50HP	4.76	10.40	15.16
0680	200V 20HP, 230V 25HP, 460V 50HP, 575V 60HP	6.75	13.10	19.85
0760	200V 25HP, 230V 30HP, 460V 60HP, 575V 75HP	7.65	13.35	21
0840	200V 30HP	8.50	13.80	22.30
0920	230V 40HP, 460V 75HP, 575V 100HP	10.80	15.10	25.90
1000	200V 40HP	12.45	16.15	28.60
1080	230V 50HP, 460V 100HP, 575V 125HP	14.95	17.80	32.75
1160	200V 50HP, 230V 60HP, 460V 125HP, 575V 150HP	17.15	18.90	36.05
1240	200V 60HP, 460V 150HP	21	22	43
1320	230V 75HP, 575V 200HP	24	23	47
1400	200V 75HP	26.50	23.50	50
1480	230V 100HP, 460V 200HP	36.50	27.50	64
1560	200V 100HP	50.50	34.50	85
1640	230V 125HP	50.50	34.50	85
1720	200V 125HP, 230V 150HP	56.50	42.50	99
1800	200V 150HP	68	46.50	114.50
1880	200V 200HP	85	49	134
1960	230V 200HP	81	51	132

D5020 Lighting and Branch Wiring

Fluorescent Fixture

Incandescent Fixture

D5020 210	Fluorescent Fixtures (by Wattage)	COST PER S.F.		
		MAT.	INST.	TOTAL
0190	Fluorescent fixtures recess mounted in ceiling			
0195	T12, standard 40 watt lamps			
0200	1 watt per S.F., 20 FC, 5 fixtures @40 watts per 1000 S.F.	.78	1.93	2.71
0240	2 watt per S.F., 40 FC, 10 fixtures @40 watt per 1000 S.F.	1.55	3.80	5.35
0280	3 watt per S.F., 60 FC, 15 fixtures @40 watt per 1000 S.F	2.30	5.75	8.05
0320	4 watt per S.F., 80 FC, 20 fixtures @40 watt per 1000 S.F.	3.07	7.60	10.67
0400	5 watt per S.F., 100 FC, 25 fixtures @40 watt per 1000 S.F.	3.84	9.55	13.39
0450	T8, energy saver 32 watt lamps			
0500	0.8 watt per S.F., 20 FC, 5 fixtures @32 watt per 1000 S.F.	.85	1.93	2.78
0520	1.6 watt per S.F., 40 FC, 10 fixtures @32 watt per 1000 S.F.	1.68	3.80	5.48
0540	2.4 watt per S.F., 60 FC, 15 fixtures @ 32 watt per 1000 S.F	2.49	5.75	8.24
0560	3.2 watt per S.F., 80 FC, 20 fixtures @32 watt per 1000 S.F.	3.33	7.60	10.93
0580	4 watt per S.F., 100 FC, 25 fixtures @32 watt per 1000 S.F.	4.17	9.55	13.72

D5020 216	Incandescent Fixture (by Wattage)	COST PER S.F.		
		MAT.	INST.	TOTAL
0190	Incandescent fixture recess mounted, type A			
0200	1 watt per S.F., 8 FC, 6 fixtures per 1000 S.F.	1	1.56	2.56
0240	2 watt per S.F., 16 FC, 12 fixtures per 1000 S.F.	2	3.12	5.12
0280	3 watt per S.F., 24 FC, 18 fixtures, per 1000 S.F.	3	4.61	7.61
0320	4 watt per S.F., 32 FC, 24 fixtures per 1000 S.F.	3.99	6.15	10.14
0400	5 watt per S.F., 40 FC, 30 fixtures per 1000 S.F.	5	7.75	12.75

D5020 Lighting and Branch Wiring

High Bay Fixture

Low Bay Fixture

D5020 226	H.I.D. Fixture, High Bay, 16' (by Wattage)	COST PER S.F.		
		MAT.	INST.	TOTAL
0190	High intensity discharge fixture, 16' above work plane			
0240	1 watt/S.F., type E, 42 FC, 1 fixture/1000 S.F.	1.12	1.63	2.75
0280	Type G, 52 FC, 1 fixture/1000 S.F.	1.12	1.63	2.75
0320	Type C, 54 FC, 2 fixture/1000 S.F.	1.28	1.82	3.10
0440	2 watt/S.F., type E, 84 FC, 2 fixture/1000 S.F.	2.24	3.32	5.56
0480	Type G, 105 FC, 2 fixture/1000 S.F.	2.24	3.32	5.56
0520	Type C, 108 FC, 4 fixture/1000 S.F.	2.56	3.64	6.20
0640	3 watt/S.F., type E, 126 FC, 3 fixture/1000 S.F.	3.34	4.94	8.28
0680	Type G, 157 FC, 3 fixture/1000 S.F.	3.34	4.94	8.28
0720	Type C, 162 FC, 6 fixture/1000 S.F.	3.84	5.45	9.29
0840	4 watt/S.F., type E, 168 FC, 4 fixture/1000 S.F.	4.49	6.60	11.09
0880	Type G, 210 FC, 4 fixture/1000 S.F.	4.49	6.60	11.09
0920	Type C, 243 FC, 9 fixture/1000 S.F.	5.55	7.60	13.15
1040	5 watt/S.F., type E, 210 FC, 5 fixture/1000 S.F.	5.60	8.25	13.85
1080	Type G, 262 FC, 5 fixture/1000 S.F.	5.60	8.25	13.85
1120	Type C, 297 FC, 11 fixture/1000 S.F.	6.85	9.40	16.25

D5020 234	H.I.D. Fixture, High Bay, 30' (by Wattage)	COST PER S.F.		
		MAT.	INST.	TOTAL
0190	High intensity discharge fixture, 30' above work plane			
0240	1 watt/S.F., type E, 37 FC, 1 fixture/1000 S.F.	1.25	2.06	3.31
0280	Type G, 45 FC., 1 fixture/1000 S.F.	1.25	2.06	3.31
0320	Type F, 50 FC, 1 fixture/1000 S.F.	1.05	1.61	2.66
0440	2 watt/S.F., type E, 74 FC, 2 fixtures/1000 S.F.	2.51	4.13	6.64
0480	Type G, 92 FC, 2 fixtures/1000 S.F.	2.51	4.13	6.64
0520	Type F, 100 FC, 2 fixtures/1000 S.F.	2.13	3.29	5.42
0640	3 watt/S.F., type E, 110 FC, 3 fixtures/1000 S.F.	3.77	6.25	10.02
0680	Type G, 138 FC, 3 fixtures/1000 S.F.	3.77	6.25	10.02
0720	Type F, 150 FC, 3 fixtures/1000 S.F.	3.16	4.90	8.06
0840	4 watt/S.F., type E, 148 FC, 4 fixtures/1000 S.F.	5.05	8.30	13.35
0880	Type G, 185 FC, 4 fixtures/1000 S.F.	5.05	8.30	13.35
0920	Type F, 200 FC, 4 fixtures/1000 S.F.	4.24	6.60	10.84
1040	5 watt/S.F., type E, 185 FC, 5 fixtures/1000 S.F.	6.30	10.40	16.70
1080	Type G, 230 FC, 5 fixtures/1000 S.F.	6.30	10.40	16.70
1120	Type F, 250 FC, 5 fixtures/1000 S.F.	5.30	8.20	13.50

D5020 238	H.I.D. Fixture, Low Bay, 8'-10' (by Wattage)	COST PER S.F.		
		MAT.	INST.	TOTAL
0190	High intensity discharge fixture, 8'-10' above work plane			
0240	1 watt/S.F., type J, 30 FC, 4 fixtures/1000 S.F.	2.51	3.36	5.87
0280	Type K, 29 FC, 5 fixtures/1000 S.F.	2.52	3.03	5.55
0400	2 watt/S.F., type J, 52 FC, 7 fixtures/1000 S.F.	4.46	6.15	10.61

D50 Electrical

D5020 Lighting and Branch Wiring

D5020 238	H.I.D. Fixture, Low Bay, 8'-10' (by Wattage)	COST PER S.F.		
		MAT.	INST.	TOTAL
0440	Type K, 63 FC, 11 fixtures/1000 S.F.	5.45	6.35	11.80
0560	3 watt/S.F., type J, 81 FC, 11 fixtures/1000 S.F.	6.90	9.30	16.20
0600	Type K, 92 FC, 16 fixtures/1000 S.F.	7.95	9.35	17.30
0720	4 watt/S.F., type J, 103 FC, 14 fixtures/1000 S.F.	8.90	12.25	21.15
0760	Type K, 127 FC, 22 fixtures/1000 S.F.	10.85	12.70	23.55
0880	5 watt/S.F., type J, 133 FC, 18 fixtures/1000 S.F.	11.35	15.45	26.80
0920	Type K, 155 FC, 27 fixtures/1000 S.F.	13.35	15.70	29.05

D5020 242	H.I.D. Fixture, Low Bay, 16' (by Wattage)	COST PER S.F.		
		MAT.	INST.	TOTAL
0190	High intensity discharge fixture, mounted 16' above work plane			
0240	1 watt/S.F., type J, 28 FC, 4 fixt./1000 S.F.	2.65	3.80	6.45
0280	Type K, 27 FC, 5 fixt./1000 S.F.	2.95	4.28	7.23
0400	2 watt/S.F., type J, 48 FC, 7 fixt/1000 S.F.	4.89	7.40	12.29
0440	Type K, 58 FC, 11 fixt/1000 S.F.	6.25	8.75	15
0560	3 watt/S.F., type J, 75 FC, 11 fixt/1000 S.F.	7.55	11.20	18.75
0600	Type K, 85 FC, 16 fixt/1000 S.F.	9.20	13	22.20
0720	4 watt/S.F., type J, 95 FC, 14 fixt/1000 S.F.	9.80	14.75	24.55
0760	Type K, 117 FC, 22 fixt/1000 S.F.	12.50	17.45	29.95
0880	5 watt/S.F., type J, 122 FC, 18 fixt/1000 S.F.	12.45	18.55	31
0920	Type K, 143 FC, 27 fixt/1000 S.F.	15.45	21.50	36.95

D5030 Communications and Security

D5030 910	Communication & Alarm Systems	COST EACH		
		MAT.	INST.	TOTAL
0200	Communication & alarm systems, includes outlets, boxes, conduit & wire			
0210	Sound system, 6 outlets	6,275	8,350	14,625
0220	12 outlets	9,075	13,300	22,375
0240	30 outlets	16,100	25,200	41,300
0280	100 outlets	49,300	84,500	133,800
0320	Fire detection systems, non-addressable, 12 detectors	3,550	6,900	10,450
0360	25 detectors	6,200	11,700	17,900
0400	50 detectors	12,200	23,200	35,400
0440	100 detectors	22,600	42,100	64,700
0450	Addressable type, 12 detectors	5,225	6,975	12,200
0452	25 detectors	9,475	11,800	21,275
0454	50 detectors	18,300	23,200	41,500
0456	100 detectors	35,300	42,500	77,800
0458	Fire alarm control panel, 8 zone, excluding wire and conduit	755	1,325	2,080
0459	12 zone	2,625	1,975	4,600
0460	Fire alarm command center, addressable without voice, excl. wire & conduit	4,925	1,150	6,075
0462	Addressable with voice	10,400	1,825	12,225
0480	Intercom systems, 6 stations	4,225	5,700	9,925
0560	25 stations	14,000	21,900	35,900
0640	100 stations	53,000	80,000	133,000
0680	Master clock systems, 6 rooms	5,575	9,300	14,875
0720	12 rooms	8,975	15,900	24,875
0760	20 rooms	12,500	22,400	34,900
0800	30 rooms	20,800	41,400	62,200
0840	50 rooms	34,000	70,000	104,000
0920	Master TV antenna systems, 6 outlets	2,775	5,900	8,675
0960	12 outlets	5,150	11,000	16,150
1000	30 outlets	12,700	25,400	38,100
1040	100 outlets	42,300	83,000	125,300

D5090   Other Electrical Systems

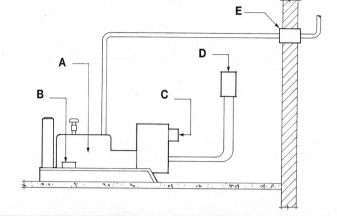

Generator System

A: Engine
B: Battery
C: Charger
D: Transfer Switch
E: Muffler

D5090 210	Generators (by kW)	COST PER kW		
		MAT.	INST.	TOTAL
0190	Generator sets, include battery, charger, muffler & transfer switch			
0200	Gas/gasoline operated, 3 phase, 4 wire, 277/480 volt, 7.5 kW	1,250	288	1,538
0240	11.5 kW	1,150	219	1,369
0280	20 kW	785	142	927
0320	35 kW	535	93	628
0360	80 kW	385	56.50	441.50
0400	100 kW	335	54.50	389.50
0440	125 kW	550	51.50	601.50
0480	185 kW	490	39	529
0560	Diesel engine with fuel tank, 30 kW	410	108	518
0600	50 kW	435	85.50	520.50
0640	75 kW	335	68.50	403.50
0680	100 kW	305	58	363
0760	150 kW	273	46	319
0840	200 kW	247	37.50	284.50
0880	250 kW	210	31	241
0960	350 kW	181	25.50	206.50
1040	500 kW	198	19.95	217.95

489

For customer support on your Square Foot Cost Data, call 877.756.2789.

E1010 Commercial Equipment

E1010 110	Security/Vault, EACH	COST EACH		
		MAT.	INST.	TOTAL
0100	Bank equipment, drive up window, drawer & mike, no glazing, economy	7,725	1,450	9,175
0110	Deluxe	10,100	2,875	12,975
0120	Night depository, economy	8,300	1,450	9,750
0130	Deluxe	11,800	2,875	14,675
0140	Pneumatic tube systems, 2 station, standard	28,800	5,100	33,900
0150	Teller, automated, 24 hour, single unit	49,700	5,100	54,800
0160	Teller window, bullet proof glazing, 44" x 60"	4,650	945	5,595
0170	Pass through, painted steel, 72" x 40"	4,800	1,800	6,600
0300	Safe, office type, 1 hr. rating, 34" x 20" x 20"	2,500		2,500
0310	4 hr. rating, 62" x 33" x 20"	9,750		9,750
0320	Data storage, 4 hr. rating, 63" x 44" x 16"	14,800		14,800
0330	Jewelers, 63" x 44" x 16"	15,800		15,800
0340	Money, "B" label, 9" x 14" x 14"	595		595
0350	Tool and torch resistive, 24" x 24" x 20"	8,650	233	8,883
0500	Security gates-scissors type, painted steel, single, 6' high, 5-1/2' wide	252	360	612
0510	Double gate, 7-1/2' high, 14' wide	695	720	1,415

E1010 510	Mercantile Equipment, EACH	COST EACH		
		MAT.	INST.	TOTAL
0015	Barber equipment, chair, hydraulic, economy	690	25	715
0020	Deluxe	4,375	37	4,412
0100	Checkout counter, single belt	3,500	93	3,593
0110	Double belt, power take-away	5,025	103	5,128
0200	Display cases, freestanding, glass and aluminum, 3'-6" x 3' x 1'-0" deep	1,425	149	1,574
0220	Wall mounted, glass and aluminum, 3' x 4' x 1'-4" deep	2,375	238	2,613
0320	Frozen food, chest type, 12 ft. long	9,050	400	9,450

E1010 610	Laundry/Dry Cleaning, EACH	COST EACH		
		MAT.	INST.	TOTAL
0100	Laundry equipment, dryers, gas fired, residential, 16 lb. capacity	755	238	993
0110	Commercial, 30 lb. capacity, single	3,700	238	3,938
0120	Dry cleaners, electric, 20 lb. capacity	37,500	6,875	44,375
0130	30 lb. capacity	57,500	9,175	66,675
0140	Ironers, commercial, 120" with canopy, 8 roll	182,500	19,600	202,100
0150	Institutional, 110", single roll	35,900	3,300	39,200
0160	Washers, residential, 4 cycle	975	238	1,213
0170	Commercial, coin operated, deluxe	3,775	238	4,013

E10 Equipment

E1020 Institutional Equipment

E1020 110	Ecclesiastical Equipment, EACH	COST EACH		
		MAT.	INST.	TOTAL
0090	Church equipment, altar, wood, custom, plain	2,875	425	3,300
0100	Granite, custom, deluxe	40,500	5,600	46,100
0110	Baptistry, fiberglass, economy	6,175	1,550	7,725
0120	Bells & carillons, keyboard operation	20,400	8,650	29,050
0130	Confessional, wood, single, economy	3,575	990	4,565
0140	Double, deluxe	20,900	2,975	23,875
0150	Steeples, translucent fiberglas, 30" square, 15' high	9,400	1,850	11,250

E1020 130	Ecclesiastical Equipment, L.F.	COST PER L.F.		
		MAT.	INST.	TOTAL
0100	Arch. equip., church equip. pews, bench type, hardwood, economy	111	29.50	140.50
0110	Deluxe	194	39.50	233.50

E1020 210	Library Equipment, EACH	COST EACH		
		MAT.	INST.	TOTAL
0110	Library equipment, carrels, metal, economy	289	119	408
0120	Hardwood, deluxe	2,250	149	2,399

E1020 230	Library Equipment, L.F.	COST PER L.F.		
		MAT.	INST.	TOTAL
0100	Library equipment, book shelf, metal, single face, 90" high x 10" shelf	134	49.50	183.50
0110	Double face, 90" high x 10" shelf	490	107	597
0120	Charging desk, built-in, with counter, plastic laminate	335	85	420

E1020 310	Theater and Stage Equipment, EACH	COST EACH		
		MAT.	INST.	TOTAL
0200	Movie equipment, changeover, economy	535		535
0210	Film transport, incl. platters and autowind, economy	5,600		5,600
0220	Lamphouses, incl. rectifiers, xenon, 1000W	7,400	330	7,730
0230	4000W	10,800	440	11,240
0240	Projector mechanisms, 35 mm, economy	12,700		12,700
0250	Deluxe	17,400		17,400
0260	Sound systems, incl. amplifier, single, economy	3,800	735	4,535
0270	Dual, Dolby/super sound	18,900	1,650	20,550
0280	Projection screens, wall hung, manual operation, 50 S.F., economy	355	119	474
0290	Electric operation, 100 S.F., deluxe	2,650	595	3,245

E1020 320	Theater and Stage Equipment, S.F.	COST PER S.F.		
		MAT.	INST.	TOTAL
0090	Movie equipment, projection screens, rigid in wall, acrylic, 1/4" thick	48	5.85	53.85
0100	1/2" thick	55.50	8.75	64.25
0110	Stage equipment, curtains, velour, medium weight	9.05	1.98	11.03
0120	Silica based yarn, fireproof	16.90	24	40.90
0130	Stages, portable with steps, folding legs, 8" high	37		37
0140	Telescoping platforms, aluminum, deluxe	55	31	86

E1020 330	Theater and Stage Equipment, L.F.	COST PER L.F.		
		MAT.	INST.	TOTAL
0100	Stage equipment, curtain track, heavy duty	69	66	135
0110	Lights, border, quartz, colored	192	33	225

E1020 610	Detention Equipment, EACH	COST EACH		
		MAT.	INST.	TOTAL
0110	Detention equipment, cell front rolling door, 7/8" bars, 5' x 7' high	5,900	1,525	7,425
0120	Cells, prefab., including front, 5' x 7' x 7' deep	10,800	2,050	12,850

E1020 Institutional Equipment

E1020 610	Detention Equipment, EACH	COST EACH		
		MAT.	INST.	TOTAL
0130	Doors and frames, 3' x 7', single plate	5,050	770	5,820
0140	Double plate	6,150	770	6,920
0150	Toilet apparatus, incl wash basin	3,775	1,025	4,800
0160	Visitor cubicle, vision panel, no intercom	3,525	1,525	5,050

E1020 710	Laboratory Equipment, EACH	COST EACH		
		MAT.	INST.	TOTAL
0110	Laboratory equipment, glassware washer, distilled water, economy	6,850	765	7,615
0120	Deluxe	15,300	1,375	16,675
0140	Radio isotope	18,600		18,600

E1020 720	Laboratory Equipment, S.F.	COST PER S.F.		
		MAT.	INST.	TOTAL
0100	Arch. equip., lab equip., counter tops, acid proof, economy	47	14.50	61.50
0110	Stainless steel	165	14.50	179.50

E1020 730	Laboratory Equipment, L.F.	COST PER L.F.		
		MAT.	INST.	TOTAL
0110	Laboratory equipment, cabinets, wall, open	203	59.50	262.50
0120	Base, drawer units	580	66	646
0130	Fume hoods, not incl. HVAC, economy	560	220	780
0140	Deluxe incl. fixtures	1,075	495	1,570

E1020 810	Medical Equipment, EACH	COST EACH		
		MAT.	INST.	TOTAL
0100	Dental equipment, central suction system, economy	1,750	595	2,345
0110	Compressor-air, deluxe	9,925	1,225	11,150
0120	Chair, hydraulic, economy	2,450	1,225	3,675
0130	Deluxe	8,600	2,450	11,050
0140	Drill console with accessories, economy	2,375	385	2,760
0150	Deluxe	5,375	385	5,760
0160	X-ray unit, portable	2,625	154	2,779
0170	Panoramic unit	17,600	1,025	18,625
0300	Medical equipment, autopsy table, standard	10,700	715	11,415
0310	Deluxe	17,900	1,200	19,100
0320	Incubators, economy	3,425		3,425
0330	Deluxe	13,100		13,100
0700	Station, scrub-surgical, single, economy	4,825	238	5,063
0710	Dietary, medium, with ice	18,400		18,400
0720	Sterilizers, general purpose, single door, 20" x 20" x 28"	14,200		14,200
0730	Floor loading, double door, 28" x 67" x 52"	227,000		227,000
0740	Surgery tables, standard	13,400	1,025	14,425
0750	Deluxe	28,900	1,450	30,350
0770	Tables, standard, with base cabinets, economy	1,250	395	1,645
0780	Deluxe	6,175	595	6,770
0790	X-ray, mobile, economy	18,500		18,500
0800	Stationary, deluxe	247,000		247,000

494

For customer support on your Square Foot Cost Data, call 877.756.2789.

E1030 Vehicular Equipment

E1030 110	Vehicular Service Equipment, EACH	COST EACH		
		MAT.	INST.	TOTAL
0110	Automotive equipment, compressors, electric, 1-1/2 H.P., std. controls	515	1,125	1,640
0120	5 H.P., dual controls	3,100	1,675	4,775
0130	Hoists, single post, 4 ton capacity, swivel arms	6,050	4,200	10,250
0140	Dual post, 12 ton capacity, adjustable frame	10,400	880	11,280
0150	Lube equipment, 3 reel type, with pumps	9,675	3,350	13,025
0160	Product dispenser, 6 nozzles, w/vapor recovery, not incl. piping, installed	27,200		27,200
0800	Scales, dial type, built in floor, 5 ton capacity, 8' x 6' platform	9,350	3,575	12,925
0810	10 ton capacity, 9' x 7' platform	12,100	5,100	17,200
0820	Truck (including weigh bridge), 20 ton capacity, 24' x 10'	13,900	5,950	19,850

E1030 210	Parking Control Equipment, EACH	COST EACH		
		MAT.	INST.	TOTAL
0110	Parking equipment, automatic gates, 8 ft. arm, one way	4,425	1,200	5,625
0120	Traffic detectors, single treadle	2,225	550	2,775
0130	Booth for attendant, economy	8,000		8,000
0140	Deluxe	29,000		29,000
0150	Ticket printer/dispenser, rate computing	9,450	940	10,390
0160	Key station on pedestal	635	320	955

E1030 310	Loading Dock Equipment, EACH	COST EACH		
		MAT.	INST.	TOTAL
0110	Dock bumpers, rubber blocks, 4-1/2" thick, 10" high, 14" long	66	23	89
0120	6" thick, 20" high, 11" long	137	45.50	182.50
0130	Dock boards, H.D., 5' x 5', aluminum, 5000 lb. capacity	1,500		1,500
0140	16,000 lb. capacity	1,725		1,725
0150	Dock levelers, hydraulic, 7' x 8', 10 ton capacity	6,325	1,500	7,825
0160	Dock lifters, platform, 6' x 6', portable, 3000 lb. capacity	10,200		10,200
0170	Dock shelters, truck, scissor arms, economy	2,050	595	2,645
0180	Deluxe	2,750	1,200	3,950

E10 Equipment

E1090   Other Equipment

E1090 210 — Solid Waste Handling Equipment, EACH

		COST EACH		
		MAT.	**INST.**	**TOTAL**
0110	Waste handling, compactors, single bag, 250 lbs./hr., hand fed	16,900	700	17,600
0120	Heavy duty industrial, 5 C.Y. capacity	36,500	3,350	39,850
0130	Incinerator, electric, 100 lbs./hr., economy	72,500	3,275	75,775
0140	Gas, 2000 lbs./hr., deluxe	419,500	27,200	446,700
0150	Shredder, no baling, 35 tons/hr.	336,500		336,500
0160	Incl. baling, 50 tons/day	672,000		672,000

E1090 350 — Food Service Equipment, EACH

		COST EACH		
		MAT.	**INST.**	**TOTAL**
0110	Kitchen equipment, bake oven, single deck	6,075	161	6,236
0120	Broiler, without oven	3,925	161	4,086
0130	Commercial dish washer, semiautomatic, 50 racks/hr.	8,100	990	9,090
0140	Automatic, 275 racks/hr.	34,600	4,175	38,775
0150	Cooler, beverage, reach-in, 6 ft. long	3,700	214	3,914
0160	Food warmer, counter, 1.65 kw	770		770
0170	Fryers, with submerger, single	1,375	184	1,559
0180	Double	2,650	257	2,907
0185	Ice maker, 1000 lb. per day, with bin	5,650	1,275	6,925
0190	Kettles, steam jacketed, 20 gallons	9,575	283	9,858
0200	Range, restaurant type, burners, 2 ovens and 24" griddle	5,225	214	5,439
0210	Range hood, incl. carbon dioxide system, economy	2,075	430	2,505
0220	Deluxe	2,075	430	2,505

E1090 360 — Food Service Equipment, S.F.

		COST PER S.F.		
		MAT.	**INST.**	**TOTAL**
0110	Refrigerators, prefab, walk-in, 7'-6" high, 6' x 6'	181	21.50	202.50
0120	12' x 20'	114	10.85	124.85

E1090 410 — Residential Equipment, EACH

		COST EACH		
		MAT.	**INST.**	**TOTAL**
0110	Arch. equip., appliances, range, cook top, 4 burner, economy	385	110	495
0120	Built in, single oven 30" wide, economy	990	110	1,100
0130	Standing, single oven-21" wide, economy	555	93	648
0135	Free standing, 30" wide, 1 oven, average	1,000	186	1,186
0140	Double oven-30" wide, deluxe	3,075	93	3,168
0150	Compactor, residential, economy	760	119	880
0160	Deluxe	1,275	198	1,473
0170	Dish washer, built-in, 2 cycles, economy	330	345	675
0180	4 or more cycles, deluxe	1,325	685	2,010
0190	Garbage disposer, sink type, economy	102	137	239
0200	Deluxe	267	137	404
0210	Refrigerator, no frost, 10 to 12 C.F., economy	520	93	613
0220	21 to 29 C.F., deluxe	4,225	310	4,535
0300	Washing machine, automatic	1,825	715	2,540

E1090 610 — School Equipment, EACH

		COST EACH		
		MAT.	**INST.**	**TOTAL**
0110	School equipment, basketball backstops, wall mounted, wood, fixed	1,650	1,050	2,700
0120	Suspended type, electrically operated	7,250	1,975	9,225
0130	Bleachers-telescoping, manual operation, 15 tier, economy (per seat)	106	37	143
0140	Power operation, 30 tier, deluxe (per seat)	435	65.50	500.50
0150	Weight lifting gym, universal, economy	385	930	1,315
0160	Deluxe	16,100	1,850	17,950
0170	Scoreboards, basketball, 1 side, economy	2,750	870	3,620
0180	4 sides, deluxe	17,300	11,900	29,200
0800	Vocational shop equipment, benches, metal	630	238	868
0810	Wood	735	238	973

E10 Equipment

E1090 Other Equipment

E1090 610	School Equipment, EACH	COST EACH		
		MAT.	INST.	TOTAL
0820	Dust collector, not incl. ductwork, 6' diam.	5,450	635	6,085
0830	Planer, 13" x 6"	1,200	297	1,497

E1090 620	School Equipment, S.F.	COST PER S.F.		
		MAT.	INST.	TOTAL
0110	School equipment, gym mats, naugahyde cover, 2" thick	5.15		5.15
0120	Wrestling, 1" thick, heavy duty	5.15		5.15

E1090 810	Athletic, Recreational, and Therapeutic Equipment, EACH	COST EACH		
		MAT.	INST.	TOTAL
0050	Bowling alley with gutters	51,000	12,500	63,500
0060	Combo. table and ball rack	1,400		1,400
0070	Bowling alley automatic scorer	11,000		11,000
0110	Sauna, prefabricated, incl. heater and controls, 7' high, 6' x 4'	5,850	900	6,750
0120	10' x 12'	14,100	1,975	16,075
0130	Heaters, wall mounted, to 200 C.F.	805		805
0140	Floor standing, to 1000 C.F., 12500 W	4,275	220	4,495
0610	Shooting range incl. bullet traps, controls, separators, ceilings, economy	40,100	7,675	47,775
0620	Deluxe	59,500	12,900	72,400
0650	Sport court, squash, regulation, in existing building, economy			40,400
0660	Deluxe			45,000
0670	Racketball, regulation, in existing building, economy	43,800	11,200	55,000
0680	Deluxe	47,400	22,500	69,900
0700	Swimming pool equipment, diving stand, stainless steel, 1 meter	10,500	440	10,940
0710	3 meter	17,200	2,975	20,175
0720	Diving boards, 16 ft. long, aluminum	4,475	440	4,915
0730	Fiberglass	3,625	440	4,065
0740	Filter system, sand, incl. pump, 6000 gal./hr.	2,225	795	3,020
0750	Lights, underwater, 12 volt with transformer, 300W	340	660	1,000
0760	Slides, fiberglass with aluminum handrails & ladder, 6' high, straight	3,975	745	4,720
0780	12' high, straight with platform	16,000	990	16,990

E1090 820	Athletic, Recreational, and Therapeutic Equipment, S.F.	COST PER S.F.		
		MAT.	INST.	TOTAL
0110	Swimming pools, residential, vinyl liner, metal sides	23.50	8.20	31.70
0120	Concrete sides	28.50	14.85	43.35
0130	Gunite shell, plaster finish, 350 S.F.	52.50	30.50	83
0140	800 S.F.	42.50	17.80	60.30
0150	Motel, gunite shell, plaster finish	65	39	104
0160	Municipal, gunite shell, tile finish, formed gutters	249	67	316

E2010 Fixed Furnishings

E2010 310	Window Treatment, EACH	COST EACH		
		MAT.	INST.	TOTAL
0110	Furnishings, blinds, exterior, aluminum, louvered, 1'-4" wide x 3'-0" long	220	59.50	279.50
0120	1'-4" wide x 6'-8" long	390	66	456
0130	Hemlock, solid raised, 1'-4" wide x 3'-0" long	79	59.50	138.50
0140	1'-4" wide x 6'-9" long	167	66	233
0150	Polystyrene, louvered, 1'-3" wide x 3'-3" long	39.50	59.50	99
0160	1'-3" wide x 6'-8" long	75	66	141
0200	Interior, wood folding panels, louvered, 7" x 20" (per pair)	90.50	35	125.50
0210	18" x 40" (per pair)	148	35	183

E2010 320	Window Treatment, S.F.	COST PER S.F.		
		MAT.	INST.	TOTAL
0110	Furnishings, blinds-interior, venetian-aluminum, stock, 2" slats	6	1.01	7.01
0120	Custom, 1" slats, deluxe	5.15	1.01	6.16
0130	Vertical, PVC or cloth, T&B track, economy	8.60	1.29	9.89
0140	Deluxe	13.80	1.49	15.29
0150	Draperies, unlined, economy	3.19		3.19
0160	Lightproof, deluxe	10.55		10.55
0510	Shades, mylar, wood roller, single layer, non-reflective	3.52	.87	4.39
0520	Metal roller, triple layer, heat reflective	8.95	.87	9.82
0530	Vinyl, light weight, 4 ga.	.69	.87	1.56
0540	Heavyweight, 6 ga.	2.12	.87	2.99
0550	Vinyl coated cotton, lightproof decorator shades	4.99	.87	5.86
0560	Woven aluminum, 3/8" thick, light and fireproof	6.65	1.70	8.35

E2010 420	Fixed Floor Grilles and Mats, S.F.	COST PER S.F.		
		MAT.	INST.	TOTAL
0110	Floor mats, recessed, inlaid black rubber, 3/8" thick, solid	25.50	3	28.50
0120	Colors, 1/2" thick, perforated	30.50	3	33.50
0130	Link-including nosings, steel-galvanized, 3/8" thick	27	3	30
0140	Vinyl, in colors	28.50	3	31.50

E2010 510	Fixed Multiple Seating, EACH	COST EACH		
		MAT.	INST.	TOTAL
0110	Seating, painted steel, upholstered, economy	144	34	178
0120	Deluxe	490	42.50	532.50
0400	Seating, lecture hall, pedestal type, economy	221	54	275
0410	Deluxe	540	82	622
0500	Auditorium chair, veneer construction	260	54	314
0510	Fully upholstered, spring seat	266	54	320

E2020 Moveable Furnishings

E2020 210	Furnishings/EACH	COST EACH		
		MAT.	INST.	TOTAL
0200	Hospital furniture, beds, manual, economy	865		865
0210	Deluxe	2,700		2,700
0220	All electric, economy	1,925		1,925
0230	Deluxe	4,175		4,175
0240	Patient wall systems, no utilities, economy, per room	1,575		1,575
0250	Deluxe, per room	2,125		2,125
0300	Hotel furnishings, standard room set, economy, per room	2,775		2,775
0310	Deluxe, per room	9,475		9,475
0500	Office furniture, standard employee set, economy, per person	705		705
0510	Deluxe, per person	2,675		2,675
0550	Posts, portable, pedestrian traffic control, economy	165		165
0560	Deluxe	465		465
0700	Restaurant furniture, booth, molded plastic, stub wall and 2 seats, economy	420	297	717
0710	Deluxe	1,550	395	1,945
0720	Upholstered seats, foursome, single-economy	860	119	979
0730	Foursome, double-deluxe	1,950	198	2,148

E2020 220	Furniture and Accessories, L.F.	COST PER L.F.		
		MAT.	INST.	TOTAL
0210	Dormitory furniture, desk top (built-in),laminated plastc, 24"deep, economy	49.50	24	73.50
0220	30" deep, deluxe	276	29.50	305.50
0230	Dressing unit, built-in, economy	204	99	303
0240	Deluxe	610	149	759
0310	Furnishings, cabinets, hospital, base, laminated plastic	300	119	419
0320	Stainless steel	555	119	674
0330	Countertop, laminated plastic, no backsplash	52	29.50	81.50
0340	Stainless steel	169	29.50	198.50
0350	Nurses station, door type, laminated plastic	350	119	469
0360	Stainless steel	670	119	789
0710	Restaurant furniture, bars, built-in, back bar	228	119	347
0720	Front bar	315	119	434
0910	Wardrobes & coatracks, standing, steel, single pedestal, 30" x 18" x 63"	120		120
0920	Double face rack, 39" x 26" x 70"	136		136
0930	Wall mounted rack, steel frame & shelves, 12" x 15" x 26"	77.50	8.45	85.95
0940	12" x 15" x 50"	42.50	4.43	46.93

F1010 Special Structures

F1010 120	Air-Supported Structures, S.F.	COST PER S.F.		
		MAT.	INST.	TOTAL
0110	Air supported struc., polyester vinyl fabric, 24oz., warehouse, 5000 S.F.	29	.37	29.37
0120	50,000 S.F.	13.55	.30	13.85
0130	Tennis, 7,200 S.F.	26	.31	26.31
0140	24,000 S.F.	18.05	.31	18.36
0150	Woven polyethylene, 6 oz., shelter, 3,000 S.F.	17.60	.62	18.22
0160	24,000 S.F.	13.20	.31	13.51
0170	Teflon coated fiberglass, stadium cover, economy	63	.16	63.16
0180	Deluxe	75	.22	75.22
0190	Air supported storage tank covers, reinf. vinyl fabric, 12 oz., 400 S.F.	25.50	.52	26.02
0200	18,000 S.F.	9.30	.47	9.77

F1010 210	Pre-Engineered Structures, EACH	COST EACH		
		MAT.	INST.	TOTAL
0600	Radio towers, guyed, 40 lb. section, 50' high, 70 MPH basic wind speed	3,075	1,450	4,525
0610	90 lb. section, 400' high, wind load 70 MPH basic wind speed	42,400	16,500	58,900
0620	Self supporting, 60' high, 70 MPH basic wind speed	5,000	2,900	7,900
0630	190' high, wind load 90 MPH basic wind speed	31,400	11,600	43,000
0700	Shelters, aluminum frame, acrylic glazing, 8' high, 3' x 9'	3,375	1,275	4,650
0710	9' x 12'	8,025	1,975	10,000

F1010 320	Other Special Structures, S.F.	COST PER S.F.		
		MAT.	INST.	TOTAL
0110	Swimming pool enclosure, translucent, freestanding, economy	37	5.95	42.95
0120	Deluxe	102	17	119
0510	Tension structures, steel frame, polyester vinyl fabric, 12,000 S.F.	14.55	2.71	17.26
0520	20,800 S.F.	14.30	2.44	16.74

F1010 330	Special Structures, EACH	COST EACH		
		MAT.	INST.	TOTAL
0110	Kiosks, round, 5' diam., 7' high, aluminum wall, illuminated	27,600		27,600
0120	Rectangular, 5' x 9', 1" insulated dbl. wall fiberglass, 7'-6" high	28,200		28,200
0220	Silos, steel prefab, 30,000 gal., painted, economy	24,300	5,800	30,100
0230	Epoxy-lined, deluxe	50,500	11,600	62,100

F1010 340	Special Structures, S.F.	COST PER S.F.		
		MAT.	INST.	TOTAL
0110	Comfort stations, prefab, mobile on steel frame, economy	196		196
0120	Permanent on concrete slab, deluxe	211	48	259
0210	Domes, bulk storage, wood framing, wood decking, 50' diam.	75.50	1.87	77.37
0220	116' diam.	36.50	2.16	38.66
0230	Steel framing, metal decking, 150' diam.	36.50	11.80	48.30
0240	400' diam.	29.50	9	38.50
0250	Geodesic, wood framing, wood panels, 30' diam.	37.50	2.14	39.64
0260	60' diam.	24.50	1.32	25.82
0270	Aluminum framing, acrylic panels, 40' diam.			70.50
0280	Aluminum panels, 400' diam.			22
0310	Garden house, prefab, wood, shell only, 48 S.F.	57	5.95	62.95
0320	200 S.F.	43.50	25	68.50
0410	Greenhouse, shell-stock, residential, lean-to, 8'-6" long x 3'-10" wide	47.50	35	82.50
0420	Freestanding, 8'-6" long x 13'-6" wide	47.50	11	58.50
0430	Commercial-truss frame, under 2000 S.F., deluxe			14.65
0440	Over 5,000 S.F., economy			13.40
0450	Institutional-rigid frame, under 500 S.F., deluxe			30
0460	Over 2,000 S.F., economy			12.25
0510	Hangar, prefab, galv. roof and walls, bottom rolling doors, economy	13.55	4.69	18.24
0520	Electric bifolding doors, deluxe	13.40	6.10	19.50

F10 Special Construction

F1020 Integrated Construction

F1020 110	Integrated Construction, EACH	COST EACH		
		MAT.	INST.	TOTAL
0110	Integrated ceilings, radiant electric, 2' x 4' panel, manila finish	101	26.50	127.50
0120	ABS plastic finish	114	50.50	164.50

F1020 120	Integrated Construction, S.F.	COST PER S.F.		
		MAT.	INST.	TOTAL
0110	Integrated ceilings, Luminaire, suspended, 5' x 5' modules, 50% lighted	4.41	14.15	18.56
0120	100% lighted	6.55	25.50	32.05
0130	Dimensionaire, 2' x 4' module tile system, no air bar	2.66	2.38	5.04
0140	With air bar, deluxe	3.90	2.38	6.28
0220	Pedestal access floor pkg., inc. stl. pnls, peds.& stringers, w/vinyl cov.	27	3.17	30.17
0230	With high pressure laminate covering	25	3.17	28.17
0240	With carpet covering	26.40	3.17	29.55
0250	Aluminum panels, no stringers, no covering	36	2.38	38.38

F1020 250	Special Purpose Room, EACH	COST EACH		
		MAT.	INST.	TOTAL
0110	Portable booth, acoustical, 27 db 1000 hz., 15 S.F. floor	4,100		4,100
0120	55 S.F. flr.	8,450		8,450

F1020 260	Special Purpose Room, S.F.	COST PER S.F.		
		MAT.	INST.	TOTAL
0110	Anechoic chambers, 7' high, 100 cps cutoff, 25 S.F.			2,650
0120	200 cps cutoff, 100 S.F.			1,375
0130	Audiometric rooms, under 500 S.F.	59	24.50	83.50
0140	Over 500 S.F.	57.50	19.80	77.30
0300	Darkrooms, shell, not including door, 240 S.F., 8' high	31	9.90	40.90
0310	64 S.F., 12' high	76.50	18.60	95.10
0510	Music practice room, modular, perforated steel, under 500 S.F.	38	17	55
0520	Over 500 S.F.	32	14.85	46.85

F1020 330	Special Construction, L.F.	COST PER L.F.		
		MAT.	INST.	TOTAL
0110	Spec. const., air curtains, shipping & receiving, 8'high x 5'wide, economy	785	140	925
0120	20' high x 8' wide, heated, deluxe	3,675	350	4,025
0130	Customer entrance, 10' high x 5' wide, economy	1,175	140	1,315
0140	12' high x 4' wide, heated, deluxe	2,125	175	2,300

For customer support on your Square Foot Cost Data, call 877.756.2789.

503

F10 Special Construction

F1030 Special Construction Systems

F1030 120	Sound, Vibration, and Seismic Construction, S.F.	COST PER S.F.		
		MAT.	INST.	TOTAL
0020	Special construction, acoustical, enclosure, 4" thick, 8 psf panels	37	25	62
0030	Reverb chamber, 4" thick, parallel walls	52	29.50	81.50
0110	Sound absorbing panels, 2'-6" x 8', painted metal	13.85	8.30	22.15
0120	Vinyl faced	10.80	7.45	18.25
0130	Flexible transparent curtain, clear	8.35	9.75	18.10
0140	With absorbing foam, 75% coverage	11.65	9.75	21.40
0150	Strip entrance, 2/3 overlap	8.35	15.55	23.90
0160	Full overlap	10.55	18.25	28.80
0200	Audio masking system, plenum mounted, over 10,000 S.F.	.81	.30	1.11
0210	Ceiling mounted, under 5,000 S.F.	1.42	.55	1.97

F1030 210	Radiation Protection, EACH	COST EACH		
		MAT.	INST.	TOTAL
0110	Shielding, lead x-ray protection, radiography room, 1/16" lead, economy	11,100	4,475	15,575
0120	Deluxe	13,400	7,450	20,850
0210	Deep therapy x-ray, 1/4" lead, economy	31,100	14,000	45,100
0220	Deluxe	38,400	18,600	57,000

F1030 220	Radiation Protection, S.F.	COST PER S.F.		
		MAT.	INST.	TOTAL
0110	Shielding, lead, gypsum board, 5/8" thick, 1/16" lead	12.30	7.10	19.40
0120	1/8" lead	25.50	8.10	33.60
0130	Lath, 1/16" thick	11.55	8.25	19.80
0140	1/8" thick	34.50	9.30	43.80
0150	Radio frequency, galvanized steel, prefab type, economy	5.65	3.17	8.82
0160	Radio frequency, door, copper/wood laminate, 4" x 7'	9.95	8.50	18.45

F1030 910	Other Special Construction Systems, EACH	COST EACH		
		MAT.	INST.	TOTAL
0110	Disappearing stairways, folding, pine, 8'-6" ceiling	194	149	343
0220	Automatic electric, wood, 8' to 9' ceiling	9,575	1,200	10,775
0300	Fireplace prefabricated, freestanding or wall hung, painted	1,625	455	2,080
0310	Stainless steel	3,350	660	4,010
0320	Woodburning stoves, cast iron, economy, less than 1500 S.F. htg. area	1,425	915	2,340
0330	Greater than 2000 S.F. htg. area	3,075	1,475	4,550

F10 Special Construction

F1040 Special Facilities

F1040 210	Ice Rinks, EACH	COAT EACH		
		MAT.	INST.	TOTAL
0100	Ice skating rink, 85' x 200', 55° system, 5 mos., 100 ton			664,000
0110	90° system, 12 mos., 135 ton			751,000
0120	Dash boards, acrylic screens, polyethylene coated plywood	156,000	40,000	196,000
0130	Fiberglass and aluminum construction	179,000	40,000	219,000

F1040 510	Liquid & Gas Storage Tanks, EACH	COST EACH		
		MAT.	INST.	TOTAL
0100	Tanks, steel, ground level, 100,000 gal.			226,500
0110	10,000,000 gal.			5,151,500
0120	Elevated water, 50,000 gal.		196,000	468,000
0130	1,000,000 gal.		913,000	2,073,000
0150	Cypress wood, ground level, 3,000 gal.	9,900	11,800	21,700
0160	Redwood, ground level, 45,000 gal.	77,500	32,100	109,600

F1040 Special Facilities

F1040 910	Special Construction, EACH	COST EACH		
		MAT.	INST.	TOTAL
0110	Special construction, bowling alley incl. pinsetter, scorer etc., economy	42,500	11,900	54,400
0120	Deluxe	58,500	13,200	71,700
0130	For automatic scorer, economy, add	6,400		6,400
0140	Deluxe, add	11,000		11,000
0300	Control tower, modular, 12' x 10', incl. instrumentation, economy			882,000
0310	Deluxe			1,386,000
0400	Garage costs, residential, prefab, wood, single car economy	6,275	1,200	7,475
0410	Two car deluxe	14,200	2,375	16,575
0500	Hangars, prefab, galv. steel, bottom rolling doors, economy (per plane)	18,000	5,200	23,200
0510	Electrical bi-folding doors, deluxe (per plane)	16,000	7,150	23,150

G1030 Site Earthwork

G1030 805	Trenching Common Earth	COST PER L.F.		
		MAT.	INST.	TOTAL
1310	Trenching, common earth, no slope, 2' wide, 2' deep, 3/8 C.Y. bucket		2.67	2.67
1330	4' deep, 3/8 C.Y. bucket		5.21	5.21
1360	10' deep, 1 C.Y. bucket		11.51	11.51
1400	4' wide, 2' deep, 3/8 C.Y. bucket		5.50	5.50
1420	4' deep, 1/2 C.Y. bucket		9.12	9.12
1450	10' deep, 1 C.Y. bucket		24.85	24.85
1480	18' deep, 2-1/2 C.Y. bucket		40.05	40.05
1520	6' wide, 6' deep, 5/8 C.Y. bucket w/trench box		24.25	24.25
1540	10' deep, 1 C.Y. bucket		33.05	33.05
1570	20' deep, 3-1/2 C.Y. bucket		56	56
1640	8' wide, 12' deep, 1-1/2 C.Y. bucket w/trench box		46.35	46.35
1680	24' deep, 3-1/2 C.Y. bucket		88.50	88.50
1730	10' wide, 20' deep, 3-1/2 C.Y. bucket w/trench box		79.50	79.50
1740	24' deep, 3-1/2 C.Y. bucket		103	103
3500	1 to 1 slope, 2' wide, 2' deep, 3/8 C.Y. bucket		5.21	5.21
3540	4' deep, 3/8 C.Y. bucket		15.49	15.49
3600	10' deep, 1 C.Y. bucket		69	69
3800	4' wide, 2' deep, 3/8 C.Y. bucket		8.06	8.06
3840	4' deep, 1/2 C.Y. bucket		18.01	18.01
3900	10' deep, 1 C.Y. bucket		87.50	87.50
4030	6' wide, 6' deep, 5/8 C.Y. bucket w/trench box		48	48
4050	10' deep, 1 C.Y. bucket		79	79
4080	20' deep, 3-1/2 C.Y. bucket		239	239
4500	8' wide, 12' deep, 1-1/2 C.Y. bucket w/trench box		118	118
4650	24' deep, 3-1/2 C.Y. bucket		361	361
4800	10' wide, 20' deep, 3-1/2 C.Y. bucket w/trench box		244.50	244.50
4850	24' deep, 3-1/2 C.Y. bucket		383	383

G1030 806	Trenching Loam & Sandy Clay	COST PER L.F.		
		MAT.	INST.	TOTAL
1310	Trenching, loam & sandy clay, no slope, 2' wide, 2' deep, 3/8 C.Y. bucket		2.61	2.61

G1030 807	Trenching Sand & Gravel	COST PER L.F.		
		MAT.	INST.	TOTAL
1310	Trenching, sand & gravel, no slope, 2' wide, 2' deep, 3/8 C.Y. bucket		2.45	2.45
1320	3' deep, 3/8 C.Y. bucket		4.05	4.05
1330	4' deep, 3/8 C.Y. bucket		4.83	4.83
1340	6' deep, 3/8 C.Y. bucket		6.05	6.05
1350	8' deep, 1/2 C.Y. bucket		8	8
1360	10' deep, 1 C.Y. bucket		8.55	8.55
1400	4' wide, 2' deep, 3/8 C.Y. bucket		5.05	5.05
1410	3' deep, 3/8 C.Y. bucket		7.45	7.45
1420	4' deep, 1/2 C.Y. bucket		8.45	8.45
1430	6' deep, 1/2 C.Y. bucket		13.20	13.20
1440	8' deep, 1/2 C.Y. bucket		18.80	18.80
1450	10' deep, 1 C.Y. bucket		19.05	19.05
1460	12' deep, 1 C.Y. bucket		24	24
1470	15' deep, 1-1/2 C.Y. bucket		28	28
1480	18' deep, 2-1/2 C.Y. bucket		31.50	31.50
1520	6' wide, 6' deep, 5/8 C.Y. bucket w/trench box		22.50	22.50
1530	8' deep, 3/4 C.Y. bucket		29.50	29.50
1540	10' deep, 1 C.Y. bucket		29.50	29.50
1550	12' deep, 1-1/2 C.Y. bucket		33	33
1560	16' deep, 2 C.Y. bucket		43	43
1570	20' deep, 3-1/2 C.Y. bucket		51.50	51.50
1580	24' deep, 3-1/2 C.Y. bucket		63	63

G1030 Site Earthwork

G1030 807	Trenching Sand & Gravel	COST PER L.F.		
		MAT.	INST.	TOTAL
1640	8' wide, 12' deep, 1-1/2 C.Y. bucket w/trench box		43.50	43.50
1650	15' deep, 1-1/2 C.Y. bucket		54.50	54.50
1660	18' deep, 2-1/2 C.Y. bucket		62.50	62.50
1680	24' deep, 3-1/2 C.Y. bucket		83	83
1730	10' wide, 20' deep, 3-1/2 C.Y. bucket w/trench box		83.50	83.50
1740	24' deep, 3-1/2 C.Y. bucket		104	104
1780	12' wide, 20' deep, 3-1/2 C.Y. bucket w/trench box		99.50	99.50
1790	25' deep, 3-1/2 C.Y. bucket		129	129
1800	1/2:1 slope, 2' wide, 2' deep, 3/8 C.Y. bucket		3.65	3.65
1810	3' deep, 3/8 C.Y. bucket		6.35	6.35
1820	4' deep, 3/8 C.Y. bucket		9.60	9.60
1840	6' deep, 3/8 C.Y. bucket		15	15
1860	8' deep, 1/2 C.Y. bucket		24	24
1880	10' deep, 1 C.Y. bucket		30	30
2300	4' wide, 2' deep, 3/8 C.Y. bucket		6.25	6.25
2310	3' deep, 3/8 C.Y. bucket		10.15	10.15
2320	4' deep, 1/2 C.Y. bucket		12.60	12.60
2340	6' deep, 1/2 C.Y. bucket		23	23
2360	8' deep, 1/2 C.Y. bucket		37.50	37.50
2380	10' deep, 1 C.Y. bucket		43	43
2400	12' deep, 1 C.Y. bucket		72	72
2430	15' deep, 1-1/2 C.Y. bucket		81.50	81.50
2460	18' deep, 2-1/2 C.Y. bucket		127	127
2840	6' wide, 6' deep, 5/8 C.Y. bucket w/trench box		33.50	33.50
2860	8' deep, 3/4 C.Y. bucket		49	49
2880	10' deep, 1 C.Y. bucket		54.50	54.50
2900	12' deep, 1-1/2 C.Y. bucket		67	67
2940	16' deep, 2 C.Y. bucket		101	101
2980	20' deep, 3-1/2 C.Y. bucket		138	138
3020	24' deep, 3-1/2 C.Y. bucket		192	192
3100	8' wide, 12' deep, 1-1/4 C.Y. bucket w/trench box		77	77
3120	15' deep, 1-1/2 C.Y. bucket		107	107
3140	18' deep, 2-1/2 C.Y. bucket		136	136
3180	24' deep, 3-1/2 C.Y. bucket		212	212
3270	10' wide, 20' deep, 3-1/2 C.Y. bucket w/trench box		171	171
3280	24' deep, 3-1/2 C.Y. bucket		232	232
3370	12' wide, 20' deep, 3-1/2 C.Y. bucket w/trench box		190	190
3380	25' deep, 3-1/2 C.Y. bucket		268	268
3500	1:1 slope, 2' wide, 2' deep, 3/8 C.Y. bucket		6.60	6.60
3520	3' deep, 3/8 C.Y. bucket		9	9
3540	4' deep, 3/8 C.Y. bucket		14.40	14.40
3560	6' deep, 3/8 C.Y. bucket		15	15
3580	8' deep, 1/2 C.Y. bucket		40	40
3600	10' deep, 1 C.Y. bucket		51.50	51.50
3800	4' wide, 2' deep, 3/8 C.Y. bucket		7.45	7.45
3820	3' deep, 3/8 C.Y. bucket		12.80	12.80
3840	4' deep, 1/2 C.Y. bucket		16.75	16.75
3860	6' deep, 1/2 C.Y. bucket		33	33
3880	8' deep, 1/2 C.Y. bucket		56	56
3900	10' deep, 1 C.Y. bucket		67.50	67.50
3920	12' deep, 1 C.Y. bucket		96	96
3940	15' deep, 1-1/2 C.Y. bucket		135	135
3960	18' deep, 2-1/2 C.Y. bucket		174	174
4030	6' wide, 6' deep, 5/8 C.Y. bucket w/trench box		44.50	44.50
4040	8' deep, 3/4 C.Y. bucket		68.50	68.50
4050	10' deep, 1 C.Y. bucket		79.50	79.50
4060	12' deep, 1-1/2 C.Y. bucket		101	101
4070	16' deep, 2 C.Y. bucket		160	160

509

G10 Site Preparation

G1030 Site Earthwork

G1030 807	Trenching Sand & Gravel	COST PER L.F.		
		MAT.	INST.	TOTAL
4080	20' deep, 3-1/2 C.Y. bucket		225	225
4090	24' deep, 3-1/2 C.Y. bucket		320	320
4500	8' wide, 12' deep, 1-1/2 C.Y. bucket w/trench box		111	111
4550	15' deep, 1-1/2 C.Y. bucket		160	160
4600	18' deep, 2-1/2 C.Y. bucket		210	210
4650	24' deep, 3-1/2 C.Y. bucket		340	340
4800	10' wide, 20' deep, 3-1/2 C.Y. bucket w/trench box		257	257
4850	24' deep, 3-1/2 C.Y. bucket		360	360
4950	12' wide, 20' deep, 3-1/2 C.Y. bucket w/trench box		274	274
4980	25' deep, 3-1/2 C.Y. bucket		410	410

G1030 815	Pipe Bedding	COST PER L.F.		
		MAT.	INST.	TOTAL
1440	Pipe bedding, side slope 0 to 1, 1' wide, pipe size 6" diameter	1.88	.93	2.81
1460	2' wide, pipe size 8" diameter	4.07	1.99	6.06
1500	Pipe size 12" diameter	4.25	2.08	6.33
1600	4' wide, pipe size 20" diameter	10.15	4.99	15.14
1660	Pipe size 30" diameter	10.65	5.25	15.90
1680	6' wide, pipe size 32" diameter	18.25	8.95	27.20
1740	8' wide, pipe size 60" diameter	36	17.75	53.75
1780	12' wide, pipe size 84" diameter	66.50	32.50	99

G2010 Roadways

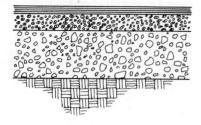

The Bituminous Roadway Systems are listed for pavement thicknesses between 3-1/2" and 7" and crushed stone bases from 3" to 22" in depth. Systems costs are expressed per linear foot for varying widths of two and multi-lane roads. Earth moving is not included. Granite curbs and line painting are added as required system components.

G2010 232	Bituminous Roadways Crushed Stone	COST PER L.F.		
		MAT.	INST.	TOTAL
1050	Bitum. roadway, two lanes, 3-1/2" th. pvmt., 3" th. crushed stone,24' wide	77.50	54	131.50
1100	28' wide	85	54.50	139.50
1150	32' wide	92.50	63	155.50
1200	4" thick crushed stone, 24' wide	80	51.50	131.50
1210	28' wide	87.50	55.50	143
1220	32' wide	95.50	59.50	155
1222	36' wide	103	63	166
1224	40' wide	111	66.50	177.50
1230	6" thick crushed stone, 24' wide	84.50	54	138.50
1240	28' wide	93	57.50	150.50
1250	32' wide	102	62	164
1252	36' wide	110	65.50	175.50
1254	40' wide	119	69.50	188.50
1256	8" thick crushed stone, 24' wide	89	55.50	144.50
1258	28' wide	98.50	60	158.50
1260	32' wide	108	64	172
1262	36' wide	117	68.50	185.50
1264	40' wide	127	73	200
1300	9" thick crushed stone, 24' wide	91.50	56.50	148
1350	28' wide	101	61.50	162.50
1400	32' wide	111	66	177
1410	10" thick crushed stone, 24' wide	93.50	57	150.50
1412	28' wide	104	61.50	165.50
1414	32' wide	114	66.50	180.50
1416	36' wide	124	71	195
1418	40' wide	135	76	211
1420	12" thick crushed stone, 24' wide	98.50	58.50	157
1422	28' wide	109	64	173
1424	32' wide	120	69	189
1426	36' wide	131	74	205
1428	40' wide	142	78.50	220.50
1430	15" thick crushed stone, 24' wide	105	61.50	166.50
1432	28' wide	117	67	184
1434	32' wide	130	72	202
1436	36' wide	142	78	220
1438	40' wide	154	83	237
1440	18" thick crushed stone, 24' wide	112	64	176
1442	28' wide	125	70	195
1444	32' wide	139	76	215
1446	36' wide	152	82	234
1448	40' wide	165	87	252
1550	4" thick pavement., 4" thick crushed stone, 24' wide	85.50	52.50	138

G2010 Roadways

G2010 232	Bituminous Roadways Crushed Stone	COST PER L.F.		
		MAT.	INST.	TOTAL
1600	28' wide	94	56	150
1650	32' wide	103	60.50	163.50
1652	36' wide	112	64	176
1654	40' wide	120	68	188
1700	6" thick crushed stone , 24' wide	90	54	144
1710	28' wide	99.50	58.50	158
1720	32' wide	109	62.50	171.50
1722	36' wide	119	66.50	185.50
1724	40' wide	128	71	199
1726	8" thick crushed stone , 24' wide	94.50	56	150.50
1728	28' wide	105	60	165
1730	32' wide	115	64.50	179.50
1732	36' wide	126	69	195
1734	40' wide	136	74	210
1800	10" thick crushed stone, 24' wide	99.50	58	157.50
1850	28' wide	111	61.50	172.50
1900	32' wide	122	67.50	189.50
1902	36' wide	132	72	204
1904	40' wide	138	75	213

G2020 Parking Lots

G2020 210	Parking Lots Gravel Base	COST PER CAR		
		MAT.	INST.	TOTAL
1500	Parking lot, 90° angle parking, 3" bituminous paving, 6" gravel base	805	440	1,245
1540	10" gravel base	915	515	1,430
1560	4" bituminous paving, 6" gravel base	1,025	470	1,495
1600	10" gravel base	1,150	540	1,690
1620	6" bituminous paving, 6" gravel base	1,375	500	1,875
1660	10" gravel base	1,500	575	2,075
1800	60° angle parking, 3" bituminous paving, 6" gravel base	805	440	1,245
1840	10" gravel base	915	515	1,430
1860	4" bituminous paving, 6" gravel base	1,025	470	1,495
1900	10" gravel base	1,150	540	1,690
1920	6" bituminous paving, 6" gravel base	1,375	500	1,875
1960	10" gravel base	1,500	575	2,075
2200	45° angle parking, 3" bituminous paving, 6" gravel base	825	455	1,280
2240	10" gravel base	935	530	1,465
2260	4" bituminous paving, 6" gravel base	1,050	480	1,530
2300	10" gravel base	1,175	555	1,730
2320	6" bituminous paving, 6" gravel base	1,425	515	1,940
2360	10" gravel base	1,525	590	2,115

G20 Site Improvements

G2040 Site Development

G2040 810	Flagpoles	COST EACH		
		MAT.	INST.	TOTAL
0110	Flagpoles, on grade, aluminum, tapered, 20' high	1,175	720	1,895
0120	70' high	9,700	1,800	11,500
0130	Fiberglass, tapered, 23' high	625	720	1,345
0140	59' high	5,050	1,600	6,650
0150	Concrete, internal halyard, 20' high	1,500	575	2,075
0160	100' high	19,900	1,450	21,350

G2040 950	Other Site Development, EACH	COST EACH		
		MAT.	INST.	TOTAL
0110	Grandstands, permanent, closed deck, steel, economy (per seat)			24
0120	Deluxe (per seat)			27
0130	Composite design, economy (per seat)			46.50
0140	Deluxe (per seat)			82.50

514

G2040 Site Development

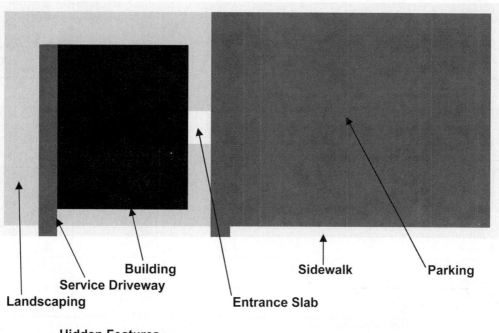

Building
Service Driveway
Landscaping
Sidewalk
Parking
Entrance Slab

Hidden Features
Utilities and Lighting
(Features do not represent actual sizes)

G2040 990	Site Development Components for Buildings	COST EACH		
		MAT.	INST.	TOTAL
0970	Assume minimal and balanced cut & fill, no rock, no demolition, no haz mat.			
0975	Lines can be adjusted linearly +/- 20% within same use & number of floors.			
0980	60,000 S.F. 2-Story Office Bldg on 3.3 Acres w/20% Green Space			
1000	Site Preparation	225	48,700	48,925
1002	Utilities	90,500	105,500	196,000
1004	Pavement	322,500	201,000	523,500
1006	Stormwater Management	297,000	59,500	356,500
1008	Sidewalks	14,600	19,100	33,700
1010	Exterior lighting	48,800	39,000	87,800
1014	Landscaping	73,500	52,000	125,500
1028	80,000 S.F. 4-Story Office Bldg on 3.5 Acres w/20% Green Space			
1030	Site preparation	230	52,500	52,730
1032	Utilities	84,000	95,500	179,500
1034	Pavement	387,000	240,500	627,500
1036	Stormwater Managmement	313,000	62,500	375,500
1038	Sidewalks	11,900	15,700	27,600
1040	Lighting	48,300	41,200	89,500
1042	Landscaping	75,500	54,000	129,500
1058	16,000 S.F. 1-Story Medical Office Bldg on 1.7 Acres w/25% Green Space			
1060	Site preparation	162	37,400	37,562
1062	Utilities	65,000	77,500	142,500
1064	Pavement	150,000	94,500	244,500
1066	Stormwater Management	145,000	29,100	174,100
1068	Sidewalks	10,600	14,000	24,600
1070	Lighting	30,000	22,600	52,600
1072	Landscaping	49,700	32,000	81,700
1098	36,000 S.F. 3-Story Apartment Bldg on 1.6 Acres w/25% Green Space			
1100	Site preparation	157	37,000	37,157

G20 Site Improvements

G2040 Site Development

G2040 990	Site Development Components for Buildings	COST EACH		
		MAT.	INST.	TOTAL
1102	Utilities	60,500	70,500	131,000
1104	Pavement	151,500	95,000	246,500
1106	Stormwater Management	135,000	27,100	162,100
1108	Sidewalks	9,225	12,100	21,325
1110	Lighting	28,000	21,500	49,500
1112	Landscaping	47,600	30,400	78,000
1128	130,000 S.F. 3-Story Hospital Bldg on 5.9 Acres w/17% Green Space			
1130	Site preparation	300	68,500	68,800
1132	Utilities	120,500	138,000	258,500
1134	Pavement	658,500	409,000	1,067,500
1136	Stormwater Management	556,000	111,500	667,500
1138	Sidewalks	17,500	23,000	40,500
1140	Lighting	79,500	69,500	149,000
1142	Landscaping	106,000	82,000	188,000
1158	60,000 S.F. 1-Story Light Manufacturing Bldg on 3.9 Acres w/18% Green Space			
1160	Site preparation	246	52,000	52,246
1162	Utilities	108,500	131,000	239,500
1164	Pavement	311,000	174,000	485,000
1166	Stormwater Management	366,000	73,500	439,500
1168	Sidewalks	20,600	27,100	47,700
1170	Lighting	59,000	44,600	103,600
1172	Landscaping	79,500	58,000	137,500
1198	8,000 S.F. 1-Story Restaurant Bldg on 1.4 Acres w/26% Green Space			
1200	Site preparation	148	36,100	36,248
1202	Utilities	54,500	61,500	116,000
1204	Pavement	143,500	91,500	235,000
1206	Stormwater Management	119,000	23,900	142,900
1208	Sidewalks	7,525	9,875	17,400
1210	Lighting	24,900	19,800	44,700
1212	Landscaping	44,700	27,900	72,600
1228	20,000 S.F. 1-Story Retail Store Bldg on 2.1 Acres w/23% Green Space			
1230	Site preparation	181	43,200	43,381
1232	Utilities	72,500	85,500	158,000
1234	Pavement	195,000	122,500	317,500
1236	Stormwater Management	185,500	37,300	222,800
1238	Sidewalks	11,900	15,700	27,600
1240	Lighting	34,400	26,200	60,600
1242	Landscaping	56,500	37,700	94,200
1258	60,000 S.F. 1-Story Warehouse on 3.1 Acres w/19% Green Space			
1260	Site preparation	219	47,800	48,019
1262	Utilities	98,000	123,000	221,000
1264	Pavement	189,500	98,000	287,500
1266	Stormwater Management	285,000	57,500	342,500
1268	Sidewalks	20,600	27,100	47,700
1270	Lighting	49,700	34,700	84,400
1272	Landscaping	68,500	48,100	116,600

G3030 Storm Sewer

G3030 210	Manholes & Catch Basins	COST PER EACH		
		MAT.	INST.	TOTAL
1920	Manhole/catch basin, brick, 4' I.D. riser, 4' deep	1,325	1,800	3,125
1980	10' deep	2,975	4,225	7,200
3200	Block, 4' I.D. riser, 4' deep	1,175	1,450	2,625
3260	10' deep	2,600	3,500	6,100
4620	Concrete, cast-in-place, 4' I.D. riser, 4' deep	1,325	2,525	3,850
4680	10' deep	3,225	6,050	9,275
5820	Concrete, precast, 4' I.D. riser, 4' deep	1,550	1,350	2,900
5880	10' deep	4,050	3,100	7,150
6200	6' I.D. riser, 4' deep	3,425	2,000	5,425
6260	10' deep	7,225	4,700	11,925

G4020 Site Lighting

Light Pole

G4020 210	Light Pole (Installed)	COST EACH		
		MAT.	INST.	TOTAL
0200	Light pole, aluminum, 20' high, 1 arm bracket	1,275	1,200	2,475
0240	2 arm brackets	1,400	1,200	2,600
0280	3 arm brackets	1,550	1,225	2,775
0320	4 arm brackets	1,675	1,225	2,900
0360	30' high, 1 arm bracket	2,225	1,500	3,725
0400	2 arm brackets	2,350	1,500	3,850
0440	3 arm brackets	2,500	1,550	4,050
0480	4 arm brackets	2,625	1,550	4,175
0680	40' high, 1 arm bracket	2,725	2,025	4,750
0720	2 arm brackets	2,850	2,025	4,875
0760	3 arm brackets	2,975	2,050	5,025
0800	4 arm brackets	3,125	2,050	5,175
0840	Steel, 20' high, 1 arm bracket	1,425	1,250	2,675
0880	2 arm brackets	1,475	1,250	2,725
0920	3 arm brackets	1,450	1,300	2,750
0960	4 arm brackets	1,550	1,300	2,850
1000	30' high, 1 arm bracket	1,750	1,625	3,375
1040	2 arm brackets	1,800	1,625	3,425
1080	3 arm brackets	1,800	1,650	3,450
1120	4 arm brackets	1,900	1,650	3,550
1320	40' high, 1 arm bracket	2,300	2,175	4,475
1360	2 arm brackets	2,350	2,175	4,525
1400	3 arm brackets	2,350	2,225	4,575
1440	4 arm brackets	2,450	2,225	4,675

Reference
Section

All the reference information is in one section making it easy to find what you need to know ... and easy to use the data set on a daily basis. This section is visually identified by a vertical black bar on the edge of pages.

In this Reference Section, we've included General Conditions; Location Factors for adjusting costs to the region you are in; Historical Cost Indexes for cost comparisons over time; a Glossary; an explanation of all Abbreviations used in the data set; and sample Estimating Forms.

Table of Contents

General Conditions, Overhead & Profit

The total building costs in the Commercial/Industrial/Institutional section include a 10% allowance for general conditions and a 15% allowance for the general contractor's overhead and profit and contingencies.

General contractor overhead includes indirect costs such as permits, Workers' Compensation, insurances, supervision and bonding fees. Overhead will vary with the size of project, the contractor's operating procedures and location. Profits will vary with economic activity and local conditions.

Contingencies provide for unforeseen construction difficulties which include material shortages and weather. In all situations, the appraiser should give consideration to possible adjustment of the 25% factor used in developing the Commercial/Industrial/Institutional models.

Architectural Fees

Tabulated below are typical percentage fees by project size, for good professional architectural service. Fees may vary from those listed depending upon degree of design difficulty and economic conditions in any particular area.

Rates can be interpolated horizontally and vertically. Various portions of the same project requiring different rates should be adjusted proportionately. For alterations, add 50% to the fee for the first $500,000 of project cost and add 25% to the fee for project cost over $500,000.

Architectural fees tabulated below include Engineering fees.

Insurance Exclusions

Many insurance companies exclude from coverage such items as architect's fees, excavation, foundations below grade, underground piping and site preparation. Since exclusions vary among insurance companies, it is recommended that for greatest accuracy each exclusion be priced separately using the unit-in-place section.

As a rule of thumb, exclusions can be calculated at 9% of total building cost plus the appropriate allowance for architect's fees.

Building Types	Total Project Size in Thousands of Dollars						
	100	250	500	1,000	5,000	10,000	50,000
Factories, garages, warehouses, repetitive housing	9.0%	8.0%	7.0%	6.2%	5.3%	4.9%	4.5%
Apartments, banks, schools, libraries, offices, municipal buildings	12.2	12.3	9.2	8.0	7.0	6.6	6.2
Churches, hospitals, homes, laboratories, museums, research	15.0	13.6	12.7	11.9	9.5	8.8	8.0
Memorials, monumental work, decorative furnishings	–	16.0	14.5	13.1	10.0	9.0	8.3

Location Factors - Residential/Commercial

Costs shown in *RSMeans Square Foot Costs* are based on national averages for materials and installation. To adjust these costs to a specific location, simply multiply the base cost by the factor for that city.

The data is arranged alphabetically by state and postal zip code numbers. For a city not listed, use the factor for a nearby city with similar economic characteristics.

STATE/ZIP	CITY	Residential	Commercial
ALABAMA			
350-352	Birmingham	.88	.90
354	Tuscaloosa	.89	.90
355	Jasper	.85	.87
356	Decatur	.88	.89
357-358	Huntsville	.87	.88
359	Gadsden	.79	.85
360-361	Montgomery	.88	.89
362	Anniston	.87	.89
363	Dothan	.88	.88
364	Evergreen	.84	.87
365-366	Mobile	.89	.89
367	Selma	.85	.88
368	Phenix City	.87	.89
369	Butler	.86	.88
ALASKA			
995-996	Anchorage	1.23	1.19
997	Fairbanks	1.27	1.21
998	Juneau	1.25	1.20
999	Ketchikan	1.27	1.27
ARIZONA			
850,853	Phoenix	.85	.88
851,852	Mesa/Tempe	.85	.88
855	Globe	.82	.87
856-857	Tucson	.84	.87
859	Show Low	.86	.88
860	Flagstaff	.87	.89
863	Prescott	.82	.86
864	Kingman	.83	.87
865	Chambers	.84	.88
ARKANSAS			
716	Pine Bluff	.79	.85
717	Camden	.71	.79
718	Texarkana	.75	.81
719	Hot Springs	.71	.79
720-722	Little Rock	.82	.85
723	West Memphis	.78	.83
724	Jonesboro	.76	.82
725	Batesville	.73	.79
726	Harrison	.75	.79
727	Fayetteville	.70	.77
728	Russellville	.76	.80
729	Fort Smith	.81	.82
CALIFORNIA			
900-902	Los Angeles	1.08	1.08
903-905	Inglewood	1.06	1.04
906-908	Long Beach	1.05	1.05
910-912	Pasadena	1.05	1.05
913-916	Van Nuys	1.08	1.07
917-918	Alhambra	1.09	1.06
919-921	San Diego	1.05	1.05
922	Palm Springs	1.05	1.05
923-924	San Bernardino	1.06	1.03
925	Riverside	1.07	1.06
926-927	Santa Ana	1.07	1.05
928	Anaheim	1.07	1.07
930	Oxnard	1.07	1.06
931	Santa Barbara	1.07	1.06
932-933	Bakersfield	1.05	1.05
934	San Luis Obispo	1.08	1.06
935	Mojave	1.07	1.04
936-938	Fresno	1.11	1.08
939	Salinas	1.14	1.10
940-941	San Francisco	1.24	1.23
942,956-958	Sacramento	1.12	1.09
943	Palo Alto	1.17	1.14
944	San Mateo	1.21	1.15
945	Vallejo	1.16	1.12
946	Oakland	1.21	1.16
947	Berkeley	1.23	1.17
948	Richmond	1.23	1.15
949	San Rafael	1.22	1.17
950	Santa Cruz	1.17	1.13
951	San Jose	1.22	1.18
952	Stockton	1.13	1.09
953	Modesto	1.11	1.08

STATE/ZIP	CITY	Residential	Commercial
CALIFORNIA (CONT'D)			
954	Santa Rosa	1.20	1.17
955	Eureka	1.16	1.10
959	Marysville	1.12	1.09
960	Redding	1.18	1.15
961	Susanville	1.16	1.14
COLORADO			
800-802	Denver	.90	.92
803	Boulder	.92	.91
804	Golden	.88	.91
805	Fort Collins	.89	.92
806	Greeley	.86	.88
807	Fort Morgan	.90	.90
808-809	Colorado Springs	.86	.91
810	Pueblo	.86	.90
811	Alamosa	.85	.91
812	Salida	.88	.91
813	Durango	.89	.92
814	Montrose	.87	.92
815	Grand Junction	.92	.93
816	Glenwood Springs	.88	.92
CONNECTICUT			
060	New Britain	1.11	1.09
061	Hartford	1.10	1.10
062	Willimantic	1.12	1.09
063	New London	1.11	1.07
064	Meriden	1.11	1.08
065	New Haven	1.11	1.10
066	Bridgeport	1.12	1.09
067	Waterbury	1.11	1.09
068	Norwalk	1.13	1.13
069	Stamford	1.13	1.13
D.C.			
200-205	Washington	.97	.99
DELAWARE			
197	Newark	1.02	1.04
198	Wilmington	1.01	1.03
199	Dover	1.01	1.04
FLORIDA			
320,322	Jacksonville	.83	.85
321	Daytona Beach	.86	.88
323	Tallahassee	.82	.84
324	Panama City	.78	.83
325	Pensacola	.88	.88
326,344	Gainesville	.82	.86
327-328,347	Orlando	.85	.87
329	Melbourne	.86	.90
330-332,340	Miami	.85	.87
333	Fort Lauderdale	.84	.86
334,349	West Palm Beach	.84	.85
335-336,346	Tampa	.85	.88
337	St. Petersburg	.84	.88
338	Lakeland	.83	.87
339,341	Fort Myers	.82	.85
342	Sarasota	.86	.89
GEORGIA			
300-303,399	Atlanta	.90	.89
304	Statesboro	.77	.82
305	Gainesville	.79	.82
306	Athens	.78	.83
307	Dalton	.78	.83
308-309	Augusta	.87	.85
310-312	Macon	.77	.83
313-314	Savannah	.82	.85
315	Waycross	.79	.83
316	Valdosta	.73	.82
317,398	Albany	.76	.82
318-319	Columbus	.78	.83
HAWAII			
967	Hilo	1.22	1.20
968	Honolulu	1.25	1.23

Location Factors - Residential/Commercial

STATE/ZIP	CITY	Residential	Commercial
STATES & POSS.			
969	Guam	.98	1.05
IDAHO			
832	Pocatello	.87	.91
833	Twin Falls	.81	.87
834	Idaho Falls	.85	.89
835	Lewiston	.97	.99
836-837	Boise	.87	.91
838	Coeur d'Alene	.98	.97
ILLINOIS			
600-603	North Suburban	1.20	1.16
604	Joliet	1.22	1.17
605	South Suburban	1.20	1.16
606-608	Chicago	1.21	1.18
609	Kankakee	1.13	1.12
610-611	Rockford	1.10	1.10
612	Rock Island	.98	.99
613	La Salle	1.10	1.10
614	Galesburg	1.03	1.02
615-616	Peoria	1.06	1.05
617	Bloomington	1.03	1.03
618-619	Champaign	1.04	1.04
620-622	East St. Louis	1.02	1.01
623	Quincy	1.02	1.00
624	Effingham	1.04	1.01
625	Decatur	1.03	1.02
626-627	Springfield	1.04	1.03
628	Centralia	1.03	1.01
629	Carbondale	1.00	1.00
INDIANA			
460	Anderson	.91	.91
461-462	Indianapolis	.93	.93
463-464	Gary	1.05	1.04
465-466	South Bend	.91	.92
467-468	Fort Wayne	.89	.89
469	Kokomo	.91	.89
470	Lawrenceburg	.86	.87
471	New Albany	.85	.86
472	Columbus	.91	.90
473	Muncie	.91	.91
474	Bloomington	.92	.91
475	Washington	.90	.90
476-477	Evansville	.90	.92
478	Terre Haute	.91	.93
479	Lafayette	.92	.90
IOWA			
500-503,509	Des Moines	.89	.92
504	Mason City	.75	.83
505	Fort Dodge	.74	.80
506-507	Waterloo	.83	.88
508	Creston	.84	.89
510-511	Sioux City	.86	.88
512	Sibley	.73	.80
513	Spencer	.75	.81
514	Carroll	.79	.86
515	Council Bluffs	.86	.91
516	Shenandoah	.81	.87
520	Dubuque	.87	.91
521	Decorah	.79	.84
522-524	Cedar Rapids	.92	.92
525	Ottumwa	.87	.88
526	Burlington	.88	.88
527-528	Davenport	.97	.97
KANSAS			
660-662	Kansas City	.98	.98
664-666	Topeka	.81	.87
667	Fort Scott	.89	.88
668	Emporia	.83	.86
669	Belleville	.81	.85
670-672	Wichita	.80	.86
673	Independence	.89	.88
674	Salina	.80	.87
675	Hutchinson	.80	.83
676	Hays	.82	.86
677	Colby	.84	.87
678	Dodge City	.81	.87
679	Liberal	.80	.86
KENTUCKY			
400-402	Louisville	.89	.90
403-405	Lexington	.86	.90

STATE/ZIP	CITY	Residential	Commercial
KENTUCKY (CONT'D)			
406	Frankfort	.86	.89
407-409	Corbin	.81	.84
410	Covington	.90	.94
411-412	Ashland	.93	.97
413-414	Campton	.86	.88
415-416	Pikeville	.90	.94
417-418	Hazard	.86	.88
420	Paducah	.88	.90
421-422	Bowling Green	.89	.91
423	Owensboro	.90	.92
424	Henderson	.89	.90
425-426	Somerset	.85	.88
427	Elizabethtown	.86	.88
LOUISIANA			
700-701	New Orleans	.86	.87
703	Thibodaux	.82	.84
704	Hammond	.77	.80
705	Lafayette	.84	.84
706	Lake Charles	.86	.85
707-708	Baton Rouge	.84	.86
710-711	Shreveport	.78	.83
712	Monroe	.75	.82
713-714	Alexandria	.78	.84
MAINE			
039	Kittery	.92	.91
040-041	Portland	.96	.96
042	Lewiston	.96	.95
043	Augusta	.92	.92
044	Bangor	.95	.94
045	Bath	.91	.91
046	Machias	.95	.92
047	Houlton	.96	.92
048	Rockland	.95	.92
049	Waterville	.90	.90
MARYLAND			
206	Waldorf	.88	.91
207-208	College Park	.86	.91
209	Silver Spring	.88	.91
210-212	Baltimore	.92	.93
214	Annapolis	.89	.93
215	Cumberland	.90	.92
216	Easton	.85	.87
217	Hagerstown	.89	.92
218	Salisbury	.83	.83
219	Elkton	.92	.91
MASSACHUSETTS			
010-011	Springfield	1.06	1.04
012	Pittsfield	1.06	1.03
013	Greenfield	1.04	1.03
014	Fitchburg	1.13	1.08
015-016	Worcester	1.15	1.10
017	Framingham	1.18	1.12
018	Lowell	1.18	1.13
019	Lawrence	1.18	1.14
020-022, 024	Boston	1.23	1.18
023	Brockton	1.16	1.10
025	Buzzards Bay	1.14	1.08
026	Hyannis	1.13	1.09
027	New Bedford	1.16	1.10
MICHIGAN			
480,483	Royal Oak	1.01	1.00
481	Ann Arbor	1.02	1.02
482	Detroit	1.05	1.03
484-485	Flint	.95	.96
486	Saginaw	.91	.94
487	Bay City	.91	.94
488-489	Lansing	.94	.97
490	Battle Creek	.90	.92
491	Kalamazoo	.90	.92
492	Jackson	.92	.95
493,495	Grand Rapids	.90	.92
494	Muskegon	.88	.91
496	Traverse City	.85	.89
497	Gaylord	.88	.91
498-499	Iron Mountain	.89	.93
MINNESOTA			
550-551	Saint Paul	1.11	1.10
553-555	Minneapolis	1.12	1.09
556-558	Duluth	1.04	1.05

Location Factors - Residential/Commercial

STATE/ZIP	CITY	Residential	Commercial
MINNESOTA (CONT'D)			
559	Rochester	1.03	1.03
560	Mankato	1.00	1.00
561	Windom	.92	.95
562	Willmar	.95	.97
563	St. Cloud	1.04	1.04
564	Brainerd	.95	.99
565	Detroit Lakes	.94	.98
566	Bemidji	.94	.99
567	Thief River Falls	.93	.96
MISSISSIPPI			
386	Clarksdale	.74	.78
387	Greenville	.82	.85
388	Tupelo	.75	.79
389	Greenwood	.77	.78
390-392	Jackson	.84	.86
393	Meridian	.82	.84
394	Laurel	.76	.80
395	Biloxi	.80	.82
396	McComb	.74	.77
397	Columbus	.75	.79
MISSOURI			
630-631	St. Louis	1.02	1.02
633	Bowling Green	.97	.98
634	Hannibal	.92	.96
635	Kirksville	.90	.96
636	Flat River	.96	.98
637	Cape Girardeau	.91	.97
638	Sikeston	.88	.95
639	Poplar Bluff	.89	.94
640-641	Kansas City	1.03	1.02
644-645	St. Joseph	.98	.98
646	Chillicothe	.96	.98
647	Harrisonville	.97	.98
648	Joplin	.90	.91
650-651	Jefferson City	.93	.99
652	Columbia	.93	1.00
653	Sedalia	.92	.98
654-655	Rolla	.97	.98
656-658	Springfield	.89	.94
MONTANA			
590-591	Billings	.90	.93
592	Wolf Point	.86	.91
593	Miles City	.87	.89
594	Great Falls	.90	.93
595	Havre	.83	.90
596	Helena	.86	.90
597	Butte	.86	.92
598	Missoula	.87	.90
599	Kalispell	.86	.90
NEBRASKA			
680-681	Omaha	.89	.91
683-685	Lincoln	.89	.90
686	Columbus	.86	.89
687	Norfolk	.88	.89
688	Grand Island	.87	.89
689	Hastings	.89	.90
690	McCook	.82	.86
691	North Platte	.87	.89
692	Valentine	.83	.88
693	Alliance	.82	.88
NEVADA			
889-891	Las Vegas	1.03	1.06
893	Ely	1.02	1.02
894-895	Reno	.91	.95
897	Carson City	.91	.95
898	Elko	.93	.92
NEW HAMPSHIRE			
030	Nashua	.98	.97
031	Manchester	.96	.97
032-033	Concord	.97	.97
034	Keene	.88	.87
035	Littleton	.91	.88
036	Charleston	.84	.85
037	Claremont	.85	.84
038	Portsmouth	.96	.95

STATE/ZIP	CITY	Residential	Commercial
NEW JERSEY			
070-071	Newark	1.15	1.13
072	Elizabeth	1.17	1.12
073	Jersey City	1.14	1.11
074-075	Paterson	1.15	1.12
076	Hackensack	1.14	1.11
077	Long Branch	1.11	1.09
078	Dover	1.14	1.11
079	Summit	1.15	1.11
080,083	Vineland	1.12	1.08
081	Camden	1.13	1.10
082,084	Atlantic City	1.17	1.10
085-086	Trenton	1.15	1.11
087	Point Pleasant	1.12	1.10
088-089	New Brunswick	1.17	1.12
NEW MEXICO			
870-872	Albuquerque	.83	.88
873	Gallup	.82	.88
874	Farmington	.83	.88
875	Santa Fe	.84	.88
877	Las Vegas	.82	.87
878	Socorro	.82	.87
879	Truth/Consequences	.81	.85
880	Las Cruces	.81	.85
881	Clovis	.82	.88
882	Roswell	.83	.89
883	Carrizozo	.82	.89
884	Tucumcari	.83	.88
NEW YORK			
100-102	New York	1.35	1.31
103	Staten Island	1.30	1.28
104	Bronx	1.30	1.26
105	Mount Vernon	1.16	1.17
106	White Plains	1.19	1.16
107	Yonkers	1.22	1.20
108	New Rochelle	1.20	1.17
109	Suffern	1.17	1.12
110	Queens	1.30	1.29
111	Long Island City	1.33	1.30
112	Brooklyn	1.36	1.31
113	Flushing	1.32	1.31
114	Jamaica	1.31	1.30
115,117,118	Hicksville	1.23	1.23
116	Far Rockaway	1.31	1.31
119	Riverhead	1.23	1.23
120-122	Albany	1.00	1.01
123	Schenectady	1.02	1.01
124	Kingston	1.04	1.09
125-126	Poughkeepsie	1.21	1.16
127	Monticello	1.04	1.07
128	Glens Falls	.95	.97
129	Plattsburgh	1.00	.98
130-132	Syracuse	.98	.98
133-135	Utica	.96	.97
136	Watertown	.94	.97
137-139	Binghamton	.98	.98
140-142	Buffalo	1.07	1.03
143	Niagara Falls	1.03	1.02
144-146	Rochester	.99	1.01
147	Jamestown	.92	.95
148-149	Elmira	.95	.98
NORTH CAROLINA			
270,272-274	Greensboro	.90	.83
271	Winston-Salem	.90	.83
275-276	Raleigh	.89	.81
277	Durham	.90	.84
278	Rocky Mount	.89	.81
279	Elizabeth City	.80	.79
280	Gastonia	.94	.83
281-282	Charlotte	.93	.84
283	Fayetteville	.90	.84
284	Wilmington	.91	.82
285	Kinston	.91	.82
286	Hickory	.89	.82
287-288	Asheville	.91	.83
289	Murphy	.83	.77
NORTH DAKOTA			
580-581	Fargo	.83	.90
582	Grand Forks	.78	.86
583	Devils Lake	.81	.86
584	Jamestown	.79	.86
585	Bismarck	.81	.88

For customer support on your Square Foot Cost Data, call 877.756.2789.

523

Location Factors - Residential/Commercial

STATE/ZIP	CITY	Residential	Commercial
NORTH DAKOTA (CONT'D)			
586	Dickinson	.79	.86
587	Minot	.86	.91
588	Williston	.79	.86
OHIO			
430-432	Columbus	.93	.94
433	Marion	.90	.91
434-436	Toledo	.98	.98
437-438	Zanesville	.90	.92
439	Steubenville	.94	.96
440	Lorain	.96	.96
441	Cleveland	1.00	.99
442-443	Akron	.98	.97
444-445	Youngstown	.95	.94
446-447	Canton	.94	.95
448-449	Mansfield	.91	.92
450	Hamilton	.92	.92
451-452	Cincinnati	.92	.92
453-454	Dayton	.93	.92
455	Springfield	.94	.92
456	Chillicothe	.97	.95
457	Athens	.94	.93
458	Lima	.92	.94
OKLAHOMA			
730-731	Oklahoma City	.84	.87
734	Ardmore	.79	.84
735	Lawton	.82	.86
736	Clinton	.80	.85
737	Enid	.79	.85
738	Woodward	.80	.85
739	Guymon	.82	.86
740-741	Tulsa	.80	.83
743	Miami	.85	.85
744	Muskogee	.77	.82
745	McAlester	.76	.82
746	Ponca City	.79	.83
747	Durant	.77	.83
748	Shawnee	.78	.83
749	Poteau	.78	.82
OREGON			
970-972	Portland	1.00	1.00
973	Salem	.99	1.00
974	Eugene	1.00	.99
975	Medford	.98	.99
976	Klamath Falls	.99	.99
977	Bend	1.01	1.00
978	Pendleton	.99	.99
979	Vale	.97	.92
PENNSYLVANIA			
150-152	Pittsburgh	1.01	1.02
153	Washington	.97	1.00
154	Uniontown	.95	1.00
155	Bedford	.92	.97
156	Greensburg	.98	1.01
157	Indiana	.96	.99
158	Dubois	.92	.98
159	Johnstown	.93	.99
160	Butler	.94	.97
161	New Castle	.94	.96
162	Kittanning	.94	.97
163	Oil City	.93	.96
164-165	Erie	.95	.96
166	Altoona	.89	.94
167	Bradford	.92	.96
168	State College	.91	.94
169	Wellsboro	.93	.96
170-171	Harrisburg	.96	.98
172	Chambersburg	.90	.94
173-174	York	.93	.96
175-176	Lancaster	.94	.96
177	Williamsport	.93	.95
178	Sunbury	.94	.96
179	Pottsville	.93	.97
180	Lehigh Valley	1.00	1.03
181	Allentown	1.05	1.03
182	Hazleton	.94	.98
183	Stroudsburg	.94	.99
184-185	Scranton	.97	.99
186-187	Wilkes-Barre	.95	.98
188	Montrose	.92	.96
189	Doylestown	1.09	1.09

STATE/ZIP	CITY	Residential	Commercial
PENNSYLVANIA (CONT'D)			
190-191	Philadelphia	1.17	1.15
193	Westchester	1.11	1.10
194	Norristown	1.11	1.11
195-196	Reading	.99	1.00
PUERTO RICO			
009	San Juan	.76	.79
RHODE ISLAND			
028	Newport	1.09	1.06
029	Providence	1.09	1.07
SOUTH CAROLINA			
290-292	Columbia	.96	.85
293	Spartanburg	.96	.85
294	Charleston	.94	.85
295	Florence	.93	.85
296	Greenville	.95	.84
297	Rock Hill	.83	.80
298	Aiken	.93	.85
299	Beaufort	.81	.79
SOUTH DAKOTA			
570-571	Sioux Falls	.77	.82
572	Watertown	.73	.78
573	Mitchell	.74	.76
574	Aberdeen	.76	.80
575	Pierre	.78	.83
576	Mobridge	.73	.77
577	Rapid City	.78	.83
TENNESSEE			
370-372	Nashville	.85	.88
373-374	Chattanooga	.84	.85
375,380-381	Memphis	.84	.87
376	Johnson City	.71	.81
377-379	Knoxville	.81	.84
382	McKenzie	.73	.81
383	Jackson	.77	.83
384	Columbia	.78	.83
385	Cookeville	.72	.81
TEXAS			
750	McKinney	.81	.84
751	Waxahackie	.81	.84
752-753	Dallas	.84	.85
754	Greenville	.82	.85
755	Texarkana	.80	.83
756	Longview	.80	.83
757	Tyler	.82	.84
758	Palestine	.79	.81
759	Lufkin	.81	.83
760-761	Fort Worth	.82	.84
762	Denton	.85	.85
763	Wichita Falls	.85	.83
764	Eastland	.83	.82
765	Temple	.81	.80
766-767	Waco	.84	.82
768	Brownwood	.79	.82
769	San Angelo	.79	.82
770-772	Houston	.84	.87
773	Huntsville	.81	.85
774	Wharton	.82	.87
775	Galveston	.83	.85
776-777	Beaumont	.86	.86
778	Bryan	.80	.84
779	Victoria	.83	.86
780	Laredo	.81	.84
781-782	San Antonio	.82	.84
783-784	Corpus Christi	.85	.85
785	McAllen	.85	.84
786-787	Austin	.80	.84
788	Del Rio	.82	.85
789	Giddings	.81	.84
790-791	Amarillo	.80	.83
792	Childress	.82	.84
793-794	Lubbock	.81	.85
795-796	Abilene	.83	.83
797	Midland	.85	.86
798-799,885	El Paso	.80	.82
UTAH			
840-841	Salt Lake City	.82	.89
842,844	Ogden	.81	.87
843	Logan	.81	.88

524

Location Factors - Residential/Commercial

STATE/ZIP	CITY	Residential	Commercial
UTAH (CONT'D)			
845	Price	.78	.86
846-847	Provo	.82	.88
VERMONT			
050	White River Jct.	.89	.89
051	Bellows Falls	.96	.95
052	Bennington	.99	.95
053	Brattleboro	.95	.95
054	Burlington	.94	.94
056	Montpelier	.92	.92
057	Rutland	.93	.92
058	St. Johnsbury	.90	.89
059	Guildhall	.90	.88
VIRGINIA			
220-221	Fairfax	1.07	.96
222	Arlington	1.10	.96
223	Alexandria	1.11	.96
224-225	Fredericksburg	1.08	.94
226	Winchester	1.02	.94
227	Culpeper	1.07	.94
228	Harrisonburg	.86	.86
229	Charlottesville	.89	.86
230-232	Richmond	.92	.86
233-235	Norfolk	.94	.86
236	Newport News	.94	.86
237	Portsmouth	.89	.84
238	Petersburg	.92	.86
239	Farmville	.83	.78
240-241	Roanoke	.95	.86
242	Bristol	.83	.80
243	Pulaski	.82	.81
244	Staunton	.86	.83
245	Lynchburg	.92	.85
246	Grundy	.80	.79
WASHINGTON			
980-981,987	Seattle	1.02	1.03
982	Everett	1.03	1.01
983-984	Tacoma	1.00	1.01
985	Olympia	.98	1.00
986	Vancouver	.96	.99
988	Wenatchee	.93	.95
989	Yakima	.98	.99
990-992	Spokane	.97	.95
993	Richland	.97	.98
994	Clarkston	.94	.94
WEST VIRGINIA			
247-248	Bluefield	.94	.96
249	Lewisburg	.93	.97
250-253	Charleston	.97	.99
254	Martinsburg	.90	.95
255-257	Huntington	.98	1.00
258-259	Beckley	.94	.97
260	Wheeling	.94	.98
261	Parkersburg	.93	.97
262	Buckhannon	.94	.97
263-264	Clarksburg	.94	.98
265	Morgantown	.94	.98
266	Gassaway	.93	.97
267	Romney	.91	.95
268	Petersburg	.91	.94
WISCONSIN			
530,532	Milwaukee	1.07	1.02
531	Kenosha	1.03	1.00
534	Racine	1.01	1.00
535	Beloit	1.00	.99
537	Madison	1.00	.99
538	Lancaster	.97	.95
539	Portage	.96	.96
540	New Richmond	.97	.96
541-543	Green Bay	1.03	1.00
544	Wausau	.97	.96
545	Rhinelander	.95	.96
546	La Crosse	.96	.96
547	Eau Claire	.99	.98
548	Superior	.96	.97
549	Oshkosh	.95	.95
WYOMING			
820	Cheyenne	.82	.87
821	Yellowstone Nat. Pk.	.82	.88
822	Wheatland	.77	.86

STATE/ZIP	CITY	Residential	Commercial
WYOMING (CONT'D)			
823	Rawlins	.85	.89
824	Worland	.80	.86
825	Riverton	.80	.86
826	Casper	.79	.86
827	Newcastle	.84	.88
828	Sheridan	.83	.87
829-831	Rock Springs	.86	.89
CANADIAN FACTORS (reflect Canadian currency)			
ALBERTA			
	Calgary	1.07	1.10
	Edmonton	1.06	1.11
	Fort McMurray	1.10	1.10
	Lethbridge	1.08	1.07
	Lloydminster	1.03	1.03
	Medicine Hat	1.03	1.02
	Red Deer	1.03	1.03
BRITISH COLUMBIA			
	Kamloops	1.01	1.04
	Prince George	1.01	1.04
	Vancouver	1.02	1.05
	Victoria	1.02	1.03
MANITOBA			
	Brandon	1.08	1.04
	Portage la Prairie	.98	.96
	Winnipeg	.95	1.00
NEW BRUNSWICK			
	Bathurst	.90	.92
	Dalhousie	.92	.92
	Fredericton	.95	.98
	Moncton	.92	.94
	Newcastle	.91	.92
	St. John	.99	.98
NEWFOUNDLAND			
	Corner Brook	1.02	1.03
	St. John's	1.03	1.06
NORTHWEST TERRITORIES			
	Yellowknife	1.12	1.13
NOVA SCOTIA			
	Bridgewater	.94	.96
	Dartmouth	1.03	1.04
	Halifax	.97	1.01
	New Glasgow	1.02	1.03
	Sydney	1.01	1.01
	Truro	.94	.96
	Yarmouth	1.02	1.03
ONTARIO			
	Barrie	1.11	1.07
	Brantford	1.10	1.07
	Cornwall	1.09	1.05
	Hamilton	1.06	1.08
	Kingston	1.09	1.06
	Kitchener	1.02	1.02
	London	1.05	1.06
	North Bay	1.17	1.12
	Oshawa	1.07	1.05
	Ottawa	1.06	1.07
	Owen Sound	1.10	1.06
	Peterborough	1.08	1.05
	Sarnia	1.10	1.07
	Sault Ste. Marie	1.04	1.02
	St. Catharines	1.04	1.02
	Sudbury	1.01	1.01
	Thunder Bay	1.06	1.02
	Timmins	1.07	1.04
	Toronto	1.08	1.10
	Windsor	1.05	1.01
PRINCE EDWARD ISLAND			
	Charlottetown	.90	.95
	Summerside	.97	.99
QUEBEC			
	Cap-de-la-Madeleine	1.08	1.01
	Charlesbourg	1.08	1.01
	Chicoutimi	1.11	1.03
	Gatineau	1.08	1.01

For customer support on your Square Foot Cost Data, call 877.756.2789.

525

Location Factors - Residential/Commercial

STATE/ZIP	CITY	Residential	Commercial
QUEBEC (CONT'D)			
	Granby	1.08	
	Hull	1.08	1.01
	Joliette	1.09	1.01
	Laval	1.08	1.01
	Montreal	1.05	1.07
	Quebec City	1.06	1.07
	Rimouski	1.11	1.03
	Rouyn-Noranda	1.08	1.01
	Saint-Hyacinthe	1.08	1.01
	Sherbrooke	1.08	1.01
	Sorel	1.09	1.01
	Saint-Jerome	1.08	1.01
	Trois-Rivieres	1.19	1.09
SASKATCHEWAN			
	Moose Jaw	.91	.92
	Prince Albert	.90	.91
	Regina	1.05	1.08
	Saskatoon	1.03	1.02
YUKON			
	Whitehorse	1.03	1.06

526

Historical Cost Indexes

The following tables are the estimated Historical Cost Indexes based on a 30-city national average with a base of 100 on January 1, 1993.

The indexes may be used to:

1. Estimate and compare construction costs for different years in the same city.

2. Estimate and compare construction costs in different cities for the same year.

3. Estimate and compare construction costs in different cities for different years.

4. Compare construction trends in any city with the national average.

EXAMPLES

1. Estimate and compare construction costs for different years in the same city.

   A. To estimate the construction cost of a building in Lexington, KY in 1970, knowing that it cost $915,000 in 2016.

   Index Lexington, KY in 1970 = 26.9

   Index Lexington, KY in 2016 = 184.9

   $$\frac{\text{Index 1970}}{\text{Index 2016}} \quad \text{x} \quad \text{Cost 2016} \quad = \quad \text{Cost 1970}$$

   $$\frac{26.9}{184.9} \quad \text{x} \quad \$915,000 \quad = \quad \$133,118$$

   Construction Cost in Lexington, KY in 1970 = $133,118

   B. To estimate the current construction cost of a building in Boston, MA that was built in 1980 for $900,000.

   Index Boston, MA in 1980 = 64.0

   Index Boston, MA in 2016 = 244.3

   $$\frac{\text{Index 2016}}{\text{Index 1980}} \quad \text{x} \quad \text{Cost 1980} \quad = \quad \text{Cost 2016}$$

   $$\frac{241.3}{64.0} \quad \text{x} \quad \$900,000 \quad = \quad \$3,393,281$$

   Construction Cost in Boston in 2016 = $3,393,281

2. Estimate and compare construction costs in different cities for the same year.

   To compare the construction cost of a building in Topeka, KS in 2016 with the known cost of $800,000 in Baltimore, MD in 2016

   Index Topeka, KS in 2016 = 176.2

   Index Baltimore, MD in 2016 = 189.7

   $$\frac{\text{Index Topeka}}{\text{Index Baltimore}} \quad \text{x} \quad \text{Cost Baltimore} = \text{Cost Topeka}$$

   $$\frac{176.2}{189.7} \quad \text{x} \quad \$800,000 \quad = \$743,068$$

   Construction Cost in Topeka in 2016 = $743,068

3. Estimate and compare construction costs in different cities for different years.

   To compare the construction cost of a building in Detroit, MI in 2016 with the known construction cost of $5,000,000 for the same building in San Francisco, CA in 1980.

   Index Detroit, MI in 2016 = 212.3

   Index San Francisco, CA in 1980 = 75.2

   $$\frac{\text{Index Detroit 2016}}{\text{Index San Francisco 1980}} \quad \text{x Cost San Francisco 1980} = \text{Cost Detroit 2016}$$

   $$\frac{212.3}{75.2} \quad \text{x} \quad \$5,000,000 \quad = \quad \$14,115,692$$

   Construction Cost in Detroit in 2016 = $14,115,692

4. Compare construction trends in any city with the national average.

   To compare the construction cost in Las Vegas, NV from 1975 to 2016 with the increase in the National Average during the same time period.

   Index Las Vegas, NV for 1975 = 42.8   For 2016 = 215.5

   Index 30 City Average for 1975 = 43.7   For 2016 = 207.2

   A. National Average escalation   = $\frac{\text{Index} - \text{30 City 2016}}{\text{Index} - \text{30 City 1975}}$
   From 1975 to 2016

   $$= \frac{207.2}{43.7}$$

   National Average escalation
   From 1975 to 2016   =   4.74 or increased by 474%

   B. Escalation for Las Vegas, NV = $\frac{\text{Index Las Vegas, NV 2016}}{\text{Index Las Vegas, NV 1975}}$
   From 1975 to 2016

   $$= \frac{215.5}{42.8}$$

   Las Vegas escalation
   From 1975 to 2016   =   5.04 or increased by 504%

   Conclusion: Construction costs in Las Vegas are higher than National average costs and increased at a greater rate from 1975 to 2016 than the National Average.

Historical Cost Indexes

Year	National 30 City Average	Birmingham	Huntsville	Mobile	Montgomery	Tuscaloosa	Anchorage	Phoenix	Tuscon	Fort Smith	Little Rock	Anaheim	Bakersfield	Fresno	Los Angeles	Oxnard
Jan 2016	207.2E	185.3E	182.6E	185.9E	183.2E	185E	245.3E	181.5E	178.4E	168E	171.5E	217.9E	215.2E	219.4E	219.1E	216.3E
2015	204.0	184.8	182.1	185.4	182.7	184.5	244.7	181.0	177.9	167.6	171.1	217.3	214.6	218.8	218.5	215.7
2014	203.0	180.3	175.8	170.6	163.1	165.9	241.0	180.0	177.0	165.1	168.4	214.2	215.2	217.0	217.3	214.8
2013	196.9	173.3	168.7	165.7	158.7	161.6	235.3	174.1	169.2	161.2	163.2	207.0	205.6	210.3	210.7	207.8
2012	194.0	169.1	162.7	163.2	153.8	154.6	232.5	172.2	166.4	158.7	161.0	204.1	202.9	207.3	207.2	204.7
2011	185.7	163.0	156.3	156.9	147.6	148.3	225.1	163.8	159.5	152.0	154.1	196.1	194.6	199.8	199.2	197.0
2010	181.6	159.6	152.9	153.1	144.4	144.9	218.3	160.7	157.3	150.3	152.5	192.8	190.8	195.0	194.9	192.8
2009	182.5	162.7	157.2	155.6	148.8	149.4	222.9	161.3	156.8	149.2	156.3	194.5	192.3	195.0	196.6	194.5
2008	171.0	150.3	146.9	143.7	138.4	138.8	210.8	152.2	148.4	138.6	145.4	182.7	179.8	183.2	184.7	182.6
2007	165.0	146.9	143.3	140.3	134.8	135.4	206.8	147.7	143.1	134.8	141.7	175.1	173.7	176.7	177.5	176.2
2006	156.2	135.7	133.6	126.7	124.0	122.2	196.4	137.9	134.8	123.2	127.2	166.8	164.9	169.8	167.3	167.4
2005	146.7	127.9	125.4	119.3	116.6	114.6	185.6	128.5	124.1	115.4	119.4	156.5	153.0	157.9	157.1	156.4
2004	132.8	115.9	112.9	107.0	104.9	102.9	167.0	116.5	113.6	103.8	108.6	142.1	139.5	143.8	142.0	140.9
2003	129.7	113.1	110.7	104.8	102.3	100.8	163.5	113.9	110.6	101.8	105.8	139.4	137.2	142.1	139.6	138.7
2002	126.7	110.0	103.4	103.6	101.7	99.7	159.5	113.3	110.2	100.4	102.1	136.5	133.4	136.0	136.4	136.3
2001	122.2	106.0	100.5	100.6	98.3	95.9	152.6	109.0	106.4	97.3	98.7	132.5	129.3	132.8	132.4	132.4
2000	118.9	104.1	98.9	99.1	94.2	94.6	148.3	106.2	104.9	94.4	95.7	129.4	125.2	129.4	129.9	129.7
1999	116.6	101.2	97.4	97.7	92.5	92.8	145.9	105.5	103.3	93.1	94.4	127.9	123.6	126.9	128.7	128.2
1998	113.6	96.2	94.0	94.8	90.3	89.8	143.8	102.1	101.0	90.6	91.4	125.2	120.3	123.9	125.8	124.7
1997	111.5	94.6	92.4	93.3	88.8	88.3	142.0	101.8	100.7	89.3	90.1	124.0	119.1	122.3	124.6	123.5
1995	105.6	87.8	88.0	88.8	84.5	83.5	138.0	96.1	95.5	85.0	85.6	120.1	115.7	117.5	120.9	120.0
1990	93.2	79.4	77.6	82.7	78.4	75.2	125.8	86.4	87.0	77.1	77.9	107.7	102.7	103.3	107.5	107.4
1985	81.8	71.1	70.5	72.7	70.9	67.9	116.0	78.1	77.7	69.6	71.0	95.4	92.2	92.6	94.6	96.6
1980	60.7	55.2	54.1	56.8	56.8	54.0	91.4	63.7	62.4	53.1	55.8	68.7	69.4	68.7	67.4	69.9
1975	43.7	40.0	40.9	41.8	40.1	37.8	57.3	44.5	45.2	39.0	38.7	47.6	46.6	47.7	48.3	47.0
1970	27.8	24.1	25.1	25.8	25.4	24.2	43.0	27.2	28.5	24.3	22.3	31.0	30.8	31.1	29.0	31.0
1965	21.5	19.6	19.3	19.6	19.5	18.7	34.9	21.8	22.0	18.7	18.5	23.9	23.7	24.0	22.7	23.9
1960	19.5	17.9	17.5	17.8	17.7	16.9	31.7	19.9	20.0	17.0	16.8	21.7	21.5	21.8	20.6	21.7
1955	16.3	14.8	14.7	14.9	14.9	14.2	26.6	16.7	16.7	14.3	14.5	18.2	18.1	18.3	17.3	18.2
1950	13.5	12.2	12.1	12.3	12.3	11.7	21.9	13.8	13.8	11.8	11.6	15.1	15.0	15.1	14.3	15.1
1945	8.6	7.8	7.7	7.8	7.8	7.5	14.0	8.8	8.8	7.5	7.4	9.6	9.5	9.6	9.1	9.6
1940	6.6	6.0	6.0	6.1	6.0	5.8	10.8	6.8	6.8	5.8	5.7	7.4	7.4	7.4	7.0	7.4

Year	National 30 City Average	Riverside	Sacramento	San Diego	San Francisco	Santa Barbara	Stockton	Vallejo	Colorado Springs	Denver	Pueblo	Bridgeport	Bristol	Hartford	New Britain	New Haven
Jan 2016	207.2E	217.7E	221.6E	213.7E	250.8E	216.3E	223.4E	230.8E	185.6E	187.1E	184E	224.7E	223.9E	224.8E	223.4E	225.6E
2015	204.0	217.1	221.0	213.1	250.2	215.7	222.8	230.2	185.1	186.6	183.5	224.1	223.3	224.2	222.8	225.0
2014	203.0	214.1	221.5	211.4	248.0	214.7	218.4	227.9	187.8	189.1	184.7	223.8	222.8	223.9	222.4	224.1
2013	196.9	207.0	215.0	203.2	241.0	207.9	211.8	222.2	180.7	183.4	180.0	217.8	217.2	218.4	216.8	218.5
2012	194.0	204.1	211.9	198.9	237.6	204.8	208.8	218.9	179.4	182.2	177.9	213.9	213.3	214.4	212.9	214.7
2011	185.7	195.8	203.1	192.3	227.9	196.2	201.3	210.3	171.7	174.1	170.7	205.2	204.2	205.1	203.9	205.8
2010	181.6	192.7	196.7	188.2	223.0	192.9	195.9	204.6	170.5	172.6	168.1	198.8	199.1	199.8	198.8	200.7
2009	182.5	193.1	198.1	191.6	224.9	193.6	194.5	202.7	168.4	172.0	166.7	199.0	197.7	198.7	197.4	199.3
2008	171.0	181.3	186.0	179.9	210.6	181.5	183.2	191.6	158.8	161.9	157.9	185.8	184.8	185.7	184.5	186.1
2007	165.0	174.7	179.1	173.6	201.1	175.2	178.2	185.9	152.6	155.9	152.3	179.6	178.3	179.6	178.0	179.5
2006	156.2	166.0	172.2	164.0	191.2	166.4	171.0	177.9	146.1	149.2	145.0	169.5	168.0	169.5	167.7	169.8
2005	146.7	155.4	161.1	153.8	179.7	155.7	160.0	167.2	138.3	141.1	136.4	160.8	159.4	159.6	159.1	160.8
2004	132.8	141.0	147.2	139.0	163.6	141.4	144.2	148.6	125.6	127.2	123.5	143.6	142.9	142.4	142.7	144.3
2003	129.7	138.7	144.9	136.6	162.1	139.2	141.8	146.4	123.1	123.7	120.7	141.3	140.2	140.3	140.0	141.6
2002	126.7	135.3	138.4	134.2	157.9	135.9	136.9	143.8	118.7	121.3	116.7	133.8	133.6	133.1	133.3	134.8
2001	122.2	131.4	135.5	129.7	151.8	131.5	134.1	140.2	113.3	117.2	113.1	128.6	128.5	128.5	128.3	128.6
2000	118.9	128.0	131.5	127.1	146.9	128.7	130.7	137.1	109.8	111.8	108.8	122.7	122.6	122.9	122.4	122.7
1999	116.6	126.5	129.4	124.9	145.1	127.2	127.9	135.3	107.0	109.1	107.1	121.1	121.3	121.2	121.1	121.3
1998	113.6	123.7	125.9	121.3	141.9	123.7	124.7	132.5	103.3	106.5	103.7	119.1	119.4	120.0	119.7	120.0
1997	111.5	122.6	124.7	120.3	139.2	122.4	123.3	130.6	101.1	104.4	102.0	119.2	119.5	119.9	119.7	120.0
1995	105.6	119.2	119.5	115.4	133.8	119.0	119.0	122.5	96.1	98.9	96.8	116.0	116.5	116.9	116.3	116.5
1990	93.2	107.0	104.9	105.6	121.8	106.4	105.4	111.4	86.9	88.8	88.8	96.3	95.9	96.6	95.9	96.5
1985	81.8	94.8	92.2	94.2	106.2	93.2	93.7	97.0	79.6	81.0	80.4	86.1	86.2	87.2	86.1	86.3
1980	60.7	68.7	71.3	68.1	75.2	71.1	71.2	71.9	60.7	60.9	59.5	61.5	60.7	61.9	60.7	61.4
1975	43.7	47.3	49.1	47.7	49.8	46.1	47.6	46.5	43.1	42.7	42.5	45.2	45.5	46.0	45.2	46.0
1970	27.8	31.0	32.1	30.7	31.6	31.3	31.8	31.8	27.6	26.1	27.3	29.2	28.2	29.6	28.2	29.3
1965	21.5	23.9	24.8	24.0	23.7	24.1	24.5	24.5	21.3	20.9	21.0	22.4	21.7	22.6	21.7	23.2
1960	19.5	21.7	22.5	21.7	21.5	21.9	22.3	22.2	19.3	19.0	19.1	20.0	19.8	20.0	19.8	20.0
1955	16.3	18.2	18.9	18.2	18.0	18.4	18.7	18.6	16.2	15.9	16.0	16.8	16.6	16.8	16.6	16.8
1950	13.5	15.0	15.6	15.0	14.9	15.2	15.4	15.4	13.4	13.2	13.2	13.9	13.7	13.8	13.7	13.9
1945	8.6	9.6	10.0	9.6	9.5	9.7	9.9	9.8	8.5	8.4	8.4	8.8	8.7	8.8	8.7	8.9
1940	6.6	7.4	7.7	7.4	7.3	7.5	7.6	7.6	6.6	6.5	6.5	6.8	6.8	6.8	6.8	6.8

For customer support on your Square Foot Cost Data, call 877.756.2789.

Historical Cost Indexes

Year	National 30 City Average	Connecticut Norwalk	Stamford	Water-bury	Delaware Wilming-ton	D.C. Washing-ton	Fort Lau-derdale	Jackson-ville	Miami	Orlando	Talla-hassee	Tampa	Albany	Atlanta	Colum-bus	Macon
Jan 2016	207.2E	230.8E	230.9E	224.4E	212.2E	197.8E	176.6E	174.7E	178.6E	178.8E	172.9E	179.4E	169E	178.8E	171.7E	170.4E
2015	204.0	230.2	230.3	223.8	211.7	197.3	176.1	174.2	178.1	178.3	172.5	178.9	168.6	178.3	171.3	170
2014	203.0	230.3	230.5	223.2	210.2	197.0	176.5	171.0	178.4	176.5	164.4	183.2	166.6	177.5	170.2	168.2
2013	196.9	223.1	224.2	217.3	203.6	191.7	173.2	168.4	175.8	174.8	160.8	180.2	163.2	173.3	166.7	164.9
2012	194.0	219.8	220.0	213.4	201.0	189.6	170.9	165.8	173.7	172.8	158.5	177.8	158.3	170.7	160.6	159.3
2011	185.7	211.0	211.2	204.6	193.3	182.3	164.4	159.4	167.0	166.1	151.7	171.4	151.8	164.0	154.3	152.7
2010	181.6	198.3	203.8	199.4	187.7	179.0	160.3	155.4	163.3	162.5	148.1	167.5	148.3	160.1	150.9	149.4
2009	182.5	198.8	204.3	198.4	188.9	181.4	160.7	152.0	165.2	163.9	145.1	165.1	152.0	164.5	155.6	153.5
2008	171.0	185.2	189.2	185.3	177.5	170.4	149.6	142.5	152.9	153.0	135.3	155.6	140.4	153.2	143.9	142.5
2007	165.0	178.6	183.3	179.1	173.4	163.1	145.6	139.0	148.9	148.1	131.7	152.0	135.8	148.0	140.1	138.1
2006	156.2	169.7	173.7	169.1	158.3	153.0	136.0	126.5	136.7	134.3	118.4	136.6	123.6	139.9	126.3	124.3
2005	146.7	160.9	164.1	160.5	149.6	142.7	126.0	119.2	127.1	126.0	110.7	127.7	115.7	131.7	118.9	116.1
2004	132.8	143.6	146.5	143.2	136.1	126.8	114.8	107.8	115.6	113.2	100.6	116.6	102.8	119.6	102.0	105.0
2003	129.7	140.6	145.1	140.8	133.0	124.6	109.0	105.6	110.5	107.8	98.0	103.5	100.5	115.8	99.1	102.7
2002	126.7	133.5	136.2	133.9	129.4	120.3	107.1	104.6	107.4	106.7	97.3	102.8	99.7	113.0	98.3	101.9
2001	122.2	128.2	131.5	128.8	124.8	115.9	104.3	100.8	104.6	103.8	94.8	100.3	96.4	109.1	95.5	99.0
2000	118.9	121.5	126.4	123.2	117.4	113.8	102.4	99.0	101.8	101.2	93.6	98.9	94.9	106.1	94.1	97.0
1999	116.6	120.0	122.0	121.2	116.6	111.5	101.4	98.0	100.8	100.1	92.5	97.9	93.5	102.9	92.8	95.8
1998	113.6	118.8	120.8	120.1	112.3	109.6	99.4	96.2	99.0	98.4	90.9	96.3	91.4	100.8	90.4	93.3
1997	111.5	118.9	120.9	120.2	110.7	106.4	98.6	95.4	98.2	97.2	89.9	95.5	89.9	98.5	88.7	91.6
1995	105.6	115.5	117.9	117.3	106.1	102.3	94.0	90.9	93.7	93.5	86.5	92.2	85.0	92.0	82.6	86.8
1990	93.2	96.3	98.9	95.1	92.5	90.4	83.9	81.1	84.0	82.9	78.4	85.0	75.0	80.4	74.0	76.9
1985	81.8	85.3	86.8	85.6	81.1	78.8	76.7	73.4	78.3	73.9	70.6	77.3	66.9	70.3	66.1	68.1
1980	60.7	60.7	60.9	62.3	58.5	59.6	55.3	55.8	56.5	56.7	53.5	57.2	51.7	54.0	51.1	51.3
1975	43.7	44.7	45.0	46.3	42.9	43.7	42.1	40.3	43.2	41.5	38.1	41.3	37.5	38.4	36.2	36.5
1970	27.8	28.1	28.2	28.9	27.0	26.3	25.7	22.8	27.0	26.2	24.5	24.2	23.8	25.2	22.8	23.4
1965	21.5	21.6	21.7	22.2	20.9	21.8	19.8	17.4	19.3	20.2	18.9	18.6	18.3	19.8	17.6	18.0
1960	19.5	19.7	19.7	20.2	18.9	19.4	18.0	15.8	17.6	18.3	17.2	16.9	16.7	17.1	16.0	16.4
1955	16.3	16.5	16.5	17.0	15.9	16.3	15.1	13.2	14.7	15.4	14.4	14.1	14.0	14.4	13.4	13.7
1950	13.5	13.6	13.7	14.0	13.1	13.4	12.5	11.0	12.2	12.7	11.9	11.7	11.5	11.9	11.0	11.3
1945	8.6	8.7	8.7	8.9	8.4	8.6	7.9	7.0	7.8	8.1	7.6	7.5	7.4	7.6	7.0	7.2
1940	6.6	6.7	6.7	6.9	6.4	6.6	6.1	5.4	6.0	6.2	5.8	5.7	5.7	5.8	5.5	5.6

Year	National 30 City Average	Georgia Savan-nah	Hawaii Hono-lulu	Boise	Poca-tello	Chicago	Decatur	Joliet	Peoria	Rock-ford	Spring-field	Ander-son	Evans-ville	Fort Wayne	Gary	Indian-apolis
Jan 2016	207.2E	170.9E	251.1E	185.9E	185.8E	239.5E	209.5E	239.3E	214.3E	225.9E	211E	186.9E	190.6E	183.5E	213E	190.1E
2015	204.0	170.5	250.5	185.4	185.3	238.9	209.0	238.7	213.7	225.3	210.5	186.4	190.1	183.0	212.4	189.6
2014	203.0	167.2	239.7	184.1	184.4	238.1	206.0	236.8	212.5	224.4	208.2	184.5	188.1	180.8	210.1	188.4
2013	196.9	163.5	231.5	179.6	179.9	231.1	200.8	228.7	207.2	217.2	203.2	177.8	182.5	175.9	203.6	182.7
2012	194.0	158.9	227.9	170.4	171.5	226.2	195.7	222.7	199.7	211.8	197.2	175.2	177.8	171.8	197.9	180.6
2011	185.7	151.7	219.0	162.8	163.6	217.6	187.9	216.8	193.4	205.8	189.4	168.4	169.7	165.0	191.0	173.0
2010	181.6	147.9	214.6	161.7	162.4	210.6	181.2	208.3	186.2	197.6	182.2	162.8	165.9	160.0	183.7	168.1
2009	182.5	152.8	218.6	162.6	163.4	209.0	182.2	205.9	186.2	196.5	184.0	164.9	166.9	163.0	185.2	170.3
2008	171.0	141.1	205.5	153.2	152.8	195.7	168.7	185.5	172.2	180.2	168.5	152.4	155.8	151.0	169.0	159.2
2007	165.0	136.9	200.4	148.0	148.1	186.5	160.5	177.7	166.0	175.0	159.2	149.0	152.6	147.7	165.4	154.1
2006	156.2	124.0	191.9	141.6	141.2	177.8	153.8	168.7	155.2	163.3	152.8	139.8	144.4	139.1	154.5	144.6
2005	146.7	116.9	181.2	133.8	132.5	164.4	145.0	158.3	146.5	151.2	145.4	131.4	136.2	131.0	146.4	137.4
2004	132.8	105.6	161.2	122.0	120.5	149.3	129.3	145.1	133.9	138.1	130.3	120.9	123.5	120.1	132.2	125.0
2003	129.7	103.0	159.4	120.2	118.7	146.2	127.4	143.5	132.4	137.3	128.4	119.7	121.4	118.2	131.4	122.5
2002	126.7	102.0	157.2	118.3	117.1	141.2	123.8	138.5	129.6	132.9	125.3	117.5	119.0	116.4	129.1	120.5
2001	122.2	99.0	150.0	114.3	113.4	135.8	120.1	133.7	124.3	127.8	119.8	113.4	115.6	112.1	123.4	116.4
2000	118.9	97.5	144.8	112.9	112.1	131.2	115.1	124.6	119.0	122.2	116.2	109.8	111.5	108.4	117.8	113.2
1999	116.6	96.0	143.0	110.9	109.6	129.6	113.0	122.6	116.4	120.7	113.8	107.1	109.3	106.8	112.8	110.6
1998	113.6	93.7	140.4	107.4	107.0	125.2	110.1	119.8	113.8	115.5	111.1	105.0	107.2	104.5	111.1	108.0
1997	111.5	92.0	139.8	104.6	104.7	121.3	107.8	117.5	111.5	113.1	108.9	101.9	104.4	101.8	110.3	105.2
1995	105.6	87.4	130.3	99.5	98.2	114.2	98.5	110.5	102.3	103.6	98.1	96.4	97.2	95.0	100.7	100.1
1990	93.2	77.9	104.7	88.2	88.1	98.4	90.9	98.4	93.7	94.0	90.1	84.6	89.3	83.4	88.4	87.1
1985	81.8	68.9	94.7	78.0	78.0	82.4	81.9	83.4	83.7	83.0	81.5	75.2	79.8	75.0	77.8	77.1
1980	60.7	52.2	68.9	60.3	59.5	62.8	62.3	63.4	64.5	61.6	61.1	56.5	59.0	56.7	59.8	57.9
1975	43.7	36.9	44.6	40.8	40.5	45.7	43.1	44.5	44.7	42.7	42.4	39.5	41.7	39.9	41.9	40.6
1970	27.8	21.0	30.4	26.7	26.6	29.1	28.0	28.6	29.0	27.5	27.6	25.4	26.4	25.5	27.1	26.2
1965	21.5	16.4	21.8	20.6	20.5	22.7	21.5	22.1	22.4	21.2	21.3	19.5	20.4	19.7	20.8	20.7
1960	19.5	14.9	19.8	18.7	18.6	20.2	19.6	20.0	20.3	19.2	19.3	17.7	18.7	17.9	18.9	18.4
1955	16.3	12.5	16.6	15.7	15.6	16.9	16.4	16.8	17.0	16.1	16.2	14.9	15.7	15.0	15.9	15.5
1950	13.5	10.3	13.7	13.0	12.9	14.0	13.6	13.9	14.0	13.3	13.4	12.3	12.9	12.4	13.1	12.8
1945	8.6	6.6	8.8	8.3	8.2	8.9	8.6	8.9	9.0	8.5	8.6	7.8	8.3	7.9	8.4	8.1
1940	6.6	5.1	6.8	6.4	6.3	6.9	6.7	6.8	6.9	6.5	6.6	6.0	6.4	6.1	6.5	6.3

529

Historical Cost Indexes

Year	National 30 City Average	Indiana Muncie	Indiana South Bend	Indiana Terre Haute	Iowa Cedar Rapids	Iowa Daven-port	Iowa Des Moines	Iowa Sioux City	Iowa Water-loo	Kansas Topeka	Kansas Wichita	Kentucky Lexing-ton	Kentucky Louis-ville	Louisiana Baton Rouge	Louisiana Lake Charles	Louisiana New Orleans
Jan 2016	207.2E	186.5E	189.1E	191.1E	189.3E	197.4E	189.5E	179.6E	178.3E	176.2E	175.6E	184.9E	186.4E	177.1E	175.9E	178.1E
2015	204.0	186.0	188.6	190.6	188.8	196.9	189.0	179.1	177.8	175.7	175.1	184.4	185.9	176.6	175.4	177.6
2014	203.0	185.0	185.5	189.2	188.3	195.9	188.5	178.9	177.4	173.6	173.7	183.7	186.1	171.3	170.9	177.5
2013	196.9	179.3	179.3	182.6	183.4	190.8	181.8	173.2	171.3	168.8	167.3	179.1	182.4	166.3	166.2	173.1
2012	194.0	175.8	174.3	178.9	180.3	184.6	179.2	168.8	169.2	162.3	161.4	174.8	177.5	164.4	163.8	171.7
2011	185.7	169.2	168.1	171.3	173.8	178.1	172.7	162.3	162.8	156.3	155.3	167.8	170.8	156.9	156.5	162.1
2010	181.6	163.1	163.5	167.8	170.4	175.6	170.2	159.0	160.2	152.8	151.3	158.7	166.9	153.7	153.4	160.3
2009	182.5	164.5	167.1	168.0	165.9	172.0	162.1	156.3	147.6	154.5	152.4	160.7	167.5	156.9	154.4	161.9
2008	171.0	153.0	153.1	155.9	157.3	163.4	152.4	147.8	139.2	144.9	143.1	152.0	156.8	144.0	141.5	149.3
2007	165.0	149.7	149.5	152.7	152.4	157.6	148.9	144.0	135.0	141.0	138.7	147.6	150.9	139.4	138.0	145.2
2006	156.2	141.2	141.3	144.0	145.8	151.1	142.9	137.3	128.4	134.7	132.9	129.0	143.0	129.4	129.8	135.8
2005	146.7	131.0	133.8	135.2	135.7	142.1	133.5	129.4	120.0	125.9	125.6	121.5	135.1	120.8	117.7	126.2
2004	132.8	120.0	120.6	123.0	122.7	128.6	121.6	116.2	108.3	112.6	113.4	110.1	120.3	105.6	109.0	115.3
2003	129.7	118.7	119.2	121.7	120.9	126.4	120.2	114.6	105.9	109.4	111.3	107.8	118.7	103.3	106.1	112.4
2002	126.7	116.7	117.1	119.9	116.0	122.1	116.7	111.4	103.7	107.8	109.6	106.2	116.4	102.5	105.0	110.6
2001	122.2	112.8	111.6	115.5	112.5	117.5	113.2	107.7	100.6	104.3	105.2	103.3	112.7	99.4	101.3	104.6
2000	118.9	109.1	106.9	110.8	108.8	112.6	108.9	99.0	98.1	101.0	101.1	101.4	109.3	97.8	99.7	102.1
1999	116.6	105.7	103.7	107.9	104.1	109.2	107.6	96.7	96.4	98.7	99.3	99.7	106.6	96.1	97.6	99.5
1998	113.6	103.6	102.3	106.0	102.5	106.8	103.8	95.1	94.8	97.4	97.4	97.4	101.5	94.7	96.1	97.7
1997	111.5	101.3	101.2	104.6	101.2	105.6	102.5	93.8	93.5	96.3	96.2	96.1	99.9	93.3	96.8	96.2
1995	105.6	95.6	94.6	97.2	95.4	96.6	94.4	88.2	88.9	91.1	91.0	91.6	94.6	89.6	92.7	91.6
1990	93.2	83.9	85.1	89.1	87.1	86.5	86.9	80.4	80.3	83.2	82.6	82.8	82.6	82.0	84.1	84.5
1985	81.8	75.2	76.0	79.2	75.9	77.6	77.1	72.1	72.3	75.0	74.7	75.5	74.5	74.9	78.5	78.2
1980	60.7	56.1	58.3	59.7	62.7	59.6	61.8	59.3	57.7	58.9	58.0	59.3	59.8	59.1	60.0	57.2
1975	43.7	39.2	40.3	41.9	43.1	41.0	43.1	41.0	40.0	42.0	43.1	42.7	42.5	40.4	40.6	41.5
1970	27.8	25.3	26.2	27.0	28.2	26.9	27.6	26.6	25.9	27.0	25.5	26.9	25.9	25.0	26.7	27.2
1965	21.5	19.5	20.2	20.9	21.7	20.8	21.7	20.5	20.0	20.8	19.6	20.7	20.3	19.4	20.6	20.4
1960	19.5	17.8	18.3	18.9	19.7	18.8	19.5	18.6	18.2	18.9	17.8	18.8	18.4	17.6	18.7	18.5
1955	16.3	14.9	15.4	15.9	16.6	15.8	16.3	15.6	15.2	15.8	15.0	15.8	15.4	14.8	15.7	15.6
1950	13.5	12.3	12.7	13.1	13.7	13.1	13.5	12.9	12.6	13.1	12.4	13.0	12.8	12.2	13.0	12.8
1945	8.6	7.8	8.1	8.4	8.7	8.3	8.6	8.2	8.0	8.3	7.9	8.3	8.1	7.8	8.3	8.2
1940	6.6	6.0	6.3	6.5	6.7	6.4	6.6	6.3	6.2	6.4	6.1	6.4	6.3	6.0	6.4	6.3

Year	National 30 City Average	Louisiana Shreve-port	Maine Lewis-ton	Maine Portland	Maryland Balti-more	Massachusetts Boston	Massachusetts Brockton	Massachusetts Fall River	Massachusetts Law-rence	Massachusetts Lowell	Massachusetts New Bedford	Massachusetts Pitts-field	Massachusetts Spring-field	Massachusetts Wor-cester	Michigan Ann Arbor	Michigan Dear-born
Jan 2016	207.2E	171.3E	194.2E	197.4E	189.7E	241.3E	228.9E	228.8E	235.1E	233.8E	228E	213.4E	215.2E	227.4E	210.3E	212E
2015	204.0	170.9	193.7	196.9	189.2	240.7	228.3	228.2	234.5	233.2	227.4	212.8	214.6	226.8	209.8	211.5
2014	203.0	167.7	192.4	195.4	187.6	239.1	229.1	229.3	234.8	233.9	228.5	213.3	214.8	227.3	206.1	208.4
2013	196.9	159.0	187.9	189.5	183.2	232.9	221.6	221.8	225.3	223.4	221.1	205.9	206.5	219.3	201.2	202.4
2012	194.0	153.2	180.7	182.4	180.8	229.3	217.2	217.8	221.6	219.3	217.0	200.2	201.9	215.2	196.1	199.2
2011	185.7	146.7	167.4	169.1	173.6	219.3	208.2	206.0	211.6	210.1	205.2	191.2	194.5	206.6	186.7	189.4
2010	181.6	145.1	162.1	163.7	168.3	214.3	202.7	200.7	204.9	204.7	200.0	185.9	189.4	201.1	183.3	186.8
2009	182.5	147.6	159.4	161.8	169.1	211.8	199.2	197.6	201.9	201.9	196.9	185.1	187.1	198.2	179.2	186.9
2008	171.0	135.9	149.1	150.9	157.7	198.6	186.4	185.1	188.7	189.1	184.5	171.3	173.7	184.5	169.6	176.4
2007	165.0	132.4	146.4	148.2	152.5	191.8	180.4	179.7	182.0	181.7	179.1	166.8	169.1	180.4	166.8	172.5
2006	156.2	125.2	140.2	139.8	144.9	180.4	171.6	170.4	172.3	173.5	170.4	157.3	159.7	171.6	161.1	165.5
2005	146.7	117.4	131.7	131.4	135.8	169.6	161.3	159.6	162.2	162.3	159.6	147.6	150.7	158.8	147.5	155.8
2004	132.8	105.7	119.8	119.4	121.4	154.1	144.4	143.6	146.4	146.7	143.6	131.9	136.5	143.2	135.9	142.1
2003	129.7	103.9	117.7	117.3	118.1	150.2	139.6	140.6	143.3	143.6	140.5	128.9	133.8	139.0	134.1	139.0
2002	126.7	102.1	117.5	117.1	115.6	145.6	136.0	135.0	136.9	137.3	134.9	124.7	128.3	134.9	131.4	134.3
2001	122.2	98.3	114.6	114.3	111.7	140.9	132.1	131.4	133.1	132.9	131.3	120.3	124.5	130.1	126.8	129.8
2000	118.9	96.2	105.7	105.4	107.7	138.9	129.4	128.4	129.7	130.3	128.3	116.7	121.0	127.3	124.3	125.8
1999	116.6	94.8	105.0	104.7	106.4	136.2	126.7	126.3	127.3	127.4	126.2	115.0	118.8	123.8	117.5	122.7
1998	113.6	92.0	102.8	102.5	104.1	132.8	125.0	124.7	125.0	125.3	124.6	114.2	117.1	122.6	116.0	119.7
1997	111.5	90.1	101.7	101.4	102.2	132.1	124.4	124.7	125.1	125.6	124.7	114.6	117.4	122.9	113.5	117.9
1995	105.6	86.7	96.8	96.5	96.1	128.6	119.6	117.7	120.5	119.6	117.2	110.7	112.7	114.8	106.1	110.5
1990	93.2	80.3	88.5	88.6	85.6	110.9	103.7	102.9	105.2	102.8	102.6	98.7	101.4	103.2	93.2	96.5
1985	81.8	73.6	76.7	77.0	72.7	92.8	88.8	88.7	89.4	88.2	88.6	85.0	85.6	86.7	80.1	82.7
1980	60.7	58.7	57.3	58.5	53.6	64.0	63.7	64.1	63.4	62.7	63.1	61.8	62.0	62.3	62.9	64.0
1975	43.7	40.5	41.9	42.1	39.8	46.6	45.8	45.7	46.2	45.7	46.1	45.5	45.8	46.0	44.1	44.9
1970	27.8	26.4	26.5	25.8	25.1	29.2	29.1	29.2	29.1	28.9	29.0	28.6	28.5	28.7	28.5	28.9
1965	21.5	20.3	20.4	19.4	20.2	23.0	22.5	22.5	22.5	22.2	22.4	22.0	22.4	22.1	22.0	22.3
1960	19.5	18.5	18.6	17.6	17.5	20.5	20.4	20.4	20.4	20.2	20.3	20.0	20.1	20.1	20.0	20.2
1955	16.3	15.5	15.6	14.7	14.7	17.2	17.1	17.1	17.1	16.9	17.1	16.8	16.9	16.8	16.7	16.9
1950	13.5	12.8	12.9	12.2	12.1	14.2	14.1	14.1	14.1	14.0	14.1	13.8	14.0	13.9	13.8	14.0
1945	8.6	8.1	8.2	7.8	7.7	9.1	9.0	9.0	9.0	8.9	9.0	8.9	8.9	8.9	8.8	8.9
1940	6.6	6.3	6.3	6.0	6.0	7.0	6.9	7.0	7.0	6.9	6.9	6.8	6.9	6.8	6.8	6.9

Historical Cost Indexes

For customer support on your Square Foot Cost Data, call 877.756.2789.

Year	National 30 City Average	Michigan Detroit	Flint	Grand Rapids	Kala-mazoo	Lansing	Sagi-naw	Minnesota Duluth	Minne-apolis	Roches-ter	Mississippi Biloxi	Jackson	Missouri Kansas City	St. Joseph	St. Louis	Spring-field
Jan 2016	207.2E	212.3E	199.4E	191.7E	190.9E	200.1E	194.6E	213.5E	222.6E	209.6E	167.7E	174.7E	210.5E	202.8E	210.6E	192.3E
2015	204.0	211.8	198.9	191.2	190.4	199.6	194.1	212.9	222.0	209.1	167.3	174.2	210.0	202.3	210.1	191.8
2014	203.0	208.7	196.0	187.9	188.4	195.7	191.7	211.1	220.7	207.5	164.6	169.8	210.8	199.3	208.6	189.5
2013	196.9	203.1	190.6	182.4	182.4	190.2	186.5	205.1	216.3	201.4	160.9	164.3	204.7	193.1	202.4	181.6
2012	194.0	200.3	186.3	176.4	178.9	185.9	182.8	203.1	214.7	199.7	157.6	160.5	200.8	189.0	198.0	177.7
2011	185.7	190.3	178.5	169.9	171.9	178.0	174.6	195.7	208.1	193.8	151.4	154.3	191.1	178.8	190.6	168.3
2010	181.6	187.5	176.2	160.6	167.3	175.8	171.9	193.1	203.8	188.9	148.0	151.0	186.1	174.2	185.9	164.3
2009	182.5	187.8	176.0	156.7	166.3	173.7	168.6	191.8	203.1	188.0	152.8	156.6	185.6	172.3	187.2	163.7
2008	171.0	177.0	164.4	140.2	155.7	161.6	158.7	177.4	190.6	175.0	141.8	147.1	175.5	164.2	176.2	152.9
2007	165.0	172.9	161.7	137.2	152.6	158.7	156.1	174.2	184.5	171.4	137.7	131.2	169.0	157.8	170.6	145.3
2006	156.2	165.8	153.8	131.1	146.0	152.5	150.6	166.2	173.9	161.9	123.3	117.1	162.0	152.4	159.3	139.4
2005	146.7	156.2	145.1	123.1	134.9	142.0	140.6	157.3	164.6	152.8	116.0	109.9	151.3	143.1	149.4	131.7
2004	132.8	141.9	129.6	112.8	122.5	129.7	127.6	139.1	150.1	137.0	105.3	99.3	135.1	126.9	135.5	116.3
2003	129.7	138.7	127.9	110.9	120.8	127.7	126.0	136.7	146.5	134.5	101.4	96.7	131.9	124.8	133.3	113.9
2002	126.7	134.2	126.9	107.4	119.9	125.3	124.4	131.9	139.5	130.0	101.0	96.3	128.4	122.7	129.6	112.5
2001	122.2	129.4	122.4	104.3	115.2	120.1	119.7	131.4	136.1	124.9	98.7	93.9	121.8	114.7	125.5	106.7
2000	118.9	125.3	119.8	102.7	111.6	117.6	115.9	124.1	131.1	120.1	97.2	92.9	118.2	111.9	122.4	104.3
1999	116.6	122.6	113.7	100.9	106.1	111.5	110.1	120.3	126.5	117.1	95.7	91.8	114.9	106.1	119.8	101.1
1998	113.6	119.5	112.3	99.7	104.9	110.1	108.7	117.7	124.6	115.5	92.1	89.5	108.2	104.2	115.9	98.9
1997	111.5	117.6	110.8	98.9	104.2	108.7	107.5	115.9	121.9	114.1	90.9	88.3	106.4	102.4	113.2	97.3
1995	105.6	110.1	104.1	91.1	96.4	98.3	102.3	100.3	111.9	102.5	84.2	83.5	99.7	96.1	106.3	89.5
1990	93.2	96.0	91.7	84.0	87.4	90.2	88.9	94.3	100.5	95.6	76.4	75.6	89.5	86.8	94.0	82.6
1985	81.8	81.6	80.7	75.8	79.4	80.0	80.5	85.2	87.9	86.5	69.4	68.5	78.2	77.3	80.8	73.2
1980	60.7	62.5	63.1	59.8	61.7	60.3	63.3	63.4	64.1	64.8	54.3	54.3	59.1	60.4	59.7	56.7
1975	43.7	45.8	44.9	41.8	43.7	44.1	44.8	45.2	46.0	45.3	37.9	37.7	42.7	45.4	44.8	41.1
1970	27.8	29.7	28.5	26.7	28.1	27.8	28.7	28.9	29.5	28.9	24.4	21.3	25.3	28.3	28.4	26.0
1965	21.5	22.1	21.9	20.6	21.7	21.5	22.2	22.3	23.4	22.3	18.8	16.4	20.6	21.8	21.8	20.0
1960	19.5	20.1	20.0	18.7	19.7	19.5	20.1	20.3	20.5	20.2	17.1	14.9	19.0	19.8	19.5	18.2
1955	16.3	16.8	16.7	15.7	16.5	16.3	16.9	17.0	17.2	17.0	14.3	12.5	16.0	16.6	16.3	15.2
1950	13.5	13.9	13.8	13.0	13.6	13.5	13.9	14.0	14.2	14.0	11.8	10.3	13.2	13.7	13.5	12.6
1945	8.6	8.9	8.8	8.3	8.7	8.6	8.9	8.9	9.0	8.9	7.6	6.6	8.4	8.8	8.6	8.0
1940	6.6	6.8	6.8	6.4	6.7	6.6	6.9	6.9	7.0	6.9	5.8	5.1	6.5	6.7	6.7	6.2

Year	National 30 City Average	Montana Billings	Great Falls	Nebraska Lincoln	Omaha	Nevada Las Vegas	Reno	New Hampshire Man-chester	Nashua	New Jersey Camden	Jersey City	Newark	Pater-son	Trenton	NM Albu-querque	NY Albany
Jan 2016	207.2E	188.6E	188.9E	184.2E	185.2E	216.1E	195.4E	199.2E	198.3E	225.1E	226.7E	230.7E	229.1E	228.2E	178.8E	208.6E
2015	204.0	188.1	188.4	183.7	184.7	215.5	194.9	198.7	197.8	224.5	226.1	230.1	228.5	227.6	178.3	208.1
2014	203.0	185.4	186.3	181.1	183.7	210.9	195.7	198.8	197.3	224.6	226.6	230.6	227.8	226.6	178.0	205.1
2013	196.9	180.5	181.1	175.5	180.6	206.8	190.4	193.8	192.5	217.7	219.3	223.6	221.2	218.7	173.6	195.3
2012	194.0	178.6	179.5	169.3	177.1	202.6	186.8	189.2	188.3	213.6	215.2	219.2	217.5	213.7	171.0	191.2
2011	185.7	168.8	170.8	162.8	169.6	195.7	179.8	176.8	176.1	205.5	207.0	210.3	209.2	206.4	163.3	180.9
2010	181.6	166.5	168.6	159.3	165.8	193.7	175.6	172.5	172.0	201.2	203.3	206.3	205.3	200.8	162.5	177.8
2009	182.5	165.3	166.0	161.7	165.0	191.5	176.5	174.0	173.5	196.8	200.5	203.6	202.0	198.6	163.0	178.1
2008	171.0	153.1	154.6	152.0	154.8	176.2	167.0	162.5	161.9	184.0	187.3	189.6	188.7	185.6	152.6	166.0
2007	165.0	147.7	147.8	147.7	150.3	166.7	161.4	159.3	158.7	179.3	183.3	185.2	184.6	181.7	146.7	159.4
2006	156.2	140.5	140.8	132.4	140.7	160.0	154.5	146.7	146.4	167.7	171.0	173.6	172.2	169.9	140.1	151.6
2005	146.7	131.9	132.4	124.1	132.8	149.5	145.4	136.8	136.5	159.0	162.4	163.9	163.0	161.5	130.4	142.2
2004	132.8	118.2	117.9	112.1	119.4	137.1	130.5	124.2	123.9	143.4	145.4	147.2	146.6	146.0	118.3	128.6
2003	129.7	115.8	115.8	110.2	117.3	133.8	128.3	122.4	122.2	142.0	144.3	145.7	145.0	143.0	116.4	126.8
2002	126.7	114.6	114.5	108.2	115.0	131.9	126.4	119.9	119.6	135.9	138.5	140.2	140.1	137.6	114.6	122.6
2001	122.2	117.5	117.7	101.3	111.6	127.8	122.5	116.2	116.2	133.4	136.7	136.9	136.8	136.0	111.4	119.2
2000	118.9	113.7	113.9	98.8	107.0	125.8	118.2	111.9	111.9	128.4	130.5	132.6	132.4	130.6	109.0	116.5
1999	116.6	112.1	112.7	96.6	104.8	121.9	114.4	109.6	109.6	125.3	128.8	131.6	129.5	129.6	106.7	114.6
1998	113.6	109.7	109.1	94.9	101.2	118.1	111.4	110.1	110.0	124.3	127.7	128.9	128.5	127.9	103.8	113.0
1997	111.5	107.4	107.6	93.3	99.5	114.6	109.8	108.6	108.6	121.9	125.1	126.4	126.4	125.2	100.8	110.0
1995	105.6	104.7	104.9	85.6	93.4	108.5	105.0	100.9	100.8	107.0	112.2	111.9	112.1	111.2	96.3	103.6
1990	93.2	92.9	93.8	78.8	85.0	96.3	94.5	86.3	86.3	93.0	93.5	93.8	97.2	94.5	84.9	93.2
1985	81.8	83.9	84.3	71.5	77.5	87.6	85.0	78.1	78.1	81.3	83.3	83.5	84.5	82.8	76.7	79.5
1980	60.7	63.9	64.6	58.5	63.5	64.6	63.0	56.6	56.0	58.6	60.6	60.1	60.0	58.9	59.0	59.5
1975	43.7	43.1	43.8	40.9	43.2	42.8	41.9	41.3	40.8	42.3	43.4	44.2	43.9	43.8	40.3	43.9
1970	27.8	28.5	28.9	26.4	26.8	29.4	28.0	26.2	25.6	27.2	27.8	29.0	27.8	27.4	26.4	28.3
1965	21.5	22.0	22.3	20.3	20.6	22.4	21.6	20.6	19.7	20.9	21.4	23.8	21.4	21.3	20.6	22.3
1960	19.5	20.0	20.3	18.5	18.7	20.2	19.6	18.0	17.9	19.0	19.4	19.4	19.4	19.2	18.5	19.3
1955	16.3	16.7	17.0	15.5	15.7	16.9	16.4	15.1	15.0	16.0	16.3	16.3	16.3	16.1	15.6	16.2
1950	13.5	13.9	14.0	12.8	13.0	14.0	13.6	12.5	12.4	13.2	13.5	13.5	13.5	13.3	12.9	13.4
1945	8.6	8.8	9.0	8.1	8.3	8.9	8.7	8.0	7.9	8.4	8.6	8.6	8.6	8.5	8.2	8.5
1940	6.6	6.8	6.9	6.3	6.4	6.9	6.7	6.2	6.1	6.5	6.6	6.6	6.6	6.6	6.3	6.6

Historical Cost Indexes

Year	National 30 City Average	New York Binghamton	Buffalo	New York	Rochester	Schenectady	Syracuse	Utica	Yonkers	North Carolina Charlotte	Durham	Greensboro	Raleigh	Winston-Salem	N. Dakota Fargo	Ohio Akron
Jan 2016	207.2E	202.9E	210.1E	268.7E	204E	207.9E	201.4E	198.6E	244E	173.5E	172.7E	170.9E	168E	171E	181.5E	199.6E
2015	204.0	202.4	209.6	268.0	203.5	207.4	200.9	198.1	243.4	173.1	172.3	170.5	167.6	170.6	181.0	199.1
2014	203.0	202.4	207.8	267.7	202.2	204.2	198.8	196.4	245.6	165.5	165.2	163.9	161.6	164.3	177.1	199.2
2013	196.9	195.5	200.7	259.4	194.3	195.5	193.6	188.5	230.0	159.6	158.4	158.7	157.5	159.1	169.6	191.8
2012	194.0	192.9	198.1	256.3	191.2	191.4	189.6	185.2	227.1	156.4	155.6	156.3	155.4	157.2	167.3	187.7
2011	185.7	176.1	188.1	245.6	181.9	181.5	180.8	173.5	218.9	142.6	143.8	142.2	142.8	140.9	160.3	180.8
2010	181.6	173.8	184.0	241.4	179.6	179.0	176.7	170.5	217.1	140.5	141.6	140.0	140.6	138.8	156.2	174.5
2009	182.5	172.9	184.4	239.9	179.6	178.8	176.9	171.3	217.5	145.1	145.8	144.0	145.1	142.7	154.3	175.7
2008	171.0	160.7	173.9	226.8	168.0	167.2	165.9	161.6	202.1	135.8	136.6	134.6	135.3	133.8	145.0	165.4
2007	165.0	155.7	168.6	215.2	163.5	159.8	159.8	155.0	196.5	132.8	133.7	131.7	132.1	131.0	139.8	159.0
2006	156.2	147.5	159.2	204.5	155.2	151.1	150.9	146.4	186.0	125.1	124.6	123.8	124.2	123.0	133.2	152.8
2005	146.7	137.5	149.4	194.0	147.3	142.2	141.9	136.9	176.1	110.5	112.1	112.1	112.1	111.1	125.3	144.4
2004	132.8	123.5	136.1	177.7	132.0	128.4	127.3	124.4	161.8	98.9	100.0	100.1	100.3	99.4	113.0	131.9
2003	129.7	121.8	132.8	173.4	130.3	126.7	124.7	121.7	160.0	96.2	97.5	97.5	97.8	96.8	110.6	129.6
2002	126.7	119.0	128.5	170.1	127.1	123.0	121.8	118.4	154.8	94.8	96.1	96.1	96.3	95.5	106.3	127.2
2001	122.2	116.0	125.2	164.4	123.1	120.1	118.6	115.3	151.4	91.5	92.9	92.9	93.3	92.3	103.1	123.5
2000	118.9	112.4	122.3	159.2	120.0	117.5	115.1	112.4	144.8	90.3	91.6	91.6	91.9	91.0	97.9	117.8
1999	116.6	108.6	120.2	155.9	116.8	114.7	113.7	108.5	140.6	89.3	90.2	90.3	90.5	90.1	96.8	116.0
1998	113.6	109.0	119.1	154.4	117.2	114.2	113.6	108.6	141.4	88.2	89.1	89.2	89.4	89.0	95.2	113.1
1997	111.5	107.2	115.7	150.3	115.2	111.2	110.6	107.0	138.8	86.8	87.7	87.8	87.9	87.6	93.5	110.6
1995	105.6	99.3	110.1	140.7	106.6	104.6	104.0	97.7	129.4	81.8	82.6	82.6	82.7	82.6	88.4	103.4
1990	93.2	87.0	94.1	118.1	94.6	93.9	91.0	85.6	111.4	74.8	75.4	75.5	75.5	75.3	81.3	94.6
1985	81.8	77.5	83.2	94.9	81.6	80.1	81.0	77.0	92.8	66.9	67.6	67.7	67.6	67.5	73.4	86.8
1980	60.7	58.0	60.6	66.0	60.5	60.3	61.6	58.5	65.9	51.1	52.2	52.5	51.7	50.8	57.4	62.3
1975	43.7	42.5	45.1	49.5	44.8	43.7	44.8	42.7	47.1	36.1	37.0	37.0	37.3	36.1	39.3	44.7
1970	27.8	27.0	28.9	32.8	29.2	27.8	28.5	26.9	30.0	20.9	23.7	23.6	23.4	23.0	25.8	28.3
1965	21.5	20.8	22.2	25.5	22.8	21.4	21.9	20.7	23.1	16.0	18.2	18.2	18.0	17.7	19.9	21.8
1960	19.5	18.9	19.9	21.5	19.7	19.5	19.9	18.8	21.0	14.4	16.6	16.6	16.4	16.1	18.1	19.9
1955	16.3	15.8	16.7	18.1	16.5	16.3	16.7	15.8	17.6	12.1	13.9	13.9	13.7	13.5	15.2	16.6
1950	13.5	13.1	13.8	14.9	13.6	13.5	13.8	13.0	14.5	10.0	11.5	11.5	11.4	11.2	12.5	13.7
1945	8.6	8.3	8.8	9.5	8.7	8.6	8.8	8.3	9.3	6.4	7.3	7.3	7.2	7.1	8.0	8.8
1940	6.6	6.4	6.8	7.4	6.7	6.7	6.8	6.4	7.2	4.9	5.6	5.6	5.6	5.5	6.2	6.8

Year	National 30 City Average	Ohio Canton	Cincinnati	Cleveland	Columbus	Dayton	Lorain	Springfield	Toledo	Youngstown	Oklahoma Lawton	Oklahoma City	Tulsa	Oregon Eugene	Portland	PA Allentown
Jan 2016	207.2E	191.6E	188.6E	203.9E	192.9E	188.4E	197.2E	188.6E	201.5E	193.8E	175.6E	177.6E	170.4E	203.6E	204.8E	210.3E
2015	204	191.1	188.1	203.4	192.4	187.9	196.7	188.1	201	193.3	175.1	177.1	170	203.1	204.3	209.8
2014	203.0	190.8	188.1	203.1	192.4	185.8	196.7	187.2	199.7	193.8	169.3	170.9	166.7	200.4	201.4	209.0
2013	196.9	183.9	181.9	195.5	187.1	180.7	187.6	181.5	193.5	186.5	162.7	165.5	161.7	194.7	195.6	203.2
2012	194.0	180.2	178.6	191.4	183.5	177.6	184.1	178.0	190.4	183.3	158.1	158.0	151.3	190.1	191.8	199.3
2011	185.7	172.1	171.3	184.0	175.7	168.4	176.8	168.4	183.1	176.5	151.7	151.6	144.8	184.9	186.6	192.2
2010	181.6	167.5	166.5	180.2	170.3	163.4	173.0	164.6	176.5	171.7	149.5	149.4	143.1	181.5	182.9	187.9
2009	182.5	168.4	168.1	181.6	171.2	165.3	174.0	166.0	178.5	172.8	152.5	153.3	146.8	181.3	183.6	188.3
2008	171.0	158.4	159.0	170.6	160.2	157.5	163.9	157.5	168.2	161.5	141.3	140.6	136.2	172.3	174.7	175.7
2007	165.0	153.2	152.4	164.9	155.1	149.4	158.1	150.3	161.6	156.7	136.2	135.6	132.5	168.5	169.5	169.4
2006	156.2	147.0	144.9	156.9	146.9	142.9	151.8	143.4	154.5	150.3	129.5	129.5	125.7	160.5	161.7	160.1
2005	146.7	137.7	136.8	147.9	138.0	134.5	143.3	134.8	145.5	141.1	121.6	121.6	117.8	150.9	152.2	150.2
2004	132.8	125.6	124.3	135.6	126.2	121.1	131.5	122.0	132.8	129.3	109.3	109.4	107.3	136.8	137.8	133.9
2003	129.7	123.7	121.3	132.7	123.7	119.4	126.1	120.0	130.4	126.2	107.5	107.8	105.0	134.2	135.9	130.3
2002	126.7	120.9	119.2	130.0	120.9	117.5	124.3	118.0	128.4	123.6	106.3	106.0	104.1	132.2	133.9	128.1
2001	122.2	117.7	115.9	125.9	117.1	114.0	120.4	114.7	123.8	119.9	102.0	102.1	99.7	129.8	131.2	123.5
2000	118.9	112.7	110.1	121.3	112.5	109.0	115.0	109.2	115.7	114.1	98.7	98.9	97.5	126.3	127.4	119.9
1999	116.6	110.6	107.9	118.7	109.5	107.1	112.0	106.4	113.6	111.9	97.5	97.4	96.2	120.9	124.3	117.8
1998	113.6	108.5	105.0	114.8	106.8	106.8	104.5	109.4	104.1	111.2	109.0	94.3	94.5	94.5	120.0	122.2	115.7
1997	111.5	106.4	102.9	112.9	104.7	102.6	107.4	102.2	108.5	107.2	92.9	93.2	93.0	118.1	119.6	113.9
1995	105.6	98.8	97.1	106.4	99.1	94.6	97.1	92.0	100.6	100.1	85.3	88.0	89.0	112.2	114.3	108.3
1990	93.2	92.1	86.7	95.1	88.1	85.9	89.4	83.4	92.9	91.8	77.1	79.9	80.0	98.0	99.4	95.0
1985	81.8	83.7	78.2	86.2	77.7	76.3	80.7	74.3	84.7	82.7	71.2	73.9	73.8	88.7	90.1	81.9
1980	60.7	60.6	59.8	61.0	58.6	56.6	60.8	56.2	64.0	61.5	52.2	55.5	57.2	68.9	68.4	58.7
1975	43.7	44.0	43.7	44.6	42.2	40.5	42.7	40.0	44.9	44.5	38.6	38.6	39.2	44.6	44.9	42.3
1970	27.8	27.8	28.2	29.6	26.7	27.0	27.5	25.4	29.1	29.6	24.1	23.4	24.1	30.4	30.0	27.3
1965	21.5	21.5	20.7	21.1	20.2	20.0	21.1	19.6	21.3	21.0	18.5	17.8	20.6	23.4	22.6	21.0
1960	19.5	19.5	19.3	19.6	18.7	18.1	19.2	17.8	19.4	19.1	16.9	16.1	18.1	21.2	21.1	19.1
1955	16.3	16.3	16.2	16.5	15.7	15.2	16.1	14.9	16.2	16.0	14.1	13.5	15.1	17.8	17.7	16.0
1950	13.5	13.5	13.3	13.6	13.0	12.5	13.3	12.3	13.4	13.2	11.7	11.2	12.5	14.7	14.6	13.2
1945	8.6	8.6	8.5	8.7	8.3	8.0	8.5	7.8	8.6	8.4	7.4	7.1	8.0	9.4	9.3	8.4
1940	6.6	6.7	6.5	6.7	6.4	6.2	6.5	6.1	6.6	6.5	5.7	5.5	6.1	7.2	7.2	6.5

For customer support on your Square Foot Cost Data, call 877.756.2789.

Historical Cost Indexes

Year	National 30 City Average	Pennsylvania						RI	South Carolina		South Dakota		Tennessee			
		Erie	Harris-burg	Phila-delphia	Pitts-burgh	Reading	Scranton	Provi-dence	Charles-ton	Colum-bia	Rapid City	Sioux Falls	Chatta-nooga	Knox-ville	Memphis	Nash-ville
Jan 2016	207.2E	195E	200.8E	235.3E	210E	204.7E	202.2E	220.5E	171.7E	172.3E	168.1E	168.4E	174.8E	171.4E	178E	180.4E
2015	204.0	194.5	200.3	234.7	209.5	204.2	201.7	219.9	171.3	171.9	167.7	168.0	174.3	171.0	177.5	179.9
2014	203.0	191.4	200.3	232.1	207.1	201.4	201.4	220.3	171.0	162.5	168.6	166.8	174.1	170.9	176.5	178.2
2013	196.9	187.2	192.5	223.6	201.4	196.1	196.0	213.4	166.3	157.9	159.6	161.2	167.4	159.3	169.3	172.8
2012	194.0	182.0	187.9	221.8	196.8	190.8	191.5	209.3	155.3	148.1	156.7	158.8	163.8	155.8	166.3	167.5
2011	185.7	176.0	180.9	212.8	187.7	184.3	185.3	196.8	149.4	142.3	149.5	151.3	157.3	149.3	158.4	160.7
2010	181.6	171.0	175.2	209.3	182.3	179.8	180.8	192.8	147.4	140.2	147.3	149.0	152.4	144.7	154.2	156.7
2009	182.5	171.9	176.5	209.5	181.7	180.7	181.3	192.9	150.8	144.3	148.2	151.1	155.7	148.0	157.2	159.8
2008	171.0	161.0	165.7	196.2	169.2	169.3	167.9	178.2	142.1	134.8	138.4	141.3	136.7	133.4	146.0	147.5
2007	165.0	156.0	159.4	188.4	163.2	164.9	163.1	174.4	139.3	131.7	129.5	133.0	133.7	130.6	143.1	144.4
2006	156.2	148.7	150.7	177.8	155.5	155.0	153.9	163.6	122.4	118.7	122.8	126.6	127.0	123.9	136.5	135.5
2005	146.7	140.6	141.5	166.4	146.6	145.5	142.9	155.6	110.2	109.6	115.2	118.5	116.0	115.1	128.6	127.9
2004	132.8	127.0	126.2	148.4	133.1	128.7	129.0	138.8	98.9	98.4	103.9	107.3	105.0	104.5	115.6	115.8
2003	129.7	124.5	123.8	145.5	130.5	126.3	126.0	135.9	96.1	95.7	101.3	104.0	102.8	102.3	111.1	110.9
2002	126.7	122.1	121.1	142.0	127.8	123.4	124.2	131.1	94.8	93.9	100.2	103.2	102.2	101.2	107.6	109.2
2001	122.2	118.9	118.4	136.9	124.1	120.4	121.1	127.8	92.1	91.3	96.8	99.7	98.2	96.5	102.6	104.3
2000	118.9	114.6	114.0	132.1	120.9	115.9	117.7	122.9	89.9	89.1	94.2	98.0	96.9	95.0	100.8	100.8
1999	116.6	113.8	112.8	129.8	119.6	114.8	116.3	121.6	89.0	88.1	92.4	95.8	95.9	93.4	99.7	98.6
1998	113.6	109.8	110.5	126.6	117.1	112.5	113.7	120.3	88.0	87.2	90.7	93.6	94.9	92.4	97.2	96.8
1997	111.5	108.5	108.8	123.3	113.9	110.8	112.3	118.9	86.5	85.6	89.3	92.1	93.6	90.8	96.1	94.9
1995	105.6	99.8	100.6	117.1	106.3	103.6	103.8	111.1	82.7	82.2	84.0	84.7	89.2	86.0	91.2	87.6
1990	93.2	88.5	89.5	98.5	91.2	90.8	91.3	94.1	73.7	74.5	77.2	78.4	79.9	75.1	81.3	77.1
1985	81.8	79.6	77.2	82.2	81.5	77.7	80.0	83.0	65.9	66.9	69.7	71.2	72.5	67.7	74.3	66.7
1980	60.7	59.4	56.9	58.7	61.3	57.7	58.4	59.2	50.6	51.1	54.8	57.1	55.0	51.6	55.9	53.1
1975	43.7	43.4	42.2	44.5	44.5	43.6	41.9	42.7	35.0	36.4	37.7	39.6	39.1	37.3	40.7	37.0
1970	27.8	27.5	25.8	27.5	28.7	27.1	27.0	27.3	22.8	22.9	24.8	25.8	24.9	22.0	23.0	22.8
1965	21.5	21.1	20.1	21.7	22.4	20.9	20.8	21.9	17.6	17.6	19.1	19.9	19.2	17.1	18.3	17.4
1960	19.5	19.1	18.6	19.4	19.7	19.0	18.9	19.1	16.0	16.0	17.3	18.1	17.4	15.5	16.6	15.8
1955	16.3	16.1	15.6	16.3	16.5	15.9	15.9	16.0	13.4	13.4	14.5	15.1	14.6	13.0	13.9	13.3
1950	13.5	13.3	12.9	13.5	13.6	13.2	13.1	13.2	11.1	11.1	12.0	12.5	12.1	10.7	11.5	10.9
1945	8.6	8.5	8.2	8.6	8.7	8.4	8.3	8.4	7.1	7.1	7.7	8.0	7.7	6.8	7.3	7.0
▼ 1940	6.6	6.5	6.3	6.6	6.7	6.4	6.5	6.5	5.4	5.5	5.9	6.2	6.0	5.3	5.6	5.4

Year	National 30 City Average	Texas														Utah
		Abi-lene	Ama-rillo	Austin	Beau-mont	Corpus Christi	Dallas	El Paso	Fort Worth	Houston	Lubbock	Odessa	San Antonio	Waco	Wichita Falls	Ogden
Jan 2016	207.2E	171.2E	171.1E	172.7E	174.5E	173.2E	174.9E	169.2E	172.9E	176.9E	173.7E	171.9E	172.2E	167.7E	169.4E	177.6E
2015	204.0	170.8	170.7	172.3	174.0	172.8	174.4	168.8	172.5	176.4	173.3	171.5	171.8	167.3	169.0	177.1
2014	203.0	160.9	167.2	167.5	165.2	167.3	172.4	156.2	169.5	176.0	166.9	159.2	169.5	161.7	160.6	174.2
2013	196.9	155.2	161.7	158.7	160.2	157.7	167.4	151.2	161.7	169.4	159.3	151.5	164.3	156.1	155.9	167.7
2012	194.0	152.5	158.6	154.3	158.2	151.8	165.2	149.1	159.4	167.8	156.6	149.1	160.7	153.8	152.9	165.6
2011	185.7	145.9	151.9	147.6	152.4	145.0	157.9	142.5	152.8	160.8	150.0	142.5	152.6	147.3	146.6	158.7
2010	181.6	144.2	150.1	144.7	150.5	142.1	155.1	140.8	149.6	157.3	148.1	140.7	147.9	145.8	145.0	157.5
2009	182.5	143.5	147.3	146.4	149.8	141.7	155.5	141.1	150.1	160.9	144.6	139.1	150.8	146.7	146.1	155.1
2008	171.0	132.4	137.0	137.5	140.3	133.2	144.1	130.7	138.3	149.1	134.7	128.9	141.0	136.1	136.1	144.8
2007	165.0	128.5	132.2	131.6	137.1	128.8	138.6	126.2	134.8	146.0	130.2	125.0	136.4	132.1	132.1	140.0
2006	156.2	121.6	125.7	125.5	130.1	122.0	131.1	120.3	127.6	138.2	123.2	118.3	129.8	125.2	125.3	133.4
2005	146.7	113.8	117.3	117.9	121.3	114.4	123.7	112.5	119.4	129.0	115.5	110.5	121.3	116.1	117.3	126.0
2004	132.8	102.9	106.3	105.7	108.5	102.9	112.0	101.5	108.4	115.9	104.7	99.9	108.4	104.8	105.1	115.2
2003	129.7	100.5	104.4	103.9	106.1	100.4	109.3	99.5	105.5	113.4	102.3	97.6	104.8	102.7	103.0	112.7
2002	126.7	99.9	101.9	102.4	104.5	98.9	107.9	98.7	104.6	111.5	100.5	95.9	103.7	100.5	101.4	110.8
2001	122.2	93.4	98.4	99.8	102.3	96.6	103.8	95.5	100.9	107.8	97.6	93.4	100.5	97.8	97.3	107.7
2000	118.9	93.4	98.1	99.1	101.6	96.9	102.7	92.4	99.9	106.0	97.6	93.4	99.4	97.3	96.9	104.6
1999	116.6	91.8	94.5	96.0	99.7	94.0	101.0	90.7	97.6	104.6	96.0	92.1	98.0	94.8	95.5	103.3
1998	113.6	89.8	92.7	94.2	97.9	91.8	97.9	88.4	94.5	101.3	93.3	90.1	94.8	92.7	92.9	98.5
1997	111.5	88.4	91.3	92.8	96.8	90.3	96.1	87.0	93.3	100.1	91.9	88.8	93.4	91.4	91.5	96.1
1995	105.6	85.2	87.4	89.3	93.7	87.4	91.4	85.2	89.5	95.9	88.4	85.6	88.9	86.4	86.8	92.2
1990	93.2	78.0	80.1	81.3	86.5	79.3	84.5	76.7	82.1	85.4	81.5	78.6	80.7	79.6	80.3	83.4
1985	81.8	71.1	72.5	74.5	79.3	72.3	77.6	69.4	75.1	79.6	74.0	71.2	73.9	71.7	73.3	75.2
1980	60.7	53.4	55.2	54.5	57.6	54.5	57.9	53.1	57.0	59.4	55.6	57.2	55.0	54.9	55.4	62.2
1975	43.7	37.6	39.0	39.0	39.6	38.1	40.7	38.0	40.4	41.2	38.9	37.9	39.0	38.6	38.0	40.0
1970	27.8	24.5	24.9	24.9	25.7	24.5	25.5	23.7	25.9	25.4	25.1	24.6	23.3	24.8	24.5	26.8
1965	21.5	18.9	19.2	19.2	19.9	18.9	19.9	19.0	19.9	20.0	19.4	19.0	18.5	19.2	18.9	20.6
1960	19.5	17.1	17.4	17.4	18.1	17.1	18.2	17.0	18.1	18.2	17.6	17.3	16.8	17.4	17.2	18.8
1955	16.3	14.4	14.6	14.6	15.1	14.4	15.3	14.3	15.2	15.2	14.8	14.5	14.1	14.6	14.4	15.7
1950	13.5	11.9	12.1	12.1	12.5	11.9	12.6	11.8	12.5	12.6	12.2	12.0	11.6	12.1	11.9	13.0
1945	8.6	7.6	7.7	7.7	8.0	7.6	8.0	7.5	8.0	8.0	7.8	7.6	7.4	7.7	7.6	8.3
▼ 1940	6.6	5.9	5.9	5.9	6.1	5.8	6.2	5.8	6.2	6.2	6.0	5.9	5.7	5.9	5.8	6.4

533

Historical Cost Indexes

Year	National 30 City Average	Utah Salt Lake City	Vermont Bur-lington	Vermont Rutland	Virginia Alex-andria	Virginia Newport News	Virginia Norfolk	Virginia Rich-mond	Virginia Roanoke	Washington Seattle	Washington Spokane	Washington Tacoma	West Virginia Charles-ton	West Virginia Hunt-ington	Wisconsin Green Bay	Wisconsin Kenosha
Jan 2016	207.2E	182.6E	192.1E	189E	193E	176.3E	177.9E	177.7E	175.8E	210.4E	193.6E	207.7E	200.4E	201.8E	201.3E	205.9E
2015	204.0	182.1	191.6	188.5	192.5	175.8	177.4	177.2	175.3	209.9	193.1	207.2	199.9	201.3	200.8	205.4
2014	203.0	177.1	191.3	188.6	190.5	174.4	175.7	176.4	173.5	209.8	190.8	205.2	197.0	199.5	198.9	205.2
2013	196.9	171.1	178.7	176.6	185.0	170.0	171.5	170.7	168.3	203.4	184.3	199.1	187.2	193.2	193.9	201.2
2012	194.0	168.8	172.4	170.5	182.4	168.3	169.7	169.4	166.4	201.9	181.6	194.3	183.4	186.4	190.0	195.5
2011	185.7	161.8	158.8	157.1	174.5	159.8	161.1	159.0	154.2	194.1	175.5	187.6	177.5	180.3	181.4	187.6
2010	181.6	160.9	156.3	154.6	170.9	156.7	158.0	156.6	151.8	191.8	171.5	186.0	171.6	176.1	179.0	185.1
2009	182.5	160.4	158.3	156.7	172.5	160.0	161.9	160.7	155.6	188.5	173.0	186.3	174.6	177.4	175.5	182.5
2008	171.0	149.7	147.2	145.8	161.8	150.8	150.8	150.9	145.9	176.9	162.4	174.9	162.8	165.7	165.6	172.2
2007	165.0	144.6	144.4	143.2	155.0	146.2	146.1	147.3	142.9	171.4	156.7	168.7	158.7	160.1	159.4	164.8
2006	156.2	137.7	131.7	130.7	146.2	133.7	134.6	134.8	129.7	162.9	150.1	160.7	149.4	151.4	152.7	157.1
2005	146.7	129.4	124.6	123.8	136.5	123.5	124.4	125.4	112.2	153.9	141.9	151.5	140.8	141.0	144.4	148.3
2004	132.8	117.8	113.0	112.3	121.5	108.9	110.2	110.9	99.7	138.0	127.6	134.5	124.8	125.6	128.9	132.4
2003	129.7	116.0	110.7	110.1	119.5	104.8	106.2	108.6	97.0	134.9	125.9	133.0	123.3	123.6	127.4	130.9
2002	126.7	113.7	109.0	108.4	115.1	102.9	104.1	106.6	95.2	132.7	123.9	131.4	121.2	120.9	123.0	127.3
2001	122.2	109.1	105.7	105.2	110.8	99.8	100.3	102.9	92.1	127.9	120.3	125.7	114.6	117.5	119.1	123.6
2000	118.9	106.5	98.9	98.3	108.1	96.5	97.6	100.2	90.7	124.6	118.3	122.9	111.5	114.4	114.6	119.1
1999	116.6	104.5	98.2	97.7	106.1	95.6	96.5	98.8	89.8	123.3	116.7	121.6	110.6	113.4	112.1	115.8
1998	113.6	99.5	97.8	97.3	104.1	93.7	93.9	97.0	88.3	119.4	114.3	118.3	106.7	109.0	109.5	112.9
1997	111.5	97.2	96.6	96.3	101.2	91.6	91.7	92.9	86.9	118.1	111.7	117.2	105.3	107.7	105.6	109.1
1995	105.6	93.1	91.1	90.8	96.3	86.0	86.4	87.8	82.8	113.7	107.4	112.8	95.8	97.2	97.6	97.9
1990	93.2	84.3	83.0	82.9	86.1	76.3	76.7	77.6	76.1	100.1	98.5	100.5	86.1	86.8	86.7	87.8
1985	81.8	75.9	74.8	74.9	75.1	68.7	68.8	69.5	67.2	88.3	89.0	91.2	77.7	77.7	76.7	77.4
1980	60.7	57.0	55.3	58.3	57.3	52.5	52.4	54.3	51.3	67.9	66.3	66.7	57.7	58.3	58.6	58.3
1975	43.7	40.1	41.8	43.9	41.7	37.2	36.9	37.1	37.1	44.9	44.4	44.5	41.0	40.0	40.9	40.5
1970	27.8	26.1	25.4	26.8	26.2	23.9	21.5	22.0	23.7	28.8	29.3	29.6	26.1	25.8	26.4	26.5
1965	21.5	20.0	19.8	20.6	20.2	18.4	17.1	17.2	18.3	22.4	22.5	22.8	20.1	19.9	20.3	20.4
1960	19.5	18.4	18.0	18.8	18.4	16.7	15.4	15.6	16.6	20.4	20.8	20.8	18.3	18.1	18.4	18.6
1955	16.3	15.4	15.1	15.7	15.4	14.0	12.9	13.1	13.9	17.1	17.4	17.4	15.4	15.2	15.5	15.6
1950	13.5	12.7	12.4	13.0	12.7	11.6	10.7	10.8	11.5	14.1	14.4	14.4	12.7	12.5	12.8	12.9
1945	8.6	8.1	7.9	8.3	8.1	7.4	6.8	6.9	7.3	9.0	9.2	9.2	8.1	8.0	8.1	8.2
1940	6.6	6.3	6.1	6.4	6.2	5.7	5.3	5.3	5.7	7.0	7.1	7.1	6.2	6.2	6.3	6.3

Year	National 30 City Average	Wisconsin Mad-ison	Wisconsin Mil-waukee	Wisconsin Racine	Wyoming Chey-enne	Canada Calgary	Canada Edmon-ton	Canada Ham-ilton	Canada London	Canada Montreal	Canada Ottawa	Canada Quebec	Canada Tor-onto	Canada Van-couver	Canada Win-nipeg
Jan 2016	207.2E	202.5E	209.9E	205.6E	175.9E	227.2E	226.9E	221.8E	217.6E	221E	221E	221E	226.5E	217.7E	204.6E
2015	204.0	202.0	209.4	205.1	175.4	226.6	226.3	221.2	217.0	220.4	220.4	220.4	225.9	217.1	204.1
2014	203.0	202.4	210.2	205.3	173.6	228.2	229.1	222.1	218.9	219.8	214.8	218.3	227.4	218.8	202.9
2013	196.9	197.8	204.6	200.5	167.6	222.7	223.3	216.0	213.7	214.5	214.1	212.6	220.9	216.2	200.2
2012	194.0	191.2	202.2	194.9	165.0	219.1	219.6	211.1	208.2	209.6	209.7	207.5	216.6	214.0	200.7
2011	185.7	183.4	191.9	187.0	155.3	212.6	212.4	204.2	200.1	202.1	202.0	200.6	209.6	207.1	192.6
2010	181.6	181.3	187.1	184.4	154.4	200.6	200.7	197.7	193.7	193.9	195.6	192.7	200.7	192.7	182.9
2009	182.5	180.0	187.3	182.8	155.2	205.4	206.6	203.0	198.9	198.8	200.1	197.4	205.6	199.6	188.5
2008	171.0	168.4	176.3	173.0	146.5	190.2	191.0	194.3	188.6	186.7	188.0	186.8	194.6	184.9	174.4
2007	165.0	160.7	168.9	164.7	141.3	183.1	184.4	186.8	182.4	181.7	182.4	182.0	187.7	180.6	170.1
2006	156.2	152.8	158.8	157.2	128.0	163.6	164.8	169.0	164.9	158.6	163.7	159.0	170.6	169.3	155.9
2005	146.7	145.2	148.4	147.9	118.9	154.4	155.7	160.0	156.2	149.2	155.1	150.1	162.6	159.4	146.9
2004	132.8	129.7	134.2	132.7	104.7	138.8	139.4	142.6	140.4	134.5	141.0	135.7	146.0	141.7	129.7
2003	129.7	128.4	131.1	131.5	102.6	133.6	134.2	139.1	137.0	130.2	137.4	131.4	142.3	137.0	124.4
2002	126.7	124.5	128.2	126.5	101.7	122.5	122.2	136.3	134.1	127.2	134.9	128.4	139.8	134.6	121.4
2001	122.2	120.6	123.9	123.3	99.0	117.5	117.4	131.5	129.1	124.4	130.2	125.5	134.7	130.2	117.2
2000	118.9	116.6	120.5	118.7	98.1	115.9	115.8	130.0	127.5	122.8	128.8	124.1	133.1	128.4	115.6
1999	116.6	115.9	117.4	115.5	96.9	115.3	115.2	128.1	125.6	120.8	126.8	121.9	131.2	127.1	115.2
1998	113.6	110.8	113.4	112.7	95.4	112.5	112.4	126.3	123.9	119.0	124.7	119.6	128.5	123.8	113.7
1997	111.5	106.1	110.1	109.3	93.2	110.7	110.6	124.4	121.9	114.6	122.8	115.4	125.7	121.9	111.4
1995	105.6	96.5	103.9	97.8	87.6	107.4	107.4	119.9	117.5	110.8	118.2	111.5	121.6	116.2	107.6
1990	93.2	84.3	88.9	87.3	79.1	98.0	97.1	103.8	101.4	99.0	102.7	96.8	104.6	103.2	95.1
1985	81.8	74.3	77.4	77.0	72.3	90.2	89.1	85.8	84.9	82.4	83.9	79.9	86.5	89.8	83.4
1980	60.7	56.8	58.8	58.1	56.9	64.9	63.3	63.7	61.9	59.2	60.9	59.3	60.9	65.0	61.7
1975	43.7	40.7	43.3	40.7	40.6	42.2	41.6	42.9	41.6	39.7	41.5	39.0	42.2	42.4	39.2
1970	27.8	26.5	29.4	26.5	26.0	28.9	28.6	28.5	27.8	25.6	27.6	26.0	25.6	26.0	23.1
1965	21.5	20.6	21.8	20.4	20.0	22.3	22.0	22.0	21.4	18.7	21.2	20.1	19.4	20.5	17.5
1960	19.5	18.1	19.0	18.6	18.2	20.2	20.0	20.0	19.5	17.0	19.3	18.2	17.6	18.6	15.8
1955	16.3	15.2	15.9	15.6	15.2	17.0	16.8	16.7	16.3	14.3	16.2	15.3	14.8	15.5	13.3
1950	13.5	12.5	13.2	12.9	12.6	14.0	13.9	13.8	13.5	11.8	13.4	12.6	12.2	12.8	10.9
1945	8.6	8.0	8.4	8.2	8.0	9.0	8.8	8.8	8.6	7.5	8.5	8.0	7.8	8.2	7.0
1940	6.6	6.2	6.5	6.3	6.2	6.9	6.8	6.8	6.6	5.8	6.6	6.2	6.0	6.3	5.4

534

Glossary

Accent lighting–Fixtures or directional beams of light arranged to bring attention to an object or area.

Acoustical material–A material fabricated for the sole purpose of absorbing sound.

Addition–An expansion to an existing structure generally in the form of a room, floor(s) or wing(s). An increase in the floor area or volume of a structure.

Aggregate–Materials such as sand, gravel, stone, vermiculite, perlite and fly ash (slag) which are essential components in the production of concrete, mortar or plaster.

Air conditioning system–An air treatment to control the temperature, humidity and cleanliness of air and to provide for its distribution throughout the structure.

Air curtain–A stream of air directed downward to prevent the loss of hot or cool air from the structure and inhibit the entrance of dust and insects. Generally installed at loading platforms.

Anodized aluminum–Aluminum treated by an electrolytic process to produce an oxide film that is corrosion resistant.

Arcade–A covered passageway between buildings often with shops and offices on one or both sides.

Ashlar–A square-cut building stone or a wall constructed of cut stone.

Asphalt paper–A sheet paper material either coated or saturated with asphalt to increase its strength and resistance to water.

Asphalt shingles–Shingles manufactured from saturated roofing felts, coated with asphalt and topped with mineral granules to prevent weathering.

Assessment ratio–The ratio between the market value and assessed valuation of a property.

Back-up–That portion of a masonry wall behind the exterior facing-usually load bearing.

Balcony–A platform projecting from a building either supported from below or cantilevered. Usually protected by railing.

Bay–Structural component consisting of beams and columns occurring consistently throughout the structure.

Beam–A horizontal structural framing member that transfers loads to vertical members (columns) or bearing walls.

Bid–An offer to perform work described in a contract at a specified price.

Board foot–A unit or measure equal in volume to a board one foot long, one foot wide and one inch thick.

Booster pump–A supplemental pump installed to increase or maintain adequate pressure in the system.

Bowstring roof–A roof supported by trusses fabricated in the shape of a bow and tied together by a straight member.

Brick veneer–A facing of brick laid against, but not bonded to, a wall.

Bridging–A method of bracing joists for stiffness, stability and load distribution.

Broom finish–A method of finishing concrete by lightly dragging a broom over freshly placed concrete.

Built-up roofing–Installed on flat or extremely low-pitched roofs, a roof covering composed of plies or laminations of saturated roofing felts alternated with layers of coal tar pitch or asphalt and surfaced with a layer of gravel or slag in a thick coat of asphalt and finished with a capping sheet.

Caisson–A watertight chamber to expedite work on foundations or structure below water level.

Cantilever–A structural member supported only at one end.

Carport–An automobile shelter having one or more sides open to the weather.

Casement window–A vertical opening window having one side fixed.

Caulking–A resilient compound generally having a silicone or rubber base used to prevent infiltration of water or outside air.

Cavity wall–An exterior wall usually of masonry having an inner and outer wall separated by a continuous air space for thermal insulation.

Cellular concrete–A lightweight concrete consisting of cement mixed with gas-producing materials which, in turn, forms bubbles resulting in good insulating qualities.

Chattel mortgage–A secured interest in a property as collateral for payment of a note.

Clapboard–A wood siding used as exterior covering in frame construction. It is applied horizontally with grain running lengthwise. The thickest section of the board is on the bottom.

Clean room–An environmentally controlled room usually found in medical facilities or precision manufacturing spaces where it is essential to eliminate dust, lint or pathogens.

Close studding–A method of construction whereby studs are spaced close together and the intervening spaces are plastered.

Cluster housing–A closely grouped series of houses resulting in a high density land use.

Cofferdam–A watertight enclosure used for foundation construction in waterfront areas. Water is pumped out of the cofferdam allowing free access to the work area.

Collar joint–Joint between the collar beam and roof rafters.

Column–A vertical structural member supporting horizontal members (beams) along the direction of its longitudinal axis.

Combination door (windows)–Door (windows) having interchangeable screens and glass for seasonal use.

Common area–Spaces either inside or outside the building designated for use by the occupant of the building but not the general public.

Common wall–A wall used jointly by two dwelling units.

Compound wall–A wall constructed of more than one material.

Concrete–A composite material consisting of sand, coarse aggregate (gravel, stone or slag), cement and water that when mixed and allowed to harden forms a hard stone-like material.

Concrete floor hardener–A mixture of chemicals applied to the surface of concrete to produce a dense, wear-resistant bearing surface.

Conduit–A tube or pipe used to protect electric wiring.

Coping–The top cover or capping on a wall.

Craneway–Steel or concrete column and beam supports and rails on which a crane travels.

Curtain wall–A non-bearing exterior wall not supported by beams or girders of the steel frame.

Dampproofing–Coating of a surface to prevent the passage of water or moisture.

Dead load–Total weight of all structural components plus permanently attached fixtures and equipment.

Decibel–Unit of acoustical measurement.

Decorative block–A concrete masonry unit having a special treatment on its face.

Deed restriction–A restriction on the use of a property as set forth in the deed.

Depreciation–A loss in property value caused by physical aging, functional or economic obsolescence.

Direct heating–Heating of spaces by means of exposed heated surfaces (stove, fire, radiators, etc.).

Distributed load–A load that acts evenly over a structural member.

Dock bumper–A resilient material attached to a loading dock to absorb the impact of trucks backing in.

Dome–A curved roof shape spanning an area.

Dormer–A structure projecting from a sloping roof.

Double-hung window–A window having two vertical sliding sashes-one covering the upper section and one covering the lower.

Downspout–A vertical pipe for carrying rain water from the roof to the ground.

Drain tile–A hollow tile used to drain water-soaked soil.

Dressed size–In lumber about 3/8" - 1/2" less than nominal size after planing and sawing.

Drop panel–The depressed surface on the bottom side of a flat concrete slab which surrounds a column.

Dry pipe sprinkler system–A sprinkler system that is activated with water only in case of fire. Used in areas susceptible to freezing or to avoid the hazards of leaking pipes.

Drywall–An interior wall constructed of gypsum board, plywood or wood paneling. No water is required for application.

Duct–Pipe for transmitting warm or cold air. A pipe containing electrical cable or wires.

Dwelling–A structure designed as living quarters for one or more families.

Easement–A right of way or free access over land owned by another.

Eave–The lower edge of a sloping roof that overhangs the side wall of the structure.

Economic rent–That rent on a property sufficient to pay all operating costs exclusive of services and utilities.

Economic life–The term during which a structure is expected to be profitable. Generally shorter than the physical life of the structure.

Economic obsolescence–Loss in value due to unfavorable economic influences occurring from outside the structure itself.

Effective age–The age a structure appears to be based on observed physical condition determined by degree of maintenance and repair.

Elevator–A mechanism for vertical transport of personnel or freight equipped with car or platform.

Ell–A secondary wing or addition to a structure at right angles to the main structure.

Eminent domain–The right of the state to take private property for public use.

Envelope–The shape of a building indicating volume.

Equity–Value of an owner's interest calculated by subtracting outstanding mortgages and expenses from the value of the property.

Escheat–The assumption of ownership by the state, of property whose owner cannot be determined.

Facade–The exterior face of a building sometimes decorated with elaborate detail.

Face brick–Brick manufactured to present an attractive appearance.

Facing block–A concrete masonry unit having a decorative exterior finish.

Feasibility study–A detailed evaluation of a proposed project to determine its financial potential.

Felt paper–Paper sheathing used on walls as an insulator and to prevent infiltration of moisture.

Fenestration–Pertaining to the density and arrangement of windows in a structure.

Fiberboard–A building material composed of wood fiber compressed with a binding agent, produced in sheet form.

Fiberglass–Fine spun filaments of glass processed into various densities to produce thermal and acoustical insulation.

Field house–A long structure used for indoor athletic events.

Finish floor–The top or wearing surface of the floor system.

Fireproofing–The use of fire-resistant materials for the protection of structural members to ensure structural integrity in the event of fire.

Fire stop–A material or member used to seal an opening to prevent the spread of fire.

Flashing–A thin, impervious material such as copper or sheet metal used to prevent air or water penetration.

Float finish–A concrete finish accomplished by using a flat tool with a handle on the back.

Floor drain–An opening installed in a floor for removing excess water into a plumbing system.

Floor load–The live load the floor system has been designed for and which may be applied safely.

Flue–A heat-resistant enclosed passage in a chimney to remove gaseous products of combustion from a fireplace or boiler.

Footing–The lowest portion of the foundation wall that transmits the load directly to the soil.

Foundation–Below grade wall system that supports the structure.

Foyer–A portion of the structure that serves as a transitional space between the interior and exterior.

Frontage–That portion of a lot that is placed facing a street, public way or body of water.

Frost action–The effects of freezing and thawing on materials and the resultant structural damage.

Functional obsolescence–An inadequacy caused by outmoded design, dated construction materials or over or undersized areas, all of which cause excessive operating costs.

Furring strips–Wood or metal channels used for attaching gypsum or metal lath to masonry walls as a finish or for leveling purposes.

Gambrel roof (mansard)–A roof system having two pitches on each side.

Garden apartment–Two or three story apartment building with common outside areas.

General contractor–The prime contractor responsible for work on the construction site.

Girder–A principal horizontal supporting member.

Girt–A horizontal framing member for adding rigid support to columns which also acts as a support for sheathing or siding.

Grade beam–That portion of the foundation system that directly supports the exterior walls of the building.

Gross area–The total enclosed floor area of a building.

Ground area–The area computed by the exterior dimensions of the structure.

Ground floor–The floor of a building in closest proximity to the ground.

Grout–A mortar containing a high water content used to fill joints and cavities in masonry work.

Gutter–A wood or sheet metal conduit set along the building eaves used to channel rainwater to leaders or downspouts.

Gunite–A concrete mix placed pneumatically.

Gypsum–Hydrated calcium sulphate used as both a retarder in Portland cement and as a major ingredient in plaster of Paris. Used in sheets as a substitute for plaster.

Hall–A passageway providing access to various parts of a building–a large room for entertainment or assembly.

Hangar–A structure for the storage or repair of aircraft.

Head room–The distance between the top of the finished floor to the bottom of the finished ceiling.

Hearth–The floor of a fireplace and the adjacent area of fireproof material.

Heat pump–A mechanical device for providing either heating or air conditioning.

Hip roof–A roof whose four sides meet at a common point with no gabled ends.

I-beam–A structural member having a cross-section resembling the letter "I".

Insulation–Material used to reduce the effects of heat, cold or sound.

Jack rafter–An unusually short rafter generally found in hip roofs.

Jalousie–Adjustable glass louvers which pivot simultaneously in a common frame.

Jamb–The vertical member on either side of a door frame or window frame.

Joist–Parallel beams of timber, concrete or steel used to support floor and ceiling.

Junction box–A box that protects splices or joints in electrical wiring.

Kalamein door–Solid core wood doors clad with galvanized sheet metal.

Kip–A unit of weight equal to 1,000 pounds.

Lally column–A concrete-filled steel pipe used as a vertical support.

Laminated beam–A beam built up by gluing together several pieces of timber.

Landing–A platform between flights of stairs.

Lath–Strips of wood or metal used as a base for plaster.

Lead-lined door, sheetrock–Doors or sheetrock internally lined with sheet lead to provide protection from X-ray radiation.

Lean-to–A small shed or building addition with a single pitched roof attached to the exterior wall of the main building.

Lintel–A horizontal framing member used to carry a load over a wall opening.

Live load–The moving or movable load on a structure composed of furnishings, equipment or personnel weight.

Load bearing partition–A partition that supports a load in addition to its own weight.

Loading dock leveler–An adjustable platform for handling off-on loading of trucks.

Loft building–A commercial/industrial type building containing large, open unpartitioned floor areas.

Louver window–A window composed of a series of sloping, overlapping blades or slats that may be adjusted to admit varying degrees of air or light.

Main beam–The principal load bearing beam used to transmit loads directly to columns.

Mansard roof–A roof having a double pitch on all four sides, the lower level having the steeper pitch.

Mansion–An extremely large and imposing residence.

Masonry–The utilization of brick, stone, concrete or block for walls and other building components.

Mastic–A sealant or adhesive compound generally used as either a binding agent or floor finish.

Membrane fireproofing–A coating of metal lath and plaster to provide resistance to fire and heat.

Mesh reinforcement–An arrangement of wire tied or welded at their intersection to provide strength and resistance to cracking.

Metal lath–A diamond-shaped metallic base for plaster.

Mezzanine–A low story situated between two main floors.

Mill construction–A heavy timber construction that achieves fire resistance by using large wood structural members, noncombustible bearing and non-bearing walls and omitting concealed spaces under floors and roof.

Mixed occupancy–Two or more classes of occupancy in a single structure.

Molded brick–A specially shaped brick used for decorative purposes.

Monolithic concrete–Concrete poured in a continuous process so there are no joints.

Movable partition–A non-load bearing demountable partition that can be relocated and can be either ceiling height or partial height.

Moving walk–A continually moving horizontal passenger carrying device.

Multi-zone system–An air conditioning system that is capable of handling several individual zones simultaneously.

Net floor area–The usable, occupied area of a structure excluding stairwells, elevator shafts and wall thicknesses.

Non-combustible construction–Construction in which the walls, partitions and the structural members are of non-combustible materials.

Nurses call system–An electrically operated system for use by patients or personnel for summoning a nurse.

Occupancy rate–The number of persons per room, per dwelling unit, etc.

One-way joist connection–A framing system for floors and roofs in a concrete building consisting of a series of parallel joists supported by girders between columns.

Open web joist–A lightweight prefabricated metal chord truss.

Parapet–A low guarding wall at the point of a drop. That portion of the exterior wall that extends above the roof line.

Parging–A thin coat of mortar applied to a masonry wall used primarily for waterproofing purposes.

Parquet–Inlaid wood flooring set in a simple design.

Peaked roof–A roof of two or more slopes that rises to a peak.

Penthouse–A structure occupying usually half the roof area and used to house elevator, HVAC equipment, or other mechanical or electrical systems.

Pier–A column designed to support a concentrated load.

Pilaster–A column usually formed of the same material and integral with, but projecting from, a wall.

Pile–A concrete, wood or steel column usually less than 2' in diameter which is driven or otherwise introduced into the soil to carry a vertical load or provide vertical support.

Pitched roof–A roof having one or more surfaces with a pitch greater than 10°.

Plenum–The space between the suspended ceiling and the floor above.

Plywood–Structural wood made of three or more layers of veneer and bonded together with glue.

Porch–A structure attached to a building to provide shelter for an entranceway or to serve as a semi-enclosed space.

Post and beam framing–A structural framing system where beams rest on posts rather than bearing walls.

Post-tensioned concrete–Concrete that has the reinforcing tendons tensioned after the concrete has set.

Pre-cast concrete–Concrete structural components fabricated at a location other than in-place.

Pre-stressed concrete–Concrete that has the reinforcing tendons tensioned prior to the concrete setting.

Purlins–A horizontal structural member supporting the roof deck and resting on the trusses, girders, beams or rafters.

Rafters–Structural members supporting the roof deck and covering.

Resilient flooring–A manufactured interior floor covering material in either sheet or tile form, that is resilient.

Ridge–The horizontal line at the junction of two sloping roof edges.

Rigid frame–A structural framing system in which all columns and beams are rigidly connected; there are no hinged joints.

Roof drain–A drain designed to accept rainwater and discharge it into a leader or downspout.

Rough floor–A layer of boards or plywood nailed to the floor joists that serves as a base for the finished floor.

Sanitary sewer–A sewer line designed to carry only liquid or waterborne waste from the structure to a central treatment plant.

Sawtooth roof–Roof shape found primarily in industrial roofs that creates the appearance of the teeth of a saw.

Semi-rigid frame–A structural system wherein the columns and beams are so attached that there is some flexibility at the joints.

Septic tank–A covered tank in which waste matter is decomposed by natural bacterial action.

Service elevator–A combination passenger and freight elevator.

Sheathing–The first covering of exterior studs by boards, plywood or particle board.

Shoring–Temporary bracing for structural components during construction.

Span–The horizontal distance between supports.

Specification–A detailed list of materials and requirements for construction of a building.

Storm sewer–A system of pipes used to carry rainwater or surface waters.

Stucco–A cement plaster used to cover exterior wall surfaces usually applied over a wood or metal lath base.

Stud–A vertical wooden structural component.

Subfloor–A system of boards, plywood or particle board laid over the floor joists to form a base.

Superstructure–That portion of the structure above the foundation or ground level.

Terrazzo–A durable floor finish made of small chips of colored stone or marble, embedded in concrete, then polished to a high sheen.

Tile–A thin piece of fired clay, stone or concrete used for floor, roof or wall finishes.

Unit cost–The cost per unit of measurement.

Vault–A room especially designed for storage.

Veneer–A thin surface layer covering a base of common material.

Vent–An opening serving as an outlet for air.

Wainscot–The lower portion of an interior wall whose surface differs from that of the upper wall.

Wall bearing construction–A structural system where the weight of the floors and roof are carried directly by the masonry walls rather than the structural framing system.

Waterproofing–Any of a number of materials applied to various surfaces to prevent the infiltration of water.

Weatherstrip–A thin strip of metal, wood or felt used to cover the joint between the door or window sash and the jamb, casing or sill to keep out air, dust, rain, etc.

Wing–A building section or addition projecting from the main structure.

X-ray protection–Lead encased in sheetrock or plaster to prevent the escape of radiation.

A	Area Square Feet; Ampere
AAFES	Army and Air Force Exchange Service
ABS	Acrylonitrile Butadiene Stryrene; Asbestos Bonded Steel
A.C., AC	Alternating Current; Air-Conditioning; Asbestos Cement; Plywood Grade A & C
ACI	American Concrete Institute
ACR	Air Conditioning Refrigeration
ADA	Americans with Disabilities Act
AD	Plywood, Grade A & D
Addit.	Additional
Adj.	Adjustable
af	Audio-frequency
AFUE	Annual Fuel Utilization Efficiency
AGA	American Gas Association
Agg.	Aggregate
A.H., Ah	Ampere Hours
A hr.	Ampere-hour
A.H.U., AHU	Air Handling Unit
A.I.A.	American Institute of Architects
AIC	Ampere Interrupting Capacity
Allow.	Allowance
alt., alt	Alternate
Alum.	Aluminum
a.m.	Ante Meridiem
Amp.	Ampere
Anod.	Anodized
ANSI	American National Standards Institute
APA	American Plywood Association
Approx.	Approximate
Apt.	Apartment
Asb.	Asbestos
A.S.B.C.	American Standard Building Code
Asbe.	Asbestos Worker
ASCE	American Society of Civil Engineers
A.S.H.R.A.E.	American Society of Heating, Refrig. & AC Engineers
ASME	American Society of Mechanical Engineers
ASTM	American Society for Testing and Materials
Attchmt.	Attachment
Avg., Ave.	Average
AWG	American Wire Gauge
AWWA	American Water Works Assoc.
Bbl.	Barrel
B&B, BB	Grade B and Better; Balled & Burlapped
B&S	Bell and Spigot
B.&W.	Black and White
b.c.c.	Body-centered Cubic
B.C.Y.	Bank Cubic Yards
BE	Bevel End
B.F.	Board Feet
Bg. cem.	Bag of Cement
BHP	Boiler Horsepower; Brake Horsepower
B.I.	Black Iron
bidir.	bidirectional
Bit., Bitum.	Bituminous
Bit., Conc.	Bituminous Concrete
Bk.	Backed
Bkrs.	Breakers
Bldg., bldg	Building
Blk.	Block
Bm.	Beam
Boil.	Boilermaker
bpm	Blows per Minute
BR	Bedroom
Brg., brng.	Bearing
Brhe.	Bricklayer Helper
Bric.	Bricklayer
Brk., brk	Brick
brkt	Bracket
Brs.	Brass
Brz.	Bronze
Bsn.	Basin
Btr.	Better
BTU	British Thermal Unit
BTUH	BTU per Hour
Bu.	Bushels
BUR	Built-up Roofing
BX	Interlocked Armored Cable
°C	Degree Centigrade
c	Conductivity, Copper Sweat
C	Hundred; Centigrade
C/C	Center to Center, Cedar on Cedar
C-C	Center to Center
Cab	Cabinet
Cair.	Air Tool Laborer
Cal.	Caliper
Calc	Calculated
Cap.	Capacity
Carp.	Carpenter
C.B.	Circuit Breaker
C.C.A.	Chromate Copper Arsenate
C.C.F.	Hundred Cubic Feet
cd	Candela
cd/sf	Candela per Square Foot
CD	Grade of Plywood Face & Back
CDX	Plywood, Grade C & D, exterior glue
Cefi.	Cement Finisher
Cem.	Cement
CF	Hundred Feet
C.F.	Cubic Feet
CFM	Cubic Feet per Minute
CFRP	Carbon Fiber Reinforced Plastic
c.g.	Center of Gravity
CHW	Chilled Water; Commercial Hot Water
C.I., CI	Cast Iron
C.I.P., CIP	Cast in Place
Circ.	Circuit
C.L.	Carload Lot
CL	Chain Link
Clab.	Common Laborer
Clam	Common Maintenance Laborer
C.L.F.	Hundred Linear Feet
CLF	Current Limiting Fuse
CLP	Cross Linked Polyethylene
cm	Centimeter
CMP	Corr. Metal Pipe
CMU	Concrete Masonry Unit
CN	Change Notice
Col.	Column
CO$_2$	Carbon Dioxide
Comb.	Combination
comm.	Commercial, Communication
Compr.	Compressor
Conc.	Concrete
Cont., cont	Continuous; Continued, Container
Corr.	Corrugated
Cos	Cosine
Cot	Cotangent
Cov.	Cover
C/P	Cedar on Paneling
CPA	Control Point Adjustment
Cplg.	Coupling
CPM	Critical Path Method
CPVC	Chlorinated Polyvinyl Chloride
C.Pr.	Hundred Pair
CRC	Cold Rolled Channel
Creos.	Creosote
Crpt.	Carpet & Linoleum Layer
CRT	Cathode-ray Tube
CS	Carbon Steel, Constant Shear Bar Joist
Csc	Cosecant
C.S.F.	Hundred Square Feet
CSI	Construction Specifications Institute
CT	Current Transformer
CTS	Copper Tube Size
Cu	Copper, Cubic
Cu. Ft.	Cubic Foot
cw	Continuous Wave
C.W.	Cool White; Cold Water
Cwt.	100 Pounds
C.W.X.	Cool White Deluxe
C.Y.	Cubic Yard (27 cubic feet)
C.Y./Hr.	Cubic Yard per Hour
Cyl.	Cylinder
d	Penny (nail size)
D	Deep; Depth; Discharge
Dis., Disch.	Discharge
Db	Decibel
Dbl.	Double
DC	Direct Current
DDC	Direct Digital Control
Demob.	Demobilization
d.f.t.	Dry Film Thickness
d.f.u.	Drainage Fixture Units
D.H.	Double Hung
DHW	Domestic Hot Water
DI	Ductile Iron
Diag.	Diagonal
Diam., Dia	Diameter
Distrib.	Distribution
Div.	Division
Dk.	Deck
D.L.	Dead Load; Diesel
DLH	Deep Long Span Bar Joist
dlx	Deluxe
Do.	Ditto
DOP	Dioctyl Phthalate Penetration Test (Air Filters)
Dp., dp	Depth
D.P.S.T.	Double Pole, Single Throw
Dr.	Drive
DR	Dimension Ratio
Drink.	Drinking
D.S.	Double Strength
D.S.A.	Double Strength A Grade
D.S.B.	Double Strength B Grade
Dty.	Duty
DWV	Drain Waste Vent
DX	Deluxe White, Direct Expansion
dyn	Dyne
e	Eccentricity
E	Equipment Only; East; Emissivity
Ea.	Each
EB	Encased Burial
Econ.	Economy
E.C.Y	Embankment Cubic Yards
EDP	Electronic Data Processing
EIFS	Exterior Insulation Finish System
E.D.R.	Equiv. Direct Radiation
Eq.	Equation
EL	Elevation
Elec.	Electrician; Electrical
Elev.	Elevator; Elevating
EMT	Electrical Metallic Conduit; Thin Wall Conduit
Eng.	Engine, Engineered
EPDM	Ethylene Propylene Diene Monomer
EPS	Expanded Polystyrene
Eqhv.	Equip. Oper., Heavy
Eqlt.	Equip. Oper., Light
Eqmd.	Equip. Oper., Medium
Eqmm.	Equip. Oper., Master Mechanic
Eqol.	Equip. Oper., Oilers
Equip.	Equipment
ERW	Electric Resistance Welded

E.S. — Energy Saver
Est. — Estimated
esu — Electrostatic Units
E.W. — Each Way
EWT — Entering Water Temperature
Excav. — Excavation
excl — Excluding
Exp., exp — Expansion, Exposure
Ext., ext — Exterior; Extension
Extru. — Extrusion
f. — Fiber Stress
F — Fahrenheit; Female; Fill
Fab., fab — Fabricated; Fabric
FBGS — Fiberglass
F.C. — Footcandles
f.c.c. — Face-centered Cubic
f'c. — Compressive Stress in Concrete; Extreme Compressive Stress
F.E. — Front End
FEP — Fluorinated Ethylene Propylene (Teflon)
F.G. — Flat Grain
F.H.A. — Federal Housing Administration
Fig. — Figure
Fin. — Finished
FIPS — Female Iron Pipe Size
Fixt. — Fixture
FJP — Finger jointed and primed
Fl. Oz. — Fluid Ounces
Flr. — Floor
FM — Frequency Modulation; Factory Mutual
Fmg. — Framing
FM/UL — Factory Mutual/Underwriters Labs
Fdn. — Foundation
FNPT — Female National Pipe Thread
Fori. — Foreman, Inside
Foro. — Foreman, Outside
Fount. — Fountain
fpm — Feet per Minute
FPT — Female Pipe Thread
Fr — Frame
F.R. — Fire Rating
FRK — Foil Reinforced Kraft
FSK — Foil/Scrim/Kraft
FRP — Fiberglass Reinforced Plastic
FS — Forged Steel
FSC — Cast Body; Cast Switch Box
Ft., ft — Foot; Feet
Ftng. — Fitting
Ftg. — Footing
Ft lb. — Foot Pound
Furn. — Furniture
FVNR — Full Voltage Non-Reversing
FVR — Full Voltage Reversing
FXM — Female by Male
Fy. — Minimum Yield Stress of Steel
g — Gram
G — Gauss
Ga. — Gauge
Gal., gal. — Gallon
Galv., galv — Galvanized
GC/MS — Gas Chromatograph/Mass Spectrometer
Gen. — General
GFI — Ground Fault Interrupter
GFRC — Glass Fiber Reinforced Concrete
Glaz. — Glazier
GPD — Gallons per Day
gpf — Gallon per Flush
GPH — Gallons per Hour
gpm, GPM — Gallons per Minute
GR — Grade
Gran. — Granular
Grnd. — Ground
GVW — Gross Vehicle Weight
GWB — Gypsum Wall Board

H — High Henry
HC — High Capacity
H.D., HD — Heavy Duty; High Density
H.D.O. — High Density Overlaid
HDPE — High Density Polyethylene Plastic
Hdr. — Header
Hdwe. — Hardware
H.I.D., HID — High Intensity Discharge
Help. — Helper Average
HEPA — High Efficiency Particulate Air Filter
Hg — Mercury
HIC — High Interrupting Capacity
HM — Hollow Metal
HMWPE — High Molecular Weight Polyethylene
HO — High Output
Horiz. — Horizontal
H.P., HP — Horsepower; High Pressure
H.P.F. — High Power Factor
Hr. — Hour
Hrs./Day — Hours per Day
HSC — High Short Circuit
Ht. — Height
Htg. — Heating
Htrs. — Heaters
HVAC — Heating, Ventilation & Air-Conditioning
Hvy. — Heavy
HW — Hot Water
Hyd.; Hydr. — Hydraulic
Hz — Hertz (cycles)
I. — Moment of Inertia
IBC — International Building Code
I.C. — Interrupting Capacity
ID — Inside Diameter
I.D. — Inside Dimension; Identification
I.F. — Inside Frosted
I.M.C. — Intermediate Metal Conduit
In. — Inch
Incan. — Incandescent
Incl. — Included; Including
Int. — Interior
Inst. — Installation
Insul., insul — Insulation/Insulated
I.P. — Iron Pipe
I.P.S., IPS — Iron Pipe Size
IPT — Iron Pipe Threaded
I.W. — Indirect Waste
J — Joule
J.I.C. — Joint Industrial Council
K — Thousand; Thousand Pounds; Heavy Wall Copper Tubing, Kelvin
K.A.H. — Thousand Amp. Hours
kcmil — Thousand Circular Mils
KD — Knock Down
K.D.A.T. — Kiln Dried After Treatment
kg — Kilogram
kG — Kilogauss
kgf — Kilogram Force
kHz — Kilohertz
Kip — 1000 Pounds
KJ — Kilojoule
K.L. — Effective Length Factor
K.L.F. — Kips per Linear Foot
Km — Kilometer
KO — Knock Out
K.S.F. — Kips per Square Foot
K.S.I. — Kips per Square Inch
kV — Kilovolt
kVA — Kilovolt Ampere
kVAR — Kilovar (Reactance)
KW — Kilowatt
KWh — Kilowatt-hour
L — Labor Only; Length; Long; Medium Wall Copper Tubing
Lab. — Labor
lat — Latitude

Lath. — Lather
Lav. — Lavatory
lb.; # — Pound
L.B., LB — Load Bearing; L Conduit Body
L. & E. — Labor & Equipment
lb./hr. — Pounds per Hour
lb./L.F. — Pounds per Linear Foot
lbf/sq.in. — Pound-force per Square Inch
L.C.L. — Less than Carload Lot
L.C.Y. — Loose Cubic Yard
Ld. — Load
LE — Lead Equivalent
LED — Light Emitting Diode
L.F. — Linear Foot
L.F. Nose — Linear Foot of Stair Nosing
L.F. Rsr — Linear Foot of Stair Riser
Lg. — Long; Length; Large
L & H — Light and Heat
LH — Long Span Bar Joist
L.H. — Labor Hours
L.L., LL — Live Load
L.L.D. — Lamp Lumen Depreciation
lm — Lumen
lm/sf — Lumen per Square Foot
lm/W — Lumen per Watt
LOA — Length Over All
log — Logarithm
L-O-L — Lateralolet
long. — Longitude
L.P., LP — Liquefied Petroleum; Low Pressure
L.P.F. — Low Power Factor
LR — Long Radius
L.S. — Lump Sum
Lt. — Light
Lt. Ga. — Light Gauge
L.T.L. — Less than Truckload Lot
Lt. Wt. — Lightweight
L.V. — Low Voltage
M — Thousand; Material; Male; Light Wall Copper Tubing
M²CA — Meters Squared Contact Area
m/hr.; M.H. — Man-hour
mA — Milliampere
Mach. — Machine
Mag. Str. — Magnetic Starter
Maint. — Maintenance
Marb. — Marble Setter
Mat; Mat'l. — Material
Max. — Maximum
MBF — Thousand Board Feet
MBH — Thousand BTU's per hr.
MC — Metal Clad Cable
MCC — Motor Control Center
M.C.F. — Thousand Cubic Feet
MCFM — Thousand Cubic Feet per Minute
M.C.M. — Thousand Circular Mils
MCP — Motor Circuit Protector
MD — Medium Duty
MDF — Medium-density fibreboard
M.D.O. — Medium Density Overlaid
Med. — Medium
MF — Thousand Feet
M.F.B.M. — Thousand Feet Board Measure
Mfg. — Manufacturing
Mfrs. — Manufacturers
mg — Milligram
MGD — Million Gallons per Day
MGPH — Million Gallons per Hour
MH, M.H. — Manhole; Metal Halide; Man-Hour
MHz — Megahertz
Mi. — Mile
MI — Malleable Iron; Mineral Insulated
MIPS — Male Iron Pipe Size
mj — Mechanical Joint
m — Meter
mm — Millimeter
Mill. — Millwright
Min., min. — Minimum, Minute

Misc.	Miscellaneous	PDCA	Painting and Decorating	SC	Screw Cover
ml	Milliliter, Mainline		Contractors of America	SCFM	Standard Cubic Feet per Minute
M.L.F.	Thousand Linear Feet	P.E., PE	Professional Engineer;	Scaf.	Scaffold
Mo.	Month		Porcelain Enamel;	Sch., Sched.	Schedule
Mobil.	Mobilization		Polyethylene; Plain End	S.C.R.	Modular Brick
Mog.	Mogul Base	P.E.C.I.	Porcelain Enamel on Cast Iron	S.D.	Sound Deadening
MPH	Miles per Hour	Perf.	Perforated	SDR	Standard Dimension Ratio
MPT	Male Pipe Thread	PEX	Cross Linked Polyethylene	S.E.	Surfaced Edge
MRGWB	Moisture Resistant Gypsum	Ph.	Phase	Sel.	Select
	Wallboard	P.I.	Pressure Injected	SER, SEU	Service Entrance Cable
MRT	Mile Round Trip	Pile.	Pile Driver	S.F.	Square Foot
ms	Millisecond	Pkg.	Package	S.F.C.A.	Square Foot Contact Area
M.S.F.	Thousand Square Feet	Pl.	Plate	S.F. Flr.	Square Foot of Floor
Mstz.	Mosaic & Terrazzo Worker	Plah.	Plasterer Helper	S.F.G.	Square Foot of Ground
M.S.Y.	Thousand Square Yards	Plas.	Plasterer	S.F. Hor.	Square Foot Horizontal
Mtd., mtd., mtd	Mounted	plf	Pounds Per Linear Foot	SFR	Square Feet of Radiation
Mthe.	Mosaic & Terrazzo Helper	Pluh.	Plumber Helper	S.F. Shlf.	Square Foot of Shelf
Mtng.	Mounting	Plum.	Plumber	S4S	Surface 4 Sides
Mult.	Multi; Multiply	Ply.	Plywood	Shee.	Sheet Metal Worker
M.V.A.	Million Volt Amperes	p.m.	Post Meridiem	Sin.	Sine
M.V.A.R.	Million Volt Amperes Reactance	Pntd.	Painted	Skwk.	Skilled Worker
MV	Megavolt	Pord.	Painter, Ordinary	SL	Saran Lined
MW	Megawatt	pp	Pages	S.L.	Slimline
MXM	Male by Male	PP, PPL	Polypropylene	Sldr.	Solder
MYD	Thousand Yards	P.P.M.	Parts per Million	SLH	Super Long Span Bar Joist
N	Natural; North	Pr.	Pair	S.N.	Solid Neutral
nA	Nanoampere	P.E.S.B.	Pre-engineered Steel Building	SO	Stranded with oil resistant inside
NA	Not Available; Not Applicable	Prefab.	Prefabricated		insulation
N.B.C.	National Building Code	Prefin.	Prefinished	S-O-L	Socketolet
NC	Normally Closed	Prop.	Propelled	sp	Standpipe
NEMA	National Electrical Manufacturers	PSF, psf	Pounds per Square Foot	S.P.	Static Pressure; Single Pole; Self-
	Assoc.	PSI, psi	Pounds per Square Inch		Propelled
NEHB	Bolted Circuit Breaker to 600V.	PSIG	Pounds per Square Inch Gauge	Spri.	Sprinkler Installer
NFPA	National Fire Protection Association	PSP	Plastic Sewer Pipe	spwg	Static Pressure Water Gauge
NLB	Non-Load-Bearing	Pspr.	Painter, Spray	S.P.D.T.	Single Pole, Double Throw
NM	Non-Metallic Cable	Psst.	Painter, Structural Steel	SPF	Spruce Pine Fir; Sprayed
nm	Nanometer	P.T.	Potential Transformer		Polyurethane Foam
No.	Number	P. & T.	Pressure & Temperature	S.P.S.T.	Single Pole, Single Throw
NO	Normally Open	Ptd.	Painted	SPT	Standard Pipe Thread
N.O.C.	Not Otherwise Classified	Ptns.	Partitions	Sq.	Square; 100 Square Feet
Nose.	Nosing	Pu	Ultimate Load	Sq. Hd.	Square Head
NPT	National Pipe Thread	PVC	Polyvinyl Chloride	Sq. In.	Square Inch
NQOD	Combination Plug-on/Bolt on	Pvmt.	Pavement	S.S.	Single Strength; Stainless Steel
	Circuit Breaker to 240V.	PRV	Pressure Relief Valve	S.S.B.	Single Strength B Grade
N.R.C., NRC	Noise Reduction Coefficient/	Pwr.	Power	sst, ss	Stainless Steel
	Nuclear Regulator Commission	Q	Quantity Heat Flow	Sswk.	Structural Steel Worker
N.R.S.	Non Rising Stem	Qt.	Quart	Sswl.	Structural Steel Welder
ns	Nanosecond	Quan., Qty.	Quantity	St.; Stl.	Steel
nW	Nanowatt	Q.C.	Quick Coupling	STC	Sound Transmission Coefficient
OB	Opposing Blade	r	Radius of Gyration	Std.	Standard
OC	On Center	R	Resistance	Stg.	Staging
OD	Outside Diameter	R.C.P.	Reinforced Concrete Pipe	STK	Select Tight Knot
O.D.	Outside Dimension	Rect.	Rectangle	STP	Standard Temperature & Pressure
ODS	Overhead Distribution System	recpt.	Receptacle	Stpi.	Steamfitter, Pipefitter
O.G.	Ogee	Reg.	Regular	Str.	Strength; Starter; Straight
O.H.	Overhead	Reinf.	Reinforced	Strd.	Stranded
O&P	Overhead and Profit	Req'd.	Required	Struct.	Structural
Oper.	Operator	Res.	Resistant	Sty.	Story
Opng.	Opening	Resi.	Residential	Subj.	Subject
Orna.	Ornamental	RF	Radio Frequency	Subs.	Subcontractors
OSB	Oriented Strand Board	RFID	Radio-frequency Identification	Surf.	Surface
OS&Y	Outside Screw and Yoke	Rgh.	Rough	Sw.	Switch
OSHA	Occupational Safety and Health	RGS	Rigid Galvanized Steel	Swbd.	Switchboard
	Act	RHW	Rubber, Heat & Water Resistant;	S.Y.	Square Yard
Ovhd.	Overhead		Residential Hot Water	Syn.	Synthetic
OWG	Oil, Water or Gas	rms	Root Mean Square	S.Y.P.	Southern Yellow Pine
Oz.	Ounce	Rnd.	Round	Sys.	System
P.	Pole; Applied Load; Projection	Rodm.	Rodman	t.	Thickness
p.	Page	Rofc.	Roofer, Composition	T	Temperature; Ton
Pape.	Paperhanger	Rofp.	Roofer, Precast	Tan	Tangent
P.A.P.R.	Powered Air Purifying Respirator	Rohe.	Roofer Helpers (Composition)	T.C.	Terra Cotta
PAR	Parabolic Reflector	Rots.	Roofer, Tile & Slate	T & C	Threaded and Coupled
P.B., PB	Push Button	R.O.W.	Right of Way	T.D.	Temperature Difference
Pc., Pcs.	Piece, Pieces	RPM	Revolutions per Minute	TDD	Telecommunications Device for
P.C.	Portland Cement; Power Connector	R.S.	Rapid Start		the Deaf
P.C.F.	Pounds per Cubic Foot	Rsr	Riser	T.E.M.	Transmission Electron Microscopy
PCM	Phase Contrast Microscopy	RT	Round Trip	temp	Temperature, Tempered, Temporary
		S.	Suction; Single Entrance; South	TFFN	Nylon Jacketed Wire
		SBS	Styrene Butadiere Styrene		

541

Abbreviations

TFE	Tetrafluoroethylene (Teflon)	U.L., UL	Underwriters Laboratory	w/	With
T. & G.	Tongue & Groove;	Uld.	Unloading	W.C., WC	Water Column; Water Closet
	Tar & Gravel	Unfin.	Unfinished	W.F.	Wide Flange
Th., Thk.	Thick	UPS	Uninterruptible Power Supply	W.G.	Water Gauge
Thn.	Thin	URD	Underground Residential	Wldg.	Welding
Thrded	Threaded		Distribution	W. Mile	Wire Mile
Tilf.	Tile Layer, Floor	US	United States	W-O-L	Weldolet
Tilh.	Tile Layer, Helper	USGBC	U.S. Green Building Council	W.R.	Water Resistant
THHN	Nylon Jacketed Wire	USP	United States Primed	Wrck.	Wrecker
THW.	Insulated Strand Wire	UTMCD	Uniform Traffic Manual For Control	WSFU	Water Supply Fixture Unit
THWN	Nylon Jacketed Wire		Devices	W.S.P.	Water, Steam, Petroleum
T.L., TL	Truckload	UTP	Unshielded Twisted Pair	WT., Wt.	Weight
T.M.	Track Mounted	V	Volt	WWF	Welded Wire Fabric
Tot.	Total	VA	Volt Amperes	XFER	Transfer
T-O-L	Threadolet	VAT	Vinyl Asbestos Tile	XFMR	Transformer
tmpd	Tempered	V.C.T.	Vinyl Composition Tile	XHD	Extra Heavy Duty
TPO	Thermoplastic Polyolefin	VAV	Variable Air Volume	XHHW	Cross-Linked Polyethylene Wire
T.S.	Trigger Start	VC	Veneer Core	XLPE	Insulation
Tr.	Trade	VDC	Volts Direct Current	XLP	Cross-linked Polyethylene
Transf.	Transformer	Vent.	Ventilation	Xport	Transport
Trhv.	Truck Driver, Heavy	Vert.	Vertical	Y	Wye
Trlr	Trailer	V.F.	Vinyl Faced	yd	Yard
Trlt.	Truck Driver, Light	V.G.	Vertical Grain	yr	Year
TTY	Teletypewriter	VHF	Very High Frequency	Δ	Delta
TV	Television	VHO	Very High Output	%	Percent
T.W.	Thermoplastic Water Resistant	Vib.	Vibrating	~	Approximately
	Wire	VLF	Vertical Linear Foot	Ø	Phase; diameter
UCI	Uniform Construction Index	VOC	Volatile Organic Compound	@	At
UF	Underground Feeder	Vol.	Volume	#	Pound; Number
UGND	Underground Feeder	VRP	Vinyl Reinforced Polyester	<	Less Than
UHF	Ultra High Frequency	W	Wire; Watt; Wide; West	>	Greater Than
U.I.	United Inch			Z	Zone

RESIDENTIAL
COST ESTIMATE

OWNER'S NAME: _____ APPRAISER: _____

RESIDENCE ADDRESS: _____ PROJECT: _____

CITY, STATE, ZIP CODE: _____ DATE: _____

CLASS OF CONSTRUCTION	RESIDENCE TYPE	CONFIGURATION	EXTERIOR WALL SYSTEM
☐ ECONOMY	☐ 1 STORY	☐ DETACHED	☐ WOOD SIDING—WOOD FRAME
☐ AVERAGE	☐ 1-1/2 STORY	☐ TOWN/ROW HOUSE	☐ BRICK VENEER—WOOD FRAME
☐ CUSTOM	☐ 2 STORY	☐ SEMI-DETACHED	☐ STUCCO ON WOOD FRAME
☐ LUXURY	☐ 2-1/2 STORY		☐ PAINTED CONCRETE BLOCK
	☐ 3 STORY	OCCUPANCY	☐ SOLID MASONRY (AVERAGE & CUSTOM)
	☐ BI-LEVEL	☐ ONE STORY	☐ STONE VENEER—WOOD FRAME
	☐ TRI-LEVEL	☐ TWO FAMILY	☐ SOLID BRICK (LUXURY)
		☐ THREE FAMILY	☐ SOLID STONE (LUXURY)
		☐ OTHER _____	

*LIVING AREA (Main Building)		*LIVING AREA (Wing or Ell) ( )		*LIVING AREA (Wing or Ell) ( )	
First Level	_____ S.F.	First Level	_____ S.F.	First Level	_____ S.F.
Second Level	_____ S.F.	Second Level	_____ S.F.	Second Level	_____ S.F.
Third Level	_____ S.F.	Third Level	_____ S.F.	Third Level	_____ S.F.
Total	_____ S.F.	Total	_____ S.F.	Total	_____ S.F.

*Basement Area is not part of living area.

MAIN BUILDING	COSTS PER S.F. LIVING AREA
Cost per Square Foot of Living Area, from Page _____	$
Basement Addition: _____ % Finished, _____ % Unfinished	+
Roof Cover Adjustment: _____ Type, Page _____ (Add or Deduct)	( )
Central Air Conditioning: ☐ Separate Ducts ☐ Heating Ducts, Page _____	+
Heating System Adjustment: _____ Type, Page _____ (Add or Deduct)	( )
Main Building: Adjusted Cost per S.F. of Living Area	$

MAIN BUILDING TOTAL COST: $ _____ /S.F. x _____ S.F. x _____ = $ _____

Cost per S.F. Living Area | Living Area | Town/Row House Multiplier (Use 1 for Detached) | TOTAL COST

WING OR ELL ( ) _____ STORY	COSTS PER S.F. LIVING AREA
Cost per Square Foot of Living Area, from Page _____	$
Basement Addition: _____ % Finished, _____ % Unfinished	+
Roof Cover Adjustment: _____ Type, Page _____ (Add or Deduct)	( )
Central Air Conditioning: ☐ Separate Ducts ☐ Heating Ducts, Page _____	+
Heating System Adjustment: _____ Type, Page _____ (Add or Deduct)	( )
Wing or Ell ( ): Adjusted Cost per S.F. of Living Area	$

WING OR ELL ( ) TOTAL COST: $ _____ /S.F. x _____ S.F. = $ _____

Cost per S.F. Living Area | Living Area | TOTAL COST

WING OR ELL ( ) _____ STORY	COSTS PER S.F. LIVING AREA
Cost per Square Foot of Living Area, from Page _____	$
Basement Addition: _____ % Finished, _____ % Unfinished	+
Roof Cover Adjustment: _____ Type, Page _____ (Add or Deduct)	( )
Central Air Conditioning: ☐ Separate Ducts ☐ Heating Ducts, Page _____	+
Heating System Adjustment: _____ Type, Page _____ (Add or Deduct)	( )
Wing or Ell ( ): Adjusted Cost per S.F. of Living Area	$

WING OR ELL ( ) TOTAL COST: $ _____ /S.F. x _____ S.F. = $ _____

Cost per S.F. Living Area | Living Area | TOTAL COST

TOTAL THIS PAGE [_____]

RESIDENTIAL
COST ESTIMATE

		QUANTITY	UNIT COST	
Total Page 1				$
Additional Bathrooms: _____ Full, _____ Half				
Finished Attic: _____ Ft. x _____ Ft.		S.F.		+
Breezeway: ☐ Open ☐ Enclosed _____ Ft. x _____ Ft.		S.F.		+
Covered Porch: ☐ Open ☐ Enclosed _____ Ft. x _____ Ft.		S.F.		+
Fireplace: ☐ Interior Chimney ☐ Exterior Chimney ☐ No. of Flues ☐ Additional Fireplaces				+
Appliances:				+
Kitchen Cabinets Adjustment: (+/−)				
☐ Garage ☐ Carport: _____ Car(s) Description _____ (+/−)				
Miscellaneous:				+

ADJUSTED TOTAL BUILDING COST $ _____

REPLACEMENT COST	
ADJUSTED TOTAL BUILDING COST	$ _____
Site Improvements	
(A) Paving & Sidewalks	$ _____
(B) Landscaping	$ _____
(C) Fences	$ _____
(D) Swimming Pool	$ _____
(E) Miscellaneous	$ _____
TOTAL	$ _____
Location Factor	x _____
Location Replacement Cost	$ _____
Depreciation	−$ _____
LOCAL DEPRECIATED COST	$ _____

INSURANCE COST	
ADJUSTED TOTAL BUILDING COST	$ _____
Insurance Exclusions	
(A) Footings, Site Work, Underground Piping	−$ _____
(B) Architects' Fees	−$ _____
Total Building Cost Less Exclusion	$ _____
Location Factor	x _____
LOCAL INSURABLE REPLACEMENT COST	$ _____

SKETCH AND ADDITIONAL CALCULATIONS

For customer support on your Square Foot Cost Data, call 877.756.2789.

CII APPRAISAL

1. SUBJECT PROPERTY: _____

2. BUILDING: _____

3. ADDRESS: _____

4. BUILDING USE: _____

5. DATE: _____

6. APPRAISER: _____

7. YEAR BUILT: _____

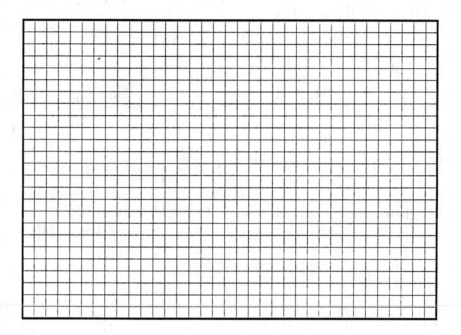

8. EXTERIOR WALL CONSTRUCTION: _____

9. FRAME: _____

10. GROUND FLOOR AREA: _____ S.F.

11. GROSS FLOOR AREA (EXCL. BASEMENT): _____ S.F.

12. NUMBER OF STORIES: _____

13. STORY HEIGHT: _____

14. PERIMETER: _____ L.F.

15. BASEMENT AREA: _____ S.F.

16. GENERAL COMMENTS: _____

_____

_____

_____

_____

_____

_____

_____

	NO.	DESCRIPTION			UNIT	UNIT COST	NEW SF COST	MODEL SF COST	+/- CHANGE
A SUBSTRUCTURE									
A	1010	Standard Foundations		Bay size:	S.F. Gnd.				
A	1030	Slab on Grade	Material:	Thickness:	S.F. Slab				
A	2010	Basement Excavation	Depth:	Area:	S.F. Gnd.				
A	2020	Basement Walls			L.F. Walls				
B SHELL									
		B10 Superstructure							
B	1010	Floor Construction	Elevated floors:		S.F. Floor				
					S.F. Floor				
B	1020	Roof Construction			S.F. Roof				
		B20 Exterior Enclosure							
B	2010	Exterior walls	Material:	Thickness: % of wall	S.F. Walls				
			Material:	Thickness: % of wall	S.F. Walls				
B	2020	Exterior Windows	Type:	% of wall	S.F. Wind.				
			Type:	% of wall Each	S.F. Wind.				
B	2030	Exterior Doors	Type:	Number:	Each				
			Type:	Number:	Each				
		B30 Roofing							
B	3010	Roof Coverings	Material:		S.F. Roof				
			Material:		S.F. Roof				
B	3020	Roof Openings			S.F. Opng.				
C INTERIORS									
C	1010	Partitions:	Material:	Density:	S.F. Part.				
			Material:	Density:					
C	1020	Interior Doors	Type:	Number:	Each				
C	1030	Fittings			Each				
C	2010	Stair Construction			Flight				
C	3010	Wall Finishes	Material:	% of Wall	S.F. Walls				
			Material:		S.F. Walls				
C	3020	Floor Finishes	Material:		S.F. Floor				
			Material:		S.F. Floor				
			Material:		S.F. Floor				
C	3030	Ceiling Finishes	Material:		S.F. Ceil.				

	NO.	SYSTEM/COMPONENT	DESCRIPTION	UNIT	UNIT COST	NEW S.F. COST	MODEL S.F. COST	+/- CHANGE
D	**SERVICES**							
	D10 Conveying							
D	1010	Elevators & Lifts	Type: Capacity: Stops:	Each				
D	1020	Escalators & Moving Walks	Type:	Each				
	D20 Plumbing							
D	2010	Plumbing		Each				
D	2020	Domestic Water Distribution		S.F. Floor				
D	2040	Rain Water Drainage		S.F. Roof				
	D30 HVAC							
D	3010	Energy Supply		Each				
D	3020	Heat Generating Systems	Type:	S.F. Floor				
D	3030	Cooling Generating Systems	Type:	S.F. Floor				
D	3090	Other HVAC Sys. & Equipment		Each				
	D40 Fire Protection							
D	4010	Sprinklers		S.F. Floor				
D	4020	Standpipes		S.F. Floor				
	D50 Electrical							
D	5010	Electrical Service/Distribution		S.F. Floor				
D	5020	Lighting & Branch Wiring		S.F. Floor				
D	5030	Communications & Security		S.F. Floor				
D	5090	Other Electrical Systems		S.F. Floor				
E	**EQUIPMENT & FURNISHINGS**							
E	1010	Commercial Equipment		Each				
E	1020	Institutional Equipment		Each				
E	1030	Vehicular Equipment		Each				
E	1090	Other Equipment		Each				
F	**SPECIAL CONSTRUCTION**							
F	1020	Integrated Construction		S.F.				
F	1040	Special Facilities		S.F.				
G	**BUILDING SITEWORK**							

ITEM		Total Change
17	Model 020 total _____ $_____	
18	Adjusted S.F. cost ____ $_____ item 17 +/- changes	
19	Building area - from item 11 _____ S.F. x adjusted S.F. cost...	$...
20	Basement area - from item 15 _____ S.F. x S.F. cost $	$..
21	Base building sub-total - item 20 + item 19	$.....
22	Miscellaneous addition (quality, etc.)	$......
23	Sub-total - item 22 + 21	$.......
24	General conditions -25 % of item 23	$......
25	Sub-total - item 24 + item 23...	$.......
26	Architects fees _____ % of item 25	$
27	Sub-total - item 26 + item 27	$......
28	Location modifier	x....
29	Local replacement cost - item 28 x item 27	$.....
30	Depreciation _____ % of item 29	$
31	Depreciated local replacement cost - item 29 less item 30	$.....
32	Exclusions	$.......
33	Net depreciated replacement cost - item 31 less item 32	$........

Base Cost per square foot floor area: *(from square foot table)*

Specify source:

| Page: _____ , Model # _____ , Area _____ S.F., |
| Exterior wall _____ , Frame _____ |

Adjustments for exterior wall variation:

Size Adjustment:
(Interpolate)

Height Adjustment: _____ + _____ = _____

Adjusted Base Cost per square foot:

Building Cost $ _____ x _____ = _____
                          Adjusted Base Cost   *Floor Area*
                          per square foot

Basement Cost $ _____ x _____ = _____
                          Basement Cost   *Basement Area*

Lump Sum Additions

TOTAL BUILDING COST *(Sum of above costs)*   _____

Modifications: *(complexity, workmanship, size)*   +/- _____ %   _____

Location Modifier: City _____ Date _____   x   _____

    Local cost of replacement   _____

Less depreciation: _____   - _____

    Local cost of replacement less depreciation $ _____

Other RSMeans Products & Services

RSMeans—a tradition of excellence in construction cost information and services since 1942

Table of Contents
Annual Cost Guides
RSMeans Online
Seminars

For more information visit the RSMeans website at www.RSMeans.com

Unit prices according to the latest MasterFormat

Cost Data Selection Guide

The following table provides definitive information on the content of each cost data publication. The number of lines of data provided in each unit price or assemblies division, as well as the number of crews, is listed for each data set. The presence of other elements such as reference tables, square foot models, equipment rental costs, historical cost indexes, and city cost indexes, is also indicated. You can use the table to help select the RSMeans data set that has the quantity and type of information you most need in your work.

Unit Cost Divisions	Building Construction	Mechanical	Electrical	Commercial Renovation	Square Foot	Site Work Landsc.	Green Building	Interior	Concrete Masonry	Open Shop	Heavy Construction	Light Commercial	Facilities Construction	Plumbing	Residential
1	562	383	400	504		498	205	302	445	561	496	245	1040	393	177
2	776	279	85	732		991	206	399	213	775	733	480	1220	286	273
3	1691	340	230	1084		1483	986	354	2037	1691	1693	482	1791	316	389
4	960	21	0	920		725	180	615	1158	928	615	533	1175	0	445
5	1893	158	155	1090		844	1799	1098	720	1893	1037	979	1909	204	746
6	2453	18	18	2111		110	589	1528	281	2449	123	2141	2125	22	2661
7	1594	215	128	1633		580	763	531	523	1591	26	1328	1695	227	1048
8	2111	81	44	2645		257	1136	1757	105	2096	0	2219	2890	0	1534
9	2004	86	45	1828		310	451	2088	391	1944	15	1668	2252	54	1437
10	1049	17	10	645		216	27	864	157	1049	29	539	1141	233	224
11	1096	206	166	551		129	54	933	28	1071	0	231	1116	169	110
12	530	0	2	285		205	133	1533	14	500	0	259	1554	23	217
13	744	149	158	253		367	122	252	78	720	244	108	759	115	103
14	273	36	0	221		0	0	257	0	273	0	12	293	16	6
21	91	0	16	37		0	0	250	0	91	0	73	535	540	220
22	1168	7548	160	1201		1572	1066	848	20	1157	1681	874	7450	9370	718
23	1178	6975	580	929		157	903	778	38	1171	110	879	5213	1922	469
26	1380	458	10150	1033		793	630	1144	55	1372	562	1310	10099	399	619
27	72	0	339	34		13	0	71	0	72	39	52	321	0	22
28	96	58	136	71		0	21	78	0	99	0	40	151	44	25
31	1510	733	610	806		3263	288	7	1217	1455	3277	604	1569	660	613
32	813	49	8	881		4409	353	405	291	784	1826	421	1687	133	468
33	529	1076	538	252		2173	41	0	236	522	2157	128	1697	1283	154
34	107	0	47	4		190	0	0	31	62	214	0	136	0	0
35	18	0	0	0		327	0	0	0	18	442	0	84	0	0
41	60	0	0	33		8	0	22	0	61	31	0	68	14	0
44	75	79	0	0		0	0	0	0	0	0	0	75	75	0
46	23	16	0	0		274	261	0	0	23	264	0	33	33	0
48	12	0	25	0		0	25	0	0	12	17	12	12	0	12
Totals	24868	18981	14050	19783		19894	10239	16114	8038	24440	15631	15617	50090	16531	12690

Assem Div	Building Construction	Mechanical	Electrical	Commercial Renovation	Square Foot	Site Work Landscape	Assemblies	Green Building	Interior	Concrete Masonry	Heavy Construction	Light Commercial	Facilities Construction	Plumbing	Asm Div	Residential
A		15	0	188	150	577	598	0	0	536	571	154	24	0	1	378
B		0	0	848	2506	0	5666	56	329	1975	368	2098	174	0	2	211
C		0	0	647	928	0	1309	0	1628	146	0	818	249	0	3	588
D		1067	941	712	1859	72	2538	330	825	0	0	1345	1105	1088	4	851
E		0	0	86	261	0	301	0	5	0	0	258	5	0	5	390
F		0	0	0	114	0	114	0	0	0	0	114	3	0	6	357
G		527	447	318	312	3364	792	0	0	534	1349	205	293	677	7	307
															8	760
															9	80
															10	0
															11	0
															12	0
Totals		1609	1388	2799	6130	4013	11318	386	2787	3191	2288	4992	1853	1765		3922

Reference Section	Building Construction Costs	Mechanical	Electrical	Commercial Renovation	Square Foot	Site Work Landscape	Assem.	Green Building	Interior	Concrete Masonry	Open Shop	Heavy Construction	Light Commercial	Facilities Construction	Plumbing	Resi.
Reference Tables	yes	yes	yes	yes	no	yes	yes	yes	yes	yes	yes	yes	yes	yes	yes	yes
Models					111			25					50			28
Crews	575	575	575	553		575		575	575	575	551	575	551	553	575	551
Equipment Rental Costs	yes	yes	yes	yes		yes		yes	yes	yes	yes	yes	yes	yes	yes	yes
Historical Cost Indexes	yes	yes	yes	yes	yes	yes	yes	yes	yes	yes	yes	yes	yes	yes	yes	no
City Cost Indexes	yes	yes	yes	yes	yes	yes	yes	yes	yes	yes	yes	yes	yes	yes	yes	yes

Online   Book   CD eBook Online add-on available for additional fee

For more information visit our website at www.RSMeans.com

Unit prices according to the latest MasterFormat

Annual Cost Guides

RSMeans Building Construction Cost Data 2016

Offers you unchallenged unit price reliability in an easy-to-use format. Whether used for verifying complete, finished estimates or for periodic checks, it supplies more cost facts better and faster than any comparable source. More than 24,700 unit prices have been updated for 2016. The City Cost Indexes and Location Factors cover more than 930 areas for indexing to any project location in North America. Order and get *RSMeans Quarterly Update Service* FREE.

RSMeans Green Building Cost Data 2016

Estimate, plan, and budget the costs of green building for both new commercial construction and renovation work with this sixth edition of *RSMeans Green Building Cost Data*. More than 10,000 unit costs for a wide array of green building products plus assemblies costs. Easily identified cross references to LEED and Green Globes building rating systems criteria.

RSMeans Mechanical Cost Data 2016

Total unit and systems price guidance for mechanical construction—materials, parts, fittings, and complete labor cost information. Includes prices for piping, heating, air conditioning, ventilation, and all related construction.

Plus new 2016 unit costs for:

- Thousands of installed HVAC/controls, sub-assemblies and assemblies
- "On-site" Location Factors for more than 930 cities and towns in the U.S. and Canada
- Crews, labor, and equipment

RSMeans Facilities Construction Cost Data 2016

For the maintenance and construction of commercial, industrial, municipal, and institutional properties. Costs are shown for new and remodeling construction and are broken down into materials, labor, equipment, and overhead and profit. Special emphasis is given to sections on mechanical, electrical, furnishings, site work, building maintenance, finish work, and demolition.

More than 49,800 unit costs, plus assemblies costs and a comprehensive Reference Section are included.

RSMeans Square Foot Costs 2016
Accurate and Easy-to-Use

- Updated price information based on nationwide figures from suppliers, estimators, labor experts, and contractors.
- Green building models
- Realistic graphics, offering true-to-life illustrations of building projects
- Extensive information on using square foot cost data, including sample estimates and alternate pricing methods

RSMeans Commercial Renovation Cost Data 2016
Commercial/Multi-family Residential

Use this valuable tool to estimate commercial and multi-family residential renovation and remodeling.

Includes: Updated costs for hundreds of unique methods, materials, and conditions that only come up in repair and remodeling, PLUS:

- Unit costs for more than 19,600 construction components
- Installed costs for more than 2,700 assemblies
- More than 930 "on-site" localization factors for the U.S. and Canada

RSMeans Electrical Cost Data 2016

Pricing information for every part of electrical cost planning. More than 13,800 unit and systems costs with design tables; clear specifications and drawings; engineering guides; illustrated estimating procedures; complete labor-hour and materials costs for better scheduling and procurement; and the latest electrical products and construction methods.

- A variety of special electrical systems, including cathodic protection
- Costs for maintenance, demolition, HVAC/mechanical, specialties, equipment, and more

RSMeans Electrical Change Order Cost Data 2016

RSMeans Electrical Change Order Cost Data provides you with electrical unit prices exclusively for pricing change orders. Analyze and check your own change order estimates against those prepared when using RSMeans cost data. It also covers productivity analysis and change order cost justifications. With useful information for calculating the effects of change orders and dealing with their administration.

RSMeans Assemblies Cost Data 2016

RSMeans Assemblies Cost Data takes the guesswork out of preliminary or conceptual estimates. Now you don't have to try to calculate the assembled cost by working up individual component costs. We've done all the work for you.

Presents detailed illustrations, descriptions, specifications, and costs for every conceivable building assembly—over 350 types in all—arranged in the easy-to-use UNIFORMAT II system. Each illustrated "assembled" cost includes a complete grouping of materials and associated installation costs, including the installing contractor's overhead and profit.

RSMeans Open Shop Building Construction Cost Data 2016

The latest costs for accurate budgeting and estimating of new commercial and residential construction, renovation work, change orders, and cost engineering.

RSMeans Open Shop "BCCD" will assist you to:

- Develop benchmark prices for change orders.
- Plug gaps in preliminary estimates and budgets.
- Estimate complex projects.
- Substantiate invoices on contracts.
- Price ADA-related renovations.

Annual Cost Guides

Unit prices according to the latest MasterFormat

RSMeans Residential Cost Data 2016

Contains square foot costs for 28 basic home models with the look of today, plus hundreds of custom additions and modifications you can quote right off the page. Includes more than 3,900 costs for 89 residential systems. Complete with blank estimating forms, sample estimates, and step-by-step instructions.

Contains line items for cultured stone and brick, PVC trim, lumber, and TPO roofing.

RSMeans Site Work & Landscape Cost Data 2016

Includes unit and assemblies costs for earthwork, sewerage, piped utilities, site improvements, drainage, paving, trees and shrubs, street openings/repairs, underground tanks, and more. Contains more than 60 types of assemblies costs for accurate conceptual estimates.

Includes:

- Estimating for infrastructure improvements
- Environmentally-oriented construction
- ADA-mandated handicapped access
- Hazardous waste line items

RSMeans Facilities Maintenance & Repair Cost Data 2016

RSMeans Facilities Maintenance & Repair Cost Data gives you a complete system to manage and plan your facility repair and maintenance costs and budget efficiently. Guidelines for auditing a facility and developing an annual maintenance plan. Budgeting is included, along with reference tables on cost and management, and information on frequency and productivity of maintenance operations.

The only nationally recognized source of maintenance and repair costs. Developed in cooperation with the Civil Engineering Research Laboratory (CERL) of the Army Corps of Engineers.

RSMeans Light Commercial Cost Data 2016

Specifically addresses the light commercial market, which is a specialized niche in the construction industry. Aids you, the owner/designer/contractor, in preparing all types of estimates—from budgets to detailed bids. Includes new advances in methods and materials.

The Assemblies section allows you to evaluate alternatives in the early stages of design/planning.

More than 15,500 unit costs ensure that you have the prices you need, when you need them.

RSMeans Concrete & Masonry Cost Data 2016

Provides you with cost facts for virtually all concrete/masonry estimating needs, from complicated formwork to various sizes and face finishes of brick and block—all in great detail. The comprehensive Unit Price Section contains more than 8,000 selected entries. Also contains an Assemblies [Cost] Section, and a detailed Reference Section that supplements the cost data.

RSMeans Labor Rates for the Construction Industry 2016

Complete information for estimating labor costs, making comparisons, and negotiating wage rates by trade for more than 300 U.S. and Canadian cities. With 46 construction trades in each city, and historical wage rates included for comparison. Each city chart lists the county and is alphabetically arranged with handy visual flip tabs for quick reference.

RSMeans Construction Cost Indexes 2016

What materials and labor costs will change unexpectedly this year? By how much?

- Breakdowns for 318 major cities
- National averages for 30 key cities
- Expanded five major city indexes
- Historical construction cost indexes

RSMeans Interior Cost Data 2016

Provides you with prices and guidance needed to make accurate interior work estimates. Contains costs on materials, equipment, hardware, custom installations, furnishings, and labor for new and remodel commercial and industrial interior construction, including updated information on office furnishings, as well as reference information.

RSMeans Heavy Construction Cost Data 2016

A comprehensive guide to heavy construction costs. Includes costs for highly specialized projects such as tunnels, dams, highways, airports, and waterways. Information on labor rates, equipment, and materials costs is included. Features unit price costs, systems costs, and numerous reference tables for costs and design.

RSMeans Plumbing Cost Data 2016

Comprehensive unit prices and assemblies for plumbing, irrigation systems, commercial and residential fire protection, point-of-use water heaters, and the latest approved materials. This publication and its companion, *RSMeans Mechanical Cost Data*, provide full-range cost estimating coverage for all the mechanical trades.

Contains updated costs for potable water, radiant heat systems, and high efficiency fixtures.

RSMeans Online

Competitive Cost Estimates Made Easy

RSMeans Online is a web-based service that provides accurate and up-to-date cost information to help you build competitive estimates or budgets in less time.

Quick, intuitive, easy to use, and automatically updated, RSMeans Online gives you instant access to hundreds of thousands of material, labor, and equipment costs from RSMeans' comprehensive database, delivering the information you need to build budgets and competitive estimates every time.

With RSMeans Online, you can perform quick searches to locate specific costs and adjust costs to reflect prices in your geographic area. Tag and store your favorites for fast access to your frequently used line items and assemblies and clone estimates to save time. System notifications will alert you as updated data becomes available. RSMeans Online is automatically updated throughout the year.

RSMeans Online's visual, interactive estimating features help you create, manage, save and share estimates with ease! It provides increased flexibility with customizable advanced reports. Easily edit custom report templates and import your company logo onto your estimates.

	Standard	Advanced	Professional
Unit Prices	✓	✓	✓
Assemblies	X	✓	✓
Sq Ft Models	X	X	✓
Editable Sq Foot Models	X	X	✓
Editable Assembly Components	X	✓	✓
Custom Cost Data	X	✓	✓
User Defined Components	X	✓	✓
Advanced Reporting & Customization	X	✓	✓
Union Labor Type	✓	✓	✓
Open Shop*	Add on for additional fee	Add on for additional fee	✓
History (3-year lookback)**	Add on for additional fee	Add on for additional fee	Add on for additional fee
Full History (2007)	Add on for additional fee	Add on for additional fee	Add on for additional fee

*Open shop included in Complete Library & Bundles
** 3-year history lookback included with complete library

Unit prices according to the latest MasterFormat

RSMeans Online

Visit RSMeans.com/Online for more details on data titles in online format.

For more information visit our website at www.RSMeans.com

Estimate with Precision

Find everything you need to develop complete, accurate estimates.

- Verified costs for construction materials
- Equipment rental costs
- Crew sizing, labor hours and labor rates
- Localized costs for U.S. and Canada

Save Time & Increase Efficiency

Make cost estimating and calculating faster and easier than ever with secure, online estimating tools.

- Quickly locate costs in the searchable database
- Create estimates in minutes with RSMeans cost lines
- Tag and store favorites for fast access to frequently used items

Improve Planning & Decision-Making

Back your estimates with complete, accurate and up-to-date cost data for informed business decisions.

- Verify construction costs from third parties
- Check validity of subcontractor proposals
- Evaluate material and assembly alternatives

Increase Profits

Use RSMeans Online to estimate projects quickly and accurately, so you can gain an edge over your competition.

- Create accurate and competitive bids
- Minimize the risk of cost overruns
- Reduce variability
- Gain control over costs

30 Day Free Trial

Register for a free trial at www.RSMeansOnline.com

- Access to unit prices, building assemblies and facilities repair and remodeling items covering every category of construction
- Powerful tools to customize, save and share cost lists, estimates and reports with ease
- Access to square foot models for quick conceptual estimates
- Allows up to 10 users to evaluate the full range of collaborative functionality used items.

2016 RSMeans Seminar Schedule

Note: call for exact dates and details.

Location	Dates	Location	Dates
Seattle, WA	January and August	Bethesda, MD	June
Dallas/Ft. Worth, TX	January	El Segundo, CA	August
Austin, TX	February	Jacksonville, FL	September
Anchorage, AK	March and September	Dallas, TX	September
Las Vegas, NV	March	Philadelphia, PA	October
New Orleans, LA	March	Houston, TX	October
Washington, DC	April and September	Salt Lake City, UT	November
Phoenix, AZ	April	Baltimore, MD	November
Kansas City, MO	April	Orlando, FL	November
Toronto	May	San Diego, CA	December
Denver, CO	May	San Antonio, TX	December
San Francisco, CA	June	Raleigh, NC	December

☎ 877-620-6245

Beginning early 2016, RSMeans will provide a suite of online, self-paced training offerings. Check our website www.RSMeans.com for more information.

Professional Development

eLearning Training Sessions

Learn how to use *RSMeans Online*® or the *RSMeans CostWorks*® CD from the convenience of your home or office. Our eLearning training sessions let you join a training conference call and share the instructors' desktops, so you can view the presentation and step-by-step instructions on your own computer screen. The live webinars are held from 9 a.m. to 4 p.m. or 11 a.m. to 6 p.m. eastern standard time, with a one-hour break for lunch.

For these sessions, you must have a computer with high speed Internet access and a compatible Web browser. Learn more at www.RSMeansOnline.com or call for a schedule: 781-422-5115. Webinars are generally held on selected Wednesdays each month.

RSMeans Online	RSMeans CostWorks CD
$299 per person	$299 per person

Seminar Schedule and Professional Development

For more information visit our website at www.RSMeans.com

Professional Development

RSMeans Online Training

Construction estimating is vital to the decision-making process at each state of every project. RSMeansOnline works the way you do. It's systematic, flexible and intuitive. In this one day class you will see how you can estimate any phase of any project faster and better.

Some of what you'll learn:
- Customizing RSMeansOnline
- Making the most of RSMeans "Circle Reference" numbers
- How to integrate your cost data
- Generate reports, exporting estimates to MS Excel, sharing, collaborating and more

Also available: RSMeans Online training webinar

Facilities Construction Estimating

In this *two-day* course, professionals working in facilities management can get help with their daily challenges to establish budgets for all phases of a project.

Some of what you'll learn:
- Determining the full scope of a project
- Identifying the scope of risks and opportunities
- Creative solutions to estimating issues
- Organizing estimates for presentation and discussion
- Special techniques for repair/remodel and maintenance projects
- Negotiating project change orders

Who should attend: facility managers, engineers, contractors, facility tradespeople, planners, and project managers.

Construction Cost Estimating: Concepts and Practice

This introductory course to improve estimating skills and effectiveness starts with the details of interpreting bid documents and ends with the summary of the estimate and bid submission.

Some of what you'll learn:
- Using the plans and specifications for creating estimates
- The takeoff process—deriving all tasks with correct quantities
- Developing pricing using various sources; how subcontractor pricing fits in
- Summarizing the estimate to arrive at the final number
- Formulas for area and cubic measure, adding waste and adjusting productivity to specific projects
- Evaluating subcontractors' proposals and prices
- Adding insurance and bonds
- Understanding how labor costs are calculated
- Submitting bids and proposals

Who should attend: project managers, architects, engineers, owners' representatives, contractors, and anyone who's responsible for budgeting or estimating construction projects.

Maintenance & Repair Estimating for Facilities

This *two-day* course teaches attendees how to plan, budget, and estimate the cost of ongoing and preventive maintenance and repair for existing buildings and grounds.

Some of what you'll learn:
- The most financially favorable maintenance, repair, and replacement scheduling and estimating
- Auditing and value engineering facilities
- Preventive planning and facilities upgrading
- Determining both in-house and contract-out service costs
- Annual, asset-protecting M&R plan

Who should attend: facility managers, maintenance supervisors, buildings and grounds superintendents, plant managers, planners, estimators, and others involved in facilities planning and budgeting.

Practical Project Management for Construction Professionals

In this *two-day* course, acquire the essential knowledge and develop the skills to effectively and efficiently execute the day-to-day responsibilities of the construction project manager.

Some of what you'll learn:
- General conditions of the construction contract
- Contract modifications: change orders and construction change directives
- Negotiations with subcontractors and vendors
- Effective writing: notification and communications
- Dispute resolution: claims and liens

Who should attend: architects, engineers, owners' representatives, and project managers.

Mechanical & Electrical Estimating

This *two-day* course teaches attendees how to prepare more accurate and complete mechanical/electrical estimates, avoid the pitfalls of omission and double-counting, and understand the composition and rationale within the RSMeans mechanical/electrical database.

Some of what you'll learn:
- The unique way mechanical and electrical systems are interrelated
- M&E estimates—conceptual, planning, budgeting, and bidding stages
- Order of magnitude, square foot, assemblies, and unit price estimating
- Comparative cost analysis of equipment and design alternatives

Who should attend: architects, engineers, facilities managers, mechanical and electrical contractors, and others who need a highly reliable method for developing, understanding, and evaluating mechanical and electrical contracts.

Visit RSMeans.com/Online for more details on data titles in online format.

For more information visit our website at www.RSMeans.com

Unit Price Estimating

This interactive *two-day* seminar teaches attendees how to interpret project information and process it into final, detailed estimates with the greatest accuracy level.

The most important credential an estimator can take to the job is the ability to visualize construction and estimate accurately.

Some of what you'll learn:
- Interpreting the design in terms of cost
- The most detailed, time-tested methodology for accurate pricing
- Key cost drivers—material, labor, equipment, staging, and subcontracts
- Understanding direct and indirect costs for accurate job cost accounting and change order management

Who should attend: corporate and government estimators and purchasers, architects, engineers, and others who need to produce accurate project estimates.

Conceptual Estimating Using the RSMeans CostWorks CD

This *two-day* class uses the leading industry data and a powerful software package to develop highly accurate conceptual estimates for your construction projects. All attendees must bring a laptop computer loaded with the current year *Square Foot Costs* and the *Assemblies Cost Data* CostWorks titles.

Some of what you'll learn:
- Introduction to conceptual estimating
- Types of conceptual estimates
- Helpful hints
- Order of magnitude estimating
- Square foot estimating
- Assemblies estimating

Who should attend: architects, engineers, contractors, construction estimators, owners' representatives, and anyone looking for an electronic method for performing square foot estimating.

RSMeans CostWorks CD Training

This *one-day* course helps users become more familiar with the functionality of the *RSMeans CostWorks* program. Each menu, icon, screen, and function found in the program is explained in depth. Time is devoted to hands-on estimating exercises.

Some of what you'll learn:
- Searching the database using all navigation methods
- Exporting RSMeans data to your preferred spreadsheet format
- Viewing crews, assembly components, and much more
- Automatically regionalizing the database

This training session requires you to bring a laptop computer to class.

When you register for this course you will receive an outline for your laptop requirements.

Also offering web training for the RSMeans CostWorks CD!

Assessing Scope of Work for Facility Construction Estimating

This *two-day* practical training program addresses the vital importance of understanding the scope of projects in order to produce accurate cost estimates for facility repair and remodeling.

Some of what you'll learn:
- Discussions of site visits, plans/specs, record drawings of facilities, and site-specific lists
- Review of CSI divisions, including means, methods, materials, and the challenges of scoping each topic
- Exercises in scope identification and scope writing for accurate estimating of projects
- Hands-on exercises that require scope, take-off, and pricing

Who should attend: corporate and government estimators, planners, facility managers, and others who need to produce accurate project estimates.

Facilities Estimating Using the RSMeans CostWorks CD

Combines hands-on skill building with best estimating practices and real-life problems. Brings you up-to-date with key concepts and provides tips, pointers, and guidelines to save time and avoid cost oversights and errors.

Some of what you'll learn:
- Estimating process concepts
- Customizing and adapting RSMeans cost data
- Establishing scope of work to account for all known variables
- Budget estimating: when, why, and how
- Site visits: what to look for—what you can't afford to overlook
- How to estimate repair and remodeling variables

This training session requires you to bring a laptop computer to class.

Who should attend: facility managers, architects, engineers, contractors, facility tradespeople, planners, project managers and anyone involved with JOC, SABRE, or IDIQ.

Unit Price Estimating Using the RSMeans CostWorks CD

Step-by-step instructions and practice problems to identify and track key cost drivers—material, labor, equipment, staging, and subcontractors—for each specific task. Learn the most detailed, time-tested methodology for accurately "pricing" these variables, their impact on each other and on total cost.

Some of what you'll learn:
- Unit price cost estimating
- Order of magnitude, square foot, and assemblies estimating
- Quantity takeoff
- Direct and indirect construction costs
- Development of contractors' bill rates
- How to use *RSMeans Building Construction Cost Data*

This training session requires you to bring a laptop computer to class.

Who should attend: architects, engineers, corporate and government estimators, facility managers, and government procurement staff.

Professional Development

For more information visit our website at www.RSMeans.com

Registration Information

Register early and save up to $100!
Register 30 days before the start date of a seminar and save $100 off your total fee. Note: This discount can be applied only once per order. It cannot be applied to team discount registrations or any other special offer.

How to register
Register by phone today! The RSMeans toll-free number for making reservations is 781-422-5115.

Two-day seminar registration fee - $935. One-day RSMeans CostWorks® or RSMeans Online training registration fee - $375.
To register by mail, complete the registration form and return, with your full fee, to: RSMeans Seminars, 1099 Hingham Street, Suite 201, Rockland, MA 02370.

One-day Construction Cost Estimating - $575.

Government pricing
All federal government employees save off the regular seminar price. Other promotional discounts cannot be combined with the government discount.

Team discount program
For over five attendee registrations. Call for pricing: 781-422-5115

Multiple course discounts
When signing up for two or more courses, call for pricing.

Refund policy
Cancellations will be accepted up to ten business days prior to the seminar start. There are no refunds for cancellations received later than ten working days prior to the first day of the seminar. A $150 processing fee will be applied for all cancellations. Written notice of the cancellation is required. Substitutions can be made at any time before the session starts. No-shows are subject to the full seminar fee.

AACE approved courses
Many seminars described and offered here have been approved for 14 hours (1.4 recertification credits) of credit by the AACE International Certification Board toward meeting the continuing education requirements for recertification as a Certified Cost Engineer/Certified Cost Consultant.

AIA Continuing Education
We are registered with the AIA Continuing Education System (AIA/CES) and are committed to developing quality learning activities in accordance with the CES criteria. Many seminars meet the AIA/CES criteria for Quality Level 2. AIA members may receive 14 learning units (LUs) for each two-day RSMeans course.

Daily course schedule
The first day of each seminar session begins at 8:30 a.m. and ends at 4:30 p.m. The second day begins at 8:00 a.m. and ends at 4:00 p.m. Participants are urged to bring a hand-held calculator since many actual problems will be worked out in each session.

Continental breakfast
Your registration includes the cost of a continental breakfast and a morning and afternoon refreshment break. These informal segments allow you to discuss topics of mutual interest with other seminar attendees. (You are free to make your own lunch and dinner arrangements.)

Hotel/transportation arrangements
RSMeans arranges to hold a block of rooms at most host hotels. To take advantage of special group rates when making your reservation, be sure to mention that you are attending the RSMeans seminar. You are, of course, free to stay at the lodging place of your choice. (Hotel reservations and transportation arrangements should be made directly by seminar attendees.)

Important
Class sizes are limited, so please register as soon as possible.

Note: Pricing subject to change.

Registration Form

ADDS-1000

Call 781-422-5115 to register or fax this form to 800-632-6732. Visit our website: www.RSMeans.com

Please register the following people for the RSMeans construction seminars as shown here. We understand that we must make our own hotel reservations if overnight stays are necessary.

☐ Full payment of $_____ enclosed.

☐ Bill me.

Please print name of registrant(s).

(To appear on certificate of completion)

_____

_____

_____

_____

P.O. #:_____
GOVERNMENT AGENCIES MUST SUPPLY PURCHASE ORDER NUMBER OR TRAINING FORM.

Please mail check to: 1099 Hingham Street, Suite 201, Rockland, MA 02370 USA

Firm name_____

Address_____

City/State/Zip_____

Telephone no._____ Fax no._____

E-mail address_____

Charge registration(s) to: ☐ MasterCard ☐ VISA ☐ American Express

Account no._____ Exp. date_____

Cardholder's signature_____

Seminar name_____

Seminar City_____

Professional Development